RODD'S CHEMISTRY OF CARBON COMPOUNDS

ELSEVIER SCIENCE PUBLISHERS B.V.
Sara Burgerhartstraat 25
P.O. Box 211, 1000 AE Amsterdam, The Netherlands

Distributors for the United States and Canada:

ELSEVIER SCIENCE PUBLISHING COMPANY INC.
655, Avenue of the Americas
New York, NY 10010, U.S.A.

ISBN 0-444-87322-8 (Vol. IV IJ)
ISBN 0-444-40664-6 (Series)

© Elsevier Science Publishers B.V., 1989

All rights reserved. No part of this publication may be reproduced, stored in a retrieval system or transmitted in any form or by any means, electronic, mechanical, photocopying, recording or otherwise, without the prior written permission of the publisher, Elsevier Science Publishers B.V./Physical Sciences & Engineering Division, P.O. Box 330, 1000 AH Amsterdam, The Netherlands.

Special regulations for readers in the USA – This publication has been registered with the Copyright Clearance Center Inc. (CCC), Salem, Massachusetts. Information can be obtained from the CCC about conditions under which photocopies of parts of this publication may be made in the USA. All other copyright questions, including photocopying outside of the USA, should be referred to the publisher.

No responsibility is assumed by the Publisher for any injury and/or damage to persons or property as a matter of products liability, negligence or otherwise, or from any use or operation of any methods, products, instructions or ideas contained in the material herein.

Printed in The Netherlands

RODD'S CHEMISTRY
OF CARBON COMPOUNDS

VOLUME I
GENERAL INTRODUCTION
ALIPHATIC COMPOUNDS

*

VOLUME II
ALICYCLIC COMPOUNDS

*

VOLUME III
AROMATIC COMPOUNDS

*

VOLUME IV
HETEROCYCLIC COMPOUNDS

*

RODD'S CHEMISTRY OF CARBON COMPOUNDS

A modern comprehensive treatise

SECOND EDITION

Edited by

S. COFFEY

M.Sc. (London), D.Sc. (Leyden), C.Chem., F.R.I.C.
formerly of
I.C.I. Dyestuffs Divison, Blackley, Manchester, Great Britain

and

MARTIN F. ANSELL

Ph.D., D.Sc. (London) F.R.S.C., C.Chem.
Reader Emeritus, Department of Chemistry, Queen Mary College,
University of London, Great Britain

VOLUME IV PART IJ

HETEROCYCLIC COMPOUNDS

Edited by

MARTIN F. ANSELL

Six-membered heterocyclic compounds with two hetero-atoms from Group V of the Periodic Table: the Pyridazine and Pyrimidine groups, the Pyrazine group. Phenoxazine, phenothiazine, phenazine and sulphur dyes. Six-membered heterocylic compounds with three or more hetero-atoms

ELSEVIER
Amsterdam–Oxford–New York–Tokyo 1989

CONTRIBUTORS TO THIS VOLUME

N. Hughes, F.R.S.C., C.CHEM.
Imperial Chemical Industries, Blackley, Manchester M9 3DA

R.N. Hunston, B.SC., PH.D.
Department of Chemistry, Brunel University, Middlesex UB8 3PH

D. Hunter, B.SC., M.SC., PH.D.
Department of Chemistry, The University, Dundee, DD1 4HN

K.J. McCullough, B.SC., PH.D.
Department of Chemistry, Heriot Watt University, Edinburgh EH14 4AS

A. McKillop, PH.D., D.SC., F.R.S.C., C.CHEM.
School of Chemical Sciences, University of East Anglia, Norwich NR4 7TJ

D.G. Neilson, B.SC., PH.D., M.R.S.C., C.CHEM.
Department of Chemistry, The University, Dundee DD1 4HN

J. Parrick, B.SC., PH.D., F.R.S.C., C.CHEM.
Department of Chemistry, Brunel University, Middlesex UB8 3PH

M. Sainsbury, D.SC., PH.D., F.R.S.C., C.CHEM.
Department of Chemistry, The University, Bath BA2 7AY

C.J.G. Shaw, B.SC., PH.D., F.R.S.C., C.CHEM.
Department of Chemistry, Brunel University, Middlesex UB8 3PH

B.P. Swann, B.SC., PH.D., C.CHEM.
Lilly Research Centre Ltd., Windlesham, Surrey GU20 6PH

R.T. Walker, B.SC., PH.D.
Department of Chemistry, The University, Birmingham B15 2TT

PREFACE TO VOLUME IV IJ

The publication of Volume IV IJ, covering six-membered mono-heterocycles with two or more hetero-atoms in the ring, completes the second edition of Rodd's Chemistry of Carbon Compounds. This edition was planned by Dr. Samuel Coffey who commenced work on it in 1962 but did not live to see it completed. An appreciation of the life and work of Dr. Coffey was published in Volume IV C (Second Edition). However, the complete second edition of "Rodd" provides a lasting tribute to the detailed planning which Coffey put into the organisation of this whole work. Indeed the contents of this volume are those outlined by Coffey. The long period over which the work has been published reflects the complexity of the task of reviewing all organic chemistry and providing a really comprehensive advanced textbook. However, this particular volume was delayed by the failure of one contributor to provide what was intended to be a major section of the book. I am therefore deeply appreciative of the work of the final contributors to this volume (see list of contributors), who in some cases I know had to overcome personal difficulties. Each contributor has provided a carefully compiled critical integration of new information with the older, yet still valuable, material recorded in the first edition. The value of "Rodd" is enhanced by the high standard of indexing. It is with much pleasure that I acknowledge the contribution of Malcolm Sainsbury who has indexed the last volume of this series. I am also indebted to the production staff at Elsevier, the artists and the printers, who have converted "authors' manuscripts" into an attractive "published work". I would like to take this opportunity to rectify an omission from Volume IV C and to acknowledge the contributions of Professor A. McKillop and Dr. B.P. Swann who initially wrote chapters 16 and 17 of Volume IV C which were up-dated by Dr. M. Sainsbury.

The objective of the Second Edition of "Rodd" as set out by Coffey in the preface to Volume I A of this series was that "... the book shall continue to serve as a work of reference for the accumulated factual knowledge of organic chemistry and to give guidance to original sources and other literature for detail beyond its scope. In the latter connection particular care has been taken to refer to the extensive literature now existing in the form of review articles and monographs".

In my view this objective set by Coffey has been achieved by the contributors to this volume.

December 1988   Martin F. Ansell

CONTENTS
VOLUME IV IJ

Heterocyclic Compounds: Six-membered heterocyclic compounds with two hetero-atoms from Group V of the Periodic Table: the Pyridazine and Pyrimidine groups, the Pyrazine group. Phenoxazine, phenothiazine, phenazine and sulphur dyes. Six-membered heterocyclic compounds with three or more hetero-atoms

PREFACE	VII
OFFICIAL PUBLICATIONS; SCIENTIFIC JOURNALS AND PERIODICALS	XVII
LIST OF ABBREVIATED NAMES OF CHEMICAL FIRMS MENTIONED IN PATENT REFERENCES	XVIII
LIST OF COMMON ABBREVIATIONS AND SYMBOLS USED	XIX

Chapter 42. Pyridazines, Cinnolines, Benzocinnolines and Phthalazines
by R.N. HUNSTON, J. PARRICK and C.J.G. SHAW

1. Pyridazines 1
   a. Introduction 1
   b. Methods of synthesis 1
   c. Pyridazine and its alkyl and aryl derivatives 6
      (*i*) Pyridazine, 6 — (*ii*) Spectra, 8 — (*iii*) Chemical reactivity, 9 — (*iv*) Alkyl- and aryl-pyridazines, 10 —
   d. Halogenopyridazines 12
   e. Nitropyridazines 14
   f. Aminopyridazines and related compounds (nitroamines, hydrazines, azides) .. 15
   g. Pyridazine-*N*-oxides 18
   h. Pyridazinones 22
   i. Pyridazinecarboxylic acids 26
   j. Pyridazine sulphur compounds 28
   k. Reduced pyridazines 30
      (*i*) Dihydropyridazines, 30 — (*ii*) Tetrahydropyridazines, 33 — (*iii*) Hexahydropyridazines (piperidazines), 34 —
2. Cinnolines 35
   a. Introduction 35
   b. Methods of synthesis 36
   c. Cinnoline and its alkyl and aryl derivatives 39
      (*i*) Cinnoline, 39 — (*ii*) Spectra, 41 — (*iii*) Chromatography, 45 — (*iv*) Reactions, 45 — (*v*) Alkylcinnolines, 47 — (*vi*) Arylcinnolines, 48 —
   d. Cinnoline-*N*-oxides 49
   e. Halocinnolines 54
   f. Nitrocinnolines 59
   g. Aminocinnolines 61
      (*i*) Hydrazinocinnolines, 68 —
   h. Hydroxycinnolines 70
      (*i*) 3- and 4-Hydroxycinnolines, 70 — (*ii*) 5-, 6-, 7- and 8-Hydroxycinnolines, 80 — (*iii*) Phenyl ethers, 82, —
   i. Cinnolinecarboxylic acids and derivatives 83

j. Reduced cinnolines 92
   (i) Reduction of cinnolines, 92 — (ii) From cyclisation reactions, 96 — (iii) Reactions and biological properties, 98 —
3. Benzocinnolines 99
   a. Introduction 99
   b. Benzo[c]cinnolines 100
      (i) Synthesis, 100 — (ii) Physical and chemical properties, 102 — (iii) Methylated benzo[c]cinnolines, 105 — (iv) Halogeno-, nitro-, and other substituted benzo[c]cinnolines, 106 — (v) Reduced benzo[c]cinnolines, 109 —
   c. Other benzocinnolines 110
4. Phthalazines 114
   a. Methods of synthesis 114
   b. Phthalazine, its alkyl and aryl derivatives 119
      (i) Phthalazine, 119 — (ii) Spectra, 120 — (iii) Reactivity, 121 — (iv) Alkyl- and aryl-phthalazines, 122 —
   c. Phthalazine-N-oxides and -N-aminoazolium salts 123
   d. Halogenophthalazines 124
   e. Aminophthalazines and related compounds 126
   f. Phthalazinones, hydroxyphthalazines and oxidophthalazinium betaines 130
      (i) 1(2H)-Phthalazinones, 130 — (ii) 3-Alkyl(or aryl)-1(2H)-phthalazinium salts and 3-alkyl(or aryl)-1-oxidophthalazinium betaines, 133 — (iii) 4-Hydroxy-1(2H)-phthalazinones, 136 — (iv) Chemiluminescence, 139 —
   g. Phthalazinethiones and mercaptophthalazines 141
   h. Phthalazine quinones 142
   i. Phthalazinecarboxylic acids and their derivatives 144
   j. Reduced phthalazines 145
      (i) 1,2-Dihydrophthalazines, 145 — (ii) 1,4-Dihydrophthalazine, 150 — (iii) 1,2,3,4-Tetrahydrophthalazines, 150 — (iv) 5,6,7,8-Tetrahydrophthalazines, 151 — (v) Other reduced phthalazines, 153 —

Chapter 43. Pyrimidines and Quinazolines
by R.T. WALKER

1. Introduction 155
2. Physical properties of pyrimidine 158
   a. Structure and geometry 158
   b. The ionisation constant of pyrimidines 159
   c. Spectroscopy 162
      (i) Nuclear magnetic resonance, 162 — (ii) Infrared and ultraviolet spectra, 163 — (iii) Mass spectra, 163 —
   d. Tautomerism 164
3. Synthesis 165
   a. The principal method 166
      (i) From β-dialdehydes, 166 — (ii) From β-aldehydo-ketones, 168 — (iii) From β-diketones, 170 — (iv) From β-aldehydo esters, 172 — (v) From β-keto esters, 173 — (vi) From β-diesters, 175 — (vii) From β-aldehydo-

nitriles, 177 — (viii) From β-keto-nitriles, 179 — (ix) From β-ester nitriles, 179 — (x) From β-dinitriles, 180 —
  b. Primary syntheses 181
    (i) Involving the formation of one bond, 181 — (ii) Involving the formation of two bonds, 181 — (iii) Involving the formation of three bonds, 184 — (iv) Involving the formation of four or more bonds, 185 —
  c. Syntheses of the pyrimidine ring from other heterocycles 186
    (i) From pyrroles, 186 — (ii) From imidazoles, 186 — (iii) From mixed 5- or 6-membered heterocycles, 187 — (iv) From purines and related heterocycles, 188 — (v) Miscellaneous examples, 189 —
  d. Some factors governing the synthesis of substituted pyrimidines 190
4. Chemical reactivity and properties of pyrimidine derivatives 193
  a. General properties 193
  b. Reactivity of the ring positions 194
    (i) To electrophiles, 194 — (ii) To nucleophiles, 197 — (iii) Photochemical reactions, 197 — (iv) Oxidation and reduction, 198 —
  c. Properties of various pyrimidine types 199
    (i) Pyrimidine, alkyl- and aryl-pyrimidines, 199 — (ii) Halogenopyrimidines, 201 — (iii) Nitro- and nitroso-pyrimidines, 208 — (iv) Pyrimidinamines, 209 — (v) Hydroxy- and alkoxy-pyrimidines, 213 — (vi) Carboxylic acids and derivatives, 218 — (vii) Sulphur-containing pyrimidines, 221 — (viii) Metallopyrimidines, 225 — (ix) Pyrimidine-N-oxides, 227 —
5. Quinazolines 227
  a. General chemical and physical properties 229
  b. Synthesis of quinazolines 231
  c. Chemical reactions of quinazolines 234
  d. Chemical reactivity of substituted quinazolines 235

Chapter 44. Pyrazines and Related Ring Structures
by K.J. McCullough

1. Pyrazines, 1,4-diazines 241
  a. Introduction 241
  b. Physical and spectroscopic properties 243
  c. General methods of synthesis 245
    (i) Self-condensation of α-aminocarbonyl compounds, 245 — (ii) Condensation of α,β-diamines with α,β-carbonyl compounds 247 — (iii) Miscellaneous condensation and dimerisation procedures, 249 — (iv) Dehydrogenation of piperazines, 251 — (v) Cleavage of quinoxalines and pteridines, 252 —

  d. Pyrazine, its homologues and derivatives 252
    (i) Pyrazine, alkyl- and aryl-pyrazines, 252 — (ii) Halopyrazines, 257 — (iii) Aminopyrazines, 259 — (iv) Hydroxypyrazines, 261 — (v) Pyrazinecarboxylic acids, 270 —
  e. Reduced pyrazines 275
    (i) Dihydropyrazines, 275 — (ii) Tetrahydropyrazines, 283 — (iii) Piperazines (hexahydropyrazines), 287 — (iv) Piperazinones (ketopiperazines), 294 —

2. Quinoxalines (benzopyrazines) 301
   a. Introduction 301
   b. Physical and spectroscopic properties 301
   c. Methods of synthesis 302
   d. Quinoxaline, its homologues and derivatives 310
      (i) Quinoxaline and its alkyl and aryl derivatives, 310 — (ii) Halogenated quinoxalines, 319 — (iii) Nitroquinoxalines, 323 — (iv) Aminoquinoxalines, 324 — (v) Hydroxyquinoxalines (quinoxalin-2(1H)-ones), 327 — (vi) Quinoxaline-N-oxides, 334 — (vii) Polyhydroxyalkylquinoxalines, 342 — (viii) Quinoxalinecarboxaldehydes, 344 — (ix) Quinoxaline-2-carboxylic acids, 345 —
   e. Reduced quinoxalines 349
      (i) Dihydroquinoxalines, 349 — (ii) Tetrahydroquinoxalines, 350 — (iii) Decahydroquinoxalines, 353 —
3. Phenazines (dibenzopyrazines) 354
   a. Introduction 354
   b. Physical and spectroscopic properties 356
   c. General methods of synthesis 357
   d. Phenazine, its homologues and derivatives 363
      (i) Phenazine, alkyl- and aryl-phenazines, 363 — (ii) Benzo- and dibenzo-phenazines, 367 — (iii) Phenazine-N-oxides, 369 — (iv) Halogenated phenazines, 375 — (v) Nitrophenazines and phenazinesulphonic acid, 378 — (vi) Aminophenazines, 381 — (vii) Hydroxyphenazines (phenazinols), 384 — (viii) Phenazinecarboxylic acids, 390 —
   e. Reduced phenazines 391
      (i) 5,10-Dihydrophenazines, 392 — (ii) 1,2,3,4-Tetrahydrophenazines, 397 — (iii) Octahydrophenazines, 398 — (iv) Decahydrophenazine, 400 — (v) Tetradecahydrophenazines, 401—

Chapter 45. Phenazine, Oxazine, Thiazine and Sulphur Dyes
by N. HUGHES

1. Introduction 403
2. Phenazine dyes 404
   a. Eurhodines and eurhodols 405
   b. Aposafranines 407
   c. Safranines 409
   d. Fluorindines 412
   e. Indulines and nigrosines 414
   f. Aniline black (C.I. Pigment Black 1, 50440) 416
3. Phenoxazine dyes 417
   a. Basic type 417
   b. Gallocyanines 420
   c. Triphendioxazines 422
4. Phenothiazine dyes 427
5. Sulphur dyes 430
   a. Constitution of the sulphur dyes 433

Chapter 46. Quinazoline Alkaloids
by A. McKillop, M. Sainsbury and B.P. Swann

1. Simple quinazolines 437
2. Vasicine and related alkaloids 440
3. Febrifugine and isofebrifugine 449
4. The indoloquinazoline alkaloids 451

Chapter 47. Compounds Containing a Six-Membered Ring with more than Two Hetero-Atoms
by D.G. Neilson and D. Hunter

1. Triazines 457
   a. 1,2,3-Triazines 457
      (i) Simple derivatives, 457 — (ii) Fused 1,2,3-triazines, 459 —
   b. 1,2,4-Triazines 461
      (i) Amino-1,2,4-triazines, 462 — (ii) 1,2,4-Triazinecarboxylic acids, 462 — (iii) Halogen-substituted 1,2,4-triazines, 463 — (iv) Hydroxy-1,2,4-triazines (triazinones), 463 — (v) 1,2,4-Triazine-N-oxides, 464 — (vi) Triazinethiones, 464 — (vii) Reduced 1,2,4-triazines, 465 — (viii) 1,2,4-Benzotriazines (α-phenotriazines), 465 — (ix) 1,2,4-Benzotriazines and related species with reduced ring systems, 467 — (x) Compounds containing 1,2,4-triazines fused to polycyclic and heterocyclic systems, 467 —
   c. 1,3,5-Triazines 468
      (i) Synthesis, 468 — (ii) 1,3,5-Triazine and its alkyl- and aryl-substituted derivatives, 471 — (iii) Amino-substituted 1,3,5-triazines (melamines), 472 — (iv) Carbonyl derivatives and carboxylic acids, 472 — (v) Cyanuric acid (2,4,6-trihydroxy-1,3,5-triazine); trithiocyanuric acid, 473 — (vi) Halogen-substituted 1,3,5-triazines — cyanuric halides, 473 — (vii) N-oxides and azides, 475 — (viii) Compounds with a fully reduced 1,3,5-triazine ring, 475 — (ix) Fused ring 1,3,5-triazines, 475 —
2. Thiadiazines 476
   (i) 1,2,3-Thiadiazines, 476 — (ii) 1,2,4-Thiadiazines, 476 — (iii) 1,2,5-Thiadiazines, 476 — (iv) 1,2,6-Thiadiazines, 476 — (v) 1,3,4-Thiadiazines, 477 — (vi) 1,3,5-Thiadiazines, 477 —
3. Oxadiazines 478
   (i) 1,2,3-Oxadiazines, 478 — (ii) 1,2,4-Oxadiazines, 479 — (iii) 1,2,5-Oxadiazines, 479 — (iv) 1,2,6-Oxadiazines, 479 — (v) 1,3,4-Oxadiazines, 479 — (vi) 1,3,5-Oxadiazines, 480 —
4. Dithiazines and dioxazines 481
   (i) Dithiazines, 481 — (ii) Dioxazines, 482 —
5. Oxathiazines 483
   (i) 1,2,3-Oxathiazines, 483 — (ii) 1,3,5-Oxathiazines, 484 — (iii) Other oxathiazine systems, 485 —
6. Trithianes, trioxanes, oxadithianes and dioxathianes 485
7. Tetrazines 486
   a. 1,2,3,4-Tetrazines 487
   b. 1,2,3,5-Tetrazines 488
   c. 1,2,4,5-Tetrazines 489

(i) General preparations and properties, 489 — (ii) Reduced 1,2,4,5-tetrazines, 491 — (iii) Properties of 1,2,4,5-tetrazines, 492 — (iv) Verdazyls, 494 —

8. Thiatriazines and oxatriazines 495
(i) 2H-1,2,3,6-Thiatriazines, 495 — (ii) 1H-1,2,4,6-Thiatriazines, 496 — (iii) 2H-1,2,4,6-Thiatriazines, 496 — (iv) Other systems, 497 — (v) Oxatriazines, 498 —
9. Dithiadiazines, oxathiadiazines, dioxadiazines and dioxathiazines 498
(i) Dithiadiazines, 498 — (ii) Oxathiadiazines, 498 — (iii) Dioxadiazines, 499 — (iv) Dioxathiazines, 500 —
10. Tetrathianes 500
11. Pentazines 501
12. Other ring systems with five hetero-atoms (N, O, S) 501

Index 503

Titles of other parts of Volume IV

HETEROCYCLIC COMPOUNDS

Vol. IV A: Three-, four- and five-membered heterocyclic compounds with a single hetero-atom in the ring

Vol. IV B: Five-membered heterocyclic compounds with a single hetero-atom in the ring: alkaloids, dyes and pigments

Vol. IV C: Five-membered heterocyclic compounds with two hetero-atoms in the ring from Groups V and/or VI of the Periodic Table

Vol. IV D: Five-membered heterocyclic compounds with more than two hetero-atoms in the ring

Vol. IV E: Six-membered monoheterocyclic compounds containing oxygen, sulphur, selenium, tellurium, silicon, germanium, tin, lead or iodine as the hetero-atom

Vol. IV F: Six-membered heterocyclic compounds with a single nitrogen atom in the ring: pyridine, polymethylenepyridines, quinoline, isoquinoline and their derivatives

Vol. IV G: Six-membered heterocyclic compounds with a single nitrogen atom in the ring to which are fused two or more carbocyclic ring systems, and six-membered ring compounds where the hetero-atom is phosphorus, arsenic, antimony or bismuth. Alkaloids containing a six-membered heterocyclic ring system

Vol. IV H: Six-membered heterocyclic compounds with (a) a nitrogen atom common to two or more fused rings; (b) one hetero-atom in each of two fused rings. Six-membered ring compounds with two hetero-atoms from Groups VI B, or V B and VI B of the Periodic Table, respectively. Isoquinoline, lupinane and quinolizidine alkaloids

Vol. IV I: Six-membered heterocyclic compounds with two hetero-atoms from Group V of the Periodic Table: the Pyridazine and Pyrimidine groups

Vol. IV J: Six-membered heterocyclic compounds with two hetero-atoms from Group V of the Periodic Table: the Pyrazine group. Phenoxazine, phenothiazine, phenazine and sulphur dyes. Six-membered heterocyclic compounds with three or more hetero-atoms

Vol. IV K: Six-membered heterocyclic compounds with two or more hetero-atoms one or more of which are from Groups II, III, IV, V or VII of the Periodic Table. Heterocyclic compounds with seven or more atoms in the ring

Vol. IV L: Fused-ring heterocyclic compounds containing three or more nitrogen atoms; purines and related ring systems, nucleosides, nucleotides and nucleic acids; pteridines, alloxazines, flavins and related compounds. The biosynthesis of plant alkaloids and nitrogenous microbial metabolites

OFFICIAL PUBLICATIONS

B.P.	British (United Kingdom) Patent
F.P.	French Patent
G.P.	German Patent
Ger. Offen.	German Patent Application, open for inspection
Sw. P.	Swiss Patent
U.S.P.	United States Patent
U.S.S.R.P.	Russian Patent
B.I.O.S.	British Intelligence Objectives Sub-Committee Reports, H.M. Stationery Office, London.
C.I.O.S.	Combined Intelligence Objectives Sub-Committee Reports
F.I.A.T.	Field Information Agency, Technical Reports of U.S. Group Control Council for Germany
B.S.	British Standards Specification
A.S.T.M.	American Society for Testing and Materials
A.P.I.	American Petroleum Institute Projects
C.I.	Colour Index Number of Dyestuffs and Pigments

SCIENTIFIC JOURNALS AND PERIODICALS

With few obvious and self-explanatory modifications the abbreviations used in references to journals and periodicals comprising the extensive literature on organic chemistry, are those used in the World List of Scientific Periodicals.

LIST OF ABBREVIATED NAMES OF CHEMICAL FIRMS MENTIONED IN PATENT REFERENCES

A.G.F.A., Agfa A.G.	Aktiengesellschaft für Anilinfabrikation (Berlin)
B.A.S.F.	Badische Anilin- und Soda-Fabrik (Ludwigshafen)
Bayer	Farbenfabriken vorm. Friedrich Bayer und Co. (Leverkusen)
Cassella	Leopold Cassella und Co. (Frankfurt am Main)
C.F.M.	Compagnie française des Matières Colorantes (Paris)
CIBA	Gesellschaft für chemische Industrie (Basel)
Du Pont	E.I. Du Pont de Nemours and Co. (U.S.A.)
G.A.F.	General Anilin and Film Corporation (U.S.A.)
Geigy A.G.	J.R. Geigy S.A. (Basel)
Hoechst	Hoechst A.G. (see M.L.B.)
I.C.I.	Imperial Chemical Industries, Ltd. (London)
I.G.	(= Interessen Gemeinschaft Farbenindustrie) of the principal dyestuffs manufacturers in Germany
Kalle	Kalle und Co., A.G. (Biebrich am Rhein)
M.L.B.	Farbwerke vormals Meister, Lucius und Brüning (Hoechst)
Sandoz	Sandoz A.G. Chemische Fabrik (Basel)

LIST OF COMMON ABBREVIATIONS AND SYMBOLS USED

A	acid
Å	Ångström units
Ac	acetyl
a	axial
as, asymm.	asymmetrical
at.	atmosphere
B	base
Bu	butyl
b.p.	boiling point
C, mC and μC	curie, millicurie and microcurie
c, C	concentration
c.d.	circular dichroism
conc.	concentrated
crit.	critical
D	Debye unit, 1×10^{-18} e.s.u.
D	dissociation energy
D	dextro-rotatory; dextro configuration
DL	optically inactive (externally compensated)
d	density
dec. or decomp.	with decomposition
deriv.	derivative
E	energy; extinction; electromeric effect
$E1$, $E2$	uni- and bi-molecular elimination mechanisms
E1cB	unimolecular elimination in conjugate base
e.s.r.	electron spin resonance
Et	ethyl
e	nuclear charge; equatorial
f	oscillator strength
f.p.	freezing point
G	free energy
g.l.c.	gas liquid chromatography
g	spectroscopic splitting factor, 2.0023
H	applied magnetic field; heat content
h	Planck's constant
Hz	hertz
I	spin quantum number; intensity; inductive effect
i.r.	infrared
J	coupling constant in n.m.r. spectra
K	dissociation constant
k	Boltzmann constant; velocity constant
kcal	kilocalories
L	laevorotatory; laevo configuration
M	molecular weight; molar; mesomeric effect
Me	methyl

LIST OF COMMON ABBREVIATIONS

m	mass; mole; molecule; *meta-*
ml	millilitre
m.p.	melting point
Ms	mesyl (methanesulphonyl)
[M]	molecular rotation
N	Avogadro number; normal
n.m.r.	nuclear magnetic resonance
N.O.E.	Nuclear Overhauser Effect
n	normal; refractive index; principal quantum number
o	*ortho-*
o.r.d.	optical rotatory dispersion
P	polarisation; probability; orbital state
Pr	propyl
Ph	phenyl
p	*para-*; orbital
p.m.r.	proton magnetic resonance
R	clockwise configuration
S	counterclockwise config.; entropy; net spin of incompleted electronic shells; orbital state
S_N1, S_N2	uni- and bi-molecular nucleophilic substitution mechanisms
S_Ni	internal nucleophilic substitution mechanisms
s	symmetrical; orbital
sec	secondary
soln.	solution
symm.	symmetrical
T	absolute temperature
Tosyl	*p*-toluenesulphonyl
Trityl	triphenylmethyl
t	time
temp.	temperature (in degrees centigrade)
tert	tertiary
U	potential energy
u.v.	ultraviolet
v	velocity
α	optical rotation (in water unless otherwise stated)
$[\alpha]$	specific optical rotation
α_A	atomic susceptibility
α_E	electronic susceptibility
ε	dielectric constant; extinction coefficient
μ	microns (10^{-4} cm); dipole moment; magnetic moment
μ_B	Bohr magneton
μg	microgram (10^{-6} g)
λ	wavelength
v	frequency; wave number

χ, χ_d, χ_μ	magnetic, diamagnetic and paramagnetic susceptibilities
~	about
(+)	dextrorotatory
(−)	laevorotatory
⊖	negative charge
⊕	positive charge

Chapter 42

Pyridazines, Cinnolines, Benzocinnolines and Phthalazines

R.N. HUNSTON, J. PARRICK and C.J.G. SHAW

1. Pyridazines

(a) Introduction

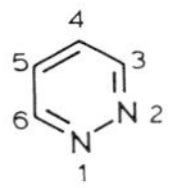

(I)

Six-membered aromatic systems with two ring nitrogen atoms are known as diazines, and the 1,2-diazine isomer is pyridazine (I). Although the first pyridazines were obtained in the late nineteenth century, interest in these compounds was not initially as great as that in the other diazines, since the pyridazines do not occur naturally. However, in recent years pyridazines have received much more attention, both from the theoretical point of view and because some pyridazine derivatives have been found to have biological activity.

The chemistry of pyridazines has been reviewed by *M. Tišler* and *B. Stanovnik*, Adv. heterocyclic Chem., 1968, **9**, 211 and 1979, **24**, 363, and in Comprehensive Heterocyclic Chemistry, Vol. 3, Part 2B, eds. *A.R. Katritzky* and *C.W. Rees*, Pergamon, Oxford, 1984, p. 1.

(b) Methods of synthesis

(1) From 1,4-keto-acids

Saturated 1,4-keto-acids and their esters react with hydrazines to form 4,5-dihydro-3(2H)-pyridazinones (II), which can be oxidised by bromine in glacial acetic acid to the corresponding 3(2H)-pyridazinones (III). Unsaturated 1,4-keto-acids yield (III) directly.

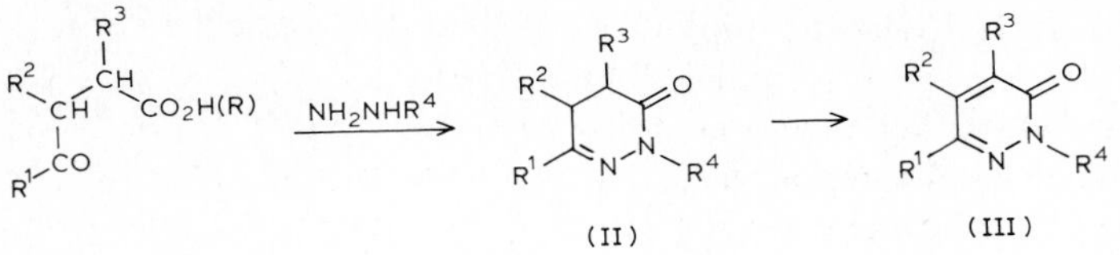

(2) From 1,4-diketones

Unsaturated 1,4-diketones and hydrazine form pyridazines. The reactions are usually performed in the presence of mineral acid, otherwise N-aminopyrroles may be formed (*G. Rio* and *A. Lecas-Nawrocka*, Bull. Soc. Chim. Fr., 1974, 2824). The *cis*-form of the diketone is required in the condensation but the *trans*-form may rearrange during the reaction to give lower yields of the pyridazine (*J.A. Hirsch* and *H.J. Szur*, J. heterocyclic Chem., 1972, **9**, 523).

Saturated 1,4-diketones and hydrazine give dihydropyridazines. Some of the latter are not particularly stable and are dehydrogenated in the presence of air or during distillation, into the more stable pyridazines: thus acetonylacetone gives a good yield of 3,6-dimethylpyridazine (*B.G. Zimmerman* and *H.L. Lochte*, J. Amer. chem. Soc., 1938, **60**, 2456).

(3) From 1,2-dicarbonyl compounds

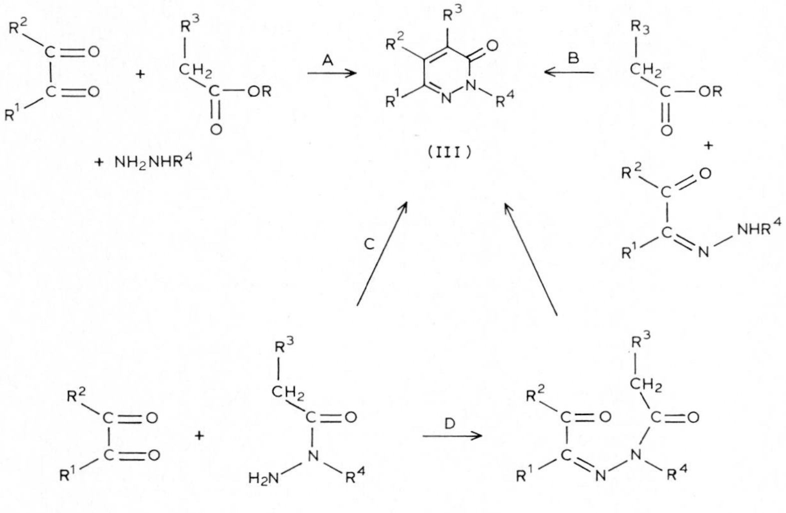

1,2-Dicarbonyl compounds (1,2-diketones, α-keto-acids, glyoxal) react with esters containing a reactive α-methylene group (malonic, acetoacetic, cyanoacetic, benzoylacetic, hippuric esters), and a hydrazine in the presence of sodium ethoxide to form 3(2H)-pyridazinones (III) (route A). However, the preferred synthetic method is either first to make the monohydrazone of the 1,2-dicarbonyl compound (particularly for aromatic diketones) and then condense this with the ester containing the reactive methylene group (route B), or prepare the acid hydrazide and condense this with the 1,2-dicarbonyl compound, when in the presence of sodium ethoxide the pyridazinone is formed directly (route C), whereas in the absence of base the hydrazone is formed (route D), which can then be cyclised in a separate step. Routes C and D are particularly useful for aliphatic dicarbonyl compounds (*P. Schmidt* and *J. Druey*, Helv., 1954, **37**, 134, 1467).

(4) From 1,2-dicarboxylic acid anhydrides

Maleic anhydride or its substituted derivatives react with hydrazines to give either the corresponding pyridazinone (V) directly or the intermediate 3-carboxyacryloylhydrazine (IV), which can then be cyclised on heating. Hydrazines with strongly electron-donating groups form the pyridazinones directly. The intermediates (IV), when dehydrated in acid media, give either *N*-aminomaleimides (VI) or pyridazinones (V). The former are isomerised in acid to pyridazinones, and the formation of (VI) can be prevented if the condensation is carried out in strongly acid solution.

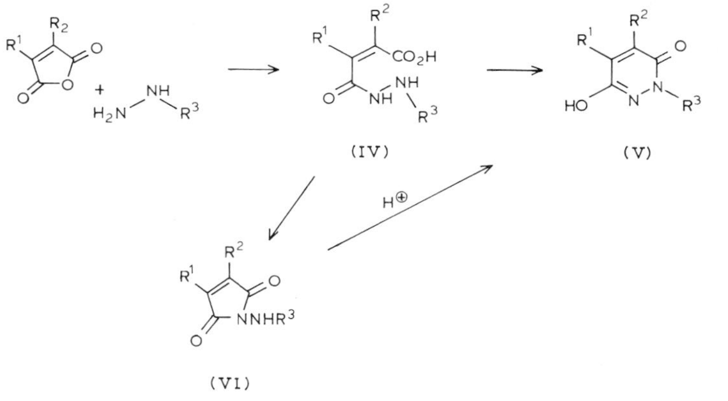

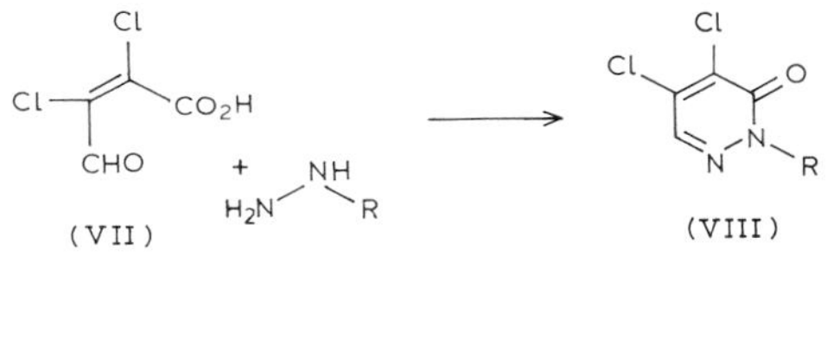

3-Formylacrylic acids or esters, such as mucochloric acid (VII) react with hydrazines to give 4,5-dichloropyridazinones (VIII) (*T. Kuraishi*, Chem. pharm. Bull. Japan, 1956, **4**, 497; *J.K. Landquist* and *S.E. Merck*, Chem. Ind., 1970, 688).

(5) Cycloaddition reactions

Pyridazines, particularly 1,2,3,6-tetrahydropyridazines (X), can be prepared by [4 + 2] cycloaddition reactions. An azo dienophile such as a dialkyl azodicarboxylate (IX) reacts with a conjugated diene to give (X).

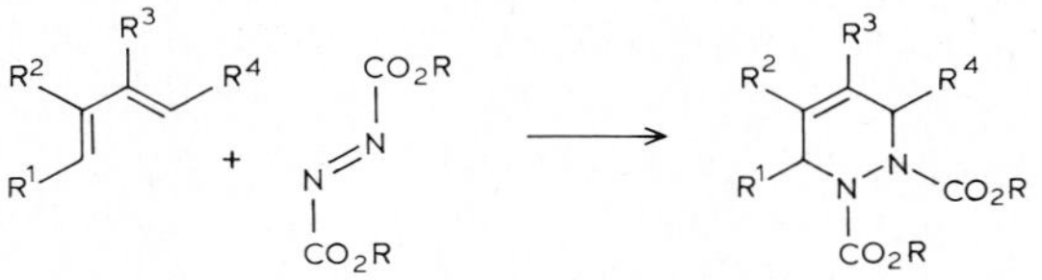

Other dienophiles employed include the tetraphenyl ester of azodiphosphoric acid (*J.L. Miesel*, Tetrahedron Letters, 1974, 3847) and hexafluoroacetone azine (*S.E. Armstrong* and *A.E. Tipping*, J. fluorine Chem., 1973, **3**, 119).

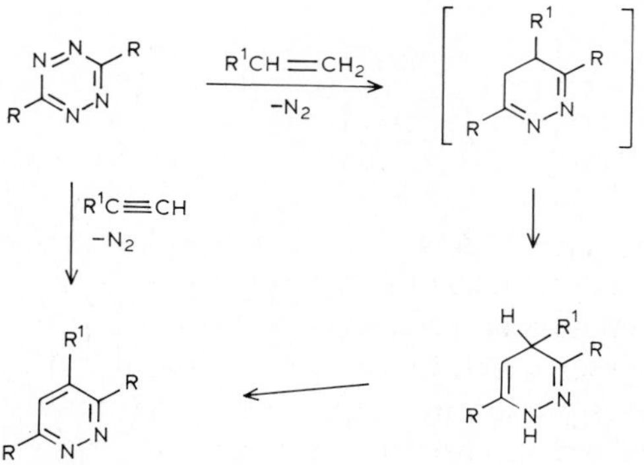

The cycloaddition reaction between 1,2,4,5-tetrazines (with strongly electrophilic substituents at positions 3 and 6) and alkenes produces 1,4-dihydropyridazines (XI) which are easily oxidised to pyridazines (XII). The latter are obtained directly from alkynes which are, however, less reactive and give lower yields (*R.A. Carboni* and *R.V. Lindsey*, J. Amer. chem. Soc., 1959, **81**, 4342; *J. Sauer et al.*, Ber., 1965, **98**, 1435). Pyridazines can also be prepared from diazo compounds and cyclopropenes by [3 + 3] cycloaddition reactions. The cycloadducts (XIII) formed first, rearrange in acid or alkali to pyridazines (*M. Franck-Neumann* and *C. Buchecker*, Tetrahedron Letters, 1969, 2659; *M.I. Komendantov*, *R.R. Beknukhametov* and *V.G. Novinskii*, Zhur. org. Khim., 1976, **12**, 801).

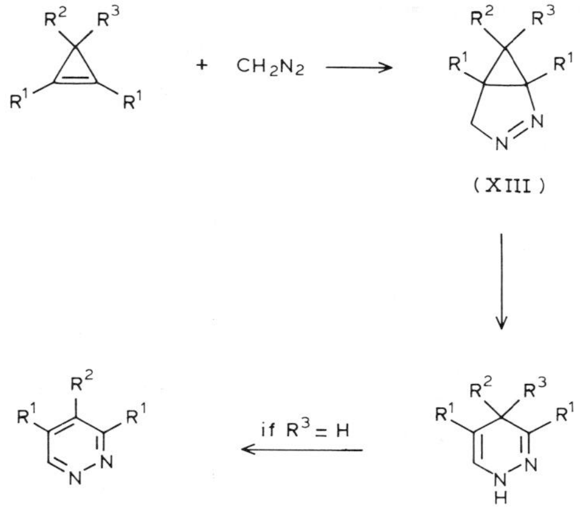

(6) From other heterocyclic compounds

Furans react with bromine in methanol to give dialkoxy-2,5-dihydrofurans (XIV). The latter undergo ring opening on treatment with acid, and subsequent reaction with hydrazine gives the corresponding pyridazine (*N. Clauson-Kaas*, *S.-O. Li* and *N. Elming*, Acta Chem. Scand., 1950, **4**, 1233).

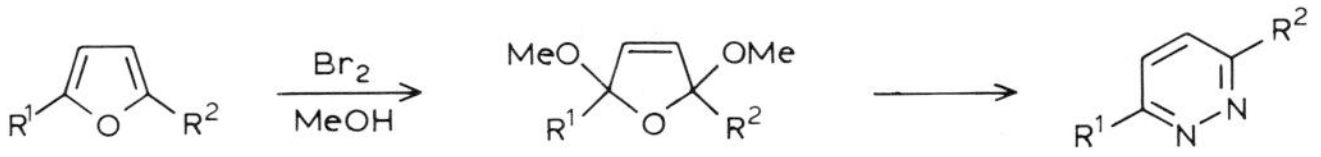

1-Aminopyrrolidines undergo ring expansion in the presence of silica gel and chloroform to give 1,4,5,6-tetrahydropyridazines (*N. Viswanathan* and *A.R. Sidhaye*, Tetrahedron Letters, 1979, 5025).

Other heterocycles which have been converted into pyridazines include pyrroles, isoxazoles, pyrans, pyridines, and 1,2-diazepines (see *H.C. Van der Plas*, Ring Transformations of Heterocycles, Academic Press, New York, NY, 1973).

(7) From carbohydrates

A number of syntheses of pyridazines from monosaccharides have been described (*K. Imada*, Chem. pharm. Bull. Japan, 1974, **22**, 1732; *P. Smit, G.A. Stork* and *H.C. Van der Plas*, J. heterocyclic Chem., 1975, **12**, 75, 957).

(c) Pyridazine and its alkyl and aryl derivatives

(i) Pyridazine

Pyridazine is a colourless liquid, m.p. $-8°$, b.p. $208°/760$ mm, $87°/14$ mm, $48°/1$ mm, prepared by (a) the acid induced ring opening of 2,5-diacetoxy- or 2,5-dimethoxy-2,5-dihydrofuran followed by reaction with hydrazine (*Clauson-Kaas, Li* and *Elming*, loc. cit.; *R. Letsinger* and *R. Lasco*, J. org. Chem., 1956, **21**, 764) (yield 60–67%); (b) the reaction of maleic dialdehyde, produced by the decomposition of the cycloaddition adduct of furan and diethyl azodicarboxylate, with hydrazine (*K.N. Zelenin* and *I.P. Bežan*, Zhur. org. Khim., 1966, **2**, 1524) (yield 55%); (c) catalytic hydrogenation of 3-chloro- or 3,6-dichloro-pyridazine (*R.H. Mizzoni* and *P.E. Spoerri*, J. Amer. chem. Soc., 1951, **73**, 1873; *T. Itai* and *H. Tgeta*, Yakugaku Zasshi, 1954, **74**, 1195) (yield 60–67%).

Pyridazine is a weak base, pK_a 2.33 (*cf*. pyridine 5.23). Among diazines, pyridazine has a relatively high pK_a value (*cf*. pyrimidine 1.30, pyrazine 0.6): this has been attributed to resonance stabilisation of the pyridazinium ion (*A. Albert, R.J. Goldacre* and *J. Phillips*, J. chem. Soc., 1948, 2240). The heterocycle acts as a mono-acidic base in forming salts: *monohydrochloride*, m.p. 161–163°, *picrate*, m.p. 170–175° (dec.), *chloroaurate*, m.p. 110° (dec.), *methiodide*, m.p. 95–96° (dec.). It has a pyridine-like smell and is completely miscible with water, alcohols, benzene and ether, but is insoluble in light-petroleum and in cyclohexane.

The pyridazine ring is planar and is best considered as a resonance hybrid of structures (I) and (Ia) with (I) making the greater contribution.

(I)   (Ia)

This is supported by evidence from electron diffraction measurements, X-ray structural analysis and microwave spectroscopy data, which all show that the N–N bond has some single bond character. Bond lengths and angles for pyridazine have been calculated (*A. Almenningen et al.*, Acta Chem. Scand., 1977, **31A**, 63). A number of theoretical calculations have been

bond lengths in Å

performed on the pyridazine molecule: these include electronic structure (*A. Toth* and *E. Dudar*, Acta chim. Acad. Sci. Hung., 1973, **77**, 69; *T. Yonezawa, H. Katô* and *H. Kato*, Theor.

TABLE 1

CHEMICAL SHIFTS AND COUPLING CONSTANTS OF PYRIDAZINE DERIVATIVES

Compound	Chemical shifts (δ, ppm)						Coupling constants (Hz)					
	H-3	H-4	H-5	H-6	Me-3	Me-4	$J_{3,5}$	$J_{3,6}$	$J_{4,5}$	$J_{4,6}$	$J_{5,6}$	$M_{Me-4,5}$
3-Methylpyridazine	–	7.38	7.40	9.06	2.74	–	–	–	8.6	1.8	4.7	–
4-Methylpyridazine	9.08	–	7.33	9.04	–	2.40	2.2	3.0	–	–	5.0	1.0

performed on the pyridazine molecule: these include electronic structure (*A. Toth* and *E. Dudar*, Acta chim. Acad. Sci. Hung., 1973, **77**, 69; *T. Yonezawa, H. Katô* and *H. Kato*, Theor. Chim. Acta, 1969, **13**, 125); polarisability (*M.A. Kovner et al.*, Opt. Spektrosk., 1970, **29**, 523); π-bond orders (*G. Häfelinger*, Ber., 1970, **103**, 3289; *G. Häfelinger*, Tetrahedron, 1971, **27**, 1635); and Dewar structure (*Z. Latajka et al.*, J. mol. Struct., 1975, **28**, 323).

The resonance energy of pyridazine has been calculated from heats of combustion to be 12.3 kcal mol^{-1} (*J. Tjebbes*, Acta chem. Scand., 1962, **16**, 916). The ionisation potential has been calculated and experimentally measured as 9.86 eV (*T. Nakajima* and *A. Pullmann*, Compt. rend., 1958, **246**, 1047). Also reported are the dipole moment, molar Kerr constant, molar Cotton–Mouton constant, and magnetic anisotropy (*M.R. Battaglia* and *G.L.D. Ritchie*, J. chem. Soc. Perkin II, 1977, 897).

(ii) Spectra

The ultraviolet absorption spectra of pyridazine and its anion and cation have been recorded in different solvents and the results discussed (*S.F. Mason*, J. chem. Soc., 1959, 1240).

The infrared and Raman spectra of pyridazine have been reported and detailed assignments made (*R.C. Lord, A.L. Marston* and *F.A. Miller*, Spectrochim. Acta, 1957, **9**, 113).

The ^{1}H-n.m.r. spectrum of pyridazine has been analysed as an A_2B_2 system. The chemical shifts are strongly concentration dependent, in a sense opposite to that normally encountered with aromatic compounds, although the coupling constants are virtually invariant (*J.A. Elvidge* and *P.D. Ralph*, J. chem. Soc. B, 1966, 249) (Tables 1 and 2).

The effects of lanthanide shift reagents on the proton resonances of pyridazine have been studied (*W.L.F. Armarego, T.J. Batterham* and *J.R. Kershaw*, Org. magn. Resonance, 1971, **3**, 575).

The ^{13}C-chemical shifts of pyridazine and its cation have been calculated (*W. Adams, A. Grimison* and *G. Rodriguez*, J. chem. Phys., 1969, **50**, 645), measured as a function of pH (*E. Breitmaier* and *K.H. Spohn*, Tetrahedron, 1973, **29**, 1145), and later improved (*G. Pouzard* and *M. Rajzmann*, Org. magn. Resonance, 1976, **8**, 271).

Also recorded are the ^{14}N-chemical shifts for pyridazine (*M. Witanowski et al.*, Tetrahedron, 1971, **27**, 3129), the ^{14}N-quadrupole resonance spectrum (*L. Guibe* and *E.A.C. Lucken*, Mol. Phys., 1968, **14**, 79), and the ^{15}N-chemical shifts (*W. Städeli et al.*, Helv., 1980, **63**, 504).

The e.s.r. spectrum of the pyridazine radical anion has been studied (*C.A. McDowell, K.F. Paulus* and *J.R. Rowlands*, Proc. chem. Soc., 1962, 60; *E.W. Stone* and *A.H. Maki*, J. chem. Phys., 1963, **39**, 1635; *R.L. Ward*, J. Amer. chem. Soc., 1962, **84**, 332; *J. Komenda* and *A. Novak*, C.A., 1972, **76**, 112509).

The mass spectrum of pyridazine shows a base peak at m/z 80 (molecular ion) with

TABLE 2

CONCENTRATION DEPENDENCE OF CHEMICAL SHIFTS AND COUPLING CONSTANTS OF PYRIDAZINE

Solvent concentration (%)		δ_{H-3}	δ_{H-4}	$J_{3,4}$	$J_{3,5}$	$J_{3,6}$	$J_{4,5}$
Neat	–	9.691	8.033	5.0	2.1	1.4	8.6
CDCl$_3$	20	9.158	7.483	5.2	1.9	1.4	8.6

fragment ions at m/z 52(20%), 51(20%), 50(10%), and 26(13%). High resolution measurements reveal that the peak at m/z 52 is a doublet composed of $C_4H_4^{+\cdot}$ (73.5%) and $C_3H_2N^{+\cdot}$ (26.5%), and the peaks at m/z 51 and 50 to be due to $C_4H_3^{\oplus}$ and $C_4H_2^{\oplus}$, respectively. The peak at m/z 26 is due to the dication $C_4H_4^{2\oplus}$, most probably the cyclobutadienyl dication (XVII). $C_4H_4^{+\cdot}$ is either the cyclobutadienyl cation radical (XV) or the tetrahedryl cation radical (XVI) (*M.H. Benn, T.S. Sorensen* and *A.M. Hogg*, Chem. Comm., 1967, 574).

$$\text{pyridazine}^+ \xrightarrow{-N_2} (XV) \text{ and/or } (XVI)$$

m/z 80 (100%)   (XV)   (XVI) m/z 52

$\downarrow -H_2CN\cdot$

$C_3H_2N^{\oplus}$ m/z 26
m/z 52
(XVII)

The ion kinetic energy spectrum of pyridazine has been studied (*J.H. Beynon, R.M. Caprioli* and *T. Ast*, Org. mass Spectrom., 1972, **6**, 273). Also reported for pyridazine are the photoelectron spectrum (*L. Asbrink et al.*, Int. J. mass Spectrom. Ion Phys., 1972, **8**, 229), fluorescence spectrum (*B.J. Cohen, H. Baba* and *L. Goodman*, J. chem. Phys., 1965, **43**, 2901), and magnetic circular dichroism spectrum (*A. Kaito, M. Hatano* and *A. Tajiri*, J. Amer. chem. Soc., 1977, **99**, 5241).

A number of analytical methods for the chromatographic separation and identification of pyridazines have been reported (*M. Tišler* and *B. Stanovnik*, Adv. heterocyclic Chem., 1979, **24**, 449).

(iii) Chemical reactivity

Pyridazine reacts readily with alkyl halides to give monoquaternary salts. Attempts to prepare diquarternary salts have only been successful with the more reactive oxonium salts such as trimethyloxonium borofluoride (*T.J. Curphey* and *K.S. Prasad*, J. org. Chem., 1972, **37**, 2259). Pyridazine is resistant to electrophilic substitution and also to oxidative attack at ring carbon. *N*-Oxidation is, however, easily achieved: treatment of pyridazine with hydrogen peroxide in glacial acetic acid or with peracids gives the 1-oxide (*T. Itai* and *S. Natsume*, Chem. pharm. Bull. Japan, 1963, **11**, 83). Pyridazine undergoes nucleophilic attack with Grignard reagents (at position 4) and organolithium compounds (generally at C-3); phenyllithium in ether for example adds to pyridazine at C-3, but in the presence of N,N,N',N'-tetramethylethylene diamine addition occurs at C-4 (*R.E. Van der Stoel* and *H.C. Van der Plas*, Rec. Trav. chim., 1978, **97**, 116).

Pyridazine is reduced by sodium in boiling ethanol to hexahydropyridazine together with 1,4-diaminobutane formed by N–N fission (*R. Marquis*, Compt. rend., 1903 **136**, 368).

(iv) Alkyl- and aryl-pyridazines
3-**Methylpyridazine**, b.p. 214.5°, 98°/7 mm, 64°/2.5 mm, hygroscopic (*hydrochloride*, m.p. 184°, *picrate*, m.p. 143–144°), is prepared by the reaction of *cis*-β-acetylacrolein with aqueous acid and hydrazine hydrate (*J.A. Hirsch* and *A.J. Szur*, J. heterocyclic Chem., 1972, **9**, 523); 4-**methylpyridazine**, b.p. 119°/27 mm, 98–100°/11 mm (*picrate*, m.p. 126–126.5°, *methiodide*, m.p. 64.5–66°), from 3,6-dichloro-4-methylpyridazine by reduction in the presence of palladium charcoal and ammonia (*R.H. Mizzoni* and *P.E. Spoerri*, J. Amer. chem. Soc., 1954, **76**, 2201); 3,4-**dimethylpyridazine**, m.p. 43–44°, b.p. 116–117°/11 mm (*picrate*, m.p. 176–177°), from 4-cyano-5,6-dimethyl-3(2*H*)-pyridazinone (*P. Schmidt* and *J. Druey*, Helv., 1954, **37**, 1467), or from 6-chloro-3,4-dimethylpyridazine (*R.H. Horning* and *E.D. Amstutz*, J. org. Chem., 1955, **20**, 707); 3,5-*dimethylpyridazine*, b.p. 106°/19 mm (*picrate*, m.p. 143–144°) (*J. Levisalles*, Bull. Soc. chim. Fr., 1957, 1004); 3,6-**dimethylpyridazine**; m.p. 34–35°, b.p. 214–215°, 55°/1 mm, hygroscopic (*hydrochloride*, m.p. 184°, *picrate*, m.p. 164°), from *cis*-hex-3-ene-2,5-dione and hydrazine hydrate (*Hirsch* and *Szur*, *loc. cit.*); 4,5-**dimethylpyridazine**, m.p. 58–59° (*picrate*, m.p. 166.5–167°), from 3,6-dichloro-4,5-dimethylpyridazine by reduction with phosphorus and hydriodic acid (*Horning* and *Amstutz*, *loc. cit.*); 3,4,6-*trimethylpyridazine*, m.p. 93–94°, b.p. 124°/16 mm (*picrate*, m.p. 135–136°) (*J. Levisalles*, Bull. Soc. chim. Fr., 1957, 1009); 3,4,5,6-*tetramethylpyridazine*, m.p. 95° (*picrolonate*, m.p. 190-191°) (*Levisalles*, *loc. cit.*); 3-*ethylpyridazine*, b.p. 103–104°/14 mm (*picrate*, m.p. 135°); 3-*n-propylpyridazine*, b.p. 108–109°/13 mm (*picrolonate*, m.p. 123–125° (dec.)); 3-*n-butylpyridazine*, b.p. 77–82°/1 mm (*picrolonate*, m.p. 134–135°) (*J. Levisalles*, Bull. Soc. chim. Fr., 1957, 997); 4-*ethyl-3-methylpyridazine*, b.p. 121°/17 mm (*picrolonate*, m.p. 182–184°); 5-*ethyl-3-methylpyridazine*, b.p. 125–126°/18 mm (*picrolonate*, m.p. 167–168°); 6-*ethyl-3-methylpyridazine*, b.p. 116°/20 mm (*picrate*, m.p. 105–106°); 5-*ethyl-4-methylpyridazine*, m.p. 33–36°, b.p. 133°/14 mm (*picrate*, m.p. 124°); 3,6-*dimethyl-4-ethylpyridazine*, b.p. 122°/16 mm (*picrate*, m.p. 133–134°); 5-*ethyl-3,4,6-trimethylpyridazine*, m.p. 91–93° (*picrolonate*, m.p. 168–169°) (*J. Levisalles*, Bull. Soc. chim. Fr., 1957, 1004, 1009); 3-*phenylpyridazine*, m.p. 105°, b.p. 330–332° (*picrate*, m.p. 130°, *methiodide*, m.p. 179°) (*R.L. Letsinger* and *R. Lasco*, J. org. Chem., 1956, **21**, 812; *N.A. Evans*, *R.B. Johns* and *K.R. Markham*, Austral. J. Chem., 1967, **20**, 713); 4-*phenylpyridazine*, m.p. 86–86.5° (*picrate*, m.p. 149–150°) (*Letsinger* and *Lasco*, *loc. cit.*; *J. Levisalles*, Bull. Soc. chim. Fr., 1957, 1004); 3,4-*diphenylpyridazine*, m.p. 106–107° (*P. Schmidt* and *J. Druey*, Helv., 1954, **37**, 134); 3,5-*diphenylpyridazine*, m.p. 144–145° (*H. Wasserman* and *J.B. Brous*, J. org. Chem., 1954, **19**, 515); 3,6-*diphenylpyridazine*, m.p. 221–222° (*F.G. Baddar*, *A. El-Haboshi* and *A.K. Fateen*, J. chem. Soc., 1965, 3342); 4,5-*diphenylpyridazine*, m.p. 164–164.5° (*R.W.H. Berry* and *A. Burawoy*, J. chem. soc. C, 1970, 1316); 3,4,6-*triphenylpyridazine*, m.p. 170° (*V. Sprio* and *P. Madonia*, Gazz., 1956, **86**, 101); 3-*methyl-6-phenylpyridazine*, m.p. 104–105° (*C. Paal* and *E. Dencks*, Ber., 1903, **36**, 491); 4-*methyl-3-phenylpyridazine*, oil (*A. Ohsawa*, *Y. Abe* and *H. Igeta*, Chem. pharm. Bull. Japan, 1978, **26**, 2550); 4-*methyl-5-phenylpyridazine*, m.p. 81–82° (*picrate*, m.p. 136–137°) (*E.J. Volker*, *M.G. Pleiss* and *J.A. Moore*, J. org. Chem., 1970, **35**, 3615); 5-*methyl-3-phenylpyridazine*, m.p. 95° (*G. Leclerc* and *C.G. Wermuth*, Bull. Soc. chim. Fr., 1971, 1752)

The ionisation constants of alkylated pyridazines depend on the number and position of the alkyl groups (*S.F. Mason*, J. chem. Soc., 1957, 1247)

The ^{1}H-n.m.r. spectrum of 3-methylpyridazine shows an ABX system, whereas that of 4-methylpyridazine has been analysed as an $ABXY_3$ system

(*K. Tori* and *M. Ogata*, Chem. pharm. Bull. Japan, 1964, **12**, 272) (see Table 1).

Loss of $N_2H^{\cdot}$ is an important fragmentation in the mass spectra of methylpyridazines (*J.H. Bowie et al.*, Austral. J. Chem., 1967, **20**, 2677; *H. Ogura et al.*, J. heterocyclic Chem., 1971, **8**, 391).

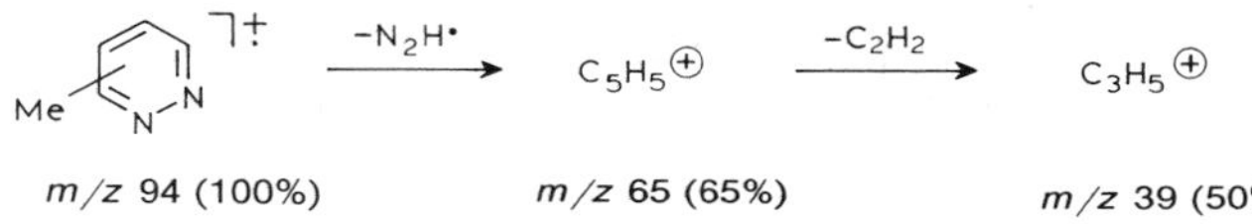

$$\underset{m/z\ 94\ (100\%)}{\text{Me}\begin{array}{c}\end{array}{\overset{\rceil \stackrel{+}{\cdot}}{\text{N-N}}}} \xrightarrow{-N_2H^{\cdot}} \underset{m/z\ 65\ (65\%)}{C_5H_5 \oplus} \xrightarrow{-C_2H_2} \underset{m/z\ 39\ (50\%)}{C_3H_5 \oplus}$$

Both 3- and 4-methylpyridazine quaternise with methyl iodide at both N-1 and N-2 (*M.S. Bale, A.B. Simmonds* and *W.F. Frager*, J. chem. Soc. B, 1966, 867).

Two isomeric *N*-oxides are usually formed by treatment of unsymmetrically substituted alkyl- or aryl-pyridazines with hydrogen peroxide in glacial acetic acid or with peracetic acid (*M. Ogata* and *H. Kano*, Chem. pharm. Bull. Japan, 1963, **11**, 29, 35). 3-Methylpyridazine is oxidised by selenium dioxide to give 3-formylpyridazine (*M. Kumagai*, Nippon Kagaku Zasshi, 1960, **81**, 492). A number of alkylpyridazines (but not 3-methylpyridazine) can be oxidised to pyridazine carboxylic acids, usually with potassium permanganate. Other substituents, such as halogen and methoxy groups, are usually unchanged. 3- And 4-methyl groups attached to pyridazine are activated by the electron-attracting ring nitrogens, and hence undergo aldol-type condensations with benzaldehyde, chloral and phthalic anhydride (*R.H. Mizzoni* and *P.E. Spoerri*, J. Amer. chem. Soc., 1954, **76**, 2201). Claisen condensations of 3- or 4-methylpyridazine with diethyl oxalate yield ethyl 3- or 4-pyridazinylpyruvate (*W.J. Haggerty, R.H. Springer* and *C.C. Cheng*, J. heterocyclic Chem., 1965, **2**, 1). 3-Methylpyridazine gives a Reissert compound (XVII) by treatment with trimethylsilyl cyanide and freshly distilled benzoyl chloride. With undistilled benzoyl chloride, however, bicyclic compounds (XIX) are formed (*S. Veeraraghavan, D. Bhattacharjee* and *F.D. Popp*, J. heterocyclic Chem., 1981, **18**, 443).

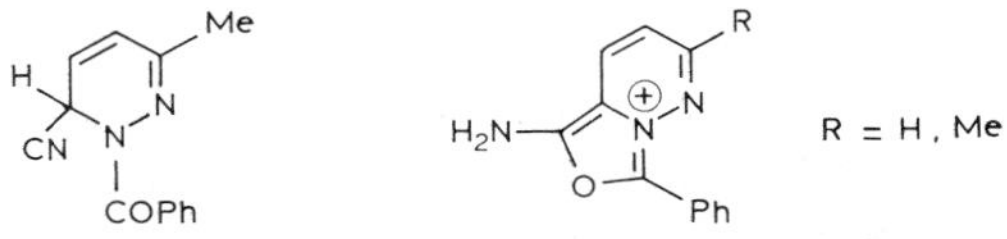

(XVIII)   (XIX)   R = H, Me

(d) Halogenopyridazines

The vast majority of compounds of this type are either 3- and/or 6-chloropyridazines, almost exclusively made from the corresponding pyridazinones by treatment with $POCl_3$ or PCl_5. The less frequently encountered 4- and 5-halopyridazines are usually prepared by the cyclisation of open-chain compounds.

3-**Chloropyridazine**, m.p. 29°, unstable at 0° even in a vacuum desiccator (*R.F. Homer et al.*, J. chem. Soc., 1948, 2195; *G. Farini* and *M. Simonetta*, Gazz., 1959, **89**, 2222); 3,6-**dichloropyridazine**, m.p. 68–69° (*P. Coad et al.*, J. org. Chem., 1963, **28**, 218); 3,4,5-**trichloropyridazine**, b.p. 117–118°/14–15 mm, prepared from diazomethane and tetrachlorocyclopropene (*H.M. Cohen*, J. heterocyclic Chem., 1967, **4**, 130); 3,4,6-**trichloropyridazine**, m.p. 57–57.5° (*Coad et al., loc. cit.*); 3,4,5,6-**tetrachloropyridazine**, m.p. 85–86°, made by chlorination of 3,6-dichloropyridazine with PCl_5 (*R.D. Chambers, J.A.H. MacBride* and *W.K.R. Musgrave*, Chem. Ind., 1966, 904); 3-*chloro-4-methylpyridazine*, m.p. 46.5–47.5° (*N. Takahayashi*, Chem. pharm. Bull. Japan, 1957, **5**, 229); 3-*chloro-5-methylpyridazine*, m.p. 139–140° (*N. Takahayashi, loc. cit.*); 3-*chloro-6-methylpyridazine*, m.p. 58° (*hydrochloride*, m.p. 250°, *hydrobromide*, m.p. 220° (dec.)); 3-*chloro-6-ethylpyridazine*, m.p. 47° (*C. Grundmann*, Ber., 1948, **81**, 1); 3-*chloro-6-phenylpyridazine*, m.p. 160° (*S. Gabriel* and *J. Colman*, Ber., 1899, **32**, 395); 3-*chloro-4,6-dimethylpyridazine*, m.p. 98–100° (*J. Levisalles*, Bull. Soc. chim. Fr., 1957, 1004); 4-*chloro-3,6-dimethylpyridazine*, m.p. 63° (*S. Sako*, Chem. pharm. Bull. Japan, 1963, **11**, 377); 6-*chloro-3,4-dimethylpyridazine*, m.p. 50–51° (*P. Schmidt* and *J. Druey*, Helv., 1954, **37**, 1467); 3-*chloro-4-ethyl-6-methylpyridazine*, m.p. 38–40°, b.p. 140–142°/12 mm (*Levisalles, loc. cit.*); 3-*Chloro-4,6-diphenylpyridazine*, m.p. 86–88° (*picrate*, m.p. 137–138°) (*G.K. Almström*, Ann., 1913, **400**, 131); 6-*chloro-3,4-diphenylpyridazine*, m.p. 111–112° (*T. Sasaki, K. Kanematsu* and *M. Murata*, J. org. Chem., 1971, **36**, 446); 3-*chloro-4-ethyl-6-phenylpyridazine*, m.p. 141° (*G. Leclerc* and *C.G. Wermuth*, Bull. Soc. chim. Fr., 1971, 1752); 3-*chloro-6-methyl-4-phenylpyridazine*, m.p. 110–112° (*Atkinson* and *Rodway, loc. cit.*); 3,6-*dichloro-4-methylpyridazine*, m.p. 83.5–84°, b.p. 149–151°/21 mm (*Coad et al., loc. cit.*); 3,6-*dichloro-4-phenylpyridazine*, m.p. 92° (*Levisalles, loc. cit.*); 3,6-*dichloro-4,5-dimethylpyridazine*, m.p. 120–121° (*E.A. Steck et al.*, J. Amer. chem. Soc., 1954, **76**, 3225); 3,6-*dichloro-5-ethyl-4-methylpyridazine*, m.p. 56–58° (*Levisalles, loc. cit.*); 3,6-*dichloro-4-methyl-5-phenylpyridazine*, m.p. 120–122° (*R.K. Bly, E.C. Zoll* and *J.A. Moore*, J. org. Chem. 1964, **29**, 2128); 3,6-*dichloro-4,5-diphenylpyridazine*, m.p. 224–225° (*R.W.H. Berry* and *A. Burawoy*, J. chem. Soc. C, 1970, 1316), 3-*bromopyridazine*, m.p. 73–74° (*Grundmann, loc. cit.*); 3-*bromo-6-methylpyridazine*, m.p. 78° (*Grundmann, loc. cit.*; 3,6-*dibromopyridazine*, m.p. 115–116° (*Coad et al., loc. cit.*); and 3,6-*dibromo-4-methylpyridazine*, m.p. 104–105° (*Coad et al., loc. cit.*).

Fluoro- and iodo-pyridazines are prepared by halogen exchange reactions. For example, 3,4,5,6-**tetrafluoropyridazine**, b.p. 117°, from tetrachloropyridazine and KF at 340° (*Chambers, MacBride* and *Musgrave, loc. cit.*); 3-**iodo-6-phenylpyridazine**, m.p. 169–170°, from the corresponding chloro-compound by short boiling with hydriodic acid and red phosphorus (*Gabriel* and *Colman, loc. cit.*); 3,6-**diiodopyridazine**, m.p. 157–158°, by treating the chloropyridazine with NaI in acetone in the presence of a catalytic amount of hydriodic acid (*Coad et al., loc. cit.*).

TABLE 3

^{1}H-N.M.R. SPECTRAL DATA

Pyridazine derivative	Chemical shifts (δ, ppm) (CDCl$_3$)							Coupling constants (Hz)				
	H-4	H-5	H-6	Me-4	Me-5	Me-6		$J_{4,5}$	$J_{4,6}$	$J_{5,6}$	$J_{Me-4,5}$	$J_{Me-5,4}$
3-Chloro-	7.59	7.55	9.17	–	–	–		8.8	1.8	4.7	–	–
3,6-Dichloro-	7.57	7.57	–	–	–	–		–	–	–	–	–
3-Chloro-4-methyl-	–	7.42	8.99	2.46	–	–		–	–	4.9	1.0	–
3-Chloro-5-methyl-	7.40	–	8.99	–	2.41	–		–	2.2	–	–	1.0
3-Chloro-6-methyl-	7.43	7.36	–	–	–	2.70		8.8	–	–	–	–

Bond lengths and angles for 3,6-dichloropyridazine have been calculated from electron diffraction and microwave data (*A. Almenningen et al.*, Acta Chem. Scand., 1977, **31A**, 63).

[Structure of 3,6-dichloropyridazine with bond lengths (Å) and angles: H−C bond 1.109; C−C bonds 1.377 and 1.373; C−Cl bond 1.717; C−N bond 1.334; N−N bond 1.339; angles 121.0°, 116.8°, 120.0°, 124.6°, 118.6°. bond lengths in Å]

The ^{1}H-n.m.r. spectra for some chloropyridazines are recorded in Table 3.

The halogenopyridazines have been converted into N-oxides (p. 18). The halogen atoms in monohalogenated pyridazines readily undergo substitution with a wide variety of nucleophiles such as ammonia, amines, hydrazines, inorganic bases, alkoxides, phenoxides, hydrogen sulphide and thiols. Halogen atoms at the 4- (or 5-) position are more reactive than those at the 3- (or 6-) position, but 3,6-dihalopyridazines are more reactive. In the latter only one halogen atom can be replaced by ammonia or primary amines, but secondary amines give 3,6-disubstituted pyridazines more easily. Unsymmetrically substituted 3,6-dihalopyridazines, when reacted with nucleophiles, give a mixture of two isomeric monosubstitution products, for example, 3,6-dichloro-4-methylpyridazine with ammonia gives a mixture of 6-amino-3-chloro-4-methyl- and 3-amino-6-chloro-4-methyl-pyridazine (p. 15).

In 3,4,5- and 3,4,6-trichloropyridazine, the chlorine at position 4 is usually the most reactive towards nucleophilic attack: exceptions occur when these compounds are refluxed with acetic acid, the 3,4,5-isomer yielding 4,6- and 5,6-dichloro-3($2H$)-pyridazinones, while the 3,4,6-compound gives 4,5-dichloro-3($2H$)-pyridazinone (p. 23).

(e) Nitropyridazines

Pyridazine is generally resistant to electrophilic substitution even in the presence of activating groups. Nevertheless some nitropyridazines have been prepared by direct nitration. For example, several 4-amino-3,6-disubstituted pyridazines have been nitrated with fuming nitric acid to give the corresponding 5-nitro derivatives in good yield (*M. Yanai et al.*, Chem. pharm. Bull. Japan, 1970, **18**, 1680). Pyridazine N-oxides on the other hand are relatively easily nitrated with mixed acids to give the 4-nitro derivatives. If the 4-position is occupied, the 5-nitro derivative is formed. Nitration of pyridazine *N*-oxides using acyl nitrates yields 3- and 5-nitro compounds (see p. 19). Nitropyridazines have been obtained from nitroamino compounds (p. 16), the latter, when heated with acid, undergoes a rearrangement in which the nitro group migrates to a neighbouring unsubstituted carbon of the pyridazine ring. In this way, 3,5-**diamino-4-nitropyridazine**, m.p. 291° (*W.D. Guither, D.G. Clark* and *R.N. Castle*, J. heterocyclic Chem., 1965, **2**, 67) and 4-**amino**-3-**methoxy**-6-**methyl**-5-**nitropyridazine**, m.p. 197–198° (*D.L. Aldous* and *R.N. Castle*, Arzneimittel-Forsch., 1963, **13**, 878) have been prepared.

4-Nitropyridazines are obtained by the reaction of either the hydrazide of nitrothioacetic acid or the corresponding amidrazone with either glyoxal or methylglyoxal (*H. Hamberger et al.*, Tetrahedron Letters, 1977, 3619).

$$R^1\text{-CO-CHO} + \underset{H_2N}{\underset{|}{CH_2(NO_2)-C(R)=N}} \longrightarrow R^1\text{-pyridazine-NO}_2, R$$

R^1 = H, Me   R = SH, NH$_2$

(f) Aminopyridazines and related compounds (nitroamines, hydrazines, azides)

The most important methods for the preparation of aminopyridazines are (a) from halopyridazines by heating with concentrated ammonia, or with anhydrous ammonia in alcohol, or with amines under pressure at 100–175°, (b) from aminohalopyridazines by hydrogenolysis over a palladium on charcoal catalyst, (c) by reduction of nitropyridazines or nitropyridazine *N*-oxides (p. 19) and (d) from hydrazinopyridazines by hydrogenolysis in the presence of Raney nickel.

3-**Aminopyridazine**, m.p. 172° (*hydrochloride*, m.p. 175.5–176.5°, *picrate*, m.p. 249–250° (dec.), *acetate*, m.p. 232°), is best prepared from 3-amino-6-chloropyridazine by method (b) (*E.A. Steck, R.P. Brundage* and *L.T. Fletcher*, J. Amer. chem. Soc., 1954, **76**, 3225); 4-**aminopyridazine**, m.p. 131–132°, from 4-amino-3,6-dichloropyridazine (*T. Itai*, Chem. pharm. Bull. Japan, 1963, **11**, 342); 3,4-*diaminopyridazine monohydrochloride*, m.p. 200.5–201.5°; 3,5-*diaminopyridazine monohydrochloride*, m.p. 268–269°; 4,5-*diaminopyridazine monohydrochloride*, m.p. 234–235° (*Guither, Clark* and *Castle, loc. cit.*), 3,6-**diaminopyridazine**, m.p. 235° (*picrate*, m.p. 277–279°) from 3-amino-6-chloropyridazine (*T.B. Gortinskaya* and *M.N. Shchukina*, Zhur. obshchei Khim., 1960, **30**, 1518); 3,4,5-*triaminopyridazine, monohydrochloride*, m.p. 214–215°, by method (c) (*Guither, Clark* and *Castle loc. cit.*); 3-*amino*-4-*methylpyridazine hydrochloride*, m.p. 264° (dec.); 3-*amino*-5-*methylpyridazine*, m.p. 194° (dec.) (*K. Mori*, Yakugaku Zasshi, 1962, **82**, 304); 3-**amino**-6-**methyl**-

pyridazine, m.p. 224–225° (*hydrochloride*, m.p. 237°, *picrate*, m.p. 220–221°, *acetate*, m.p. 214–215°), from 3-chloro-6-methylpyridazine and anhydrous ammonia (*C. Grundmann*, Ber., 1948, **81**, 1); 4-*amino*-3-*methylpyridazine*, m.p. 162–163° (*T. Nakagome*, Yakugaku Zasshi, 1962, **82**, 253); 3-*amino*-6-*ethylpyridazine*, m.p. 150°, 3-*amino*-6-*phenylpyridazine*, m.p. 152° (*Grundmann, loc. cit.*); 4-*amino*-5-*phenylpyridazine*, m.p. 154–156° (*benzoate*, m.p. 202–204°) (*C.M. Atkinson* and *R.E. Rodway*, J. Chem. Soc., 1959, 1); 4-*amino*-3,6-*dimethylpyridazine*, m.p. 162–163° (*Nakagome, loc. cit.*); 3-*amino*-6-*methyl*-4-*phenylpyridazine*, m.p. 191–193° (*benzoate*, m.p. 208–210°) (*Atkinson* and *Rodway,loc. cit.*); 3-**amino**-6-**chloropyridazine**, m.p. 213–214° (dec.), by heating 3,6-dichloropyridazine with concentrated ammonia at 100° for 6 h, 3-*amino*-6-*bromopyridazine*, m.p. 205–206.5° (dec.), similarly (*J. Druey, K. Meier* and *K. Eichenberger*, Helv., 1954, **37**, 121); 4-*amino*-5-*chloropyridazine*, m.p. 70–73° (*T. Kuraishi* and *R.N. Castle*, J. heterocyclic Chem., 1964, **1**, 42), m.p. 120–122° (*G.B. Barlin*, J. chem. Soc. Perkin I, 1976, 1424); 3-*amino*-6-*chloro*-4-*methylpyridazine*, m.p. 111–113°; 6-*amino*-3-*chloro*-4-*methylpyridazine*, m.p. 190–192° (*Mori, loc. cit.*); 4-*amino*-6-*chloro*-3-*methylpyridazine*, m.p. 158–159° (*Nakagome, loc. cit.*); 3-*amino*-6-*chloro*-4,5-*dimethylpyridazine*, m.p. 198–201° (*I. Satoda, F. Kusada* and *K. Mori*, Yakugaku Zasshi, 1962, **82**, 233); 4-**amino**-3,5-**dichloropyridazine**, m.p. 151°, and 5-**amino**-3,4-**dichloropyridazine**, m.p. 178° (*acetate*, m.p. 258° (dec.)), by heating 3,4,5-trichloropyridazine in a sealed tube with ethanolic ammonia (*Kuraishi* and *Castle, loc. cit.*); 4(5)-**amino**-3,6-**dichloropyridazine**, m.p. 203°, by heating 5-amino-6-chloro-3(2*H*)-pyridazinone with POCl$_3$ (*T. Kuraishi*, Chem. pharm. Bull. Japan, 1958, **6**, 641); 5-*chloro*-3,4-*diaminopyridazine*, (*hemihydrate*, m.p. 194–196°, *anhydrous*, m.p. 205°, *picrate*, m.p. 266°); 6-*chloro*-3,4-*diaminopyridazine*, m.p. 186–187° (*Kuraishi* and *Castle, loc. cit.*; *G.A. Gerhardt* and *R.N. Castle*, J. heterocyclic Chem., 1964, **1**, 247); and 4-*chloro*-3,5-*diaminopyridazine*, m.p. 198° (*Guither, Clark* and *Castle, loc. cit.*).

Infrared spectroscopy shows that the aminopyridazines exist in the amino form (*Y. Nitta, R. Tomii* and *F. Yoneda*, Chem. pharm. Bull. Japan, 1963, **11**, 744). They are monobasic and exhibit many of the typical properties of aromatic amines: thus they are formylated with formic acid; acetylated with acetic anhydride; and form sulphonyl derivatives with methane- or *p*-toluene-sulphonylchloride. Many pyridazinesulphonamides have been prepared by the condensation of aminopyridazines with *p*-acylaminobenzonesulphonyl chlorides (see p. 30). Stable diazonium salts are produced only from aminopyridazine *N*-oxides (p. 18).

N-Amination of various pyridazines takes place when they are treated with hydroxylamine-*O*-sulphonic acid or *O*-mesitylenesulphonylhydroxylamine: thus 3-aminopyridazines give the corresponding 3-amino-2-pyridazinium salts (*Y. Tamura, J.H. Kim* and *M. Ikeda*, J. heterocyclic Chem., 1975, **12**, 107; *A. Tomažič, M. Tišler* and *B. Stanovnik*, Tetrahedron, 1981, **37**, 1787). These compounds are intermediates for the preparation of bicyclic heterocyclic compounds with a bridgehead nitrogen atom.

3-**Nitroaminopyridazine**, m.p. 184° (dec.), is formed by the reaction of 3-aminopyridazine with concentrated sulphuric acid and potassium nitrate at room temperature. Its *potassium* *salt*, m.p. 175° (dec.), with methyl iodide forms 3-*N-methylnitraminopyridazine*, m.p. 98°. The 3-nitramino compound heated in xylene with benzylamine yields 3-*benzylaminopyridazine*, m.p. 110°. Similarly, 3-amino-6-methylpyridazine gives a *nitramine*, m.p. 178° (*potassium salt*, m.p. 188°, 3-*N-methylnitramino* compound, m.p. 148°). It is reduced in the presence of palladium on charcoal to the starting aminopyridazine (not to the hydrazino compound) and with benzylamine it yields 3-*benzylamino-6-methylpyridazine*, m.p. 138° (*picrate*, m.p. 165°), identical with the product formed from 3-chloro-6-methylpyridazine and benzylamine (*S. Dixon* and *L.F. Wiggins*, J. chem. Soc., 1950, 3236). Others include 4-*nitraminopyridazine*, m.p. 185° (dec.); 3,4-*dinitraminopyridazine*, m.p. 144° (dec.); and 5-*amino-4-nitraminopyridazine*, m.p. > 400° (dec.) (*Guither, Clark* and *Castle, loc. cit.*). The nitramines undergo a rearrangement to nitropyridazines when heated in acid (p. 15).

Hydrazinopyridazines are generally prepared by the reaction of halopyridazines with hydrazine at about 100°. Examples include: 3-*hydrazinopyridazine*, oil (*picrate*, m.p. 169° (dec.)) (*T. Itai* and *S. Kamiya*, Chem. pharm. Bull. Japan, 1963, **11**, 348); 4-*hydrazinopyridazine*, hydrochloride, m.p. 240–242° (dec.) (*T. Kuraishi*, Chem. pharm. Bull. Japan, 1957, **5**, 376); 3,6-*dihydrazinopyridazine*, m.p. 193–195° (*nitrate*, m.p. 191–192°, *acid sulphate*, m.p. 215° (dec.), *monohydrochloride*, m.p. 232–233°, *dihydrochloride*, m.p. 221–222°) (*Druey, Meier* and *Eichenberger*, loc. cit.; *T.B. Gortinskaya* and *M.N. Shchukina*, Zhur. obshchei Khim., 1960, **30**, 1518); 3-*hydrazino-6-methylpyridazine*, m.p. 116° (*monohydrate*, m.p. 75–76°, hydrochloride, m.p. 221–222° (dec.)) (*H.H. Stroh, H. Hempel* and *R. Apel*, Ber., 1965, **98**, 2500); 3-*hydrazino-6-phenylpyridazine*, m.p. 143–144° (*hydrochloride*, m.p. 231–233°) (*J. Druey* and *B.H. Ringier*, Helv., 1951, **34**, 195; *F. Gross et al.*, Experientia, 1952, **8**, 229; *D. Libermann* and *A. Rouaix*, Bull. Soc. chim. Fr., 1959, 1793); 3,4-*diphenyl-6-hydrazinopyridazine*, hygroscopic, no m.p. given (*T. Sasaki, K. Kanematsu* and *M. Murata*, J. org. Chem., 1971, **36**, 446); 4,6-*diphenyl-3-hydrazinopyridazine*, hydrochloride, m.p. 205–208° (dec.) (*Druey* and *Ringier*, loc. cit.); 6-*chloro-3-hydrazinopyridazine*, m.p. 137–138° (hydrochloride, m.p. > 250° (dec.)) (*Druey, Meier* and *Eichenberger*, loc. cit.); 6-*chloro-3-hydrazino-4-methylpyridazine*, m.p. 149°; 3-*chloro-6-hydrazino-4-methylpyridazine*, m.p. 193° (*N. Takahayashi*, Chem. pharm. Bull. Japan, 1957, **5**, 229); 3,6-*dichloro-4-hydrazinopyridazine*, m.p. 195–196° (dec.) (*Kuraishi*, loc. cit.); 4-*amino-5-hydrazinopyridazine*, m.p. 150–150.5° (*Guither, Clark* and *Castle*, loc. cit.); 4-*amino-5-chloro-3-hydrazinopyridazine*, m.p. 201–202° (dec.) (*Kuraishi* and *Castle*, loc. cit.); 4-*amino-6-chloro-3-hydrazinopyridazine*, m.p. 209° (*Gerhardt* and *Castle*, loc. cit.).

Hydrazinopyridazines are formylated with formic acid or ethyl formate and acetylated with acetic anhydride. They condense with aldehydes and ketones to give hydrazones and with 1,3-dicarbonyl compounds to give *N*-pyridazinylpyrazoles (*H. Jahine et al.*, Indian J. Chem., 1977, **15B**, 250). With nitrous acid they form azides (*K. Dury*, Angew. Chem., 1965, **77**, 282; *T. Itai* and *S. Kamiya*, Chem. pharm. Bull. Japan, 1963, **11** 1073).

3-Azidopyridazines cyclise with the adjacent nitrogen to form tetrazolo[1,5-*b*]pyridazines. The same compounds are formed when chloro compounds are treated with sodium azide in ethanol. The hydrazino group can be replaced by hydrogen by heating in a solution of sodium carbonate in the presence of cupric oxide (*Dury*, loc. cit.; *Kuraishi* and *Castle*, loc.

cit., or by halogen by treatment with strongly acidic sodium hypohalite solution (*S. Linholter, R. Rosenoern* and *L. Vincents*, Acta chem. Scand., 1963, **17**, 960).

(g) *Pyridazine-N-oxides*

Pyridazines undergo *N*-oxidation when treated with peracids such as peracetic, perbenzoic or monoperphthalic, or with hydrogen peroxide in acetic, sulphuric or polyphosphoric acids. Generally mono-*N*-oxides are formed, but in some cases 1,2-dioxides can also be obtained, usually in low yield. Unsymmetrically substituted pyridazines can produce two isomeric mono-1- and mono-2-oxides.

Pyridazine-1-**oxide**, m.p. 38–40°, b.p. 138–140°/4 mm, is formed in 89% yield when pyridazine reacts with hydrogen peroxide in glacial acetic acid (*T. Itai* and *S. Natsume*, Chem. pharm. Bull. Japan, 1963, **11**, 83). 3-Methylpyridazine gives a mixture of the 1-*oxide* (XX), m.p. 68–69° and the 2-*oxide* (XXI), m.p. 85–86° (*T. Nakagome*, Yakugaku Zasshi, 1962, **82**, 249; *M. Ogata* and *H. Kano*, Chem. pharm. Bull. Japan, 1963, **11**, 29).

Other *N*-oxides similarly prepared include: 4-*methylpyridazine*-1-*oxide*, m.p. 83–84°, and -2-*oxide*, b.p. 135°/4 mm (*M. Ogata* and *H. Kano*, Chem. pharm. Bull. Japan, 1963, **11**, 35); 3,4-*dimethylpyridazine*-1-*oxide*, m.p. 149–150°, and -2-*oxide*, m.p. 110–111° (*T. Nakagome, ibid.*,, 1963, **11**, 721); 3,6-*dimethylpyridazine*-1-*oxide*, m.p. 113–114° (*T. Itai* and *S. Sako, ibid.*, 1961, **9**, 149). 3-Chloropyridazine forms the 1-*oxide*, m.p. 93°, with perbenzoic acid (*H. Igeta, ibid.*, 1960, **8**, 559); the isomeric 2-*oxide*, m.p. 157–158°, is prepared by carrying out the Gattermann reaction on diazotised 3-aminopyridazine-2-oxide (*S. Sako, ibid.*, 1963, **11**, 261). 4-*Chloropyridazine*-1-*oxide*, m.p. 119–121° (*Itai* and *Natsume, loc. cit.*); 3,6-*dichloropyridazine*-1-*oxide*, m.p. 110–112° (*T. Itai* and *S. Sako*, Chem. pharm. Bull. Japan, 1962, **10**, 989; *T. Nakagome*, Yakugaku Zasshi, 1962, **82**, 244); 5-*chloro-3-methylpyridazine*-1-*oxide*, m.p. 148° (*T. Itai* and *S. Natsum*, Chem. pharm. Bull. Japan, 1964, **12**, 228); 3-*chloro-4-methylpyridazine*-1-*oxide*, m.p. 148–149°; 3-*chloro-5-methylpyridazine*-2-*oxide*, m.p. 127–128°; 3-*chloro-6-methylpyridazine*-1-*oxide*, m.p. 163–164°; 5-*chloro-3-methylpyridazine*-2-*oxide*, m.p. 166–167° (*M. Ogata* and *H. Kano, ibid.*, 1963, **11**, 29, 35); 4-*chloro-3-methylpyridazine*-1-*oxide*, m.p. 132.5–133°; 4-*chloro-5-methylpyridazine*-1-*oxide*, m.p. 61–62° (*M. Ogata, ibid.*, 1963, **11**, 1511); 4-*chloro-3,6-dimethylpyridazine*-1-*oxide*, m.p. 132–133°, and -2-*oxide*, m.p. 126–127° (*S. Sako, ibid.*, 1963, **11**, 337); 6-*chloro-3,4-dimethylpyridazine*-1-*oxide*, m.p. 184–184.5°, and -2-*oxide*, m.p. 109–110° (*T. Nakagome, ibid.*, 1963, **11**, 721); 6-*amino-3-chloropyridazine*-1-*oxide*, m.p. 248° (dec.) (*T. Itai* and *T. Nakashima, ibid.*, 1962, **10**, 347); 6-*chloro-3-methoxypyridazine*-1-*oxide*, m.p. 157–158° (*T. Nakagome*, Yakugaku Zasshi, 1962, **82**, 224); 3,6-*di*-

chloro-4-*methoxypyridazine*-1-*oxide*, m.p. 174–175°, and -2-*oxide*, m.p. 162.5–164° (*T. Itai* and *S. Natsume*, Chem. pharm. Bull. Japan, 1962, **10**, 643); 6-*methyl*-3(2*H*)-*pyridazinone*-1-*oxide*, m.p. 201–202° (*T. Nakagome*, Yakugaku Zasshi, 1962, **82**, 249); 3-*methoxypyridazine*-1-*oxide*, m.p. 79–80° (*T. Itai* and *S. Sako*, Chem. pharm. Bull, Japan, 1962, **10**, 989); 4-*methoxypyridazine*-1-*oxide*, m.p. 124–124.5°, and -2-*oxide*, m.p. 111° (*Itai* and *Natsume*, *loc. cit.*; *T. Nakagome*, Yakugaku Zasshi, 1962, **82**, 253); 3,6-*dimethoxypyridazine*-1-*oxide*, m.p. 152° (*M. Yanai* and *T. Kinoshita*, Yakugaku Zasshi, 1965, **85**, 344); 3-*methoxy*-6-*methylpyridazine*-1-*oxide*, m.p. 98–99° (*T. Nakagome*, *ibid.*, 1961, **81**, 1048); 4-*methoxy*-3-*methylpyridazine*-1-*oxide*, m.p. 105–106°; 4,6-*dimethoxy*-3-*methylpyridazine*-1-*oxide*, m.p. 148–149° (*idem*, *ibid.*, 1962, **82**, 253).

The nitration of pyridazine *N*-oxides (p. 18) with mixed acids gives the 4-nitro derivatives, for example, 4-*nitropyridazine*-1-*oxide*, m.p. 150–151° (*picrate*, m.p. 81–82°); 3-*methyl*-5-*nitropyridazine*-2-*oxide*, m.p. 120–121°; 3,6-*dimethyl*-4-*nitropyridazine*-1-*oxide*, m.p. 117–118°; 3-*methoxy*-6-*methyl*-4-*nitropyridazine*-1-*oxide*, m.p. 114–115°; 3-*chloro*-6-*methyl*-4-*nitropyridazine*-1-*oxide*, m.p. 103° (*Nakagome*, *loc. cit.*).

If the 4-position is occupied, *e.g.* as in 3,6-dimethyl-4-hydroxypyridazine-1-oxide, the 5-*nitro* derivative, m.p. 184–185° (dec.), is formed (*S. Kamiya* and *M. Tanno*, Chem. pharm. Bull. Japan, 1975, **23**, 1879). The nitration of pyridazine-*N*-oxides using acyl nitrates yields 3- and 5-nitro compounds: for example, pyridazine-1-oxide with benzoyl nitrate gives 3-*nitropyridazine*-1-*oxide*, m.p. 160°, and the 5-*nitro* isomer, m.p. 142–143° (*T. Itai* and *S. Natsume*, Chem. pharm. Bull. Japan, 1963, **11**, 342). 3-Nitropyridazine-1-oxides are also obtained by the oxidation of 3-aminopyridazines with hydrogen peroxide in polyphosphoric acid (*A. Pollak*, *B. Stanovnik* and *M. Tišler*, J. org. Chem., 1970, **35**, 2478). Oxidation of 3-aminopyridazine with hydrogen peroxide in acetic acid gives 3-*aminopyridazine*-2-*oxide*, m.p. 210–211° (*T. Itai* and *T. Nakashima*, Chem. pharm. Bull. Japan, 1962, **10**, 347, 936). Aminopyridazine-*N*-oxides are also produced by the catalytic reduction of nitropyridazine-*N*-oxides using palladium on charcoal in neutral solution, for example, 3-*aminopyridazine*-1-*oxide*, m.p. 139–141° (*Itai* and *Natsume*, *loc. cit.*). However, when the hydrogenation is carried out in aqueous or alcoholic acid solution with palladium on charcoal or Raney nickel, then simultaneous reduction of the nitro group and deoxygenation of the *N*-oxide takes place (*T. Nakagome*, Chem. pharm. Bull. Japan, 1963, **11**, 726). In all these reaction, if a halogen group is present, then dehalogenation usually occurs simultaneously (*H. Igeta*, *ibid.*, 1960, **8**, 368).

5-*Amino*-4-*nitropyridazine*-1-*oxide*, m.p. 287–288°; 5-*amino*-3-*methoxy*-4-*nitropyridazine*-1-*oxide*, m.p. 205–206°; 5-*amino*-3,6-*dimethoxy*-4-*nitropyridazine*-1-*oxide*, m.p. 187–188°, by amination of the appropriate 4-nitropyridazine-1-oxide with potassium permanganate in liquid ammonia (*H. Tondys* and *H.C. Van der Plas*, J. heterocyclic Chem., 1986, **23**, 621).

Pyridazine-1,2-**dioxides** include: *pyridazine*-1,2-*dioxide*, m.p. 222° (dec.); 3-*methylpyridazine*-1,2-*dioxide*, m.p. 156–157°; 3-*methylpyridazine*-1,2-*dioxide*, m.p. 150–150.5°; and 3,6-*dimethylpyridazine*-1,2-*dioxide*, m.p. 216° (dec.) (*M. Nakadate*, *S. Sueyoshi* and *I. Suzuki*, Chem. pharm. Bull. Japan, 1970, **18**, 1211). The dioxides can be reduced with hydrogen in the presence of palladium on charcoal to monoxides (*S. Sueyoshi* and *I. Suzuki*, *ibid.*, 1975, **23**, 2767).

The electronic structures of the pyridazine-*N*-oxides have been calculated using an LCAO-MO method (*T. Kubota* and *H. Watanabe*, Bull. chem. Soc. Japan, 1963, **36**, 1093). They give simple first-order ^{1}H-n.m.r. spectra (*Y. Kawazoe* and *S. Natsume*, Yakugaku Zasshi, 1963, **83**, 523; *K. Tori*,

TABLE 4

¹H-N.M.R. SPECTRAL DATA FOR PYRIDAZINE-N-OXIDES

1-Oxide	Chemical shifts (δ, ppm)						Coupling constants (Hz)					
	H-3	H-4	H-5	H-6	Me	MeO	$J_{3,4}$	$J_{3,5}$	$J_{3,6}$	$J_{4,5}$	$J_{4,6}$	$J_{5,6}$
Pyridazine	8.54	7.22	7.83	8.26	–	–	5.3	2.5	1.0	8.0	1.0	6.5
3-Cl	–	7.23	7.75	8.18	–	–	–	–	–	8.3	1.0	6.5
3-Me	–	6.98	7.58	8.10	2.52	–	–	–	–	8.2	0.5	6.1
4-Me	8.35	–	7.57	8.17	2.36	–	0.2[a]	2.8	0.5	0.5	0.2	6.2
5-Me (4-Me, 2-oxide)	8.38	6.99	–	8.08	2.37	–	5.6	0.2[a]	0.5	0.7[a]	0.7	0.7[a]
6-Me (3-Me, 2-oxide)	8.37	7.12	7.73	–	2.51	–	5.6	2.5	–	8.2	–	–
3,6-DiCl	–	7.22	7.90	–	–	–	–	–	–	8.4	–	–
3-Cl, 4-Me	–	–	7.59	8.10	2.39	–	–	–	–	0.7[a]	0.2	6.2
3-Cl, 5-Me	–	7.05	–	7.99	2.36	–	–	–	–	0.7[a]	0.7	0.7[a]
3-Cl, 6-Me	–	7.17	7.73	–	2.49	–	–	–	–	8.3	–	–
4-Cl, 6-Me (5-Cl, 3-Me, 2-oxide)	8.39	–	7.68	–	2.51	–	–	3.0	–	–	–	–
3-OMe	–	6.69	7.53	7.95	–	4.02	–	–	–	8.6	0.7	5.8
3-OMe, 6-Me	–	6.67	7.52	–	2.45	4.02	–	–	–	8.5	–	–
3-OMe, 6-Cl	–	6.72	7.71	–	–	4.04	–	–	–	8.8	–	–
3,6-DiOMe	–	6.73	7.36	–	–	4.08	–	–	–	8.7	–	–
4-NO₂	9.30	–	8.65	8.16	–	–	–	2.3	0.5	–	–	7.0
4-NO₂, 6-Me (5-NO₂, 3-Me, 2-oxide)	9.20	–	8.45	–	2.59	–	–	3.3	–	–	–	–

[a] CH_3–H coupling.

TABLE 5
¹H-N.M.R. SPECTRAL DATA FOR PYRIDAZINE-1,2-DIOXIDES

1,2-Dioxide	Chemical shifts (δ, ppm)					Coupling constants (Hz)					
	H-2	H-4	H-5	H-6	Me	$J_{3,4}$	$J_{3,5}$	$J_{3,6}$	$J_{4,5}$	$J_{4,6}$	$J_{5,6}$
Pyridazine	8.15	7.08	7.08	8.15	—	—	—	—	—	—	—
3-Me	—	7.10	7.05	8.14	2.56	~0.4 [a]	—	~0.6 [a]	8.1	2.4	6.1
4-Me	8.07	—	6.98	8.11	2.35	~0.8 [a]	2.4	0.7	~0.7 [a]	—	6.8
3,6-DiMe	—	6.96	6.96	—	2.53	<0.2 [a]	—	—	—	—	<0.2 [a]

[a] CH_3–H coupling.
(*Y. Kawazoe* and *M. Ohnishi*, Chem. pharm. Bull. Japan, 1963, **11**, 243; *I. Suyuki, M. Nakadate* and *S. Sueyoshi*, Tetrahedron Letters, 1968, 1855.)

M. Ogata and H. Kano, Chem. pharm. Bull. Japan, 1963, **11**, 235) (see Tables 4 and 5).

The mass spectrum of pyridazine-1-oxide shows loss of NO˙ followed by HCN (H. Ogura et al., J. heterocyclic Chem., 1971, **8**, 391).

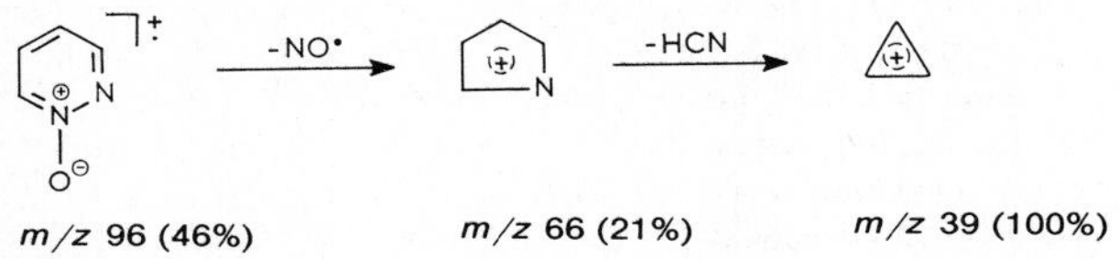

[pyridazine-1-oxide]⁺˙ —NO˙→ [C₄H₄N]⁺ —HCN→ [C₃H₃]⁺

m/z 96 (46%)   m/z 66 (21%)   m/z 39 (100%)

Nucleophilic substitution of substituted pyridazine-N-oxides has been widely studied. Chlorine atoms can be replaced by hydroxy, alkoxy, aryloxy, amino, piperidino, hydrazino, azido, hydroxylamino, mercapto and methyl-sulphonyl groups. Nitro groups can be replaced by chlorine, bromine, alkoxy or aryloxy groups and amino groups by halogens or hydroxyl groups.

Pyridazine-1-oxides without a substituent at position 6 rearrange when heated with acetic anhydride to give 3(2H)-pyridazinones: thus 3-aminopyridazine-1-oxide yields 6-amino-3(2H)-pyridazinone (T. Hone and T. Uedda, Chem. pharm. Bull. Japan, 1963, **11**, 114). If the 6-position is occupied by methyl, alkoxyl or halogen, the latter may be attacked: for example, 3-methoxy-6-methylpyridazine-1-oxide gives 6-*acetoxymethyl-3-methoxypyridazine*, m.p. 59–60° (M. Ogata and H. Kano, ibid., 1963, **11**, 29).

Methylpyridazine-N-oxides condense with aromatic aldehydes: thus with benzaldehyde 3,6-dimethylpyridazine-1-oxide gives 3,6-distyrylpyridazine-1-oxide (T. Itai, S. Sako and G. Okusa, ibid., 1963, **11**, 1146).

Phosphorus oxychloride with pyridazine-N-oxides causes simultaneous deoxygenation and chlorination at positions α or γ to the N-oxide group: thus, 3-methoxypyridazine-1-oxide gives 3-chloro-6-methoxypyridazine and 3,6-dimethoxypyridazine-1-oxide yields 4-chloro-3,6-dimethoxypyridazine (H. Igeta, ibid., 1959, **7**, 985; 1960, **8**, 368).

(h) Pyridazinones

These are prepared: (a) from 1,4-keto-acids (method 1, p. 1), (b) from 1,2-dicarbonyl compounds (method 3, p. 2), (c) from 1,2-dicarboxylic acid anhydrides (method 4, p. 3), (d) by the catalytic dehalogenation of halo-pyridazinones, (e) by the decarboxylation of pyridazinone carboxylic acids, and (f) by the acid or base hydrolysis of chloro- or alkoxy-pyridazines.

3(2H)-**Pyridazinone**, m.p. 104–105° (*monohydrate*, m.p. 70–71°), is prepared by method (a) (*C. Grundmann*, Ber., 1948, **81**, 1), (d) (*P. Coad* and *R.A. Coad*, J. org. Chem., 1963, **28**, 1919), or (e) (*P. Schmidt* and *J. Druey*, Helv., 1954, **37**, 134); 4-*methyl*-3(2H)-*pyridazinone*, m.p. 134°, method (c) (*Schmidt* and *Druey, loc. cit.*; *F.H. McMillan et al.*, J. Amer. chem. Soc., 1956, **78**, 407), or (f) (*S. Linhotter et al.*, Acta chem. Scand., 1961, **15**, 1660); 5-*methyl*-3(2H)-*pyridazinone*, m.p. 151–153° (*P. Schmidt* and *J. Druey*, Helv., 1954, **37** 1467; *N. Takahayashi*, Chem. pharm. Bull. Japan, 1957, **5**, 229); 6-*methyl*-3(2H)-*pyridazinone* (*monohydrate*, m.p. 119–123°, *anhydrous*, m.p. 138–140°, *hydrochloride*, m.p. 176–176.5°, *hydrobromide*, m.p. 184.5–185°), by method (f) (*R.F. Homer et al.*, J. chem. Soc., 1948, 2195); 4,6-*dimethyl*-3(2H)-*pyridazinone*, m.p. 124–125° (*J. Levisalles*, Bull. Soc. chim. Fr., 1957, 1004); 3,4-*dimethyl*-6(1H)-*pyridazinone*, m.p. 221–222° (*Schmidt* and *Druey, loc. cit.*); 4-*ethyl*-6-*methyl*-3(2H)-*pyridazinone*, m.p. 111–112° (*hydrobromide*, m.p. 237°) (*Levisalles, loc. cit.*); 4-*methyl*-6-*phenyl*-3(2H)-*pyridazinone*, m.p. 189–190° (*A. Oppenheim*, Ber., 1901, **34**, 4227); 6-*methyl*-4-*phenyl*-3(2H)-*pyridazinone*, m.p. 121–123° (*hydrobromide*, m.p. 251–254° (dec.) (*C.M. Atkinson* and *R.E. Rodway*, J. chem. Soc., 1959, 6); 6-*ethyl*-3(2H)-*pyridazinone*, m.p. 43°, b.p. 140–142°/11 mm, (*hydrobromide*, m.p. 107–112°) (*Grundmann, loc. cit.*); 6-*phenyl*-3(2H)-*pyridazinone*, m.p. 201–202° (*S. Gabriel* and *J. Colman*, Ber., 1899, **32**, 395); 4,6-*diphenyl*-3(2H)-*pyridazinone*, m.p. 183–184°; 3,4-*diphenyl*-6(1H)-*pyridazinone*, m.p. 177–178° (*G.K. Almström*, Ann., 1913, **400**, 131); 4(1H)-*pyridazinone*, m.p. 245–246° (*A.F. Thomas* and *A. Marxer*, Helv., 1958, **41**, 1898); 6-*chloro*-3(2H)-*pyridazinone*, m.p. 138–140°, method (f) (*J. Druey, K. Meier* and *K. Eichenberger*, Helv., 1954, **37**, 121); 4,5-*dichloro*-3(2H)-*pyridazinone*, m.p. 199–200° (*T. Kuraishi*, Chem. pharm. Bull. Japan, 1956, **4**, 497); 4,6-*dichloro*-3(2H)-*pyridazinone*, m.p. 203–204° (*T. Kuraishi, ibid.*, 1957, **5**, 376); 3,4-*dichloro*-6(1H)-*pyridazinone*, m.p. 203° (*T. Kuraishi, ibid.*, 1958, **6**, 641); 6-*chloro*-4-*methyl*-3(2H)-*pyridazinone*, m.p. 148°; 3-*chloro*-4-*methyl*-6(1H)-*pyridazinone*, m.p. 225° (*Takahayashi, loc. cit.*; *T. Kuraishi, ibid.*, 1957, **5**, 587); 5-*bromo*-4-*chloro*-3(2H)-*pyridazinone*, m.p. 208° (*T. Kuraishi, ibid.*, 1958, **6**, 641); 5-*amino*-4-*chloro*-3(2H)-*pyridazinone*, m.p. >300° (*acetate*, m.p. 277–279°); 4-*amino*-6-*chloro*-3(2H)-*pyridazinone*, m.p. 285° (*acetate* m.p. 255–256°; *benzoate*, m.p. 235°) (*T. Kuraishi, ibid.*, 1958, **6**, 331); 4-*amino*-3-*chloro*-6(1H)-*pyridazinone*, m.p. 278–280° (*T. Kuraishi ibid.*, 1958, **6**, 641); 4-*amino*-3(2H)-*pyridazinone*, m.p. 228–229° (*acetate*, m.p. 272°); 5-*amino*-3(2H)-*pyridazinone*, m.p. 286–287° (*T. Kuraishi, ibid.*, 1958, **6**, 331); 6-*amino*-4-*methyl*-3(2H)-*pyridazinone*, m.p. 213° (*K. Mori*, Yakugaku Zasshi, 1962, **82**, 304).

3-**Methoxypyridazine**, b.p. 86–87°/13 mm; 4-*methoxypyridazine*, m.p. 43–44° (*K. Eichenberger, R. Rometsch* and *J. Druey*, Helv., 1956, **39**, 1755); 4-*ethoxypyridazine*, b.p. 118–119°/5 mm; 3-*methoxy*-4-*methylpyridazine* (*hydrochloride*, m.p. 150° (dec.)) (*Mori, loc. cit.*); 3-*methoxy*-6-*methylpyridazine*, b.p. 210° (*hydrochloride*, m.p. 137–138°) (*W.G. Overend* and *L.F. Wiggins*, J. chem. Soc., 1947, 239); 3-*methoxy*-6-*phenylpyridazine*, m.p. 116–117° (*Gabriel* and *Colman, loc. cit.*); 3-*methoxy*-4-*methyl*-6-*phenylpyridazine*, m.p. 53° (*G. Leclerc* and *C.G. Wermuth*, Bull. Soc. chim. Fr., 1971, 1752); 6-*methyl*-3-*phenoxy*-4-*phenylpyridazine*, m.p. 96–98° (*Mori, loc. cit.*); 6-*chloro*-3-*methoxypyridazine*, m.p. 90° (*Druey, Meier* and *Eichenberger, loc. cit.*); 3-*chloro*-6-*methoxy*-4-*methylpyridazine* m.p. 112–116° (*Mori, loc. cit.*); 6-*chloro*-3-*methoxy*-4-*methylpyridazine*, m.p. 68–70° (*T. Nakagome*, Yakugaku Zasshi, 1962, **82**, 244); 4-*amino*-3-*methoxypyridazine*, m.p. 127–128°; 6-*amino*-3-*methoxypyridazine*, m.p. 107–108°

(*picrate,* m.p. 222°) (*T. Nakagome, ibid.*, 1961, **81**, 554); 6-*amino*-3-*methoxy*-4-*methylpyridazine*, m.p. 83–85° (*Mori, loc. cit.*); 4-*amino*-3-*methoxy*-6-*methylpyridazine* m.p. 159–161°; 4,5-*diamino*-3-*methoxy*-6-*methylpyridazine*, m.p. 210–211° (*D. L. Aldous* and *R. N. Castle*, Arzneimittel-Forsch., 1963, **13**, 878).

6-**Hydroxy**-3(2*H*)-**pyridazinone** (maleic hydrazide), m.p. 306–308°, method (c) (*R. H. Mizzoni* and *P. E. Spoerri*, J. Amer. chem. Soc., 1951, **73**, 1873); 6-*hydroxy*-4-*methyl*-5-*phenyl*-3(2*H*)-*pyridazinone*, m.p. 322–326° (*R. K. Bly, E. C. Zoll* and *J. A. Moore*, J. org. Chem., 1964, **29**, 2128); 6-*hydroxy*-4,5-*diphenyl*-3(2*H*)-*pyridazinone*, m.p. 350° (*R. W. H. Berry* and *A. Burawoy*, J. chem. Soc. C, 1970, 1316); 6-*methoxy*-3(2*H*)-*pyridazinone*, m.p. 162–163° (*T. Nakagome*, Yakugaki Zasshi, 1962, **82**, 244); 3,4-*dimethoxypyridazine*, m.p. 55–57° (*picrate*, m.p. 151°); 6-*chloro*-3,4-*dimethoxypyridazine*, m.p. 126° (*T. Itai* and *S. Natsume*, Chem. pharm. Bull. Japan, 1962, **10**, 643); 3,6-*dimethoxypyridazine*, m.p. 106–107°, and many other 3,6-dialkoxypyridazines (*Druey, Meier* and *Eichenberger, loc. cit.*).

3- and 4-Hydroxypyridazines exist predominantly in the keto forms (XXII) and (XXII) respectively. This has been established from u.v. and i.r.

(XXII)   (XXIII)   (XXIV)

spectroscopic evidence and from X-ray analysis. Maleic hydrazide exists in the monohydroxy-monoketo form (XXIV). Unsubstituted 3(2*H*)-pyridazinones behave both as weak acids and weak bases. They form salts with bases and their salts with mineral acids tend to dissociate in water. Maleic hydrazide is more acidic and forms monometallic salts with bases. Data from pK_a measurements and ^{1}H-n.m.r. spectra for pyridazinones are given in Table 6.

The mass spectrum of 3(2*H*)-pyridazinone shows loss of CO followed by $N_2H^{\cdot}$ (*J. H. Bowie et al.*, Austral. J. Chem., 1967, **20**, 2677).

m/z 96 (33%)   m/z 68 (40%)   m/z 39 (100%)

6-Hydroxy-3(2*H*)-pyridazinone gives a base peak at $[M - 1]^{\oplus}$. The molecular ion can also lose N_2H_2 (*H. Ogura et al.*, J. heterocyclic Chem., 1971, **8**, 371).

Alkylation of a pyridazinone can give either an *N*- or an *O*-substituted product. Methylation with methyl iodide or dimethyl sulphate usually produces the *N*-methyl derivative. Thus 3(2*H*)-pyridazinone gives 2-*methyl-*

TABLE 6

pK_a VALUES AND ^{1}H-N.M.R. SPECTRAL DATA FOR PYRIDAZINONES

Compound	pK_a Values	
	Proton gain	Proton loss
3(2H)-Pyridazinone	−1.8	10.46
4(2H)-Pyridazinone	1.07	8.68
6-Hydroxy-3(2H)-pyridazinone (Maleic hydrazide)	5.5	13

3(2H)-Pyridazinone	^{1}H-n.m.r. spectral data (DMSO)				
	Chemical shifts (δ, ppm)				Coupling constants
	H-4	H-5	N-Me	O-Me	$J_{4,5}$
6-Cl	7.00	7.51	–	–	9.9
2-Me, 6-Cl	7.04	7.54	3.64	–	9.8
2-Me, 6-OH	6.87	7.04	3.49	–	9.4
2-Me, 6-OMe	6.91	7.16	3.53	3.78	9.0
Maleic hydrazide	6.96	6.96	–	–	–
3,6-Dimethoxypyridazine	7.16	7.16	–	3.99	–

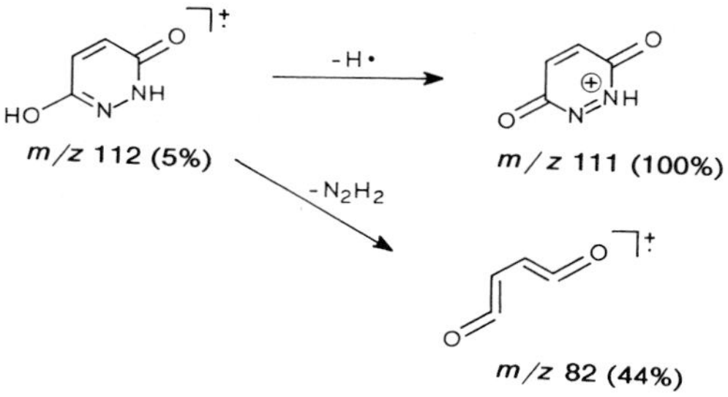

3(2H)-*pyridazinone*, m.p. 42–43° (*S. Hünig* and *K.-H. Oette*, Ann., 1961, **640**, 98). Diazomethane gives both the *N*- and the *O*-methyl compound: for example, 6-methyl-3(2H)-pyridazinone yields a mixture of 2,6-*dimethyl-3(2H)-pyridazinone*, m.p. 50–51°, and 3-*methoxy-6-methylpyridazine*, m.p. 210–215° (*hydrochloride*, m.p. 131–132°). Methylation of maleic hydrazide (XXIV) with diazomethane gives first the mono-*O*-substituted derivative (XXV) and then on continued treatment the *N,O*-dimethyl compound (XXVI), m.p. 64–65° (*F. Arndt*, Angew. Chem., 1949, **61**, 397). Methylation with dimethyl sulphate at 100° gives the *mono-N* derivative (XXVII), m.p. 210–211°, and, on further treatment at 150°, a mixture of the

N,N'-*dimethyl* compound (XXVIII), m.p. 137–138°, and the 2-methyl-6-methoxy derivative (XXVI). Prolonged heating favours the formation of (XXVIII) as the result of $O \to N$-methyl-group migration (*K. Eichenberger, A. Staehelin* and *J. Druey*, Helv., 1954, **37**, 837).

(XXV)   (XXVI)   (XXVII)   (XXVIII)

The $O \to N$ rearrangement of alkyl groups in alkoxypyridazine derivatives is well known. Thus, 3,6-dimethoxypyridazine when heated with methyl iodide or dimethyl sulphate or with strong acids or aluminium trichloride rearranges to (XXVI) and on prolonged heating to (XXVIII) (*Eichenberger, Staehelin* and *Druey, loc. cit.*).

Acylations of pyridazones and maleic hydrazide generally give *O*-acylated derivatives: thus maleic hydrazide gives an *O-acetate*, m.p. 122–123°, with acetic anhydride or acetyl chloride, an *O-benzenesulphonate*, m.p. 144° (*monohydrate*, m.p. 134–135°) with benzenesulphonyl chloride (*H. Feuer* and *H. Rubinstein*, J. Amer. chem. Soc., 1958, **80**, 5873), and an *O-benzoate*, m.p. 162.5–164.5° with benzoyl chloride (*D. Stefanye* and *W.L. Howard*, J. org. Chem., 1954, **19**, 115). *N*-Acylated maleic hydrazides are formed from the rearrangement of the corresponding acylated *N*-aminomaleimides (*Feuer* and *Rubinstein, loc. cit.*; *H. Feuer* and *J.P. Asunskis*, J. org. Chem., 1962, **27**, 4684).

The pyridazinones are converted into 3- and/or 6-chloropyridazines by heating with phosphorus oxychloride or phosphorus pentachloride (p. 12).

Maleic hydrazide and related compounds have been used as selective plant growth regulators (*D.L. Schoene* and *O.L. Hoffmann*, Science, 1949, **109**, 588; *E. Parups, I. Hoffmann* and *H.V. Morley*, Canad. J. Biochem. Physiol., 1962, **40**, 1159).

(i) Pyridazinecarboxylic acids

Pyridazinecarboxylic acids are prepared by (a) oxidation with potassium permanganate of benzopyridazines such as cinnolines, benzo[*c*]cinnolines, or phthalazines; (b) oxidation of pyridazines having a suitable side-chain; (c) method (3) (p. 2) using diethyl malonate $R^3 = CO_2H$), leading to substituted 4-carboxylic acids; (d) method (5) (p. 4) using the cycloaddition of diethyl azodicarboxylate to conjugated dienes to form tetrahydropyridazine-1,2-dicarboxylic acids; (e) hydrolysis of cyanopyridazines with concentrated sulphuric acid at about 150°; and (f) partial decarboxylation of pyridazinepolycarboxylic acids.

Pyridazine-3-carboxylic acid, m.p. 200–201° (dec.) (*methyl ester*, m.p. 139°, *ethyl ester*, m.p. 68°, *amide*, m.p. 182–182.5°, *hydrazide*, m.p. 151–152°) is prepared by the oxidation of 3-hydroxymethylpyridazine, m.p. 66° (*N. Clauson-Kaas* and *F. Limborg*, Acta chem. Scand., 1947, **1**, 619; *R. Delaby, R. Damiens* and *M. Robba*, Compt. rend., 1958, **247**, 1739; *W.J. Leanza, H.J. Becker* and *E.F. Rogers*, J. Amer. chem. Soc., 1953, **75**, 4086), or by the oxidation of 3-*n*-butylpyridazine (*R.L. Letsinger* and *R. Lasco*, J. org. Chem., 1956, **21**, 812); **pyridazine-4-carboxylic acid**, m.p. 239–240° (dec.). (*methyl ester*, m.p. 63°, *ethyl ester*, b.p. 125°/13 mm, *amide*, m.p. 191–192°, *hydrazide*, m.p. 124–125°) is obtained by partial decarboxylation of the 4,5-dicarboxylic acid by heating it in aqueous solution at 200° for 2 h (*Leanza, Becker* and *Rogers, loc. cit.*), or by the oxidation of 4-*n*-butylpyridazine (*Letsinger* and *Lasco, loc. cit.*), or from 3,6-dichloro-4-methylpyridazine by dichromate oxidation to 3,6-*dichloropyridazine-4-carboxylic acid*, m.p. 144° (dec.), and then catalytic hydrogenation (*G. Heinisch*, Monatsch., 1973, **104**, 953); **pyridazine-4,5-dicarboxylic acid**, m.p. 213° (dec.) by the oxidation of phthalazine (*S. Gabriel*, Ber., 1903, **36**, 3373; *H. Raistrick* and *P. Rudman*, Biochem. J., 1956, **63**, 395), or by the partial decarboxylation of the dipotassium salt of the tetracarboxylic acid with 15% HCl at 100° (*E. Täuber*, Ber, 1895, **28**, 451); **pyridazine-3,4,5,6-tetracarboxylic acid**, m.p. not given, is prepared by the oxidation of benzo[*c*]cinnoline (*Täuber, loc. cit.*); 5-**phenylpyridazine-4-carboxylic acid**, m.p. 200–201° (dec.) (*R. Stoermer* and *H. Fincke*, Ber., 1909, **42**, 3115), m.p. 222–224° (dec.) (*methyl ester*, m.p. 103–105°) (*C.M. Atkinson* and *R.E. Rodway*, J. chem. Soc., 1959, 1) prepared by heating the 3,4-dicarboxylic acid (below) at 125°; 6-**phenylpyridazine-3-carboxylic acid**, m.p. 130–131°, by the oxidation of 3-methyl-6-phenylpyridazine (*C. Paal* and *E. Dencks*, Ber., 1903, **36**, 491); 4-*methyl-5-phenylpyridazine-3-carboxylic acid*, m.p. 135–136° (dec.) (*methyl ester*, m.p. 104°) (*R.T. Bly, E.C. Zoll* and *J.A. Moore*, J. org. Chem., 1964, **29**, 2128); 5-**phenylpyridazine-3,4-dicarboxylic acid**, m.p. 225–227° (dec.) (*dimethyl ester*, m.p. 131–132°, 3-*monomethyl ester*, m.p. 178–180°, 4-*monomethyl ester*, m.p. 105–112°), by the permanganate oxidation of 4-phenylcinnoline (*Stoermer* and *Fincke, loc. cit.; Bly, Zoll* and *Moore, loc. cit.; Atkinson* and *Rodway, loc. cit.*, give m.p. 148–150° (efferv.), remelting at 214° (dec.)); 3,6-*dimethylpyridazine-4,5-dicarboxylic acid*, m.p. 225–226° (*C. Paal* and *J. Ubber*, Ber., 1903, **36**, 497); 6-**methyl-5-phenylpyridazine-3,4-dicarboxylic acid**, m.p. 228–229°, by permanganate oxidation of 3-methyl-4-phenylcinnoline (*Stoermer* and *Fincke, loc. cit.*), 6-**chloropyridazine-3-carboxylic acid**, m.p. 146°, by oxidation of 6-chloro-3-methylpyridazine (*R.F. Homer et al.*, J. chem. Soc., 1948, 2195). With hydrochloric acid in ethanol it yields 6-*carbethoxy-3*(2*H*)-*pyridazinone*, m.p. 121–122° (*Homer et al., loc. cit.*).

Substituted **3(2***H***)-pyridazinone-4-carboxylic acids** are readily available from diethyl malonate by method (c) and include: 5-*methyl*, m.p. 193–194°, 6-*methyl*, m.p. 182–183°, 5,6-*dimethyl*, m.p. 172–173°, 5,6-*diphenyl*, m.p. 243–244° (dec.), and 6-*chloro-3*(2*H*)-*pyridazinone-4-carboxylic acid*, m.p. 210° (dec.) (*methyl ester*, m.p. 132°) (*R.F. Homer, H. Gregory* and *L.F. Wiggins*, J. chem. Soc. 1948, 2191; *P. Schmidt* and *J. Druey*, Helv., 1954, **37**, 1467; *K. Mori*, Yakugaku Zasshi, 1962, **82**, 304).

Diethyl tetrahydropyridazine-1,2-dicarboxylates formed by method (d) include **diethyl 3-methyl-6-phenyl-1,2,3,6-tetrahydropyridazine-1,2-dicarboxylate**, m.p. 75–76°, from 1-methyl-4-phenylbutadiene, and **diethyl 3,6-diphenyl-1,2,3,6-tetrahydropyridazine-1,2-dicarboxylate**, m.p. 132°, from 1,4-diphenylbutadiene (*K. Alder et al.*, Ann., 1954, **585**, 81; *J. Levisalles* and *P. Baranger*, Compt. rend., 1954, **238**, 592).

4-Methylpyridazine and chloral give an aldol product, m.p. 117–118°, which on hydrolysis gives β-4-*pyridazinylacrylic acid*, m.p. 224–225° (*R.H. Mizzoni* and *P.E. Spoerri*, J. Amer. chem. Soc., 1954, **76**, 2201; *R.G. Jones, E.C. Kornfeld* and *K.C. McLaughlin*, J. Amer. chem. Soc., 1950, **72**, 3539).

Cyanopyridazines are prepared by (a) method (3) (p. 2) using ethyl cyanoacetate (R^3 = CN), leading to 4-cyano-3(2H)-pyridazinones, (b) the dehydration of amides, and (c) the Reissert reaction of pyridazine-N-oxides leading to 3-cyanopyridazines. Examples include; 3-**cyanopyridazine**, m.p. 43–44°; 4-*cyanopyridazine*, m.p. 79–80° (*M. Robba*, Ann. Chim., 1960, **5**, 351; *R. Delaby, R. Damiens* and *M. Robba*, Compt. rend., 1958, **247**, 1739); 3-*chloro-4-cyanopyridazine*, m.p. 41–42°; 3-*chloro-6-cyanopyridazine*, m.p. 90–91.5°; 3-*chloro-4-cyano-6-methylpyridazine*, m.p. 106° (dec.) (*U.M. Teotino* and *G. Cignarella*, Gazz., 1959, **89**, 1200: *A. Dornow* and *W. Abele*, Ber., 1964, **97**, 3349).

For **4-cyano**-3(2H)-**pyridazinone**, m.p. 184–185°, and its 5-*methyl*, m.p. 228–230°, 6-*methyl*, m.p. 169–170°, 5,6-*dimethyl*, m.p. 210–211°, and 5,6-*diphenyl*, m.p. 272–273°, derivatives see *P. Schmidt* and *J. Druey*, Helv., 1954, **37**, 134, 1467.

(j) Pyridazine sulphur compounds

Pyridazine mercaptans (thiones) are prepared from the corresponding halopyridazines by treatment with either alcoholic sodium (potassium) hydrogen sulphide or thiourea in alcohol or phosphorus pentasulphide in pyridine. The latter also converts pyridazinones into pyridazinethiones.

3(2H)-**Pyridazinethione**, m.p. 170° (*G.F. Duffin* and *J.D. Kendall*, J. chem. Soc., 1959, 3789); 4(1H)-*pyridazinethione*, m.p. 206–210° (dec.) (*A. Albert* and *G.B. Barlin*, ibid., 1962, 3129); 6-*methyl*-3(2H)-*pyridazinethione*, m.p. 203.5–205° (dec.) (*H. Gregory, W.G. Overend* and *L.F. Wiggins*, ibid., 1948, 2199; *Duffin* and *Kendall*, loc. cit.; *R.N. Castle et al.*, J. heterocyclic Chem., 1966, **3**, 79); 6-*chloro*-3(2H)-*pyridazinethione*, m.p. 150° (*A. Pollak, B. Stanovnik* and *M. Tišler*, Canad. J. Chem., 1966, **44**, 829); 6-*bromo*-3(2H)-*pyridazinethione*, m.p. 140–145° (dec.) (*J. Kinugawa, M. Ochiai* and *H. Yamamoto*, Yakugaku Zasshi, 1960, **80**, 1559); 6-*amino*-3(2H)-*pyridazinethione*, m.p. 281–282° (*Castle et al.*, loc. cit.); 6-*hydroxy*-3(2H)-*pyridazinethione*, m.p. 157–158° (*J. Druey, K. Meier* and *K. Eichenberger*, Helv., 1954, **37**, 121); 6-*mercapto*-3(2H)-*pyridazinethione*, m.p. 245–246° (dec.) (*Druey, Meier* and *Eichenberger*, loc. cit.; *Castle et al.*, loc. cit.; *Pollak, Stanovnik* and *Tišler*, loc. cit.); 6-*mercapto*-4-*methyl*-3(2H)-*pyridazinethione*, m.p. 187° (dec.) (*N. Takahayashi*, Chem. pharm. Bull. Japan, 1957, **5**, 229); 4-*amino*-6-*chloro*-3(2H)-*pyridazinethione*, m.p. 185–195° (dec.) (*Kinugawa, Ochiai* and *Yamamoto*, loc. cit.); 6-*methoxy*-3(2H)-*pyridazinethione*, m.p. 205° (dec.) (*B. Stanovnik* and *M. Tišler*, Croat. Chem. Acta, 1964, **36**, 81), m.p. 191–193° (*M. Fujisaka et al.*, Bull. chem. Soc. Japan, 1964, **37**, 1107).

X-ray structural analysis of 3(2H)-pyridazinethione (XXIX) confirms that this compound exists in the thione form in the solid state, with the planar molecules arranged in layers forming dimers by hydrogen bonding between the NH-group of one molecule and the sulphur atom of the next (*C.H. Carlisle* and *M.B. Hossain*. Acta Cryst., 1966, **21**, 249).

Using u.v. and i.r. spectroscopy, 6-mercapto-3(2H)-pyridazinethione has been shown to exist predominantly in the mono-thiol-mono-thione form (XXX) in the solid state and in aquous solution (*Stanovnik* and *Tišler*, loc. cit.; *Fujisaka et al.*, loc. cit.) Similarly, 6-hydroxy-

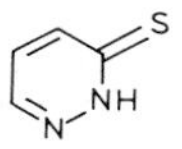

(XXIX) 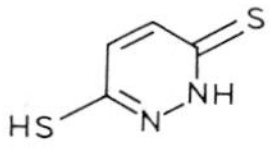 (XXX) 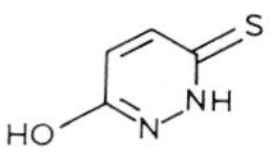 (XXXI) 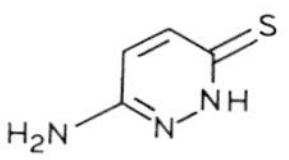 (XXXII)

3(2H)-pyridazinethione exists predominantly in the hydroxythione form (XXXI) and 6-amino-3(2H)-pyridazinethione in the aminothione form (XXXV).

Pyridazinethiones are weaker bases than the corresponding pyridazinones; pK_a values are given in Table 7.

Alkylation of pyridazinethiones with alkyl halides yields exclusively S-alkylated derivatives (cf. pyridazinones, p. 24).

3-Methylthiopyridazine, m.p. 37–38°, b.p. 138°/15 mm; 4-methylthiopyridazine (picrate, m.p. 149–150.5°, hydrochloride, m.p. 190–191°); 6-methyl-3-methylthiopyridazine, b.p. 135–141°/20 mm; 3-ethylthio-6-methylpyridazine, m.p. 41°; 6-chloro-3-methylthiopyridazine, m.p. 103–104°, 6-methoxy-3-methylthiopyridazine, m.p. 87°; 4-methyl-6-methylthio-3(2H)-pyridazinone, m.p. 104°; 4-methyl-3-methoxy-6-methylthiopyridazine, m.p. 58°; 4-methyl-6-methoxy-3-methylthiopyridazine, m.p. 100° (Duffin and Kendall, loc. cit.; N. Takahayashi, Chem. pharm. Bull. Japan, 1957, **5**, 229).

N-Substituted pyridazinethiones are only available by the action of phosphorus pentasulphide on the corresponding pyridazinones.

6-Mercapto-3(2H)-pyridazinethiones are also S-alkylated to give 3,6-dialkylthiopyridazines, as are 6-hydroxy-3(2H)-pyridazinethiones which yield 6-alkylthio-3(2H)-pyridazinones.

Acylation of pyridazinethiones gives only S-acylated products. Pyridazinethiones are oxidised to disulphides by iodine, iron (III) chloride, acid permanganate, and upon long exposure to air. With conc. nitric acid they

TABLE 7

pK_a VALUES FOR PYRIDAZINETHIONES

Compound	Proton gain	Proton loss	
		First	Second
3(2H)-Pyridazinethione	−2.68	8.30	—
4(1H)-Pyridazinethione	−0.75	6.54	—
6-Hydroxy-3(2H)-pyridazinethione	−1.7 (−1.4)	3.6 (3.3)	—
6-Mercapto-3(2H)-pyridazinethione	−0.5	2.1	10.4
6-Amino-3(2H)-pyridazinethione	−0.14	9.05	—
6-Methoxy-3(2H)-pyridazinethione	−2.36 (−2.3)	6.95 (8.5)	—

yield sulphonic acids. Thioethers are oxidised by both hydrogen peroxide and peracids to either sulphoxides or sulphones (*H. Gregory, W.G. Overend* and *L.F. Wiggins*, J. chem. Soc., 1949, 2066; *N. Takahayashi*, J. pharm. Soc. Japan, 1956, **76**, 1293).

A large number of pyridazine "*sulpha drugs*" have been synthesised: for example many 6-substituted 3-sulphanilamidopyridazines have been made by the reaction of an arylsulphonyl chloride with an aminopyridazine in the presence of pyridine, or by heating an arylsulphonamide with a halopyridazine in the presence of base.

3-*Sulphanilamidopyridazine*, m.p. 207°; 6-*methyl*, m.p. 190–191°, 6-*ethyl*, m.p. 160°, and 6-*phenyl*, m.p. 225°, analogues (*C. Grundmann*, Ber., 1948, **81**, 1: *W.G. Overend* and *L.F. Wiggins*, J. chem. Soc., 1947, 239).

(k) Reduced pyridazines

(i) Dihydropyridazines

Dihydropyridazines have been reviewed (*A.L. Weis*, Adv. heterocyclic Chem., 1985, **38**, 23). Derivatives of five dihydropyridazines, 1,2-, 1,4-, 1,6-, 4,5-, and 5,6-dihydropyridazines, are known.

(*A*) 1,2-*Dihydropyridazines* have been prepared by direct cyclisation, reduction of pyridazines and by oxidation of reduced pyridazines.

Ethyl α-cinnamoylacetoacetate with benzenediazonium chloride gives a hydrazone (XXIII), m.p. 86–87°, which rearranges in alcohol forming 3-*carbethoxy*-1,6-*diphenyl*-4-*hydroxy*-1,2-*dihydropyridazine* (XXXIV), m.p. 104–105° (*D. Shapiro et al.*, J. Amer. chem. Soc., 1956, **78**, 2144).

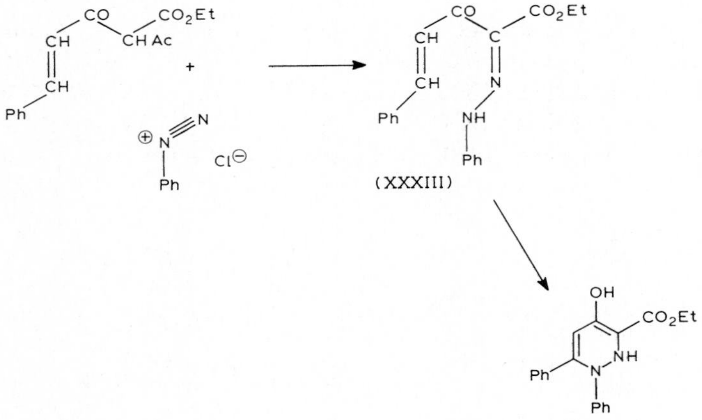

Hydrazobenzene and α-alkylsuccinyl chloride in benzene–pyridine solution form 4-alkyl-3,6-dihydroxy-1,2-diphenyl-1,2-dihydropyridazines, *e.g.* 4-*methyl*, m.p. 183–184°, 4-*ethyl*, m.p. 144–144.5°, and 4-*n-propyl*, m.p. 122.5–124.5° (*F.H. McMillan et al.*, J. Amer. chem. Soc., 1955, **77**, 2900).

Reduction of 3,6-diphenylpyridazine with sodium and ethanol gives 1,2-*dihydro*-3,6-*diphenylpyridazine*, m.p. 170–175° (*G. Rosseels*, Ing. Chim. Brussels, 1964, **46**, 7). This compound is unstable and is oxidised back to the parent pyridazine in air or on attempted acetylation. In alkali it isomerises to the corresponding 4,5-dihydro compound (*K. Alder et al.*, Ann., 1954, **585**, 81).

1,2-**Dicarbethoxy**-1,2-**dihydro**-3,6-**diphenylpyridazine**, m.p. 198°, is formed by selenium dioxide oxidation of the corresponding 1,2,3,6-tetrahydro compound (*Alder et al., loc. cit.*). Treatment of 1,2-dicarbethoxyhexahydropyridazine with alkali gives 1-*carbethoxy*-, m.p. 33–34°, b.p. 68–69°/0.02 mm, and 1,2-*dicarbethoxy*-1,2-*dihydropyridazine*, b.p. 107–109°/0.02–0.03 mm, (*M. Rink, S. Mehta* and *K. Grabowski*, Arch. Pharm., 1959, **292**, 225).

(*B*) 1,4-*Dihydropyridazines* are widely available from 1,4-dicarbonyl compounds and monosubstituted hydrazines (method (2), p. 2) and from the cycloaddition reaction between 1,2,4,5-tetrazines and alkenes (method (5), p. 4), and between diazo compounds and cyclopropenes (p. 5).

Polarographic reduction of pyridazine in alkaline solution yields 1,4-dihydropyridazine (*S. Millefiori*, Ann. Chim., 1969, **59**, 15). The addition of *t*-butylmagnesium chloride to 3,6-disubstituted pyridazines gives the corresponding 4- and 5-*t*-butyl-3,6-disubstituted-1,4-dihydropyridazines. Thus 4,5-di-*t*-butyl-3,6-dimethoxy-1,4-dihydropyridazine, characterised only by its n.m.r. spectrum, is prepared from 4-*t*-butyl-3,6-dimethoxypyridazine and *t*-butylmagnesium chloride. Acids rapidly convert this into *cis*-4,5-di-*t*-butyl-3,6-dimethoxy-4,5-dihydropyridazine, and more slowly into the *trans* isomer, m.p. 68–70° (*I. Crossland*, Acta Chem. Scand., 1972, **26**, 3257).

(*C*) 1,6-*Dihydropyridazines* are obtained by the peroxide-induced rearrangement of 1-phenyliminopyridines. Peroxide attack at either C-2 or C-5 of the pyridine ring (XXXV), followed by ring opening and recyclisation leads to the isomers (XXXVI) and (XXXVII) (*V. Snieckus* and *G. Kan*, Tetrahedron Letters, 1970, 2267).

Simple 1,6-dihydropyridazines have been prepared from 1-methylpyridazinium salts and sodium borohydride (*C. Kaneko, T. Tsuchiya* and *H. Igeta*, Chem. pharm. Bull. Japan, 1974, **22**, 2844). They slowly decompose in air at room temperature.

(*D*) 4,5-*Dihydropyridazines* are formed from the condensation of di- and poly-substituted saturated 1,4-diketones with hydrazines. They also result from the addition of Grignard reagents or organolithium compounds to 3,6-disubstituted pyridazines. Thus 3-methoxy-6-phenylpyridazine reacts with *t*-butylmagnesium chloride to give a mixture of 4- and 5-*t*-butyl-3-methoxy-6-phenyl-4,5-dihydropyridazines (not separated, but characterised by hydrolysis). Bromination in cold hydrochloric acid gives 5-*bromo*-4-*t*-*butyl*-3-*methoxy*-6-*phenyl*-4,5-*dihydropyridazine*, m.p. 134–136° (dec.) (*I. Crossland* and *L.K. Rasmussen*, Acta Chem. Scand., 1965, **19**, 1652).

3,6-**Diphenyl**-4,5-**dihydropyridazine**, m.p. 149°, is prepared from acetophenone hydrazone *via* the 1,6-dilithio salt (XXXVIII) which is cyclised by 2,3-dibromo-2,3-dimethylbutane *via* the bromolithio salt (XXXIX) (*F.E. Henoch, K.G. Hampton* and *C.R. Hauser*, J. Amer. chem. Soc., 1969, **91**, 676). It is also obtained by isomerisation in alkali of the corresponding 1,2-dihydro compound (*Alder et al., loc. cit.*) (p. 31). Although this compound is reasonably stable, other 4,5-dihydropyridazines readily dimerise or trimerise (*B.K. Bandlish et al.*, J. org. Chem., 1973, **38**, 1102).

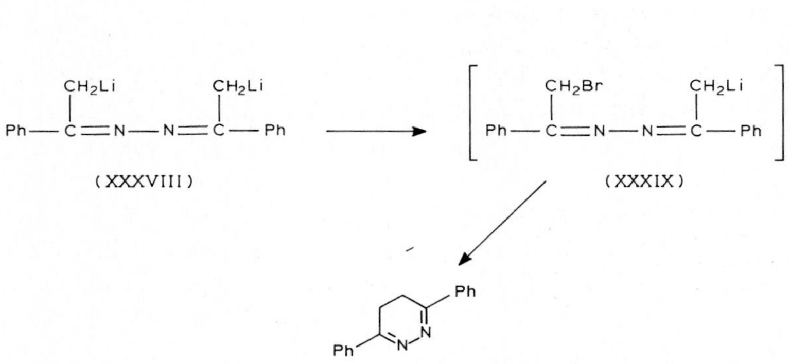

(*E*) 5,6-*Dihydropyridazines* are rare. 6-*n*-*Butyl*-3,6-*dimethyl*-5,6-*dihydropyridazine*, no b.p. given, is obtained by halogenation and subsequent dehydrohalogenation of the corresponding 1,4,5,6-tetrahydro compound. On heating it rearranges to the 1,6-dihydro derivative (*P. De Mayo* and *M.C. Usselham*, Canad. J. Chem., 1973, **51**, 1724).

(*F*) 3,6-*Dihydropyridazines* are known only as labile intermediates and rapidly lose nitrogen even at $-78°$. However, the corresponding 2-oxides are stable and lose N_2O only at $300°$ (*J.F.M. Oth, H. Olsen* and *J.P. Snyder*, J. Amer. chem. Soc., 1977, **99**, 8505).

(ii) Tetrahydropyridazines

(*A*) 1,2,3,6-*Tetrahydropyridazines* can be prepared by the cycloaddition of an azo compound and a conjugated diene (method (5), p. 4).

Dimethyl azodicarboxylate with butadiene gives 1,2-*dicarbomethoxy*-1,2,3,6-*tetrahydropyridazine*, m.p. 38°, b.p. 149.5°/12 mm, (*dibromide*, m.p. 125°), and with *trans,trans*-1,4-diphenylbutadiene, it yields 1,2-*dicarbomethoxy*-3,6-*diphenyl*-1,2,3,6-*tetrahydropyridazine*, m.p. 132° (*Alder et al., loc. cit.*). With *trans,trans*-hexa-2,4-diene, the azo ester gives a quantitative yield of 1,2-dicarbomethoxy-3,6-dimethyl-1,2,3,6-tetrahydropyridazine. The reaction with *cis,trans*-hex-2,4-diene is slower and yields a 4:1 mixture of the *trans*- and the *cis*-1,2-dicarbomethoxy-3,6-dimethyl-1,2,3,6-tetrahydropyridazine. Hydrolysis and decarboxylation yield the *trans*- and the *cis*-3,6-dimethyl-1,2,3,6-tetrahydropyridazine, respectively (*J.A. Berson* and *S.S. Olin*, J. Amer. chem. Soc., 1969, **91**, 777; *R. Daniels* and *K.A. Roseman*, Tetrahedron Letters, 1966, 1335). In some cases, however, removal of the carboxyl groups proceeds with migration of the double bond to give 1,4,5,6-tetrahydropyridazines (see below).

The conformation of 1,2,3,6-tetrahydropyridazines has been studied by variable temperature ^{1}H-n.m.r. spectroscopy. Ring inversion (XL) as well as nitrogen inversion is frequently observed. With *N*-carboxyalkyl groups present, restricted rotation about the N–COOR bonds is also involved and interaction between the *N*-acyl substituents produces unusually high energy barriers for ring inversion. (75–78 kJ mol^{-1}).

(XL)

1,2,3,6-Tetrahydropyridazines are oxidised to pyridazines by bromine and reduced to hexahydropyridazines. Reduction of 1,2-carbethoxy-1,2,3,6-tetrahydropyridazines by lithium aluminium hydride yields 1,2-dimethyl-hexahydropyridazines (see below).

(*B*) 1,4,5,6-*Tetrahydropyridazines* are sometimes produced on decarboxylation of the adducts formed from azo compounds and conjugated dienes (see above). Thus the adduct (XLI) of diethyl azodicarboxylate and 1,4-di-

(XLI)

phenylbutadiene, on hydrolysis followed by decarboxylation gives 3,6-*diphenyl*-1,4,5,6-*tetrahydropyridazine*, m.p. 157–159°. The same compound is also obtained when the product from the reaction of 1,2-dibenzoylethane and hydrazine is hydrogenated, and also by heating α-chloracetophenone semicarbazone with sodium bicarbonate in aqueous solution (*S.G. Cohen et al.*, J. Amer. chem. Soc., 1957, **79**, 4400).

1,4,5,6-Tetrahydropyridazines are produced by ring enlargement of 1-aminopyrrolidines (p. 5).

1,4,5,6-Tetrahydropyridazines are oxidised to pyridazines by lead dioxide and reduced to hexahydropyridazines either catalytically or with lithium aluminium hydride.

1,3-**Diphenyl**-1,4,5,6-**tetrahydropyridazine**, m.p. 138–139°, is prepared from 3-chloropropyl phenyl ketone and phenylhydrazine (*J.B. Conant et al.*, J. Amer. chem. Soc., 1924, **46**, 1882).

(C) 3,4,5,6-*Tetrahydropyridazines*

3,6-**Diphenyl**-3,4,5,6-**tetrahydropyridazine**, m.p. 120° (dec.) is obtained by reduction of the adduct (XXXIII), followed by hydrolysis, decarboxylation and autoxidation of the product in ether. It isomerises to 3,6-diphenyl-1,4,5,6-tetrahydropyridazine on heating or in acetic acid (*C.H. Wang et al.*, J. Amer. chem. Soc., 1957, **79**, 2661; *Cohen et al.*, loc. cit.). *meso*-(*cis*)- and *dl*-(*trans*)-3,6-Diethyl-3,6-dimethyl-3,4,5,6-tetrahydropyridazine are obtained by the oxidation of *meso*- and *dl*-3,6-diamino-3,6-dimethyloctane, respectively, with iodine pentafluoride at −20° (*P.D. Bartlett* and *N.A. Porter*, J. Amer. chem. Soc., 1968, **90**, 5317).

(D) 2,3,4,5-*Tetrahydropyridazines*

2,3,4,5-Tetrahydropyridazines have been prepared from 4,5-dihydropyridazines, for example, 3-*n*-**butyl**-3,6-**dimethyl**-2,3,4,5-**tetrahydropyridazine**, b.p. 52°/0.2 mm, (*P. De Mayo, J.B. Stothers* and *M.C. Usselman*, Canad. J. Chem., 1972, **50**, 612), and 3,6-**dimethyl**-3-**phenyl**-2,3,4,5-**tetrahydropyridazine**, m.p. 84.5–87.5° (*C.G. Overberger* and *G. Kesslin*, J. org. Chem., 1962, **27**, 3898) are made by the addition of *n*-butyllithium or phenyllithium to 3,6-dimethyl-4,5-dihydropyridazine.

(iii) Hexahydropyridazines (piperidazines)

Piperidazine, b.p. 149–150°, 57–58°/15 mm, (*picrolonate*, m.p. 190°), is prepared by acid hydrolysis of its *N,N*-di-*iso*-butyryl derivative, b.p. 121–124°/2 mm, which is obtained from *N,N'*-di-*iso*-butyrylhydrazine and trimethylene dibromide in DMF (*H. Stetter* and *H. Spangenberger*, Ber., 1958, **91**, 1982). The reduction of pyridazine with sodium in boiling ethanol yields piperidazine together with 1,4-diaminobutane formed by N–N bond fission (*R. Marquis*, Compt. rend., 1903, **136**, 368). Another method involves the hydrogenation and

subsequent hydrolysis and decarboxylation of 1,2-dicarbethoxy-1,2,3,6-tetrahydropyridazine (*M. Rink, S. Mehta* and *K. Grabowski,* Arch. Pharm., 1959, **292**, 225).
Piperidazine reacts with succinic anhydride at 160° and the resulting 1,2-*succinylpiperidazine* (XLII), m.p. 179–180°, with lithium aluminium hydride in tetrahydrofuran gives 10-azaquinolizidine (XLIII) (*Stetter* and *Spangenberger, loc. cit.*).

(XLII)   (XLIII)   (XLIV)

Compound (XLII) is also obtained directly from piperidazine by reaction with 1,4-dihalobutanes. Piperidazine reacts with formaldehyde and with benzaldehyde to give (XLIV) and its diphenyl analogue respectively (*M. Rink* and *S. Mehta,* Naturwiss., 1958, **45**, 313).
Hydrogenation of the double bond of a number of 1,2,3,6-tetrahydropyridazines has resulted in the preparation of the following hexahydropyridazines: 3-*methyl,* b.p. 73°/29 mm (*Y.S. Shabarov, M.G. Kuz'min* and *R.Y. Levina,* Zhur. obshchei Khim., 1960, **30**, 2473); 4-*methyl,* b.p. 165° (*picrolonate,* m.p. 199°) (*P. Baranger* and *J. Levisalles,* Bull. Soc. chim. Fr., 1957, 704); 3,6-*dimethyl,* b.p. 65°/7 mm (*S.G. Cohen* and *R. Zand,* J. Amer. chem. soc., 1962, **84**, 586); 4,5-*dimethyl,* b.p. 90°/30 mm (*B.T. Gillis* and *P.E. Beck,* J. org. Chem., 1962, **27**, 1947); and 3-*phenyl,* b.p. 158–160°/10 mm, (*Y.S. Shabarov et al.,* Zhur. obshchei Khim., 1963, **33**, 1206).
1,2-**Dimethylhexahydropyridazine**, b.p. 140–141°, is obtained by lithium aluminium hydride reduction of 1,2-dicarbethoxy-1,2,3,6-tetrahydropyridazine (*H.R. Snyder* and *J.G. Michels,* J. org. Chem., 1963, **28**, 1144).
Conformational studies of hexahydropyridazines have been made using n.m.r. spectroscopy and photoelectron spectroscopy (see *S.F. Nelson,* Acc. chem. Res. 1978, **11**, 14).

2. Cinnolines

(a) Introduction

The chemistry of the two benzopyridazines*, cinnoline and phthalazine, has been extensively studied although none of their derivatives occur naturally. However, research interest in this area has been stimulated due to the medicinal applications of some of their derivatives.

---

* Condensed Pyridazine and Pyrazine Rings by *J.C.E. Simpson,* in The Chemistry of Heterocyclic Compounds, Vol. 5, ed. *A. Weissberger,* Interscience, New York, 1953, reviews the literature up to 1949, and *G.M. Singerman, ibid.,* Vol. 27, ed. *R.N. Castle,* Interscience, New York, 1973, covers work up to 1971.

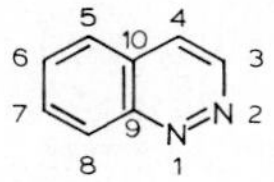

Cinnoline
(1,2-diazanaphthalene, benzo[c]pyridazine)

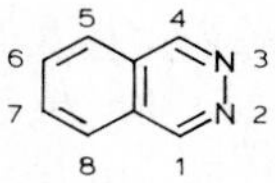

Phthalazine
(2,3-diazanaphthalene, benzo[d]pyridazine)

(b) Methods of synthesis

The general methods of synthesis depend on the cyclisation of diazonium salts.

(*1*) The cinnoline ring system was discovered in 1883 by von Richter and his original synthesis involves the intramolecular cyclisation of the diazonium salt of *o*-aminophenylpropiolic acid, which on heating yields 4-hydroxycinnoline-3-carboxylic acid (*V. von Richter*, Ber., 1883, **16**, 677)*.

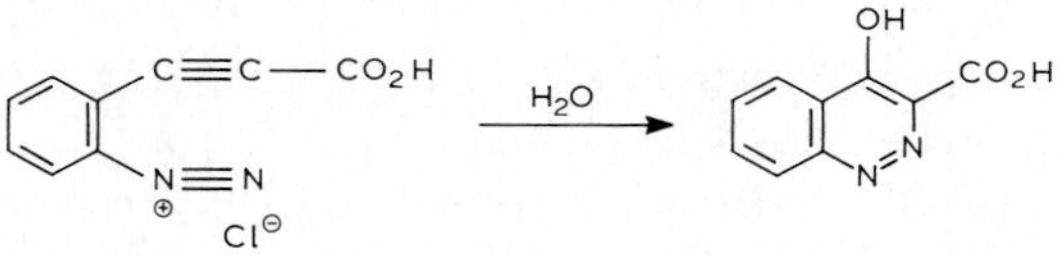

The von Richter reaction has been extended and represents a general method for the synthesis of 4-hydroxycinnolines from diazotised *o*-aminophenylacetylenes (*K. Schofield* and *J.C.E. Simpson*, J. chem. Soc., 1945, 512). However, the reaction is of quite limited application and 4-hydroxycinnolines are more conveniently prepared *via* the Borsche synthesis.

(*2*) *The Widman–Stoermer synthesis.* One of the most widely used methods of synthesis of 4-substituted and/or 3,4-disubstituted derivatives involves the cyclisation of diazotised *o*-aminoarylethylenes (*O. Widman*, Ber., 1884, **17**, 722; *R. Stoermer* and *H. Fincke*, ibid., 1909, **42**, 3115). The

---

* Throughout this section, such names as 4-hydroxycinnoline will be used without implying that the tautomer with a hydroxyl group is necessarily present in more than a trace quantity at equilibrium. 4-Hydroxycinnolines are known to exist predominantly in the oxo form, *i.e.* 4(1*H*)-cinnolinone.

synthesis is quite general and R^1 can be alkyl, aryl, or heteroaryl but not hydrogen, and R^2 can be hydrogen, alkyl and aryl. The cyclisation is facilitated when R^1 is an electron donor, e.g. Me or Ph, and the presence of electron-withdrawing groups in the side-chain prevents the cyclisation step.

(*3*) *The Borsche synthesis.* Related to the Widman–Stoermer reaction the Borsche synthesis utilises diazotised *o*-aminoacetophenone derivatives to yield 4-hydroxycinnolines (*W. Borsche* and *A. Herbert*, Ann., 1941, **546**, 293). A mechanism involving the cyclisation of the enol form of acetophenone appears probable. This reaction provides a general method

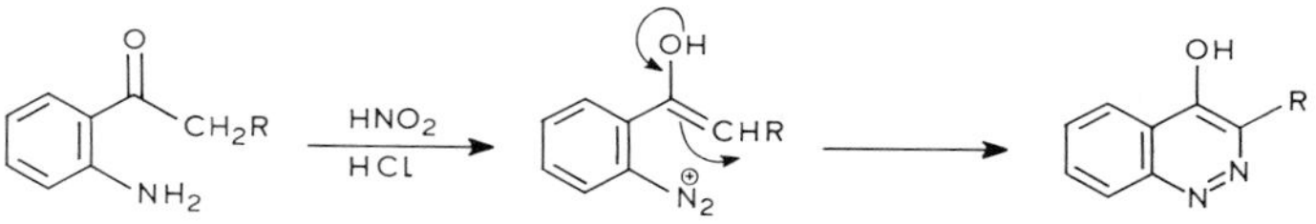

for obtaining cinnoline derivatives with alkyl, aryl, or other substituents in the benzene ring of the bicyclic system. The cyclisation is slower than in method (2) but is facilitated by electron-withdrawing groups *meta* to the acetyl group. Yields are generally good but heat must be avoided to prevent decomposition of the diazonium salt to the phenol. Concentrated hydrochloric acid assists ring closure but can lead to the replacement of a nitro group by chlorine, e.g. 2-amino-4-chloro-5-nitroacetophenone yields 6,7-dichloro-4-hydroxycinnoline (*C.M. Atkinson* and *J.C.E. Simpson*, J. chem. Soc., 1947, 232).

Baumgarten devised an important modification of the Borsche synthesis which involves coupling of the diazotised *o*-aminobenzaldehyde or *o*-aminoacetophenone with nitromethane in a dilute, basic solution to give nitroformaldehyde 2-formyl/acetyl-phenylhydrazone in excellent yields. The hydrazone is then cyclised in the presence of aluminium oxide, sodium hydroxide, or an anion exchange resin to give a nitrocinnoline (*H.E. Baumgarten* and *M.R. Debrunner*, J. Amer. chem. Soc., 1954, **76**, 3489).

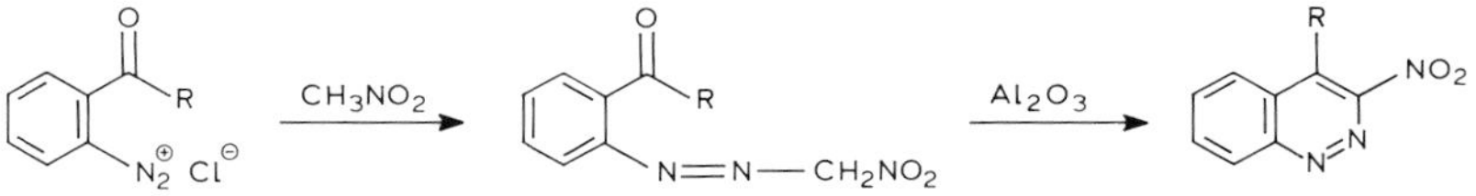

In an improvement to the synthesis of cinnolines from *o*-aminoacetophenone, the methyl group has been activated by substitution with a methylsulphinyl group. Diazotisation of *o*-amino-α-(methylsulphinyl)acetophenones gives the corresponding 3-methylsulphinylcinnoline-4(1H)-ones in good yield. Removal of the methylsulphinyl group is achieved by treatment with Raney nickel (*M. von Strandtmann et al.*, J. heterocyclic Chem., 1972, **9**, 173). By this method various 5- (R = Me, H, Cl, MeO), 6- (H, Cl, MeO) substituted 3-methylsulphinylcinnolinones and derivatives have been prepared and shown to be useful immunosuppressants (*M. von Strandtmann et al.*, U.S. Pat. 3,798,219; 3,937,704; C.A., 1974, **80**, 146187; 1976, **84**, 150650). The preparation of 3-methanesulphonyl-4-hydroxycinnoline derivatives has been described (*R. Albrecht*, Ann., 1978, 617). The 3-methanesulphonyl group may be replaced by cyano by heating it with KCN in DMF at 120° for 24 h.

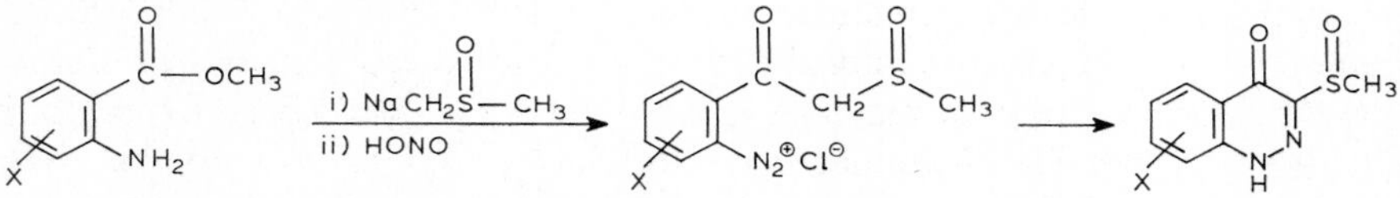

(*4*) Barber and co-workers devised an alternative approach to the synthesis of 4-hydroxycinnolines starting from readily accessible diethyl mesoxalate phenylhydrazones. Phenylhydrazine reacts with diethyl mesoxalate to give the corresponding hydrazone. Saponification and reaction with thionyl chloride gives the acid chloride which undergoes Friedel–Crafts catalysed cyclisation to give 4-hydroxycinnoline-3-carboxylic acids, which on heating decarboxylate to give 4-hydroxycinnolines (*H.J. Barber et al.*, J. chem. Soc., 1961, 2828).

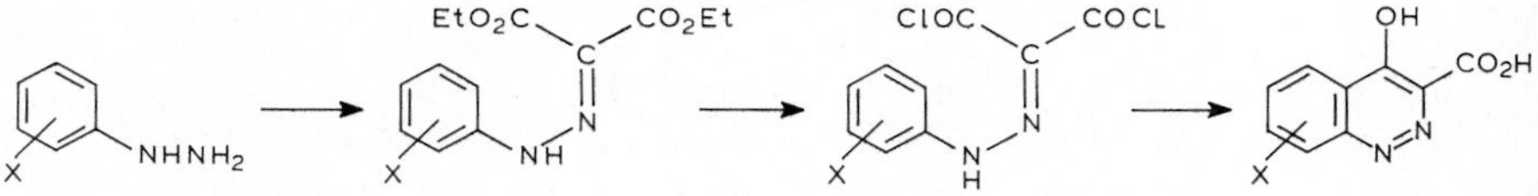

In a synthesis based on the work of Barber, 4-methoxy-2-nitroaniline has been converted in seven steps into 8-amino-6-methoxycinnoline in an overall yield of 33% (*R.A. Coburn* and *D. Gala*, J. heterocyclic Chem., 1982, **19**, 757). In a similar procedure benzaldehyde phenylhydrazone is treated with oxalylchloride, followed by Friedel–Crafts cyclisation of the resulting *N*-benzylideneamino-*N*-phenyloxamoyl chloride to give *N*-benzylideneaminoisatin. The latter with hot aqueous potassium hydroxide gives the cinnoline derivative in good yield.

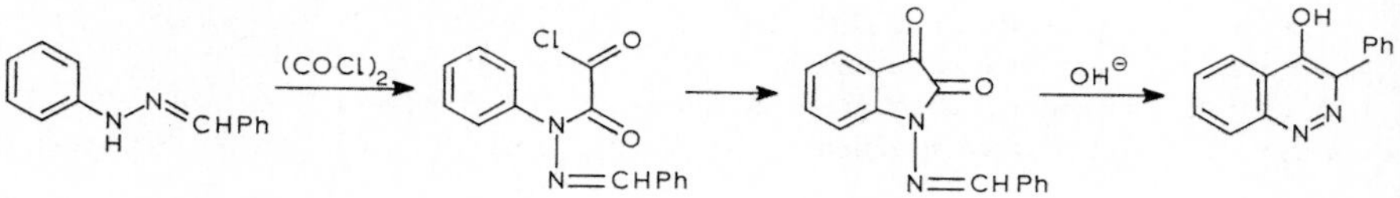

(c) Cinnoline and its alkyl and aryl derivatives

(i) Cinnoline

Cinnoline is most conveniently prepared by the thermal decarboxylation of cinnoline-4-carboxylic acid in benzophenone at about 160°, a small amount of 4,4′-dicinnolinyl, m.p. 237–238°, being formed as a by-product (*T.L. Jacobs et al.*, J. Amer. chem. Soc., 1946, **68**, 1310). An alternative route is the reduction of 4-chlorocinnoline, with iron and sulphuric acid, to give 1,4-dihydrocinnoline which can then be oxidised to cinnoline with mercury(II) oxide (*M. Busch* with *M. Klett*, Ber., 1892, **25**, 2847; with *A. Rast*, Ber., 1897, **30**, 521). The most common method used for obtaining 4-chlorocinnoline is by treatment of the readily prepared 4-hydroxycinnoline with phosphorus oxychloride. Reduction either with lithium aluminium hydride or by catalytic methods in methanol with palladium on calcium carbonate gives 4,4′-dicinnolinyl as the main product. However, catalytic reduction of 4,7-dichlorocinnoline using the above conditions gives 7-chlorocinnoline (*J.S. Morley*, J. chem. Soc., 1951, 1971). Heterocyclic halogens are often best replaced by a method introduced by *M.J.S. Dewar* (*ibid.*, 1944, 619) and widely used by *A. Albert* and his co-workers (*e.g., ibid.*, 1948, 1284; 1949, 1148). The reactive halocinnoline is first condensed with toluene-*p*-sulphonylhydrazide and the product decomposed by aqueous sodium carbonate to give cinnoline in good yield (*E.J. Alford* and *K. Schofield*, *ibid.*, 1953, 609).

Direct reduction of 4-hydroxycinnoline with lithium aluminium hydride in refluxing tetrahydrofuran for 8 h followed by mercury(II) oxide oxidation of the partially reduced cinnoline gives cinnoline in 74% yield (*C.M. Atkinson* and *C.J. Sharpe*, *ibid.*, 1959, 2858). A similar reduction of 3-hydroxycinnoline using lithium aluminium hydride in refluxing 1,2-dimethoxyethane also gives 1,2,3,4-tetrahydrocinnoline as one of the partially reduced products (*D.E. Ames* and *H.Z. Kucharska*, *ibid.*, 1962, 1509).

Cinnoline, m.p. 40–41°, b.p. 114°/0.35 mm, is a pale yellow solid which is soluble in water and organic solvents. It may be stored as such under nitrogen at 0° but rapidly liquifies and turns green on standing in air (*T.L. Jacobs et al.*, J. Amer. chem. Soc., 1946, **68**, 1310), apparently with little decomposition (*J.S. Morley*, J. chem. Soc., 1951, 1971). It crystallises from ether as a colourless *etherate complex* m.p. 24–25° (*M. Busch* and *A. Rast*, Ber., 1897, **30**, 521). It is a weak base, pK_a 2.70 (*A. Albert et al.*, J. chem. Soc., 1948, 2240) or 2.29 (*A.R. Osborn et al., ibid.*, 1956, 4191) at 20° in water and forms the following stable salts: *hydrochloride*, m.p. 154–156°, sublimes at 110–115°/3 mm; *picrate*, m.p. 196°; *methiodide*, dark reddish-brown crystals, m.p. 168° (*Jacobs et al., loc. cit.*); *chloroplatinate*, m.p. 280° (dec.) and *aurichloride*, m.p. 146° (*Busch* and *Rast, loc. cit.*).

Cinnoline does not yield crystals of sufficient quality for a structure determination by X-ray crystallography, however the structure of 4-methylcinnoline has been determined. The planar molecules crystallise with monoclinic P2/C symmetry (*G.J. von Hummel et al.*, Acta Cryst., 1979, **35B**, 516). Several molecular orbital calculations of the π-electron density distribution in cinnoline have been made by the Hückel method (*H.C. Longuet-Higgins* and *C.A. Coulson*, J. chem. Soc., 1949, 971; *S.C. Wait Jr.* and *J.W. Wesley*, J. mol. Spectrosc., 1966, **19**, 25). In a study of various heterocyclic systems including cinnoline, a linear correlation between chemical shifts and electron densities has been used to establish the validity of the Hückel MO method in predicting the order of chemical shifts (*M.K. Mahanti*, Indian J. Chem., 1977, **15B**, 168). The CNDO method has also been used (*R.J. Pugmire et al.*, J. Amer. chem. Soc., 1969, **91**, 6381), and in each case the highest electron density for the ring carbon atoms was shown to be at positions 5 and 8. On the basis of molecular orbital calculations it has been predicted that the preference for electrophilic substitution at different positions of the cinnoline ring system is in the order $5 = 8 > 6 = 7 > 3 \gg 4$ (*M.J.S. Dewar* and *P.M. Maitlis*, J. chem. Soc., 1957, 2521). These calculations are in good qualitative agreement with the observed reactivity of cinnoline with simple electrophiles (*J.S. Morley, ibid.*, 1951, 1971; *E.J. Alford* and *K. Schofield, ibid.*, 1953, 609). Nitration in sulphuric acid results in the formation of 33% 5-nitrocinnoline and 28% 8-nitrocinnoline as the sole nitration products. The actual species which undergoes nitration is in fact the 2-cinnolinium cation (*R.B. Moodie et al.*, J. Chem. Soc. B, 1968, 312). A higher π-electron density has been assigned to N-1 than to N-2, although it has been shown experimentally that cinnoline undergoes *N*-oxidation (*I. Suzuki et al.*, Tetrahedron Letters, 1966, 2899), protonation (*D.E. Ames et al.*, Chem. Ind., 1966, 458), and alkylation (*D.E. Ames* and *H.Z. Kucharska*, J. chem. Soc., 1964, 283) preferentially at N-2. Proof was obtained by showing that the u.v. spectra of protonated 3-, 4-, and 8-methylcinnoline are very similar to the spectra of the corresponding 2,3-, 2,4- and 2,8-dimethylcinnolinium perchlorates, the spectra of the 1,3-, 1,4- and 1,8-dimethylcinnolinium salts being quite different (*D.E. Ames et al.*, J. chem. Soc., B, 1967, 748). The relative rates of methylation of cinnoline at N-1 and N-2 with methyl iodide in dimethyl sulphoxide at 23° have been studied by n.m.r. spectroscopy. The rates were faster than predicted and the reaction led to a mixture containing N-2 and N-1 methylated products in a ratio of 9:1 (*J.A. Zoltewicz* and *L.W. Deady*, J. Amer. chem. Soc., 1972, **94**, 2765). However molecular orbital calculations have shown the electron densities N-1 and N-2 to be similar and the difference in protonation may be due to steric hindrance arising from the proximity of C-8 (*M.H. Palmer et al.*, Tetrahedron, 1971, **27**, 2921). MO calculations using the CNDO/2 method confirms that the electron densities are essentially equal at N-1 and N-2 in cinnoline (*M.K. Mahanti*, Indian J. Chem., 1979, **18B**, 359). The preferential N-2 reactivity due to steric hindrance of N-1 reactivity by the peri C-8 proton can be balanced by a substituent in the 3-position. STO-3G minimal basis set *ab initio* calculations have been used to determine the charge distribution of various nitrogen-containing heterocycles including cinnoline and other diazanaphthalenes. The electron density is slightly higher at N-1 and the results are in good agreement with ^{15}N-n.m.r. measurements (*F. Escudero et al.*, J. chem. Soc., Perkin II, 1983, 1735). Calculations of the most probable protonation sites for the four asymmetric diazanaphthalenes have been carried out (*P. Van de Weijer et al.*, Theor. Chim. Acta, 1975, **38**, 223) using the molecular-potential method (*R. Bonaccorsi et al., ibid.*, 1971, **20**, 331). Only the results for cinnoline disagreed with the protonation studies done by n.m.r. spectroscopy (*P. Van de Weijer et al.*, Org. magn. Resonance, 1976, **8**, 187). This discrepancy is attributed to the fact that the molecular potential is calculated on the basis of a naphthalene-like geometry. Calculation of the molecular potential on the basis of geometries extrapolated from the crystal structure of 4-methylcinnoline indicate that protonation should occur preferentially at the β-site, in agreement with the results of the n.m.r. experiment. The first and second nitrogen protonation in cinnoline and other diazanaphthalenes has been investigated using ^{13}C-n.m.r. spectroscopy (*P. van de Weijer et al., ibid.*, 1977, **9**, 281). The ^{13}C chemical shifts were reported as a function of pH to give classical titration curves from which pK_1 and pK_2 were determined.

The enthalpy and free energy of proton transfer between cinnoline and reference bases in the gas phase has been determined from measurements taken in the ion source of a pulsed high-pressure mass spectrometer, thus enabling evaluation of intrinsic basicity (*M. Meot-Ner*, J. Amer. chem. Soc., 1979, **101**, 2396). A comparison is given of the trends in the relation between gas-phase and solution basicities. Lone pair charges have been calculated by using *ab initio* molecular orbital theory at the STO-3G level and a good correlation is obtained between the charge values and experimental gas-phase basicities mentioned above (*J. Catalan et al., ibid.*, 1984, **106**, 6552), and measurements for 4-methyl- and 4-amino-cinnoline indicate N-1 to be the more basic centre (*C.M. Atkinson* and *J.C.E. Simpson*, J. chem. Soc., 1949, 1354).

Values for molar polarisation at infinite dilution, molar refractions and dipole moments have been given for cinnoline and its 4-methyl analogue (*M.R. Rogers* and *T.W. Campbell*, J. Amer. chem. Soc., 1953, **73**, 1209). The dipole moment for cinnoline in benzene solution (4.1 D) is slightly lower than that observed for pyridazine (4.32 D).

The first and second ionisation potentials of cinnoline have been determined by photoelectron spectroscopy (*M.J.S. Dewar* and *S.D. Warley*, J. chem. Phys., 1969, **51**, 263). The first ionisation potential of 8.51 eV corresponds to loss of the nitrogen lone-pair electrons (phthalazine 8.68 eV), and the second is a π ionisation (9.03 eV). The photoelectron spectroscopy of the azabenzenes and azanaphthalenes has been reviewed (*Warley*, Chem. Reviews., 1971, **71**, 295). However, an assignment taking into account the large through-bond interaction (*R. Hoffmann*, Acc. chem. Res., 1971, **4**, 1) between nitrogen lone-pair orbitals has been reported for bands with $I_v < 13$ eV in the photoelectron spectra of cinnoline and other azanaphthalenes (*F. Brogli*, Helv., 1972, **55**, 274). The high-resolution photoelectron spectra (He, 584 Å) and calculated ionisation potentials of cinnoline and nine other diazanaphthalenes give further support to the interaction of non-bonding electrons of equivalently placed atoms as suggested by Hoffmann. The first ionisation in cinnoline arises from an *n*-orbital whereas the first ionisation in phthalazine proceeds from a π-orbital (*D.M.W. Van der Ham* and *D. Van der Meer*, Chem. phys. Letters, 1972, **2**, 247).

The heat of atomisation of cinnoline has been calculated to be 79.167 eV, similar to the value obtained for phthalazine (79.215 eV) (*M.J.S. Dewar* and *T. Morita*, J. Amer. chem. Soc., 1969, **91**, 769).

(ii) Spectra

The u.v.-absorption spectrum of cinnoline has been measured in several different solvents including ethanol (*J.M. Hearn et al.*, J. chem. Soc., 1951, 3318), cyclohexane (*G. Farini et al.*, Nuovo Cimento, 1959, **8**, 60), and in neutral and acidic aqueous solution (*A.R. Osborn et al.*, J. chem. Soc., 1956, 4191). Theoretical calculations of transition energies and band intensities have also been made (*G. Farini et al.*, Theor. Chim. Acta, 1965, **3**, 45, 418; *L. Goodman* and *R.W. Harrell*, J. chem. Phys., 1959, **30**, 1131). The spectrum displays three to six absorption maxima in the 200–380 nm range which are attributable to $\pi - \pi$ * transitions. The lower energy $n - \pi$ * transition arising from the nitrogen lone-pair gives a weak absorption at approximately 390 nm. A comparison of the u.v. spectra of cinnoline, phthalazine and naphthalene has been reported (*E.D. Amstutz*, J. org. Chem., 1952, **17**, 1508). A modified INDO method has been used to calculate the electronic spectrum of various diazanaphthalenes including cinnoline, and the better classified bands of the series can be reproduced (*J.E. Ridley* and *M.S. Zerner*, J. mol. Spectrosc., 1974, **50**, 457). A reported temperature variation in the electronic spectrum of cinnoline in non-polar solvents appears to be explicable in terms of various molecular interactions (*J.R. Honner et al.*, Austral. J. Chem., 1974, **27**, 1613). In studies of the protonation site of 3-, 4-, and 8-methylcinnoline, u.v. spectra of the above compounds in acidic solution were compared with those of the 2,3-, 2,4-, and 2,8-dimethylcinnolinium perchlorates (*D.E. Ames et al.*, J. chem. Soc. B, 1967, 748). U.v. spectra of various other alkyl- and aryl-cinnolines have been reported (*A.R. Osborn et al.*, J. chem. Soc., 1956, 4191; *J.M. Bruce, ibid.*, 1959, 2366; *H.S. Lowrie*, J. med. Chem., 1966, **9**, 784). U.v. spectra have been reported for various hydroxy-, alkoxy-, amino-, nitro- and halo-cinnolines, and the data is discussed in terms of the possible tautomers present. (*J.C.E. Simpson et al.*, J. chem. Soc., 1951, 1971). The u.v. spectrum of cinnoline in the vapour state has been recorded (*S.C. Wait, Jr.* and *F.M. Grogan*, J. mol. Spectrosc., 1967, **24**, 383; *R.W. Glass et al.*, J. chem. Phys., 1970, **53**, 3857). The first electronic absorption region ($n - \pi$ *) of cinnoline has been measured in the solid state at 4 K using durene and naphthalene as host crystals (*A.D. Jordan* and *G.I. Ross*, J. mol. Spectrosc., 1973, **46**, 316). It was analysed as a single transition in agreement with vapour-phase studies. The salient properties of the first $\pi - \pi$ * transition are also reported.

A mint-green emission at 22° from liquid solutions of cinnoline, as well as its 4-methyl derivative, has been observed and assigned as fluorescence from the lowest excited singlet (n, π * state). The fluorescences have lifetimes less than 2 ns with quantum yields of 1.8×10^{-3} for cinnoline and 1.4×10^{-3} for 4-methylcinnoline (*J.A. Stikeleather*, Chem. phys. Letters, 1973, **21**, 326). Cinnoline and its 4-methyl derivative are non-phosphorescent in hydrocarbon rigid glass solvents, but undergo phosphorescence activation in hydroxylic rigid glass solvents. These yellow phosphorescences (o,o bands at 515 nm) are of (π, π *) character having lifetimes of 0.17 and 0.23 s, respectively (*Stikeleather, ibid.*, 1974, **24**, 253). The energies of the low-lying triplet states of azanaphthalenes have been calculated both by the CNDO/S CI method and by a direct SCF calculation of the appropriate states by a variant of the CNDO/S method (*H.H. Jaffe* and *C.A. Masmanidis, ibid.*, 1974, **24**, 416). A pulsed-frequency doubled-dye laser has been used to study the time and wavelength resolved emission from cinnoline vapours (*J. McDonald* and *L.E. Bru*, J. chem. Phys., 1974, **61**, 3895). Cinnoline did not exhibit any detectable long-lived phosphorescence or intermediate strong coupling emission. Excited state absorption spectra and intersystem crossing kinetics in cinnoline and other diazanaphthalenes have been obtained for their solutions at room temperature, and transient bands have been assigned to $Sn - S_1$ and $Tn - T_1$ absorptions (*G.W. Scott et al., ibid.* 1980, **72**, 5002).

The infrared absorption spectra of cinnoline and other diazanaphthalenes have been assigned where possible (*W.L.F. Armarego et al.*, Spectrochim. Acta, 1966, **22**, 117, *R.W. Mitchell et al.*, J. mol. Spectrosc., 1970, **36**, 310). Infrared spectra of alkyl- and aryl-cinnolines have been recorded (*H.E. Baumgarten* and *J.C. Furnas*, J. org. Chem., 1961, **26**, 1536). The Raman spectrum of an aqueous solution of cinnoline has been obtained using an Ar laser source (*Mitchell et al.*, *loc. cit.*). Experimental studies of the variation of intensity with excitation wave number of some a_1 vibrational bands in the Stokes Raman spectra of cinnoline and other diazanaphthalenes have been reported (*A. Aminzadeh et al.*, J. Raman Spectrosc., 1980, **9**, 214).

The ^{1}H-n.m.r. spectrum of cinnolines has been recorded in deuterated acetone and carbon tetrachloride, and the chemical shifts and coupling constants are included in Table 8 (*P.J. Black* and *M.L. Hefferman*, Austral. J. Chem., 1965, **18**, 707; *L.S. Besford et al.*, J. chem. Soc., 1963, 2867). The ^{1}H-n.m.r. spectra of cinnolines are quite complex and unequivocal assignments are based on the complete interpretation of the spectra of many derivatives including deuteration studies (see Table 8; *T.J. Batterham*, NMR Spectra of Simple Heterocycles, Wiley, New York, 1971). Proton magnetic resonance spectra have been reported for various alkyl- and aryl-cinnolines (*H.S. Lowrie*, J. med. Chem., 1966, **9**, 670; *A.W. Ellis* and *A.C. Lovesey*, J. chem. Soc. B, 1967, 1285; *M.H. Palmer* and *E.R.R. Russell*, J. chem. Soc. C, 1968, 2621). The high field (250 mHz) ^{1}H-n.m.r. spectrum of cinnoline in deuterated chloroform has been reported (*G.C. Pappalardo*, Org. magn. Resonance, 1975, **7**, 504). The inter-ring H-H couplings have been determined experimentally, and theoretically by INDO calculations. The ^{1}H-n.m.r. spectra of cinnoline and various 8-substituted derivatives in deuterated dimethylsulphoxide have been reported and the observed chemical shifts are included in Table 8 (*R.G. Guy et al.*, Tetrahedron, 1978, **34**, 941). The previous assignment of

TABLE 8

^{1}H-N.M.R. SPECTRAL DATA FOR CINNOLINES (δ, ppm)

Substituent	Solvent	H-3	H-4	H-5	H-6	H-7	H-8
None [a]	Me$_2$CO	9.32	8.08	8.01	7.84	7.93	8.49
None	CCl$_4$	9.10	7.73	7.57	7.57	7.57	8.30
None	DMSO	9.44	8.25	8.09	7.88	7.98	8.48
3-NH$_2$	DMSO		7.00	7.53	7.53	7.53	8.17
3-NO$_2$	DMSO		9.23	8.30	8.07	8.07	8.60
4-Me	CDCl$_3$	8.96	(2.53)	7.68	7.68	7.68	8.30
4-CO$_2$H	DMSO	9.85		8.50	7.93	7.93	8.63
4-NH$_2$	DMSO	8.73	(7.37)	7.65	7.65	8.17	8.17
5-NH$_2$	DMSO	9.13	8.25		6.82	7.52	7.52
5-NO$_2$	CDCl$_3$	9.40	8.80		8.83	7.90	8.62
8-NH$_2$	CDCl$_3$	9.03	6.85	7.55	6.85	7.30	(5.38)
8-NH$_2$	DMSO	9.23	7.97	7.05	7.53	7.01	
8-NO$_2$	CDCl$_3$	9.40	7.95	7.20	7.95	7.95	
8-NO$_2$	DMSO	9.68	8.50	8.44	8.09	8.53	
8-Me	DMSO	9.43	8.18	7.75	7.88	7.75	
8-OH	DMSO	9.39	8.16	7.47	7.75	7.26	
8-OMe	DMSO	9.43	8.17	7.58	7.81	7.36	
8-Cl	DMSO	9.59	8.35	8.11	7.88	8.15	

[a] Coupling constants (Hz): $J_{3,4}$ 5.7, $J_{4,8}$ 0.8, $J_{5,6}$ 7.8, $J_{5,7}$ 1.5, $J_{5,8}$ 0.8, $J_{6,7}$ 6.9, $J_{6,8}$ 1.3, $J_{7,8}$ 8.6.

the chemical shifts of the heterocyclic ring protons H-3 and H-4 by *P.J. Black* and *M.L. Hefferman* (1965) was confirmed by deuteration studies. Variations in chemical shift due to solvent effects, perturbation effects of the heterocyclic ring, and electronic and proximity effects of the 8-substituents have been discussed.

The ^{13}C n.m.r. chemical shifts for various diazanaphthalenes including cinnoline (see Table 9) have been measured and quantum mechanical treatment of the chemical shift data has been used to rationalise shift values in terms of the paramagnetic screening mechanisms (*R.J. Pugmire et al.*, J. Amer. chem. Soc., 1969, **91**, 6381). *Van de Weijer et al.* have reported a ^{13}C-n.m.r.-spectroscopic investigation of the site of protonation and pK_a values of various diazanaphthalenes (Org. magn. Resonance, 1976, **8**, 187). The relationship between ^{13}C-substituent chemical shifts and proton coupling constants in aza-aromatic systems, including cinnoline, has been reported (see Table 9) (*I.B. Cook et al.*, Austral. J. Chem., 1984, **37**, 311). ^{13}C-n.m.r. spectroscopy has been used to distinguish between *O*-methyl and nuclear *N*-methyl groups in various nitrogen heterocycles including cinnoline (*G.B. Barlin, D.J. Brown* and *M.D. Fenn, ibid.*, 1984, **37**, 2391). Spectral data have been reported for 1-methoxycinnoline and 2-methyl-1-oxocinnoline in CDCl$_3$.

In an ^{15}N-n.m.r.-spectroscopic study of various nitrogen containing heterocycles, *W. Städeli* and co-workers have reported the ^{15}N chemical shifts of several diazines including cinnoline (Helv., 1980, **63**, 504). The assignment of the chemical shifts of cinnoline in DMSO was based on the *N*-screening constants (calculated by CNDO methods), which showed N-1 to be less shielded than N-2. The values (ppm) of +44.6 (N-1) and +41.3 (N-2) are relative to external nitromethane. Similarly, INDO/S parameterised screening calculations have been used to assign signals in the ^{15}N-n.m.r. spectra of cinnoline and various polycyclic diazines and azines (*M. Witanowski et al.*, Org. magn. Resonance, 1981, **16**, 309; *L. Stefaniak et al., ibid.*, 1984, **22**, 201). ^{14}N- and ^{15}N-n.m.r. data have been reported for cinnoline-*N*-oxides, and discussed on the basis of ^{14}N-line-widths and electric-field gradients calculated from INDO results. The *N*-oxide signal was observed to be the sharpest (*L. Stefaniak, M. Witanowski* and *G.A. Webb*, Pol. J. chem., 1981, **55**, 1431; Bull. Acad. Pol. Sci., 1983, **29**, 489).

The ^{14}N-nuclear-quadrupole resonance in cinnoline, and other compounds containing N–N bonds has been obtained and analysed to yield the average σ- and π-electron charge densities as a function of the C–N–N angle (*E. Sauer et al.*, J. chem. Phys., 1972, **57**, 3807).

The u.v. irradiation of cinnoline, phthalazine, and various of their derivatives in methanol gives the corresponding N–H semiquinone radicals by a 2-proton process. The radicals have

TABLE 9

^{13}C-N.M.R. SPECTRAL DATA FOR CINNOLINES (δ, ppm)

Solvent	C-3	C-4	C-5	C-6	C-7	C-8	C-9	C-10 [a]
H$_2$O (pH 6) [b]	145.4	126.3	127.8	133.1	133.1	128.2	150.1	127.8
H$_2$O (pH 1) [b]	141.4	135.5	129.0	138.9	138.1	128.5	149.1	133.2
CHCl$_3$ [c]	146.3	124.8	128.1	132.5	132.3	129.7	151.2	126.1

[a] For numbering of C-atoms see p. 36.
[b] *P. van de Weijer et al.*, Org. magn. Resonance, 1976, **8**, 187.
[c] *R.J. Pugmire et al.*, J. Amer. chem. Soc., 1969, **91**, 6381.

been characterised by their second moment, and by a comparison of their electron spin resonance spectra with the simulated spectra of their semiquinones in methanol (*A. Castellano et al.*, J. chem. Res., Synop., 1979, 70). Diazanaphthalenes, including cinnoline have been used in their excited states to form radicals by hydrogen abstraction from suitable quenchers (*S. Basu et al.*, Chem. Phys., 1983, **79**, 95). Cinnoline is considered to react through both its singlet and polarised triplet states.

The mass spectrum of cinnoline was first reported by *J.R. Elkins* and *E.V. Brown* (J. heterocyclic Chem., 1968, **5**, 639). Fragmentation upon electron impact occurs with the consecutive loss of N_2 and C_2H_2, the molecular ion (m/z 130) giving the base peak. In a subsequent study, the fragmentation pattern was investigated in the light of results obtained with D- and ^{15}N-labelled derivatives (*M.H. Palmer et al.*, Org. Mass Spectrom., 1969, **2**, 1265), however, the structure of the C_8H_6 cation resulting from initial loss of N_2 could not be elucidated. The mass spectra of various alkyl- and aryl-cinnolines have been reported (*Palmer et al.*, loc. cit.; *S.N. Bannore et al.*, Indian J. Chem., 1967, **7**, 654; *J.R. Elkins* and *I.V. Brown*, J. heterocyclic Chem., 1968, **5**, 639).

The magnetic circular dichroism (c.d.) spectra of cinnoline and other diazanaphthalenes has been measured in the 20,000–50,000 cm^{-1} wavenumber region and agree with calculated values (*A. Kaito* and *M. Hatano*, J. Amer. chem. Soc., 1978, **100**, 4037; *M. Vasak et al.*, ibid., 1978, **100**, 6838).

(iii) Chromatography

Various aza-arenes, including cinnoline have been separated on Florisil using ether saturated with ammonia as eluant. Retention is influenced by the Brønsted basicity of the heterocyclic N atom(s), and, in contrast with adsorption on silica and alumina, steric interactions have minimal effects on retention (*J. Adams* and *C.S. Giam*, Int. J. environ. anal. Chem., 1984, **18**, 195).

(iv) Reactions

As mentioned above, cinnoline reacts with simple electrophiles to give a mixture of 5- and 8-substituted products. Similar results have been observed for 3- and 4-alkylcinnolines (*E.A. Hobday et al.*, J. chem. Soc., 1962, 4914; *M.H. Palmer* and *E.R.R. Russell*, J. chem. Soc. C, 1962, 2621). The nitration of cinnoline to 5- and 8-nitrocinnoline, their separation, and their conversion to cinnoline carbonitriles has been described (*G. Heinisch*, Sci. Pharm., 1982, **50**, 246). Reduction of cinnoline with lithium aluminium hydride in boiling ether gives 1,4-dihydrocinnoline in good yield. Similar reduction of various 3- and 4-alkylated cinnolines also results in formation of the corresponding 1,4-dihydro compound (*L.S. Besford et al.*, J. chem. Soc., 1968, 2867; *D.I. Haddesley et al.*, ibid., 1964, 5269). Treatment of cinnoline, 4-methyl-, and 4-phenyl-cinnoline with amalgamated zinc in refluxing 33% aqueous acetic acid for 2 h gives indole, skatole, and 3-phenylindole, respectively, in very good yield (*L.S. Besford* and *J.M.*

Bruce, ibid., 1964, 4037; J.M. Bruce, ibid., 1959, 2366). Very short (4–10 min) reaction time gives the 1,4-dihydro compound, which is then converted into the indole ring *via* elimination of the nitrogen atom at N-2. The

catalytic hydrogenation of cinnoline has been shown to give several products at varying stages of reduction, including 1,4-dihydro- and 1,2,3,4-tetrahydro-cinnoline, indole and its dihydro and octahydro derivatives, and 2-(*o*-aminophenyl)-ethylamine (J.D. *Westover*, Ph.D. Dissertation, Brigham Young University, Provo, Utah, 1965). The reduction of 4-methylcinnoline under similar conditions also gives a wide range of reduced products (W.E. *Maycock*, M.S. Thesis, Brigham Young University, Provo, Utah, 1964).

The catalytic deuteration of cinnoline has been carried out by heating a mixture of cinnoline, D_2O, and platinised asbestos in a sealed tube at approximately 200°. Cinnoline showed a preferential order of exchange of H-3 ≫ H-4, H-6, H-7, H-8 (G. *Fischer* and M. *Puza*, Synthesis, 1973, 219).

The electrolytic reduction of 4-methylcinnoline at -0.4 V in 1 N hydrochloric acid gives 1,4-dihydro-4-methylcinnoline. At -1.0 V the reduction consumes an additional 2 electrons to give skatole (H. *Lund*, Acta Chem. Scand., 1967, **21**, 2525). In alkaline solution however the reduction stops after the consumption of 2 electrons to give 1,4-dihydro-4-methylcinnoline. 3-Phenylcinnoline is also reduced under similar conditions to give the 1,4-dihydro compound. The electrochemical reduction of cinnoline in DMF–water mixtures at a Hg electrode has been studied (M. *Muruyama* and K. *Murakami* C.A., 1977, **87**, 124592). The electrochemical behaviour of cinnoline at $-40°$ in liquid ammonia has been studied by cyclic voltammetry at a smooth Pt electrode (M. *Herlem* and G. *Van Amerongen*, Anal. Letters., 1980, **13**, 549).

The cinnoline ring system is quite susceptible to oxidation, the nitrogen-containing ring being more stable than the benzenoid ring. Oxidation of 4-phenylcinnoline with hot, aqueous potassium permanganate gives 5-phenylpyridazine-3,4-dicarboxylic acid in excellent yield (R. *Stoermer* and H. *Fincke*, Ber., 1909, **42**, 3115).

When cinnoline, or its alkyl and aryl derivatives are treated with peracids, or hydrogen peroxide, then a mixture of the 1- and 2-oxide is obtained (*C.M. Atkinson* and *J.C.E. Simpson*, J. chem. Soc., 1947, 1649; *I. Suzuki et al.*, Tetrahedron Letters, 1966, 2899; *H.S. Lowrie*, J. med. Chem., 1966, **9**, 670). Treatment of cinnoline with hydrogen peroxide in acetic acid results in the formation of a mixture of cinnoline-1- and -2-oxides (*vide infra*).

Cinnoline is toxic and has anti-bacterial action against *Escherichia coli*. Very few pharmacological studies have been reported on the alkyl- and aryl-cinnolines. 3-(*p*-Dialkylaminoalkoxyphenyl)-cinnolines are both anti-ulcer and anti-inflammatory agents. They are also said to be appetite inhibitors and central nervous system stimulants (*H.S. Lowrie*, U.S.P. 3,239,523, 1966). The anti-inflammatory and anti-microbial activities of pyridazine and benzopyridazine compounds, including cinnoline derivatives, along with structure-activity relationships have been reviewed (*M. Ungurearu et al.*, Rev. Med.-Chir., 1984, **88**, 709; C.A., 1985, **103**, 171461). Various 3- and 4-alkyl- and aryl-cinnoline derivatives have also been reported as active acaricides and insecticides (*J.S. Badmin* and *R.F. Jones*, C.A., 1979, **90**, 49667). 3-Methylcinnoline is effective against *Megoura vicial* and *Tetranychus urticae* on broad bean leaves.

(v) Alkylcinnolines

3-**Methylcinnoline** is a red oil obtained from 4-chloro-3-methylcinnoline *via* the 4-toluene-*p*-sulphonylhydrazide (*E.J. Alford* and *K. Schofield*, J. chem. Soc., 1953, 1811). 3-Methylcinnoline has been prepared by a general procedure for the synthesis of cinnolines involving oxidation of an aminoindole derivative by nitrobenzene (*M. Somei* and *Y. Kurizuka*, Chem. Letters, 1979, 127). Thus when 1-amino-2-methylindole and nitrobenzene in 3% methanolic hydrochloric acid is refluxed for 42 h 3-methylcinnoline is formed in 92% yield. Reductive ring closure of a dinitro-alcohol yields the corresponding dihydroxyamino-alcohol which in the presence of oxygen condenses in a base-catalysed reaction to give a cinnoline. 3-Methylcinnoline has been prepared in good yield by this route (*H. Lund* and *N.N. Nilsson*, Acta Chem. Scand., 1976, **30B**, 5). The 3-methyl group lacks the reactivity of that of the 4-methyl group, since there is no reaction with chloral, but the methiodide condenses with *p*-dimethylaminobenzaldehyde. 3-Methylcinnoline is characterised as its *picrate*, m.p. 175–176.5° (*Alford* and

R = 1-methyl-2-quinolylidene

Schofield, loc. cit.). 3-Methylcinnoline 2-methiodide has also been used to prepare cyanine dyes (*D.R. Sinha* and *A.B. Lal*, Indian J. Chem., 1973, **11**, 126). Reaction of the methiodide with diphenylformamidine gives 3-*β-anilinovinylcinnoline 2-methiodide*, m.p. 212–214°, which on treatment with quinaldine methiodide gives (I) (m.p. 275–276°).

4-Methylcinnoline has been prepared by the Widman–Stoermer synthesis involving cyclisation of diazotised *o*-amino-(propen-2-yl)benzene. Condensations of 4-chlorocinnoline with ethyl cyanoacetate, followed by hydrolysis of the resulting ethyl 4-cinnolinylcyanoacetate provides an indirect synthesis of 4-methylcinnoline (*Y. Mizuno et al.*, Chem. pharm. Bull. Japan, 1954, **2**, 225).

It has also been synthesised by oxidation of 1-amino-3-methylindole (*M. Somai* and *Y. Kurizuka, loc. cit.*). It crystallises as yellow needles, m.p. 74°, readily forms an *ethiodide*, red needles, m.p. 152–154° (*Jacobs et al., loc. cit.*; *C.M. Atkinson* and *J.C.E. Simpson*, J. chem. Soc., 1947, 808), a *picrate* m.p. 177–179°, *2-methiodide*, m.p. 201–203°, and a *perchlorate*, m.p. 148–149°. Nitration gives the 8-*nitro* derivative, m.p. 138–139° (dec.) which is reduced, at 50° by tin(II) chloride in hydrochloric acid solution, to the 8-*amino*- compound, m.p. 126–127° (*K. Schofield* and *T. Swain, ibid.*, 1949, 1367). The acidity of the 4-methyl substituent allows condensation with refluxing benzaldehyde in the presence of anhydrous zinc chloride to give **4-styrylcinnoline**, m.p. 121.5–122.5°, in good yield. This compound on oxidation with permanganate in aqueous pyridine gives cinnoline-4-carboxylic acid (*Jacobs et al., loc. cit.*). Various other substituted benzaldehydes have also been reacted with 4-methylcinnoline in the presence of zinc chloride (*R.N. Castle* and *D.B. Cox*, J. org. Chem., 1953, **18**, 1706). As described for 3-methylcinnoline, the reactivity of the methyl group is greatly increased with quaternisation and quaternised 4-methylcinnoline will condense with *p*-dimethylaminobenzaldehyde to give the styryl derivative in the absence of catalyst (*Atkinson* and *Simpson, loc. cit.*).

Treatment of 4-methylcinnoline with potassium amide and an alkyl nitrite in liquid ammonia results in nitration to give 4-(*nitromethyl*)*cinnoline*, m.p. 152° (dec.) (*H. Feuer* and *J.P. Lawrence*, J. org. Chem., 1972, **37**, 3662). Spectral data indicate that the primary nitro compound is in equilibrium with its dipolar structure.

Reaction of 4-methylcinnoline with ethyl nitrite in ethanolic hydrochloric acid gives 4-*cinnolinecarboxaldehyde oxime*, m.p. 195–198°, in good yield (*H. Bredrick et al.*, Ann., 1970, **737**, 39). Various cyanine dyes have been prepared from quaternary salts of 4-methylcinnoline. 4-Anilinovinylcinnoline methiodide is prepared in a similar manner to that described for the 3-derivative (*A.B. Lal*, J. Indian. chem. Soc., 1959, **36**, 64; Ber., 1961, 1723). Treatment of 4-methylcinnoline with iodine in refluxing pyridine solution gives 4-*cinnolinylmethylpyridinium iodide*, m.p. 185–186° (dec.), in good yield (*R.N. Castle* and *M. Onda*, J. org. Chem., 1961, **26**, 4465). On reduction with sodium and alcohol, 4-methylcinnoline and its 6- and 7-chloro derivatives yield skatole, a change which is characteristic of cinnolines carrying a hydrocarbon substituent at C-4.

(vi) Arylcinnolines

Many arylcinnolines have been prepared by the Widman–Stoermer reaction, *e.g.* **4-phenyl-**, **4-*p*-tolyl-**, and **3-methyl-4-phenyl-cinnoline**, m.ps.

67–67.5°, 58–59° and 135°, respectively, as well as a number of 3-pyridyl- and 3-quinolyl-cinnolines (*A.J. Nunn* and *K. Schofield*, J. chem. Soc., 1953, 3700). Several 3-phenylcinnolines have been prepared by cyclisation of phenylhydrazone derivatives (method 4) (*H.E. Baumgarten* and *J.L. Furnas*, J. org. Chem., 1961, **26**, 1536). Similarly, 6-methyl-3-phenylcinnoline-4-carboxylic acid is prepared from benzaldehyde-*p*-tolylhydrazone and undergoes thermal decarboxylation to give *6-methyl-3-phenylcinnoline*, m.p. 138.5–139.5°, in good yield. However, failure with other phenylhydrazones suggests that this does not present a general method for the synthesis of 3-arylcinnolines.

3-Phenylcinnoline has been prepared by oxidation of an aminoindole derivative by nitrobenzene (*M. Somei* and *Y. Kurizuka*, Chem. Letters, 1979, 127). The reductive metalation of 3-phenylcinnoline by treatment with sodium in THF gives the dianion which shows reactivity at the 1- and the 4-position (*S. Kaban* and *J.G. Smith*. Organometallics, 1983, **2**, 1351). After initial reaction at the 4-position, the resulting delocalised anion is a weak nucleophile, and reacts at the 1-position only under favourable conditions. The mono-*N*-amino salt of 3-phenylcinnoline has been prepared in high yield by direct *N*-amination using *O*-mesitylenesulphonylhydroxylamine (*Y. Tamura et al.*, J. heterocyclic Chem., 1974, **11**, 675). Cyclodehydration of *cis*-phenylglyoxal 2-phenylhydrazone at 150–160° with an anhydrous AlCl$_3$–NaCl melt gives a mixture of 4-phenylcinnoline and 2-phenylquinoxaline (in poor yield) (*S.N. Banmore* and *J.L. Bose*, Indian J. Chem., 1973, **11**, 631). The 4-phenylcinnolines on reduction yield indoles. Thus 4-phenylcinnoline with amalgamated zinc in acetic acid, and 3-methyl-4-phenylcinnoline with sodium and alcohol, yield 3-phenylindole and its 2-methyl derivative, respectively (*C.M. Atkinson* and *J.C.E. Simpson*, J. chem. Soc., 1947, 1649). Oxidation of 3-methyl-4-phenylcinnoline with hydrogen peroxide in acetic acid yields an *N*-oxide, considered to be at position 1, but the product shows no peroxidic properties. *Lund* and *Nilsson* have also described the formation of 3-phenylcinnoline-1-oxide in appreciable amounts during their electrosynthesis of cinnolines by ring closure of dinitro compounds (*loc. cit.*).

(d) Cinnoline-N-oxides

Treatment of 3,4-disubstituted cinnolines with hydrogen peroxide in acetic acid gives the corresponding 1-oxides in excellent yield (*Atkinson* and *Simpson* J. chem. Soc., 1947, 1649).

Oxidation at N-1 occurs due to the steric hindrance caused by the bulky 3-substituent. *N*-Oxidation of cinnoline under similar conditions results in the formation of four products with the 2-oxide being predominant. The products are: cinnoline-1-oxide (II; 26%), cinnoline-2-oxide (III; 50%), cinnoline-1,2-dioxide (IV; 0.3%, m.p. 235° (dec.)) and indazole (V; 3%) (*M. Ogata, H. Kano* and *K. Tori*, Chem. pharm. Bull. Japan, 1962, **10**, 1123; 1963, **11**, 1527; *I. Suzuki et al.*, ibid., 1967, **15**, 1088; Tetrahedron Letters,

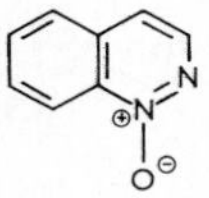

(II) 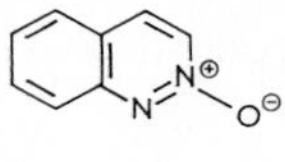   (III) 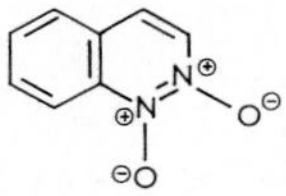   (IV) 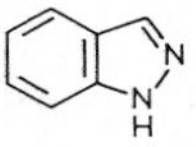   (V)

1966, 2899). Under forcing conditions the yield of the dioxide can be increased to 13% and treatment of the mono-*N*-oxides under similar conditions gives the dioxide in low yields. Oxidation of 3-phenylcinnoline with hydrogen peroxide in acetic acid gives the unsubstituted indazole. It is suggested that ring contraction occurs *via* loss of C-3 (*H.S. Lowrie*, J. med. Chem., 1966, **9**, 670). 4-Methylcinnoline gives the corresponding 1-oxide and 2-oxide in the ratio of 1:2, along with 3% of the 1,2-dioxide (m.p. 168–169°), and similar results are obtained for the *N*-oxidation of 4-methoxycinnoline. Although 4-ethyl-3-methylcinnoline, 3,4-dimethylcinnoline and cinnoline-4-carboxylic acid give no 1-oxide, 4-methyl-3-nitrocinnoline and 3-chlorocinnoline give only their respective 1-oxides. (ref. 3, Table 10). *N*-Oxidation of 8-nitrocinnoline, as would be expected because of the close proximity of the bulky 8-substituent to N-1, gives 8-nitrocinnoline-2-oxide (VI; 20%), 7-nitroindazole (VII; 45%), and 4-hydroxy-8-nitrocinnoline (VIII; 3%) (ref. 8, Table 10).

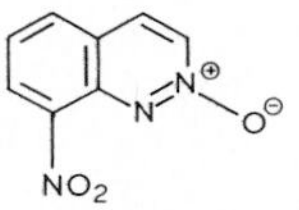

(VI) 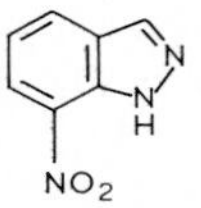   (VII) 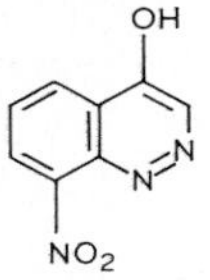   (VIII)

Besides hydrogen peroxide, phthalic monoperacid in ether has also been used to prepare cinnoline-1- and -2-oxides (ref. 2, Table 10). Treatment of 2-phenylisatogen with ethanolic ammonia at 145° for 6 h gives 3-phenyl-4-hydroxycinnoline-1-oxide *via* ring expansion and nitrogen insertion (*W.E. Noland* and *D.A. Jones*, J. org. Chem., 1962, **27**, 341).

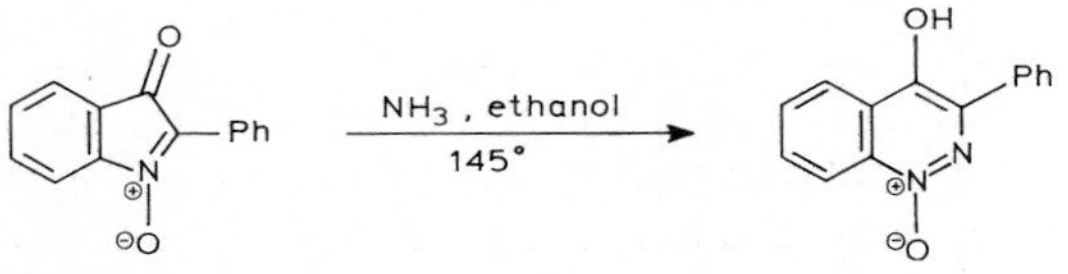

3-Phenylcinnoline-1-oxides are formed in appreciable amounts during the electrosynthesis of cinnolines by ring closure of dinitro compounds (*H.*

TABLE 10

CINNOLINE-N-OXIDES

Compound	m.p. (°C)	Compound	m.p. (°C)
Cinnoline-1-oxide	110.5–111.5 [1]	Cinnoline-2-oxide	125–126 [1]
	111–112 [2]		122–123 [2]
4-Methyl-	94–95 [1]	4-Methyl-	151–152 [1]
4,6-Dimethyl-	160–161 [3]	4,6-Dimethyl-	230–231 [3]
3-Phenyl-	138–139 [4]	3,4-Dimethyl-	160–161 [3]
3-Nitro-	214–215 [5,6]	3-Phenyl-	181–182 [4]
4-Nitro-	161–162 [5,6]	3-Nitro-	227–228 [7]
5-Nitro-	182–183 [5,8]	8-Nitro-	228 (dec.) [8,9,10]
3-Chloro-	168–169 [1]	5-Nitro-	215–217 [8,9,10]
4-Chloro-	94–94.5 [2]	4-Chloro-	150–151 [2]
4-Hydroxy-	153 (dec.) [2]		257 (dec.) [2]

1 M. Ogato, H. Kano and K. Tori, Chem. pharm. Bull. Japan, 1963, **11**, 1527.
2 I. Suzuki and T. Nakashima, ibid., 1964, **12**, 619.
3 M.H. Palmer and E.R.R. Russell, J. chem. Soc. C, 1968, 2621.
4 H.S. Lowrie, J. med. Chem., 1966, **9**, 670.
5 I. Suzuki et al., Chem. pharm. Bull. Japan, 1964, **12**, 1090.
6 I. Suzuki, T. Nakashima and T. Itai, ibid., 1963, **11**, 268.
7 T. Itai, I. Suzuki and T. Nakajima, C.A., 1967, **66**, 95067.
8 I. Suzuki, T. Nakashima and N. Nagasawa, Chem. pharm. Bull. Japan, 1965, **13**, 713.
9 Idem, ibid., 1966, **14**, 816.
10 Idem, ibid., 1963, **11**, 1326.

Lund and N.H. Nilsson, Acta Chem. Scand., 1976, **30B**, 5). The cinnoline-N-oxides and some of their derivatives are listed in Table 10.

The N-oxidation of 3-chloro-5,6,7,8-tetrahydrocinnoline (IX) by hydro-

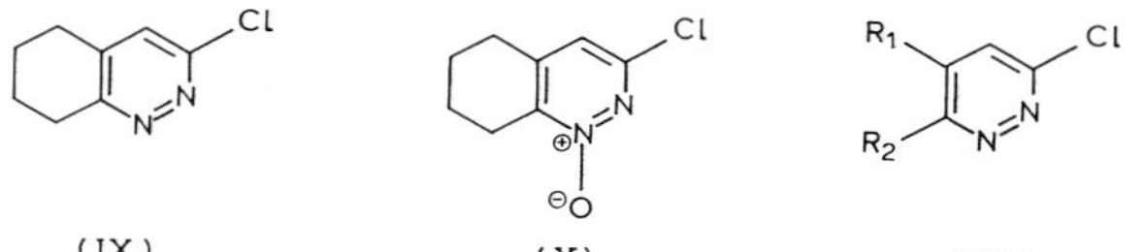

gen peroxide in acetic acid gives the expected 1-oxide (X) by analogy with the known reactions of 5-, and 6-methyl-3-chloropyridazines (XI: $R_1 = CH_3$, $R_2 = H$; $R_1 = H$, $R_2 = CH_3$, respectively) which give their corresponding 1-oxides (ref. 1, Table 10).

Removal of the chloro group by hydrogenation over palladium–charcoal, and aromatisation by bromination and dehydrobromination gives cinnoline-1-oxide.

Cinnoline-N-oxides have been identified by comparison of their u.v.- and n.m.r.-spectral data with those of the known cinnoline-1-, and -2-oxides. The above aromatisation procedure has been used as a general procedure for preparing cinnoline-1-oxides, albeit in only poor yields. The position of N-oxide groups in diazine ring systems have been determined by additivity effects in the nitrogen chemical shifts and a simple theoretical explanation in molecular orbital terms (*L. Stefaniak*, Spectrochim. Acta, 1976, **32A**, 345; *M. Witanowski et al.*, Org. magn. Resonance, 1980, **14**, 305; *L. Stefaniak, M. Witanowski* and *G.A. Webb*, Bull. Acad. Pol. Sci., 1981, **29**, 489).

^{1}H-Nuclear magnetic resonance spectroscopic studies of cinnoline-N-oxides have been reported. The magnetic anisotropy effect of the N-oxide results in a low field shift from H-8 in 1-oxides but not in the 2-oxide derivatives (*K. Tori, M. Ogato* and *H. Kano*, Chem. pharm. Bull. Japan, 1963, **11**, 268). The mass spectra of cinnoline-N-oxides have also been studied (*M.H. Palmer* and *E.R.R. Russell*, Org. Mass. Spectrom., 1969, **2**, 1265).

The anti-tumour activity of several cinnoline-N-oxides has been investigated and 4-nitrocinnoline-1-oxide found to be the most effective (*S. Kamiya et al.*, Chem. pharm. Bull. Japan, 1977, **25**, 504). 3-Nitrocinnoline-1-oxide is reported to be weakly mutagenic. 4-Azidocinnoline-1-oxide and -2-oxide have been prepared and shown to be devoid of anti-tumour or mutagenic activity (*idem, ibid.*, 1980, **28**, 1485).

Cinnolyloxyphenoxypropionate esters have been reported as effective herbicides (Jap. Pat., 45167/1982; C.A., 1982, **97**, 72380). Thus, 3,7-dichlorocinnoline with hydrogen peroxide gives the N-1-oxide, which with sodium carbonate and methyl 2-(4-hydroxyphenoxy)propionate gives give the 7-chloro-3-phenoxy derivative.

The reactivity of the cinnoline ring towards electrophiles is changed when cinnoline is N-oxidised, thereby giving access to different substitution patterns of the ring. For example, nitration of cinnoline-1-oxide no longer gives the 5-, and the 8-nitro derivatives observed for cinnoline, but a mixture of 4-nitro- and 4,5-dinitro-cinnoline-1-oxide, the relative yields of which depend upon the conditions used (Table 10, ref. 5, 6).

Treatment of cinnoline-1-oxide with benzoyl nitrate in chloroform gives 3-nitrocinnoline-1-oxide in good yield. Nitration of cinnoline-2-oxides with a nitric acid–sulphuric acid mixture produces a mixture of 5-, 6-, and 8-nitrocinnoline-2-oxide. Increasing the acidity of the nitrating mixture results in an increased yield of the 5-, and the 8-nitro isomer, whereas that of 6-nitrocinnoline-2-oxide decreases (*J.T. Gleghorn et al.*, J. chem. Soc. B, 1968, 316).

Substituents such as nitro, chloro, or methoxy at the 4-position of cinnoline-1-oxide can be replaced by a variety of nucleophilic reagents. For example, treatment of 4-chloro-, or 4-nitro-cinnoline-1-oxides with sodium methoxide yields the corresponding 4-methoxy compound. Subsequent treatment with sodium hydroxide gives 4-hydroxycinnoline-1-oxide (Table 10, ref. 2, 5, 6). 4-Azidocinnoline-1-oxide and -2-oxide, prepared by a similar procedure, are devoid of anti-tumour activity (*Kamiya et al., loc. cit.*).

When cinnoline-1-oxide is reacted with phosphorus oxychloride, 4-chlorocinnoline is obtained in good yield. The N–O bond of cinnoline-1-, and -2-oxide can also be reduced by treatment with phosphorus trichloride in refluxing chloroform to give cinnoline in 27 and 13% yields, respectively, the majority of the reaction product consisting of unreacted starting material (*I. Suzuki et al.*, Chem. pharm. Bull. Japan, 1967, **15**, 1088).

Catalytic reduction has been used to convert N-oxides to the corresponding cinnolines. However, reduction of substituents such as nitro or chloro may also occur. For example, the Raney nickel catalysed hydrogenation of 4-nitrocinnoline-1-oxide reduces both the N-oxide and nitro group to give 4-aminocinnoline in good yield (Table 10, ref. 5, 6). Similarly, 6-, and 8-nitrocinnoline-2-oxide are reduced by hydrogen over palladium–charcoal to give the corresponding aminocinnolines. Hydrogenation of the 4-chlorocinnoline-N-oxides over palladium–charcoal results in preferred removal of the chloro group and the N-oxide bond does not react (Table 10, ref. 2).

Cinnoline-1,2-dioxide consumes three molar equivalents of hydrogen when reacted over palladium–charcoal to give 1,4-dihydrocinnoline (*Suzuki et al., loc. cit.*).

As mentioned earlier, the acidity of the methyl group of 4-methylcinnoline is considerably increased on N-oxidation. When 4-methylcinnoline-2-oxide is treated with 4-dimethylamino-, or 4-nitro-benzaldehyde in the presence of potassium ethoxide, 4-styryl derivatives are obtained. Excess of the aldehyde may result in deoxygenation, to give, for example, 4-styrylcinnoline (*R.N. Castle, K. Adachi* and *W.D. Guither*, J. heterocyclic Chem., 1965, **2**, 459).

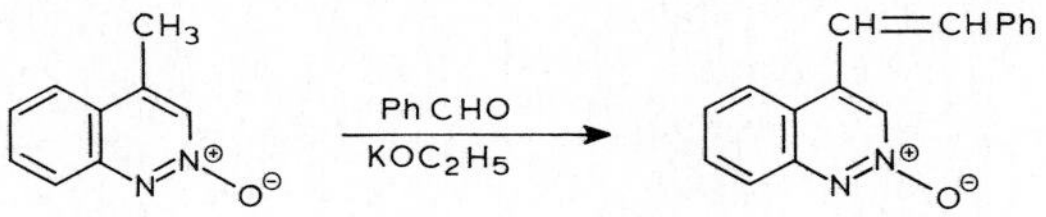

[Reaction: 4-methylcinnoline-2-oxide + PhCHO / KOC₂H₅ → 4-styrylcinnoline-2-oxide]

The photolysis of 4-methylcinnoline-2-oxide gives 58% 4-methylcinnoline, 25% 3-methylindazole and 8% 3-methylindole. 4-Methylcinnoline-1-oxide upon irradiation gives 42% 4-methylcinnoline, 11% 3-methylanthranil, and 4% 2-aminoacetophenone (*W.M. Horspool, J.R. Kershaw* and *A.W. Murray*, Chem. Comm., 1973, 345).

Cinnoline-2-oxide reacts with phenylmagnesium bromide to give phenanthrene (XII), *cis*- and *trans*-stilbene (XIII), 1,2-dihydro-2,3-diphenylcinnoline (XIV), and 2-styrylazobenzene (XV), in yields of 1–15% (*H. Igeta et al.*, Chem. pharm. Bull. Japan, 1970, **18**, 1497).

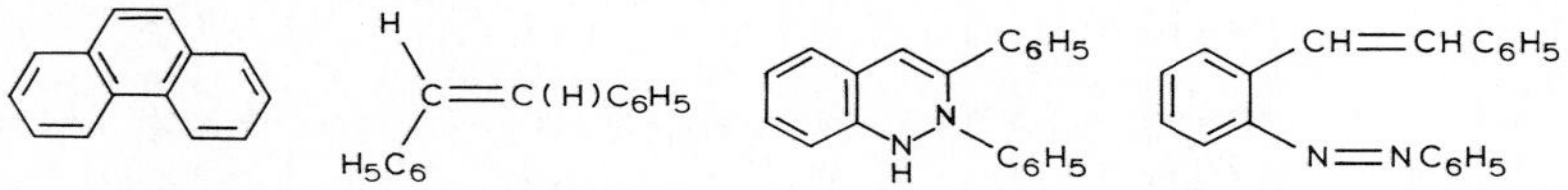

(XII)   (XIII)   (XIV)   (XV)

Similar products are obtained from the reaction of 4-methylcinnoline-2-oxide and phenylmagnesium bromide. However, the carbon atom β to the *N*-oxide group in cinnoline-1-oxide is relatively insensitive to this addition (*idem*, Chem. Comm., 1973, 622).

(e) Halocinnolines

As in pyridazine, the halogen substituents in the heterocyclic ring are reactive and are therefore useful as intermediates in the synthesis of various cinnoline derivaties. 4-Chlorocinnolines are readily available from the corresponding hydroxy compounds by heating them with phosphorus oxychloride, generally without phosphorus pentachloride which may cause further substitution or replacement. There has been little interest in the synthesis of 4-bromo derivatives since the 4-chloro compounds are exceptionally reactive, even more so than the 4-chloroquinolines. However, 4-bromocinnoline has been prepared and found to be active, both as an acaracide and insecticide. 4-Chlorocinnoline can also be prepared in good yield by treatment of cinnoline-1- or -2-oxides with phosphorus oxychloride in chloroform solution (Table 10, ref. 4, 5).

3-Chloro- and 3-bromo-cinnoline have been prepared from 3-halo-4-hydroxy compounds by treatment with phosphoryl chloride and selective removal of the chlorine atoms in the 4-position by the *p*-toluenesulphonylhydrazide process. Direct halogenation of 4-hydroxycinnoline at C-3 can be achieved using sulphuryl chloride but poor yields (20%) are usually obtained (Table 11, ref. 5). However, in a synthesis of hexafluorocinnoline, 7,8-dichloro-4-hydroxycinnoline is reacted with sulphuryl chloride and acetic anhydride in acetic acid to give 3,7,8-*trichloro*-4-*hydroxycinnoline*, m.p.

TABLE 11

HALOCINNOLINES

Compound	m.p. (°C)	Compound	m.p. (°C)
3-Bromo	92–93 [1]	3,4-Dichloro	128–129 [11]
3-Chloro	90–91 [1]	4,6-Dichloro	112–113 [12]
3-Iodo	102–103 [2]	4,8-Dichloro	146–147 [12]
4-Chloro	78–79, 79 [3,4]	4,7-Dichloro	143–144
6-Bromo	129–130 [5]	3-Bromo-4-chloro	153–154 [1]
6-Chloro	119–120 [5]	3-Iodo-4-chloro	165–166
	131–131.5 [6]	4-Chloro-6-fluoro	89 [13]
7-Chloro	89–90 [6]	3,4,6-Trichloro	141–142 [5]
	92–93 [7]	4,6,7-Trichloro	140–141 [12]
8-Chloro	88–89 [6]	4,7,8-Trichloro	223–225 [12]
3-Chloro-4-methyl	145–146 [8]	3,4,7,8-Tetrachloro	253–254.5 [14]
3-Bromo-4-methyl	158.5–159 [8]	5,6,7,8-Tetrafluoro	107–108 [15]
6-Chloro-4-methyl	176–177	3,4,5,6,7,8-Hexachloro	188–190 [14]
8-Chloro-4-methyl	124–125 [9]	3,4,5,6,7,8-Hexafluoro	100–102 [14]
	126–127 [10]		

1 E.J. Alford and K. Schofield, J. chem. Soc., 1953, 609; 1811.
2 D.E. Ames and D. Bull, Tetrahedron, 1982, **38**, 383.
3 J.R. Keneford and J.C.E. Simpson, J. chem. Soc., 1947, 917.
4 N.J. Leonard and S.N. Boyd, J. org. Chem., 1946, **11**, 419.
5 K. Schofield and T. Swain, J. chem. Soc., 1950, 384; 392.
6 A.R. Osborn and K. Schofield, ibid., 1956, 4207.
7 J.S. Morley, ibid., 1951, 1971.
8 H.E. Baumgarten, W.F. Murdock and J.E. Dirks, J. org. Chem., 1961, **26**, 803.
9 L. McKenzie and C.S. Hamilton, ibid., 1951, **16**, 1414.
10 K. Schofield and T. Swain, J. chem. Soc., 1949, 1367.
11 H.E. Baumgarten, J. Amer. chem. Soc., 1955, **77**, 5109.
12 E. Lunt, K. Washbourn and W.R. Wragg, J. chem. Soc. C, 1968, 1152.
13 R.N. Castle et al., J. heterocyclic Chem., 1964, **1**, 98.
14 R.D. Chamber, J.A.H. McBride and W.K.R. Musgrave, J. chem. Soc. D, 1970, 739.
15 V.P. Petrov and V.A. Barkhash, C.A., 1970, **73**, 25229.

323–325°, which is converted into 3,4,7,8-*tetrachlorocinnoline*, m.p. 253–254.5°, with phosphorus oxychloride and phosphorus pentachloride (*R.D. Chambers, J.A.H. McBride* and *W.K.R. Musgrave*, Chem. Comm., 1970, 739). A better route to the 3-halo-4-hydroxy compounds involves the introduction of the halogen atom prior to the formation of the heterocyclic ring, *e.g.* by diazotisation of the halo-*o*-aminoacetophenones (*K. Schofield* and *J.C.E. Simpson*, J. chem. Soc., 1948, 1170). 3-Chlorocinnoline has been prepared by direct chlorination of 3-hydroxycinnoline with phosphorus oxychloride but the yield is quite low (Table 11, ref. 1). The direct replacement of a nitro group at C-3 with a chlorine atom has been used to prepare 3-chlorocinnolines (Table 11, ref. 11). The presence of a hydroxy group at C-4 will also lead to the formation of some 4-chloro compound. Thus, 3,4-dichlorocinnoline has been prepared from both 4-chloro-3-nitrocinnoline and 4-hydroxy-3-nitrocinnoline. Dechlorination at the 4-position on treatment of 3,4,7-trichlorocinnoline with sodium carbonate in water at 95° has been used to prepare 3,7-dichlorocinnoline. Subsequent reaction with 30% hydrogen peroxide gives the N-1-oxide (Jap. Pat., 45167, 1982; C.A., 1982, **97**, 72380). Bromination of 4-hydroxycinnolines at the 3-position occurs quite readily using an aqueous alkaline bromine solution and subsequent removal of the 4-hydroxy group *via* the *p*-toluenesulphonylhydrazide as described above, has been reported to give 3-bromocinnoline in an overall yield of 80% (*E.J. Alford* and *K. Schofield*, J. chem. Soc., 1953, 609; *D.E. Ames, R.F. Chapman* and *D. Waite*, J. Chem. Soc. C, 1966, 470). Treatment of 3-bromo-4-hydroxycinnoline with phosphorus tribromide in bromobenzene at 170° has been reported to give 3,4-dibromocinnoline (*S. Kanoktanaporn* and *J.A.H. McBride*, J. chem. Res. Synop., 1980, 206). Iodination of 4-hydroxycinnoline using iodine monochloride and sodium acetate in acetic acid yields 4-hydroxy-3-iodocinnoline in excellent yield (Table 11, ref. 2). The latter has been converted to the 4-chloro compound using phosphorus oxychloride and dechlorinated *via* the 4-toluenesulphonylhydrazone to give 3-iodocinnoline. The synthesis of cinnolines containing halo substituents on the benzene ring moiety has generally been achieved by diazotisation of halo-substituted aromatic amines. For example, the Widman–Stoermer synthesis has been used to prepare the 6-chloro derivative, and the Borsche synthesis gives 8-iodo- and 6-fluoro-4-hydroxycinnolines (*A.J. Nunn* and *K. Schofield*, J. chem. Soc., 1953, 3700; *K. Schofield* and *R.S. Theobald, ibid.*, 1950, 395).

The Barber synthesis (method 4) of 4-hydroxycinnolines which involves the cyclisation of mesoxalyl chloride phenylhydrazones has also been used to prepare various 5-, 6-, 7- and 8-halogenated cinnolines. Cyclisation of the acid chloride using titanium tetrachloride in nitrobenzene gives the halogen-substituted 3-carboxy-4-hydroxycinnoline in yields depending on the particular substituent. (Table 11, ref. 13). Halogenated cinnolines which have been prepared by this route include: 5-, 6-, 7- and 8-chloro-; 6- and 8-bromo-; 6-, 7- and 8-fluoro-; 5,6-, 5,8-, 6,8- and 7,8-dichloro-cinnoline. As previously mentioned, metathesis of nitro to chloro derivatives on conversion of 4-hydroxy-nitrocinnolines into their respective 4-chloro derivatives provides an alternative approach to the synthesis of chlorocinnolines which does not involve diazotisation (*H.J. Barber et al.*, J. chem. Soc. C, 1967, 1957). All four 8-halocinnolines have been prepared and their proton magnetic resonance spectra studied (*R.G. Guy, F.J. Swinbourne* and *T.McD. McClymont*, Tetrahedron, 1978, **34**, 941). The fluoro and chloro derivatives are prepared *via* cyclisation of the mesoxalyl chloride halophenylhydrazones. 8-Bromo- and 8-iodo-cinnoline are prepared by diazotisation of 8-aminocinnoline and subsequent treatment with CuBr/Cu powder and NaI, respectively. 4-Chlorocinnolines are hydrolysed to the corresponding 4-hydroxy compounds on heating with dilute aqueous acid or even with water alone (*J.R. Keneford et al.*, J. chem. Soc., 1950, 1104). The halogen is also readily replaced by reaction with an alkoxide, aryloxide, thioalkoxide, amine or hydrazine. 4-Phenoxycinnoline may be obtained by heating the 4-chloro- derivative with a mixture of ammonium carbonate and phenol. The 4-phenoxycinnoline can be readily converted into 4-aminocinnoline by heating it with ammonium acetate at 160° (*C.M. Atkinson, J.C.E. Simpson* and *A. Taylor*, ibid., 1954, 1381). Direct conversion of 4-chlorocinnoline into the 4-amino compound by its reaction with alcoholic ammonia is less efficient (*K. Baker*, ibid., 1948, 1713). The chloro group is quite effectively displaced by the sodium derivatives of active hydrogen compounds, such as phenylacetonitrile, ethyl acetoacetate and ethyl cyanoacetate. For example, dimethyl (4-cinnolinyl)malonate is obtained in good yield by treatment of 4-chlorocinnoline with dimethyl malonate, and sodium amide in boiling benzene (*Y. Mizuno, K. Adachi* and *K. Ikeda*, Chem. pharm. Bull. Japan 1954, **2**, 225). Treatment of 4-chlorocinnoline with phosphorus pentasulphide in refluxing pyridine gives 4-cinnolinethiol (*R.N. Castle et al.*, J. heterocyclic Chem., 1966, **3**, 79). Other methods for the replacement of chlorine have been discussed in the preparation of cinnoline and hydrazinocinnoline.

The nucleophilic displacement of chlorine from 4-chlorocinnoline is reported to occur by a one-stage bimolecular mechanism (*N.B. Chapman* and *D.Q. Russell-Hill*, J. chem. Soc., 1956, 1563). Halogens substituted at the 3-position are less susceptible to nucleophilic substitution than the

4-chloro compounds. This is clearly demonstrated by the selective amination of 3,4-dichlorocinnoline to 4-amino-3-chlorocinnoline, and by the hydrolysis of 3,4,6-trichlorocinnoline to give 3,6-dichloro-4-hydroxycinnoline (Table 11, ref. 5; *E. Lunt, K. Washburn* and *W.R. Wragg*, J. chem. Soc. C., 1968, 687).

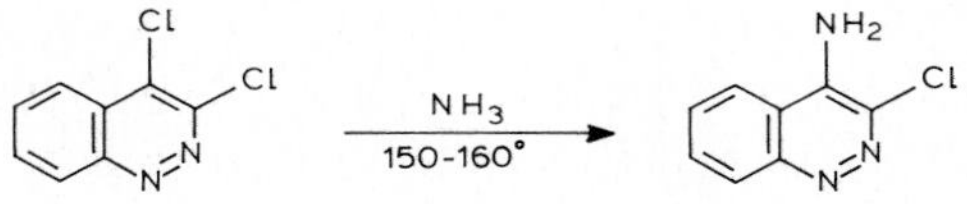

However, 3-bromocinnolines do undergo nucleophilic displacement reactions under forcing conditions (Table 11, ref. 1). 3-Bromocinnoline is converted by standard methods into 3-amino- and 3-methoxy-cinnoline and is reduced in the presence of methanolic potassium hydroxide by palladium–charcoal and hydrogen to cinnoline.

Halogen exchange is possible, usually under somewhat drastic conditions, and the following example (Table 11, ref. 12), is illustrative.

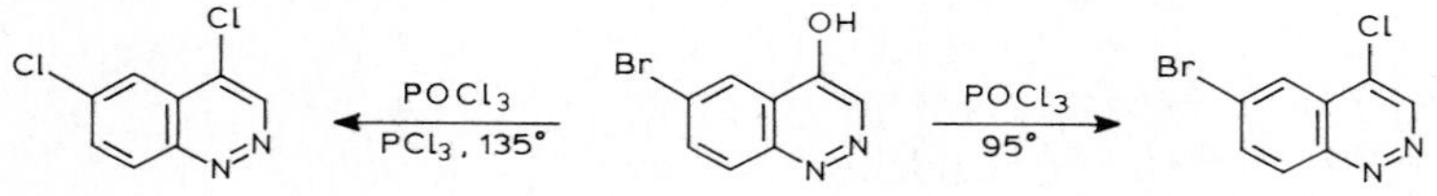

3,4-Dicyanocinnoline is prepared in very good yield by the sequential treatment of 3-bromo-4-chlorocinnoline with sodium *p*-toluenesulphonate and potassium cyanide in DMF (Table II, ref. 2). Condensation of 3,4-dibromocinnoline with *o*-phenylenediamine gives a dark blue hydrobromide, which reacts with alkali and air to give the fully aromatic system, *quinoxalino*[2,3-*c*]*cinnoline* (m.p. 228–230°) (*S. Kanoktanaporn* and *J.A.H. McBride*, J. chem. Res. Synop., 1980, 206).

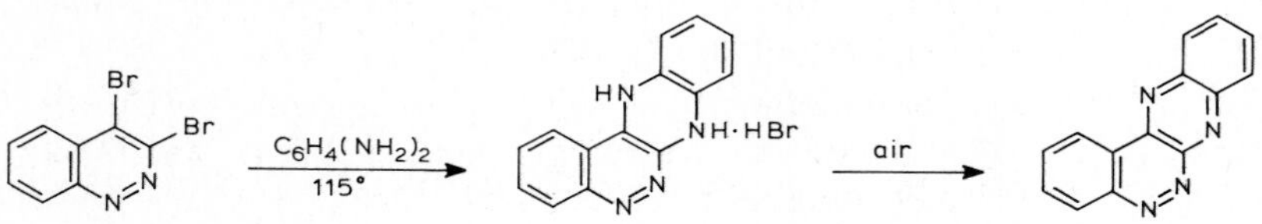

Condensation of 3-bromo- or 3-iodo-cinnoline with a terminal alkyne in the presence of a palladium or a copper compound as a catalyst gives a 3-alkynylcinnoline. For example, 3-iodocinnoline with phenylacetylene in the presence of copper(I) iodide and bis(triphenylphosphine) palladium (II) chloride in trimethylamine gives 3-(phenylethynyl)-cinnoline (Table 11, ref. 2). 4-Chloro-3-(phenylethynyl)-cinnoline, prepared in a similar way, reacts with hydrazine in refluxing ethanol to give pyrazolocinnoline.

(f) Nitrocinnolines

Nitrocinnolines with a nitro group in the benzene ring are readily prepared from 3-, 4-, 5- and 6-nitro-2-aminoacetophenones by the Borsch synthesis which involves diazotisation and cyclisation and yields a nitro-substituted hydroxycinnoline as the initial product (Table 11, ref. 5). In the case of 6-nitrocinnoline this involves the oxidation of *acetone 6-nitro-4-cinnolinyl hydrazone*, m.p. 179–180°, by copper(II) sulphate, but decomposition of the appropriate 4-*p*-toluenesulphonylhydrazino derivative is used for the others. The Baumgarten modification of the Borsch synthesis (*vide supra*) provides a new approach to 3-nitrocinnolines. The diazonium salt from an *o*-aminobenzaldehyde or acetophenone is treated with nitromethane, and the resulting nitroformaldehyde-*o*-formyl- or -*o*-acetyl-phenylhydrazone is cyclised with aluminium oxide to give 3-nitrocinnoline or its 4-methyl derivative (Table 11, ref. 1). In the corresponding reaction from ethyl anthranilate the cyclisation stage fails.

In good agreement with molecular orbital calculations (*M.J.S. Dewar* and *P.M. Maitlis*, J. chem. Soc., 1957, 2421), the nitration of cinnoline yields 33% 5-nitrocinnoline and 28% 8-nitrocinnoline as the sole nitration products (*R.B. Moodie et al.*, J. chem. Soc. B, 1968, 312). Comparison of the rate of nitration of cinnoline with that of 2-methylcinnolinium perchlorate leads to the conclusion that the nitration of cinnoline proceeds *via* the 2-cinnolinium cation. Nitration of cinnoline-1-oxide with nitric acid and sulphuric acid gives a mixture of 4- and 5-nitrocinnoline (*vide supra*). Nitration of cinnoline-2-oxide yields a mixture of 5-, 6- and 8-nitrocinnoline-2-oxide. 4-Hydroxycinnoline gives the 6-nitro derivative as the main nitration product (Table 12, ref. 6).

The cyclisation of mesoxalyl chloride phenylhydrazones has also been used to prepare 4-hydroxycinnolines which contain nitro substituents in the benzenoid ring (*A.J. Barber et al.*, J. chem. Soc. C, 1967, 1657). The nitrocinnolines and some of their 4-substituted derivatives are listed in Table 12.

TABLE 12

NITROCINNOLINES

Compound	m.p. (°C)	4-Chloro deriv., m.p. (°C)	4-Toluene-p-sulphonyl hydrazino deriv., m.p. (°C)
3-Nitro	205–206.5 [1]	169–170 [2]	
3-Nitro-4-methyl	188–189 [1]		
5-Nitro	151–152 [3]		
	147–148.5 [5]	170–171 [4]	
6-Nitro	205–206 [3]	135–137 [6]	190–192 (dec.) [5]
			212–213 (dec.) [3]
7-Nitro	153–154.5 [5]	148–149 [4]	195–196 (dec.) [5]
8-Nitro	133–134 [3]	180 (dec.) [7]	195–196 (dec.) [5]
	132–132.5 [5]	167–169 [4]	
8-Nitro-4-methyl	138–139 (dec.) [8]		
8-Nitro-3,4-dimethyl	150–151 [9]		
8-Nitro-4-t-butyl	195 [9]		
8-Nitro-7-methyl		210–211 (dec.) [7]	

1 H.E. Baumgarten and M.R. DeBrunner, J. Amer. chem. Soc., 1954, **76**, 3489.
2 H.E. Baumgarten, ibid., 1955, **77**, 5109.
3 J.S. Morley, J. chem. Soc., 1951, 1971.
4 K. Schofield and R.S. Theobald, ibid., 1949, 2404.
5 E.J. Alford and K. Schofield, ibid., 1953, 1609.
6 K. Schofield and J.C.E. Simpson, ibid., 1945, 512.
7 J.R. Keneford, J.S. Morley and J.C.E. Simpson, ibid., 1948, 1702.
8 K. Schofield and T. Swain, ibid., 1949, 1367.
9 M.H. Palmer and E.R.R. Russell, J. Chem. Soc. C, 1968, 2621.

Reduction of nitrocinnolines provides an easy route to the corresponding amino derivatives. 3-Nitrocinnoline has been reduced using tin(II) chloride in hydrochloric acid (Table 12, ref. 1) and 5-, 6-, and 8-aminocinnolines have been prepared from the corresponding nitro compounds by catalytic hydrogenation in the presence of palladium–charcoal in ethanol (*E.J. Alford* and *K. Schofield*, J. chem. Soc., 1953, 1811).

Hydrogenation or reduction with tin(II) chloride of 5-, and 8-nitrocinnoline yields the corresponding amine derivative (*G. Heinisch*, Sci. Pharm., 1982, **50**, 246). The use of aqueous titanium(III) chloride for the reduction of 5-, and 8-nitrocinnoline has been described (*M. Somei*, *K. Kato* and *S. Inone*, Chem. pharm. Bull. Japan, 1980, **28**, 2515). This procedure avoids the use of a prolonged reaction time at high temperature and handling material under reduced pressure. The use of excess titanium(III) chloride effects reduction of the ring. 1,4-Dihydro-5-nitrocinnoline has been reduced by stirring it with titanium(III) chloride in acetic acid–water to give 5-amino-1,4-dihydrocinnoline in excellent yield (Jap. Pat., 24368, 1982; C.A., 1982, **96**, 162724).

The reduction of 5- and 8-nitrocinnoline by zinc amalgam to give 4- and 7-aminoindole, respectively, in 70% yields has been reported (Jap. Pat., 67667, 1981, C.A., 1982, **96**, 122635). Reduction of 4-nitrocinnoline-1-oxide by hydrogen over Raney nickel gives 4-aminocinnoline in good yield (*A. R. Osborn* and *K. Schofield*, J. chem. Soc., 1955, 2100; *E. J. Alford et al.*, ibid. 1952, 3009).

As mentioned earlier, chlorination under forcing conditions using phosphorus oxychloride and phosphorus pentachloride may result in replacement of a nitro group, particularly in the case of 6-nitro compounds. For example, 4-hydroxy-6-nitrocinnoline gives a mixture of 4-chloro-6-nitrocinnoline and 4,6-dichlorocinnoline (*H. J. Barber et al.*, J. Chem. Soc. C, 1967, 1657). Similar observations have also been made with 3-nitrocinnolines (Table 12, ref. 2). Although 5- and 8-nitrocinnoline are both quite stable in the presence of weak bases and mineral acids, they decompose in warm, dilute aqueous sodium hydroxide solution to give high melting amorphous solids (Table 12, ref. 3).

(g) Aminocinnolines*

The reduction of nitrocinnolines, either chemically or catalytically, is a general procedure for the preparation of aminocinnolines, and in particular their 4-hydroxy derivatives. Examples of this method have been discussed in the preceding section under the reactions of nitrocinnolines.

3-Aminocinnoline and its 1-methyl homologue can be prepared by the action of aqueous ammonia, at 130–140° for 20 h in the presence of a little copper(II) sulphate, on the corresponding 3-bromo compounds. 3-Bromocinnoline reacts with anhydrous dimethylamine in a sealed tube at 65° for 24 h to give 3-dimethylaminocinnoline in moderate yield (Table 13, ref. 18).

Although the halogen in 4-chlorocinnoline is reactive, some difficulty is experienced in its replacement by the amino group. The use of aqueous ammonia is partly successful (Table 13, ref. 4) but with alcoholic ammonia,

---

* The aminocinnolines can also be found in the literature by their alternative name, cinnolinamines.

liquid ammonia, or phenol and ammonium carbonate introduction of the amino group is not achieved. The last method, however, gives 4-phenoxycinnoline which on fusion with ammonium acetate at about 200° is converted smoothly into the amino compound, and thus provides a convenient method for its preparation. A wide variety of 4-dialkyl- and 4-alkyl-aminocinnolines can be prepared from the appropriate amine and either 4-phenoxycinnoline (*J.R. Keneford* and *J.C.E. Simpson*, J. chem. Soc., 1947, 917) or 4-chlorocinnoline (Table 13, ref. 14). Of the compounds shown to have anti-malarial activity (against *Plasmodium gallimaceum* in chicks), 7-chloro-4-(4'-dimethylamino-1'-methyl-*n*-butylamino)-cinnoline (XVI) is the most effective.

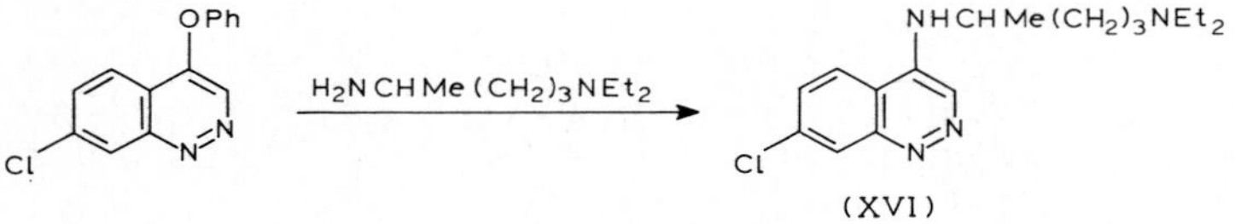

A methanesulphonyl group in the 4-position of the cinnoline ring is labile and is displaced with greater ease than the chloro group. Compounds containing such a group react with aniline, ammonia, alkyl amines and hydrazine to give the corresponding 4-aminated compound (Table 13, ref. 16, 19). Various 5,6,7,8-halo- and -nitro-4-aziridinocinnolines (XVII) are prepared in moderate to excellent yields by the reaction of the corresponding 4-chlorocinnoline with aziridine in dry benzene at 45–55° (Table 13, ref. 20). 6-Nitro-4-aziridinocinnoline shows anti-tumour activity against K B cells. Aromatic amines can be readily prepared from the corresponding hydroxy compound by heating it with aqueous ammonium sulphite or bisulphite, and this method has been used to convert, 5-, 6-, 7-, and 8-hydroxycinnoline into the corresponding aminocinnolines (Table 13, ref. 5, 6). 7-Acetyl-4-hydroxycinnoline (XVIII) is converted by a Schmidt reaction ($H_2SO_4/CHCl_3/NaN_3$) into 7-acetamido-4-hydroxycinnoline (XIX) which may be hydrolysed to the 7-amino compound (XX) (Table 13,

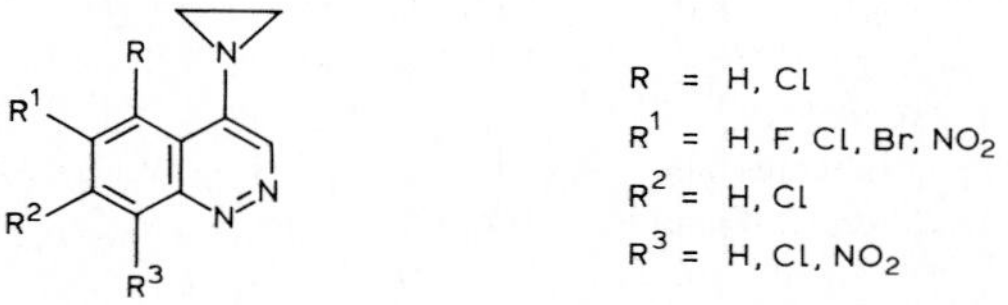

TABLE 13
AMINOCINNOLINES

Compound	m.p. (°C)	Compound	m.p. (°C)
3-Amino	165–166 [1,2]	6-Amino-4-hydroxy	275–276 [14]
4-Amino	212–213 [3]	7-Amino-4-hydroxy	276–277 [13]
	209.5–210.5 [4]	8-Amino-4-hydroxy	290–291 [13]
5-Amino	160–161 [5]	3,4-diamino	220–220.5 [11]
6-Amino	203–204 [5]	4,6-diamino	262–264 (dec.) [12]
7-Amino	191–192 [5]	4,7-diamino	179–180 [15]
8-Amino	hemihydrate 89–92 [6,7]	4,8-diamino	167–168 [13]
3-Amino-4-methyl	159.5–160 [2]	4-methylamino	228–230 [16,17]
3-Amino-6-chloro	215 (dec.) [8]	4-dimethylamino	brownoil [18]
4-Amino-3-chloro	228–229 [9]	3-dimethylamino	107–108 [18]
4-Amino-6-chloro	277–278 [3]	4-anilino	240 [19]
4-Amino-7-chloro	209–210 [3]	4-aziridino	84 [20]
4-Amino-8-methyl	142–152 [10]	4-aziridino-6-nitro	175–177 [20]
4-Amino-3-nitro	308–308.5 [11]	5-amino-6,7-dimethyl-3-phenyl	209–211 [21]
4-Amino-6-nitro	288–289 (dec.) [3,12]	5-amino-3,6,7-trimethyl	235–236 [21]
4-Amino-7-nitro	300–301 (dec.) [13,14]	5-amino-3-phenyl	219–221 [21]
	235–236 (dec.) [13]		
4-Amino-8-nitro	242–243 (dec.) [10]		
4-Amino-3-methyl-6-nitro	320 (dec.) [3,12]		
4-Amino-3-methyl-8-nitro	283–285 (dec.) [3]		

1 *E.J. Alford* and *K. Schofield*, J. chem. Soc., 1953, 1811.
2 *H.E. Baumgarten* and *M.R. DeBrunner*, J. Amer. chem. Soc., 1954, **76**, 3489.
3 *J.R. Keneford, K. Schofield* and *J.C.E. Simpson*, J. chem. Soc., 1948, 358.
4 *K. Baker*, ibid., 1948, 1713.
5 *A.R. Osborn* and *K. Schofield*, ibid., 1955, 2100.
6 *E.J. Alford et al.*, ibid., 1952, 3009.
7 *J.S. Morley*, ibid., 1951, 1971.
8 *H.E. Baumgarten, D.L. Pedesen* and *M.W. Hunt*, J. Amer. chem. Soc., 1958, **80**, 1977.
9 *K. Schofield* and *T. Swain*, J. chem. Soc., 1950, 392.
10 *J.R. Keneford, J.S. Morley* and *J.C.E. Simpson*, ibid., 1948, 1702.
11 *H.E. Baumgarten*, J. Amer. chem. Soc., 1955, **77**, 5109.
12 *J.R. Keneford et al.*, J. chem. Soc., 1952, 2595.
13 *K. Schofield* and *R.S. Theobald*, ibid., 1949, 2404.
14 *N.J. Leonard* and *S.N. Boyd*, J. org. Chem., 1946, **11**, 419.
15 *C.M. Atkinson, J.C.E. Simpson* and *A. Taylor*, J. chem. Soc., 1954, 1381.
16 *G.B. Barlin* and *W.V. Brown*, J. chem. Soc. C, 1967, 2473.
17 *C.M. Atkinson* and *A. Taylor* J. chem. Soc., 1955, 4236.
18 *A.J. Boulton, I.J. Fletcher* and *A.R. Katritzky*, J. Chem. Soc. B, 1971, 2344.
19 *E. Hayashi* and *T. Watanabe*, C.A., 1968, **69**, 19106.
20 *T. Yamazaki, R.E. Draper* and *R.N. Castle*, J. heterocyclic Chem., 1978, **15**, 1039.
21 *K. Nagarajan* and *R.K. Shah*, Chem. Comm., 1973, 926.

ref. 13). 2-Phenacylcyclohexane-1,3-diones may be cyclised with hydrazine to 3-substituted 5-oxo-5,6,7,8-tetrahydrocinnolines (XXI). The latter compounds react with hydrazoic acid in sulphuric acid, but instead of undergoing ring enlargement they form 5-aminocinnolines (Table 13, ref. 21).

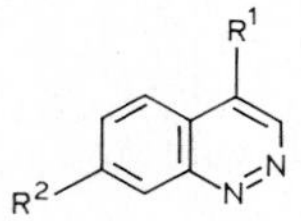

(XVIII) R^1 = OH, R^2 = Ac
(XIX) R^1 = OH, R^2 = NHAc
(XX) R^1 = OH, R^2 = NH_2

Diaminocinnolines can be prepared by amination and reduction of nitro-4-phenoxycinnolines. For example, 6-nitro-4-phenoxycinnoline gives 4-amino-6-nitrocinnoline on fusion with sodium amide, and subsequent reduction gives 4,6-diaminocinnoline. 3-Nitrocinnoline reacts with hydroxylamine to give a good yield of 4-amino-3-nitrocinnoline. The latter on reduction with tin(II) chloride in hydrochloric acid gives 3,4-diaminocinnoline (Table 13, ref. 11, 12). In polyphosphoric acid the oximes of compounds (XXI) are similarly transformed. Rearrangement of the gem dimethyl group in 7,7-dimethyl-5-oxo-3-phenyl-5,6,7,8-tetrahydrocinnoline (XXI; R = Me) to 5-amino-6,7-dimethyl-3-phenylcinnoline (XXII) takes place under these conditions.

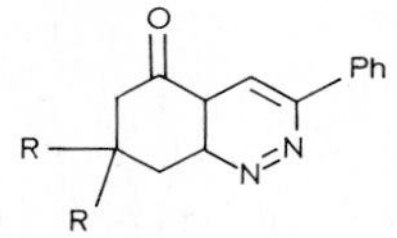

(XXI) 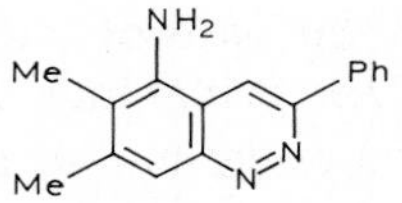   (XXII)
R = Me, H

5-Amino-3,6,7-trimethylcinnoline and 5-amino-3-phenylcinnoline have also been prepared by this route. The latter when diazotised and treated with hypophosphorous acid yields 3-phenylcinnoline.

Flash pyrolysis of 3-methyl-4-methylene-1,2,3-benzotriazine (XXIII) gives 4-methylaminocinnoline (XXIV) in 40% yield together with 2-ethylbenzonitrile (XXV) in 30% yield.

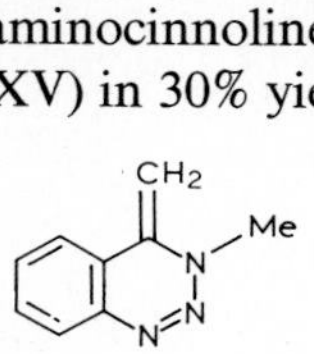

(XXIII) 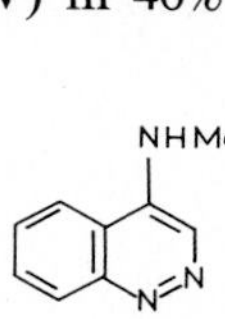   (XXIV) 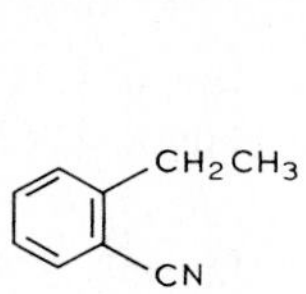   (XXV)

The aminocinnolines are in general yellow crystalline materials with moderately high melting points. They are basic substances, dissolving in acetic acid to give deeply coloured aqueous acidic solutions, and are reprecipitated by ammonia; many can be crystallised from water (Table 13, ref. 5, 6). Acetyl derivatives are readily prepared and methiodides may be formed at N-1, without methylation of the amino group. Quaternisation of 3-aminocinnoline with methyl iodide gives two salts, each of which retains the characteristic NH_2-adsorption in its i.r. spectrum (*D.E. Ames et al.*, J. chem. Soc. B, 1967, 748). The more basic of the two ring nitrogen atoms has been determined by protonation studies and by comparison of the u.v. spectrum of the derived salt with those of the known 1-, and 2-methyl salts. The mono-aminocinnolines are stronger bases than cinnoline itself. The pK_a values of the aminocinnolines have been determined, and the order of decreasing basic strength is $4 > 6 > 7 > 3,8 > 5$-aminocinnoline (*A.R. Osborn, K. Schofield* and *L.N. Short*, J. chem. Soc., 1956, 4191). The increased basicity of the aminocinnolines is due to resonance interaction between the amino group and the protonated ring nitrogen atom, thereby stabilising the aminocinnolinium cation. Resonance in the case of 4-aminocinnoline stabilises N-1 and this has been observed to be the most basic centre. The only exception to this approach occurs for the weakest base of the monoaminocinnolines, 5-aminocinnoline.

As for most *N*-heteroaromatic systems, the amino groups of 5-, 6-, 7-, and 8-aminocinnoline exist in the amino ($-NH_2$), and not the imino ($=NH$) form. However, on the basis of conflicting i.r.-spectral studies, the situation with respect to 4-aminocinnoline is less clear (*R.N. Castle, D.B. Cox* and *J.F. Suttle*, J. Amer. pharm. Assoc., 1959, **48**, 135; *Osborn, Schofield* and *Short*, loc. cit.). Subsequently, u.v.-spectroscopic studies and basicity measurements by Katritzky and co-workers (Table 13, ref. 18) have shown that the amino form is generally favoured and the results indicate that the inductive effect of N-2 is the dominant factor in affecting the equilibrium positions. ^1H-n.m.r. spectroscopy has been used to show that 4-alkylamino-3-phenylcinnolines exist in the amino form (Table 13, ref. 4).

The mass spectra of 3-, 4-, 5-, and 8-aminocinnolines have been reported to show initial fragmentation by loss of molecular nitrogen to give an m/z 117 radical cation, which then loses HCN and H˙. Initial loss of HCN is an alternative pathway for 4-aminocinnoline (*J.R. Elkins* and *E.V. Brown*, J. heterocyclic Chem., 1968, **5**, 639).

3-**Aminocinnoline**, pK_a 3.63, gives a greenish-yellow aqueous solution with a greenish-blue fluorescence. The diazonium salt is quite unstable, and unlike 5-, 6-, 7-, and 8-aminocinnoline diazonium salts which couple with β-naphthol, it does not couple with either α-, or β-naphthol, most probably reacting with the solvent system to give 3-hydroxycinnoline (*H.E.*

Baumgarten, W.F. Murdock and *J.E. Dirks,* J. org. Chem., 1961, **26**, 803; *E.J. Alford et al.,* Table 13, ref. 1).

4-**Aminocinnoline**, pK_a 6.26, unlike cinnoline, is stable in air and can be boiled with 2.5 N sodium hydroxide without loss of ammonia (Table 13, ref. 4). As for the 3-amino compound, diazotisation is possible but the diazonium salt is not stable. 4-*Anilinocinnoline*, m.p. 236–237°, and other 4-arylaminocinnolines are readily available from 4-chlorocinnoline and appropriate arylamines. On nitration the 4-phenylamino compound gives the 7-*nitro* derivative, m.p. 229° (*K. Schofield* and *J.C.E. Simpson*, J. chem. Soc., 1945, 512).

4-Dimethylamino-2-methylcinnolinium halides have been prepared by quaternisation, and their ionisation constants, u.v. and ^{1}H-n.m.r. spectra reported (*G.B. Barlin* and *A.C. Young*, J. chem. Soc. B, 1971, 2323).

4-Amino-6-chlorocinnoline is of interest since its *methiodide*, m.p. 225–226°, on boiling with sodium hydroxide loses the amino group as ammonia. 6-Chloro-1-methylcinnol-4-one (XXVI) is recovered and thereby establishes that the methiodide forms at N-1 (*J.C.E. Simpson, ibid.,* 1947, 1653).

5- and 8-Aminocinnoline each give a violet-red solution in acid whilst the 6- and 7-amine each form a yellow solution. Each of these amines can be diazotised and coupled with alkaline β-naphthol to form a red dye (Table 13, ref. 5). The *picrate* of 8-*aminocinnoline* is reported as jet black needles, m.p. 236–238°. 8-Amino-4-hydroxycinnoline reacts with nitrous acid to give 1,8-azo-1,4-dihydrocinnol-4-one (XXVII). This is comparable to the behavior of 8-amino-4-hydroxyquinaldine (*B.A. Halcrow* and *W.O. Kermack*, J. chem. Soc., 1945, 415).

In an attempt to synthesise 2-aza analogues of primaquine (XXVIII), 8-amino-6-methoxycinnoline has been prepared by the Barber method. However, introduction of a 5-aminopentan-2-yl side chain in the amino group to give 2-azaprimaquine (XXIX) could not be achieved (*R.A. Coburn* and *D. Gala*, J. heterocyclic Chem., 1982, **19**, 757).

8-Aminocinnoline does not undergo the Sandmeyer reaction with copper(I) cyanide, but 5-aminocinnoline has been converted into the corresponding 5-cyanocinnoline (*G. Heinisch* and *M. Koch*, Sci. Pharm., 1982, **50**, 246).

Nucleophilic displacement of a 4-amino group is greatly enhanced by the presence of either a nitro group in the ring, or by alkylation at N-1 or N-2. 4-Amino-3-nitrocinnoline is readily hydrolysed on warming with dilute potassium hydroxide to give 4-hydroxy-3-nitrocinnoline (Table 13, ref. 11), and both N-1 and N-2-methyl 4-aminocinnolinium iodides give the corresponding hydroxy derivative on hydrolysis (*D.E. Ames, R.F. Chapman* and *H.Z. Kucharska*, J. chem. Soc., 1964, 5659). The 4-methylamino group can be similarly displaced (Table 13, ref. 17; *H.J. Barber* and *E. Lunt, ibid.*, 1965, 1468).

The aminocinnolines react with aldehydes and ketones to form Schiff bases (Table 13, ref. 11). For example, 6-chloro-3,4-diaminocinnoline reacts with diacetyl to give 6-chloro-2,3-dimethyl-1,4,9,10-tetrazaphenanthrene (XXX) (*E. Lunt, K. Washbourn* and *W.R. Wragg*, J. chem. Soc. C, 1967, 1657).

3,4-Diaminocinnoline has been reported to undergo cyclocondensations with glyoxal, biacetyl and benzyl in good yields (*S. Kanaktanaporn* and *J.A.H. McBride*, J. chem. Res. Synop. 1980, 206; Tetrahedron Letters, 1977, 1817).

The diaminocinnolines became specially important when it was found that crude quaternary salts prepared from the reduction of nitro compounds were active against trypanosome infections in mice (Table 13, ref. 12). An extensive systematic search led to the discovery of a number of active compounds, in which two quaternised amino heterocyclic compounds are linked in various ways. *J. McIntyre* and *J.C.E. Simpson* (J. chem. Soc., 1952, 2606, 2615) prepared 4,4'-diamino- and 4,4'-bis-methylamino-6,6'-azocinnoline (XXXI); each gives two dimethochlorides with one member of each pair being active against *Trypanosome congolense*. The preparation involves the diazotisation of 3,3'-diacetyl-4,4'-diaminoazobenzene in formic acid to build up the two cinnoline nuclei required. 6-Benzeneazo-4-aminocinnoline is, however, inactive.

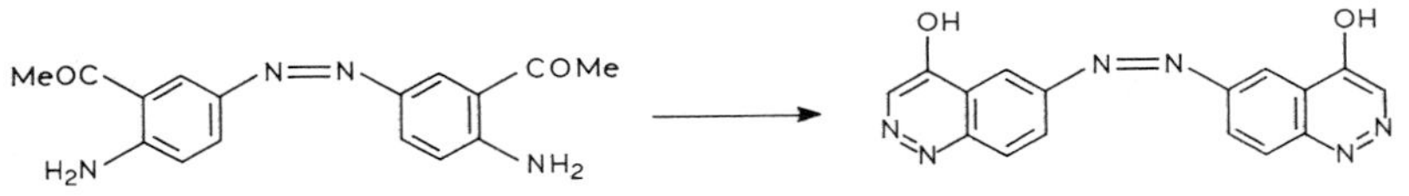

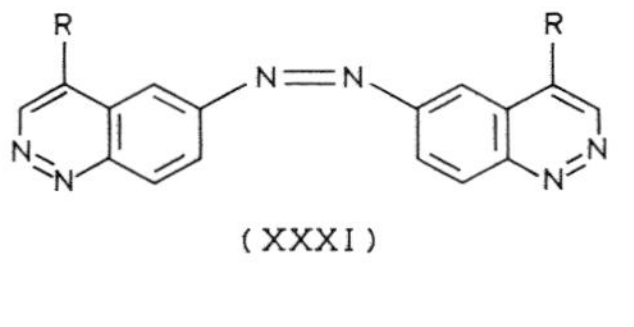

Other ways of linking the two heterocyclic nuclei have been described (*J.S. Morley* and *J.C.E. Simpson*, *ibid.*, 1952, 2617) in the synthesis of quaternary salts of dicinnolyl-ureas, -thioureas and -guanidines from 4,6-diaminocinnoline hydrochloride and its 3-methyl homologue.

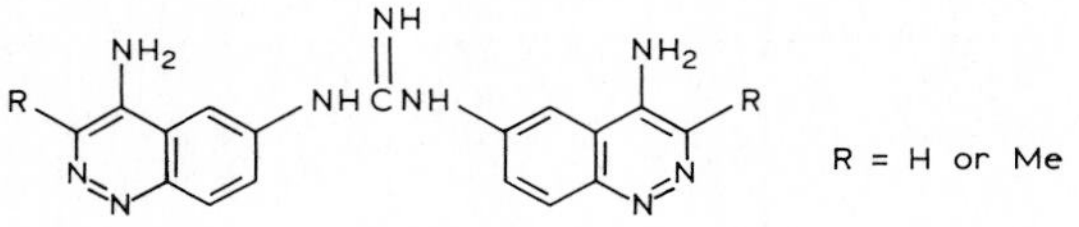

(XXXII)   R = H or Me

N^1,N^3-Di(4-amino-6-cinnolyl)guanidine (XXXII, R = H) gives a dimethiodide highly active against *T. congolense* infections in mice (*E.M. Lourie et al.*, Brit. J. Pharmacol., 1951, **6**, 643) and exhibits a chemotherapeutic index of the same order as that of antrycide methosulphate.

The quaternary salts of pyrimidylaminocinnolines have trypanocidal properties. The compounds are prepared by combining 4,6-diaminocinnoline methiodide or its 3-methyl homologue with the methiodides of various 4-chloropyrimidines (B.P. 663,095/6, 1951).

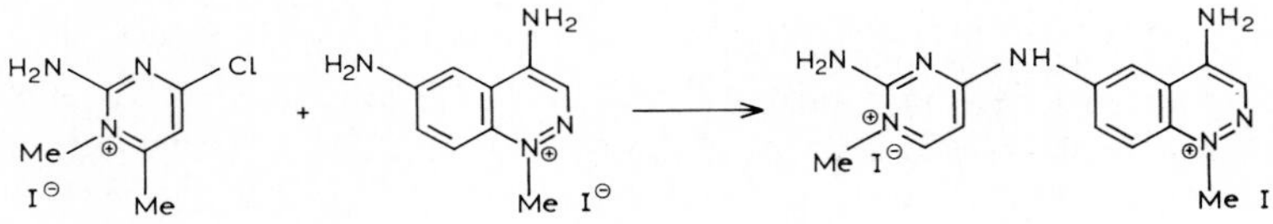

(i) Hydrazinocinnolines

Many hydrazinocinnolines have been prepared by the direct action of hydrazine on the 4-chloro compounds. The replacement proceeds slowly, so that the reaction of hydrazine with 4-chlorocinnoline requires some days at room temperature whilst with hydrazine hydrate it is necessary to keep the mixture at 150–160° for 20 h. Phenylhydrazine and toluene-*p*-sulphonylhydrazide react similarly (Table 14, ref. 1). The formation of these substituted amino derivatives is to be contrasted with the difficulty of forming 4-aminocinnolines from 4-chloro compounds and ammonia. The 4-methanesulphonyl group is displaced more readily than the 4-chloro atom and 4-methanesulphonylcinnoline reacts with hydrazine to give the corresponding 4-hydrazino compound (Table 13, ref. 16, 19).

The 4-hydrazinocinnolines are used for the preparation of cinnoline unsubstituted in the 4-position. The routes are simple and effective since 4-hydrazinocinnolines undergo oxidative deamination when heated with a

TABLE 14
HYDRAZINOCINNOLINES

Additional substituent	4-Hydrazino deriv., m.p. (°C)	4-Phenylhydrazino deriv., m.p. (°C)	4-p-Toluene-sulphonylhydrazino deriv., m.p. (°C)
none	two forms [1] see text	238 [1]	hydrochloride 224–226 [2]
3-methyl			hydrochloride 187–188 [2]
3-phenyl			hydrochloride 209–210 [3]
3-chloro	300 [1]	134–135 [1]	167–169 [2]
3-bromo			187–189 (dec.) [2]
6-chloro	320 [1]		
6-bromo	300 [1]		
6-nitro	330 [4]		190–192 [2] 212–213 [4]
6-methoxy			199–201 (dec.) [5] (as hydrate)
7-nitro			195–196 [2]

1 K. Schofield and T. Swain, J. chem. Soc., 1950, 392.
2 E.J. Alford and K. Schofield, ibid., 1953, 609.
3 H.S. Lawrie, J. med. Chem., 1966, **9**, 670.
4 J.S. Morley, J. chem. Soc., 1951, 1971.
5 A.R. Osborn and K. Schofield, ibid., 1955, 2100.

10% aqueous solution of copper(II) sulphate, and the 4-(p-toluene-sulphonylhydrazino)cinnolines are similarly decomposed by sodium carbonate solution. For example, 6-halo-4-hydrazinocinnolines are deaminated using boiling aqueous copper(II) sulphate solution to give the corresponding 6-halocinnoline (Table 14, ref. 1).

3-Hydrazino-5,6,7,8-tetrahydrocinnolines, prepared by treatment of the corresponding 3-chloro compound with hydrazine, are anti-hypertensive agents (U.S.P., C.A., 1985, **103**, 37488).

4-Hydrazinocinnoline occurs in two crystalline forms, m.ps. 226–227° and 296–297°. Both give the same *hydrochloride*, m.p. 244–245°, the same *diacetyl* derivative, m.p. 208–209°, and with anhydrous formic acid, the same *monoformyl* derivative, m.p. 229–230°. It has not been possible to moderate the oxidation of 3-chloro-4-hydrazinocinnoline so as to prepare 3-chlorocinnoline. The corresponding 6-nitro-4-hydrazinocinnoline is converted into 6-nitrocinnoline if first condensed with acetone and the resulting *acetone 6-nitro-4-cinnolylhydrazone*, m.p. 176–177°, oxidised with copper(II) sulphate. For other hydrazinocinnolines see Table 14).

(h) Hydroxycinnolines

The hydroxycinnolines are all more acidic than phenol, the strongest acid being 5-hydroxycinnoline (pK_a 7.40), followed by the 6-, 7-, and 8-derivatives, and the weakest being 4-hydroxycinnoline (pK_a 9.27). The addition of a nitrogen atom in the hydroxyquinoline ring system has a base-weakening effect. 6-Hydroxycinnoline (pK_a 3.65) is the strongest base and 4-hydroxycinnoline (pK_a −0.35) the weakest.

(i) 3- and 4-Hydroxycinnolines

Both 3-hydroxycinnoline and 4-hydroxycinnoline exist principally in the keto form: 3(2H)-cinnolinone (XXXIII) and 4(1H)-cinnolinone (XXXIV) respectively.

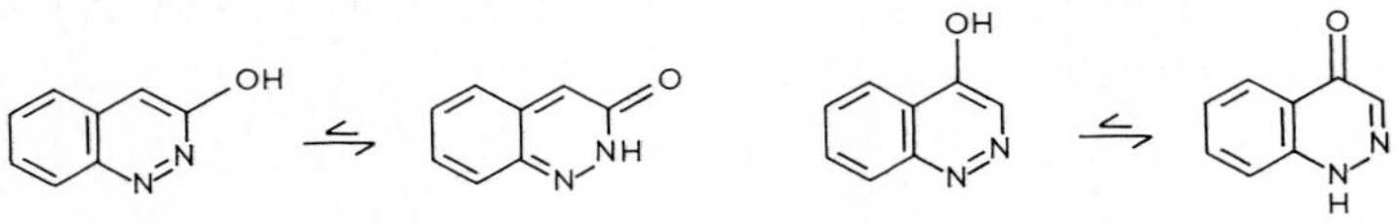

(XXXIII)   (XXXIV)

The amide-enol ratio has been reported to be 4000:1 for 4-hydroxycinnoline (*A. Albert* and *G.B. Barlin*, J. chem. Soc., 1962, 3129) as compared to 380:1 for 3-hydroxycinnoline (*A.J. Boulton*, *L.J. Fletcher* and *A.R. Katritzky*, J. chem. Soc. B, 1971, 2344). The lower value for the latter may be due to disruption of the benzenoid character of the homocyclic ring by the *ortho*-quinoid resonance form (XXXV) of the amide structure.

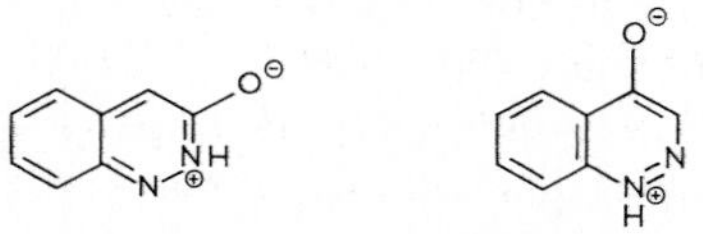

(XXXV)   (XXXVI)

The site of protonation in (XXXV) and (XXXVI) for 3- and 4-hydroxyinnoline, respectively, has been determined by study of their ionisation constants and u.v. spectra (*G.B. Barlin*, ibid., 1965, 2260; *D.E. Ames*, *R.E. Chapman* and *D. Waite*, J. chem. Soc. C, 1966, 470). In addition, their i.r. spectra, in both solution and solid state, show absorption due to N–H and C=O groups, but no O–H, stretching vibrations. Molecular orbital calculations of the π-charge density, free valence index, and self polarisability have recently been reported (*O.N. Chupakhin*, *E.O. Sidorov* and *E.G. Kovalev*, Zhur. org. Khim., 1977, **13**, 204).

3-Hydroxycinnoline (XXXVII) was first prepared (*P.W. Neber* and *G. Bossel*, Ann., 1929, **471**, 113) from *o*-nitromandelonitrile. The method has been elaborated by *E.J. Alford* and *K. Schofield* (Table 15, ref. 3).

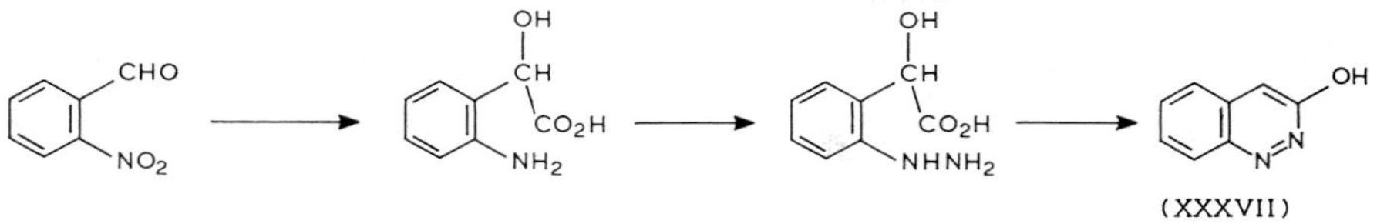

2-Amino-5-chloromandelic acid and 2-amino-5-methoxymandelic acid have similarly been converted into 6-*chloro*-3-*hydroxycinnoline*, m.p. 262–265°, and 3,6-dihydroxycinnoline respectively. α-Substituted mandelates (XXXVIII) give initially 3-substituted 1-aminodioxindoles (XXXIX) which may be converted into the 4-substituted 3-hydroxycinnoline *via* the *o*-hydrazinomandelate.

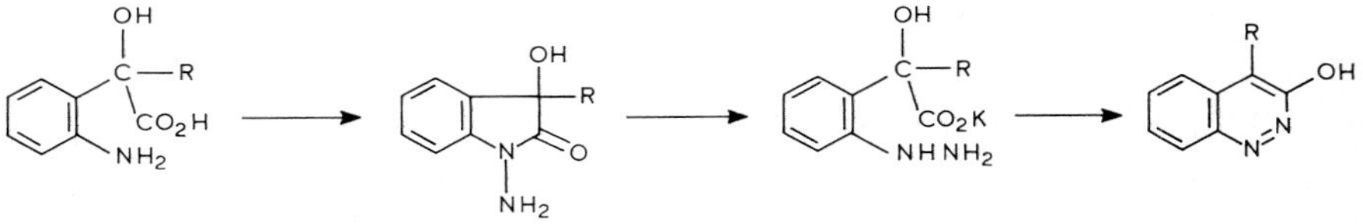

Oxidation of 1-amino-oxindole (XL) with lead(IV) acetate in benzene gives 3-hydroxycinnoline in good yield (*H.E. Baumgarten, P.L. Creger* and *R.L. Zey*, J. Amer. chem. Soc., 1960, **82**, 3977; *M. Lora-Tamayo, B. Marco* and *P. Navarro*, An. Quim., 1976, **72**, 914). Oxidation of (XL) in dimethylsulphoxide gives the adduct (XLI) (*Lora-Tamayo et al., loc. cit.*), whereas one equivalent of *t*-butyl hypochlorite has been shown to give 3-hydroxycinnoline in near quantitative yield (Table 15, ref. 4). As substituted 1-amino-oxindoles are not readily available, it is doubtful whether the above method will prove to be a general procedure.

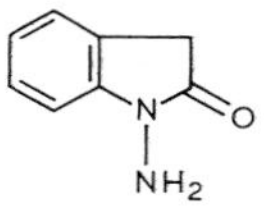

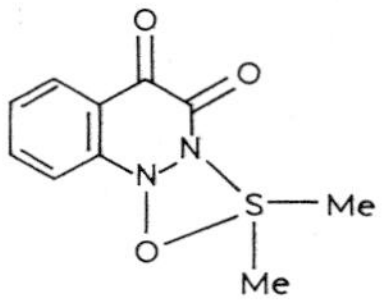

3,4-Dihydroxycinnolines (XLII) have been prepared from 4,5,6-substituted-isatins (XLIII). Thus, 5,7-dimethyl-3,4-dihydroxycinnoline and various 6-substituted-3,4-dihydroxycinnolines are obtained by treatment of the corresponding isatin with sodium nitrite (*M. Lora-Tamayo, B. Marco* and *C. Sender*, Org. Prep. Proced. Int., 1978, **10**, 298).

(XLII)   (XLIII)

Basic hydrolysis of isatin, followed by catalytic reduction of the resulting o-$H_2NC_6H_4COCO_2^\ominus Na^\oplus$ gives sodium o-aminomandelate which, in a similar fashion to that described by Bossel, may be diazotised, reduced with tin(II) chloride and cyclised to give 3-hydroxycinnoline (*R. Zey*, J. heterocyclic Chem., 1972, **9**, 1177). A modification of Zey's method has been reported to give 3-hydroxycinnoline in 80% yield (*M. Lora-Tamayo, B. Marco* and *P. Navarro*, Org. Prep. Proced. Int., 1976, **8**, 45).

Diazotisation of 3-aminocinnolines in dilute mineral acid has also been used to prepare 3-hydroxycinnolines in moderate to good yields. Stronger acids, for example concentrated hydrochloric, result in halogenation of the 3-position (Table 15, ref. 1). 3-Hydroxycinnoline and its 4-methyl derivative have been prepared by treating the 4,5,6,7-tetrahydro compounds with N-bromosuccinimide followed by dehydrobromination of the dibromo compounds (*J. Daunis, M. Gherret-Rigail* and *R. Jacquier*, Bull. Soc. chim. Fr., 1972, 3198).

3-**Hydroxycinnoline**, bright yellow needles, m.p. 201–203°, *benzoyl* derivative, m.p. 148–149°. 3-**Methoxycinnoline**, m.p. 40–42° (*picrate*, m.p. 155–157.5°) is prepared from the 3-bromo compound (*E.J. Alford* and *K. Schofield*, J. chem. Soc., 1953, 1811).

3-Hydroxycinnoline, pK_a 8.64, is soluble in aqueous sodium hydroxide, and is methylated by diazomethane or dimethyl sulphate to give *2-methylcinnolin-3(2H)-one* (XLIV), m.p. 132–134° (*Alford* and *Schofield, loc. cit.*; *D.E. Ames* and *H.Z. Kucharska*, *ibid.*, 1963, 4924), which is reduced with zinc dust and alcohol to *2-methyl*-1,2,3,4-*tetrahydrocinnolin-3-one*, m.p. 91–92.5°. In a similar manner, benzylation with benzyl chloride (*idem., ibid.*, 1964, 283) and cyanoethylation with acrylonitrile in the presence of benzyltrimethylammonium hydroxide (*D.E. Ames et al., ibid.*, 1965, 5391) affords the corresponding 2-substituted-cinnolin-3(2*H*)-ones.

(XLIV)

Glucosidation with tetra-*O*-acetyl-α-D-glucopyranosyl bromide in the

TABLE 15

SUBSTITUTED 3-HYDROXYCINNOLINES

Substituent	m.p. (°C)	Substituent	m.p. (°C)
4-methyl	232–234 (dec.) [1]	4-chloro	220 [4]
4-phenyl	300–302 (dec.) [2]	5-chloro	268 (dec.) [4]
		6-chloro	262–265 [3]
6-hydroxy	> 300 [3]	7-chloro	263–264.5 (dec.) [1]

1 H.E. Baumgarten, W.F. Murdock and J.E. Dirks, J. org. Chem., 1961, **26**, 803.
2 H.E. Baumgarten and P.L. Creger, J. Amer. chem. Soc., 1960, **82**, 4634.
3 E.J. Alford and K. Schofield, J. chem. Soc., 1952, 2102.
4 H.E. Baumgarten, W.F. Whitman and G.J. Lehmann, J. heterocyclic Chem., 1969, 6, 333.

presence of potassium hydroxide in aqueous acetone gives the intensely yellow 2-(*tetra-O-acetyl-α-D-glucopyranosyl*)-*cinnoline*-3(2H)-*one*, m.p. 160–165°. However, glucosidation of the silver salt of cinnolin-3(2H)-one produces the corresponding O-substituted compound (G. Wagner and D. Heller, Z. Chem., 1964, **4**, 349).

3-Hydroxycinnoline is reported (P.W. Neber and G. Bossel, Ann., 1929, **471**, 113) to give oxindole on reduction for one hour with red phosphorus and hydriodic acid, a rearrangement which does not occur with 4-hydroxycinnoline. Reduction with zinc dust and sulphuric acid gives 1-*amino-oxindole* (XL), m.p. 126–127°, in good yield (H.E. Baumgarten, P.L. Creger and R.L. Zey, J. Amer. chem. Soc., 1960, **82**, 3977), contrary to an earlier report that *tetrahydrocinnolin-3-one* (XLV), m.p. 160°, is formed by this method. However, it has been shown that the reduction to 1-amino-oxindole and its subsequent rearrangement to (XLV) is dependent on the reaction conditions (G. Winters, V. Aresi and G. Nathansohn, J. heterocyclic Chem., 1974, **11**, 997). Thus, reduction with zinc and sulphuric acid in a two-phase system at room temperature gives (XLV) in 78% yield. Further treatment with not acid gives a near quantitative conversion into (XL).

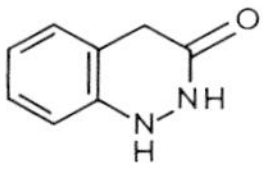

(XLV) 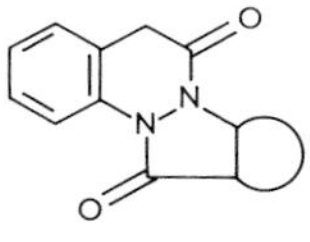   (XLVI) 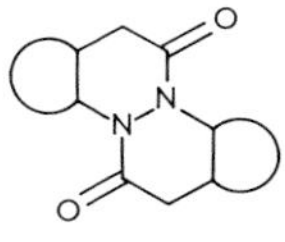   (XLVII) 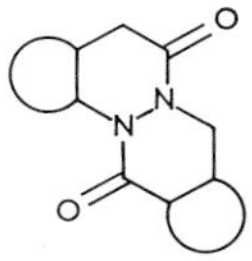   (XLVIII)

Tetrahydrocinnolin-3-one (XLV) was used to prepare tetracyclic compounds (XLVI–XLVIII) as potential anti-inflammatory agents.

3-Hydroxycinnoline can be converted, albeit in low yield, into the corresponding 3-chloro compound on treatment with phosphorus oxychloride (*E.J. Alford* and *K. Schofield*, J. chem. Soc., 1953, 1811). Direct thiation with phosphorus pentasulphide has proved difficult but N-2-alkylated derivatives (*e.g.* XLIV) react more readily (*D.E. Ames* and *R.F. Chapman*, ibid., 1967, 40).

4-Hydroxycinnolines. The most useful method for the preparation of 4-hydroxycinnolines is by the Borsche reaction (general method 3). In particular, *o*-aminophenyl benzyl ketone on diazotisation in concentrated hydrochloric acid and cyclisation of the diazonium compound rapidly gives 4-hydroxy-3-phenylcinnoline (XLIX). There is no evidence of the formation of 9-phenanthrol (L) by a competing Pschorr reaction (*P.W. Ockenden* and *K. Schofield*, ibid., 1953, 370). Other substituted *o*-aminoacetophenones have also been used to give various 3-substituted-4-hydroxycinnolines (*Ockenden* and *Schofield*, loc. cit., Table 16, ref 4).

(XLIX)   (L)

3-Cyclohexyl-4-hydroxycinnolines with methoxy substituents at positions 6 and 7 have been similarly prepared (*I. Malba* and *R.N. Castle*, J. heterocyclic Chem., 1980, **17**, 407). Treatment with *N*-bromosuccinimide gives the 8-bromo derivative. 3-Methanesulphinyl and sulphonyl derivatives have been previously discussed in general method (3).

The Richter synthesis (general method 1) leads to 4-hydroxycinnoline-3-carboxylic acid which is then decarboxylated by heating in benzophenone at 210° (*K. Schofield* and *T. Swain*, J. chem. Soc., 1949, 2392). Another route, for which there is only a single example, is the preparation of 5-*chloro*-4-*hydroxy*-3-*phenylcinnoline*, m.p. > 300°, by condensation of 6-chloro-2-hydrazinobenzoic acid with benzaldehyde (*K. Pfannstiel* and *J. Janecke*, Ber., 1942, **75**, 1096).

Barber and co-workers have described the Friedel–Crafts cyclisation of mesoxalyl chloride phenylhydrazones (general method 4) to give 4-hydroxycinnoline-3-carboxylic acids, which on decarboxylation give 4-hydroxycinnolines (Table 16, ref. 7). The precursor diethyl mesoxalate phenylhy-

TABLE 16
SUBSTITUTED 4-HYDROXYCINNOLINES

Substituent	m.p. (°C)	Substituent	m.p. (°C)
3-Methyl	241–242 [1,2]	7-Methyl	243.5–244.5 [1,3]
acetyl deriv.	117–117.5	acetyl deriv.	117–118
3-Chloro	278–279 [3]	7-Chloro	276–277 [2,14]
acetyl deriv.	125–126	7-Nitro	295–296 [5]
3-Nitro	276.5–277.5 [4–6]	acetyl deriv.	140–141
5-Chloro	330–332 [7]	7-Amino	276–277 [5]
5-Nitro	304–305 [5]	diacetyl deriv.	330
acetyl deriv.	185–186	7-Methoxy	255–257 [8]
5-Methoxy	275 [8]	8-Methyl	220–221 [7]
6-Methyl	271 [7]	8-Chloro	198–199 [9]
acetyl deriv.	117–118	8-Nitro	185–187 [4,15]
6-Chloro	294–295 [2,7,9–11]	acetyl deriv.	not formed
acetyl deriv.	159–160	8-Amino	290–291 [5]
6-Nitro	330–331, 336 [2,4,7,12]	diacetyl deriv.	282–283
acetyl deriv.	147–148	8-Methoxy	162–163 [7]
6-Amino	275–276 [2]	5,8-Dichloro	222–224 [7]
6-Methoxy	252, 254–255 [7,8]	6,8-Dichloro	221–223 [7]
		7,8-Dichloro	261–262 [7]

1 J.R. Keneford and J.C.E. Simpson, J. chem. Soc., 1448, 354.
2 N.J. Leonard and S.N. Boyd, J. org. Chem., 1946, **11**, 419.
3 K. Schofield and J.C.E. Simpson, J. chem. Soc., 1948, 1170.
4 idem, ibid., 1945, 512.
5 K. Schofield and R.S. Theobald, ibid., 1949, 2404.
6 K. Schofield and T. Swain, ibid., 1949, 1367.
7 H. J. Barber et al., ibid., 1961, 2828.
8 A.R. Osborn and K. Schofield, ibid., 1955, 2100.
9 K. Schofield and J.C.E. Simpson, ibid., 1945, 520.
10 J.R. Keneford et al., ibid., 1950, 1104.
11 K. Schofield and T. Swain, ibid., 1949, 2393.
12 W. Borsche and A. Herbert, Ann., 1941, **546**, 293.
13 J.R. Keneford, J.S. Marley and J.C.E. Simpson, J. chem. Soc., 1948, 1702.
14 C.M. Atkinson and J.C.E. Simpson, ibid., 1947, 232.
15 J.C.E. Simpson, ibid., 1947, 237.

drazones have been known to decarboxylate after hydrolysis to the diacids (LI) in certain cases (R = p-OH and p-$C_6H_5CH_2O$-). The corresponding mesoxalyl chlorides are moderately stable and are cyclised by heating with titanium tetrachloride in nitrobenzene at 100°. Decarboxylation has been

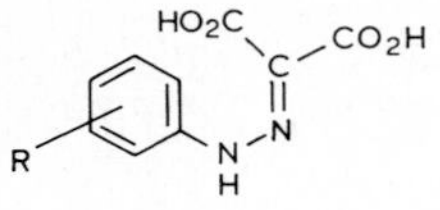

(LI) 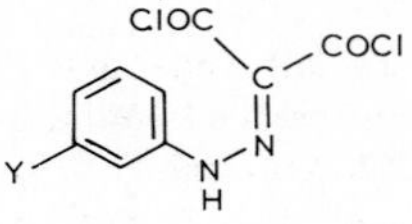 (LII)

effected by heating in benzophenone at 200–215°. Although a variety of substituted 4-hydroxycinnoline-3-carboxylic acids have been prepared by this method, problems arise for the cyclisation of 3-substituted mesoxalyl chloride phenylhydrazone (LII) which may cyclise on to either adjacent *ortho*-carbon to give a mixture of the 5- and the 7-substituted-4-hydroxycinnoline. For example, cyclisation of LII, (Y = Cl), gives a mixture of 5-chloro- and 7-chloro-4-hydroxycinnoline. (*R.R. Sharp* and *R.N. Castle*, J. heterocyclic Chem., 1965, **2**, 63).

The synthesis of 6-nitro-4-hydroxy- and 8-nitro-4-hydroxy-cinnolines from *o*- and *p*-nitroaniline has been described (*P.S. Fernandes et al.*, J. Indian chem. Soc., 1975, **52**, 546). *p*-Nitroaniline is diazotised and treated with diethyl malonate and the product hydrolysed and decarboxylated to give (LIII) which with polyphosphoric acid cyclises to 6-nitro-4-hydroxycinnoline (LIV).

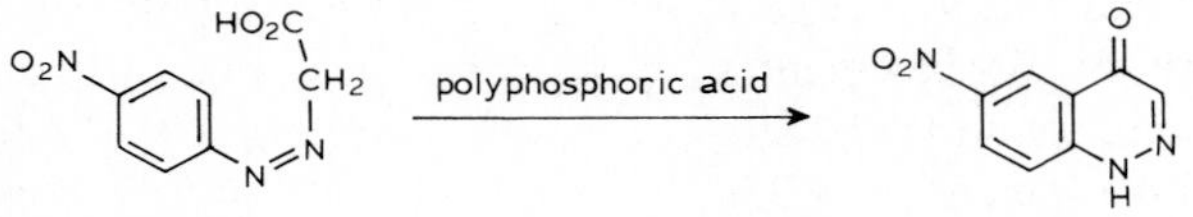

(LIII) → polyphosphoric acid → (LIV)

A heterocyclisation reaction leading to cinnolin-4(1*H*)-one derivatives has been reported (*A.A. Sandison* and *G. Tennant*, Chem. Comm., 1974, 752). 2-Nitrophenacylidene phenylhydrazones (LV) undergo base-catalysed cyclisation by intramolecular nucleophilic displacement of the nitro group by the *ortho*-side-chain, providing an efficient general route to 3-substituted-1-phenylcinnolin-4(1*H*)-ones (LVI) (*e.g.* R = acetyl, yield = 92%).

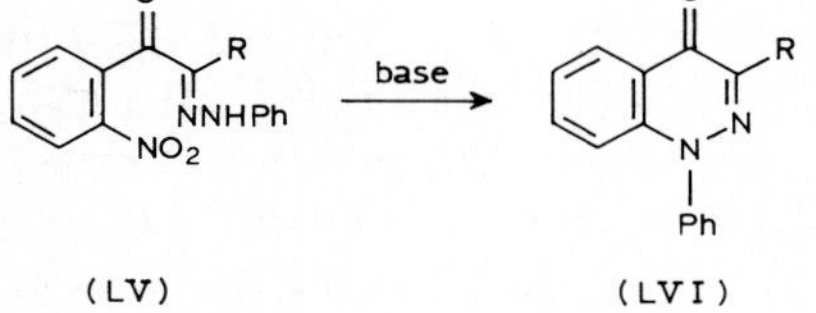

(LV) → base → (LVI) (LVII)

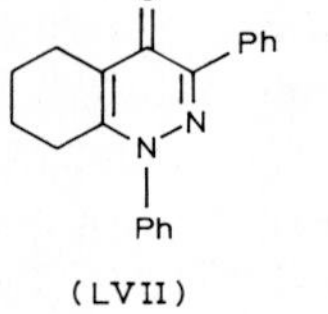

1,3-Diphenylcinnoline-4(1H)-one, m.p. 183°, (LVI, R = Ph) has been prepared from the 5,6,7,8-tetrahydro derivative (LVII) by dehydrogenation with 10% palladium on charcoal in decalin (*V. Sprio, O. Migliara* and *S. Plescia*, Ann. Chim. (Rome), 1971, **61**, 648).

Hydrolysis of a 4-substituted-cinnoline such as 4-amino (*H.E. Baumgarten*, J. Amer. chem. Soc., 1955, **77**, 5109), 4-chloro (*K. Hayashi* and *T. Watanabe*, Yakugaku Zasshi, 1968, **88**, 94), and 4-methanesulphonyl (*G. Barlin* and *W.V. Brown*, J. chem. Soc. C, 1967, 2473) gives the corresponding 4-hydroxycinnoline.

4-Hydroxy-3-phenylcinnoline, m.p. 260–261°, has been prepared in good yield from the high temperature reaction of 3-phenylcinnoline-4-carboxylic acid with potassium hydroxide and copper(II) oxide.

4-Hydroxycinnoline, m.p. 234°, pK_a (as acid) 9.53 in 50% aqueous alcohol, forms an *acetyl* derivative, m.p. 127–128° (Table 16, ref. 2, 14). 4-*Methoxy-*, 4-*ethoxy-* and 4-*methoxy-6-nitrocinnoline* have m.ps. 127–128, 101–102° and 194–194.5° respectively (Table 16, ref. 14) and the first named has pK_a 2.7 in 50% aqueous alcohol (*J.R. Keneford et al.*, J. chem. Soc., 1949, 1356). 6-*Chloro-4-methoxycinnoline* has m.p. 169.5–170° (*J.C.E. Simpson*, ibid., 1947, 1653). Data for other substituted 4-hydroxycinnolines are presented in Table 16.

4-Hydroxycinnolines are soluble in dilute sodium hydroxide and even in aqueous sodium carbonate when electronegative groups, such as nitro, are substituents in the benzene ring (Table 16, ref. 4, 7). Of greatest preparative value has been their conversion in to the corresponding 4-chloro derivatives in which the halogen is readily replaced by hydrogen. Treatment with phosphorus pentasulphide in refluxing pyridine gives the 4-thiol derivative in excellent yield (*R.N. Castle et al.*, J. org. Chem., 1960, **25**, 570), often with concomitant displacement of halogen substituents on the homocycle.

4-Hydroxycinnoline is reduced by lithium aluminium hydride in refluxing tetrahydrofuran to a 1,2,3,4-tetrahydrocinnoline/cinnoline mixture. Gentle oxidation of the mixture using red mercury(II) oxide gives cinnoline in good yield (*C.M. Atkinson* and *J.C.E. Sharpe*, J. chem. Soc., 1959, 2858; *D.E. Ames* and *H.Z. Kucharska*, ibid., 1962, 1509). Similar reduction has been reported for phenyl- and methoxy-substituted cinnolines (*N.B. Chapman* and *D.Q. Russell-Hill*, ibid., 1956, 1563). Treatment of 5-chloro- and 8-chloro-4-hydroxycinnoline with fuming sulphuric acid at 170–180° gives the 8- and 5-sulphonic acid derivatives, respectively (*E. Lunt, K. Washbourn* and *W.R. Wragg*, J. chem. Soc. C, 1968, 687). Bromination at the 3-position of various 4-hydroxycinnolines has been effected by addition of bromine to an alkaline solution of the cinnoline (*D.E. Ames* and *R.F. Chapman*, ibid., 1967, 40). Sulphuryl chloride gives the corresponding 3-chloro derivative in poor yield (*K. Schofield* and *T. Swain*, ibid., 1950, 384). Nitration of 4-hydroxycinnoline in sulphuric acid solution gives various nitro-4-hydroxycinnolines, of which the 6-nitro (58%) and the 8-nitro (30%) derivative are most abundant (*R.B. Moodie, J.R.*

Penton and *K. Schofield*, J. chem. Soc. B, 1971, 1493). The species undergoing nitration is considered to be the 4-hydroxycinnolinium cation. The nitration of various 4-hydroxycinnolines including the 5-, 6-, 7-, and 8-chloro derivatives has also been reported (*D.E. Ames, R.F. Chapman* and *D. Waite*, J. chem. Soc. C, 1966, 470; *Lunt et al. loc. cit.*).

In contrast to the protonation of 4-hydroxycinnoline, methylation with dimethyl sulphate, diazomethane, or methyl iodide gives principally the anhydro base (LVIII) of 2-methyl-4-hydroxycinnolinium hydroxide, together with a small amount of 1-methyl-4-cinnolone (LIX). Similarly, other alkylating agents yield the 2-alkyl derivative as the major product (*D.E. Ames et al.*, J. chem. Soc., 1963, 4924; 1964, 283; 1965, 5391). Substitution at the 3-position results in alkylation at N-1 whereas alkylation of a 6-substituted compound gives a mixture of N-1 and N-2 alkylated products*. The presence of an 8-substituent results in N-2 alkylation only (*idem*, J. chem. Soc. C, 1966, 470). The reversible base-catalysed cyanoethylation of 4-hydroxycinnoline with acrylonitrile gives exclusively 1-cyanoethyl-4-cinnolone (LX).

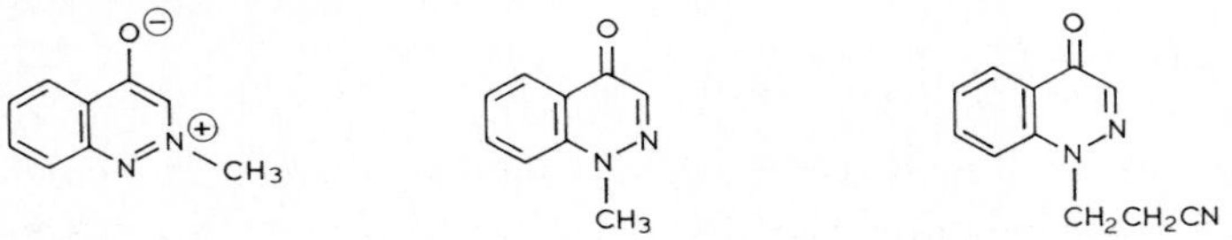

(LVIII)   (LIX)   (LX)

A study of the alkylation of 6-bromo-1,4-dihydro-4-oxo-3-cinnolinecarboxylic acid (LXI) in basic solution has shown that N-2 is alkylated to the same extent as N-1 (*R.P. Brundage* and *G.Y. Lesher*, J. heterocyclic Chem., 1976, **13**, 1085). The intermediate inner salt, 2-methyl-6-bromo-3-carboxy-4-hydroxycinnolinium hydroxide (LXIII) has been isolated and its facile decarboxylation to (LXII) studied.

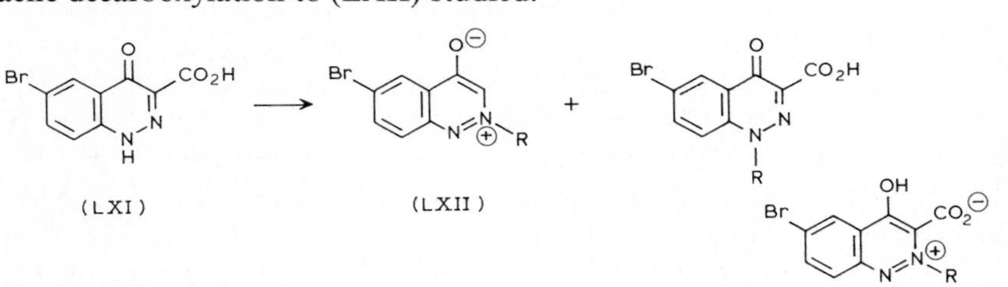

(LXI)   (LXII)

(LXIII)

---

* The 2-methyl isomer of 6-nitro-4-hydroxycinnoline was wrongly formulated as the methyl nitronate (Table 16, ref. 10).

Irradiation of various 2-alkylcinnolinium-4-olates has been shown to cause their photochemical rearrangement to 3-alkyl-4(3*H*)-quinzolones (*D.E. Ames, S. Chandrasekhar* and *R. Simpson*, J. chem. Soc. Perkin I, 1975, 2035).

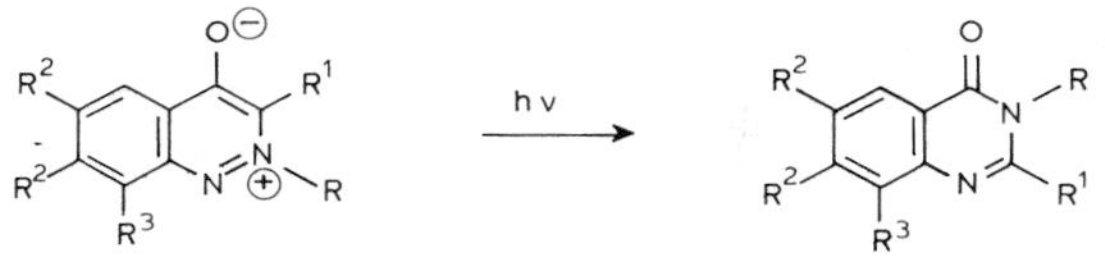

Quaternisation of 6-chlorocinnolin-4(1*H*)-one with methyl tosylate gives (LXIV) as the predominant tautomer. Treatment of the latter with alkali gives the betaine (LXV) which with dimethyl acetylenedicarboxylate gives the crystalline cycloadduct (LXVI), m.p. 89° (*N. Dennis, A.R. Katrikzky* and *M. Ramaiah, ibid.*, 1975, 1506).

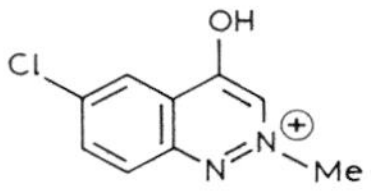

(LXIV) 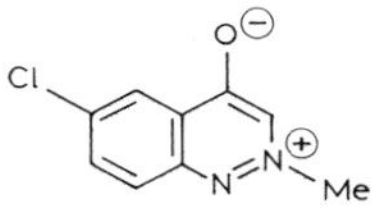   (LXV) 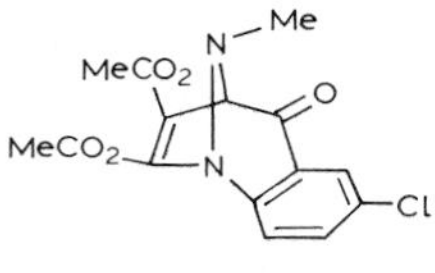   (LXVI)

Acetyl derivatives, however, are considered to be 4-acetoxy compounds and not 1-acetyl-4-cinnolones. The 8-substituted derivatives appear to resist acetylation and, in certain cases, *e.g.* 8-nitro-4-hydroxycinnolines, this can be explained in terms of intramolecular hydrogen bonding as in (LXVII) (Table 16, ref. 1, 4, 14).

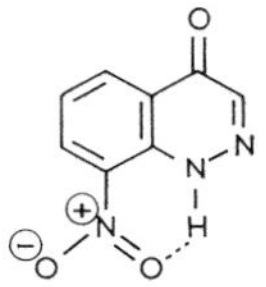

(LXVII) 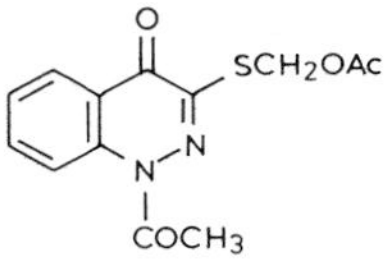   (LXVIII) 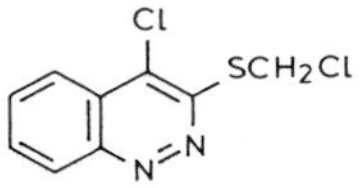   (LXIX)

3-Methylsulphinyl-4(1*H*)-cinnolone gives the diacetate (LXVIII) and the dichloro derivative (LXIX) on treatment with acetic anhydride and thionyl chloride (*D.T. Connor, P. Young* and *M. Von Strandtmann*, J. heterocyclic Chem., 1978, **15**, 115).

A study of the mass spectral fragmentation of 4-hydroxycinnoline shows that consecutive loss of two hydrogen cyanide molecules, carbon monoxide, and a hydrogen atom occurs, giving ions at m/z 119, 92, 64 and 63 (*J.R. Elkins* and *E.V. Brown, ibid.*, 1965, **5**, 639).

(ii) 5-, 6-, 7- and 8-Hydroxycinnolines

Appropriate methoxy-*o*-nitroacetophenones give, on reduction followed by diazotisation, 5-, 6-, 7- and 8-methoxy-4-hydroxycinnolines, from which 5-, 6-, 7- and 8-methoxycinnolines are obtained by dehalogenation of the corresponding chloro compounds. The 8-methoxy derivative quickly liquifies and becomes green in air. Subsequent demethylation with hydrobromic acid gives the corresponding hydroxycinnolines in good overall yield (*K. Schofield* and co-workers: refs. given with Table 17). The 5-, 6- and 7-hydroxycinnolines are high melting crystalline solids.

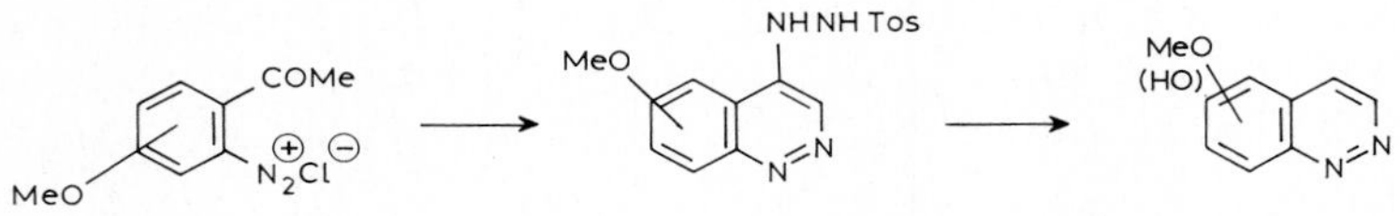

An alternative synthesis of 8-hydroxycinnoline (*Albert* and *Hampton, loc. cit.*) is from methyl 3-methoxyanthranilate by a method based on the synthesis of cinnoline. The intermediate 4-*methyl*-8-*methoxycinnoline*, m.p. 129–130°, is demethylated and the resulting 8-*hydroxy*-4-*methylcinnoline*, m.p. 176.7–177.5°, converted into its hydrochloride which condenses better

TABLE 17

Bz-HYDROXYCINNOLINES AND DERIVATIVES

Compound	Position of substituent, m.p. (°C)			
	5[1]	6[1]	7[1]	8[2]
x-Hydroxy	285 (dec.)	> 300	> 300	185–186 sublimes at 120°/0.5 mm
x-Methoxy	92–93.5	87–88	monohydrate 109–100	67–70.5
4-Hydroxy-*x*-methoxy	275	252[3]	255–257	monohydrate 164–165
4-Chloro-*x*-methoxy	141–142	149–151	178–179	142–143
x-Methoxy-4-toluene-*p*-sulphonylhydrazino-cinnoline hydrochloride	221 (dec.)	199–201 (dec.)	169–172 (dec.)	169–172 (dec.)

1 5-, 6- and 7-Series by *A.R. Osborn* and *K. Schofield*, J. chem. Soc., 1955, 2100.
2 8-Series, *K. Schofield et al., ibid.*, 1952, 3009; *A. Albert* and *A. Hampton, ibid.*, 1952, 4985.
3 Identical with the product from the Richter synthesis (*K. Schofield* and *J.C.E. Simpson, ibid.*, 1945, 512).

with benzaldehyde than does the free base to give 8-*hydroxy-4-styrylcinnoline*, m.p. 200–201°. This, as its *benzoyl* derivative, m.p. 212–213°, is oxidised by potassium permanganate to 8-*hydroxycinnoline-4-carboxylic acid*, m.p. 200° (dec.), which is readily decarboxylated to 8-hydroxycinnoline on heating in ethylene glycol.

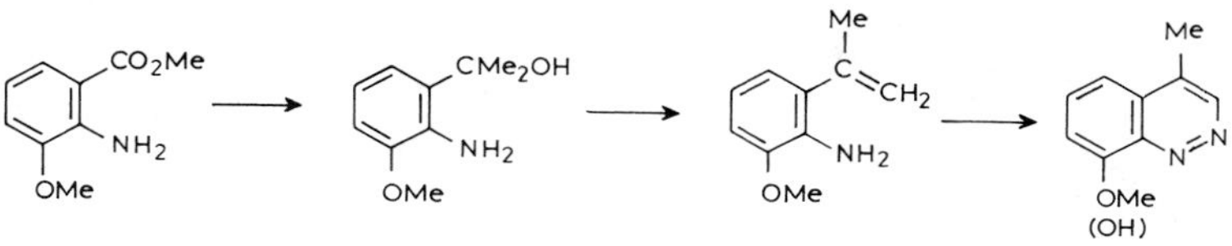

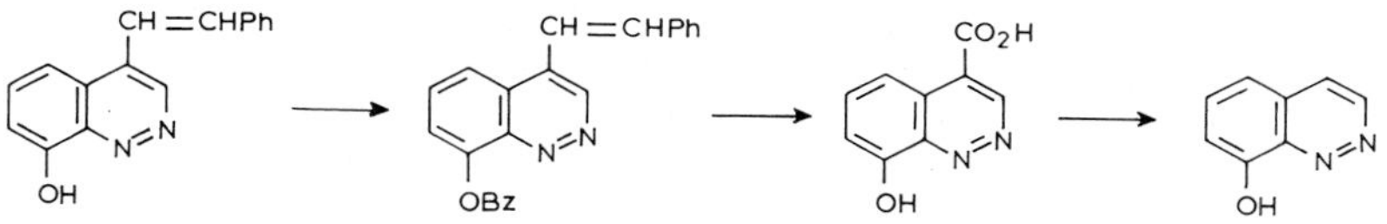

Alternatively, the reaction of 2-amino-3-methoxyacetophenone with either methylmagnesium iodide, or phenylmagnesium bromide gives the carbinols (LXX; R = CH$_3$ and C$_6$H$_5$) which are readily dehydrated. Diazotisation and cyclisation of the alkenes (LXXI) gives the corresponding 4-*methyl-* and 4-*phenyl-4-hydroxycinnolines* (m.ps. 177–178.5° and 142–143.5°, respectively) (*E.J. Alford et al., J. chem. Soc.*, 1952, 3009).

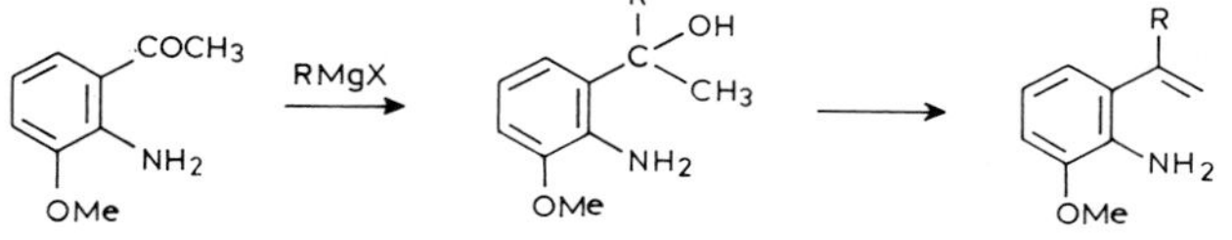

Cyclisation of mesoxalyl chloride *p*-methoxyphenylhydrazone in dry nitrobenzene at 95° in the presence of titanium tetrachloride affords 4-hydroxy-6-methoxycinnoline-3-carboxylic acid in good yield (*H.J. Barber et al., J. chem. Soc.*, 1961, 2828).

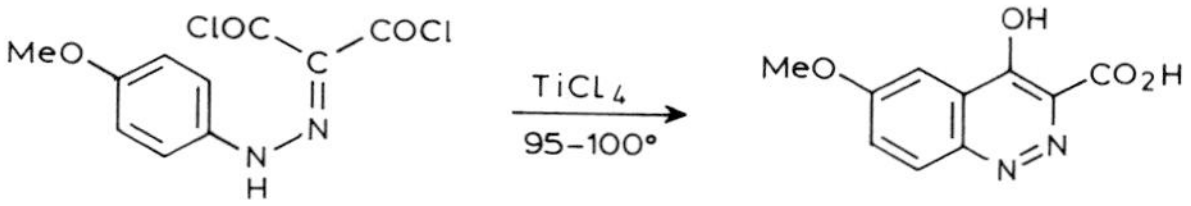

The 5-, 6-, 7-, and 8-hydroxycinnolines show phenolic properties but give little colour with iron(III) chloride. 8-Hydroxycinnoline forms chelates with various metal ions but seems to offer no advantages over the corresponding more stable 8-hydroxyquinoline complexes (*Albert* and *Hampton, loc. cit.*; *H. Irving* and *H.S. Rosotti*, Analyst, 1955, **80**, 245). The properties of 8-hydroxycinnoline have been extensively studied including its solubility as determined by the quantitative formation of its dibromo derivative. Absorption spectra and dissociation constants are reported for 8-hydroxycinnoline and its 4-methyl derivative in comparison with analogous quinoline derivatives.

The 5- and 8-hydroxy isomers couple with benzenediazonium chloride, while the 6- and 7-compounds do not, but all give very good yields of the corresponding amines, as yellow crystalline substances, in the Bucherer reaction.

Their u.v. spectra closely resemble the spectra of the corresponding methoxy derivatives except for the presence of maxima above 400 nm in the spectra of the hydroxy derivatives suggests some zwitterionic structure (*e.g.* LXXII and LXXIII) (*A.R. Osborn* and *K. Schofield*, J. chem. Soc., 1956, 4207).

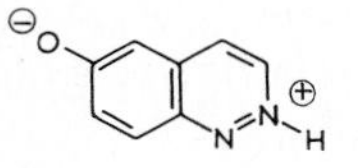

(LXXII) 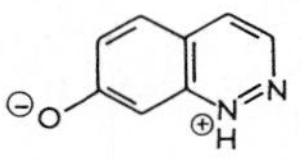   (LXXIII)

Their i.r. spectra show absorption due to O–H stretching vibration and give no indication of C=O stretching (*S.F. Mason, ibid.*, 1957, 4874).

(iii) Phenyl ethers

Many 4-phenoxycinnolines have been made because of their usefulness for the preparation of 4-aminocinnolines. The 4-phenoxycinnolines given in Table 18 are prepared by heating the chloro compounds in phenol for about one hour at 95° with either one equivalent of potassium hydroxide or excess of ammonium carbonate. The 4-phenoxycinnolines, like the chloro compounds, are stable and not hydrolysed by boiling water, However, with very dilute acid or on long boiling with alcohol they yield the hydroxy compounds.

4-*Phenoxycinnoline*, pK_a 2.27 in 50% aqueous ethyl alcohol, is converted by boiling acetic anhydride to 4-*acetoxycinnoline*, m.p. 127–128° (Table 18, ref. 7; Table 17, ref. 3). In addition to their value in the preparation of simple 4-aminocinnolines, 4-phenoxycinnolines can easily be converted to substituted amino compounds.

As for the 4-hydroxycinnolines, absence of a bulky substituent at posi-

TABLE 18

SUBSTITUTED 4-PHENOXYCINNOLINES

Substituent	m.p. (°C)	Substituent	m.p. (°C)
None	94–95 [1]	3-Chloro	120–121 [2]
5-Chloro	118–119 [3]	5,6-Dichloro	189 [3]
6-Chloro	128–129 [1]	7-Chloro	127–128 [1]
8-Chloro	158–159 [4]	3-Nitro	108–109 [5]
5-Nitro	123 [6]	6-Nitro	117–118 [5,7]
7-Nitro	172–173 [8]	8-Nitro	166–167 [4,8]
5-Amino	199–200 [6]	7-Amino	179–180 [6]
8-Amino	130 [6]	3-Methyl-6-nitro	129–130 [9]
3-Methyl-8-nitro	137.5–138 [8]		

1 J.R. Keneford and J.C.E. Simpson, J. chem. Soc., 1947, 917.
2 K. Schofield and T. Swain, ibid., 1950, 384.
3 H.J. Barber et al., ibid., 1961, 2828.
4 J.R. Keneford, J.S. Morley and J.S.E. Simpson, ibid., 1948, 1702.
5 K. Schofield and T. Swain, ibid., 1949, 1367.
6 C.M. Atkinson, J.C.E. Simpson and A. Taylor, ibid., 1954, 1381.
7 J.R. Keneford, K. Schofield and J.C.E. Simpson, ibid., 1948, 358.
8 K. Schofield and T. Swain, ibid., 1949, 2393.
9 J.R. Keneford and J.C.E. Simpson, ibid., 1948, 354.

tion 3 results in methylation at N-2 on quaternisation with methyl iodide in ethanol (*D.E. Ames et al.*, J. chem. Soc., 1965, 5391).

(i) Cinnolinecarboxylic acids and derivatives

Of the simple cinnolinecarboxylic acids, the 3-, 4-, and 7-carboxylic acids have been most studied. The von Richter synthesis (general method 1), which involves diazotisation and cyclisation of an *o*-aminophenylpropiolic acid leads to the formation of 4-hydroxy-3-cinnolinecarboxylic acids. The synthesis of fifty-one 4-hydroxycinnoline-3-carboxylic acids and their esters by the method of Barber and co-workers (general method 4) has been reported in a patent (*D.J. Gilman*, C.A., 1972, **77**, 34545). Diazotised substituted anilines were treated with diethyl malonate, the product saponified, and the derived acid chlorides cyclised to give the required compounds which are reported to be useful for the treatment of asthma, hay fever, and urticaria. In a similar way, the preparation of ethyl 6-ethyl-4-hydroxy-3-cinnolylcarboxylate-3-^{14}C using $^{14}CH_2(CO_2Et)_2$ has been described (*S.W. Longworth et al.*, J. labelled Compd., 1974, **10**, 423). Numerous 6-substituted (amino, acyl, cyano and carboxy)-4-hydroxycinnoline-3-carboxylic acids and their esters have been prepared and reported to have anti-anaphylactic activity (*J. Preston* and *A.J. Reeve*, C.A., 1976, **85**, 78150). Similarly, thirty 6-substituted-cinnoline-3-carboxylic acids and derivatives have been prepared by reduction of the corresponding 4-chlorocinnolinecarboxylates and reported to have antigen-antibody reaction-inhibiting activity (Pat., C.A., 1976, **84**, 180256). Ethyl 4-hydroxy-6-phenylcinnoline-3-carboxylate, an anti-allergic compound, is prepared from the 6-amino-compound by refluxing it for two hours with amyl nitrite and benzene (*J. Morton* and *J. Preston*, B.P., 1,472,767, 1977).

Heating of 1-ethyl-4-hydroxy-3-methanesulphonylcinnoline (LXXIV), prepared by a modification of the Borsch synthesis (general method 3), with potassium cyanide in DMF for 24 h at 120° gives the 3-cyano compound, which on hydrolysis by refluxing acetic acid–hydrochloric acid gives the 3-carboxylic acid (LXXV) (*R. Albrecht*, Ann., 1978, 617).

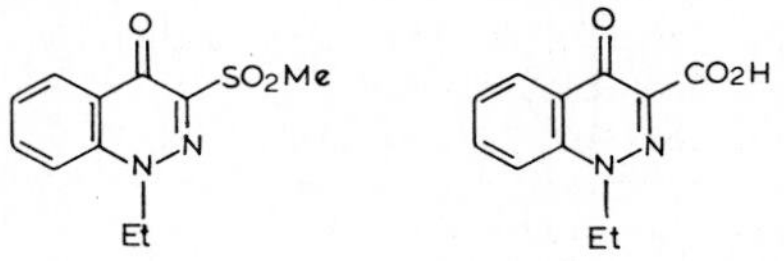

(LXXIV)   (LXXV)

Acetylation of 4-hydroxycinnoline-3-carboxylic acid with acetic anhydride is accompanied by decarboxylation and the product is 4-acetoxycinnoline. The hydroxy-acid displays a characteristic and unusual behavior towards heating with a mixture of pyridine and acetic anhydride at 95° (*K. Schofield* and *J.C.E. Simpson*, J. chem. Soc., 1946, 472). A crystalline product, m.p. 217°, is formed which is a zwitterion based on a resonance hybrid associated with structures (LXXVI) and (LXXVII). When this substance is refluxed with either an alcohol or aniline in benzene, an acidic addition product of type (LXXVIII) (in which R = OMe, OEt, NHPh, etc.) is formed.

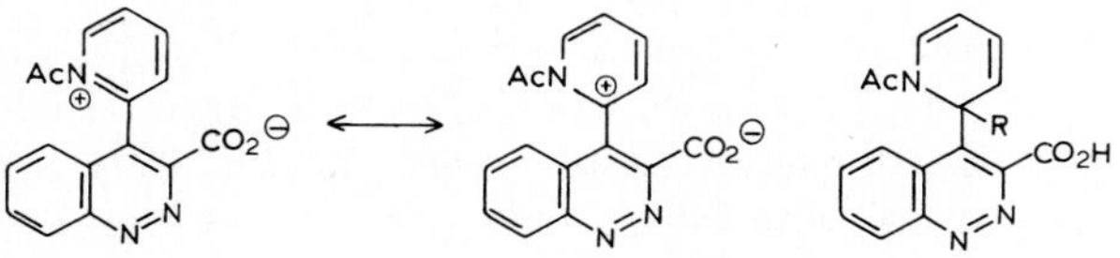

(LXXVI)   (LXXVII)   (LXXVIII)

6-Chloro-1-methyl-4-oxocinnoline-3-carboxylic acid is readily decarboxylated on heating to give 6-chloro-1-methyl-4-cinnolone (Table 19, ref. 12).

TABLE 19

CINNOLINECARBOXYLIC ACIDS AND DERIVATIVES

Compound	m.p. (°C)	Compound	m.p. (°C)
3-Carboxylic acid	206 (dec.) [1]	4-Carboxylic acid	195–196 (dec.) [2]
ethyl ester	97–97.5 [3]	ethyl ester	48.5–49.5 [2]
4-hydroxy-	268–268.5 [4]	3-phenyl-	224–225 [5,6]
4-hydroxy-, ethyl ester	191–192 [7]	3-phenyl-,	
	194–195 (dec.) [8]	ethyl ester	92–93 [6]
(acetyl derivative)	82–83 [7]	4-Carbaldehyde	147–149 [11]
4-hydroxy-6-chloro-	263–264 (dec.) [9]	4-Acetyl-	100–101 [12]
	267 (dec.) [10]	4-Carbonylchloride [2]	–
4-hydroxy-6-bromo-	264 (dec.) [9]	3-phenyl-	139–142 [6]
	256–259 (dec.) [10]	4-Carboxamide	
4-hydroxy-6-methoxy-	268 (dec.) [9]	3-phenyl-	272–273 [6]
	258–259 (dec.) [10]	4-Carbonitrile	139–140 [13]
3-Carbaldehyde	119–120 [1]		146.5 [14]
3-Acetyl-	155–156 [3]	4,6-Dicarbonitrile	179 [13]
3-Carboxamide, 4-hydroxy-	354–355 [8]		
3,4-Dibenzoyl-	162–163 [15]		

1 H.J. Haas and A. Seeliger, Ber., 1963, **96**, 2427.
2 T.L. Jacobs et al., J. Amer. chem. Soc., 1946, **68**, 1310.
3 H.E. Baumgarten and C.H. Anderson, ibid., 1958, **80**, 1981.
4 K. Schofield and J.C.E. Simpson, J. chem. Soc., 1945, 512.
5 H.E. Baumgarten and J.L. Furnas, J. org. Chem., 1961, **26**, 1536.
6 H.S. Lowrie, J. med. Chem., 1966, **9**, 664.
7 K. Schofield and J.C.E. Simpson, J. chem. Soc., 1946, 1035.
8 E. Lunt, K. Washbourn and W.R. Wragg, J. chem. Soc. C, 1968, 687.
9 K. Schofield and T. Swain, J. chem. Soc., 1949, 2393.
10 H.J. Barber et al., ibid., 1961, 2828.
11 R.N. Castle and M. Onda, J. org. Chem., 1961, **26**, 4465.
12 D.E. Ames et al., J. chem. Soc., 1965, 5391.
13 E. Hayashi, Y. Akahori and T. Watanabe, C.A., 1968, **68**, 49538.
14 G.B. Barlin and W.V. Brown, J. chem. Soc. C, 1967, 2473.
15 D.E. Ames, H.R. Ansari and A.W. Ellis, ibid., 1969, 1795.

Esterification of 6-chloro-4-hydroxycinnoline-3-carboxylic acid has been reported using ethanol with boron trifluoride etherate as catalyst (*D.E. Ames. R.F. Chapman* and *H.Z. Kucharska, ibid.*, Suppl. 1, 1964, 5659). However, this method has been reported as being unsuccessful for the parent compound, 4-hydroxycinnoline-3-carboxylic acid (*J. Daunis, M. Guerret-Rigail* and *R. Jacquier*, Bull. Soc. chim. Fr., 1972, 3198). Other methods tried include: ethanolic sulphuric acid giving the desired product in 10% yield; methyl iodide, which gave the N-1-methyl derivative; and a mixture of thionyl chloride and phosphorus pentachloride followed by ethanol which gave the ester in 70% yield if performed on a small scale. Esterification of the N-1 alkylated cinnoline-3-carboxylic acids is much easier than that of the parent acid. Diazotisation of *o*-aminobenzaldehyde followed by coupling with diethyl malonate gives (LXXIX, R = OEt) which cyclises spontaneously to give ethyl cinnoline-3-carboxylate albeit in poor yield (Table 19, ref. 3). Replacement of diethyl malonate by ethyl acetoacetate produces (LXXIX, R = CH_3) which undergoes spontaneous cyclisation to give 3-cinnolinyl methyl ketone. The 6-chloro and the 7-chloro derivatives have been made similarly.

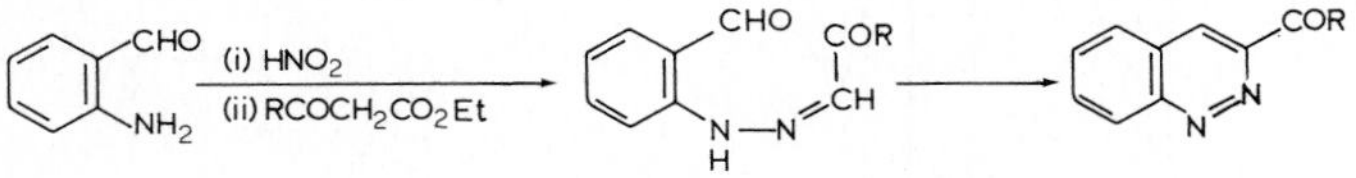

Certain 1a,7b-dihydro-1*H*-cyclopropa[*c*]cinnolines give, on thermal rearrangement in boiling xylene, a mixture of (LXXX; 28%) and (LXXXI; 7%) (*L. Garanti* and *G. Zecchi*, J. heterocyclic Chem., 1978, **15**, 509).

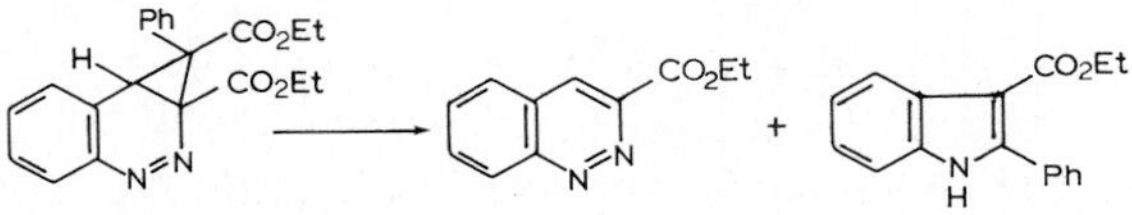

A convenient synthesis of cinnoline-3-carboxylates involves treatment of a functionalised enamine (LXXXII) with an aryldiazonium tetrafluoroborate followed by cyclisation-deamination of the resulting hydrazone. Structural factors affecting the cyclisation have been discussed (*C. B. Kanner* and *U. K. Pandit*, Heterocycles, 1978, **9**, 1381).

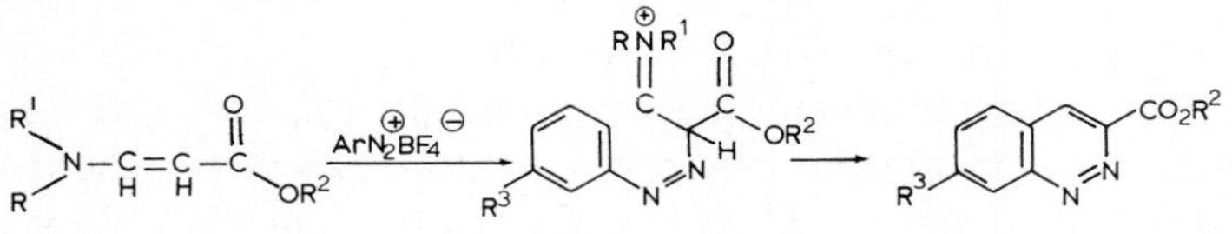

3-Methylsulphonylcinnolines react with potassium cyanide to give the corresponding 3-cyano derivative (*R. Albrecht, loc. cit.*). 3-Bromocinnolines (*e.g.* LXXXIII) undergo a similar displacement using copper(I) cyanide (*W.A. White*, G.P., C.A., 1970, **73**, 77269; *D.E. Ames* and *C.J.A. Byrne*, J. chem. Soc. Perkin I, 1976, 592).

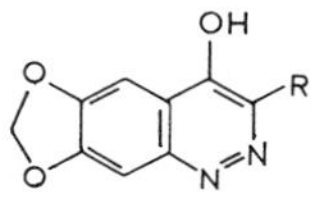

(LXXXIII) 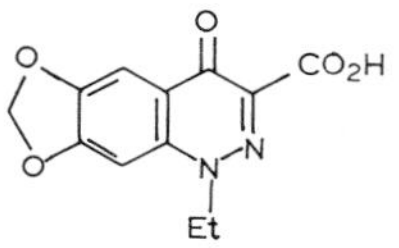 (LXXXIV) 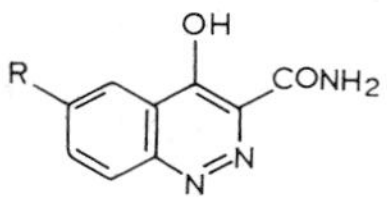 (LXXXV)

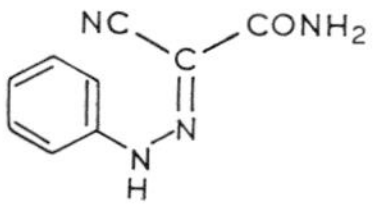

(LXXXVI)

The reaction (LXXXIII, R = Br) with copper(I) cyanide in DMF and treatment with iron(III) chloride/hydrochloric acid at 60° gives (LXXXIII, R = CN). The latter is alkylated by treatment with sodium hydride in DMF followed by ethyl iodide to give the 1-ethyl derivative. Reflux with acetic acid in hydrochloric acid gives an excellent yield of cinoxacin (LXXXIV) which shows anti-microbial and anti-cancer activity (*White, loc. cit.*; *T. Sakano, S. Masuda* and *S. Amano*, Chemotherapy, 1980, **28**, 139).

5-Substituted-3-halo-3-cinnolinecarbonitriles have also been prepared by treating the corresponding 4-hydroxycinnoline-3-carboxamide (LXXXV) with a phosphorus oxyhalide (Pat., C.A., 1972, **77**, 19666).

Cyclisation of cyano hydrazone derivatives has been reported to give cinnoline-3-carboxamides in good yield. Thus, when (LXXXVI) is refluxed for 1 h in chlorobenzene in the presence of aluminium trichloride 4-aminocinnoline-3-carboxamide is obtained in 86% yield (Pat., C.A., 1977, **87**, 117893).

4-Hydroxycinnoline-3-carboxamide derivatives of penicillins and cephalosporins have been synthesised as potential bacteriocides. For example (LXXXVII) is prepared by the reaction of the penicillamines with either the cinnolinecarbonyl chloride or the *N*-hydroxysuccinimide ester (Pat., C.A., 1977, **86**, 189920; *H. Tobiki et al.*, Yakugaku Zasshi, 1980, **100**, 38). Similarly, the cephalosporin derivative (LXXXVIII, R = 4-hydroxy-3-cinnolinecarbonyl-) is prepared by acylation of the free amine (Pat., C.A., 1978, **88**, 62402).

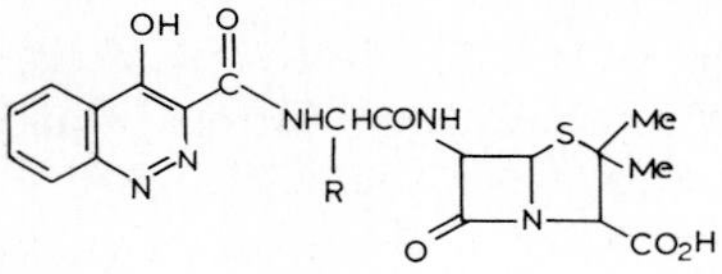

(LXXXVII) 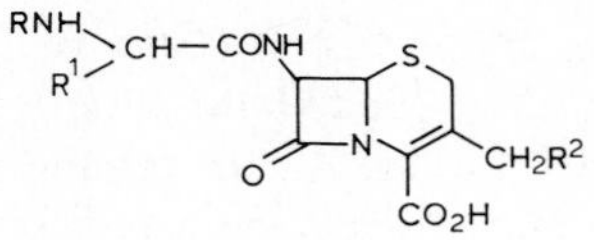   (LXXXVIII)

Surprisingly, when one of the natural sugars D-glucose, D-mannose, D-galactose or D-xylose is treated with phenylhydrazine in aqueous acetic acid at 115–120° for 3 h, there is obtained, in addition to the expected phenylosazones, a moderate yield of a 3-substituted cinnoline. For the first two, the compound obtained is 3-(D-arabo-tetrahydroxybutyl)cinnoline (LXXXIX) which, when treated with aqueous sodium periodate in the dark, gives 3-cinnolinecarbaldehyde in good yield (*H.J. Haas* and *A. Seeliger*, Ber., 1963, **96**, 2427). The aldehyde is readily oxidised to the acid by basic hydrogen peroxide. Reduction using hydrazine hydrate gives 3-methylcinnoline. Reaction of the aldehyde with catalytic amounts of ethanolic potassium cyanide gives the diol (XC; 32%) which is oxidised by air at 80° to give the corresponding dione (*E. Lippmann* and *S. Ungethuem*, Z. Chem., 1973, **13**, 343).

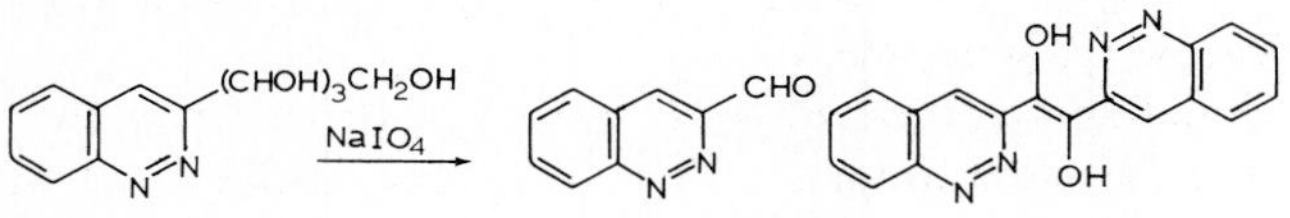

(LXXXIX)   (XC)

Cinnoline-4-carboxylic acids are important in that they provide a facile route to cinnolines *via* decarboxylation. They have been prepared by condensation of 4-methylcinnoline with benzaldehyde to give 4-styrylcinnoline which is readily oxidised to cinnoline-4-carboxylic acid. Cyclisation of benzaldehyde phenylhydrazone derivatives yields *N*-benzylideneaminoisation which gives 3-phenylcinnoline-4-carboxylic acid on rearrangement (Table 19, ref. 5, 6). Thermal decarboxylation at 150–200° results in formation of 3-phenylcinnoline. However, heating at 200° in the presence of potassium hydroxide, copper(II) oxide and metallic copper gives 4-hydroxy-3-phenylcinnoline in good yield (*H.S. Lowrie*, J. med. Chem., 1966, **9**, 670).

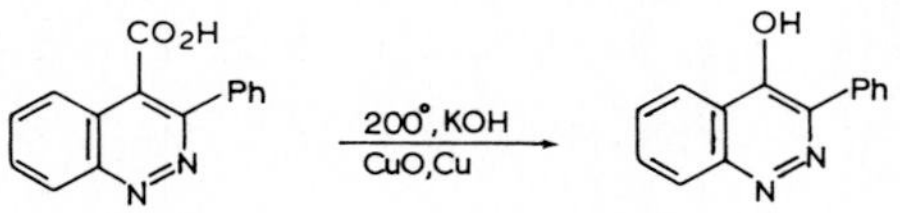

Polarographic reduction of 3-phenylcinnoline-4-carboxylic acid gives 1,4-dihydro-3-phenylcinnoline, which may undergo further reduction as described earlier. As for cinnoline, cinnoline-4-carboxylic acid gives indole on reduction with zinc amalgam in acetic acid, the decarboxylation most probably occurring at the intermediary dihydro stage. N-Oxidation of cinnoline-4-carboxylic acid by hydrogen peroxide in acetic acid gives 4-carboxycinnoline-2-oxide as the only isolable product (*M.H. Palmer* and *E.R.R. Russell*, J. chem. Soc. C, 1968, 2621).

Esterification of cinnoline-4-carboxylic acid can be achieved either directly using ethanol and sulphuric acid at reflux, or *via* the carbonyl chloride (Table 19, ref. 5, 11). The ethyl ester undergoes a Claisen condensation with ethyl acetate and hydrolysis of the resulting β-keto ester yields 4-*acetylcinnoline*, oxime m.p. 165–165.5°.

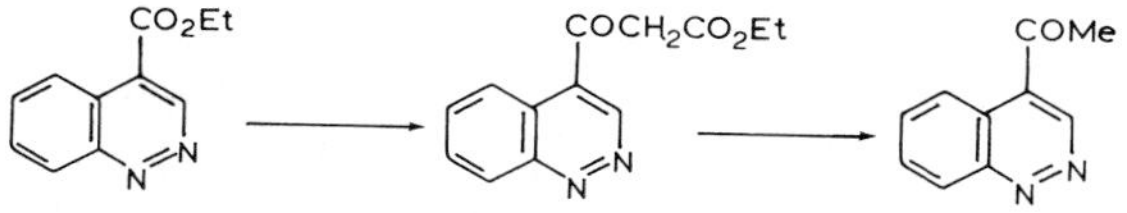

Ethyl 3-phenylcinnoline-4-carboxylate reacts with phenylmagnesium bromide to give the 1-phenyl derivative and not the expected Grignard product. Methyl hydrazine also gives the unexpected 1,4-dihydro ester and not the hydrazide (*Lowrie, loc. cit.*).

Various cinnoline-4-carboxamides have been prepared from the carbonyl chloride and the desired amine, and their pharmacological activities have been evaluated (*Lowrie, loc. cit.*).

Although cinnoline-4-carbaldehyde has been prepared by reduction of the ester with lithium aluminium hydride and of the carbonyl chloride with tri-t-butoxyaluminium hydride followed by partial reoxidation, in each case the aldehyde was isolated as its semicarbazone (Table 19, ref. 11). However, it has been isolated as yellow needles in a synthesis starting from 4-methylcinnoline. Reaction with pyridine and iodine gives the methylpyridinium iodide (XCI), which reacts with *p*-nitrosodimethylaniline to give (XCII). Conversion to the hydrazone (XCIII) and treatment with nitrous acid gives the aldehyde in good yield.

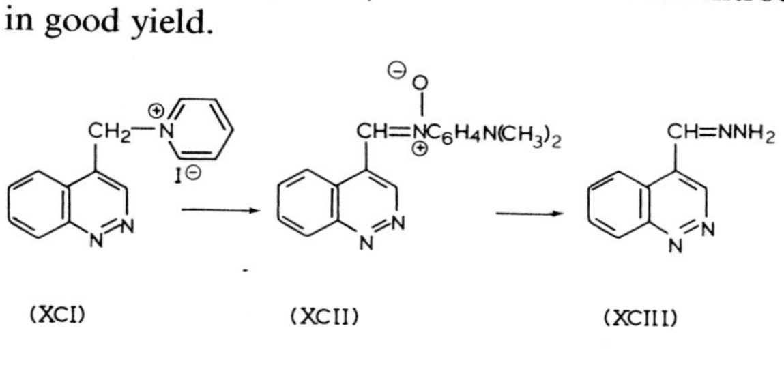

(XCI) (XCII) (XCIII)

Cinnoline-4-carbonylchloride reacts with diazomethane to give a diazoketone (XCIV) which is converted into 4-cinnolinyl chloromethyl ketone with dry hydrogen chloride, the ketone being stable as its hydrochloride salt (Table 19, ref. 2).

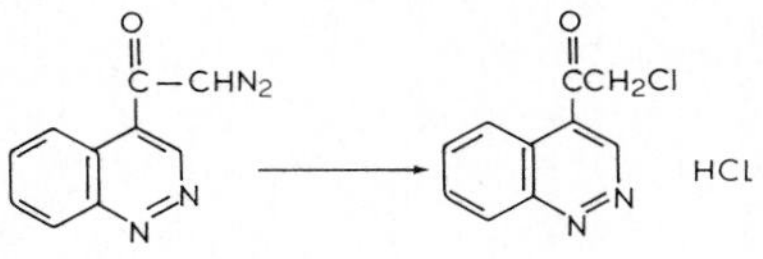

(XCIV)

3-Phenyl-4-cinnolinecarbonylchloride undergoes an aluminium trichloride Friedel–Crafts cyclisation to give 11-indeno[1,2-c]cinnoline-11-one (XCV) in good yield (*H.S. Lowrie*, J. med. Chem., 1966, **9**, 670).

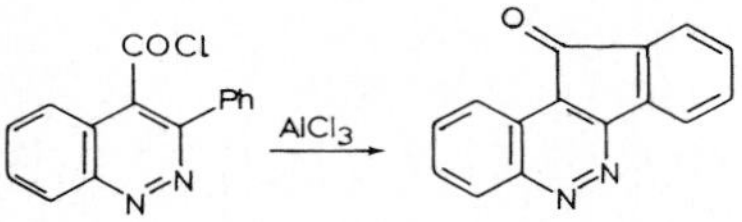

(XCV)

Oxidation of 1,3-diphenyl-2H-pyrrolo[3,4-c]cinnoline (XCVI) with nitric acid gives 3,4-dibenzoylcinnoline (XCVII) which is also produced in small amounts from the selenium dioxide oxidation of 3,4-dibenzylcinnoline (*F. Angelico*, C.A., 1910, **4**, 1618).

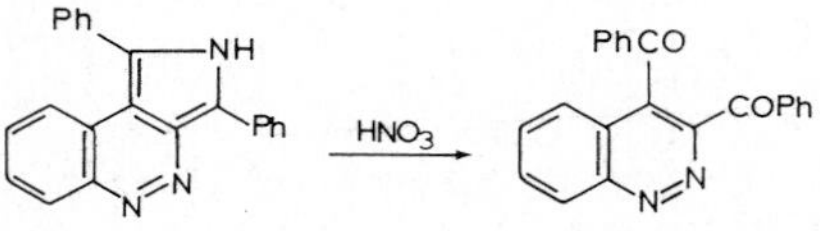

(XCVI)   (XCVII)

As for cinnoline-3-carbonitrile, the 4-derivative is readily prepared by displacement of the 4-methanesulphonyl group by cyanide, some cinnoline-4,6-dicarbonitrile being formed with excess cyanide (Table 19, ref. 13).

The mass spectra of 3-alkyl-4-cinnolinecarbonitriles have been reported (*M. Uchida, T. Higashimo* and *E. Hayashi*, Chem. pharm. Bull. Japan, 1977, **25**, 2225). The nitrile group is readily displaced by various nucleophiles (*T. Watanabe*, C.A., 1970, **72**, 3452) and has also been converted into the corresponding acid and carboxamide. Cinnoline-4-carbonitrile reacts with MeCOR (R = Me, Et, Ph) in benzene in the presence of sodium amide to give 4-cinnolinyl methyl ketones (XCVIII). Reaction with benzyl cyanide under similar conditions gives α-phenyl-4-cinnolinylacetonitrile (XCIX) in 60% yield (*E. Hayashi* and *I. Utsunomiya*, C.A., 1974, **81**, 136080; 1975, **82**, 4191). The reaction of 4-cinnolinecarbonitrile with Grignard reagents RMgX gives the 4-R, 4-RCO, and (unexpectedly) N-R (C) derivatives (*E. Hayashi et al.*, Chem. pharm. Bull. Japan, 1977, **25**, 579).

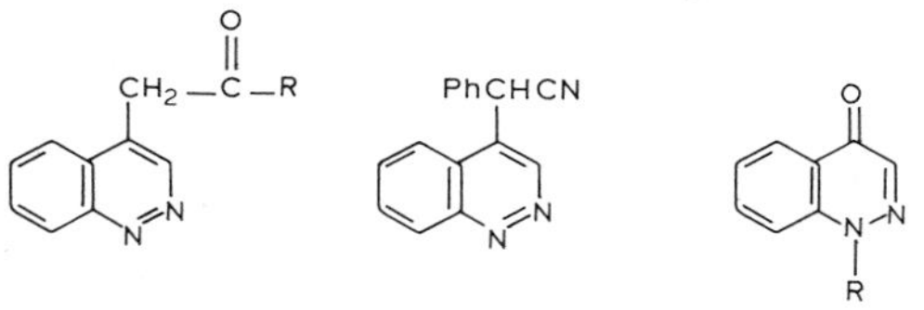

(XCVIII)   (XCIX)   (C)

4-Alkylcinnoline-7-carboxylic acids can be prepared by the Widman–Stoermer synthesis (general method 2) from 4-carboxy-2-aminoarylalkenes (CI). Diazotisation and cyclisation of 2-amino-5-cyanoacetophenone yields 4-hydroxycinnoline-6-carbonitrile, and ethyl β-(2-amino-4-carbethoxybenzoyl)propionate (CII) gives the diester (CIII).

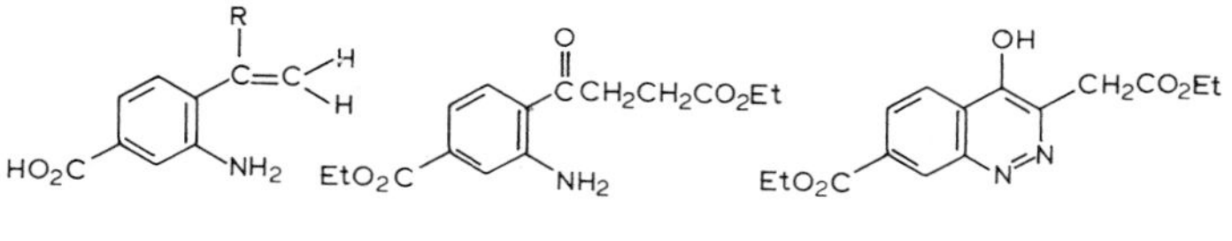

(CI)   (CII)   (CIII)

The reaction between styrene and diethyl azodicarboxylate in the presence of a radical inhibitor gives the cinnoline-1,2-dicarboxylate derivative (CIV) (*G. Ahlgren et al.*, Acta Chem. Scand., 1975, **29B**, 524). Treatment of cinnoline with benzoyl chloride and trimethylsilyl cyanide in anhydrous methylene chloride gives 1,2-dibenzoylcinnoline-4-carbonitrile (CV) in reasonable yield. Hydrolysis of (CV) gives rise to 4,4′-bicinnolyl (*D. Bhattacharjee* and *F. D. Popp*, J. heterocyclic Chem., 1980, **17**, 1211).

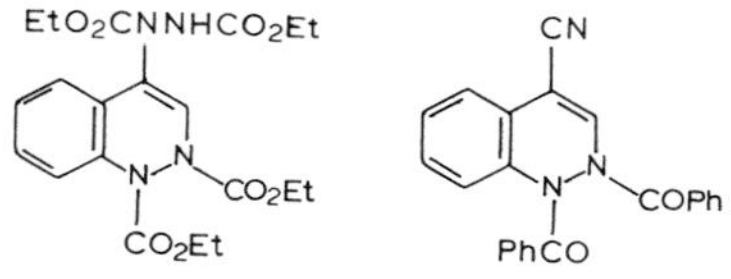

(CIV)   (CV)

(j) Reduced cinnolines

The reduction of cinnoline and its alkyl and aryl derivatives, halocinnolines, nitrocinnolines, etc. has already been discussed in some detail in their respective sections. In general terms, the chemical, catalytic and electrolytic reduction of cinnolines produces various products derived from reduction of the hetero-ring, but leaves the benzenoid ring intact. Cinnoline derivatives which contain the fully or partially reduced benzenoid ring are generally obtained from numerous and diverse cyclisation reactions.

(i) Reduction of cinnolines

Contrary to earlier reports, reduction of cinnoline and its derivatives produces, in the first instance, 1,4-dihydro compounds and not the 1,2-dihydro derivatives as first proposed. However, it has been shown that under certain conditions the 1,4-dihydro compound may rearrange to its 1,2-dihydro form. For example, 1,4-dihydro-4-phenylcinnoline gives the 1,2-*phthalyl* derivative (CVI), mp. 178–178.5°, in the presence of phthalic anhydride and pyridine, and the 1,2-*butylmalonyl* derivative (CVII), m.p. 160–161°, with diethyl butylmalonate and sodium methoxide (Table 20, ref. 1; *D.E. Ames, R.F. Chapman* and *H.Z. Kucharska*, J. chem. Soc., 1964, 5659). Alkyl- and aryl-substituted 1,2-malonyl-1,2-dihydrocinnolines show anti-inflammatory, analgesic and anti-pyretic properties, similar to the structurally related drug phenylbutazone (CVIII) (*F. Schatz* and *T. Wagner-Jauregg*, Helv., 1968, **51**, 1919).

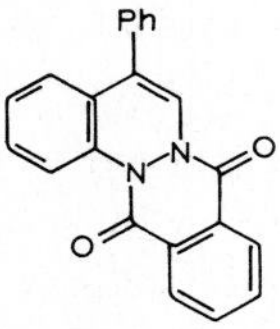

(CVI) 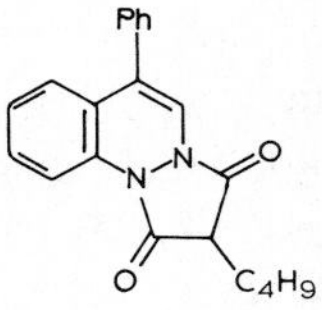   (CVII) 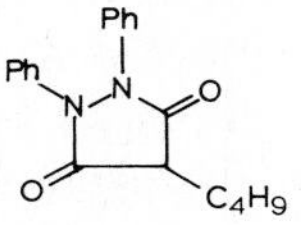   (CVIII)

Reduction of cinnoline and various 4-alkyl and 4-aryl derivatives with lithium aluminium hydride yields the corresponding 1,4-dihydro derivatives, which, in hot dilute acid, are in equilibrium with 1-aminoindoles. The 1,4-dihydro-4-methyl derivative undergoes further reduction with hydrogen and platinum in acetic acid to give 1,2,3,4-tetrahydrocinnoline which also forms 1,2-phthaloyl derivatives (Table 20, ref. 1, 3; *D.I. Haddlesey, P.A. Mayor* and *S.S. Szinai*, J. chem. Soc., 1964, 5269). Lithium aluminium

TABLE 20

HETERO-REDUCED CINNOLINES

Compound	m.p. (°C)	Compound	m.p. (°C), b.p./mm
1,4-Dihydrocinnoline	81–82.5 [1]	1,2,3,4-Tetrahydrocinnoline	82–83/0.2 [7]
	87–88 [2]	Hydrochloride	171–173 [7]
4-Methyl-	63–65 [1]	Picrate	122–123 [7]
4-Phenyl-	112.5–113 [1]	4-Methyl-	109–110 (dec.) [3]
	115–116 [3]	2-Methyl-	70–71/0.2 [8]
3-Phenyl-	152–153 [4]	Hydrochloride	160–163 (dec.) [8]
1-Methyl-3,4-diphenyl-	120 [5]	Picrate	118–119 (dec.) [8]
1,3,4-Triphenyl-	162 [5]	3-Oxo-	160–164 [9]
1,4,4-Triphenyl-	143 [6]	4-Oxo-, hydrochloride	216–220 [9]

1 L.S. Besford, G. Allen and J.M. Bruce, J. chem. Soc., 1963, 2867.
2 M. Busch and A. Rast, Ber., 1897, 30, 521.
3 R.N. Castle and M. Onda, J. org. Chem. 1961, 26, 4465.
4 H.S. Lowrie, J. med. Chem., 1966, 9, 664.
5 G. Cauquis, B. Chaband and M. Genies, Bull. Soc. Chem. Fr., 1973, 3487.
6 Idem, ibid., 1975, 583.
7 D.E. Ames and H.Z. Kucharska, J. chem. Soc., 1962, 1509.
8 Idem, ibid., 1963, 4924.
9 H. Lund, Acta Chem. Scand., 1967, 21, 2525.

hydride reduction of 3-hydroxy- and 4-hydroxy-cinnoline, and their 2-methyl and 1-methyl derivatives, respectively, gives 1,2,3,4-tetrahydrocinnoline, and its 2-methyl and 1-methyl derivative respectively (Table 20, ref. 7, 8). Under similar conditions 4-methoxycinnoline yields a mixture of cinnoline and 1,2,3,4-tetrahydrocinnoline. Reduction of cinnoline and its 4-carboxylic acid by zinc amalgam in acetic acid gives, in the first instance, 1,4-dihydrocinnoline which is further reduced to indole (L.S. Besford and J.M. Bruce, J. chem. Soc., 1964, 4037). Similarly, 4-alkyl and 4-aryl derivatives give 3-substituted-indoles as final products. Reduction with sodium in ethanol gives a mixture of the 1,4-dihydro compound and the corresponding indole (Haddlesey, Mayor and Szinai, loc. cit.). Cinnoline and 4-methylcinnoline give complex mixtures containing hydrocinnolines and hydroindoles on catalytic hydrogenation. As observed for the lithium aluminium hydride reduction of 4-phenylcinnoline, the catalytic hydrogenation stops at the 1,4-dihydro stage (Schatz and Wagner-Jauregg, loc. cit.). Reduction of 3-phenylcinnoline, either electrolytically or with zinc dust and barium hydroxide, also gives the 1,4-dihydro compound (Table 20, ref. 4). The latter forms a 1,2-phthalyl derivative (CIX) which can be catalytically reduced to give 1,2-phthalyl-1,2,3,4-tetrahydro-3-phenylcinnoline (CX).

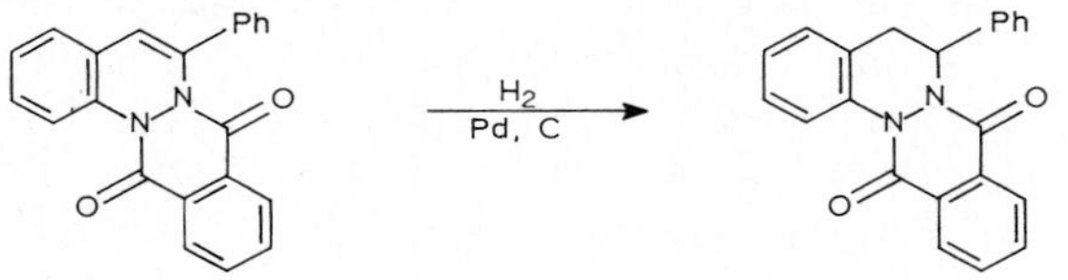

(CIX)   →[H₂, Pd, C]   (CX)

Reduction of 4-chlorocinnoline, either by lithium aluminium hydride, electrolytically or catalytically over palladium–charcoal on calcium carbonate, gives 4,4′-bicinnolinyl (*J. Morley*, J. chem. Soc., 1951, 1971). However, catalytic reduction of 3-chlorocinnoline over palladium on charcoal yields 1,4-dihydrocinnoline (*D. E. Ames, R. F. Chapman* and *H. Z. Kucharska*, ibid., 1964, 5659).

Electrolytic reduction of 4-methylcinnoline produces 4-methyl-1,4-dihydrocinnoline which can be further reduced to skatole. 4-Hydroxy- and 4-methoxy-cinnoline give 4-oxo-1,2,3,4-tetrahydrocinnoline (CXI) in the first instance, which may undergo further reduction to yield 4-hydroxy-1,2,3,4-tetrahydrocinnoline (CXII). 4-Hydroxycinnoline gives 3-oxo-1,2,3,4-tetrahydrocinnoline and 1-aminooxindole (CXIII) depending on the acidity of the reaction (Table 20, ref. 9). The mechanism of the electrochemical reduction of cinnoline in DMF–water at a mercury electrode has been reported (*M. Maruyana* and *K. Murakami*, Nippon Kagaku Kaishi, 1977, 990).

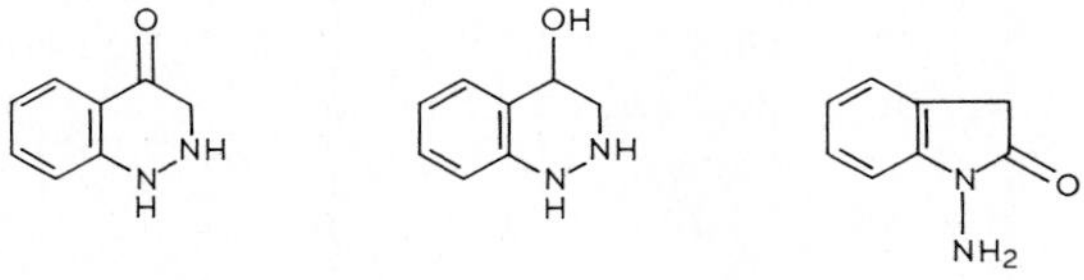

(CXI)   (CXII)   (CXIII)

Various cinnoline derivatives, reduced in the hetero-ring, have been prepared by condensation of alkenes with diazenium cations obtained by electrochemical oxidation of phenyl derivatives of hydrazine. Thus, electrochemical oxidation of Ph_2NNH_2 gives (CXIV), which on treatment with styrene gives 1,4-diphenyl-1,2,3,4-tetrahydrocinnoline (CXV) in excellent yield (*G. Cauquis* and *M. Genies*, Tetrahedron Letters, 1971, 3959). Oxidation of (XCV) gives 1,4-dihydro-1,4-diphenylcinnoline (CXVI, R = H) and reaction of (CXIV) with *trans*-stilbene gives directly 1,4-dihydro-1,3,4-triphenylcinnoline (CXVI, R = Ph). *cis*-Stilbene, however, does not react. (Table 20, ref. 5). The electrochemical oxidation of diverse 1,2,3,4-tetrahydrocinnolines in acetonitrile to give 1,4-dihydrocinnolines and cinnolinium cations has also been reported (Table 20, ref. 6).

The stereochemistry of the condensation of aryldiazenium salts and alkenes to give 1-alkyl/aryl-4-phenyl-1,2,3,4-tetrahydrocinnolines has been studied using *cis-* and *trans*-deuterostyrene. Vinyl esters react with the triphenyldiazenium cation to give 4-alkoxy-1,2-diphenyl-1,2,3,4-tetrahydrocinnolines (*K.N. Zelenin, V.N. Verbov* and *Z.M. Matveeva*. Khim. Geterot. Soedin., 1977, 1658; *K.N. Zelenin* and *V.N. Verbov, ibid.*, 1979, 1547). Similarly, the cycloaddition of diethyl azodicarboxylate to styrenes gives tetraethyl 4-(1,2-dicarboxyhydrazino)-1,2,3,4-tetrahydrocinnoline-1,2-dicarboxylate (CXVII) (*L. Horner* and *W. Naumann*, Ann., 1954, **587**, 81; *K. Alder* and *H. Niklas, ibid.*, 1954, **585**, 97) and addition to 1-vinylcyclohexene gives diethyl 1,2,3,5,6,7,8,8a-octahydrocinnoline-1,2-dicarboxylate (CXVIII) which can be reduced and decarboxylated to give decahydrocinnoline (CXIV) (*J.M. Bruce* and *P. Knowles*, J. chem. Soc., 1964, 4046).

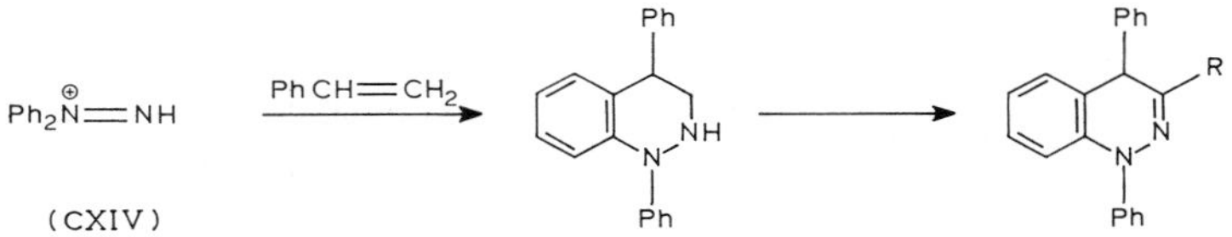

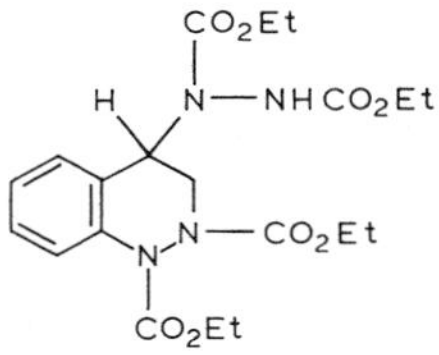

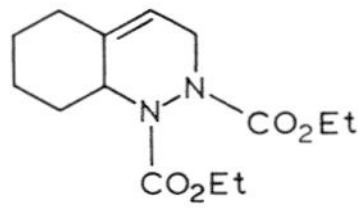

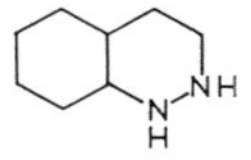

Diazotised arylamines add to 1,1-bis(*p*-alkoxyphenyl)ethylenes to yield 2-arylazo derivatives which cyclise on heating in acetic acid to give 1-aryl-7-alkoxy-4-(*p*-alkoxyphenyl)-1,2-dihydrocinnolines (CXX) (*A.B. Sakla et al.*, Indian J. Chem., 1976, **14B**, 742). The electrochemical oxidation of 2-(*o*-hydroxyaminophenyl)ethylamine (CXXI, Y = NHOH) leads, by way of the corresponding nitroso derivatives (CXXI, Y = NO), to 1,4-dihydrocinnoline (*R. Hazard* and *A. Tallec*, Bull. Soc chim. Fr., 1976, 433).

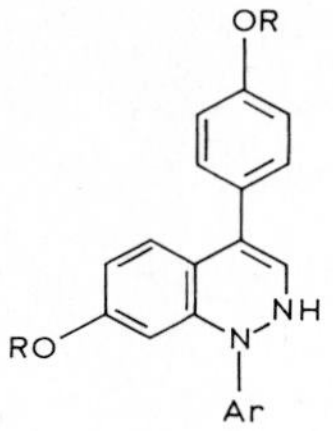

(CXX) 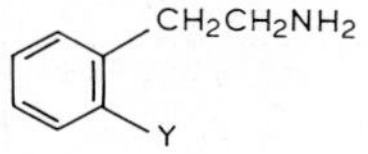   (CXXI)

Ring enlargement of 1-aminoindoles on treatment with acid gives 1,4-dihydrocinolines. Thus 1-aminoindole, 2-methylaminoindole and 3-methylaminoindole give 1,4-dihydrocinnoline, and its 3-methyl and 4-methyl derivatives respectively. The corresponding cinnoline is also isolated from the reaction mixture (*M. Somei* and *K. Ura*, Chem. Letters, 1978, 707).

The mass spectral fragmentation of various tetrahydrocinnolines has been reported and shown to occur by a retro-addition process to give a diazenium cation and an alkene (*G. Cauquis, B. Chaband* and *J. Ulrich*, Org. Mass Spectrom., 1977, **12**, 717).

(ii) From cyclisation reactions

Cyclisation reactions are particularly useful in preparing cinnolines with a reduced benzenoid ring. Perhaps the most used method involves the condensation of hydrazine with a 2-oxocyclohexylmethyl ketone. Thus, 2-oxocyclohexylacetone and α-(2-oxocyclohexyl)acetophenone react with hydrazine to give the hexahydrocinnoline (CXXII) which is readily oxidised to give 3-methyl- and 3-phenyl-5,6,7,8-tetrahydrocinnoline (CXXIII; R = CH$_3$, R = C$_6$H$_5$) respectively in good yield (Table 21, ref. 1).

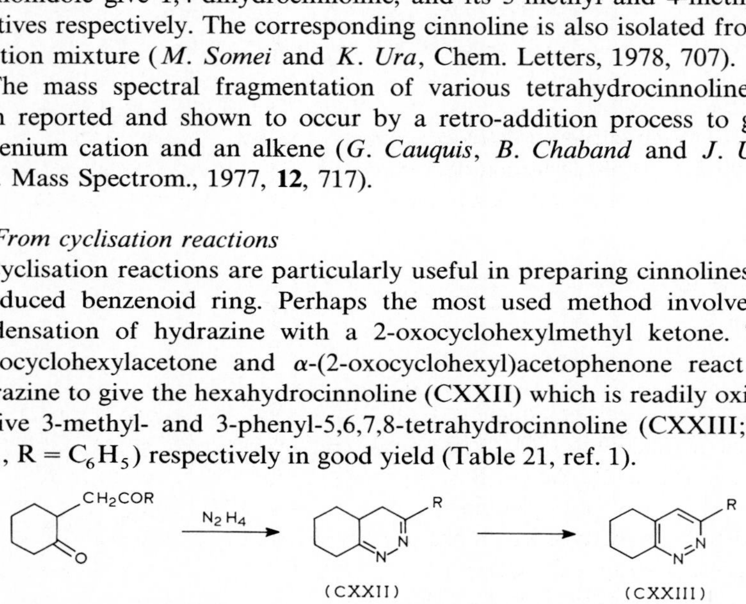

(CXXII)   (CXXIII)

The presence of a hydroxyl group either at position 1 (*e.g.*, CXXIV) of the cyclohexane ring, or α to the carbonyl function (*e.g.*, CXXV) results in the direct formation of a 5,6,7,8-tetrahydrocinnoline. The 4-oxoaldehyde (CXXV; R,R^1 = H) leads to unsubstituted 5,6,7,8-tetrahydrocinnoline (*J. Levisalles* and *P. Baranger*, Compt. rend., 1956, **242**, 1336; *J. Levisalles*, Bull. Soc. chim. Fr., 1957, 1009).

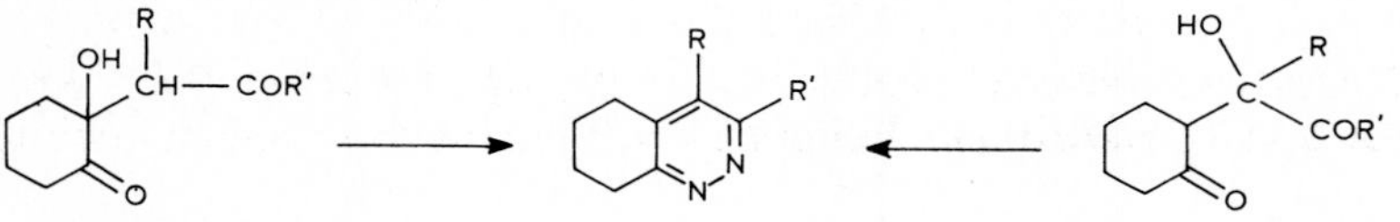

(CXXIV)   (CXXV)

TABLE 21

5,6,7,8-TETRAHYDROCINNOLINES

Substituent	*m.p.* (°*C*), *b.p./mm*	*Substituent*	*m.p.* (°*C*)
3-Methyl-	53.5–54.5 [1]	3-Oxo-	192–194 [1,3]
3,5-Dimethyl-	92/0.005 [2]	hydrobromide	193–197 [1,3]
3-Phenyl-	86–87.5 [1]	5-Oxo-7,7-dimethyl-3-phenyl	121–123 [4]
3-Phenyl-5-methyl-	64 [2]		
3-Chloro-	123–127/0.5 [1,3]		
3-Methoxy-	94–98/0.4 [1]		

1 *H.E. Baumgarten, P.L. Creger* and *C.E. Villars*, J. Amer. chem. Soc., 1958, **80**, 6609.
2 *H. Stetter* and *A. Mertens*, Ber., 1981, **114**, 2479.
3 *R.H. Horning* and *E.D. Amstutz*, J. org. Chem., 1955, **20**, 707.
4 *K. Nagarajan* and *R.K. Shah*, Chem. Comm., 1973, 926.

The 1,4,8-triketone (CXXVI) condenses with hydrazine to give 3,5-dimethyl-5,6,7,8-tetrahydrocinnoline, (CXXVIII), probably *via* the intramolecular aldol condensation product (CXXVII) (Table 21, ref. 2).

(CXXVI) (CXXVII) (CXXVIII)

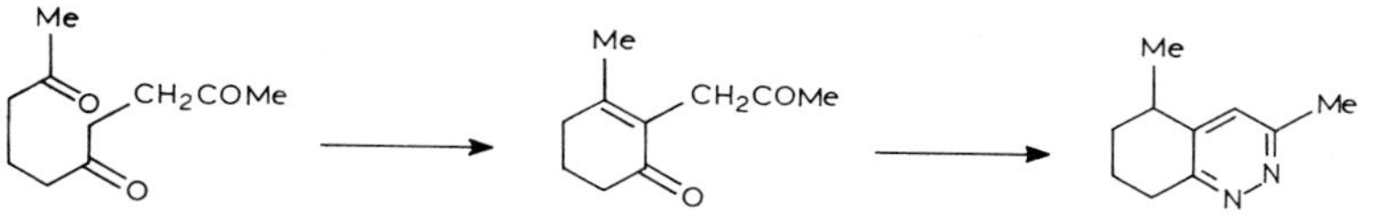

The 2-oxocyclohexylacetic acid derivative (CXXIX) yields the 4,4a,5,6,7,8-hexahydro-3(3*H*)cinnolinone (CXXX; R = Me, Ph) (*E. Buchta, G. Wolfrum* and *H. Ziener*, Ber., 1958, **91**, 1552). The reaction of (CXXIX; R = 3-methoxyphenyl) with hydrazine gives the reduced cinnoline (CXXX; R = 3-methoxyphenyl), which has been transformed in a number of steps to the azamorphinan (CXXXII) (*T. Kametani et al.*, Chem. pharm. Bull. Japan, 1968, **16**, 296; 1969, **17**, 1096).

(CXXIX) (CXXX) (CXXXI) (CXXXII)

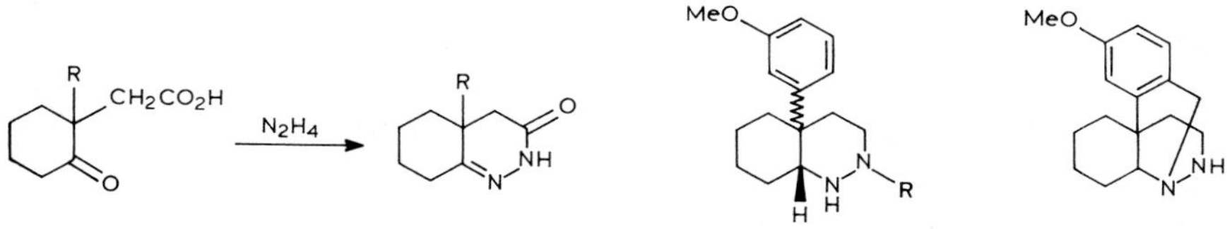

2-Alkyl derivatives of (CXXX; R = 3-benzyloxyphenoxy) have been prepared and converted *via* lithium aluminium hydride reduction into the decahydrocinnolines (CXXXI; R = Me, PhCH$_2$CH$_2$, cyclopropylmethyl), the 2-methyl derivative has analgesic activity 27 times that of morphine (*T. Kametani et al.*, Heterocycles, 1980, **14**, 449; Yakugaku Zasshi, 1980, **100**, 641). Condensation with esters of (CXXIX) also gives the hexahydrocinnolines (CXXX) (*Baumgarten, Creger* and *Villars, loc. cit.*; *W. Reid* and *A. Draisbach*, Ber., 1959, **92**, 949; Table 21, ref. 3). The ethyl ester of CXXIX, R = H) may be cyclised with a (halophenyl)hydrazine to give a 2-arylhexahydrocinnoline which after bromination-dehydrobromination gives a cinnolinone such as the herbicide 2-(4-chloro-2-fluorophenyl)-5,6,7,8-tetrahydro-3-cinnolinone (CXXXIII) (U.S.P., C.A., 1977, **87**, 135377). The 3-oxo group can be chlorinated to give a 3-chloro-5,6,7,8-tetrahydrocinnoline (CXXXIV) which can be converted into other 3-substituted derivatives or reduced to give the parent 5,6,7,8-tetrahydrocinnolinone (*Baumgarten, Creger* and *Villars, loc. cit., Horning* and *Amstutz, loc. cit., M. Ogata, H. Kano* and *K. Tori*, Chem. pharm. Bull. Japan, 1963, **11**, 1527).

(CXXXIII)   (CXXXIV)

Lactonisation of (CXXIX, R = OH) followed by hydrolysis of the acetoxy-lactone (CXXXV) and condensation with a hydrazine has been used to prepare 5,6,7,8-tetrahydro-3(2*H*) cinnolinone and its 2-aryl derivatives similar to (CXXXIII) (*W. Yu* and *H. Jing-Jain*, Acta chim. Sin., 1962, **28**, 351; *A. Mondon, H. Menz* and *J. Zander*, Ber., 1963, **96**, 826).

(iii) Reactions and biological properties

Bromination of 5,6,7,8-tetrahydrocinnolin-3(2*H*)-one with *N*-bromosuccinimide gives cinnolin-3(2*H*)-one, whereas its 2-methyl derivative gives the 5,8-dibromo-5,6,7,8-tetrahydrocinnolin-3(2*H*)-one (CXXXVI) which shows no tendency to aromatise. Raney nickel reduces 3-ethoxycarbonylcinnoline-4(1*H*)-thione to 3-ethoxycarbonyl-1,4-dihydrocinnoline (CXXXVII) which is aromatised to 4-bromo-3-ethoxycarbonylcinnoline on bromination (*J. Daunis, M. Guerret-Rigail* and *R. Jacquier*. Bull. Soc. chim. Fr., 1972, 3198). A study of the reduction of 5,6,7,8-tetrahydrocinnolin-3(2*H*)-one, including its 2-methyl and 4-ethoxycarbonyl derivatives with Raney nickel, has shown that a variety of hexahydro- and octahydro-cinnolines are formed depending on the conditions used (*idem, ibid.*, 1972, 1994).

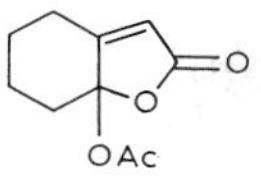

(CXXXV) 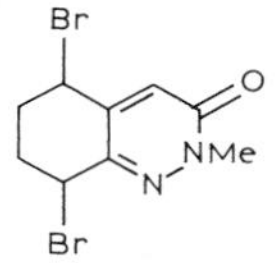   (CXXXVI) 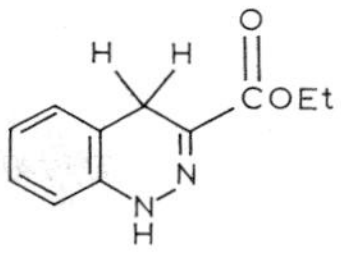   (CXXXVII)

Certain 4-cyano-5,6,7,8-tetrahydro-3(2H)cinnolinones are claimed to have analgesic properties and have been prepared by condensation of cyanoacetates and hydrazines with cyclohexane-1,2-dione (*P. Schmidt* and *J. Druey*, Helv. 1957, **40**, 1749; *E. Jucker* and *R. Suess*, ibid., 1959, **42**, 2506).

5-Oxo-1,4,5,6,7,8-hexahydrocinnolines (CXXXVIII) are prepared by condensation of 2-acetonyl- or 2-phenacyl-cyclohexane-1,3-diones with hydrazines. The 1-aminoalkyl compounds have been shown to be potent but toxic CNS depressants (*K. Nagarajan et al.*, J. med. Chem., 1976, **19**, 508). They aromatise on heating in pyridine in the presence of air to give e.g. (CXXXIX), while with acid (CXL) is formed (*K. Nagarajan* and *R.K. Shah*, Chem. Comm., 1973, 926).

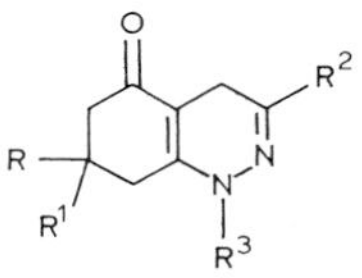

(CXXXVIII) 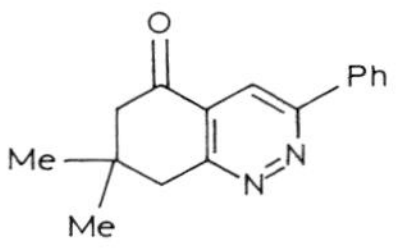   (CXXXIX) 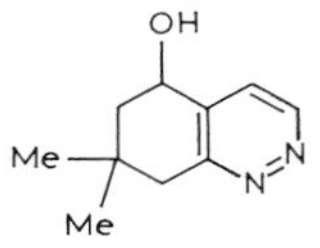   (CXL)

3. Benzocinnolines*

(a) Introduction

Of the possible benzocinnolines only benzo[c]cinnoline (I, previously called 3,4-benzocinnoline) has received much attention. The numbering system currently in use is shown on structure (I). The alternative numbering, analogous to that used for phenanthrene, and the name phenazone have been used in Beilstein's Handbuch and other publications but are now obsolete.

---

* Benzo[c]cinnolines have been the subject of a review of work published up to 1977; *J.W. Barton*, Adv. heterocyclic Chem., 1979, **24**, 151.

(b) Benzo[c]cinnolines

(i) Synthesis

The parent compound was originally synthesised by sodium amalgam reduction of 2,2'-dinitrobiphenyl (II) and this has been the most widely used preparative method (*E. Taüber*, Ber., 1891, **24**, 3081). Other workers have subsequently used Taüber's conditions, in addition to reduction with sodium sulphide (*J. Radell, L. Spialter* and *J. Hollander*, J. org. chem., 1956, **21**, 1051; *J.C. Arcos, M. Arcos* and *J.A. Miller*, ibid., 1956, **21**, 651; *R.S.W. Braithwaite, P.F. Holt* and *A.N. Hughes*, J. chem. Soc., 1958, 4073) and hydrosulphide (*J.F. Corbett* and *P.F. Holt*, ibid., 1961, 5029; *J.W. Barton* and *J.F. Thomas*, ibid., 1964, 1265). Catalytic reduction, using either hydrogen (*A. Etienne* and *G. Izoret*, Bull. Soc. chim. Fr., 1964, 2897; *F.E. Kempter* and *R.N. Castle*, J. heterocyclic Chem., 1964, **6**, 523) or hydrazine (*J.W. Barton* and *M.A. Cockett*, J. chem. Soc., 1962, 2454), has been used. Particularly effective methods include electrolytic reduction (*G. Wittig* and *O. Stichnoth*, Ber., 1935, **68**, 928) and lithium aluminium hydride (*G.M. Badger, J.H. Seidler* and *B. Thomson*, J. chem. Soc., 1951, 3207; *Corbett* and *Holt*, loc. cit.; ibid., 1960, 3646) and lithium bis(2-methoxyethoxy)aluminium hydride (*J.F. Corbett*, Chem. Comm., 1968, 1257) in non-protic solvents. Simultaneous reduction of other groups may limit use, e.g., 2,2'-dimethyl-4,4',6,6'-tetranitrobiphenyl does not give a benzocinnoline.

Cyclisation reactions of 2,2'-disubstituted biphenyls have been reviewed (*R.E. Buntrock* and *E.C. Taylor*, Chem. Rev., 1968, **68**, 209).

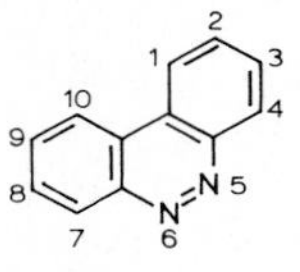

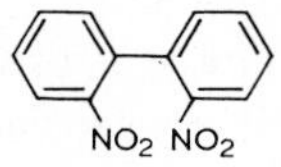

Benzo[c]cinnoline-5-oxide and -5,6-dioxide have been isolated as intermediates in the above reduction when suitable reaction conditions are chosen. The end product is 5,6-dihydrobenzo[c]cinnoline, a hydrazo compound, which is easily oxidised back to the fully aromatic (I) and is not, in general, isolated.

2-Amino-2'-nitrobiphenyls, which may be intermediates in some of the reductions, cyclise to benzo[c]cinnolines on treatment with base (*C.W. Muth et al.*, J. org. Chem., 1960, **25**, 736). When reducible substituents are present, the less efficient oxidation of 2,2'-diaminobiphenyl with, for example, manganese dioxide (*I. Bhatnagar* and *M.V. George, ibid.*, 1968, **33**, 2407) may be used. Bis-diazotisation of 2,2'-diaminobiphenyls in the presence of sodium arsenite or hypophosphorous acid gives benzo[*c*]cinnolines in poor yield (*R.B. Sandin* and *T.L. Cairns*, J. Amer. chem. Soc., 1936, **58**, 2019; *Barton* and *Crockett, loc. cit.*). Subsequent to his original report, Taüber also described the synthesis of benzo[*c*]cinnoline by heating 2,2'-dihydrazinobiphenyl with 20% hydrochloric acid at 150° (Ber., 1896, **29**, 2270). 2,2'-Diazidobiphenyl is quantitatively converted into benzo[*c*]cinnoline by low temperature photolysis (*A. Yabe* and *K. Honda*, Bull. chem. Soc. Japan, 1976, **49**, 2495, *A. Yabe, ibid.*, 1980, **53**, 2933).

Diazotisation of 2-amino-3'-methoxybiphenyl, and the more reactive 2-amino-3',4'-dimethoxy compound yields 2-methoxybenzo[*c*]cinnoline and 2,3-dimethoxybenzo[*c*]cinnoline, respectively, on cyclisation (*J.S. Swenton, J.J. Ikeler* and *B.H. Williams*, J. Amer. chem. Soc., 1970, **92**, 3103; *J.M. Blatchly, J.F.W. McOmie* and *M.L. Watts*, J. chem. Soc., 1962, 5085).

PhN≡NPh

(III)

Ph C⟨N—N—Ph / N=N—Ph⟩ X⊖

(IV)

Dehydrogenation of azo compounds, such as (III) in a melt of aluminium chloride and sodium chloride at about 100° results in the formation of benzo[*c*]cinnolines (Pat., C.A., 1931, **25**, 1266). Although this reaction does not appear to be general, better yields have been obtained using a solution of aluminium trichloride in refluxing methylene chloride (*P.F. Holt* and *C.W. Went*, J. chem. Soc., 1963, 4099). Two reports appeared almost simultaneously concerning the photochemical cyclisation of azobenzene to benzo[*c*]cinnoline (*G.E. Lewis*, Tetrahedron Letters, 1960, 12; *P. Hugelshofer, J. Kalroda* and *K. Shaffner*, Helv., 1960, **43**, 1329); the mechanism of the reaction has been investigated (*G.M. Badger, R.J. Drewer* and *G.E. Lewis*. Austral. J. Chem., 1966, **19**, 643). Various substituted azobenzenes have been reported to cyclise although the presence of a *p*-acetyl or *p*-nitro group slows the reaction, and the presence of a *p*-amino group prevents cyclisation (*idem, ibid.*, 1965, **18**, 190; *G.M. Badger, C.P. Joshua* and *G.E. Lewis, ibid.*, 1965, **18**, 1639; *G.E. Lewis, R.H. Prager* and *R.H.M. Ross, ibid.*, 1975, **28**, 2459; *C.P. Joshua* and *V.N.R. Pillai*, Indian J. Chem., 1976, **14B**, 525, and references cited therein).

Benzo[c]cinnoline derivatives with alkoxyl, carboxyl, and halogen substituents have been obtained from the photochemical bridging and ring cleavage of 2,3,5-triphenyltetrazolium salts (IV) (*F. Weygand* and *I. Frank*, Z. Naturforsch., 1948, **3B**, 377; *D. Jerchel* and *H. Frocher*, Ann., 1954, **590**, 216). Ring contraction of diazepines (*R.J. Dubois* and *F.D. Popp*, J. heterocyclic Chem., 1969, **6**, 113), triazepines (*S.F. Gait et al.*, J. chem. Soc. Perkin I, 1975, 19) and thiadiazepines (*G.R. Collins*, Ph.D. Indiana University (1965) , Diss. Abstr., 1966, **27(B)**, 403) have also been reported.

(ii) Physical and chemical properties

Benzo[c]cinnoline, m.p. 156°, is weakly basic, pK_a 2.2 in water at 20° (compare cinnoline, pK_a 2.7) (*P.H. Gore* and *J.N. Phillips*, Nature, 1949, **163**, 690). It is very soluble in alcohols, ether, etc., and is recovered unchanged and not as the hydrochloride after evaporation with hydrochloric acid; *dipicrate*, m.p. 192°. The *methiodide*, m.p. 185-187°, is decomposed by ammonia and re-forms benzocinnoline (*T. Wohlfahrt*, J. pr. Chem., 1902, [ii], **65**, 295). X-ray crystallography has shown the molecule to be only approximately planar, forming dimers similar to those observed in solution at low temperature (*D.N. de Vries Reilingh et al.*, J. chem. Phys., 1971, **54**, 2722). The bond lengths indicate that some bond fixation occurs with (I) the preferred Kekulé structure. The u.v. spectrum shows three $\pi-\pi^*$ transitions at 250, 300, and 350 nm, similar to phenanthrene, and a weak absorption at 400 nm due to an $n-\pi^*$ transition. *N*-Oxidation, protonation and substituent effects have also been reported (*G.M. Badger* and *I.S. Walker*, J. chem. Soc., 1956, 122; *P.F. Holt et al.*, ibid., 1961, 5029; 1962, 1812; 1966, 1306). The dipole movements of benzo[c]cinnoline and its *N*-oxide have been reported (*K.E. Calderbank* and *R.J.W. Lefevre*, ibid., 1948, 1949).

Despite differing reports concerning the assignment of the signals in the proton magnetic resonance spectrum of benzo[c]cinnoline the accepted interpretation of the three multiplets is δ (ppm) 8.64 ($H_{4,7}$), 8.43 ($H_{1,10}$) and 7.82 ($H_{2,3,8,9}$) (*R.H. Martin et al.*, Tetrahedron, 1966, Suppl. **8**, 181; *J.W. Barton*, Adv. heterocyclic Chem., 1979, **24**, 151). The ^{13}C-n.m.r. spectrum of benzo[c]cinnoline has been determined and the chemical shifts compared with the calculated molecular orbital charge densities (*A. Koennecke et al.*, Org. magn. Resonance, 1979, **12**, 696). Other workers have also reported the ^{13}C-n.m.r. spectra of benzocinnolines (*R.P. Bennet*, Inorg. Chem., 1970, **9**, 2184; *S.R. Challand et al.*, J. chem. Soc. Perkin I, 1975, 26). ^{13}C-^{15}N Nuclear spin coupling constants in neutral and acidic solutions have been reported (*Y. Kuroda* and *Y. Fujiwara*, J. phys. Chem., 1981, **85**, 2655).

The mass spectra of benzo[c]cinnoline and various *C*-monosubstituted derivatives have been examined. In most cases, a biphenyl radical ion is observed due to loss of N_2, however, certain alkoxy and carboxy substituents may be eliminated before N_2 loss, and benzo[c]cinnoline-5-oxide loses NO (*J.H. Bowie*, *G.E. Lewis* and *J.A. Reiss*, Austral. J. Chem., 1968, **21**, 1233; *J.H. Bowie et al.*, ibid., 1967, **20** 2545).

The magnetic circular dichroism of benzo[c]cinnoline has been reported and interpreted for the 4 lowest $\pi-\pi^*$ bands and the $n-\pi^*$ transition (*M. Vasak*, *M.R. Whipple* and *J. Michl*, J. Amer. chem. Soc., 1978, **100**, 6867). E.s.r. and luminescence spectra at low temperature in solution have been investigated (*R. Schaaf* and *H. Perkampus*, Tetrahedron, 1981, **37**, 341).

Benzo[c]cinnoline undergoes quaternisation when treated with 1,3-dibromopropane leading on elimination of hydrogen bromide to the cyclic imminium salt (V). The latter in boiling ethanol gives the dehydrogenated, fully aromatic benzo[c]pyridazo[1,2-a]cinnolinium salt (VI) (*D.G. Farnum, R.J. Alaimo* and *J.M. Dunston*, J. org. Chem., 1967, **32**, 1130). Treatment of related quaternary salts (VII, R = carboxyl derivative) with base produces ylides (VIII) (*E. Carp, M. Dorneaunu* and *I. Zugnarescu*, Rev. Roum. Chim., 1974, **19**, 1507; *S.F. Gait et al.*, J. chem. Soc. Perkin I, 1975, 556).

Benzo[c]cinnoline is reduced by zinc in acid or alkaline solution, and hydrogen over a palladium/alumina catalyst at one atmosphere to give the unstable 5,6-dihydrobenzo[c]cinnoline which may be isolated as its hydrochloride (*H. Kuhne* and *H. Erlenmeyer*, Helv., 1955, **38**, 531; *H. Duval*, Bull. Soc. chim. Fr., 1910, 485; *A. Etienne* and *R. Piat*, ibid., 1962, 292). More drastic catalytic reduction cleaves the heterocyclic ring with the formation of 2,2′-diaminobiphenyl, as does reduction with hydrazine in the presence of Raney nickel (*J. Radell, L. Spialter* and *J. Hollander*, J. org. Chem., 1956, **21**, 1051; *R.E. Moore* and *A. Furst*, ibid., 1958, **23**, 1504). Exposure of a solution of benzo[c]cinnoline to u.v. light gives firstly the 5,6-dihydro compound, which loses ammonia to give carbazole (*H. Inoue* and *Y. Matsuda*, Chem. Letters, 1972, **8**, 713). Alkali metals add to benzocinnoline in ether in the absence of oxygen to give 5,6-dimetallo derivatives which yield the 5,6-dihydro compound on treatment with a proton source. The dihydro compounds may be benzoylated to give, after disproportionation, benzo[c]cinnoline and 2,2′-dibenzamidobiphenyl or may be used to form 5,6-dimethyl derivatives with dimethyl sulphate, (*G. Wittig, M.A. Jesaitus* and *M. Glos*, Ann., 1952, **577**, 1; *G. Wittig* and *O. Stichnoth*, Ber., 1935, **68**, 928). Phenyl lithium also adds at the nitrogen atoms to give the air-sensitive 5,6-dihydro-5-phenylbenzo[c]cinnoline, which on oxidation with lead tetraacetate gives the quinonimine (IX) (*G. Wittig* and *A. Schumacher*, ibid., 1955, **88**, 234).

On oxidation of (I) by permanganate the heterocyclic ring survives and pyridazinetetracarboxylic acid (X) is formed (*E. Taüber*, Ber., 1895, **28**,

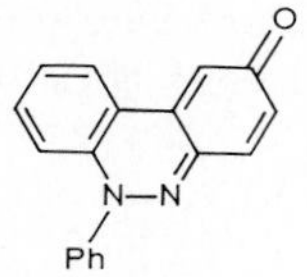

451). Hydrogen peroxide at 0° gives the 5-oxide (*R.S.W. Braithwaite* and *P.F. Holt*, J. chem. Soc., 1959, 3025), whereas heating at 110–120° gives the 5,6-dioxide (*I. Suzuki, M. Nakadate* and *T. Nakashima*, Tetrahedron Letters, 1966, 2899). However, the main method of preparation is the controlled reduction of 2,2′-dinitrobiphenyls (*E. Taüber*, Ber., 1891, **24**, 3083). Cyclisation of 2-amino-2′-nitrobiphenyls provides a route to certain benzo[*c*]cinnoline monoxides of known orientation (*J.F. Corbett* and *P.F. Holt*, J. chem. Soc., 1961, 5029; *J.W. Barton* and *J.F. Thomas*, ibid., 1964 1265).

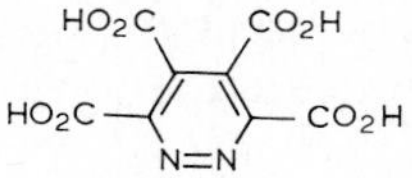

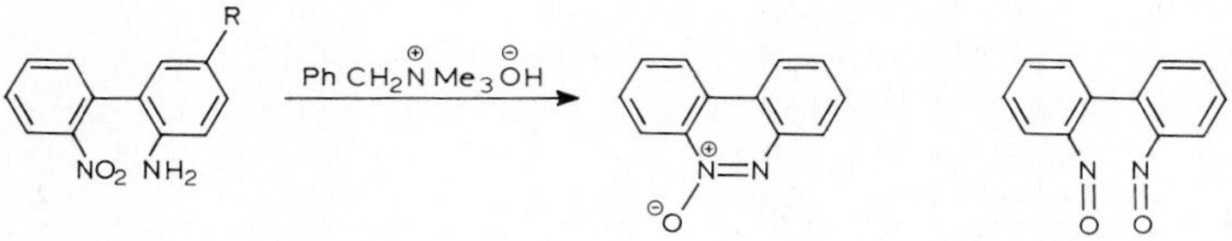

Benzo[c]cinnoline-5-oxide, bright yellow needles, has m.p. 139°. The 5,6-*dioxide*, m.p. 233–236° (dec.), 243°, is colourless, non-basic and sparingly soluble in organic solvents. It has been referred to as the dinitroso compound (XI), but the lack of nitroso-type reactions does not support this structure. Spectroscopic arguments have also been examined (*S.D. Ross, G.H. Kahan* and *W.A. Leach*, J. Amer. chem. Soc., 1952, **74**, 4122; *W. Lüttke*, Z. Electrochem., 1957, **61**, 976). Nitration with sulphuric and nitric acids at 70–80° yields 1-nitrobenzo[*c*]cinnoline-6-oxide (43%) and the 4-nitro isomer (26%) (*J.W. Barton* and *M.A. Cockett*, J. chem. Soc., 1962, 2454; *J.F. Corbett, P.F. Holt* and *M.L. Vickery*, ibid., 1962, 4384), whereas with fuming nitric acid alone 2-nitrobenzo[*c*]cinnoline-6-oxide is obtained in good yield.

Benzo[*c*]cinnoline is not readily attacked by electrophilic reagents but nitration occurs at position 1 (56%) and position 4 (15%) at temperatures between 0 and 100°. A large excess of acid at room temperature has been reported to give the 1,10-dinitro derivative (*Idem, ibid.*, 1962, 4860; *W.T. Smith* and *P.R. Ruby*, J. Amer. chem. Soc., 1954, **76**, 5807), and nitration of 1,10-dimethylbenzo[*c*]cinnoline gives the 4- and the 7-derivative (*W. Theilacker* and *F. Baxmann*, Ann., 1953, **581**, 117). Theoretical studies of the reactivity of both neutral and monoprotonated benzo[c]cinnoline predicts that position 1 is the most reactive but there is uncertainty as to whether the next most reactive position is 3 or 4 (*A. Pullman*, Rev. Sci, 1948, **86**, 219; *H.C. Longuet-Higgins* and *C.A. Coulson*, J. chem. Soc., 1949, 971, *M.J.S. Dewar* and *P.M. Maitlis*, ibid., 1957, 2521). Molecular bromine forms a complex with benzo[c]cinnoline (*Smith* and *Ruby*, loc. cit.), but bromine with silver sulphate in sulphuric acid gives a mixture of the 1- and the 4-bromo derivative in a ratio of 2.3 : 1, and a small amount of the 1,4-dibromo compound (*Barton*, loc. cit.). Chlorination of benzo[c]cinnoline with chlorine in the presence of aluminium trichloride gives the octachloro derivative (*S. Konaktanoporn* and *J.A.H.McBride*, J. chem. Res. Synop., 1980, 203). Sulphonation has been reported to give the 3-mono- and the 3,8-di-sulphonic acid (Pat., C.A., 1934, **28**, 654).

(iii) Methylated benzo[c]cinnolines

The photochemical cyclisation of 3-methylazobenzene gives **1-methylbenzo[c]cinnoline**, m.p. 117.5°, in 13% yield and *3-methylbenzo[c]cinnoline*, m.p. 125–125.5°, in 27% yield. The *2-methyl*, m.p. 137–138°, and the *4-methyl* derivative, m.p. 129°, are obtained from 4-methylazobenzene and 2-methylazobenzene, respectively. The 1,8-, (m.p. 118.5–119°), 1.10-, (m.p. 114.5°), 2,9-, (m.p. 190–191°), 3,8-, (m.p. 188°), 2,4-, (m.p. 121.5°) and 4,7-*dimethylbenzo[c]cinnolines* (m.p. 169–170°), and the 1,2,4-*trimethyl* derivative (m.p. 146.5–147.5°) have been similarly prepared (*G.M. Badger, R.J. Drewer* and *G.E. Lewis*, Austral. J. Chem., 1963, **16**, 1042; 1964, **17**, 1036). The following benzo[c]cinnolines have been prepared by the reductive cyclisation of the appropriate 2,2′-dinitrobiphenyl: 1,10-*dimethyl*- (*J. Kenner* and *W.V. Stubbings*, J. chem. Soc., 1921, 593); 2,9-*dimethyl*-, m.p. 187°, and its *dioxide*, m.p. 128° (dec.) (*L. Meyer*, Ber., 1893, **26**, 2239) and 3,8-*dimethyl*- and its *monoxide*, m.p. 209° (*F. Ullman* and *P. Dierterle*, ibid., 1904, **37**, 23).

As with the corresponding phenanthrene derivatives, benzo[c]cinnolines with bulky substitutents at positions 1 and 10 should be nonplanar and optically resolvable. Attempts to resolve 1,10-dimethylbenzo[c]cinnoline, its 5,6-dihydro derivative or the corresponding 1,5,6,10-tetramethyl-5,6-dihydrobenzo[c]cinnoline have been unsuccessful (*Wittig* and *Stichnoth*, loc. cit.). However, 4,7-diamino-1,10-dimethylbenzo[c]cinnoline has been resolved but the active forms $[\alpha]_D^{25} \pm 58°$, racemise readily in boiling methanol (*W. Theilacker* and *F. Baxmann*, Ann., 1953, **581**, 117).

2,3,8,9-**Tetramethyl**- and 2,3,4,7,8,9-**hexamethyl-benzo**[c]**cinnolines** have been made by cyclisation of biaryls (*I. Puskas, E.K. Fields* and *E.M. Banas*, Amer. chem. Soc., Div. Pet. Chem., Prepr., 1972, **17**, B6). Several aryl derivatives of benzo[c]cinnoline including 1-*phenyl*-, m.p. 165–166°, 2-*phenyl*-, m.p. 161–162°, and 3-*phenyl-benzo*[c]*cinnoline*, m.p. 153–154°, and *naphtho*[1,2-c]*cinnoline*, m.p. 195–196°, have been prepared by photocyclisation of azobenzenes (*G.M. Badger, R.J. Drewer* and *G.E. Lewis*, Austral. J. Chem., 1965, **18**, 190; *N.C. Jamieson* and *G.E. Lewis*, ibid., 1967, **20**, 321).

(iv) Halogeno-, nitro-, and other substituted benzo[c]cinnolines

Bromination of benzo[c]cinnoline by bromine in sulphuric acid in the presence of silver sulphate gives the 1-bromo, 4-bromo and 1,4-dibromo derivative in order of decreasing yield. Chlorination has also been shown to give the same isomers in a similar ratio (Table 22, ref. 1, 4, 7). The 1-, 2-, and 4-bromo compounds have also been obtained by the Sandmeyer reaction of their corresponding aminobenzo[c]cinnolines (Table 22, ref. 1, 3, 7). Mono- and di-bromobenzo[c]cinnolines have been made by the cyclisa-

TABLE 22

SUBSTITUTED BENZO[c]CINNOLINES

Substituent	m.p. (°C)	Substituent	m.p. (°C)
1-Bromo-	121–122.5 [1]	1-Nitro-	160–161 [8]
-6-oxide	202–204 [1]		162–163 [3]
-5-oxide	156–158 [1]	4-Nitro-	237–238 [3]
4-Bromo-	198–199 [1,2]		232 [9]
-6-oxide	235–236 [1,2]	1,4-Dinitro-	270 [10]
2-Bromo-	221–222 [3,4]	2-Nitro-	269 [3,11]
-6-oxide	249–250 [3,4]	3-Nitro-	259 [3,11]
3-Bromo-	194–195 [4,7]	1-Amino-	167 [3,9]
1,4-Dibromo-	196–198 [4,7]	2-Amino-	244–245 [3,9]
-6-oxide	241 [4]	3-Amino-	162–163 [3,12]
1-Chloro-	145–146 [5]	4-Amino-	198–200 [8,9]
2-Chloro-	215.5–216 [5]	2-Hydroxy-	316–318 [13]
3-Chloro-	189.5–190.5 [5]	3-Hydroxy-	275 [14]
4-Chloro-	191–192 [5]	2-Methoxy-	148 [15,16]
2,4-Dichloro-	236–237 [6]	4-Methoxy-	179–189 [15]

1 P.F. *Holt* and R. *Oakland*, J. chem. Soc. C, 1966, 1306.
2 J.W. *Barton* and J.F. *Thomas*, J. chem. Soc., 1964, 1265.
3 J.W. *Barton* and M.A. *Cockett, ibid.*, 1962, 2454.
4 J.F. *Corbett* and P.F. *Holt, ibid.*, 1961, 5029.
5 G.M. *Badger*, R.J. *Drewer* and G.E. *Lewis*, Austral. J. Chem., 1964, **17**, 1036.
6 G.E. *Lewis*, R.H. *Prager* and R.H.M. *Ross, ibid.*, 1975, **28**, 2459.
7 J.W. *Barton* and D.J. *Lapham*, J. chem. Soc. Perkin I, 1979, 1503.
8 W.T. *Smith* and P.R. *Ruby*, J. Amer. chem. Soc., 1954, **76**, 5807.
9 J.F. *Corbett*, P.F. *Holt* and M.L. *Vickery*, J. chem. Soc., 1962, 4860.
10 *Idem, ibid.*, 1962, 4384.
11 G.M. *Badger*, C.P. *Joshua* and G.E. *Lewis*, Austral. J. Chem., 1965, **18**, 1639.
12 J.C. *Arcos*, M. *Arcos* and J.A. *Miller*, J. org. Chem., 1956, **21**, 1051.
13 G.E. *Lewis et al.*, Austral. J. Chem., 1970, **23**, 619.
14 O. *Goll*, Pat., C.A., 1934, **28**, 654.
15 G.E. *Lewis* and J.A. *Reiss*, Austral. J. Chem., 1968, **21**, 1097.
16 J.S. *Swenton*, T.J. *Ikeler* and B.H. *Williams*, J. Amer. chem. Soc., 1970, **92**, 3103.

tion of halogenobiphenyls (Table 22, ref. 3, 4; *Corbett* and *Holt, ibid.*, 1961, 3695; *Corbett et al., ibid.*, 1962, 1812; *S.D. Ross et al.*, J. Amer. chem. Soc., 1952, **74**, 1297) and others, including all the mono-chloro and mono-iodo isomers, by the photocyclisation of halogenoazobenzenes (Table 22, ref. 5, 6). Reduction of halobenzo[c]cinnoline oxides has also been used to obtain the corresponding halobenzo[c]cinnolines. The halogenation of benzo[c]cinnoline has recently been reported, along with an n.m.r.-spectroscopic investigation of the various bromo- and chloro-benzo[c]cinnolines obtained (Table 22, ref. 7). Halobenzo[c]cinnolines undergo nucleophilic displacement reactions with alkoxides, dialkylamines and metallic amides. The reactions of the amides proceed *via* aryne intermediates rather than by direct nucleophilic displacement (*G.E. Lewis* and *J.A. Reiss*, Austral. J. Chem., 1967, **20**, 2217; *idem, ibid.*, 1975, **28**, 2057; Table 22, ref. 6).

The nitration of benzo[c]cinnoline gives the 1-nitro, 4-nitro (not 3-nitro as first reported), and 1,10-dinitro derivatives (Table 22, ref. 3, 8, 9, 10). Nonreductive cyclisation methods have been used to prepare 2- and 3-nitrobenzo[c]cinnolines (Table 22, ref. 3, 9, 11). Both chemical and catalytic reduction of nitro derivatives and their N-oxides have been used to prepare the corresponding amines (see Table 22). Reactions of halobenzo[c]cinnolines with potassium amide have given the 2-, the 3-, and the 4-amino compound. The mono-aminobenzo[c]cinnolines show important colour changes in acid solution (*Lewis* and *Reiss, loc. cit.*). Their solutions in alcohol or ether exhibit an intense green fluorescence, especially the 2-compound. As for other aromatic amines they undergo acylation and diazotisation.

A pressure reduction of 1-*benzenesulphonamidobenzocinnoline*, m.p. 211–213°, with Raney nickel as catalyst, cleaves the heterocyclic ring. The resulting benzenesulphonamido-*o*-diaminobiphenyl is deaminated by diazotisation in the presence of hypophosphorous acid with the formation of the known 2-benzene-sulphonamidobiphenyl. This proves that the nitro group in the main product from the nitration of benzocinnoline is at the 1-position. The second product gives an amine with m.p. 198–200°, identical to that of the 4-amino compound (*Smith* and *Ruby, loc. cit.*).

3,8-**Diamino**-, m.p. 265–267°, 3,8-**dimethylamino**-, m.p. 276°, (*hydrochloride*, m.p. 236°) and 3,8-**diethylamino-benzo[c]cinnoline** are prepared from the appropriate diphenyl compounds (*Ullmann* and *Dierterle, loc. cit.*).

4,7-**Diamino**-1,10-**dimethylbenzo[c]cinnoline**, m.p. 178–179°; *diacetyl* deriv., m.p. 221°, *dibenzoyl* deriv., m.p. 279°.

3-Hydroxy-, and 3,8-dihydroxy-benzo[c]cinnoline are obtained from the fusion of the corresponding sulphonic acids with alkali (*Goll, loc. cit.*). The former, and its 6-oxide, are obtained from the base-catalysed rearrangement of 3′-hydroxy-2-nitrobenzensulphonanilide (*E. Waldan* and *R. Pütter*, Angew. Chem., intern. Edn., 1972, **11**, 826; *S.F. Gait et al.*, J. chem. Soc. Perkin I, 1975, 19). Methoxybenzo[c]cinnolines are obtained by cyclisation of methoxylated biaryls and also by the reaction of halobenzo[c]cinnolines with sodium methoxide. 3,8-**Dimethoxybenzo[c]cinnoline**, m.p. 197°, has been synthesised by reductive cyclisation of 2,2′-dinitro-4,4′-dimethoxy-biphenyl followed by reduction of the 5,6-dioxide formed, and by oxidative cyclisation of the 2,2′-diamino compound (*F.E. Kempter* and *R.N. Castle*, J. heterocyclic Chem., 1969, **6**, 523; *K. Hata, K. Tatematsu* and *B. Kubota*, Bull. chem. Soc. Japan, 1935, **10**, 425). They can be demethylated to the hydroxy derivatives by heating with hydrobromic acid or with aluminium chloride (see Table 22). Various benzo[c]cinnolines including the 2-, and 3-methoxy derivatives have been reported to be active as herbicides (Pat., C.A., 1981, **95**, 182265).

Benzo[c]cinnoline-2-carboxylic acid. m.p. 363–364° (*methyl ester*, m.p. 185.5°) and -4-**carboxylic acid**, m.p. 283.5–285°, have been prepared by the photocyclisation of the corresponding azobenzene-carboxylic acids. Irradiation of azobenzene-3-carboxylic acid gives, in addition to the expected **benzo[c]cinnoline-3-carboxylic acid** (*methyl ester*, m.p. 177°) none of the alternative 1-carboxylic acid, but 1-**hydroxybenzo[c]cinnoline-10-carboxylic acid lactone** (XII), m.p. 329–330°, is obtained (Table 22, ref. 5; *C.P. Joshua* and *G.E. Lewis, ibid.*, 1967, **20**, 929). The synthesis and n.m.r. spectra of the methyl and ethyl esters of the 2- and the 4-carboxylic acid have been reported (Table 22, ref. 15).

Reductive ring cleavage of bridged tetrazolium salts (XIII) has also been used to prepare the 2-carboxylic acid and its methyl ester (*D. Jerchel* and *H. Fischer*, Ann., 1954, **590**, 216). 2,9-Dicyanobenzo[c]cinnoline has been prepared in a similar way (no details given) (*F.A. Neugebauer*, Tetrahedron, 1970, **26**, 4843). **Benzo[c]cinnoline-2,9-dicarboxylic acid**, m.p. 323°, -3,8-**dicarboxylic acid**, m.p. 335°, and -4,7-**dicarboxylic acid**, m.p. 312°, and their methyl and ethyl esters have been prepared by photochemical cyclisation of the appropriate azobenzene dicarboxylic acid (*C.P. Joshua* and *V.N.R. Pillai*, Indian J. Chem., 1974, **12**, 60). 2-**Methylbenzo[c]cinnoline-9-carboxylic acid**, m.p. > 290°, is prepared by oxidation of the 2,9-dimethyl compound by chromic acid at 100° (*Meyer, loc. cit.*). **Benzo[c]cinnoline-3,8-dicarboxylic acid chloride**, m.p. 208°, is obtained by reducing 2,2′-dinitrodiphenyl-4,4′-dicarboxylic acid with zinc dust and ammonia and treating the resulting acid with phosphorus pentachloride (Pat., C.A., 1932, **26**, 5124).

(XII)   (XIII)   (XIV)

Reductive electrochemical carboxylation has been used to prepare diethyl 1,4-dihydrobenzo[c]cinnoline-1,2-dicarboxylate (U. Hess et al., Z. chem., 1980, **20**, 64). 6,6'-Di-(hydroxymethyl)-2,2'-dinitrobiphenyl gives, on electrolytic reduction, **benzo[c]cinnoline-1,10-dimethanol monoxide**, m.p. 216°. However, an attempt to synthesise the 1,10-dicarbaldehyde in a similar way gave 4,9-diazapyrene (XIV) via its 4,9-dioxide.

(v) Reduced benzo[c]cinnolines

5,6-Dihydrobenzo[c]cinnoline is the unstable reduction product of benzo[c]cinnoline and it readily re-forms the parent. It is isolated as its hydrochloride and the corresponding 1,10-dimethyl homologue as the sulphate, m.p. 134° (*Wittig* and *Stichnoth, loc. cit.*). By reaction with diethyl malonate in the presence of sodium ethoxide 5,6-dihydrobenzo[c]cinnoline gives the compound (XV), m.p. 210–212° (*H. Kuhne* and *H. Erlenmeyer*, Helv., 1955, **38**, 531).

(XV)　　　(XVI)　　　(XVII)

Other methods of synthesis and reactions have been discussed under the properties of benzo[c]cinnoline and its alkyl derivatives. The photochemical reduction of benzo[c]cinnoline in strongly acidic propanol gives 5,6-dihydrocinnoline, which shows significant phosphorescence (*H. Inone, Y. Hiroshima* and *N. Makita*, Bull. chem. Soc. Japan, 1979, **52**, 351). Photolysis of the triphenyldiazenium cation gives the dihydrobenzocinnolinium cation (XVI) which on dehydrogenation gives (XVII) (*G. Cauquis* and *G. Reverdy*, Tetrahedron Letters, 1977, 3267). Electron spin resonance studies of the radical cations from various 5,6-dihydrobenzo[c]cinnolines have been reported (*F.A. Neugebauer* and *H. Weger*, J. phys. Chem., 1978, **82**, 1152).

1,2,3,4-**Tetrahydrobenzo[c]cinnoline** (XIX, R = R^1 = H), m.p. 98°, and its 2,4-*dimethyl* derivative, m.p. 111°, are prepared from the monophenyl hydrazone of cyclohexane-1,2-dione and its 4,6-dimethyl derivative, respectively, by the action of either sulphuric or phosphoric acids. Only poor yields of the cinnolines (XIX; R = R^1 = H and R = H, R^1 = OMe) are obtained from the corresponding hydrazones (XVIII) (*B.P. Moore*, Nature, 1949, **163**, 918; *R.A. Soutter* and *M. Tomlinson*, J. chem. Soc., 1961, 4256; *R.S.W. Braithwaite* and *G.K. Robinson*, ibid., 1962, 3671; *M.J.M. Pollmann*, Rec. Trav. chim., 1970, **89**, 929).

(XVIII)　　　(XIX)　　　(XX)

1,2,3,4-Tetrahydrobenzo[c]cinnoline has also been prepared by the oxidative rearrangement of the aminoindole (XX) (*M. Somei* and *Y. Kurizuka*, Chem. Letters, 1979, 127). Cycloadditions of 1,1'-bicyclohexene (XXI) with acylazo compounds give the tetrahydropyridazines (XXII) (*Y.S. Shabarov et al.*, J. gen. Chem. U.S.S.R. 1962, **32**, 2806; *B.T. Gillis* and *P.E. Beck*, J. org. Chem., 1963, **28**, 3177).

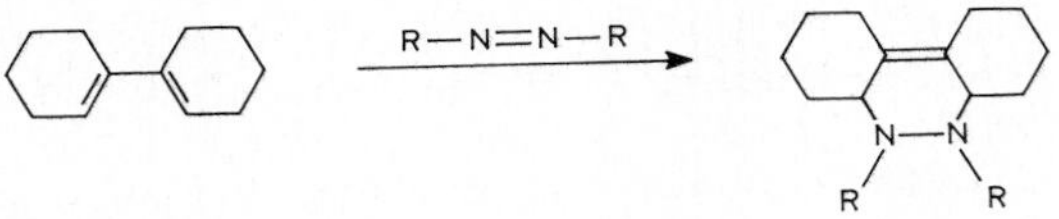

(XXI)   R—N=N—R   (XXII)

(c) Other benzocinnolines

Other benzocinnolines include the naphthopyridazines: benzo[f]cinnoline (XXIII), benzo[g]cinnoline (XXIV), and benzo[h]cinnoline (XXV), and also 1H-benzo[d,e]cinnoline (XXVI).

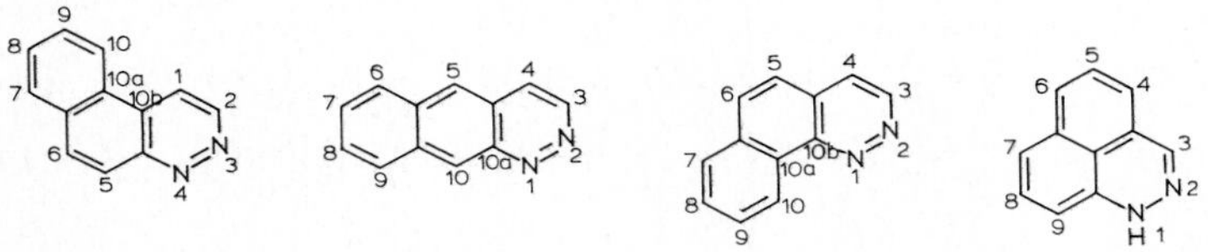

(XXIII)   (XXIV)   (XXV)   (XXVI)

Alkylation of 2-tetralone, and its 6-methoxy derivative *via* the corresponding pyrrolidine enamine leads to the substituted methyl acetate (XXVII, R = H, OMe). The latter with hydrazine hydrate gives the 2-oxo-1,10b,2,3,5,6-hexahydrobenzo[f]cinnoline (XXVIII, R = H, OMe). Reduction with lithium aluminium hydride produces 1,2,3,4,5,6-hexahydrobenzo[f]cinnolines (XXIX, R = H, OMe) (*U.K. Pandit, K. De Jonge* and *H.O. Huisman*, Rec. Trav. chim., 1969, **88**, 149).

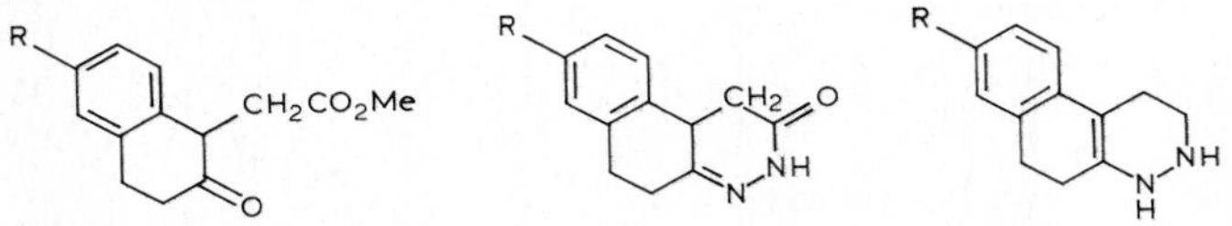

(XXVII)   (XXVIII)   (XXIX)

The parent compound **benzo[f]cinnoline**, m.p. 116.5–118°, has been prepared in moderate yield by the photocyclodehydrogenation of *cis*-3-styrylpyridazine, which is obtained by isomerisation of the corresponding *trans* derivative (*H.-H. Perkampus* and *Th. Bluhm*, Tetrahedron, 1972, **28**, 2099). **Benzo[h]cinnoline**, m.p. 100–106°, is prepared in an similar manner.

Their electron density and proton chemical shifts (*H.-H. Perkampus, T. Bluhm* and *J.V.*

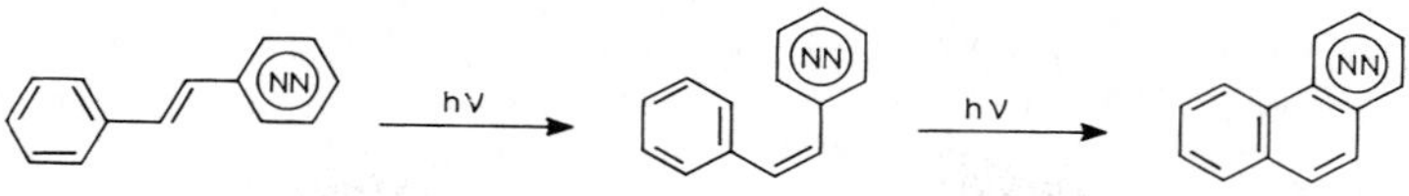

Knop, Z. Naturforsch. Teil A, 1972, **27**, 310), electronic absorption spectra (*idem*, Spectrochim. Acta, 1972, **28A**, 2179), luminescence behaviour (*idem*, C.A., 1973, **78**, 64606), pK_a and basicity (*T. Bluhm et al.*, *ibid.*, 1979, **90**, 5690), and mass spectra (*G. Schmidtberg* and *Th. Bluhm*, Org. Mass Spectrom., 1974, **9**, 453) have been reported. Their photochemical instability has been studied by e.s.r. and luminescence spectroscopy (*R. Schaaf* and *H.-H. Perkampus*, Tetrahedron, 1981, **37**, 341).

1-Aminobenzo[*f*]cinnoline-2-carbonitrile and -2-carboxamide have been prepared by the aluminium trichloride catalysed cyclisation of (XXX, R = CN, $CONH_2$) (Pat., C.A., 1977, **87**, 117893).

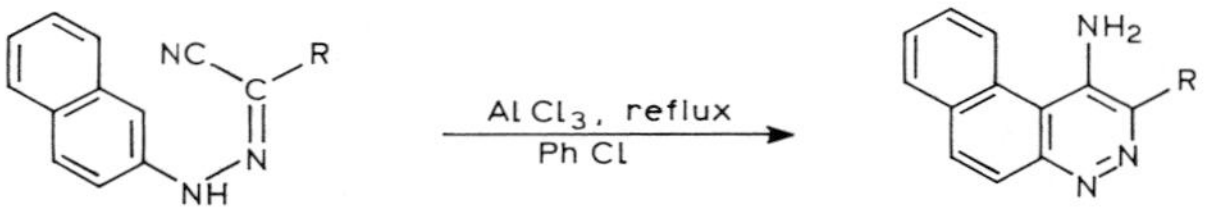

The natural product, methyl communate (XXXI) has been converted, *via* potassium permanganate oxidation, methylation with diazomethane and reaction with hydrazine into (XXXII), m.p. 199–201° (*S. Braun* and *H. Breitenbach*, Tetrahedron, 1977, **33**, 145).

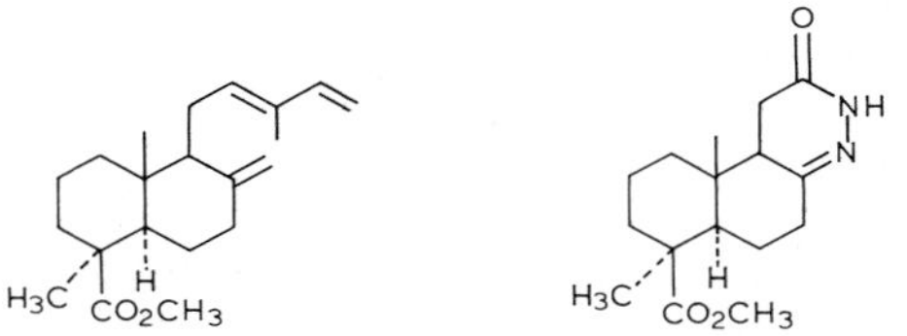

Hydrazine and 1-oxo-3-phenyl-1,2,3,4-tetrahydronaphthalene-2-acetic acid condense to form 2,3,4,4a,5,6-**hexahydro**-3-**oxo**-5-**phenylbenzo**[*h*]**cinnoline** (XXXIII, R = Ph), m.p. 191° (*W. Borsche* and *F. Sinn*, Ann., 1943, **555**, 70). Similarly, 1-tetralone-2-acetic acid with hydrazine gives the unsubstituted compound (XXXIII, R = H), m.p. 199–200°, which with bromine is converted into 3-*oxo*-2,3,5,6-*tetrahydrobenzo*[*h*]*cinnoline* (XXXIV), m.p. 257–261°, and hence with phosphorus oxychloride into 3-*chloro*-5,6-*dihydrobenzo*[*h*]*cinnoline* (XXXV, R = Cl), m.p. 151–154°. The latter with hydrazine gives the corresponding 3-*hydrazinohydrochloride* (XXXV, R = $NHNH_2 \cdot HCl$), m.p. 196° (dec.) (*H.M. Holava* and *R.A. Partyka*, J. med. Chem., 1971, **14**, 262).

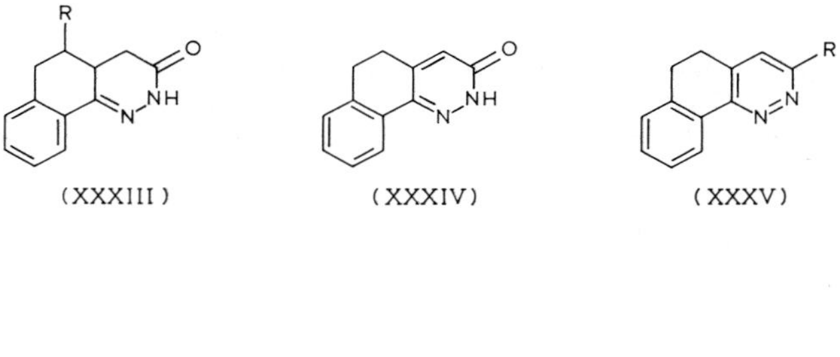

Following on from the above work, Bluhm dechlorinated (XXXV, R = Cl) to give 5,6-*dihydrobenzo*[*h*]*cinnoline*, m.p. 58–60°, which was dehydrogenated to give (XXV), m.p. 67–68° (*Th. Bluhm*, J. heterocyclic Chem., 1981, **18**, 189).

Cyclisation of 6-benzyl-3-phenylpyridazine-5-carboxylic chloride in the presence of aluminium trichloride gives 5,10-**dihydro**-5-**oxo**-3-**phenylbenzo**[*g*]**cinnoline**, m.p. 236° (*dinitrophenylhydrazone*, m.p. 244°), an early example of the [*g*] class, previously named as derivatives of naphthol[2,3-*c*]pyridazine (*W. Borsche* and *A. Klein*, Ann., 1941, **548**, 74).

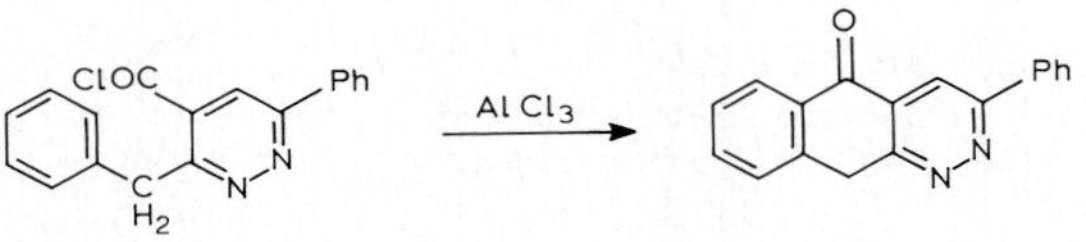

As part of the structure determination of atractylon, its hydroxylated and oxidised derivative (XXXVI) was treated with phenylhydrazine to give 5,5a,6,7,8,9,9a,10-*octahydro*-4,9a-*dimethyl*-6-*methylene*-2-*phenylbenzo*[*g*]*cinnoline*-3[2*H*]-*one*, m.p. 210–212° (*H. Hikino, Y. Hikino* and *I. Yoshioka*, Chem. pharm. Bull. Japan, 1962, **10**, 641; 1964, **12**, 755).

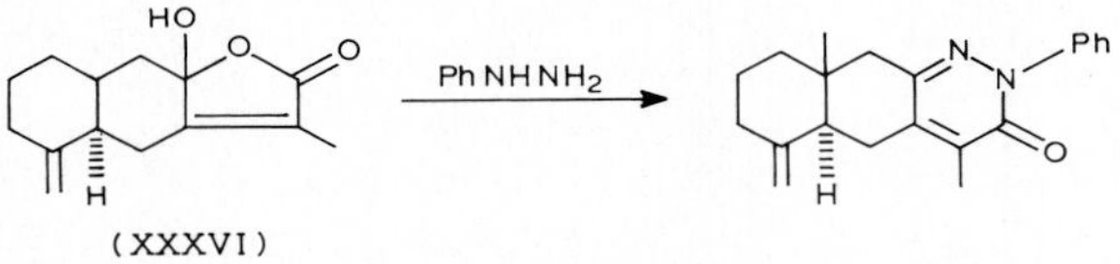

(XXXVI)

Treatment of either dinaphthoquinone (XXXVII) or 1,4-naphthoquinone with hydrazine has been shown to yield the black *benzo*[*g*]*cinnoline* derivative (XXXVIII), m.p. 300° (dec.) (*R. Pummerer et al.*, Ber., 1929, **62**, 2135; *E.S. Hand* and *T. Cohen*, Tetrahedron, 1967, **23**, 2911).

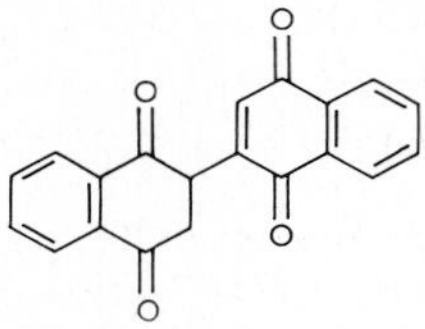

(XXXVII) 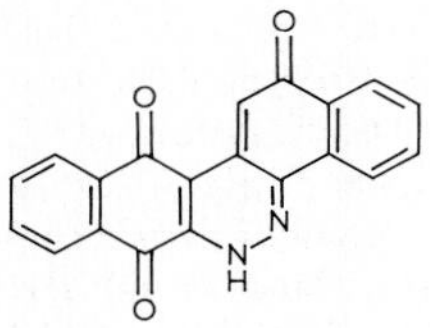 (XXXVIII)

The reaction of the lactone (XXXIX) with hydrazine gives 1,4-*dihydro*-5,10-*dihydroxy*-4,4-*dimethylbenzo*[*g*]*cinnoline*-3(2*H*)-*one*, m.p. 192° (dec.) (*S. Petersen* and *H. Heizer*, Ann., 1972, **764**, 36).

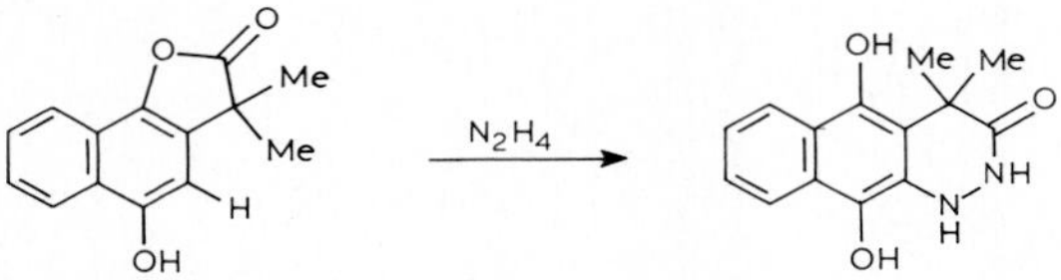

(XXXIX)

Hydrogenation of 2,3,4,4a,5,10-hexahydrobenzo[*g*]cinnoline derivative (XL) over platinum oxide gives the *cis*- (70%) and the *trans*- (30%) octahydro derivative (XLI) (*T. Tanaka et al.*, C.A., 1979, **91**, 211355).

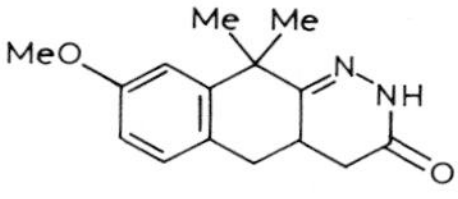

(XL) 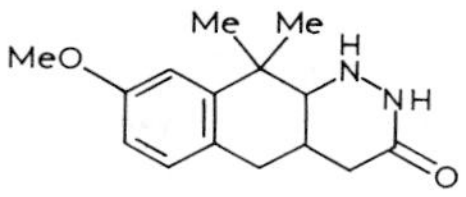   (XLI)

Cyclisation of 1-benzeneazo-2-hydroxynaphthalene-8-carboxylic acid by means of acid is reported to give 3,9-**dioxo-2-phenyl-2,3-dihydrobenzo**[*d,e*]**cinnoline** (XLII), m.p. 201–202°, the transformation being reversed by alkali. An isomeric product is formed from 1-benzeneazo-4-hydroxynaphthalene-8-carboxylic acid (*K. Dziewonski* and *T. Stolyhwo*, Ber., 1924, **57**, 1540).

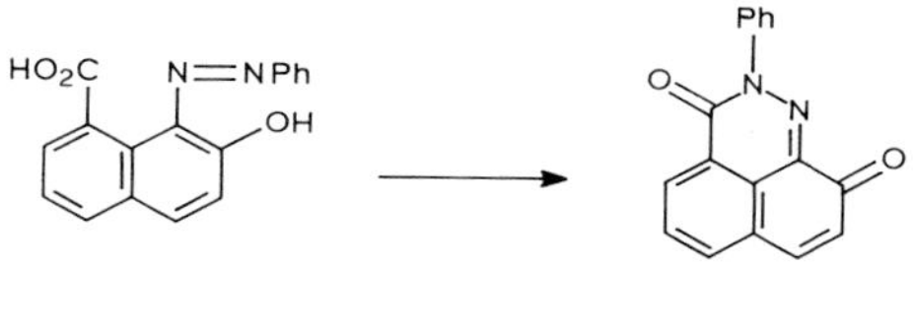

(XLII) 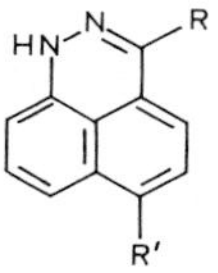   (XLIII)

The calculated orbital energies and electron densities have been reported for the benzo[*d,e*]cinnoline ring system (*A.F. Pozharskii* and *E.N. Malysheva*, Khim. Geterot. Soedin., 1970, 103).

8-Formyl-, 8-acetyl-, and 8-benzoyl-1-naphthol have been condensed with hydrazine to give respectively 1*H*-*benzo*[*d,e*]*cinnoline* (XLIII, R = R′ = H), m.p. 148–151°, its 3-*methyl* (XLIII, R = CH$_3$ R′ = H), m.p. 153–155°, and 3-*phenyl* (XLIII, R = C$_6$H$_5$, R′ = H), m.p. 206–208°C, derivative in good to excellent yield. Reaction with methyl iodide in the presence of base gives the corresponding 1-methyl derivative (*P.H. Lacy* and *D.C.C. Smith*, J. chem. Soc. C, 1971, 747; *idem*, J. chem. Soc. Perkin I, 1975, 419).

5-Methoxy-1-oxotetralin-8-carboxylic acid with hydrazine gives 3-hydroxy-6-methoxy-8,9-dihydro-1*H*-benzo[*d,e*]cinnoline (XLIII, R = OH, R′ = OCH$_3$), which with phosphorus oxychloride gives the 3-chloro derivative. The latter with hydrazine forms (XLIII, R = NHNH$_2$, R′ = OCH$_3$) (Pat., C.A., 1972, **77**, 164733).

1-Acyl-4,8-dimethoxynaphthalenes could not be cyclised with hydrazine although 4-methoxy-8-hydroxynaphthalene-1-carboxaldehyde gives (XLIII, R = H, R′ = OCH$_3$) with excess hydrazine (*V.V. Mezheritskii et al.*, Zhur. org. Khim., 1981, **17**, 627).

4. Phthalazines*

(a) Methods of synthesis

Phthalazine (I) was first obtained in 1893 from $\alpha,\alpha,\alpha',\alpha'$-tetrachloro-*o*-xylene and aqueous hydrazine at 150° under pressure (*S. Gabriel* and *G. Pinkus*, Ber., 1893, **26**, 2210). A more convenient process was later found to be the reaction of the tetrachloro-xylene with hydrazine sulphate in sulphuric acid (F.P., 1,438,827, 1966). Alternatively, *o*-phthalaldehyde may be treated with either aqueous hydrazine sulphate or alcoholic hydrazine hydrate to give (I) in high yield (*R.F. Smith* and *E.D. Otrempa*, J. org. Chem., 1962, **22**, 879; Pat., C.A., 1966, **65**, 726).

(*1*) The use of substituted *o*-phthalaldehydes in a reaction with hydrazine produces phthalazines substituted in the benzene nucleus and *o*-diacylbenzenes similarly yield 1,4-disubstituted phthalazines (*F.M. Dean, D.R. Randell* and *G. Winfield*, J. chem. Soc., 1959, 1071; *C.R. Warner, E.J. Walsh* and *R.F. Smith*, ibid., 1962, 1232; *J.M. Van der Zanden* and *G. De Vries*, Rec. Trav. chim., 1955, **74**, 52).

(*2*) *o*-Acylbenzoic acids and phthalide and its derivatives are important precursors of phthalazines, again by reaction with hydrazines. Phthaldehydic acid is converted into phthalazine-1(2*H*)-one (II) in quantitative yield by reaction with hydrazine (*S. Biniecki* and *B. Gutkowska*, Acta Polon. Pharm., 1955, **11**, 27). In a similar way *o*-acylbenzoic acids yield 4-substituted phthalazin-1(2*H*)-ones (III).

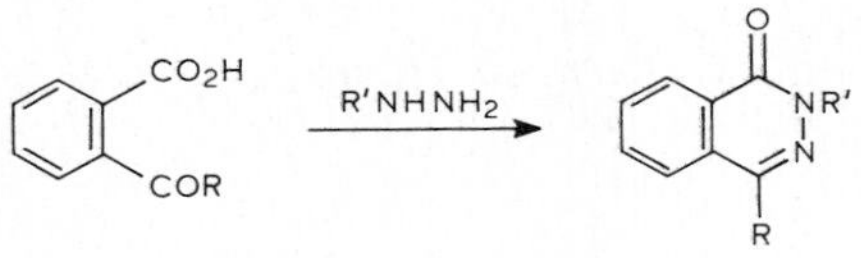

(II) R = H
(III) R ≠ H

---

* *N.R. Patel*, in Heterocyclic Chemistry, Vol. 27, Series ed. A. Weissberger and E.C. Taylor; Condensed Pyridazines including Cinnoline and Phthalazine, ed. *R.N. Castle*, J. Wiley and Sons Ltd., New York, 1973, pp. 323–760.

Alternative starting materials are phthalide and its derivatives substituted in the 3-, 4-, 5-, 6- or 7-position (*S.S. Berg* and *E.W. Parnell*, J. chem. Soc., 1961, 5275) and phthalimide and its derivatives (*W. Triebs et al.*, Ann., 1951, **574**, 54).

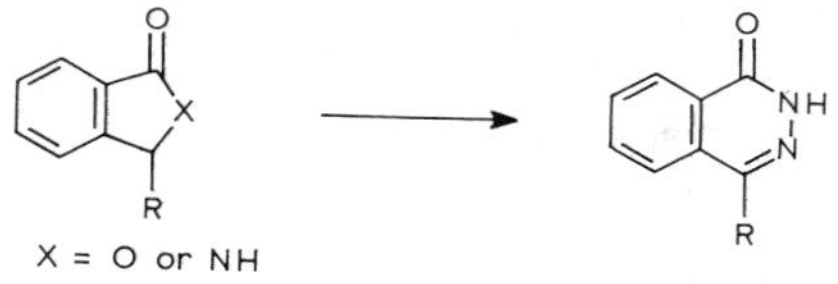

X = O or NH

3-Alkylidenephthalides may be used and decarboxylation of (IV) provides a good yield of 4-methylphthalazine-1(2*H*)-one (*S. Foldeak et al.*, C.A., 1970, **72**, 100626).

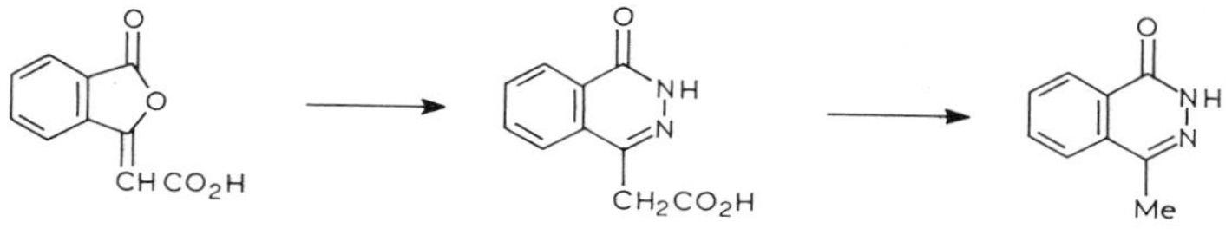

(IV)

(*3*) As might be expected, phthalic acid derivatives are readily converted into phthalazines. Thus, in general diesters or disodium salts of phthalic acid, phthalic anhydride or phthaloyl chloride and their substituted derivatives react with hydrazine or substituted hydrazines to give phthalazine derivatives. For instance, phthalic anhydride and hydrazine hydrate in ethanol afford 4-hydroxyphthalazin-1(2*H*)-one (V) in 85% yield (*G. Rosseels*, Bull. Soc. chim. Belg., 1965, **74**, 91) and phthalic anhydride with methylhydrazine in acetic acid yields 2-methyl-1-hydroxyphthalazine-1(2*H*)-one (VI) (*E.H. White*, *D.F. Roswell* and *O.G. Zafirion*, J. org. Chem., 1969, **34**, 2462). However, the reaction of phenylhydrazine with

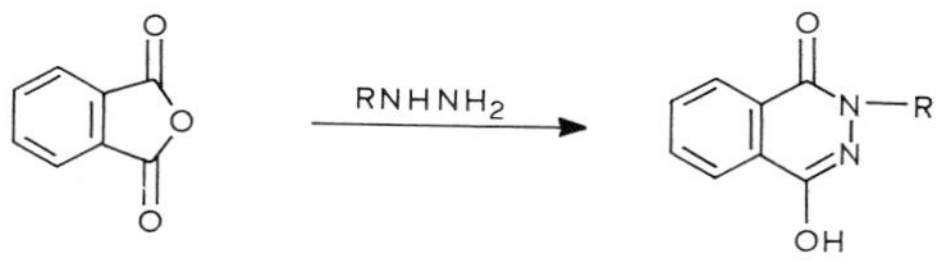

(V) R = H
(VI) R = Me

phthalic anhydride gives mainly 2-anilinophthalimide (VII) (*D.J. Drain* and *D.E. Seymour*, J. chem. Soc., 1955, 852), but this is readily converted into 4-hydroxy-2-phenylphthalazine-1(2*H*)-one (IX) by treatment with base.

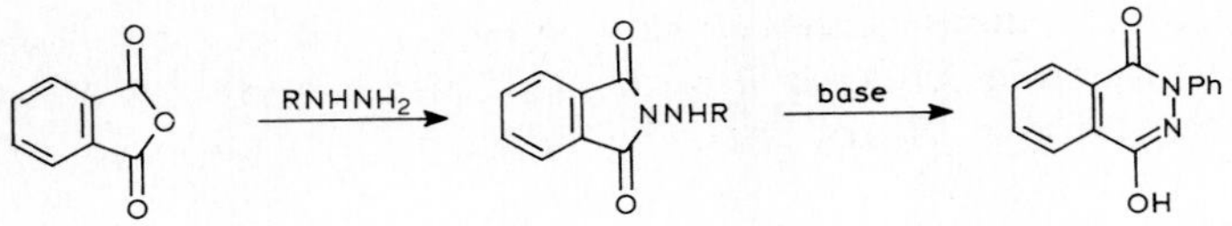

(VII) R = Ph
(VIII) R = H

(IX)

In a similar way, 2-aminophthalimide (VIII) on treatment with aqueous base gives (V) in quantitative yield (*J. Nishe* and *S. Kamimoto*, Nippon Kagaku Zasshi, 1958, **79**, 1403).

(*4*) Reductive removal of chlorine from 1-chlorophthalazine (XI) by catalytic reduction in the presence of palladium–charcoal (*E.F.M. Stephenson*, Chem. Ind. (London), 1957, 174) or from 1,4-dichlorophthalazine (XII) by chemical reduction with hydriodic acid and red phosphorus (*C.M. Atkinson* and *C.J. Sharpe*, J. chem. Soc., 1959, 3040) occurs smoothly to give phthalazine (I). This approach is useful because the monochloro and dichloro compounds are readily obtained from the corresponding phthalazinones, (X) and (V), respectively.

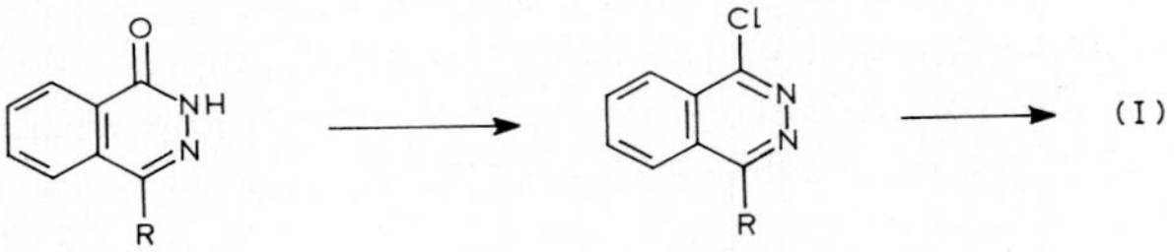

(X) R = H
(V) R = OH

(XI) R = H
(XII) R = Cl

→ (I)

Alternatively, the chlorine in (XI) may be removed by a two-step process by conversion of (XI) to the corresponding hydrazine (XIII) followed by oxidative removal of the substituent with copper sulphate (*A. Albert* and *G. Catterall*, ibid., 1967, 1533). The readily accessible *p*-toluenesulphonyl derivative (XIV) may be converted into (I) by the action of hot alkali (*Atkinson* and *Sharpe*, loc. cit.).

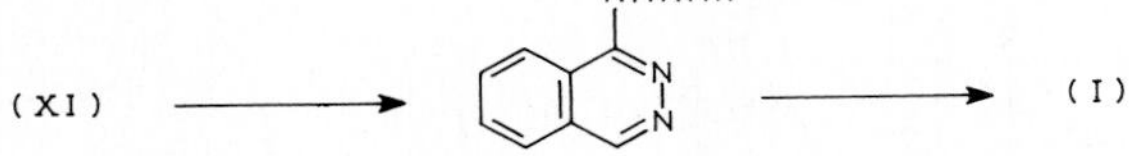

(XI) →

(XIII) R = H
(XIV) R = $O_2SC_6H_4Me$-*p*

→ (I)

These dehalogenation methods may be used to obtain alkyl- or aryl-sub-

These dehalogenation methods may be used to obtain alkyl- or aryl-substituted phthalazines from the corresponding chloro compounds (*E. Hayashi* and *E. Oishi*, C.A., 196, **65**, 15373).

(*5*) Phthalazines having a good leaving group in the 1-position, *e.g.* chlorine or methylsulphonyl, undergo nucleophilic attack with active methylene compounds in the presence of base. Thus diethyl malonate reacts with 1-chlorophthalazine (XI) to give (XV) which undergoes acidic hydrolysis and decarboxylation to give (XVI) (*Y. Mizuno, K. Adachi* and *K. Ikeda*, Chem. pharm. Bull. Japan, 1954, **2**, 225). In a similar way 1-methylsulphonylphthalazine (*E. Oishi*, Yakagaku Zasshi, 1969, **89**, 959) and its 4-aryl derivatives (*E. Hayashi* and *E. Oishi*, *ibid.*, 1968, **88**, 83) yield 1-alkyl- and 1-alkyl-4-aryl-phthalazines.

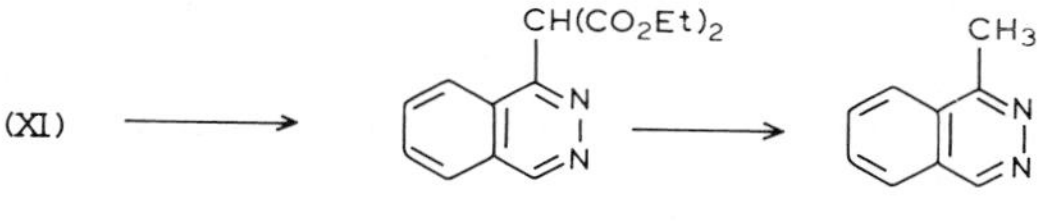

(XI) ⟶ (XV) ⟶ (XVI)

(*6*) Phthalazine undergoes addition of Grignard or organo-lithium reagents across the C=N bond to yield the 1-substituted-1,2-dihydrophthalazine (XVII) which may be oxidised to the 1-alkyl- or 1-aryl-phthalazine (*A. Marxer, F. Hofer* and *U. Salzman*, Helv., 1969, **52**, 1376).

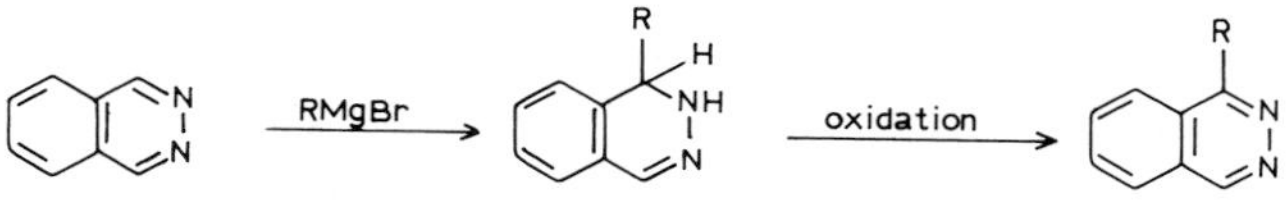

⟶ RMgBr ⟶ (XVIII) ⟶ oxidation ⟶

1-Methyl-1,2-dihydrophthalazine (XVIII), obtained from methyllithium, yields the isolable biphthalazine peroxide (XIX) which may be quantitatively reduced to 1-methylphthalazine (XX) with bisulphite (*A. Hirsch* and *D.G. Orphanos*, Canad. J. chem., 1966, **44**, 2109).

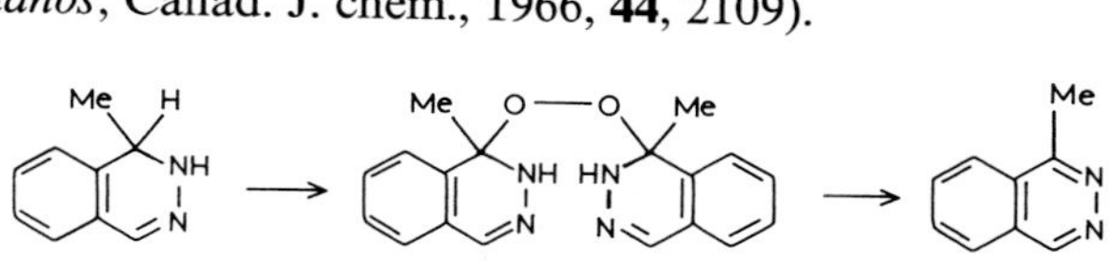

(XVIII) ⟶ (XIX) ⟶ (XX)

1-Alkyl- or 1-aryl-phthalazines will undergo attack by Grignard reagents at the unsubstituted C-atom of the heterocyclic moiety to give 1,4-disubstituted derivatives (*Marxer, Hofer* and *Salzman, loc. cit.*).

(*7*) The Reissert compound (XXI) on treatment with methyl iodide and sodium hydride in dimethylformamide yields (XXII) which undergoes hydrolysis under alkaline conditions to afford 1-methylphthalazine (XVI) (*F.D. Popp, J.M. Wefer* and *C.W. Klinowski*, J. heterocyclic Chem., 1968, **5**, 879).

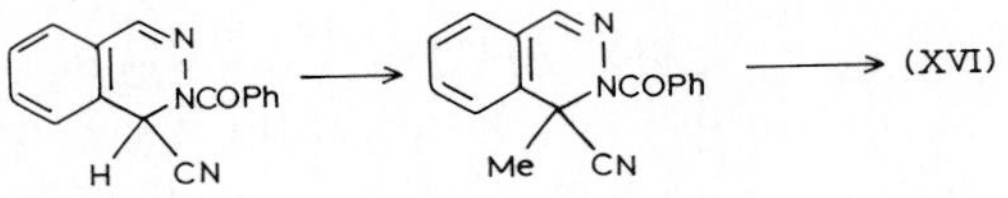

(*8*) Miscellaneous methods of synthesis include the formation of 1,4-diphenylphthalazine (XXIV) by the reaction of 1,4-diphenyl-*sym*-tetrazine (XXIII) with benzyne generated from anthranilic acid diazonium betaine (*J. Sauer* and *G. Heinrichs*, Tetrahedron Letters, 1966, 4979). Phthalazine-N-oxides are reduced to the corresponding phthalazine by the usual methods (*E. Hayashi* and *E. Oishi*, Yakugaku Zasshi, 1966, **86**, 576; C.A., 1966, **65**, 15373). It is noteworthy that a previously described route (C.C.C., 1st Edn., Vol. IVB, p. 1239) to 1-arylphthalazines has been shown to yield azines and not phthalazines (*H.J. Rodda* and *P.E. Rogash*, J. chem. Soc., 1956, 3927). The azines are formed by decomposition of the starting material under the conditions described.

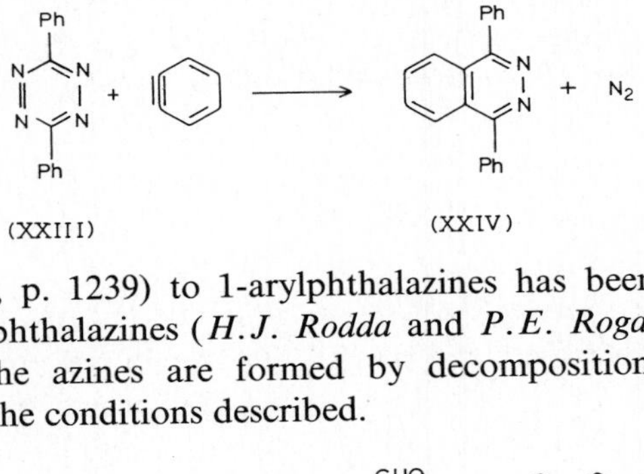

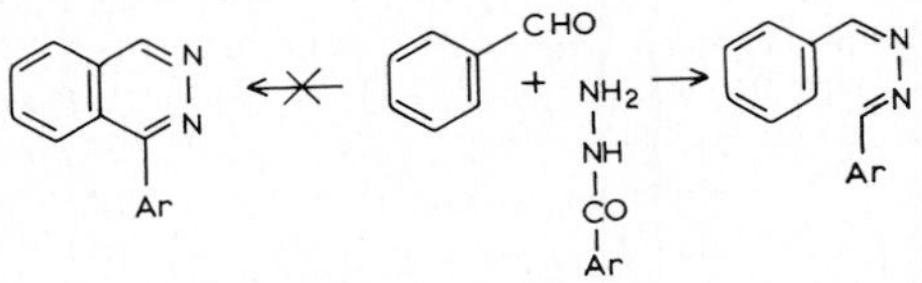

However, 1-benzoyl-2-benzylidenehydrazines containing electron releasing substitutents on the benzylidene nucleus give substituted 1-phenylphthalazines upon treatment with base (*A.M. Barghash et al.*, Pharmazie, 1972, **27**,

796). Although the photolysis of *N*-alkyl-*N*-acylhydrazones of benzaldehyde generally gives products derived from benzyne, the presence of an *o*-nitro group alters the course of the reaction to produce 1-substituted 5-nitrophthalazine by a photochemical 6π pericyclic reaction process (*Y. Maki* and *T. Furuta*, Synthesis, 1976, 263).

(b) Phthalazine, its alkyl and aryl derivatives

(i) Phthalazine

Phthalazine forms colourless prisms, m.p. 90–91°, and boils (315–317°) with decomposition at atmospheric pressure but can be distilled at 175°/17 mm. It may be purified by sublimation at 125°/0.3 mm. The parent heterocycle is easily soluble in water and most organic solvents. Phthalazine has pK_a of 3.5 in water at 20° (*A. Albert et al.*, J. chem. Soc., 1961, 2689) and so is a stronger base than either pyridazine or cinnoline. The heterocycle acts as a monoacidic base in forming salts: *monohydrochloride*, m.p. 235–236° (dec.). *monohydrobromide*, m.p. 245–255°, *monohydriodide*, m.p. 203°, *picrate*, m.p. 208–210°. Aryl sulphonic acid salts of phthalazine may be obtained directly by oxidative cyclisation of *o*-phthalaldehyde bis(arylsulphonylhydrazone) (*R.N. Butler* and *J.P. Jones*, J. chem. Res. Synop., 1982, 348).

The base forms a stable complex with bromine, m.p. 122–123°, which may be used as a convenient source of bromine in addition or substitution reactions (*A. Hirsch* and *D. Orphanos*, Canad. J. Chem., 1968, **46**, 1455).

Quaternisation of the heterocycle with the lower alkyl halides and benzyl halides occurs readily (*R.C. Smith* and *E.D. Otremba*, J. org. Chem., 1962, **27**, 897): 2-*methylphthalazinium iodide*, m.p. 240–243°. Higher alkyl halides require more vigorous conditions (*H. Daeniker* and *J. Druey*, Helv., 1957, **40**, 918) and the quaternary salts may be converted into the isolable pseudo-bases, 1-hydroxy-2-alkyl-1,2-dihydrophthalazine. 1-Substituted phthalazines are quaternised mainly at position 3 while the relative steric size of substituents in 1,4-disubstituted phthalazines determines the quaternisation pattern.

Several theoretical studies of the phthalazine molecule have been made (*A.H. Gawer* and *B.P. Dailey*, J. chem. Phys., 1965, **42**, 2568; *S.C. Wait* and *J.W. Wesley*, J. mol. Spectrosc., 1966, **19**, 25) and use of extended Hückel theory with CNDO has given the σ and π charge densities and bond orders (*R.J. Pugmire*, J. Amer. chem. Soc., 1969, **91**, 6381).

Charge density (Pugmire)

Bond lengths (Huiszoon) in Å

X-ray diffraction studies have been made (C. *Huiszoon et al.*, Acta Crystal., 1972, **28B**, 3415).

The ionisation potential for phthalazine is calculated to be 9.57 eV (*T. Nakajima* and *A. Pullman*, J. chim. Phys., 1958, **55**, 793). The behaviour of phthalazine and certain derivatives under polarographic reduction conditions have been investigated (*H. Lund* and *E.T. Jansen*, Acta Chem. Scand., 1970, **24**, 1867). The calculated value for the dipole moment, 4.56 D (*H.F. Hameka* and *A.M. Laquori*, Mol. Phys., 1958, **1**, 9), and the experimental value, 4.88 D (*J. Crossley* and *S. Walker*, Canad. J. Chem., 1968, **46**, 2369), are in reasonable agreement.

(ii) Spectra

The u.v. spectrum of phthalazine has been recorded in various solvents including cyclohexane, ethanol, water (*R.C. Hirt et al.*, J. chem. Phys., 1956, **25**, 574), methanol and aqueous acid (*R.M. Acheson* and *F.W. Foxton*, J. chem. Soc., 1966, 2218). The spectra of phthalazine and naphthalene are similar with the latter showing the greater absorption at slightly longer wavelengths (*E.D. Amstutz*,, J. org. Chem., 1952, **17**, 1508). Calculations of transition energies and band intensities have been made (*G. Favini, I. Vandoni* and *M. Simonetta*, Theoret. Chim. Acta, 1965, **3**, 418).

The i.r. spectrum of phthalazine has been reported (*A. Hirsch* and *D. Orphanos*, J. heterocyclic Chem., 1965, **2**, 206) but detailed study has been neglected. The variation of intensity with excitation wave number of various vibrational bands in the Raman spectrum of phthalazine has been investigated (*A. Aminzadeh, V. Fawcett* and *D.A. Long*, J. Raman Spectrosc., 1980, **9**, 219).

The observed p.m.r. spectrum has been compared with the computed spectrum and the absorption bands assigned (see Table 23) (*P.J. Black* and *M.L. Heffernan*, Austral. J. Chem., 1965, **18**, 707). The effects of europium shift reagents on the spectrum have been studied (*W.L.F. Armarego, T.J. Batterham* and *J.R. Kershaw*, Org. magn. Resonance, 1971, **3**, 575). The H–H, C–H, N–H, C–C and N–C spin–spin coupling constants have been calculated using the FP-INDO method (*S.A.T. Long* and *J.D. Memory*, J. magn. Resonance, 1981, **44**, 355).

The ^{13}C-n.m.r. spectrum of phthalazine has been determined in chloroform relative to benzene (*Pugmire, loc. cit.*). The effect of protonation at various values of pH has been investigated by a study of the ^{13}C-n.m.r.-spectra (*P. Van de Weijer* and *D.M.W. Van der*

TABLE 23

P.M.R. SPECTRUM OF PHTHALAZINE (IN ACETONE)

Proton	δ (*ppm*)	Coupling constants (*Hz*)			
1 and 4	9.6	$J_{4,8}$	0.4		
5 and 8	8.13	$J_{5,8}$	0.6	$J_{5,6}$	8.2
6 and 7	8.0	$J_{6,7}$	6.8	$J_{5,7}$	1.2

Ham, Org. Magn. Resonance, 1977, **9**, 281). The INDO method has been used to calculate the electronic charge distribution in the ground and lowest excited singlet and triplet states of neutral and protonated phthalazines, and experimentally observed changes in pK_a upon excitation have been correlated with the theoretical results (*J. Waluk*, *A. Grabowska* and *J. Lipinski*, Chem. phys. Letters, 1980, **70**, 175).

The signals in the ^{15}N-n.m.r. spectrum of phthalazine (*M. Witanowski*, *L. Stefaniak* and *G.A. Webb*, J. org. Res., 1981, **16**, 309) and in that of its *N*-oxide (*W. Staedelli et al.*, Helv., 1980, **63**, 504) have been discussed. In DMSO, the phthalazine nitrogen exhibits δ -10.3 ppm relative to nitromethane, whereas phthalazine-2-oxide has 2-N with δ -68.9 and 3-N δ -53.2 ppm.

Irradiation of phthalazine in methanol with u.v. light produces semiquinone radicals by a process involving 2 photons. In methanol containing hydrochloric acid a similar process yields the radical cation (*A. Castellano et al.*, J. chem. Res. Synop., 1979, 70). These two species have been studied by e.s.r. and optical spectroscopy (*T. Kato* and *T. Shida*, J. Amer. chem. Soc., 1979, **101**, 6869). The hyperfine coupling in the e.s.r. spectrum of the radical anion has been studied by means of spin density calculations (*M.H. Palmer* and *I. Simpson*, Z. Naturforsch. Teil A, 1983, **38**, 415).

The electron impact mass spectrum of phthalazine shows loss of N_2 from the base peak molecular ion to be a relatively minor pathway. This is quite different from the cases of pyridazine and cinnoline. The phthalazine molecular ion fragments by sequential loss of two HCN units followed by ethyne (*J.H. Bowie et al.*, Austral. J. Chem., 1967, **20**, 2677).

$$M^{+\cdot}(m/z\ 130) \xrightarrow{-HCN} m/z\ 103 \xrightarrow{-HCN} m/z\ 76 \xrightarrow{-C_2H_2} m/z\ 50$$

1-Methyl- and 1-phenyl-phthalazine lose H˙ as an important first step. This 'proximity to nitrogen effect' is also shown by 1-methoxy- and 1,4-diamino-phthalazine.

The presence of a carbonyl or thiocarbonyl group in the heterocyclic system causes an increase in the number of major fragmentation pathways from the molecular ion. Thus, phthalazinone loses CO, N_2 and $HN_2^\cdot$ from the molecular ion in an approximately equal extent. Phthalazine-*N*-oxide molecular ion decomposes by loss of NO and then HCN (*J.H. Bowie et al.*, ibid., 1967, **20**, 2545).

Phosphorescence, excitation and absorption spectra of phthalazine have been determined (*H. Baba*, *I. Yamazaki* and *T. Takemura*, Bull. chem. Soc. Japan, 1969, **42**, 276) and it has been shown that intersystem crossing from second and third excited singlet states competes favourably with internal conversion (*Y.H. Li* and *E.C. Lim*, J. chem. Phys., 1972, **56**, 1004). The fluorescence spectrum has been studied in some detail (*R.W. Anderson* and *W. Knox*, J. Lumin., 1981, **24/25**, 647). The magnetic circular dichroism spectrum has been investigated (*M. Vasak*, *M.R. Whipple* and *J. Michl*, J. Amer. chem. Soc., 1978, **100**, 6838).

(iii) Reactivity

The oxidation of phthalazine with alkaline permanganate yields pyridazine-4,5-dicarboxylic acid (*S. Gabriel*, Ber., 1903, **36**, 3373) while treatment with monoperphthalic acid in ether affords *phthalazine-N-oxide*, m.p. 143° (*E. Hayashi et al.*, Yakugaku Zasshi, 1962, **82**, 584). Nitration of phthalazine gives mainly 5-*nitrophthalazine*, m.p. 187–188° (*S. Kanahara*, ibid., 1964, **84**, 489). Phthalazine yields a Reissert compound by treatment with benzoyl chloride and potassium cyanide (*F.D. Popp*, *J.M. Wefer* and *C.W. Klinowski*, J. heterocyclic Chem., 1968,

5, 879). The reaction with maleic anhydride (2 molar equivalents) yields the spiro products (XXV) (*M.B. Hocking, ibid.*, 1977, **14**, 829): two stereoisomers have been isolated.

(XXV)   (XXVI)

Ylides of the type (XXVI) may be obtained by treatment of phthalazine with phenacyl bromide and subsequent basification of the salt (*M. Petrovanu et al.*, Rev. Roum. Chem., 1969, **14**, 1153).

The reduction of phthalazine occurs readily in the heterocyclic moiety: 1,2-dihydrophthalazine may be obtained upon reduction with lithium aluminium hydride (*C.C. Leznoff*, Canad. J. Chem., 1968, **46**, 1152) and 1,2,3,4-tetrahydrophthalazine with sodium amalgam. Ring-opening reactions occur on reduction with zinc and hydrochloric acid to give *o*-aminomethylbenzylamide and a similar result is obtained on catalytic reduction with hydrogen (*E.F. Elslager et al.*, J. heterocyclic Chem., 1968, **5**, 609). Reduced phthalazines are discussed in more detail in Section *j*.

(iv) Alkyl- and aryl-phthalazines

1-**Methylphthalazine**, m.p. 74°, is obtained from 4-chloro-1-methylphthalazine by reduction with hydriodic acid and red phosphorus. It gives a *monohydrochloride*, m.p. 222–223°, *monohydriodide*, m.p. 287° (dec.), and a *picrate*, m.p. 206°. Nitration yields the 5-*nitro* derivative, m.p. 195–197°, with some 4-*methyl*-8-*nitro*-1(2*H*)-*phthalazinone*, m.p. 293–294°. Quaternisation of the 5-nitro compound with methyl iodide gives both 1,2-*dimethyl*-5-*nitrophthalazinium iodide*, m.p. 208–213° (dec.) and 1,3-*dimethyl*-5-*nitrophthalazinium iodide*, m.p. 167–170° (dec.) (*Kanahara, loc. cit.*).

1-**Ethylphthalazine**, m.p. 23.5°, b.p. 196°/16 mm, undergoes oxidation in air. The *monohydrochloride*, *monohydriodide* and *picrate* have m.ps. 216°, 203°, and 175° (dec.), respectively (*V. Paul, Ber.*, 1899, **32**, 2014). 1-*Benzylphthalazine* m.p. 81°, on oxidation with permanganate is converted into 1-*benzoylphthalazine*, m.p. 123–124°, *oxime*, m.p. 243–244° (*A. Lieck, ibid.*, 1905, **38**, 3918). 1-**Phenylphthalazine** is a yellow solid, m.p. 138–141° [though a m.p. 144° has been reported (*A. Hirsch* and *D.G. Orphanos*, J. heterocyclic Chem., 1966, **3**, 38)], *picrate*, m.p. 180°. Substituted 1-phenylphthalazines have been described (*Aggrawal et al., loc. cit.*).

1,4-**Dimethylphthalazine**, m.p. 108°, is obtained as a yellow solid from 1-chloro-4-methylphthalazine by the action of diethyl malonate in the presence of base followed by hydrolysis and decarboxylation of the initial reaction product. Dimethylketen reacts with 1,4-dimethylphthalazine to give tricyclic compounds having an oxazinophthalazine skeleton (*M.A. Shah* and *G.A. Taylor*, J. chem. Soc. C, 1970, 1651).

1,4-**Diphenylphthalazine**, m.p. 197.5–199.5°, is a colourless solid soluble in dilute mineral acids (*T.H. Regan* and *J.B. Miller*, J. org. Chem., 1966, **31**, 3053) and is prepared by the action of hydrazine on *o*-dibenzoylbenzene.

1-**Methyl-4-phenylphthalazine**, m.p. 125–126°, is obtained by hydrolytic decarboxylation of 1-bis(diethoxycarbonyl)methyl-4-phenylphthalazine (XXVII) (*M.A. Shah* and *G.A. Taylor, loc. cit.*). Compounds containing a 1-substituted alkyl group such as (XXVII) are conveniently obtained by nucleophilic displacement of a methylsulphonyl substituent (*E. Hayashi* and *E. Oishi*, Yagugaku Zasshi, 1968, **88**, 83).

(XXVII)

(c) Phthalazine-N-oxides and -N-aminoazolium salts

The phthalazine nucleus is readily oxidised to the mono-N-oxide by the action of the peracids. Phthalazine-2-oxide was first obtained by the oxidation of phthalazine with monoperphthalic acid in ethereal solution at 0° (*E. Hayashi et al.*, Yagugaku Zasshi, 1962, **82**, 584). Unsymmetrically substituted phthalazines can, and often do, give two isomeric mono-N-oxides. The N-oxide function is readily removed by standard reagents such as phosphorus trihalides in chloroform (*E. Hayashi* and *E. Oishi, ibid.*, 1966, **86**, 576). The N-oxide function provides a route for substitution at the adjacent carbon atom by use of standard reactions.

The ultraviolet spectra of N-oxides and that of the corresponding phthalazine have been compared (*Hayashi* and *Oishi, loc. cit.*) as have their p.m.r spectra (*K. Tori, M. Ogata* and *H. Kano*, Chem. pharm. Bull Japan, 1963, **11**, 681) and their ^{15}N-n.m.r. spectra (*Staedeli et al., loc. cit.*).

Phthalazine 2-**oxide**, m.p. 143°, forms a *hydrate*, m.p. 94°, and a *picrate*, m.p. 152–155°.

1-**Methylphthalazine-3-oxide**, m.p. 192–193°, is formed as the major component (29% yield) in the oxidation of 1-methylphthalazine with monoperphthalic acid in chloroform–ether solution (*E. Hayashi* and *E. Oishi*, Yakugaku Zasshi, 1968, **88**, 1333). The minor component in the reaction product is 1-*methylphthalazine-2-oxide*, m.p. 100–110° and 138° (double m.p.), formed in 16% yield. 1-Benzylphthalazine gives 1-*benzylphthalazine-2-oxide*, m.p. 140–141°, and 1-*benzylphthalazine-3-oxide*, m.p. 172–174°, in better total yield and a ratio of isomers of 3:1, respectively.

1-**Phenylphthalazine-3-oxide**, m.p. 181–183°, is a pale yellow solid; the only product upon N-oxidation of 1-phenylphthalazine (*E. Hayashi* and *E. Oishi, ibid.*, 1966, **89**, 576). The N-oxide reacts with Grignard reagents to give 1-phenyl-4-substituted-phthalazine-3-oxides and with phosphorus oxychloride to yield 1-chloro-4-phenylphthalazine. 1-Phenylphthalazine-3-

oxide reacts with phenyl isocyanate to produce 1-anilino-4-phenylphthalazine (*Hayashi* and *Oishi, loc. cit.*).

N-Oxidation of 1-alkoxyphthalazines gives the corresponding 3-oxide as the sole product, while 1,4-dialkoxyphthalazines appear not to undergo *N*-oxidation.

1,4-Diphenylphthalazine-2-oxide is a yellow solid, m.p. 188–190°, which on photolysis in the absence of oxygen gives 1,3-diphenylisobenzofuran (*K.B. Tomer et al.*, J. Amer. chem. Soc., 1973, **95**, 7402).

Treatment of phthalazine and 1-phenylphthalazine with *O*-mesitylenesulphonylhydroxylamine gives the *N*-aminoazolium salts, (XXVIII) and (XXIX), respectively. The salts are converted into the betaines (XXX) and (XXXI), respectively, by the action of benzoyl chloride.

(XXVIII) R = H
(XXIX)   R = Ph

(XXX)   R = H
(XXXI) R = Ph

2-Aminophthalazinium mesitylenesulphonate, m.p. 157–158°, gives the 2-*benzoylimino* derivative, m.p. 203–204° (*Y. Tamura et al.*, J. heterocyclic Chem., 1974, **11**, 675). The salt (XXVIII) reacts with dimethyl acetylenedicarboxylate to give the *tricyclic azaindolizine* derivative (XXXII), m.p. 118–119° (*Y. Tamura, Y. Miki* and *M. Ikeda, ibid.*, 1975, **12**, 119).

(XXVIII)

(XXXII)

2-Amino-4-phenylphthalazinium mesitylenesulphonate, m.p. 144–145°, gives the 3-*benzoylimino* derivative, m.p. 204–206° on treatment with benzoyl chloride. Mass spectrometry and p.m.r. spectra show the substitution pattern of the benzoyl derivative. This compound, (XXXI), has been used to introduce substituents into the 1-position of the phthalazine nucleus. Thus, 1-**cyano-4-phenylphthalazine**, m.p. 185–185.5°, is obtained on treatment of (XXXI) with cyanide ion (*Y. Tamura et al., ibid.*, 1976, **13**, 23).

(d) Halogenophthalazines

The 1-chloro- and 1,4-dichloro-phthalazines are readily obtained from phthalazin-1-one and 4-hydroxyphthalazin-1(2*H*)-one, respectively, by treatment with phosphorus oxychloride (*E. Hayashi et al.*, Yagugaku Zasshi, 1962, **82**, 584), although the formation of the dichloro compound requires the addition of phosphorus pentachloride to the reaction mixture. Alternatively, the 1,4-dichloro compound may be obtained by the action of phosphorus pentachloride alone (*A. Hirsch* and *D. Orphanos*, Canad. J.

Chem., 1954, **43**, 1551). Similarly, the 1-bromo- and 1,4-dibromo-phthalazines are obtained by use of the appropriate inorganic bromine derivatives (*idem, ibid.*, 1965, **43**, 2708). On the other hand, the iodo (*idem, ibid.*, 1965, **43**, 1551) and fluoro derivatives of phthalazine (*R.D. Chambers et al.*, Tetrahedron Letters, 1970, 57) are best obtained by halogen-exchange reactions from the corresponding chloro compound.

The halogen atoms show characteristic reactivity, including nucleophilic displacement of the 5-, 6-, 7-, and 8-fluorine atoms in hexafluorophthalazine. Generally, the readily available 1-chloro- or 1,4-dichloro-phthalazines are used as intermediates when it is desired to introduce other substituents into these positions. However, sometimes, as for example in the preparation of 1-cyano- or 1,4-dicyano-phthalazine, the greater reactivity of the iodo compounds is necessary (*A. Hirsch* and *D. Orphanos*, Canad. J. Chem., 1965, **43**, 1551). 1-Chlorophthalazine is hydrolysed in boiling aqueous alkali to phthalazin-1(2*H*)-one. This same compound is a minor product from the hydrolysis under aqueous acidic conditions; the major product is 2-(1-phthalazinyl)-1(2*H*)-phthalazinone (XXXIII) (*C.M. Atkinson, C.W. Brown* and *J.C.E. Simpson*, J. chem. Soc., 1956, 1081). However, 1,4-dichlorophthalazine is hydrolysed smoothly in dilute aqueous acid to 4-chlorophthalazin-1(2*H*)-one. Nucleophilic substitution reactions of 1-halogeno- and 1,4-dihalogeno-phthalazines have been used extensively for the preparation of the other derivatives. The kinetics of the reaction of 1-chlorophthalazine with ethoxide ion and with piperidine have been studied and it has been shown that 1-chlorophthalazine and 4-chlorocinnoline have approximately the same reactivity but are less reactive than 3-chloroquinazoline (*N.B. Chapman* and *D.Q. Russell-Hill*, *ibid.*, 1956, 1563). Some further examples of the usefulness of the reactions of chlorophthalazines are given in the sections on aminophthalazines and phthalazinones.

(XXXIII)   (XXXIV)

Catalytic hydrogenolysis of a 1-chlorine atom appears to proceed smoothly to give the corresponding phthalazine as does reduction of 1,4-dichlorophthalazine or 1-chloro-4-phenylphthalazine with hydriodic acid and red phosphorus. However, reduction of 1-chloro-4-methylphthalazine with hydriodic acid and red phosphorus is said to give 1-methylisoindole

(XXXIV) and reduction of 1-chlorophthalazine and 1-chloro-4-alkylphthalazines with zinc and hydrochloric acid give 1,3-dihydroisoindoles.

1-Chlorophthalazine is an unstable compound and, perhaps because of this, the melting point quoted ranges from 109° to 121°, but is most commonly given at about 114°. The *picrate* has m.p. 135°. The u.v. spectrum of 1-chlorophthalazine has been studied in isooctane [λ_{max} 307.5 (log ϵ 2.97) and 265 nm (log ϵ 3.70)] and in methanol [λ_{max} 307 (log ϵ 3.02) and 267.5 nm (log ϵ 3.65)] (*G. Favini* and *M. Simonetta, Gazz.*, 1960, **90**, 369). 4-Substituted 1-chlorophthalazines include the 4-*methyl*-, m.p. 133°, and 4-*phenyl*-, m.p. 160°, and 4-*chlorophthalazin*-1(2*H*)-*one*, m.p. 274°, and 4-*chloro-2-methylphthalazin*-1(2*H*)-*one*, m.p. 129°.

1,4-**Dichlorophthalazine**, m.p. 164°, forms a *stable complex* with *bromine*, m.p. 155°, and undergoes methanolysis to afford 1-*chloro-4-methoxyphthalazine*, m.p. 108–109°. 1,4,5,6,7,8-**Hexachlorophthalazine**, 194–195.5°, is obtained from 1,4-dichlorophthalazine by the action of chlorine in the presence of anhydrous aluminium chloride at 200° (*R.D. Chambers et al.*, Tetrahedron Letters, 1970, 57).

1-**Bromophthalazine**, m.p. 175°, and **1,4-dibromophthalazine**, m.p. 160–161°, are obtained in good yield by the action of phosphorus oxybromide in ether on 1-phthalazinone and phosphorus pentabromide on 4-hydroxyphthalazin-1(2*H*)-one, respectively.

1-**Iodo-** and 1,4-**diiodo-phthalazine**, m.p. 78 (dec.) and 161° respectively, are obtained by the action of sodium iodide in acetone on the corresponding chlorophthalazine in the presence of hydriodic acid at room temperature. These iodo compounds yield the corresponding *cyano* compound on treatment with copper(I) cyanide, m.p. 156–157° and 204–205°, respectively (*A. Hirsch* and *D. Orphanos*, Canad. J. Chem., 1965, **43**, 1551).

1,4,5,6,7,8-**Hexafluorophthalazine**, m.p. 91–93°, is obtained by treatment of hexachlorophthalazine with potassium fluoride at 290°. Hydrolysis with aqueous acid or treatment with sodium methoxide in methanol at −15° give 4,5,6,7,8-*pentafluorophthalazin*-1(2*H*)-*one*, m.p. 244–246°, and 1,5,6,7,8-*pentafluoro-4-methoxyphthalazine*, m.p. 67.5–69.5°. Under more vigorous conditions, all six fluorines can be replaced by methoxy groups to give 1,4,5,6,7,8-*hexamethoxyphthalazine*, m.p. 104–106° (*Chambers et al., loc. cit.*).

(e) Aminophthalazines and related compounds

The preparation of a 1-aminophthalazine and related compounds is usually achieved by nucleophilic displacement of a substituent from the 1-position by the appropriate amine. The readily displaced substituent is commonly halogen (especially chlorine), alkyl or aryl esters, thiol or alkyl thiol, methylsulphoxide or methylsulphone groups (*G.B. Barlin* and *W.V. Brown*, J. chem. Soc. C, 1969, 921). The use of 1,4-dichlorophthalazine in this reaction generally gives a 1-amino-4-chlorophthalazine with strongly basic amines such as ammonia or alkylamines unless forcing conditions are used. In contrast, weakly basic amines, such as anilines, have to be used under carefully controlled conditions in order to obtain the mono-substitution product, and the diamines are readily obtained (*R.D. Haworth* and *S. Robinson*, J. chem. Soc., 1948, 777). Useful routes to 1,4-diaminophthalazine are the reactions of hydrazine with either phthalonitrile or 1,4-diphenoxyphthalazine.

Useful routes to 4-aminophthalazinones include the ring expansion of phthalamide derivatives (*W. Koehler, M. Bubner* and *G. Ulbricht*, Ber., 1967, **100**, 1073). 4-Chloro-2-substituted phthalazinones can be converted to the corresponding 4-aminophthalazinones (*L. Rylski et al.*, Acta Polon. Pharm., 1965, **22**, 111).

Reduction of the corresponding nitro compound is the usually employed route to phthalazines carrying an amino substituent on the carbocycle.

1-**Aminophthalazine**, m.p. 212–213° (*C.M. Atkinson, C.W. Brown* and *C.J.E. Simpson*, J. chem. Soc. 1956, 1081) is a cream solid which forms a *hydrochloride*, m.p. 205–206°, *picrate*, m.p. 301° (*H.J. Rodda*, ibid., 1956, 3509), and a quaternary *methiodide*, m.p. 251–252°. The exocyclic nitrogen atom is readily acylated and sulphonated (*Rodda, loc. cit.*).

1,4-**Diaminophthalazine** is a colourless solid, m.p. 254°, which forms a *hydrochloride*, m.p. 230–231°. 1-*Amino-4-chlorophthalazine* has m.p. 202° (*I. Satoda, F. Kusuda* and *K. Mori*, Yakugaku Zasshi, 1962, **82**, 233) and 4-*aminophthalazinone* has m.p. 265–266° (*Koehler, Bubner* and *Ulbricht, loc. cit.*).

1-Hydroxyaminophthalazines are obtained by the action of hydroxylamine on either 1-alkoxy- or 1,4-dialkoxy-phthalazine. The hydroxyaminophthalazines (XXXV) give amido-esters of carbonic acid (XXXVI) with ethyl chloroformate at room temperature and these cyclise to the oxadiazolophthalazines (XXXVII) upon brief warming in DMF solution (Pat., C.A., 1968, **69**, 10456).

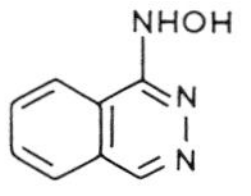

(XXXV) 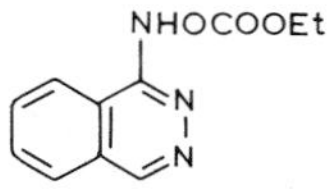   (XXXVI) 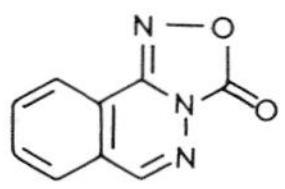   (XXXVII)

Hydrazinophthalazines have been obtained by nucleophilic replacement reactions with 1-substituted phthalazines and hydrazine in reactions similar to those described for aminophthalazines. Monoalkylhydrazines react at the secondary nitrogen atom (*K.T. Potts* and *C. Lovelette*, J. org. Chem., 1969, **34**, 3221). Thus, methylhydrazine and 1-chlorophthalazine afford (XXXVIII), whereas *p*-toluenesulphonylhydrazide gives (XXXIX) (*C.M. Atkinson* and *C.J. Sharpe*, J. chem. Soc., 1959, 3040).

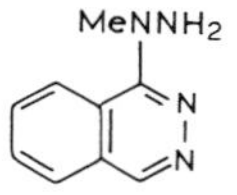

(XXXVIII) 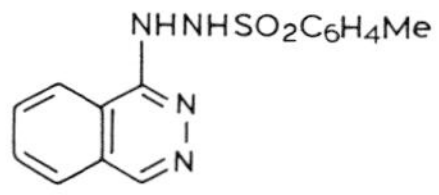   (XXXIX)

1,4-Dihydrazinophthalazines are readily obtained by the action of hydrazine or hydrazine hydrate upon suitably symmetrically or unsymmetrically 1,4-disubstituted phthalazines. Alternatively, 1,4-dihydrazinophthalazine (XL) is readily obtained by the reaction of hydrazine hydrate with phthalonitrile (*G.A. Reynolds, J.A. Van Allan* and *J.F. Tinker*, J. org. Chem., 1959, **24**, 1205).

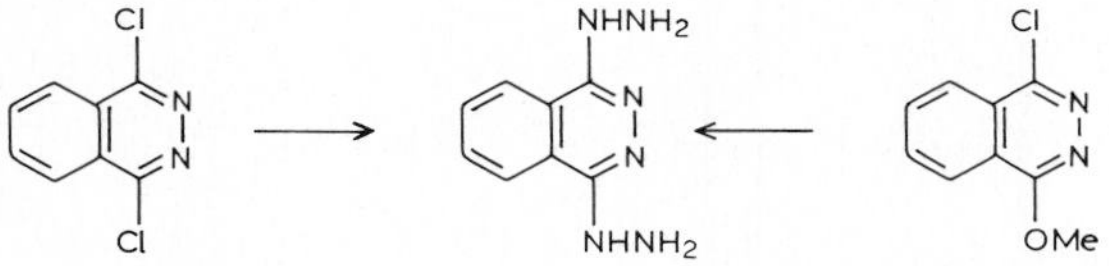

(XL)

4-Hydrazinophthalazinones may be obtained by replacement of a chlorine atom from 4-chlorophthalazinone (*Koehler, Bubner* and *Ulbricht, loc. cit.*) or by the action of hydrazine hydrate upon phthalimide derivatives, for example 3-iminophthalimide.

1-Hydrazinophthalazine and 1,4-dihydrazinophthalazine are clinically useful anti-hypertensive agents, hydralazine and dihydralazine, respectively. They are thought to act as vasodilators (*J. Druey* and *B.H. Ringier*, Helv., 1951, **34**, 195).

1-**Hydrazinophthalazine**, m.p. 172–173°, is a somewhat unstable yellow solid which gives a stable *hydrochloride*, m.p. 273° (dec.), having a u.v.-absorption spectrum with λ_{max} (water) 240 (ϵ 11,100), 260 (ϵ 10,650) and 303 nm (ϵ 5,150). 4-Substituted derivatives include 4-*methyl*, m.p. 310° (dec.), *hydrochloride*, m.p. 285° and 4-*phenyl*, m.p. 135°, *hydrochloride*, m.p. 290–291°, 1,4-**Dihydrazinophthalazine** is an orange solid whose m.p. has been variously quoted between 180° (dec.) and 191–193° (dec.). The *dihydrochloride* has m.p. 257° (dec.) and the *sulphate*, m.p. 233° (dec.). The sulphate in water has λ_{max} 321 nm (ϵ 5910). 4-**Hydrazino-**1(2*H*)-**phthalazinone**, m.p. 269° (dec.) affords a *hydrochloride*, m.p. 267° (dec.).

Aldehydes and ketones form hydrazones with hydrazinophthalazines and similarly 1,4-dihydrazinophthalazine (XLI) forms dihydrazones. The hydrazine group can be incorporated into a pyrazole nucleus using standard reactions (*R. Ruggieri*, C.A., 1964, **61**, 4344).

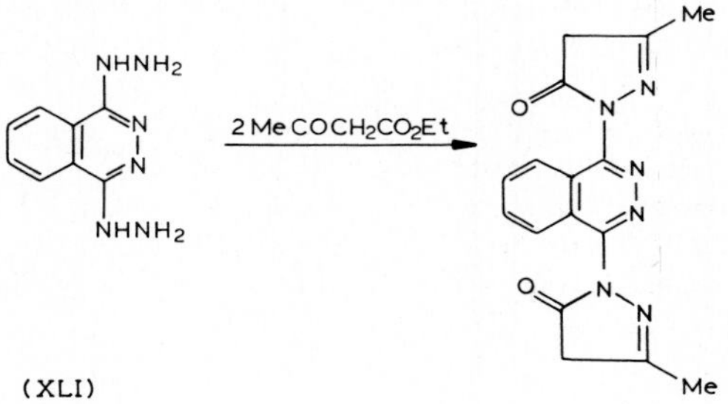

(XLI)

Reactions in which the hydrazine moiety is used in an annulation process usually occur easily so that a range of tri- and tetra-cyclic systems are readily produced. 1-Hydrazinophthalazine with nitrous acid yields *tetrazolo*[4,5-*a*] *phthalazine* (XLIII), m.p. 209–210°, while 1,4-dihydrazinophthalazine gives 6-*azidotetrazolo*[4,5-*a*] *phthalazine* (XLIV), m.p. 195° (*Reynolds, Van Allan* and *Tinker*, loc. cit.).

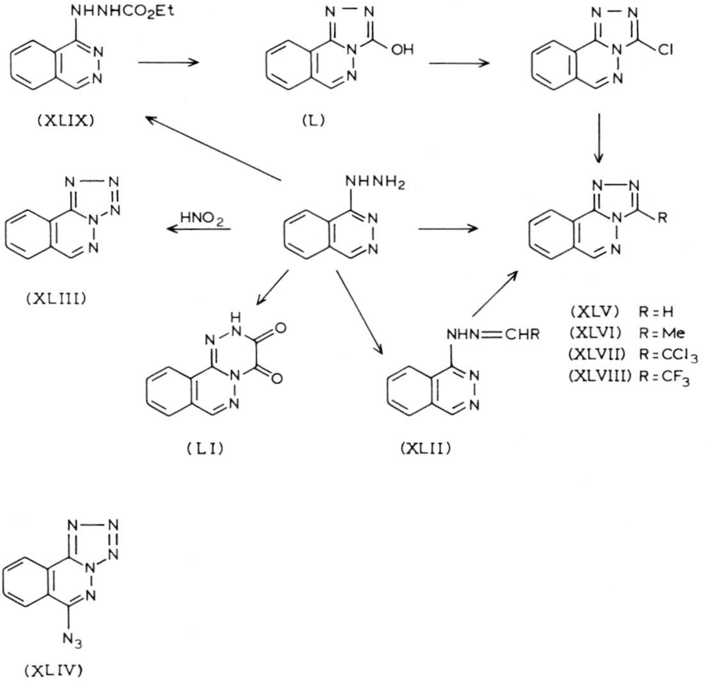

1-Hydrazinophthalazine with carboxylic acids, acid chloride, or esters gives *s*-triazolo[3,4-*a*]phthalazines in high yields (*Druey* and *Ringier*, loc. cit.). The *p*-nitrophenyl esters of carboxylic acids are very useful reagents and the five-membered triazole ring forms particularly easily (*H. Zimmer, J.M. Kokosa* and *K.J. Shah*, J. org. Chem., 1975, **40**, 2901).

1,2,4-**Triazolo**[4,5-*a*]**phthalazine** (XLV), m.p. 190–191°, and its 3-*methyl* derivative (XLVI), m.p. 171–172°, are formed by the action of 85% formic acid and acetic anhydride, respectively. Similarly, the 3-*trichloromethyl* derivative (XLVII) m.p. 252–253°, and the 3-*trifluoromethyl* derivative (XLVIII), m.p. 201–203°, are produced with trichloroacetonitrile and trifluoroacetic anhydride, respectively. The *ethoxycarbonyl* derivative (XLIX), m.p. 198–200°, obtained by the action of ethyl chloroformate on 1-hydrazinophthalazine, affords the 3-*hydroxy* derivative (L), m.p. 275–277°, on boiling with dilute aqueous sodium hydroxide. The reaction of 1-hydrazinophthalazine with an excess of oxalic acid gives (XLV), presumably by decarboxylation of the originally formed 3-carboxylic acid at the temperature of the reaction (160°), but use of the diethyl oxalate furnishes the all six-membered annulated system, 2*H-as*-**triazino**[3,4-*a*]**phthalazin**-3,4-**dione**, m.p. 355° (*Zimmer et al., loc. cit.*).

(f) Phthalazinones, hydroxyphthalazines and oxidophthalazinium betaines

(i) 1(2H)-Phthalazinones

A phthalazine bearing a hydroxyl group in the 5-, 6-, 7-, or 8-position is accurately described as a hydroxyphthalazine but when the oxygen function is in the 1-position the compound is correctly described as a 1(2H)-phthalazinone. Phthalazinone may be obtained from *o*-formylbenzoic acid and hydrazine while 4-substituted phthalazinones are obtained in a similar way from *o*-acylbenzoic acids or phthalides (section *a*, method 2). When a substituted hydrazine is used the corresponding 2-substituted 1(2H)-phthalazinone is obtained (*W.R. Vaughan*, Chem. Rev., 1948, **43**, 447).

Certain 2-substituted derivatives are also available by alkylation of 1(2H)-phthalazinone with diazomethane (*R. Gomper*, Ber., 1960, **93**, 187) or by the action of an alkyl halide or sulphate in the presence of base (*C.M. Atkinson, C.W. Brown* and *J.C.E. Simpson*, J. chem. Soc., 1956, 1081).

Alternative routes to 1(2H)-phthalazinone include the reaction of hydrazine with phthalimidine (*W. Treibs et al.*, Ann., 1951, **574**, 54) or by decarboxylation of the 4-carboxylic acid derivative (*K. Frankel*, Ber., 1900, **33**, 2808). 4-Substituted 1(2H)-phthalazinones are obtained from the action of hydrazine on 4-aryl-2,3-benzoxazin-1-ones (*F.G. Baddar, A.F.M. Fahmy* and *N.F. Aly*, J. chem. Soc. Perkin I, 1973, 2448). *N*-Aminophthalimide in the presence of benzene or its simple derivatives and anhydrous aluminium chloride gives 4-aryl-1(2H)-phthalazinones (*M.F. Ismail, F.A. El-Bassiouny* and *H.A. Younes*, Tetrahedron, 1984, **40**, 2983). Certain 4-substituted 2-alkylphthalazinium hexacyanatoferrate iodides undergo oxidation in the presence of potassium to give 4-substituted 2-alkyl-1(2H)-phthazinones (*T. Ikeda, S. Kanahara* and *K. Aoki*, Yakugaku Zasshi, 1968, **88**, 521).

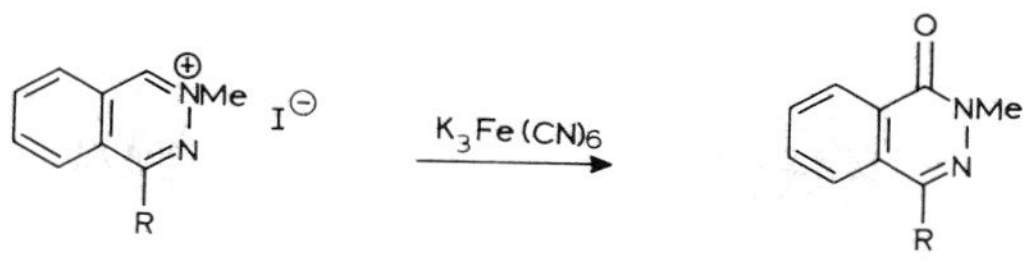

1(2H)-Phthalazinone is potentially capable of existing in either the carbonyl (LI) or hydroxy (LIII) form but studies using a variety of physical techniques have shown the carbonyl form to be the major tautomer. The u.v. spectra of (LI), (LII) and (LIII) have been recorded in water at pH 7.0

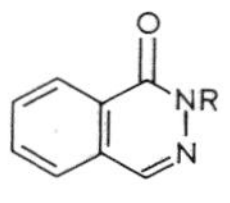

(LI) R=H
(LII) R=Me

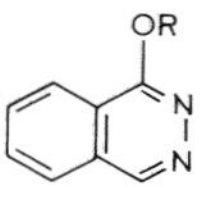

(LIII) R=H
(LIV) R=Me

and as cationic species, and the close similarity between the spectra of (LI) and (LII) has been demonstrated (*A. Albert* and *G.B. Barlin*, J. chem. Soc., 1962, 3129). The i.r. spectrum of (LI) has been studied in solution in carbon tetrachloride and in chloroform, as a nujol mull, and in the solid state (*S.F. Mason*, ibid., 1957, 4874; *K. Mori*, Yakugaku Zasshi, 1962, **82**, 1161). On the basis of i.r.-spectroscopic evidence, it has been suggested that (LI) exists as a hydrogen-bonded dimer in solution and this conclusion has been supported by molecular-weight measurements. Protonation of 2-methylphthalazinone (LII) occurs at N-3 (*D. Cook*,, Canad. J. Chem., 1964, **42**, 2292).

The mass spectra of phthalazinones has been discussed in Section (*b*)*ii* (*vide supra*).

As mentioned earlier, alkylation of (LI) with an alkyl halide or sulphate in the presence of alkali gives the 2-substituted phthalazinone, and acetylation also occurs at N-2 but O-alkylated derivatives may be obtained by use of the silver salt (*G. Wagner* and *D. Heller*, Arch. Pharm., 1966, **299**, 768). Addition of organolithium reagents to the 3,4-double bond in phthalazinones leads to 4-substituted derivatives (*A. Hirsch* and *D.G. Orphanos*, J. heterocyclic Chem., 1966, **3**, 38). However, phenyl Grignard reagents react with (LI) to give 1,4-diphenylphthalazine. See Table 24 for melting points of substituted phthalazinones.

Catalytic reduction of (LI) in the presence of platinum gives 3,4-dihydro-1(2H)-phthalazinone but 4-methyl-8-nitro-1(2H)-phthalazinone in the presence of Raney nickel affords 4-methyl-8-amino-1(2H)-phthalazinone

TABLE 24
1(2H)-PHTHALAZINONES

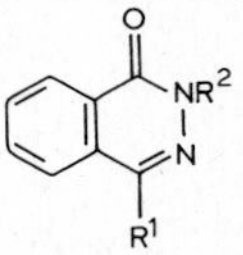

R^1	R^2	m.p. (°C)	R^1	R^2	m.p. (°C)
H	H	183–184 [1]	H	Me	113–114 [2]
Me	H	222–224 [3]	H	Et	59 [4]
Et	H	170–171 [5]	Me	Me	112 [3,6]
CH$_2$Ph	H	196 [7]	Et	Me	78–79 [8]
Ph	H	236 [2]	Et	Et	49–50 [5]
Cl	H	274 [9]			

1-ALKOXY- AND 1-ARYLOXY-PHTHALAZINES

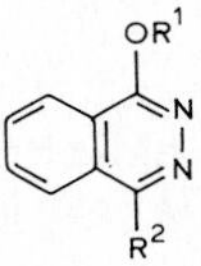

R^1	R^2	m.p. (°C)	R^1	R^2	m.p. (°C)
Me	H	60–61 [10]	Me	Me	53–54 [3]
Et	H	29–31 [10]	Me	Et	56–57 [7]
Ph	H	107 [11]	Et	Me	49 [5]

1 S. Gabriel and A. Neumann, Ber., 1893, **26**, 521.
2 R. von Rothenburg, J. pr. Chem., 1895, **51**, 140.
3 F.M. Rowe and A.T. Peters, J. chem. Soc., 1933, 1331.
4 A.N. Kost, S. Foldeak and K. Grabliauskas, C.A., 1967, **67**, 100090.
5 A. Daube, Ber., 1905, **38**, 206.
6 S. Gabriel and G. Eschenbach, ibid., 1897, **30**, 3022.
7 S. Gabriel and A. Neumann., ibid., 1893, **26**, 705.
8 V. Paul, ibid., 1899, **32**, 2014.
9 R.D. Haworth and S. Robinson, J. chem. Soc., 1948, 777.
10 A. Lieck, Ber., 1905, **38**, 3918.
11 M. Hartmann and J. Druey, U.S.P. 2,484,029, 1949.

(*J. Finkelstein* and *J.A. Romano*, J. med. Chem., 1968, **11**, 398). Lithium aluminium hydride reduces the lactam system to give 1,2-dihydrophthalazine but reduction with dissolving metals in acid gives the ring-contracted product phthalimidine (*Vaughan, loc. cit.*).

1(2H)-Phthalazinones form quaternary salts by reaction at N-3 with alkylating agents, and these salts are significant as intermediates for the formation of important 3-alkyl-1-oxidophthalazinium betaines (see below).

1(2H)-**Phthalazinone**, m.p. 183–184°, has pK_a in water at 20° of 11.99. The m.ps. for other phthalazinones and 1-alkoxyphthalazines are given in Table 24. Nitration of phthalazinone with strong sulphuric acid and potassium nitrate at 90° affords the 5-*nitro* derivative, m.p. 263–265° as the main product, although in poor yield, together with the 8-*nitro* isomer, m.p. 252–253° (*S. Kanaha*, Yakugaku Zasshi, 1964, **84**, 489). Nitration of 4-methylphthalazinone gives mainly 4-*methyl*-8-*nitrophthalazinone*, m.p. 293°, together with the 5-*nitro* isomer. m.p. 273–275°, while a similar reaction with 1-methoxyphthalazine affords mainly 1-*methoxy*-5-*nitrophthalazine*, m.p. 207–210°, with a small amount of 5-*nitro*-1(2H)-*phthalazinone*, m.p. 260°. *N*-Alkyl-nitrophthalazinones are obtained by oxidation of nitrophthalazinium quaternary salts. For, example, 5-nitrophthalazine (obtained by nitration of phthalazine) on treatment with methyl iodide yields yellow 2-*methyl*-5-*nitrophthalazinium iodide*, m.p. 195–202.5° (dec.) and red 3-*methyl*-5-*nitrophthalazinium iodide*, m.p. 170–175° (dec.). Oxidation of the former with potassium hexacyanatoferrate affords 2-*methyl*-5-*nitro*-1(2H)-*phthalazinone*, m.p. 125–126°, while the latter yields 2-*methyl*-8-*nitro*-1(2H)-*phthalazinone*, m.p. 226–227° (*Kanahara, loc. cit.*).

(ii) 3-Alkyl(or aryl)-1(2H)-phthalazinium salts and 3-alkyl(or aryl)-1-oxidophthalazinium betaines

Treatment of phthalazinone (LI) with methyl *p*-toluenesulphonate gives the salt (LV), m.p. 196°, which is converted into the water soluble 3-*methyl*-1-*oxidophthalazinium betaine* (LVI), m.p. 235°, by the action of base (*N. Dennis, A.R. Katritzky* and *M. Ramaiah*, J. chem. Soc. Perkin I, 1975, 1506). The betaine (LVI) undergoes addition with diphenylacetylene across the 2,4-positions to give (LVII). There is no physical evidence for the

(LV)   (LVI)   (LVII)

presence of the tautomer (LVIII) in phthalazinone but unexpectedly (LI) undergoes addition reactions similar to the betaine (LVI). Thus, two molecules of dimethyl acetylenedicarboxylate add to (LI) form (LIX), which then undergoes further reaction to give the product (LX).

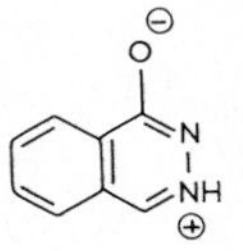

Betaines like (LVI) have been known for some time and earlier were termed pseudo-phthalazinones but their reactions and use in synthesis as illustrated above have been explored only recently. An unexpected route to the corresponding phenyl betaines starts from 2-naphthol-1-sulphonic acid and was uncovered in the extensive investigations of the action of diazonium salts on the naphthalene derivative by F.M. Rowe. The first product of the reaction is the benzenediazosulphonate (LXI) which is converted by aqueous sodium carbonate into the 1-benzeneazo-2-oxo-1,2-dihydronaphthalene-1-sulphonate (LXII). The non-aromatic ring is cleaved by the action of sodium hydroxide and the intermediate cyclises to give a phthalazine derivative (LXIII) which, on hydrolysis yields 1-hydroxy-3-phenyl-3,4-dihydrophthalazinyl-4-acetic acid (LXIV). The reaction is known as the Rowe rearrangement (*F.M. Rowe et al.*, J. chem. Soc., 1948, 1249); yields are good and *W.R. Vaughan et al.* (J. Amer. chem. Soc., 1951, **73**, 2298) have shown that no migration of the aryl group occurs. The phthalazinylacetic acid (LXIV) may be oxidised with hot aqueous permanganate to 4-hydroxy-2-phenyl-1(2*H*)-phthalazinone (LXV) or converted into the betaines (LXVI) and (LXVII). Treatment of (LXIV) with aqueous sulphuric acid or hot alkali gives (LXVI) while the action of cold aqueous dichromate causes decarboxylation and oxidation to yield (LXVII).

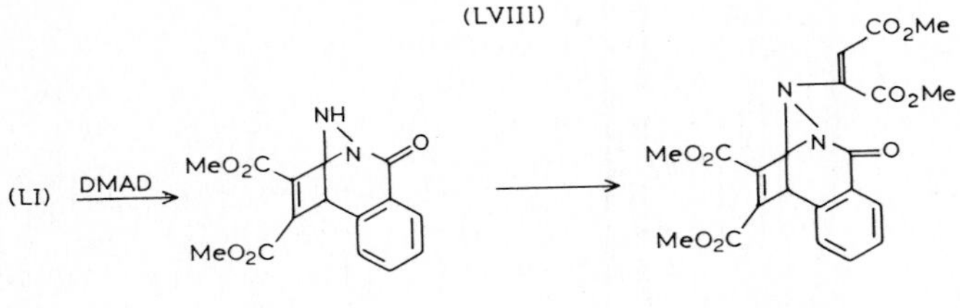

The ^{13}C-n.m.r. spectrum and the mass spectrum of oxidobetaines have been studied (*Y. Takeuchi* and *N. Dennis*, Org. magn. Resonance, 1976, **8**, 21; *N. Dennis, A.R. Katritzky* and *M. Ramaiah*, J. Indian chem. Soc., 1978, **55**, 1235).

(LXI) (LXII) (LXIII) (LXIV) (LXV) (LXVI) (LXVII)

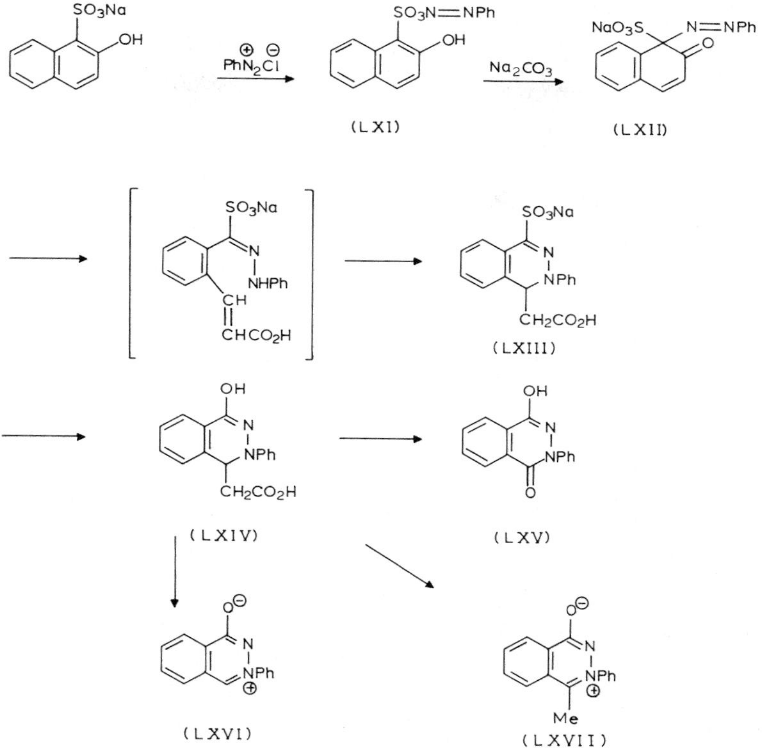

3-**Phenyl**-1-**oxidophthalazinium betaine** (LXVI), m.p. 208 °, is obtained from 1-*hydroxy*-3-*phenyl*-2,3-*dihydrophthalazinyl*-4-*acetic acid* (LXIV), m.p. 264 ° (dec.), and yields a *picrate*, m.p. 226 °. Methylation with dimethyl sulphate or diazomethane yields a pseudo-base (LXVIII) which may be characterised by its product (LXIX) formed by reaction with alkoxide.

(LXVIII) (LXIX)

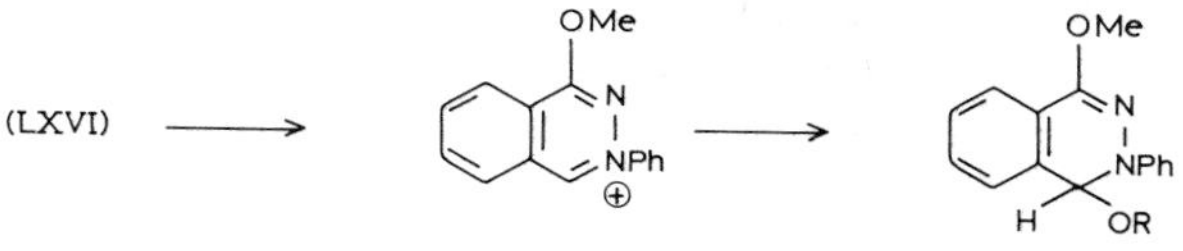

4-**Methyl**-3-**phenyl**-1-**oxidophthalazinium betaine** (LXVII), m.p. 235 °, forms a *picrate*, m.p. 197 °.

4-**Methyl**-3-*p*-**nitrophenyl**-1-**oxidophthalazinium betaine** (LXX) m.p. 251 °, forms a *picrate*, m.p. 208 °, but more interestingly upon attempted methylation gives a 4-*methylene* compound (LXXI), m.p. 134 °, which yields a *perchlorate*, m.p. 199 ° (*F.M. Rowe* and *H.J. Twitchett*, J. chem. Soc., 1936, 1704).

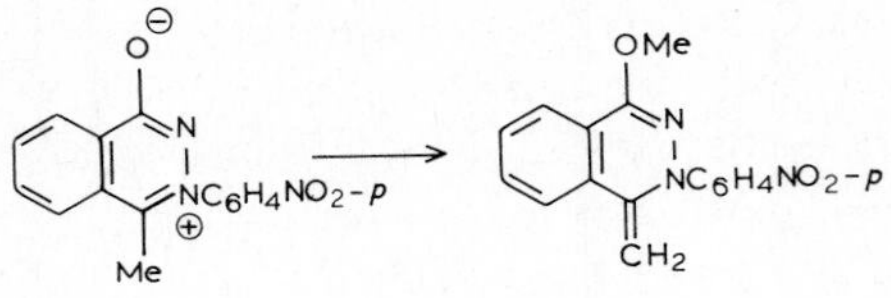

(LXX) → (LXXI)

The 3-phenyl betaine (LXVI) gives the expected adducts with styrene, diphenylacetylene or dimethyl acetylenedicarboxylate in xylene. However, the last with (LXVI) in chloroform gives the unexpected adduct (LXXII). Both the normal adduct and (LXXII) rearrange to (LXXIII) on heating in the absence of a solvent (*N. Dennis, A.R. Katritzky* and *M. Ramaiah*, J. chem. Soc. Perkin I, 1976, 2281). Oxidophthalazinium betaines (LXXIV)

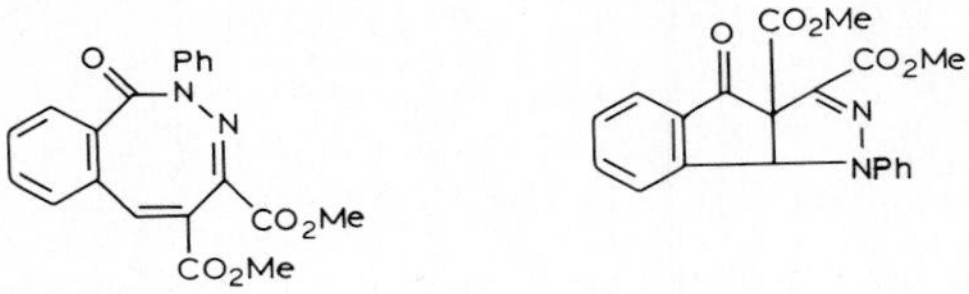

(LXXII)   (LXXIII)

undergo photochemical intramolecular cyclisation to (LXXV) and ring opening to give (LXXVII) or (LXXVIII) in the presence of methanol or ethylamine, respectively (*Y. Maki et al.*, Chem. Letters, 1977, 1005). The diaziridine (LXXV) is converted by water into the 2-methylphthalazinone (LXXVI) (*Y. Maki et al.*, J. chem. Soc. Perkin I, 1979, 1199).

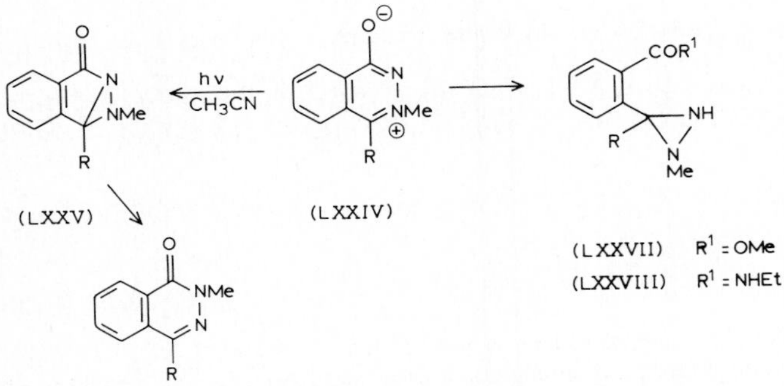

(LXXV) ⇐ hν/CH₃CN (LXXIV) →

(LXXVII)   R¹ = OMe
(LXXVIII) R¹ = NHEt

(LXXVI)

(iii) 4-Hydroxy-1(2H)-phthalazinones

These compounds have been known as 1,4-dihydroxyphthalazines and phthalazin-1,4-diones. They are most readily obtained from diethyl phthalate or phthalic anhydride by treatment wit hydrazine hydrate, or from N-aminophthalimide by the action of either aqueous acid or alkali, or from N-benzoylphthalimide and hydrazine hydrate. Derivatives having substituents in the 2-position may be obtained by use of substituted hydrazines in all but the last of the above reactions or by the methylation of 4-hydroxy-1(2H)-phthalazinone (LXXIX) with diazomethane. On the other hand, the action of methyl iodide on the silver salt of (LXXIX) gives 1-methoxy-1(2H)-phthalazinone. 2,3-Disubstituted phthalazin-1,4-diones

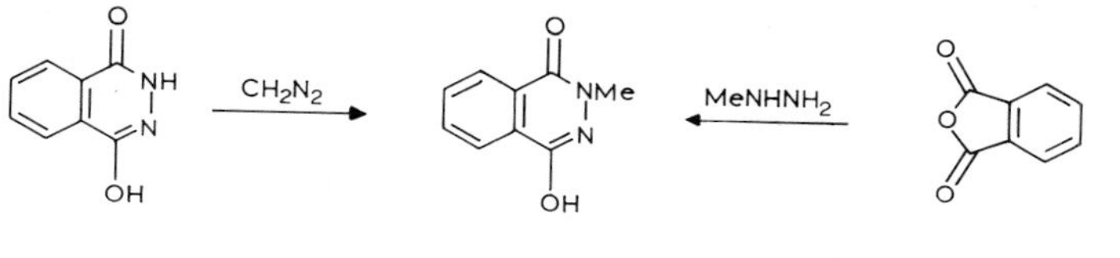

are obtained by the action of phthalic anhydride or phthaloyl chloride on 1,2-disubstituted hydrazines.

Expansion of a five-membered ring to a six-membered ring may be used to form 1-hydroxyphthalazinones. For instance phthalides (LXXX) and (LXXXI) give (LXXIX) upon reaction with hydrazine. 1-Hydroxyphthalazinone can be formed from N-aminophthalimide (H.D.K. Drew and

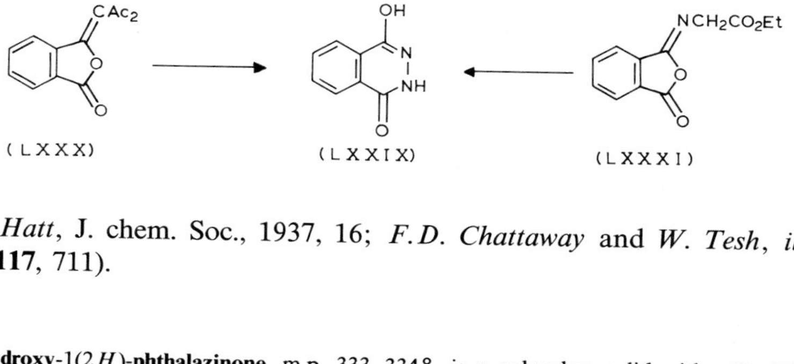

H.H. Hatt, J. chem. Soc., 1937, 16; F.D. Chattaway and W. Tesh, ibid., 1920, **117**, 711).

4-**Hydroxy-1(2H)-phthalazinone**, m.p. 333–334°, is a colourless solid with pK_a 5.95 in aqueous ethanol at 24°. It is amphoteric and forms stable salts with ammonia and hydrazine or aryl sulphonic acids. The u.v. spectrum in ethanol shows λ_{max} 268 (ϵ 3020) and 296 nm (ϵ 5370) and the i.r. spectrum (KBr disc) shows ν_{max} 3130, m (NH) and 1655 cm^{-1}, s (CO). Comparison of these and other spectral data with methyl derivatives in which the tautomeric form is fixed indicate that (LXXIX) is the major tautomer (J.A. Elvidge and A.P. Redman, J. chem. Soc., 1960, 1710). The ionisation constants have been measured by spectroscopic methods (*I.E. Kalinichenko*, *A.T. Pilipenko* and *V.A. Barovskii*, Zhur. obshchei Khim., 1978, **48**, 334) and the e.s.r. spectrum of the radical formed by oxidation of (LXXIX) with cerium(II) sulphate in sulphuric acid has been studied (*D.M. Hotton*, *P.M. Hoyle* and *D. Murphy*, Tetrahedron Letters, 1979, 2821).

The oxidation of (LXXIX) is considered under phthalazin-1,4-dione (see section *h*). The heterocyclic ring in hydroxyphthalazinone is moderately resistant to catalytic hydrogenation and this allows unsaturated substituents on the benzene moiety (*e.g.* nitro groups) to be reduced without reduction of the ring system (see luminol, p. 140). Lithium aluminium hydride reduces the carbonyl groups when one or more *N*-methyl groups are present. Thus, 2-methyl-4-hydroxy-1(2*H*)-phthalazinone and 2,3-dimethylphthalazin-1,4-dione give 1-hydroxy-3-methyl-3,4-dihydrophthalazine and 2,3-dimethyl-1,2,3,4-tetrahydrophthalazine, respectively.

All the methyl, (LXXXII) and (LXXXIII), and dimethyl derivatives, (LXXXIV). (LXXXVIII) and (LXXXIX), and two monoacetyl, (LXXXV)

(LXXXII) $R^1 = H$, $R^2 = Me$
(LXXXIII) $R^1 = Me$, $R^2 = H$
(LXXXIV) $R^1 = R^2 = Me$
(LXXXV) $R^1 = H$, $R^2 = Ac$
(LXXXVI) $R^1 = Ac$, $R^2 = H$
(LXXXVII) $R^1 = R^2 = Ac$

(LXXXVIII) $R^1 = R^2 = Me$
(XC) $R^1 = Me$, $R^2 = Ac$

(LXXXIX)

and (LXXXVI), and one diacetyl (LXXXVII) derivative are known. The physical data for these compounds are given in Table 25.

The earlier confusion over the structure of the acetyl compounds has been resolved by studies of their p.m.r. spectra. *ortho*-Effects are seen in the mass spectra of phthalazines having a methoxyl (or other) substituent in the 1- or 4-position and this produces an enhanced $[M-1]^+$ ion (*J.H. Bowie et al.*, Austral. J. Chem., 1967, **20**, 2677).

Elvidge and *Redman* (*loc. cit.*) have shown that the action of morpholine on (LXXXVIII) is to give *N*-methylmorpholine and (LXXIX). Similarly, aniline is methylated by the action of (LXXXIII). On the other hand, 1,4-diphenoxyphthalazine and morpholine gives 1,4-dimorpholinophthalazine.

4-Hydroxy-1(2*H*)-phthalazinones (LXXIX) add to activated double bonds (see diagrams below). The reaction is carried out by heating the

TABLE 25

METHYL AND ACETYL DERIVATIVES OF 4-HYDROXY-1(2H)-PHTHALAZINONE

Compound	m.p. (°C)	Compound	m.p. (°C)
(LXXXII)	188 [1]	(LXXXV)	172–173 [2,4]
(LXXXIII)	238 [1,2]		200 [5]
(LXXXIV)	93 [3]	(LXXXVI)	175–176 [2,4]
(LXXXVIII)	93 [3]		191 [5]
(LXXXIX)	175–176 [2,3]	(LXXXVII)	133 [5]
		(XC)	137 [1]

1 F.M. Rowe and A.T. Peters, J. chem. Soc., 1933, 1331.
2 H.D.K. Drew, H.H. Hatt and F.A. Hobart, ibid., 1937, 33.
3 J.A. Elvidge and A.P. Redman, ibid., 1960, 1710.
4 P.G. Parsons and H.J. Rodda, Austral. J. Chem., 1964, 17, 491.
5 A. Le Berre, J. Godin and R. Garreau, Compt. rend., 1967, 265, 570; C.A., 1968, 68, 68948.

reactants in refluxing ethanol or acetic acid solution and yields a 2-substituted-4-hydroxy-1(2H)-phthalazinone (J. Godin and A. Le Berre, Bull. Soc. chim. Fr., 1968, 4222). Substitution at the same position can be achieved by Mannich alkylation (H. Hellman and I. Loschmann, Ber., 1956, 89, 594) and by use of phthalazine-1,4-dione (vide infra).

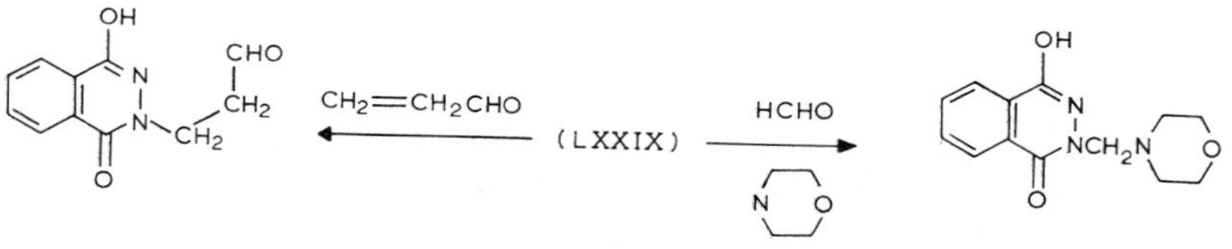

4-Mercapto-1(2H)-phthalazinethione, m.p. 262–264°, is obtained by the action of phosphorus pentasulphide on (LXXIX) and the same compound is obtained from 1,4-dichlorophthalazine and sodium hydrogen sulphide (H.L. Yale, J. Amer. chem. Soc., 1953, 75, 675) (see section g).

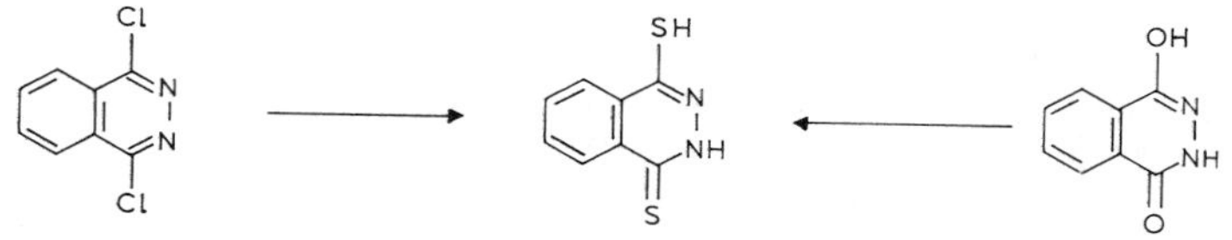

(iv) Chemiluminescence

5-Amino-4-hydroxy-1(2H)-phthalazinone (XCI) (luminol), m.p. 332–333°, is prepared by reduction of the corresponding nitro compound (C.T. Redemann and C.E. Redemann, Org.

Synth., 1949, **29**, 8). Luminol in the presence of acetyl chloride under various reaction conditions affords the 5-*acetamido* derivative (XCII), m.p. 315° (dec.) or the *N,O-diacetyl* derivative (XCIII), m.p. 251° (dec.), and the *N,N'-diacetyl* compound (XCIV), m.p. 260°

(XCI) $R^1 = R^2 = R^3 = H$
(XCII) $R^1 = Ac, R^2 = R^3 = H$
(XCIII) $R^1 = R^2 = Ac, R^3 = H$
(XCIV) $R^1 = R^3 = Ac, R^2 = H$

(XCV)

(dec.). In the presence of acetic anhydride the *tricyclic* compound (XCV), m.p. 195°, is formed (*Y. Omote et al.*, Bull. chem. Soc. Japan, 1966, **39**, 932; 1967, **40**, 899).

The action of sodium hypochlorite or potassium hexacyanatoferrate(III) (*H.O. Albrecht*, Z. physik. Chem., 1928, **135**, 321) on luminol produces chemiluminescence; hence the trivial name of the compound. The glow is enhanced in the presence of hydrogen peroxide and is then visible even at concentrations as low as 10^{-8} M. This is an example of the 'aqueous system' for the production of chemiluminescence from cyclic hydrazides. An 'aprotic system' in which the hydrazide in an aprotic solvent is treated with oxygen and a strong base also produces chemiluminescence. The increase in luminescence in the presence of certain materials (*e.g.* blood) may be used as the basis of sensitive test systems.

The detailed chemistry of the reactions which produce light have been the subject of extended and detailed investigations, and is still the subject of mechanistic investigations and speculation (*F. McCapra*, Pure appl. Chem., 1970, **24**, 611; *E.H. White* and *D.F. Roswell*, Acc. chem. Res., 1970, **3**, 54). The presence of both hydrogen atoms in the heterocyclic ring is essential for chemiluminescence. Easily oxidised compounds decrease the intensity of the chemiluminescence and, significantly, the addition of cyclopentadiene prevents light emission. The only organic product from the chemiluminescence reaction of luminol is 3-aminophthalate. Energy considerations indicate that chemiluminescence is favoured by exothermic processes which lead to small product molecules by a concerted reaction and where the product molecules have few bonds capable of absorbing vibrational energy (*M.M. Rauhut*, Acc. chem. Res., 1969, **2**, 80). The aminophthalate anion has been identified as the light emitting species formed from luminol (*E.H. White* and *M.M. Bursey*, J. Amer. chem. Soc., 1964, **86**, 941).

Recent ideas about the chemiluminescent reactions of luminol have involved the diazaquinone (XCVI) as an intermediate, though the specific role assigned to this intermediate has varied (*D.B. Paul*, Austral. J. Chem.,

1984, **37**, 1001). The pH dependence of the chemiluminescence from (XCVI) and hydroperoxide has been studied (*T.E. Eriksen, J. Lind* and *G. Merenyi*, J. chem. Soc. Faraday I, 1981, **77**, 2137). This investigation indicates that attack of (XCVI) by a hydroperoxide anion yields (XCVII) which then rearranges to the *endo*-peroxide (XCVIII). The latter then undergoes a two-step decomposition through the bicyclic peroxide (XCIX) to the excited phthalate anion. A related mechanism but involving an electron transfer process involving a diradical cation has been suggested (*F. McCapra* and *P.D. Leeson*, Chem. Comm., 1979, 114).

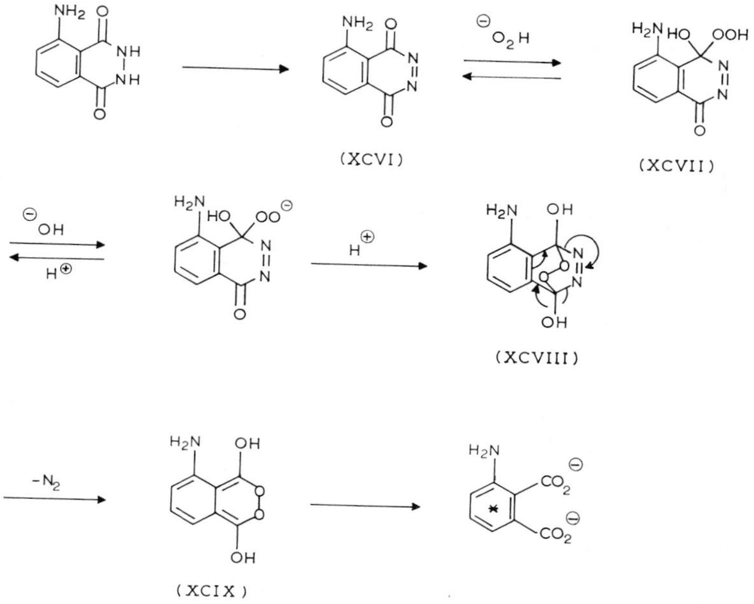

(XCVI)

(XCVII)

(XCVIII)

(XCIX)

(g) Phthalazinethiones and mercaptophthalazines

1(2H)-Phthalazinethiones are usually obtained from the corresponding phthalazinone by treatment with phosphorus pentasulphide in dry pyridine (*A. Albert* and *G. Barlin*, J. chem. Soc., 1962, 3129) or toluene (*A. Mustafa, A.H. Harhash* and *A.A.S. Saleh*, J. Amer. chem. Soc., 1960, **82**, 2735). The reaction is successful in the absence of a 4-substituent or in the presence of 4-alkyl or 4-aryl substituents. An alternative route is from the

1-chlorophthalazine by the action of potassium hydrogen sulphide or thiourea followed by decomposition of the thiouronium salt with alkali. This method has been used in the presence of a 4-substituent such as phenyl or chlorine (*E. Hayashi* and *E. Oishi*, Yakugaku Zasshi, 1966, **86**, 576) and hydrazine.

The action of phosphorus pentasulphide on phthalazinones is versatile and is used to obtain 4-mercapto-1(2*H*)-phthalazinethione from 4-hydroxy-1(2*H*)-phthalazinone (*H.L. Yale*, J. Amer. chem. Soc., 1953, **75**, 675). Also the action of P_2S_5 on 2-substituted-1(2*H*)-phthalazinones yields the corresponding thione (*D.J. Drain* and *D.E. Seymour*, J. chem. Soc., 1955, 852).

Alkylation of phthalazinethiones occurs on sulphur. Thus, the action of methyl iodide or dimethyl sulphate on phthalazinethione and 4-mercaptophthalazinethione is to yield 1-methylthiophthalazine and 1,4-bis(methylthio)phthalazine, respectively. (*K. Asano* and *S. Asai*, Yakugaku Zasshi, 1958, **78**, 450). The arylthio derivatives are obtained readily from a 1-chlorophthalazine and an aryl or heteroaryl thiol in the presence of a base (*N.B. Bednyagina* and *I.Y. Postovski*, J. gen. Chem. U.S.S.R., 1956, **26**, 2549).

(2*H*)-**Phthalazinethione**, m.p. 169–170°, is a yellow solid which exists mainly as the thione tautomer (*Albert* and *Barlin*, *loc. cit.*). The u.v. spectrum of 1(2*H*)-phthalazinethione and that of 2-*methyl*-1(2*H*)-*phthalazinethione*, m.p. 128–129°, in neutral and acidic medium are almost identical. 1(2*H*)-Phthalazinethione is soluble in aqueous alkali and the u.v. spectrum of the cation has been reported. 4-*Chloro*-, m.p. 222–224°, and 4-*mercapto*-1(2*H*)-*phthalazinethione*, m.p. 262–264°, are useful derivatives since the 4-substituent in the former is easily replaced and in the latter readily alkylated. 1-*Methylthio*-, m.p. 75–77°, and *bis*(*methylthio*)-*phthalazine*, m.p. 163°, are examples of alkylation products. The oxidation of the former to the corresponding 1-*methylsulphinyl*-, m.p. 105°, and 1-*methylsulphonyl-phthalazine*, m.p. 156°, has been studied (*G.B. Barlin* and *W.V. Brown*, J. chem. Soc. B, 1968, 1435). The nucleophilic displacement of the methylthio, methylsulphinyl and methylsulphonyl groups have been investigated in some detail (*idem*, J. chem. Soc. C, 1969, 921; *E. Hayashi et al.*, Yakugaku Zasshi, 1967, **87**, 687; *E. Oishi, ibid.*, 1969, **89**, 959).

(h) Phthalazine quinones

Potentially there are two types of quinone: one where the carbonyl groups are in the heterocyclic ring, *i.e.* the parent phthalazine-1,4-dione (C), and the group of quinones having the carbonyl groups in the carbocyclic system. The parent systems of this latter type are (CI), (CII) and (CIII). None of these last three are known.

Phthalazine-1,4-diones are reactive intermediates formed in the reaction mixture by the oxidation of 4-hydroxy-1(2*H*)-phthalazinones with reagents

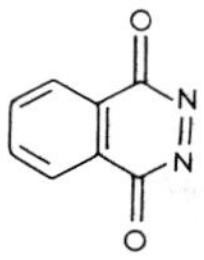

(C) 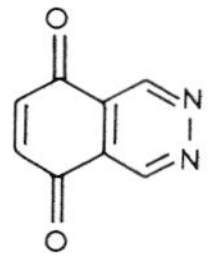   (CI) 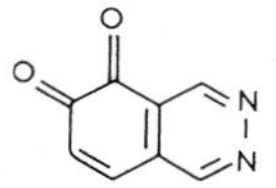   (CII)

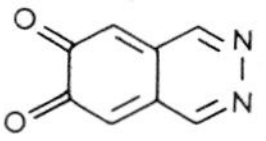

(CIII)

such as *tert*-butyl hypochlorite or lead tetracetate (*T. J. Kealy*, J. Amer. chem. Soc., 1962, **84**, 966; *R. A. Clement*, J. org. Chem., 1962, **27**, 1115). The involvement of the 5-amino derivative of (C) in the process leading to the chemiluminescence of 5-amino-4-hydroxy-1(2*H*)-phthalazinone has been mentioned in a previous section. The one electron reduction potential of this phthalazine-1,4-dione has been determined (*G. Merenyi, J. Lind* and *T. E. Eriksen*, J. phys. Chem., 1984, **88**, 2320).

Phthalazine-1,4-dione (C) is an important synthetic intermediate since it is a powerful dienophile that reacts with a variety of dienes to give [4 + 2] adducts which are difficulty accessible by other routes. For example, the sodium salt of 4-hydroxyphthalazinone on treatment with *t*-butylhypochlorite and *N*-methylpyridone at low temperature gives the adduct (CIV) (*V. V. Kane, H. Werblood* and *S. D. Levine*, J. heterocyclic Chem., 1976, **13**, 673). The addition of the quinone imine (CV) or 1,3-cyclohexadiene to (C) affords (CVI) and (CVII) respectively (*M. F. Brana* and *J. L. Soto*, An. Quim., 1974, **70**, 970; C.A., 1975, **83**, 178966 w).

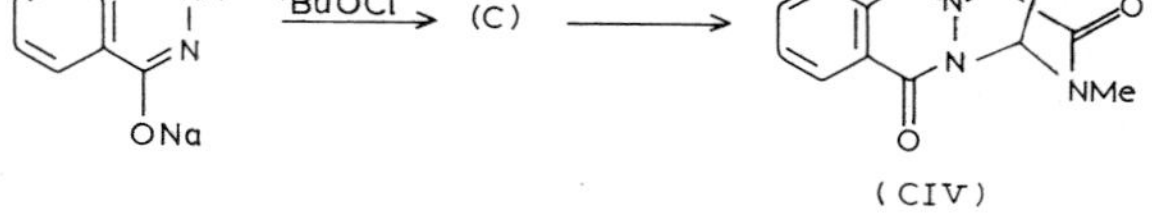

(CIV)

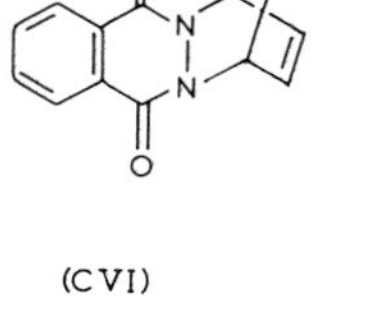

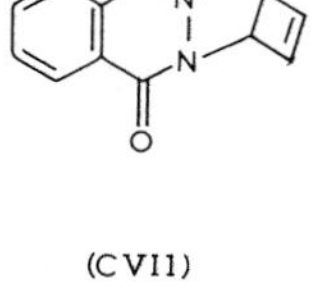

(CV)   (CVI)   (CVII)

Enamines react at one of the diazaquinone nitrogen atoms to give 2-substituted 4-hydroxy-1(2*H*)-phthalazinones. For instance, the enamine (CVIII) yields (CIX) (*H. Warnhoff* and *K. Wald*, Ber., 1977, **110**, 1716). Styrenes undergo successively a [4 + 2] cycloaddition and electrophilic attack of the intermediate on another molecule of the quinone. Thus, α-methylstyrene gives the product (CX) (*C. Seoane* and *J.L. Soto*, An. Quim., 1977, **73**, 1035).

(CVIII) (CIX)

(CX)

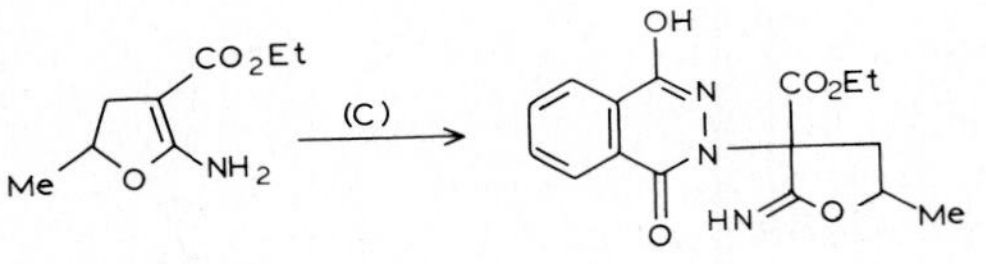

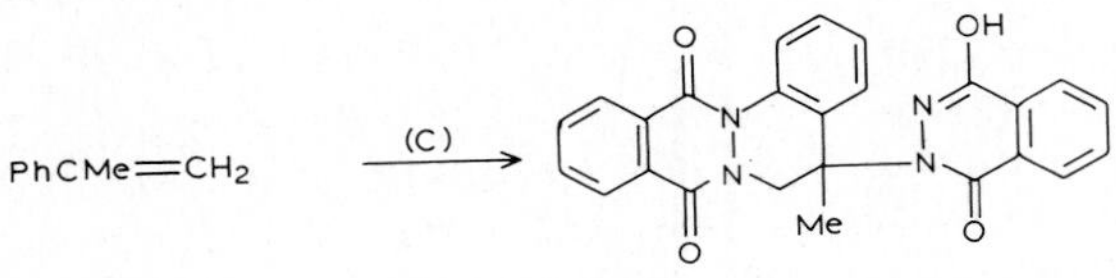

(*i*) *Phthalazinecarboxylic acids and their derivatives*

Phthalazine-1-carboxylic acid (CXII) has been obtained as its hydrobromide by hydrolysis of the Reissert compound (CXI) (*F.D. Popp*, *J.M. Wefer* and *C.W. Khinowski*, J. heterocyclic Chem., 1968, **5**, 879).

(CXI) (CXII)

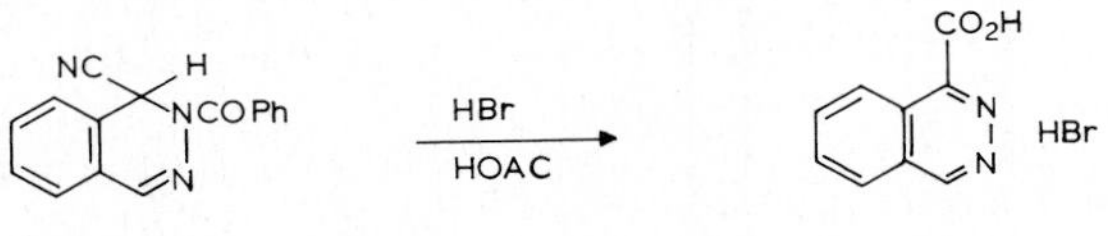

Diethyl phthalazine-1,4-dicarboxylate (CXIV) has been obtained (*J. Sauer* and *G. Heinrichs*, Tetrahedron Letters, 1966, 4979) by a one-pot process involving a Diels–Alder reaction of diethyl tetrazine-1,4-dicarboxylate and benzyne followed by extrusion of dinitrogen from the adduct (CXIII).

(CXIII) (CXIV)

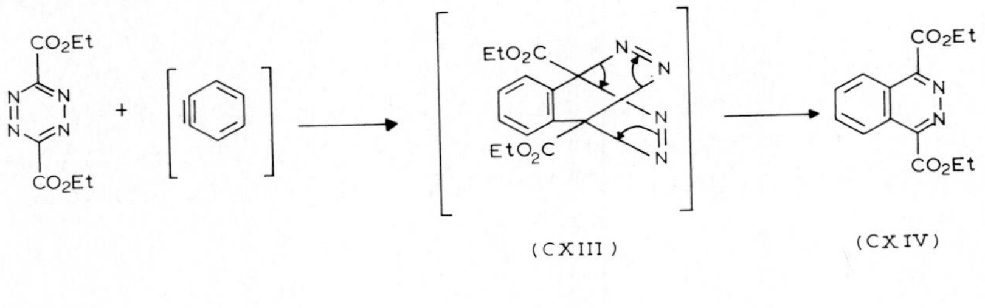

Although 1,4-dicyanophthalazine is known (*A. Hirsch* and *D.G. Orphanos*, Canad. J. Chem., 1965, **43**, 1551) (see section *d*), the corresponding diamide and diacid do not appear to have been reported. Alkaline hydrolysis of 1,4-dicyanophthalazine gave 4-cyanophthalazinone (CX) and then phthalazinone-4-carboxylic acid (CXVI).

(CXV)   (CXVI)

Phthalazine-1-carboxylic acid (CXII) forms a *hydrobromide*, m.p. 198–200°.

1(2*H*)-**Phthalazinone-4-carboxylic acid** (CXVI), m.p. 230–232° (dec.), is most conveniently obtained from phthalonic acid and hydrazine (*I. Satoda, N. Yoshida* and *K. Mori*, Yakugaku Zasshi, 1956, **28**, 613) and 2-*methyl*-, m.p. 236° (dec.), and 2-*phenyl-phthalazinone-4-carboxylic acid*, m.p. 219°, are obtained similarly using appropriately substituted hydrazines (*T.C. Bruice* and *F.M. Ridhard*, J. org. Chem., 1958, **23**, 145). The carboxylic acid function in (CXVI) may be modified by reaction with thionyl chloride to give the acid chloride. The esters can be obtained from the acid (CXVI) and an alcohol in the presence of strong mineral acid. Thus, the *methyl* and *ethyl esters*, m.ps. 208–209° and 170–171°, respectively, are obtained with the corresponding alcohols and sulphuric acid (*V.R. Brasyunas* and *K. Grabliauskas*, C.A., 1961, **55**, 4517). The *amide* and *hydrazide*, m.ps. 318° and 235–236° (dec.), respectively, are readily obtained from the esters. Nitration of (CXVI) with sulphuric acid and potassium nitrate mixture in the cold gives 5-*nitro*, m.p. 265–266° (dec.), and 7-*nitro-1(2H)-phthalazinone-4-carboxylic acid*, m.p. 252–253° (dec.). At higher temperature the 7-nitro derivative and *4,5-dinitro-1(2H)-phthalazinone*, m.p. 233–234°, are obtained (*S. Kanahara*, Yakugaku Zasshi, 1964, **84**, 483). Alkyl esters on (CXVI) on treatment with phosphorus pentasulphide yield the corresponding phthalazinethione esters which undergo hydrolysis in acid to give the thione acids.

1(2*H*)-**Phthalazinethione-4-carboxylic acid**, m.p. 212° (dec.) (*V. Brasyunas* and *K. Grabliauskas*, C.A., 1967, **67**, 100089) is obtained by hydrolysis of *ethyl phthalazinethione-1-carboxylate*, m.p. 176°. The action of methyl iodide on the ester is to provide *ethyl 1-methylthiophthalazine-4-carboxylate*, m.p. 127–128°, from a reaction in acetone in the presence of base or 1-*methylthiophthalazine-4-carboxylic acid*, m.p. 106–107°, from ethanolic sodium hydroxide reaction medium. On the other hand, 2-**methylphthalazinethione**-4-**carboxylic acid**, m.p. 209–219°, is obtained by the action of phosphorus pentasulphide on 2-methylphthalazinone-4-carboxylic acid (*A.N. Kost* and *K. Grabliauskas*, ibid., 1969, **70**, 4001).

(j) Reduced phthalazines

(i) 1,2-Dihydrophthalazines

The parent compound (CXVII), 3,4-dihydro-1(2*H*)-phthalazinones (CXVIII) and their *N*-substituted derivatives are usually obtained by partial reduction of the aromatic compounds. More highly substituted derivatives are obtained by addition reactions with organometallic reagents or from the Reissert reaction.

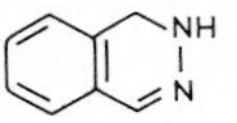

(CXVII) 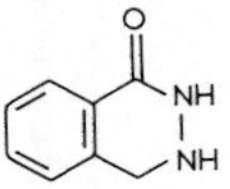 (CXVIII)

Reduction of phthalazine or phthalazinone with lithium aluminium hydride yields 1,2-dihydrophthalazine (*Y.S. Shabarov, N.I. Vasilev* and *R.Y. Levina*, Zhur. obshchei Khim., 1961, **31**, 2478; *B.K. Diep* and *B. Cauvin*, Compt. rend., 1966, **262C**, 1010). Unexpectedly, 4-methyl-1(2*H*)-phthalazinone on reduction with lithium aluminium hydride was found to yield a mixture of 4-methyl-1,2-dihydrophthalazine (57%) and 1-methyl-1,2,3,4-tetrahydrophthalazine (41%) (*Y.S. Shabarov et al.*, Zhur. obshchei Khim., 1963, **33**, 1206) although the same group of workers had found only the dihydro compound upon reduction of 4-phenylphthalazinone under similar conditions. Electrochemical reduction of 2- or 4-substituted 1(2*H*)-phthalazinones causes reduction of the 3,4-double bond (*H. Lund*, Tetrahedron Letters, 1965, 3973). Selective reduction of 6,7-disubstituted-2-substituted phthalazinones has been achieved using sodium borohydride in the presence of aluminium chloride in diglyme. For instance, the sulphonamide (CXIX) is reduced in this way to the dihydrophthalazinone (CXX). Both (CXIX) and (CXX) showed interesting diuretic and peripheral vasodilating effects (*S. Cherkez, J. Herzig* and *H. Yellin*, J. med. Chem., 1986, **29**, 947).

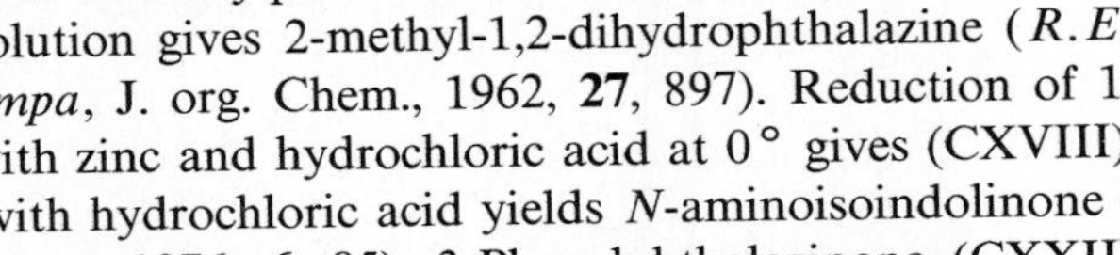

(CXIX) $\xrightarrow{\text{NaBH}_4, \text{AlCl}_3}$ (CXX)

Reduction of 2-methylphthalazinium iodide with sodium borohydride in aqueous solution gives 2-methyl-1,2-dihydrophthalazine (*R.E. Smith* and *E.D. Otrempa*, J. org. Chem., 1962, **27**, 897). Reduction of 1(2*H*)-phthalazinone with zinc and hydrochloric acid at 0° gives (CXVIII) and this on refluxing with hydrochloric acid yields *N*-aminoisoindolinone (*E. Bellasio*, Synth. Comm., 1976, **6**, 85). 3-Phenylphthalazinone (CXXII) (*Diep* and *Cauvin, loc. cit.*) is obtained by reduction of 4-hydroxy-2-phenyl-1(2*H*)-phthalazinone (CXXI). Phthalazine in an isopropanol solution undergoes photoreduction to 1,2-dihydrophthalazine and the dimer (CXXIII) upon

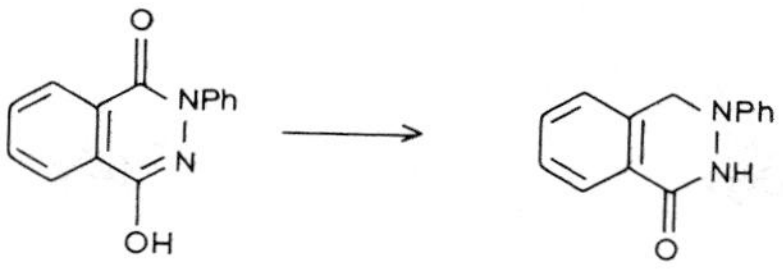

(CXXI) → (CXXII)

irradiation with ultraviolet light (*R. Sano* and *H. Inoue* Chem. Letters 1984, 1901).

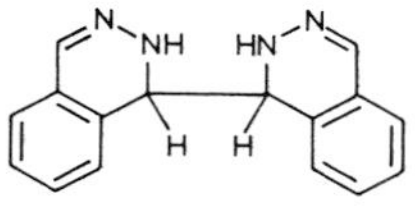

(CXXIII)

2,4-Diphenyl-1,2-dihydrophthalazine (CXXV) is obtained by cyclisation of the substituted hydrazone (CXXIV) in polyphosphoric acid (*L. Baiocchi* and *M. Giannangeli*, J. heterocyclic Chem., 1983, **20**, 225).

PhN N:CClPh
|
CH₂Ph

(CXXIV) —PPA→ (CXXV)

The carbonyl group of 2-substituted 1(2H)-phthalazinones is susceptible to nucleophilic attack by a Grignard reagent. Thus (CXXVI) yields the dihydrophthalazine (CXXVII). However, certain 4-substituents may cause

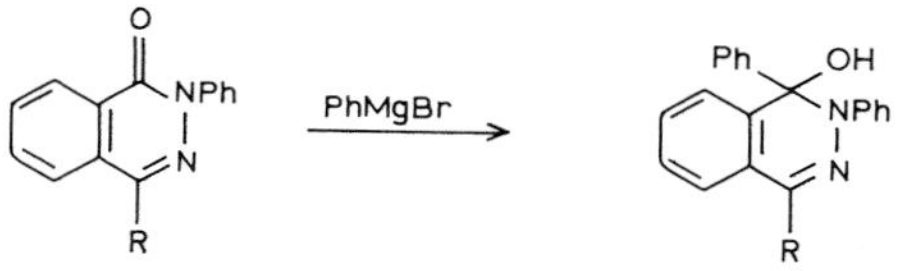

(CXXVI) R = H
(CXXVIII) R = OH

PhMgBr →

(CXXVII) R = H
(CXXIX) R = Ph

further reactions to occur. Thus, the 4-hydroxy derivative (CXXVIII) gives the triphenyldihydrophthalazine (CXXIX) (*Diep* and *Cauvin*, Compt. rend., 1966, **262C**, 1010). The product from the reaction of methylmagnesium iodide upon the 4-phenyl derivative (CXXX) is the *gem*-dimethyl dihydrophthalazine (CXXXI), although phenylmagnesium bromide with (CXXX) gives the expected 1-hydroxy-1,2,4-triphenyl-1,2-dihydrophthalazine. The abnormal behaviour in the reaction of the methyl Grignard reagent, which is also shown by other alkyl and arylalkyl Grignard reagents, has been explained by the intermediacy of ring-opened organometallic species (*A. Mustafa, A.H. Harhash* and *A.A.S. Saleh*, J. Amer. chem. Soc., 1960, **82**, 2735).

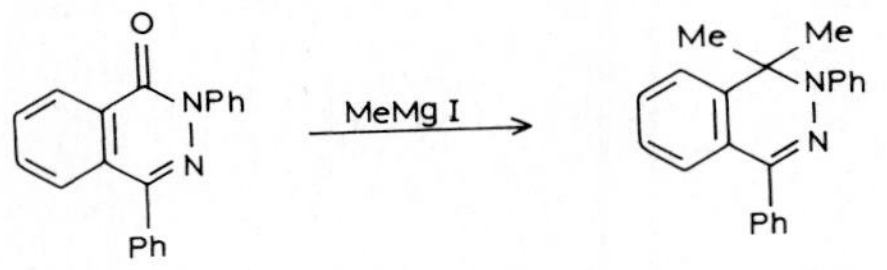

(CXXX) → MeMgI → (CXXXI)

In contrast to the reactions of Grignard reagents, vinyllithium adds to the carbon–nitrogen bond in both phthalazine and phthalazinone giving (CXXXII) from the latter (*A. Hirsch* and *D.G. Orphanos*, J. heterocyclic Chem., 1966, **3**, 38). Addition to the 1,2-bond of phthalazine also occurs in the Reissert reaction (*F.D. Popp, J.M. Weger* and *C.W. Klinowski*, *ibid.*, 1968, **5**, 879), for instance to give (CXXXIII), and in reactions with dimethylketene to produce the tricyclic (CXXIV) (*M.A. Shah* and *G.A. Taylor*, J. chem. Soc. C, 1970, 1651).

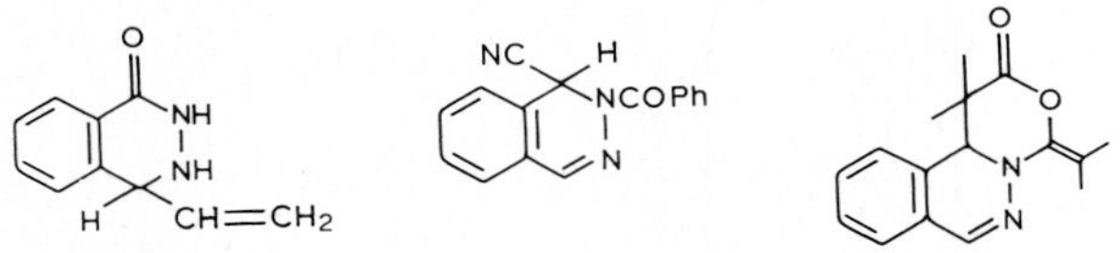

(CXXXII)   (CXXXIII)   (CXXXIV)

1,2-Dihydrophthalazine and 1-substituted 1,2-dihydrophthalazines are easily oxidised to the corresponding phthalazines. 2-Substituted 1,2-dihydrophthalazines give 2-substituted phthalazinones. The Reissert compounds such as (CXXXIII) are important intermediates in the formation of phthalazine-1-carboxylic acids (*F.D. Popp*, Heterocycles, 1980, **14**, 1033). In 1,2-dihydrophthalazines and 3,4-dihydrophthalazinones the nitrogen atom which is in the reduced part of the molecule is both the more basic and more nucleophilic nitrogen atom. Compounds such as (CXVIII) react readily with reagents having suitably disposed alkylating and acylating functions to give tricyclic compounds having two bridgehead nitrogen atoms produced by acylation at the amine nitrogen and alkylation at the

amide nitrogen atom (*E. Bellasio* and *E. Testa*, C.A., 1970, **73**, 25836).
≠ 4,4-Diphenyl-3,4-dihydrophthalazinone (CXXXVI), obtained from Crystal Violet lactone (CXXXV), is oxidised by lead tetraacetate at low temperature in the presence of triethylamine to 4,4-diphenylphthalazin-1(4*H*)-one (CXXXVII) (*M. Kuzuya, F. Miyake* and *T. Okuda*, J. chem. Soc. Perkin II, 1984, 1465). The compound is stable only below $-60°$ and at higher temperature ($> -50°$) in methylene dichloride solution a re-

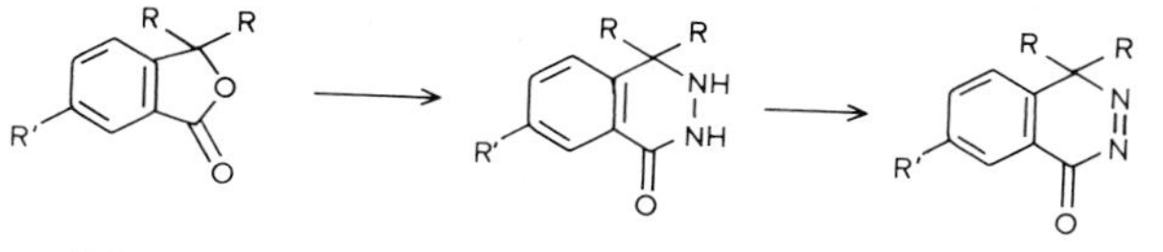

(CXXXV)
R = $C_6H_4NMe_2$-*p*
R' = NMe_2

(CXXXVI)

(CXXXVII)

arrangement occurs to give the azulenone (CXXXVIII). A mechanism for this transformation has been proposed (*idem, ibid.*, 1984, 1471). In acidic methylene dichloride solution, (CXXXVI) is converted into 3,4-diphenyl 1-oxidophthalazine betaine (CXXXIX).

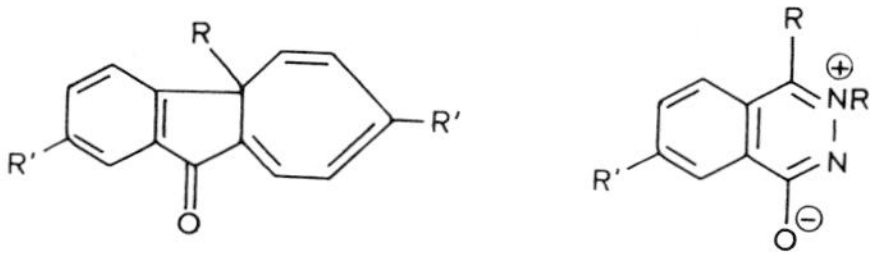

(CXXXVIII)
R = $C_6H_4NMe_2$-*p*
R' = NMe_2

(CXXXIX)

1,2-Dihydrophthalazine, m.p. 47–48°, forms a 2-*p-toluenesulphonyl* derivative, m.p. 172–174°. 2-*Methyl-* and 2-*phenyl-1,2-dihydrophthalazine*, m.ps. 34–35° and 67–68°, respectively, react with phenyl isocyanate to give *phenylthiourea* derivatives, m.ps. 235–236° and 174–175°, respectively. 2-**Methyl-1,2-dihydrophthalazine**, formed by reduction of 2-methylphthalazinium iodide with sodium borohydride, is a yellow oil, b.p. 129–130°/17 mm, and gives a *hydrochloride*, m.p. 133–135°, and a *methiodide*, m.p. 174–176°. 2-**Benzoyl-2-cyano-1,2-dihydrophthalazine**, m.p. 163–164°, is obtained by the action of potassium cyanide and benzoyl chloride in dichloromethane–water mixture at room temperature. Treatment of the Reissert compound with sodium hydride and methyl iodide in DMF gives 2-*benzoyl-1-cyano-1-methyl-1,2-dihydrophthalazine*, m.p. 143–145°, in quantitative yield, and alkaline hydrolysis of this gives 1-methylphthalazine.

3,4-Dihydro-1(2*H*)-phthalazinone, m.p. 170°, obtained by polarographic reduction, gives a *hydrochloride*, m.p. 222–224°, and with acetyl chloride yields the 3-*acetyl* derivative, m.p. 164–166°. Treatment of the acetyl compound with triethyloxonium tetrafluoroborate gives

2-**acetyl**-4-**ethoxy**-1,2-**dihydrophthalazine**, m.p. 70–72° (*E. Bellasio* and *E. Testa*, Ann. Chim. (Rome), 1969, **59**, 443, 451). The latter on alkaline hydrolysis affords 4-*ethoxy*-1,2-*dihydrophthalazine*, b.p. 110°/0.2 mm; *hydrochloride*, m.p. 166–168°. The action of triethyloxonium tetrafluoroborate on 3,4-dihydro-1(2*H*)-phthalazinone is to produce the 3-*ethyl* derivative, m.p. 179–180°.

(ii) 1,4-Dihydrophthalazine

Unstable 1,4-**dihydrophthalazine** (CXL) has been obtained by low temperature oxidation of 1,2,3,4-tetrahydrophthalazine (*vide infra*). The low temperature u.v. spectrum, p.m.r. spectrum, and melting point (−9 to −7°) have been determined. Loss of dinitrogen occurs readily from (CXL) and photolysis provides a useful route to *o*-xylylene (CXLI) (*C.R. Flynn* and *J. Michl*, J. Amer. chem. Soc., 1974, **96**, 3280).

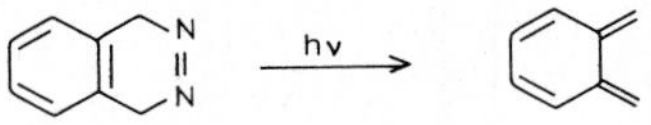

(CXL) $\xrightarrow{h\nu}$ (CXLI)

(iii) 1,2,3,4-Tetrahydrophthalazines

The hydrochloride of 1,2,3,4-tetrahydrophthalazine (CXLII) has been isolated from the reduction of phthalazine with sodium amalgam, and 2,3-dimethyl-1,2,3,4-tetrahydrophthalazine (CXLIII) is the product obtained when 2,3-dimethylphthalazine-1,4-dione is reduced with lithium aluminium hydride (*B. Junge* and *H.A. Staab*, Tetrahedron Letters, 1967, 709).

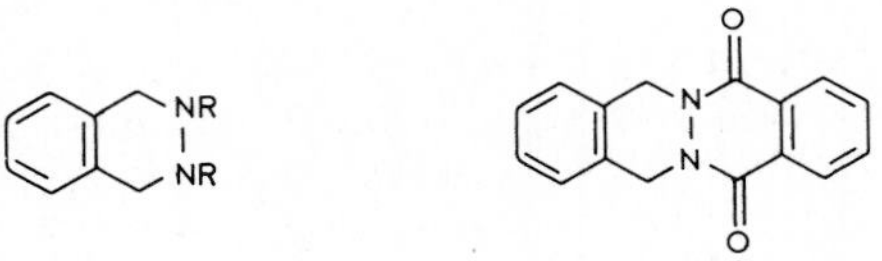

(CXLII) R = H	(CXLVI)
(CXLIII) R = Me	
(CXLIV) R = Ac	
(CXLV) R = Ph	

Less direct but generally better syntheses of 1,2,3,4-tetrahydrophthalazine (CXLII) and its derivatives include the reaction of 1,2-bis(bromomethyl)benzene with the potassium salt of 1,2-diacetylhydrazine to give (CXLIV) (*S. Groozkowski* and *B. Wesolowska*, Arch. Pharm., 1981, **314**, 880). Hydrolysis of diacetyltetrahydrophthalazine leads to (CXLII). This is essentially the route used in the synthesis of (CXLVI) from 4-hydroxy-1(2*H*)-phthalazinone and α,α′-dichloro-*o*-xylene, and subsequent hydrolysis of (CXLVI) to (CXLII) (*B. Junge* and *H.A. Staab*, Tetrahedron Letters, 1967, 709).

1,2,3,4-Tetrahydrophthalazine behaves like a *sym*-dialkylhydrazine: alkylation and acylation at nitrogen occurs readily. The 2,3-diphenyl derivative (CXLV) adopts the *anti* arrangement of phenyl groups in the gas phase (*P. Rademacher et al.*, Ber., 1977, **10**, 1939). The e.s.r. spectrum from the radical cation of (CXLII) has been studied (*F.A. Neugebauer* and *H. Weger*, J. phys. Chem., 1978, **82**, 1152). Probably the most interesting reactions in this group of compounds are the rearrangements which occur on reduction and the use of the tetrahydro compound (CXLII) as a source of *o*-xylylene. Treatment of (CXLV) with tin(II) chloride and hydrochloric acid affords 1,2-bis(phenylaminomethyl)benzene by cleavage of the 2,3-bond (*G. Wittig, W. Joss* and *P. Rathfelder*, Ann., 1957, **610**, 180). Oxidation of (CXLII) with mercury(II) oxide gives products derived from *o*-xylylene (*L.A. Carpino*, J. Amer. chem. Soc., 1963, **85**, 2144), but when the oxidation is performed at -20 to $20°$ in the presence of sulphur dioxide the *o*-quinodimethane is trapped as (mainly) the sulphinate (CXLVII) and the sulphone (CXLVIII) (*T. Durst* and *L. Tetreault-Ryan*, Tetrahedron Letters, 1978, 2353).

$$(CXLII) \longrightarrow \left[\begin{array}{c} CH_2 \\ CH_2 \end{array} \right] \xrightarrow{SO_2} \underset{9}{\underset{(CXLVII)}{\bigodot\!\!{\overset{O}{\underset{SO_2}{|}}}}} + \underset{1}{\underset{(CXLVIII)}{\bigodot\!\!SO_2}}$$

1,2,3,4-**Tetrahydrophthalazine** is an unstable solid which yields a stable *hydrochloride*, m.p. 236–238°. The 2-*benzoyl* and 2,3-*dibenzoyl* derivatives have m.ps. 90–91° and 210–212°, respectively. Other derivatives of the hydrazine NH groups are readily obtained by the action of electrophilic reagents. For instance, ammonium isothiocyanate yields the 2-*thiocarboxamide*, m.p. 191–192°. 2,3-**Diphenyl**-1,2,3,4-**tetrahydrophthalazine**, m.p. 92–92.5°, is obtained from 1,2-diphenylhydrazine.

(iv) 5,6,7,8-Tetrahydrophthalazines

These compounds are equally well considered as 4,5-dialkylpyridazines. The parent compound (CL) is prepared from 3,4,5,6-tetrahydrophthalic anhydride by routes which correspond to the synthesis of phthalazine from phthalic anhydride. The 1,4-dichloro compound (CXLIX) is a useful source of 5,6,7,8-tetrahydrophthalazine (CL) and its 1,4-disubstituted derivatives (*T. Kametani et al.*, Chem. pharm. Bull. Japan, 1971, **19**, 1794). The reduction of (CXLIX) with red phosphorus and hydriodic acid gives (CL) in moderate yield together with a smaller quantity of 1-iodo-5,6,7,8-tetrahydrophthalazine (CLI) (*R.H. Horning* and *E.D. Amstutz*, J. org. Chem.,

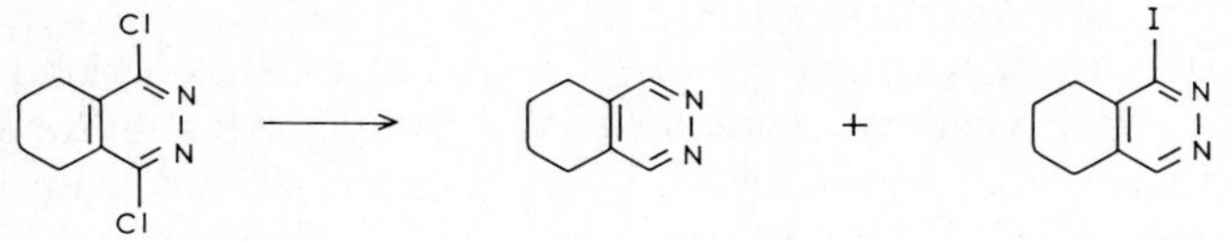

1955, **20**, 707, 1069). Catalytic reduction of 2-acetyl-1(2H)-phthalazinone with platinium and hydrogen under pressure gives a mixture of 3,4,4a,5,6,7,8,8a-octahydro- (CLII) and 5,6,7,8-tetrahydro-1(2H)-phthalazinone (CLIII) (E. Bellasio, C.A., 1974, **80**, 146097).

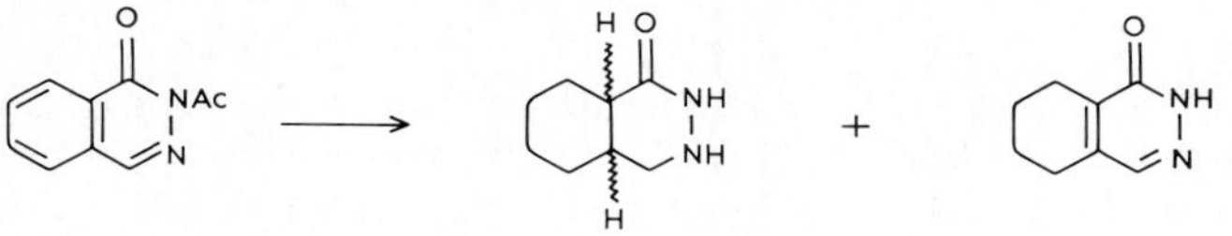

Dimethyl 1,2,4,5-tetrazine-3,6-dicarboxylate (CLIV) undergoes inverse electron demand Diels–Alder reactions with electron-rich olefins such as enamines. In the case of cyclohexenamines the product is dimethyl 5,6,7,8-tetrahydrophthalazine-1,4-dicarboxylate (CLV) (P.L. Boger et al., J. org. Chem., 1984, **49**, 4405). Hydrolytic decarboxylation then affords the parent heterocycle. Reduction of (CLV) with zinc and acetic acid causes rearrangement and formation of dimethyl 4,5,6,7-tetrahydroisoindole-1,3-dicarboxylate.

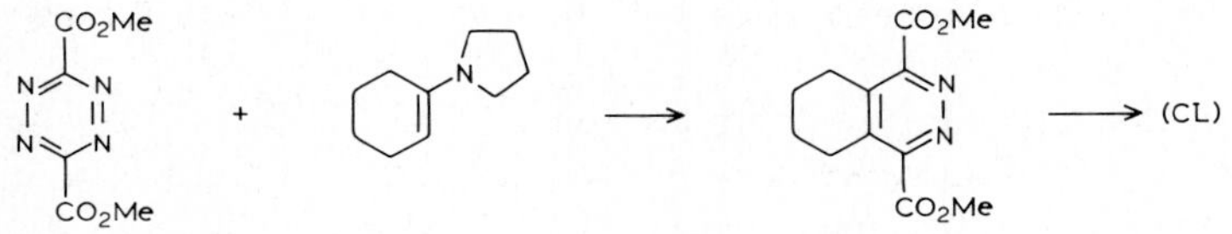

5,6,7,8-**Tetrahydrophthalazine**, m.p. 88.5–89°, is obtained by reduction of its 1,4-*dichloro* derivative, m.p. 156°, or by hydrolysis and decarboxylation of its 1,4-*dimethyl ester*, m.p. 131–132°. The ^{1}H-n.m.r. spectrum of the tetrahydrophthalazine (CL) shows δ (CDCl$_3$) 8.80 (1- and 4-H), 2.5–2.8 (5- and 8-H) and 1.6–1.9 ppm (6- and 7-H). 5,6,7,8-*Tetrahydrophthalazinone* and its 4-*hydroxy* derivative have m.ps. 144–145° and 295–298°, respectively.

(v) Other reduced phthalazines

4a,5,8,8a-Tetrahydrophthalazine derivatives have been obtained either by reaction of 4,5-disubstituted cyclohexenes with hydrazine or its derivatives (*S. Dixon* and *L.F. Wiggins*, J. chem. Soc., 1954, 594) or by condensation of 1,2-disubstituted pyridazine-3,6-dione with 1,3-dienes in a Diels–Alder reaction (*J. Druey, K. Meier* and *A. Staehlin*, Helv., 1962, **45**, 1485).

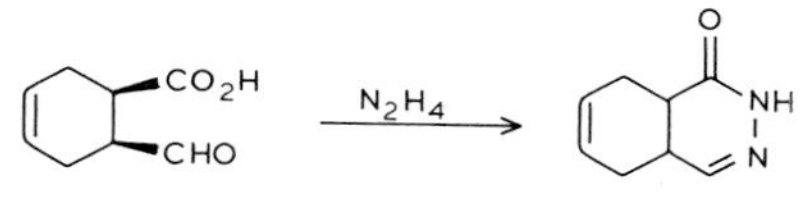

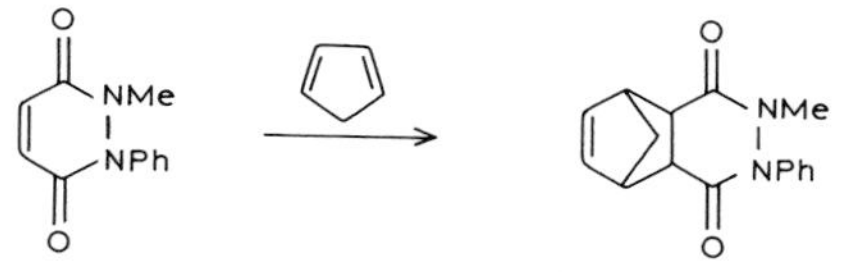

4a,5,8,8a-Tetrahydrophthalazin-1(2H)-one, m.p. 218–219° is obtained from *cis*-1,2,3,6-phthalaldehydic acid with hydrazine. The stereochemistry about the ring junction in this and other compounds of the series has not been established.

4a,5,6,7,8,8a-Hexahydrophthalazines are probably the best represented of the remaining reduced phthalazines. Suitably 1,2-disubstituted derivatives of cyclohexane react with hydrazine and its derivatives to give 4a,5,6,7,8,8a-hexahydrophthalazines. Both the *cis*- and the *trans*-isomers can be oxidised by bromine in acetic acid to give 5,6,7,8-tetrahydrophthalazines (*E. Jucker* and *R. Suess*, Helv., 1959, **42**, 2506). A Diels–Alder reaction between a cyclohexene and a 3,6-bis(fluoroalkyl)-*sym*-tetrazine yields a 1,4-bis(fluoroalkyl)hexahydrophthalazine (*R.A. Carboni* and *R.V. Lindsey, Jr.*, J. Amer. chem. Soc., 1959, **81**, 4342). In this way the vapour phase photochemical reaction of the cyclohexene and 3,6-bis(trifluoromethyl)-*sym*-tetrazine yields 1,4-bis(trifluoromethyl)-4a,5,6,7,8,8a-hexahydrophthalazine (*M.G. Barlow, R.N. Haszeldine* and *J.A. Pickett*, J. chem. Soc. Perkin I, 1978, 378).

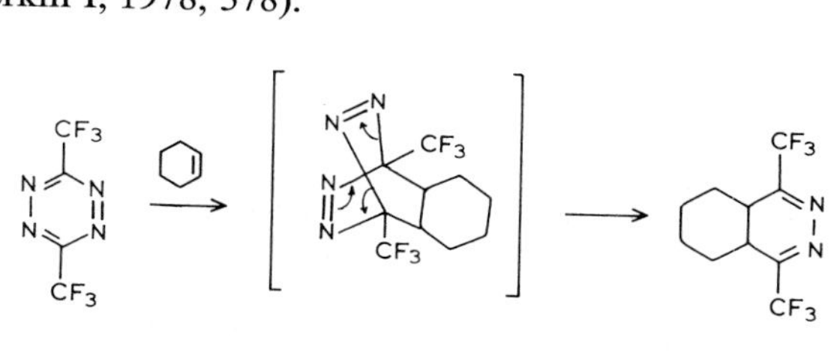

Chapter 43

Pyrimidines and Quinazolines

R.T. WALKER

1. Introduction

Pyrimidines, which normally account for 50% of the heterocyclic bases in every molecule of DNA would thereby justify extensive coverage in a review of heterocyclic chemistry even if they occurred nowhere else. However, in addition, the chemistry of pyrimidines can be traced back to the earliest days of organic chemistry when uric acid (I) was oxidised to alloxan (II) in 1818.

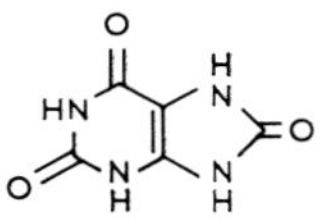

(I) 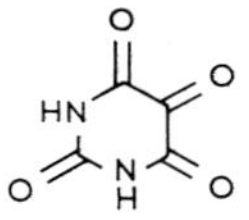   (II)

Pyrimidines are also found as the soporific and hypnotic drugs; the barbiturates, *e.g.* Luminal (III). Some derivatives show useful antibacterial (*e.g.* sulphadiazine, IV) and antimalarial properties (*e.g.* pyrimethamine, 2,4-diamino-5-*p*-chlorophenyl-6-ethylpyrimidine, V). Vitamin B1 (VI) and several coenzymes (*e.g.* CDP-ethanolamine, VII) also contain the pyrimidine ring.

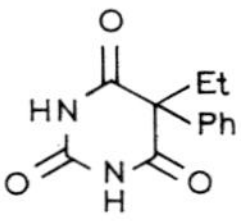

(III) 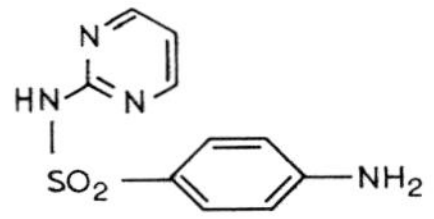   (IV) 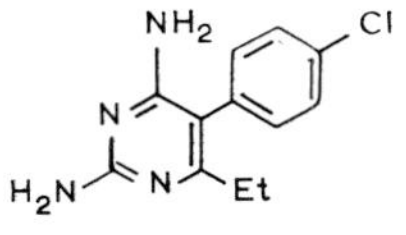   (V)

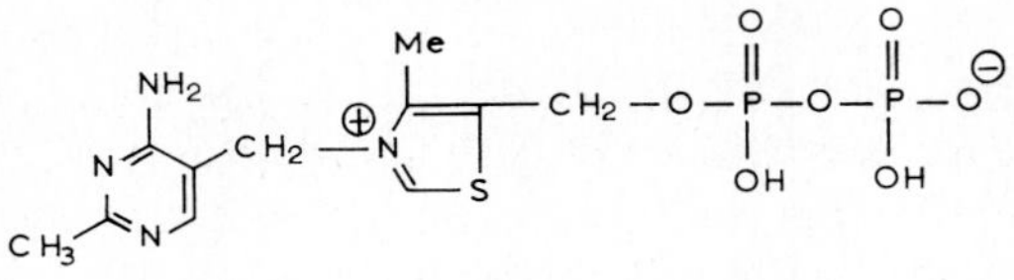

(VI)

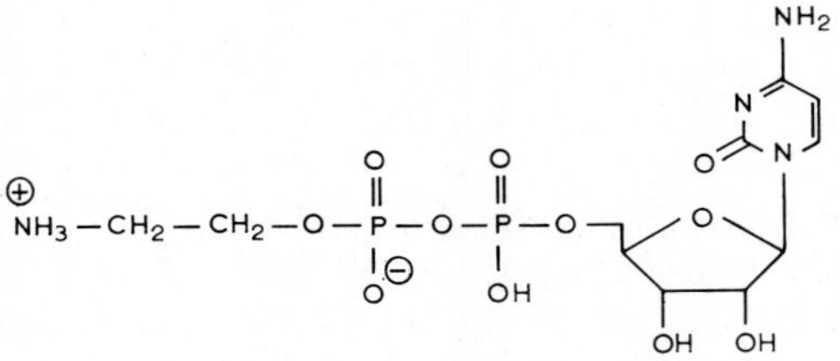

(VII)

Some idea of the complexity of the chemistry of the pyrimidine ring system can however be gained from a glance at the contents page of a general organic chemistry text book. If mentioned at all, the chemistry is dismissed usually in less than one page and even in specialised books on heterocyclic chemistry, it is rare to find more than a few pages devoted to chemistry of pyrimidines. One might be forgiven for thinking that there was very little recorded chemistry of pyrimidines and/or that what there was, was incapable of a rational explanation but neither of these conclusions would be correct.

Over 7,000 references dealing with pyrimidine chemistry have been catalogued and scientists working in this field are privileged to have available one of the most thorough and scholarly reviews of any area in organic chemistry. Reviews of pyrimidine chemistry are inextricably linked with the name of D.J. Brown who, since the appearance of the review on pyrimidine chemistry by G.R. Ramage in the first edition of this work nearly 30 years ago, has published a book and two supplements which cover the literature in a comprehensive and comprehensible fashion from the earliest reports through to 1983 (*D.J. Brown*, The Pyrimidines, in The Chemistry of Heterocyclic Compounds, Vol. 16, ed. *A. Weissberger*, Wiley, New York, 1962; *D.J. Brown*, The Pyrimidines, Supplement I, in The Chemistry of Heterocyclic Compounds, Vol. 16, eds. *A. Weissberger* and *E.C. Taylor*, Wiley, New York, 1970; *D.J. Brown*, The Pyrimidines, Supplement II, in The Chemistry of Heterocyclic Compounds, Vol. 16, eds. *A. Weissberger* and *E.C. Taylor*, Wiley, New York, 1985).

1 INTRODUCTION

The aim of this chapter is to bridge the gap between the totally inadequate coverage found elsewhere and the totally comprehensive approach of Brown. Emphasis will be placed more on general principles rather than specific examples in the knowledge that the reader can either always find the specific example in "The Pyrimidines" or they can be fairly sure that it has not been reported. Interestingly, Brown's survey shows that whereas until 1960, the U.S.A., the British Commonwealth and Germany accounted for nearly 80% of the relevant publications, in the last decade only 46% came from these countries with a further 30% from Japan and the Soviet Union.

The nomenclature of pyrimidine and its derivatives has caused considerable confusion over the years. The problem arises from the fact that since pyrimidine (VIII) is symmetrical about the line joining C-2 to C-5, positions 1 and 3 are equivalent, as are positions 4 and 6. In this review, the pyrimidine ring will be written in the form shown (VIII) which at least is familiar to biochemists and the numbering system is as shown. Thus cytosine (IX) is 4-amino-2-hydroxypyrimidine. The nomenclature problem

(VIII)   (IX)   (X)

becomes more complex when one nitrogen is methylated so that the two nitrogen atoms are no longer equivalent and all symmetry is lost. Thus, methylation of cytosine gives 1-methylcytosine (X) (sometimes known as 3-methylcytosine!) also known as 4-amino-1,2-dihydro-1-methyl-2-oxopyrimidine, a name which (wrongly) implies that the nucleus is in a reduced state. Anyway the correct name for cytosine is 4-aminopyrimidin-2(1H)-one. When the first comprehensive review of the pyrimidines appeared (*Brown, loc.cit.*), the decision was taken to use the prefixes hydroxy, mercapto or amino whatever the evidence for the actual major tautomeric form. Understandably, (but unfortunately) subsequent supplements in order to achieve consistency, have perpetuated this nomenclature which is totally at variance with current recommendations and thus to those conversant with the field and also to many working in it, the naming of pyrimidines is confused. The only way to be absolutely sure of locating a compound is to use a formula index. This chapter is certainly not the place to try to bring order to chaos and a mixture of all systems will be used but usually the actual formula will also be given.

2. Physical properties of pyrimidine

(a) Structure and geometry

Pyrimidine (VIII) only retains symmetry about the 2,5-axis and when compared with benzene, the replacement of two CH units by nitrogen atoms, which are electron attracting and in mutually reinforcing positions, means that the reactivities of the remaining carbon atoms and substituents attached to them vary considerably. Substituents at positions 4 and/or 6 are often found in naturally occurring pyrimidines. Such substitution removes the remaining symmetry.

The electron-distribution diagram (XI), the result of calculations by the VESCF method, shows that as expected, there is depletion of electron density at positions 2, 4 and 6, a slight depletion at the 5-position and a greatly enhanced density at the nitrogen atoms. The pyrimidine ring itself is flat and the bond lengths (XII) and bond angles (XIII) are shown. However, pyrimidine itself is of little interest or use; it is not found in nature and has found no application in general organic synthesis or as a precursor in the synthesis of its derivatives.

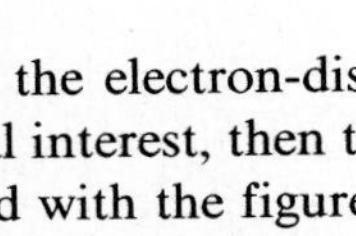

(XI) 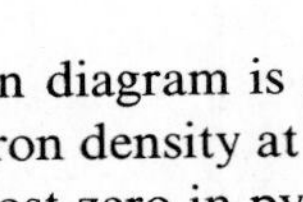   (XII) 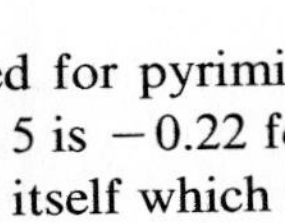   (XIII)

When the electron-distribution diagram is calculated for pyrimidines of biological interest, then the electron density at position 5 is -0.22 for uracil compared with the figure of almost zero in pyrimidine itself which explains the ease of electrophilic attack at this position.

By 1965, more than 100 diffraction studies on pyrimidines had been reported and reviewed (*G.H. Jeffrey, D. Mootz* and *D. Mootz*, Acta Cryst., 1965, **19**, 691) and the crystallographic data published by 1970 has been summarised (*D. Voet* and *A. Rich*, Prog. Nucleic Acid Res. Mol. Biol., 1970, **10**, 183). Data for uracils and uridines have also been separately reviewed (*A. Albert*, Phys. Meth. heterocyclic Chem., 1962, **1**, 1) and the more recent references containing crystallographic data of some 5-substituted 2′-deoxyuridines of biological interest have been listed (*E. De Clercq* and *R.T. Walker*, Pharmac. Ther., 1984, **26**, 1). The dipole moment of pyrimidine as determined experimentally (2.10–2.40D) is in good agreement with calculated values of 2.13–2.25D.

(b) *The ionisation constant of pyrimidines*

Pyrimidine (pK_a 1.31 and *ca.* −6.3) in common with the other diazines, pyrazine and pyridazine, is essentially a monobasic compound and is a much weaker base than pyridine which has a pK_a of 5.2. The drop in basicity is the consequence of destabilisation of the N-1 protonated species by inductive electron withdrawal caused by the second nitrogen atom and thus the system approximates to 3-nitropyridine (pK_a 0.8). However, substituents such as methyl or methoxyl can overcome this electron deficiency so that the following pK_a values are found: 4-methylpyrimidine, 1.98; 4-methoxypyrimidine, 2.5; 4,6-dimethylpyrimidine, 2.8; 4-methoxy-6-methylpyrimidine, 3.65. The basic strength of uracil (XIV) however again falls with a pK_a of −3.4 as both ring nitrogens are now involved in cyclic amide formation and therefore the oxygen atoms are the basic centres of interest.

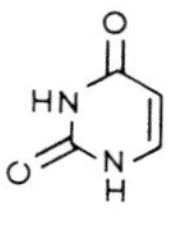

(XIV)

Pyrimidines of biological interest such as the pyrimidinones (*e.g.* XV), pyrimidinecarboxylic acids, the thiones, sulphonic acids and pyrimidin-5-ols all ionise as acids and because pyrimidine is such a weak base, it is not usual for zwitterionic species to form. Pyrimidin-2(1*H*)-one (XV; R = H) is a stronger acid (pK_a 9.17) than the corresponding pyridone (pK_a 11.7) because of the electron-withdrawing effect of the second nitrogen atom but again, addition of a methyl group reduces the acid strength so that 4-methylpyrimidine-2(1*H*)-one (XV; R = CH$_3$) has a pK_a of 9.8. Uracil

(pK_a 9.4 and *ca.* 12) is slightly less acidic than the pyrimidinone as now both nitrogen atoms are involved as cyclic amides.

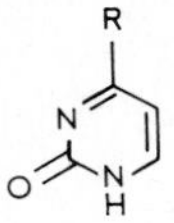

(XV)

Barbituric acid (XVI) and dialuric acid (XVII), as their names suggest, have considerable acidity (pK_a 2.8) because of their potential hydroxyl groups. Thiones are stronger acids than the corresponding pyrimidinones by a factor of about 100. The presence of electron-withdrawing groups causes an increase in acidity such that the pK_a of 5-nitrouracil (XVIII, R = NO_2) is 5.6 and that of 5-bromouracil (XVIII; R = Br) is 8.0. The pyrimidine-2-carboxylic acids and pyrimidine-2-sulphonic acids are stronger then the corresponding benzene derivatives.

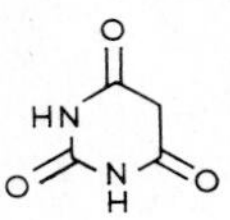

(XVI) 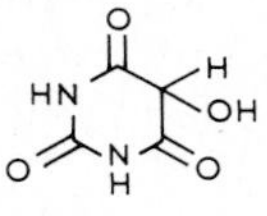   (XVII) 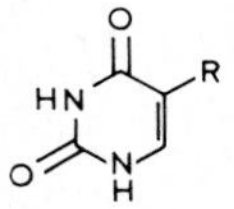   (XVIII)

Simple pyrimidinamines, *e.g.* XIX, XX, are bases of moderate strength with pK_a 3.54 and 5.71, respectively. When compared with the pK_a of pyrimidine itself (1.31), the increase in basicity in this series is much higher than that in the corresponding series from aniline to the phenylenediamines where the increase is only a factor of ten. This can be attributed to the increased stability of the cations (XXI) when compared with the neutral molecule. As expected, pyrimidin-5-amine (pK_a 2.60) does not show this

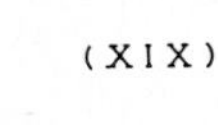

(XIX) 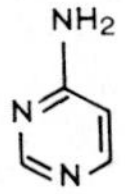   (XX)

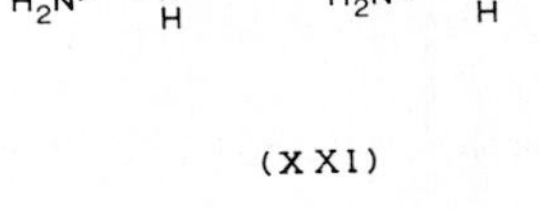

(XXI)

effect. As usual, C-alkylation results in an increase in basicity as does alkylation of the amino group. Other electron-releasing groups have the same effect and for similar reasons. An electron-withdrawing group such as bromo, cyano or nitro, decreases the basicity such that the following pK_a range can be observed: pyrimidine-2,4-diamine, 7.3; pyrimidine-4,5-diamine, 6.03; 4,6-dimethylpyrimidin-2-amine, 4.85; 4-methylpyrimidin-2-amine, 4.15; pyrimidine-2-N-methylamine, 3.82; pyrimidin-2-amine, 3.54; 5-bromopyrimidin-2-amine, 1.95; 5-nitropyrimidin-2-amine, 0.35. As expected, alkylation of the ring-nitrogen causes even greater stabilisation of the cation and compound (XXII) has a pK_a of 10.7. The effects seen with

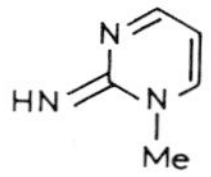

(XXII)

the aminopyrimidinones are much more difficult to rationalise except to say that, in general, the pK_a values of aminopyrimidinones show that the amino group has weakened the acid function and the oxo-substituent has reduced the basic strength. However, it is among the biologically important members of this group that some of the more inexplicable effects are seen; thus 2-aminopyrimidin-4(3H)-one (isocytosine) (XXIII) has an acidic pK_a of 9.59 compared with the corresponding pyrimidinone of 8.59, whereas 4-aminopyrimidin-2(1H)-one (cytosine) (XXIV) has an acidic pK_a of 12.15 compared with the corresponding pyrimidinone of 9.17. Also neither the acidic nor the basic strengths of cytosine or isocytosine are effected by additional amino groups. Pyrimidine-N-oxides show that upon N-oxidation one gets a weaker base and a stronger acid than the parent compound although N-3 oxidation of cytosine results in a slight increase in basic strength.

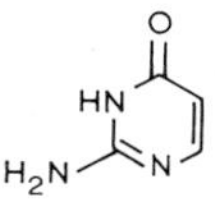

(XXIII) 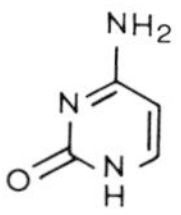   (XXIV)

Ionisation constants have usually been determined by traditional methods but more recently ^{1}H- or ^{13}C-n.m.r. spectroscopy has been used to determine both the ionisation constant and the site of protonation. pK_a

Values, both acidic and basic, for several hundred pyrimidines have been listed (*Brown*, *loc.cit*., p. 472; Suppl. I, p. 368; Suppl. II, p. 481).

(c) *Spectroscopy*

(i) *Nuclear magnetic resonance*

Since the first edition of this chapter, the use of n.m.r. spectroscopy for the identification of organic compounds has become one of the most widely used techniques but any detailed discussion is beyond the scope of this chapter and only one or two general points are mentioned here.

^{1}H-N.m.r. spectra have been reviewed (*T.J. Batterham*, N.M.R. Spectra of Simple Heterocycles, Wiley, New York, 1973; *T.J. Batterham*, The Pyrimidines, in Chemistry of Heterocyclic Compounds, Vol. 16, Suppl. 1, ed. *A. Weissberger* and *E.C. Taylor*, Wiley, New York, 1970). The first reported spectrum of pyrimidine appeared in 1960 (*S. Gronotwitz* and *R.A. Hoffman*, Ark. Kemi, 1960, **16**, 459), who measured the chemical shifts at 40 Mc/s in a number of solvents with respect to the solvent peak or an external water standard. These results were later confirmed and quoted in more useful values with respect to TMS (*G.S. Reddy*, *R.T. Hobgood* and *J.H. Goldstein*, J. Amer. Chem. Soc., 1962, **84**, 336). The results show that as expected, the relative deshielding of the four protons is H-2 > H-4 = H-6 > H-5. Methyl substituents show long-range coupling between the methyl protons and *ortho*- or *para*-ring protons. In general, the n.m.r. spectra of most monosubstituted pyrimidines, which do not have tautomeric groups, are straight forward. Thus 2-substituted pyrimidines show a doublet and a triplet typical of an A_2X system and 4-substituted pyrimidines have no symmetry and show typical ABX patterns. 5-Substituted pyrimidines show two broad singlets with no *meta*-coupling across the ring nitrogen atoms.

^{13}C-N.m.r. spectra can be measured using natural abundance of the isotope. The values for the ^{13}C-chemical shifts for pyrimidines have been listed (*R.J. Pugmire* and *D.M. Grant*, J. Amer. Chem. Soc., 1968, **90**, 697; *F.J. Weigert*, *S. Husar* and *J.D. Roberts*, J. Org. Chem., 1973 **38**, 1313). Both ^{15}N- and ^{14}N-isotopes can be used for chemical shift determination. ^{14}N has a relative sensitivity of 1×10^{-3} which is twenty times better than carbon with a spin of $I = 1$ and a moderate quadrupole moment. ^{15}N has a very low sensitivity of 1% that of ^{13}C but has a spin of $I = 0.5$. It is desirable to use ^{15}N-enriched samples and the spectra are complicated by both ^{15}N–^{1}H and ^{15}N–^{13}C coupling. The nitrogen chemical shifts for over 50 pyrimidines have been listed (*Brown*, *loc.cit*., Suppl. 2, p. 494).

(ii) Infrared and ultraviolet spectra

Very little work of general importance has been reported since the early 1960s; in fact since the technique of n.m.r. arrived to replace i.r.- and u.v.-assisted identification of compounds. Clearly i.r. spectroscopy can be used to identify the presence (or indicate the absence) of certain substituents but while giving corroborative evidence, it is rarely used nowadays as the primary method of identification. A brief but now outdated review is available (*Brown, loc.cit.*, p. 494).

While u.v. absorption is undoubtedly the most widely used technique for the location of pyrimidines during synthesis and separation procedures (TLC, HPLC, *etc.*) and it is usual to quote the λ_{max} and molar extinction coefficient for compounds so isolated, again it is rare nowadays to rely exclusively on this one technique (or any one other for that matter) for identification purposes. Some relatively recent compilations of data and reviews are available (*S.F. Mason*, Phys. Meth. heterocyclic Chem., 1963, **2**, 1; *W.L.F. Armarego*, ibid., 1971, **3**, 67; *A. Albert*; in Synthetic Procedures in Nucleic Acid Chemistry, Vol. 2, eds. *W. Zorbach* and *R.S. Tipson*, Wiley, New York, 1973).

Pyrimidine itself shows two absorption maxima centred at 243 and 298 nm. The former, which is the more intense, is due to a $\pi-\pi^*$ transition and is hence unaffected by solvent changes. The latter is an $n-\pi^*$ transition and shows a hypsochromic shift on moving from cyclohexane to water as the solvent. As might be expected, electron-releasing substituents cause a bathochromic shift in the $n-\pi^*$ transition while electron-withdrawing substituents do the reverse. The $\pi-\pi^*$ transition undergoes a bathochromic shift with either type of substituent and the intensity is increased. If the substituent groups affect the fundamental structure of the molecule (that is, allow a decrease or increase in conjugation), then other considerations apply.

(iii) Mass spectra

This technique is routinely used for helping in the identification of pyrimidines but it is rare to get a review and explanation of the fragmentation pattern observed. Rather the presence of molecular ion and/or an accurately mass-measured peak is all that is quoted. The fragmentation of pyrimidine is the loss of HCN twice. Uracil and thymine undergo an initial retro-Diels-Alder reaction by losing HCNO to give (XXV, R = H or CH_3).

(XXV)

Cytosine appears to fragment by at least three pathways. Some of the patterns of pyrimidine fragmentation have been summarised. (*Q.N. Porter* and *J. Baldas*, Mass Spectroscopy of Heterocyclic Compounds, Wiley, New York, 1971).

(d) Tautomerism

Over the years there has been much controversy about the tautomeric species present in various pyrimidinones and pyrimidinamines, particularly those naturally occurring in the nucleic acids such as cytosine (XXIV), uracil (XIV) and thymine (XVIII, R = CH$_3$). Even simple pyrimidinones were originally suggested, from i.r. evidence, to exist as pyrimidinols but this suggestion was quickly shown to be incorrect. Uracil and thymine each exist as the dioxo tautomer as confirmed by an X-ray crystallographic study for the solid state and many spectral determinations in solution. Barbituric acid (XVI) adopts the trioxo form and loses its first proton from carbon and the subsequent one from nitrogen to give the dianion. Cytosine, despite several claims to the contrary (*G.C.Y. Lee, J.H. Prestegard* and *S.J. Chan*, J. Amer. chem. Soc., 1972, **94**, 951) which have been refuted (*Y.P. Wong*, ibid., 1973, **95**, 3511), exists in the amino-oxo form, a fact which has been confirmed by X-ray crystallographic and u.v.-, n.m.r.- and Raman-spectroscopic studies. Pyrimidine tautomerism has been reviewed (*A.R. Katritzky* and *J.M. Lagowski*, Adv. heterocyclic Chem., 1963, **1**, 339; *J. Elguero et al.*, ibid., 1976 Suppl., **1**, 71) as has the tautomerism of the biologically important pyrimidines, uracil, thymine and cytosine (*J.S. Kwiatkowski* and *B. Pullman*, ibid., 1975, **18**, 189).

However, the wheel has now turned full circle and once everything was thought to have been satisfactorily settled, recent i.r.-spectroscopic studies on 2-hydroxy- and 4-hydroxy-pyrimidines have shown the existence of the hydroxy form in the gas phase. Even under these conditions uracil is still present in the dioxo form (*D. Shugar* and *S. Szczepanick*, Int. J. quantum Chem., 1981, **20**, 573).

Studies on the monoanions of uracil and its 5-alkylated derivatives have indicated that the structure present in the equilibrium mixture are depen-

(XXVIa) ⇌ (XXVIb)

dent upon the substituents. Thus 5-butyluracil forms an equilibrium strongly in favour of (XXVIa; R^1 = Bu, R^2 = H) whereas 5-ethyl-6-propyluracil has isomer (XXVIb; R^1 = Et, R^2 = Pr) predominating (*R. L. Lipnick* and *J. Fissekis*, J. org. Chem., 1979, **44**, 1627).

3. Synthesis

The pyrimidine ring has sufficient aromatic character to show great stability and this in large measure accounts for the wide variety of synthetic methods which have been used. These include, apart from primary syntheses, breakdown or modification of other heterocyclic systems and also transformations reactions on preformed pyrimidine rings. Here we are concerned with primary syntheses which, unlike benzene derivatives most of which come from preformed aromatic rings, account for the vast majority of pyrimidine ring syntheses. An early attempt to classify these synthetic reactions (*Brown*, loc.cit., 1959) has proved to stand the test of time and will be followed here. The initial classification depends upon the number of bonds being formed which can range from one to five. Within these classes, the individual precursors used can be further classified. The most useful and most-used method of pyrimidine ring synthesis — the Principal Synthesis — which accounts for about 80% of all the synthetic routes, involves the condensation of two three-atom units. This is therefore an example of a synthesis involving the formation of two bonds from two fragments each containing three atoms. One fragment contains the final pyrimidine N-1 + C-2 + N-3 and the other the C-3 + C-4 + C-5. The three-carbon fragment can be one of a number of ten classes of compounds ranging from β-dialdehydes to β-dinitriles and the other fragment is either urea, thiourea, an amidine or a guanidine.

A classical example of this type of synthesis is the condensation of acetyl acetone (XXVIII) with benzamidine (XXVII) to give 4,6-dimethyl-2-phenyl-pyrimidine (XXIX) (*A. Pinner*, Ber., 1893, **26**, 2122).

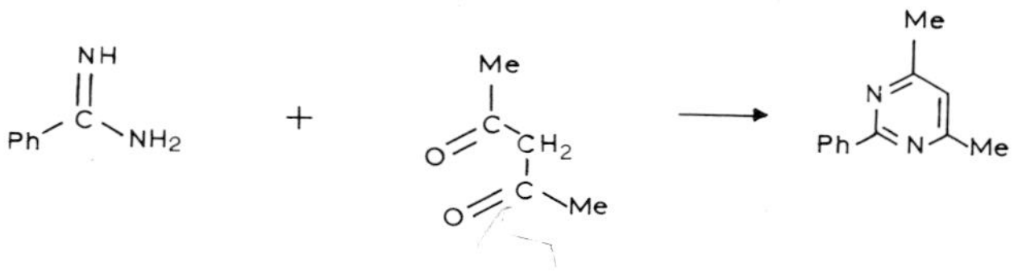

(XXVII) (XXVIII) (XXIX)

Pyrimidine itself can also be synthesised by a method in this class, although the primary synthesis is of uracil (XXXII). Thus the reaction of α-formylacetic acid (XXXI), which is produced *in situ* by decarbonylation of malic acid, with the diamino component urea (XXX) gives a reasonable yield (55%) of uracil. The latter reacts with phosphorus oxychloride to give 2,4-dichloropyrimidine (XXXIII) which can be reduced to pyrimidine (XXXIV) in overall 40% yield.

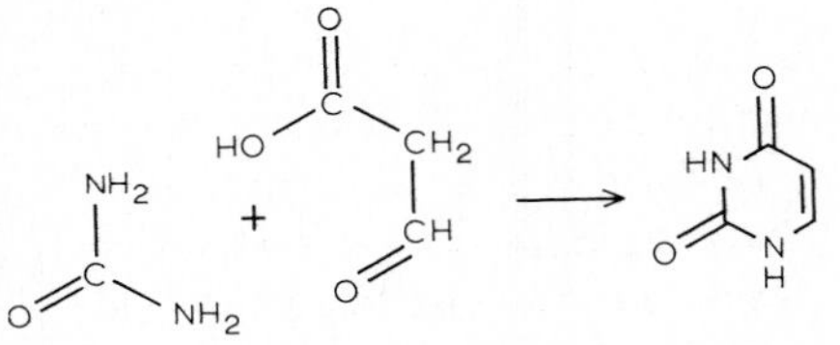

(XXX)   (XXXI)   (XXXII)

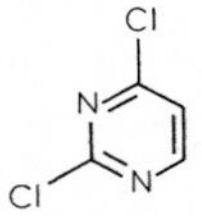

(XXXIII) 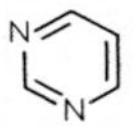   (XXXIV)

In this chapter, the Principal Synthesis will be discussed first and at some length. Thereafter the other methods available will be discussed in a logical order but in much less detail. The reviews by Brown contain a detailed discussion of all the methods available which had been reported by the end of 1983.

(a) The principal method

As stated above, this method which involves the condensation of two three-atom fragments, accounts for the large majority of pyrimidine syntheses. The method has been subdivided into ten classes, each of which utilises a different type of three-carbon fragment and each of these will now be examined in turn.

(i) From β-dialdehydes

It is not usual to use the free aldehyde but rather the acetal is used as typified by the reaction of 1,1,3,3-tetraethoxypropane (XXXVI) (which is the tetraethyl acetal of malondialdehyde) with thiourea (XXXV) in alcoholic hydrochloric acid to give 2-mercaptopyrimidine or pyrimidine-

2(1H)-thione (XXXVII) (*R.R. Hunt, I.F.W. McOmie* and *E.R. Sayer*, J. chem. Soc., 1959, 525).

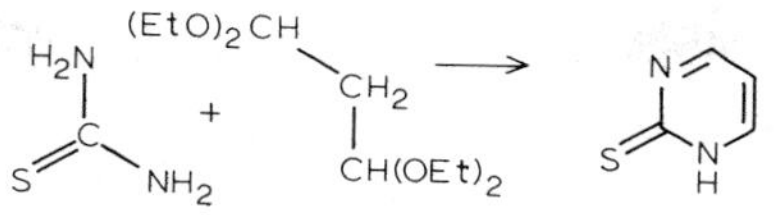

(XXXV)   (XXXVI)   (XXXVII)

Alternatively, the heteroatoms can be provided by urea or substituted ureas so that *N*-methylurea (XXXVIII) under similar conditions gives 1-methylpyrimidin-2(1H)-one (XXXIX). Amidines are a further source of the heteroatoms when sodionitromalondialdehyde (XL) reacts with the hydrochloride of furan-2-carboxamidine (XLI; R = fur-2-yl) in aqueous piperidine to give 2-fur-2'-yl-5-nitropyrimidine (XLII; R = fur-2-yl) in high yield (*D.T. Hurst* and *J. Christophides*, Heterocycles, 1977, **6**, 1999).

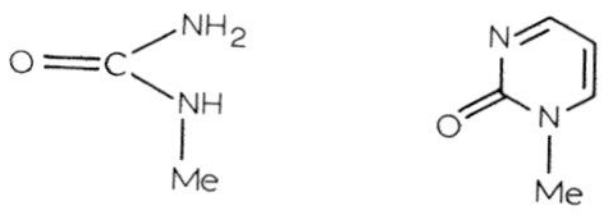

(XXXVIII)   (XXXIX)

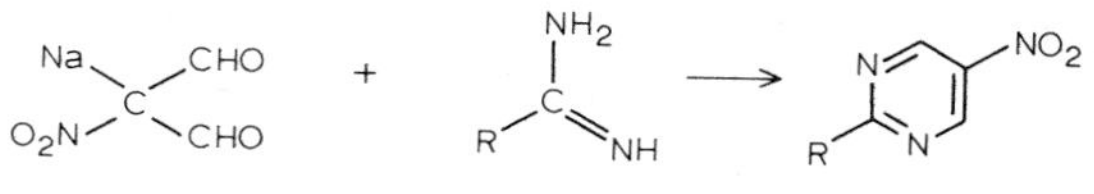

(XL)   (XLI)   (XLII)

Another source of the heteroatoms is guanidine (XLIII) which has been condensed in alcoholic acid with 1,3,3-triethoxypropene (XLIV) to give 2-aminopyrimidine (XLV). 3-Alkoxyacroleins, in place of a β-dialdehyde,

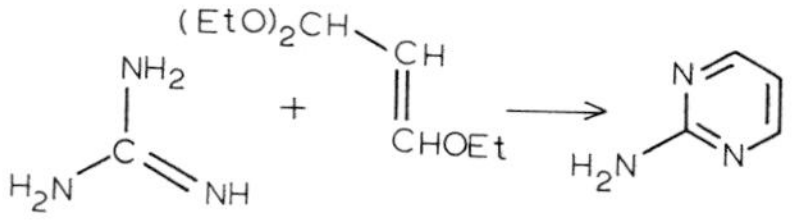

(XLIII)   (XLIV)   (XLV)

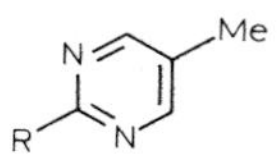

(XLVI)

have widespread use in pyrimidine synthesis and thus 3-ethoxy-2-methyl-acrolein can react with acetamidine, thiourea or N-carboxymethyl-N-methylguanidine (in base) or urea (in acid) to give 2,5-dimethyl- (XLVI; R = Me), 2-mercapto-5-methyl (XLVI; R = SH), 2N-carboxymethyl-N-methylamino-5-methyl- (XLVI; R = NMeCH$_2$COOH) and 2-hydroxy-5-methyl-pyrimidine (XLVI; R = OH), respectively (*C. Kruse* and *W. Breitmaier*, Chem. Ztg., 1977, **101**, 305).

A further β-dialdehyde equivalent which has been used successfully in pyrimidine synthesis is the bisdimethylaminotrimethinium salt (XLVII). Thus reaction of this salt with benzamidine (XLI; R = Ph) in ethanolic sodium methoxide gives 2-phenylpyrimidine (XLVIII) (*R.M. Wagner* and *C. Jutz*, Ber., 1971, **104**, 2975).

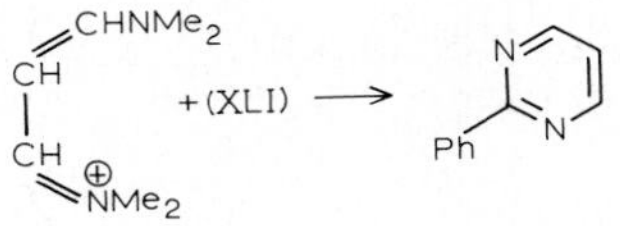

(ii) From β-aldehydo-ketones

Once again it is usual to use the acetal rather than the free aldehydo-ketone but one example of the latter is the reaction of ethyl α-formylacetoacetate (XLIX) with benzoylguanidine in ethanol to give 2-benzamido-5-ethoxycarbonyl-4-methylpyrimidine (L). An example of an acetal reaction is that of 4,4-dimethoxybutan-2-one (LI) with N,N'-dimethylurea under acidic conditions to give the pyrimidinium salt (LII).

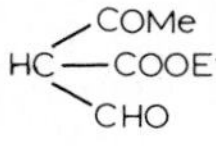

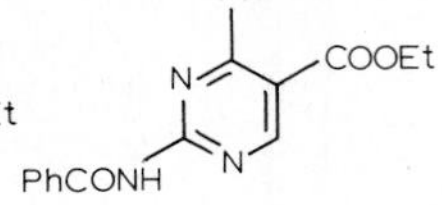

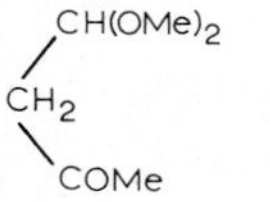

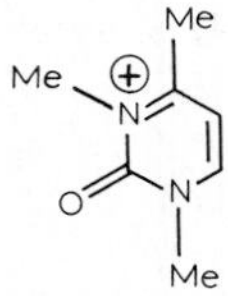

α-Alkoxymethylene ketones such as α-(ethoxymethylene)acetophenone (LIII) can react as β-aldehydo-ketones. Thus the reaction of compound (LIII) with urea in ethanolic sodium ethoxide gives 2-hydroxy-4-phenyl-pyrimidine (LIV) (*A.P. Kroon, H.C. van der Plas*, Rec Trav. chim., 1973, **92**, 1020).

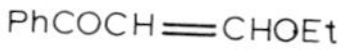

PhCOCH=CHOEt

(LIII) 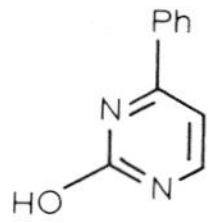   (LIV)

Bromomucic acid (LV) has also been used as an indirect intermediate of formylbromopyruvic acid (LVI). When condensed with acetamidine, 5-bromo-4-carboxy-2-methylpyrimidine (LVII) is produced (*A. Budesinsky*, Coll. Czech Chem. Comm., 1949, **14**, 223).

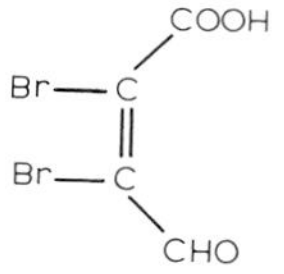

(LV) 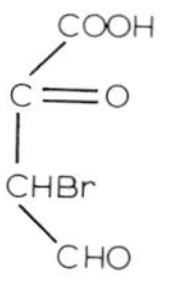   (LVI) 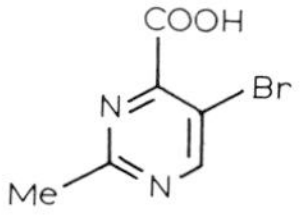   (LVIII)

Some acetylenes also serve as suitable intermediates so that ethoxymethylenepropyne (LVIII) reacts with thiourea to give 2-mercapto-4-methylpyrimidine (LIX).

(LVIII) 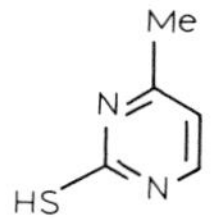   (LIX)

Other widely used aldehydo-ketone equivalents are compounds such as 4-dimethylamino-1,1-dimethoxybut-3-en-2-one (LX) which reacts with formamidine acetate to give 4-dimethoxymethylpyrimidine (LXI) (*G. Maury, J.-P. Pangan* and *R. Pangan*, J. heterocyclic Chem., 1978, **15**, 1041). 1-Benzoyl-1,2-bisdimethylaminomethylene (LXII) and benzamidine give 5-dimethylamino-2,4-diphenylpyrimidine (LXIII) (*H. Bredereck, G.*

Simchen and *W. Griebenow*, Ber., 1974, **107**, 1545). Bisdimethylaminotrimethinium salts (*e.g.* LXIV) with guanidine give 2-amino-4-phenylpyrimidine (LXV) (*R.M. Wagner* and *C. Jutz*, Ber., 1971, **104**, 2975).

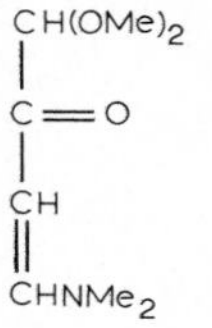

(LX), 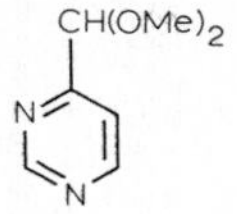 (LXI), 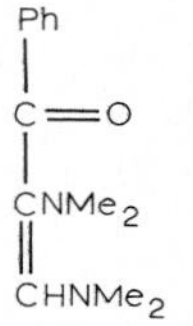 (LXII), 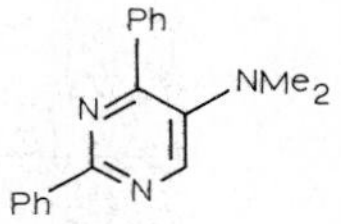 (LXIII), 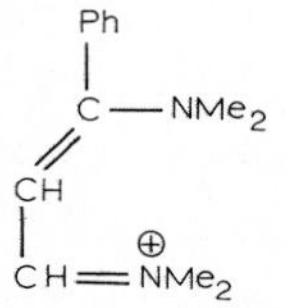 (LXIV), 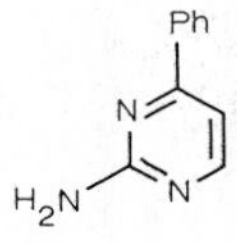 (LXV)

(iii) From β-diketones

One example of this method, the Pinner synthesis, has already been mentioned. Acetylacetone (LXVI) has also been condensed with other amidines, *e.g.* with acetamidine to give 2,4,6-trimethylpyrimidine (LXVII, R = Me), with urea to give 2-hydroxy-4,6-dimethylpyrimidine (LXVII; R = OH), with thiourea to give 2-mercapto-4,6-dimethylpyrimidine (LXVII; R = SH) and with guanidine to give 2-amino-4,6-dimethylpyrimidine (LXVII; R = NH$_2$). Guanidines react particularly easily with β-diketones or suitable intermediates and thus guanidine carbonate reacts with 1-benzoyl-3,3,3-trifluoro-acetone at 150° to give 2-amino-4-phenyl-6-trifluoromethylpyrimidine (LXVIII), (*T. Nishiwaki*, Bull. chem. Soc. Japan, 1969, **42**, 3024) and ethyl guanidine gives 2-ethylamino-4,6-dimethylpyrimidine (LXVII; R = NHEt) with acetylacetone (LXVI).

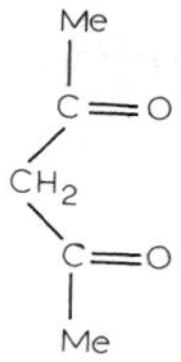

(LXVI) 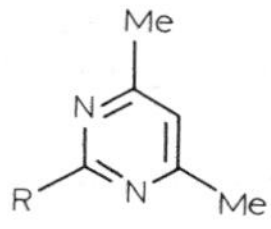 (LXVII) 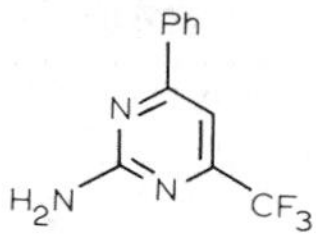 (LXVIII)

Condensation of *N*-alkylureas with unsymmetrical β-diketones can lead to the production of two isomers. The predominant isomer is controlled by the enol form of the diketone and the basicity of the ureas (*C. Kashima, Y. Yamamoto* and *Y. Omote*, Heterocycles, 1976, **4**, 1387). Thus heptan-2,4-dione (LXIX) and methylurea give equal amounts of 1,2-dihydro-1,6-dimethyl-2-oxo-4-propylpyrimidine (LXX) and 1,2-dihydro-1,4-dimethyl-2-oxo-6-propylpyrimidine (LXXI) whereas the diketone equivalent 2-aminohept-2-en-4-one (LXXII) gives them in the ratio 2:1. Benzoylacetone (LXXIII) and *N*-phenylurea give predominantly 1,2-dihydro-6-methyl-2-oxo-1,4-diphenylpyrimidine (LXXIV) whereas if *N*-phenylthiourea is used, the corresponding alternative isomer, 1,2-dihydro-4-methyl-1,6-diphenyl-2-thiopyrimidine (LXXV; X = S) is formed. This latter can easily be converted into the oxo-analogue (LXXV; X = O), thus providing practical preparation of both isomers.

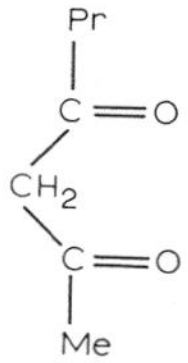

(LXIX) 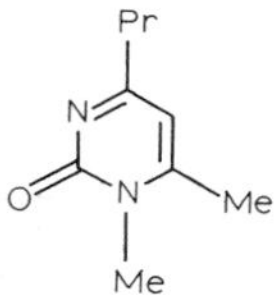 (LXX) 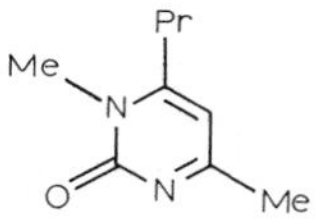 (LXXI)

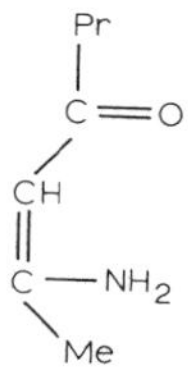

(LXXII)

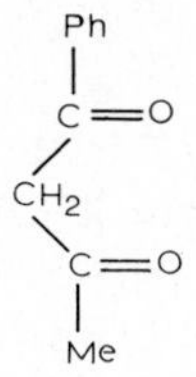

(LXXIII) 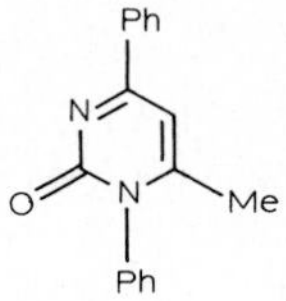   (LXXIV) 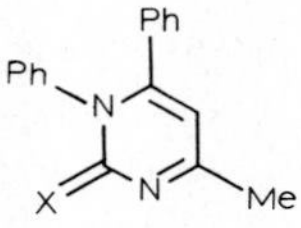   (LXXV)

(iv) From β-aldehydo esters

These compounds are the preferred starting materials for most of the syntheses leading to the pyrimidine bases which occur in nucleic acids. Ethyl sodioformylacetate (LXXVI) reacts in aqueous solution with thiourea to give 2-thiouracil (LXXVII; X = S). Initial attempts to replace thiourea with urea failed, but eventually (*J.N. Davidson* and *O. Baudisch*, J. Amer. chem. Soc., 1926, **48**, 2379) use was made of the transient production of formylacetic acid from malic acid in fuming sulphuric acid to produce a reasonable (50%) yield of uracil (LXXVII; X = O) following the addition of urea. When guanidine is substituted for urea, isocytosine (2-amino-4-hydroxypyrimidine or 2-aminopyrimdin-4(3*H*)-one) (LXXVIII) is produced. β-Methylmalic acid can be used to give pyrimidines with a methyl group in the 5-position and thus its reaction with urea gives 5-methyluracil (otherwise known as thymine or 2,4-dihydroxy-5-methylpyrimidine or 5-methylpyrimidine-2,4(1*H*,3*H*)-dione) (LXXIX).

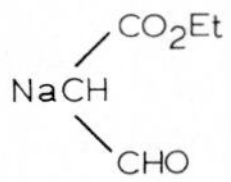

(LXXVI) 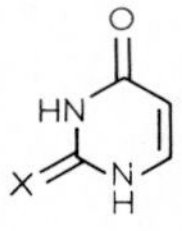   (LXXVII)

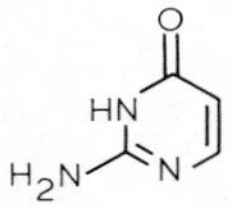

(LXXVIII) 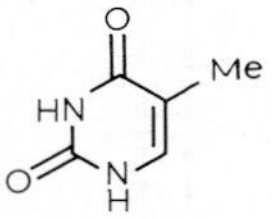   (LXXIX) 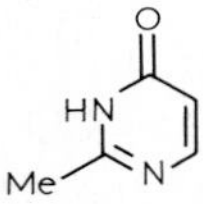   (LXXX)

As usual, amidines can be used as the source of the hetero-atoms and acetamidine with ethyl formylacetate in aqueous solution gives 4-hydroxy-2-methylpyrimidine (LXXX). Diethyl ethoxymethylenemalonate (LXXXI) has been used as a precursor of an aldehydo ester (LXXXII) whereas it could potentially react as a diester. However, only the product from the aldehydo ester is found showing that the aldehydo group is the more reactive. The reaction of compound (LXXXI) with acetamidine thus gives 5-ethoxycarbonyl-4-hydroxy-2-methylpyrimidine (LXXXIII). Other aldehyde ester equivalents which have been used are ethyl propynoate (LXXXIV) (*K. Gupta, K. Saxena* and *P.C. Jain*, Synthesis, 1981, 905), 2,3-dichloroethylpropenoate (LXXXV), dimethyl ethoxymethylenemalonate (LXXXVI) (*S. Yurgi et al.*, Chem. Pharm. Bull. Japan, 1971, 19, 2354) and methyl 3-methoxy-2-methoxymethylenepropionate (LXXXVII) (*T. Nishino, Y. Miichi* and *K. Tokuyama*, Tetrahedron Letters, 1970, 4335).

EtOHC=C(CO$_2$Et)$_2$ OHCCH(CO$_2$Et)$_2$

(LXXXI)   (LXXXII)   (LXXXIII)

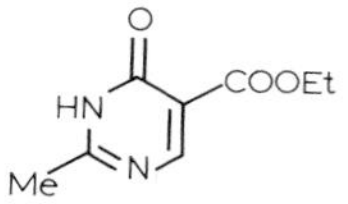

CH≡CH—COOEt   ClCH=CCl—COOEt   EtOCH=C(COOMe)$_2$

(LXXXIV)   (LXXXV)   (LXXXVI)

MeOCH=C—COOMe
          |
          CH$_2$OMe

(LXXXVII)

(*v*) *From β-keto esters*

The reactions of β-keto esters are very similar to those of β-aldehydo esters and occur in a predictable manner with amidines, guanidines, ureas and thioureas. Thus ethyl acetoacetate condenses with thiourea to give 4-hydroxy-2-mercapto-6-methylpyrimidine (LXXXVIII; R = SH), with acetamidine to give 4-hydroxy-2,6-dimethylpyrimidine (LXXXVIII; R = CH$_3$), with guanidine to give 2-amino-4-hydroxy-6-methylpyrimidine (LXXXVIII; R = NH$_2$) and with urea to give 2,4-dihydroxy-6-methylpyrimidine (LXXXVIII; R = OH) *via* an isolatable intermediate, ethyl β-ureidocrotonic acid (LXXXIX). Many β-keto ester equivalents have also been used. Diphenylcyclopropenone (XC) and benzamide oxime (XCI) give 4-hydroxy-2,5,6-triphenylpyrimidine (XCII), and α-acetyl-γ-butyrolactone

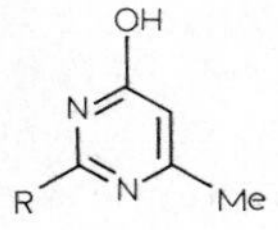

(LXXXVIII) 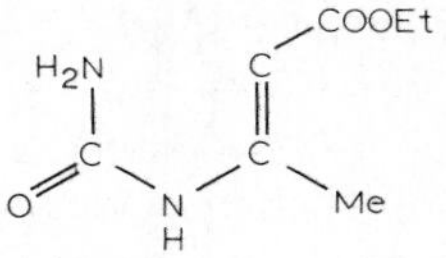   (LXXXIX)

(XCIII) and 3-ethoxymethylene-2,3-dihydro-5-methyl-2-oxofuran (XCIV) have also been used. Diketene (XCV) acts as ethyl acetoacetate and the thiol-lactone (XCVI) as ethyl 2-mercaptoethylacetoacetate (XCVII). Thus diketene gives 2,4-dihydroxy-6-methylpyrimidine (XCVIII) when reacted with urea and compound (XCVI) gives the thiol (XCIX) when it reacts with guanidine.

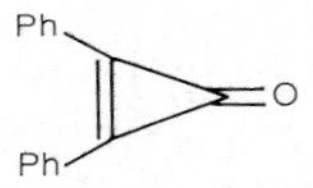

(XC) 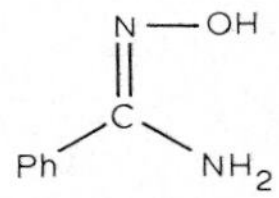   (XCI) 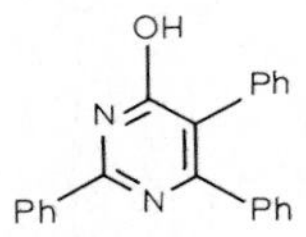   (XCII)

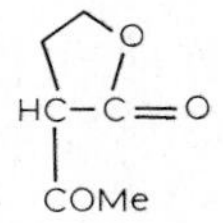

(XCIII) 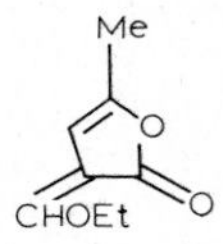   (XCIV) 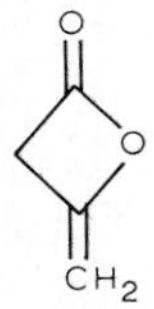   (XCV)

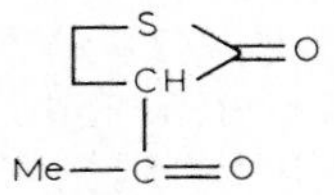

(XCVI) 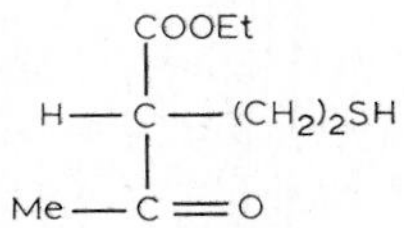   (XCVII) 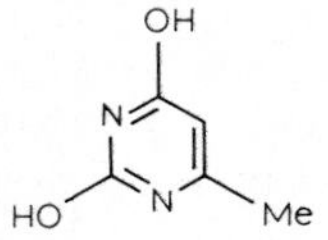   (XCVIII)

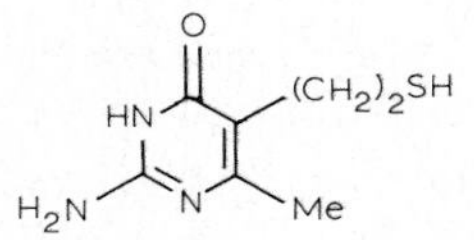

(XCIX)

(vi) From β-diesters

All the barbituric acid derivatives have been synthesised from β-diesters and hence there are thousands of examples of this type of synthesis recorded in the literature. As usual, the nitrogen atoms are provided by either an amidine, a urea, a thiourea or a guanidine and only a few representative examples can be given here as Tables listing many individual examples can be found in the reviews already cited.

Thus, even formamidine reacts at room temperature with diethyl malonate to give 4,6-dihydroxypyrimidine (C) in 80% yield. (*G.W. Kenner et al.*, J. chem. Soc., 1943, 388).

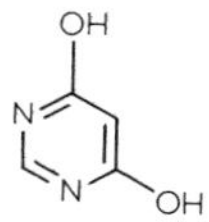

(C)

The reaction of ureas is rather more difficult and usually requires elevated temperatures in a sealed tube. Dimethyl α,α-dimethoxymalonate reacts with urea in methanolic sodium methoxide under reflux to give a 48% yield of 5,5-dimethoxybarbituric acid (CI) after 8 h (*Y. Otsuji, S. Wake* and *E. Imoto*, Tetrahedron, 1970, **26**, 4293).

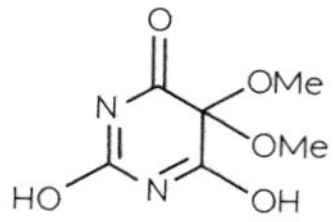

(CI)

In order to reduce these rather drastic conditions, condensation is often carried out in acetic anhydride or phosphorus oxychloride. N,N'-Dimethylurea can be condensed with diethyl malonate in the presence of sodium ethoxide or with malonyl chloride in phosphorus oxychloride to give 1,3-dimethylbarbituric acid (CII; $R = CH_3$). Malonic acid reacts readily with NN'-dicyclohexylcarbodiimide (CIII) in tetrahydrofuran to give 1,3-dicyclohexylbarbituric acid (CII; $R = C_6H_{11}$) in 65% yield which is much superior to the 5.5% yield obtained from the normal condensation of NN'-dicyclohexylurea and malonyl dichloride (*A.K. Bose* and *S. Garrett*, Tetrahedron, 1963, **19**, 85).

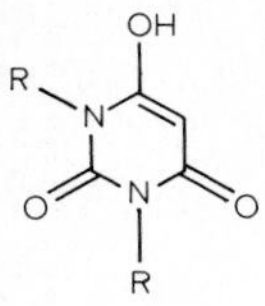

(CII)

(CIII)

Thioureas have not been widely used although thiourea itself condenses with diethyl malonate to give 4,6-dihydroxy-2-mercaptopyrimidine (CIV). Guanidine reacts with diethyl malonate by heating for 30 min in refluxing alcoholic sodium ethoxide to give a reasonable yield of 2-amino-4,6-dihydroxypyrimidine (CV).

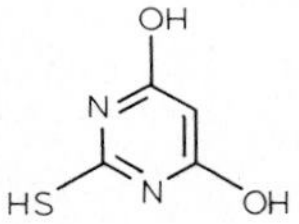

(CIV)

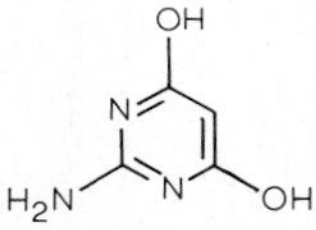

(CV)

Other diester analogues that have been used include carbon suboxide (CVI) which reacts with N,N'-diphenylbenzamidine in ether initially at $-20°$ and then at $25°$ to give anhydro-4-hydroxy-6-oxo-1,2,3-triphenylpyrimidine hydroxide (CVII); methyl 2-ethoxycarbonyldithioacetate (CVIII) which reacts with acetamidine to give a separable mixture of 4-hydroxy-6-mercapto-2-methylpyrimidine (CIX); R = SH) and the ethoxy analogue (CIX; R = OEt); and ethyl 2-methoxy(thiocarbonyl)acetate (CX) which reacts with O-methylurea to give 4-hydroxy-2,6-dimethoxypyrimidine (CXI).

$OC\!=\!C\!=\!CO$

(CVI)

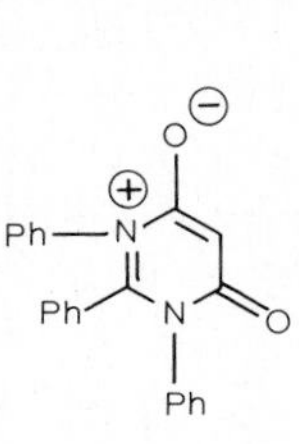

(CVII)

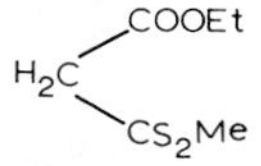

(CVIII) 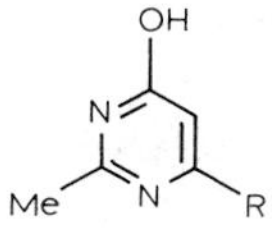 (CIX)

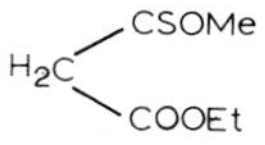

(CX) 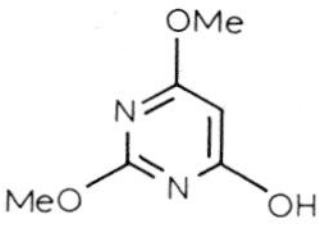 (CXI)

(vii) From β-aldehydo-nitriles

The product from this starting material is a 4- or 6-aminopyrimidine, the nitrogen atoms are provided by the usual amidine, urea, thiourea or guanidine derivatives (*A. Bendich, H. Getler* and *G.B. Brown*, J. biol. Chem., 1949, **177**, 565). A typical example is the condensation of the diethyl acetal of cyanoacetaldehyde (CXII) with urea to give 4-amino-2-hydroxy-pyrimidine (CXIII) otherwise known as cytosine. In refluxing butanol containing sodium butoxide the intermediate (CXIV) is formed which can be ring closed under acidic conditions.

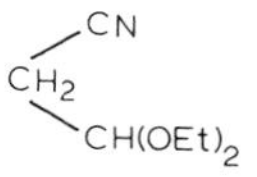

(CXII) 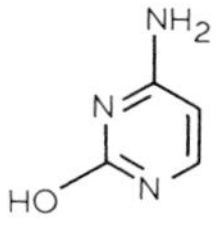 (CXIII) 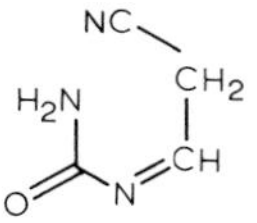 (CXIV)

Condensation of urea with ethyl ethoxymethylenecyanoacetate and related aldehydo-nitriles yields cytosine derivatives substituted at N-3 and C-5. This synthesis is often referred to as the Whitehead synthesis (*S. Senda, K. Hirota* and *J. Notani*, Chem. Pharm. Bull. Japan, 1972, **20**, 1380) and involves the intermediate synthesis of the ureidomethylene compound (CXV) from a mixture of nitrile, orthoformate and *N*-alkylurea. This intermediate can be ring closed in ethanolic base to give the cytosine (CXVI). Thus 3-methyl-5-nitrocytosine (XVI; $R^1 = NO_2$, $R^2 = CH_3$) and 5-cyanocytosine (CXVI; $R^1 = CN$, $R^2 = H$) have been synthesised. Attempted decarboxylation of 3-methylcytosine-5-carboxylic acid (CXVI; $R^1 = COOH$, $R^2 = CH_3$) leads to the methylamino compound (CXVII) following a Dimroth rearrangement.

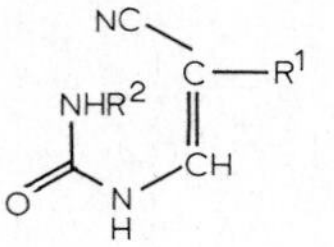

(CXV) 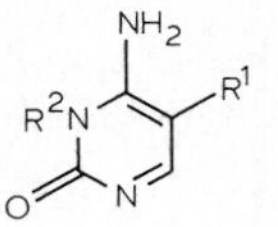   (CXVI) 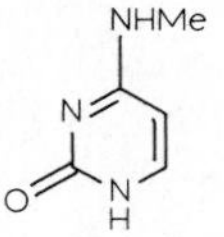   (CXVII)

Reactions of aldehydo-nitriles and guanidine have largely been aimed at the synthesis of analogues of trimethoprim (CXVIII), an antibacterial compound. Trimethoprim itself (*W. Schliemann*, Pharmazie, 1976, **31**, 140) can be synthesised by the reaction of guanidine is ethanol with 2-di-ethoxymethyl-2-formyl-3-(3′,4′,5′-trimethoxyphenyl)proprionitrile (CXIX). The reaction goes *via* the intermediate enol ether (CXX). *N*-Methylthiourea reacts with α-ethoxymethyleneacetonitrile (CXXI) to give the unexpected product (CXXII).

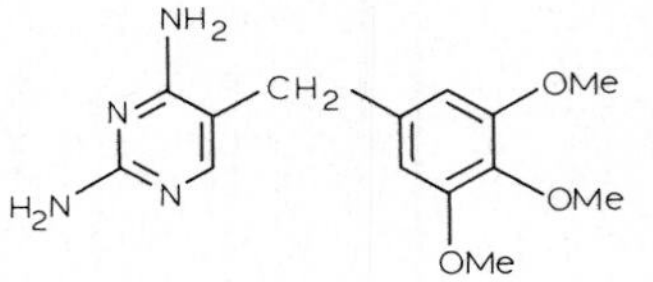

(CXVIII)

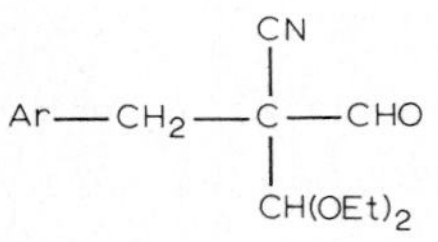

(Ar = 3,4,5-trimethoxyphenyl)
(CXIX)   (CXX)

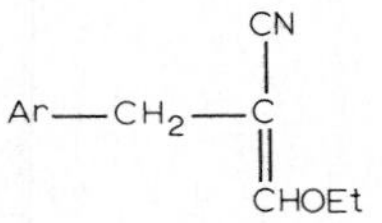

(CXXI) 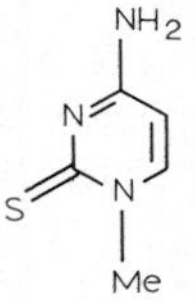   (CXXII)

(viii) From β-keto-nitriles

The synthesis of pyrimidines from β-keto-nitriles has not been used or investigated as extensively as the other syntheses, although some straightforward reactions are reported. Thus the imine of cyanoacetone (β-iminobutyronitrile (CXXIII) reacts with thiourea to give a reasonable yield of 4-amino-2-mercapto-6-methylpyrimidine (CXXIV) but guanidine, urea or acetamidine do not react (*R. Peereboon* and *H.C. Van der Plas*, Rec. Trav. Chim., 1974, **93**, 284). Several 5-aryl-2,4-diaminopyrimidines with potential biological activity have been synthesised from β-keto-nitriles and guanidine, for example 3-butyloxy-3-isopropyl-2-phenylacrylonitrile (CXXV) reacts with guanidine to give 2,4-diamino-6-isopropyl-5-phenylpyrimidine (CXXVI).

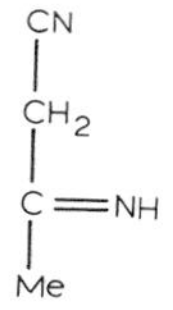

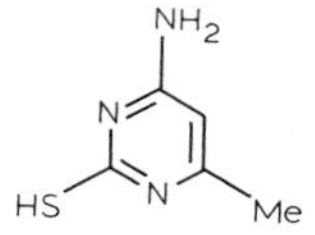

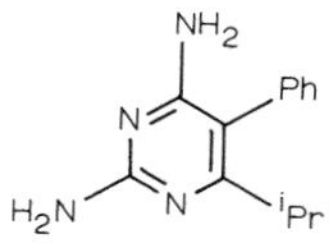

(ix) From β-ester nitriles

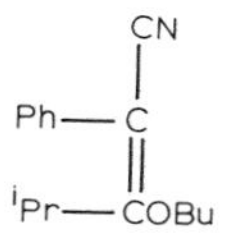

The product of this reaction is a derivative of 4-amino-6-hydroxypyrimidine (*A. Lefebre, J.L. Bernies* and *Ch. Lespagnol*, J. heterocyclic Chem., 1976, **13**, 167). Aliphatic amidines give poor yields but the aryl derivatives, for example, benzamidine, react normally in the presence of ethoxide with ethyl cyanoacetate to give (CXXVII). Ethyl cyanoacetate reacts with urea under basic conditions to give 4-amino-2,6-dihydroxypyrimidine (CXXVIII) (or 6-hydroxycytosine) and *N*-methylurea reacts under similar conditions to give 6-amino-1-methyluracil (CXXIX). With thiourea, ethyl cyanoacetate gives 4-amino-6-hydroxy-2-mercaptopyrimidine (CXXX) and with guanidine 2,4-diamino-6-hydroxypyrimidine (CXXXI) is produced. A variant on the use of an ester nitrile is the use of 2-cyano-γ-butyrolactone (CXXXII) which with urea gives 4-amino-2,6-dihydroxy-5-(β-hydroxyethyl)pyrimidine (CXXXIII).

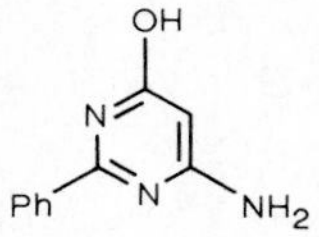

(CXXVII) 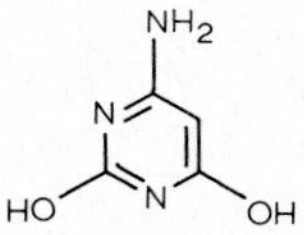 (CXXVIII) 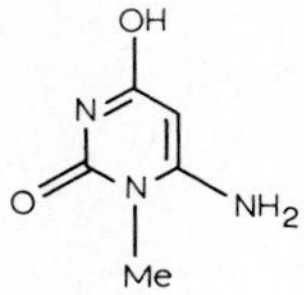 (CXXIX)

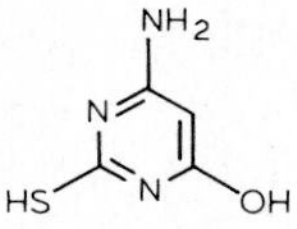

(CXXX) 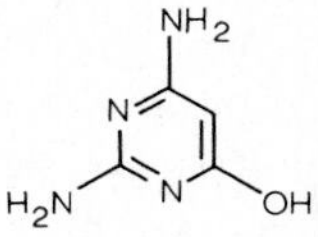 (CXXXI)

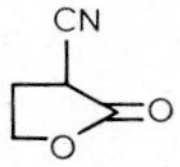

(CXXXII) 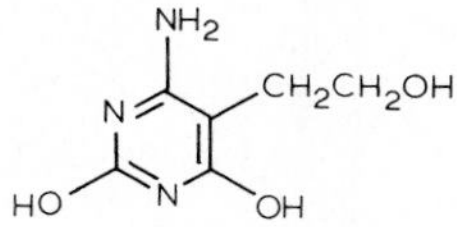 (CXXXIII)

(x) From β-dinitriles

With thiourea and guanidines, malononitriles react normally (*A. Bendich, J.F. Tinker* and *G.B. Brown*, J. Amer. chem. Soc., 1948, **70**, 3109). Thus malononitrile itself (CXXXIV) under basic conditions with thiourea gives 4,6-diamino-2-mercaptopyrimidine (4,6-diaminopyrimidine-2(1*H*)-thione) (CXXXV) and with guanidine gives 2,4,6-triaminopyrimidine (CXXVI). However with amidines and ureas, problems arise but substituted dinitriles such as tricyanomethane (CXXXVII) react normally with benzamidine to finally yield 4,6-diamino-5-cyano-2-phenylpyrimidine (CXXVIII).

$CH_2(CN)_2$

(CXXXIV) (CXXXV) 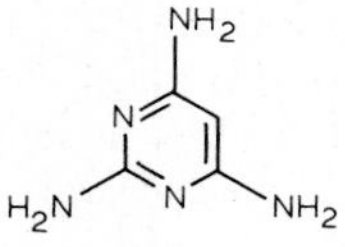 (CXXXVI)

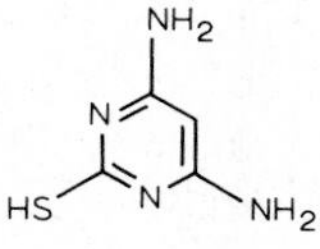

$HC(CN)_3$

(CXXXVII) (CXXXVIII)

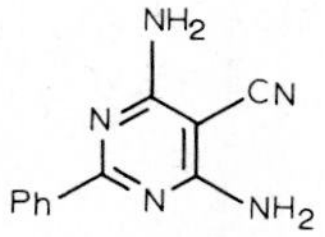

(b) Primary syntheses

(i) Involving the formation of one bond

Some examples of this type of synthesis have already been dealt with in the previous sections, as sometimes an intermediate is isolable which then just requires the final ring closure by forming one bond. However, there are very few pyrimidine syntheses which involve the formation of only one bond and the only one of any importance is the Rinkes synthesis (*I. J. Rinkes*, Rec. Trav. chim., 1927, 268) of uracil which requires the treatment of malediamide (CXXXIX) with sodium hypochlorite to give the intermediate (CXL) which cyclises to give uracil (CXLI).

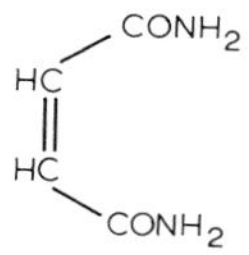

(CXXXIX) 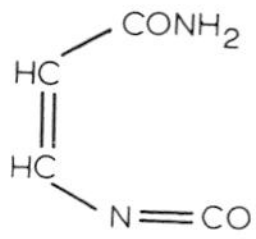   (CXL) 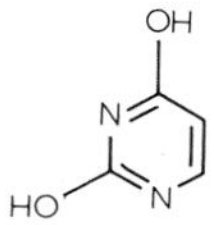   (CXLI)

(ii) Involving the formation of two bonds

(a) [1 + 5] Fragments

(1) *The one-atom fragment provides C-2*. There are several general syntheses which come under this classification starting from malondiamides or malondiamidines (*K. Sasse*, Ann., 1976, 768). A simple example of the former is the reaction of malondiamide (CXLII) with ethyl formate to give 4,6-dihydroxypyrimidine (CXLIII). Various α-substituted malondiamides and *N*-substituted derivatives can be used to give a wide variety of substituted pyrimidines. The ester can also be replaced by an amide or an acid chloride. Thus malondiamide with formamide gives 4,6-dihydroxypyrimidine (CXLIII) and with oxalyl chloride in refluxing benzene gives barbituric acid (CXLIV).

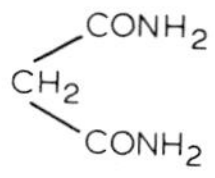

(CXLII) 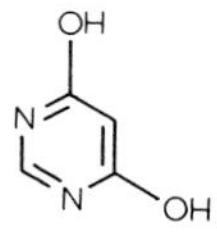   (CXLIII) 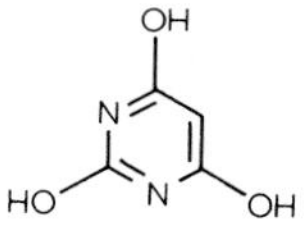   (CXLIV)

Malondiamidine (CXLV) is more reactive than the corresponding amide and reacts in the cold with ethyl formate to give 4,6-diaminopyrimidine (CXLVI). *N*-Substituted diamidines can be used and the ester can be replaced by an amide which increases the versatility of the reaction.

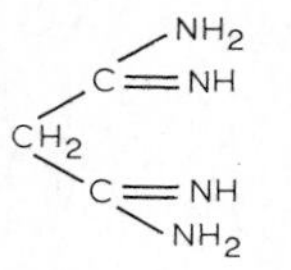

(CXLV) 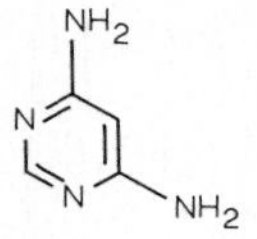   (CXLVI)

(2) *The one-atom fragment provides N-1 or N-3.* The best-known synthesis of this type is due to *R.K. Ralph, G. Shaw* and *R.N. Taylor* (J. chem. Soc., 1959, 1169) which leads to the production of uracil and thiouracil derivatives. The reaction conditions are mild and involve the cyclisation of an aminomethyleneacylurethane (CXLVII) under basic conditions to give a pyrimidine (CXLVIII). The preparation of the intermediate (CXLVII) depends upon whether or not substitution is required in the final pyrimidine at C-5. Thus condensation of cyanoacetic acid with *N*-methylurethane gives compound (CXLIX) which reacts with triethylorthoformate and acetic anhydride to give the intermediate (CL). This is then treated with the

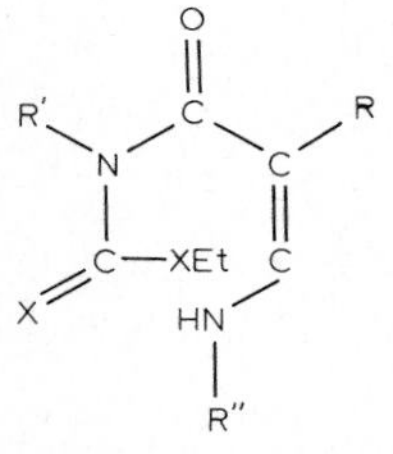

X = O or S, R = H or CN   R = H or CN, X = O or S

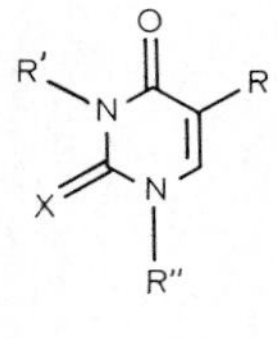

(CXLVII)   (CXLVIII)

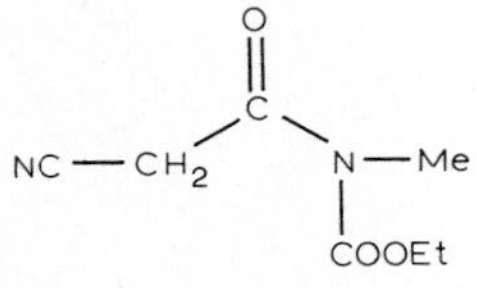

(CXLIX) 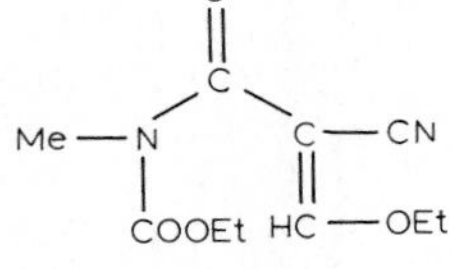   (CL) 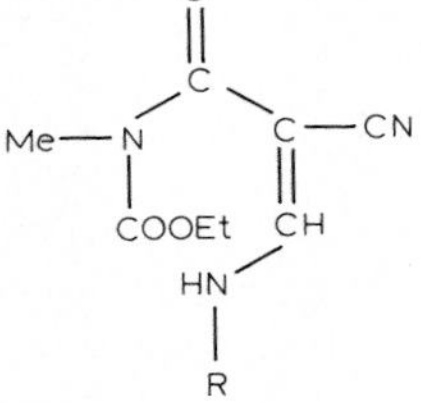   (CLI)

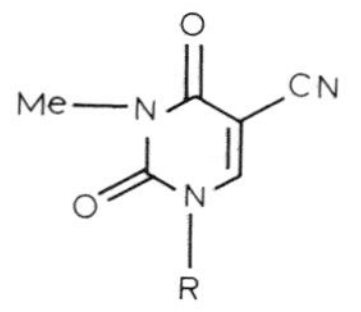

(CLII)

one-nitrogen atom fragment, such as aniline, ammonia or in the case of nucleoside synthesis tri-*O*-benzoylribosylamine, to give compound (CLI) which can be cyclised to the pyrimidine (CLII).

With no substitution or an alkyl group at C-5, the route can be exemplified by the reaction of propiolic anhydride and urethane to give (CLIII) which on treatment with a one-nitrogen atom fragment gives compound (CLIV) which can be cyclised to the uracil derivative (CLV). Many variations on this method are known.

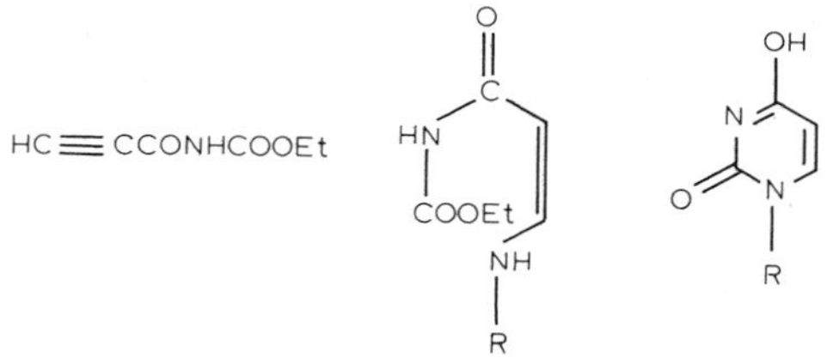

HC≡CCONHCOOEt

(CLIII)   (CLIV)   (CLV)

In a similar reaction sequence, N-3 can be provided in the cyclisation of a β-acylaminovinyl alkyl (or aryl) ketone. β-Acetamidovinyl phenyl ketone (CLVI) reacts with ammonia at 200° to give 2-methyl-4-phenylpyrimidine (CLVII).

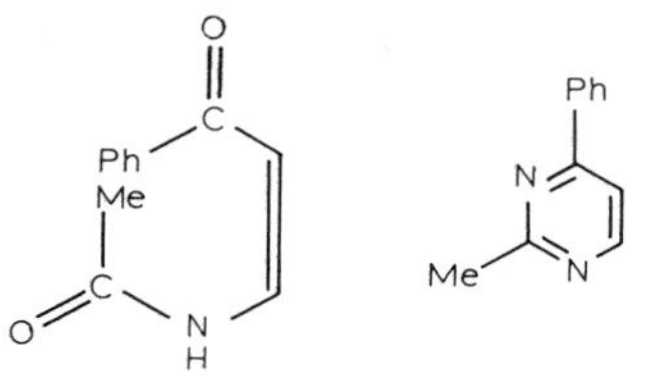

(CLVI)   (CLVII)

(b) [2 + 4] Fragments

The two-atom fragment provides C-2 and N-3. The reaction of methyl isocyanate and ethyl 3-aminocrotonate (CLVIII) gives a substituted urea (CLIX) which can be easily cyclised to 3,6-dimethyluracil (CLX).

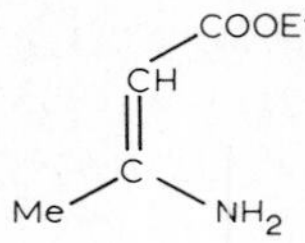

(CLVIII) 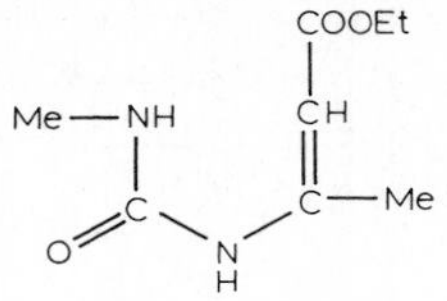   (CLIX) 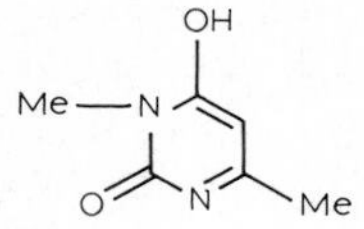   (CLX)

There are a few examples of the condensation of amides or thioamides with aminomethylene derivatives; for example ethyl 3-amino-2-cyanoacrylate (CLXI) and thioacetamide will condense under alkaline conditions to give ethyl 4-amino-2-methylpyrimidine-5-carboxylate (CLXII).

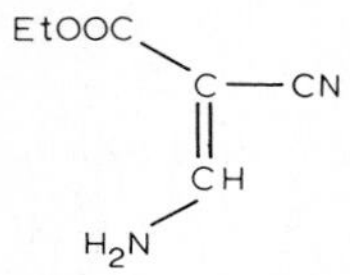

(CLXI) 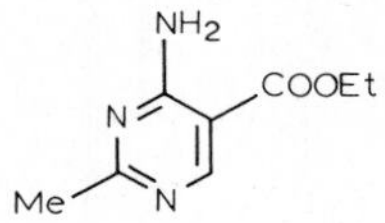   (CLXII)

Methods involving [3 + 3] atom fragments have already been covered and only involve the different methods involved in the Principal Synthesis.

(iii) Involving the formation of three bonds

These reactions involve the reaction of [2 + 2 + 2] or [3 + 2 + 1] atom fragments and have little in common. The original examples of the type were described many years ago and little further development of the methods have occurred because of the subsequent versatility of the Principal Synthesis. A few examples only of some of the 'named' syntheses are given here.

The Frankland-Kolbé synthesis dates from 1848 and involves the trimerisation of a nitrile by heating it with molten potassium or an alkoxide (*G.W. Miller* and *F.L. Rose*, J. chem. Soc., 1965, 3357). Thus the substituted acetonitrile (CLXIII) and potassium ethoxide gives the substituted 4-amino-2,6-dimethylpyrimidine (CLXIV).

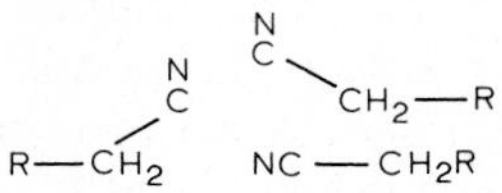

(CLXIII) 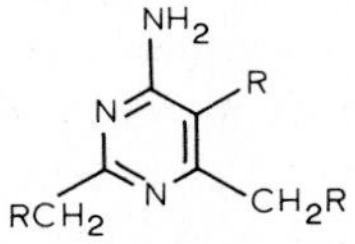   (CLXIV)

The Bredereck synthesis (*H. Bredereck, R. Gomper* and *G. Morlock*, Ber., 1957, **90**, 942) requires the heating of a β-dicarbonyl compound at 200° in an excess of formamide. Thus benzoylacetone (CLXV) when heated in formamide at > 220° gives 4-methyl-6-phenylpyrimidine (CLXVI).

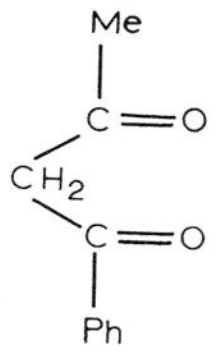

(CLXV) 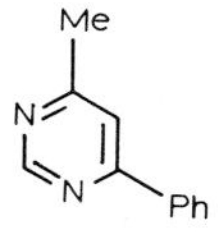   (CLXVI)

The Biginelli reaction (*K. Folker* and *T.B. Johnson*, J. Amer. chem. Soc., 1933, **55**, 3784) was first reported in 1893 and requires the condensation of a ketone, having an unsubstituted methylene group adjacent to the carbonyl group, with an aromatic aldehyde and urea under acidic conditions to give a 2-hydroxydihydropyrimidine. Thus benzaldehyde, urea and ethyl acetoacetate react through several intermediates to give the pyrimidine (CLXVII).

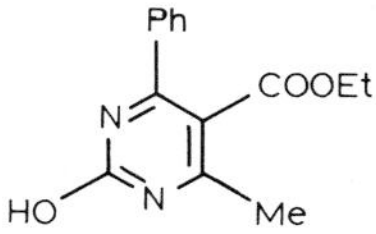

(CLXVII)

(iv) Involving the formation of four or more bonds

There are no generally applicable reactions here and the only method worth mentioning in a general review of this type involves the reaction of an aldehyde, ammonia and a β-dicarbonyl compound which is capable of being oxidised so that a pyrimidine is the final product. Thus α-benzylideneacetylacetone (CLXVIII), benzaldehyde and ammonia (which provides both nitrogen atoms) react to give the intermediate benzylidenepyrimidine (CLXIX) which rearranges to the final product (CLXX) (*F. Krohnke, E. Schmidt* and *W. Zecher*, Ber., 1964, **97**, 1163).

(CLXVIII) PhCH=C(COMe)₂

(CLXIX) [pyrimidine with Me, CHPh, Ph, Me substituents]

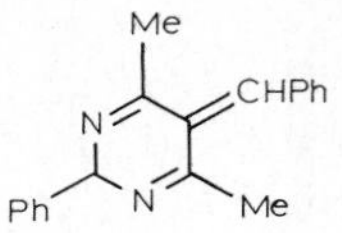

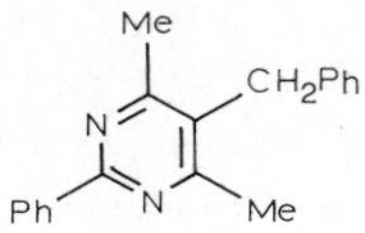

(CLXX) [pyrimidine with Me, CH₂Ph, Ph, Me substituents]

(c) Syntheses of the pyrimidine ring from other heterocycles

Only those syntheses of general use are described here and it is presupposed that the heterocycle used as the starting material has not been formed from a pyrimidine in the first place.

(i) From pyrroles

A pyrrole oxime (CLXXI) when treated with phosphorus pentachloride in ether gives two products (CLXXII and CLXXIII). The first can be converted into the pyrimidine (CLXXIV) on heating and the second can be reduced with zinc/acetic acid to give the pyrimidine (CLXXV).

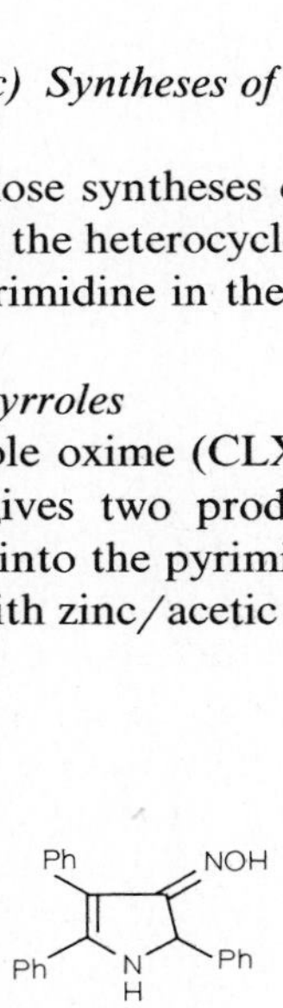

(CLXXI) 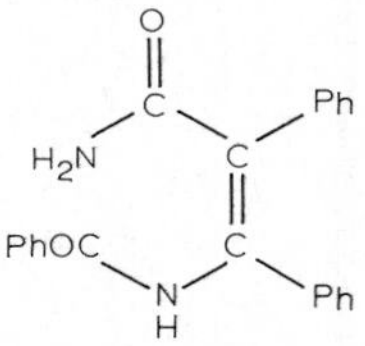   (CLXXII) 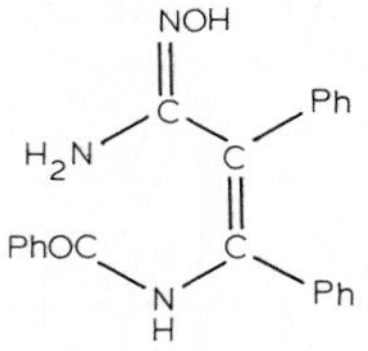   (CLXXIII)

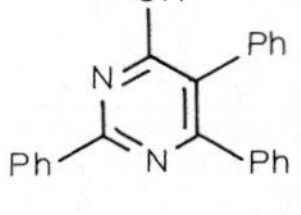

(CLXXIV) 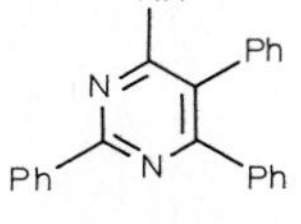   (CLXXV)

(ii) From imidazoles

Although imidazole (CLXXVI) reacts with chloroform at 550° to give a 35% yield of 5-chloropyrimidine (CLXXVII) this reaction is of little preparative value. The only useful example of this type of reaction is the ring opening and subsequent ring closure of hydantoins to give orotic acid products (*H.K. Mitchell* and *J.F. Nye*, J. Amer. chem. Soc., 1947, **69**, 674). Normally the hydantoins have only been suggested as non-isolable intermediates in the reaction of diethyl oxalacetate and ureas but if the reaction is performed in 96% sulphuric acid, the intermediate hydantoin (CLXXVIII) can be isolated and converted into orotic acid (CLXXIX) by the addition of alkali.

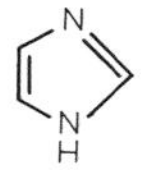

(CLXXVI) 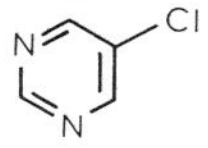   (CLXXVII)

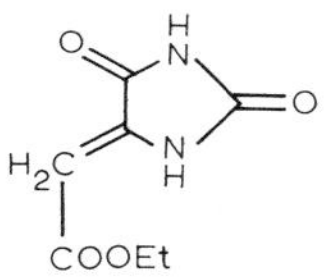

(CLXXVIII) 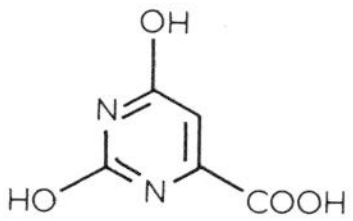   (CLXXIX)

(iii) From mixed 5- or 6-membered heterocycles

Because of the stability of the pyrimidine nucleus, it is often possible to use conditions which cause ring opening of other heterocyclic systems which can then ring close to form pyrimidines. Many of these reactions have limited application. Thus simple oxazines (CLXXX) rearrange in boiling glacial acetic acid to give the corresponding pyrimidine (CLXXXI). Ring opening occurs in a different manner when the oxazine (CLXXXII) is treated with thiopropionamide to give 5-acetyl-4-ethyl-6-hydroxy-2-phenyl-pyrimidine (CLXXXIII).

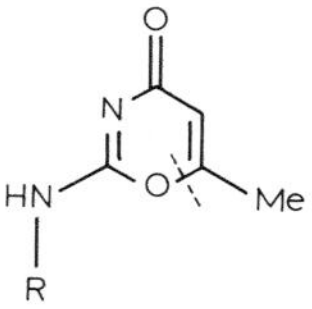

(CLXXX) 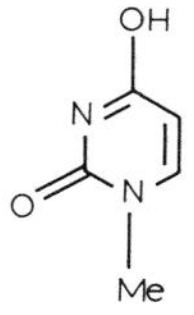   (CLXXXI)

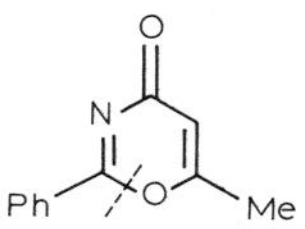

(CLXXXII) 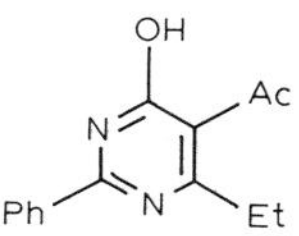   (CLXXXIII)

The oxadiazine ring can also be converted into a pyrimidine ring system. The oxadiazine (CLXXXIV) when heated with malononitrile in triethylamine gives the pyrimidine (CLXXXV).

(CLXXXIV)   (CLXXXV)

Thiazines can also be converted into pyrimidines and this often occurs spontaneously during their preparation. However, the thiazine (CLXXXVI) can be isolated and converted into the pyrimidine (CLXXXVII) by treatment with pyrrolidine at 20°.

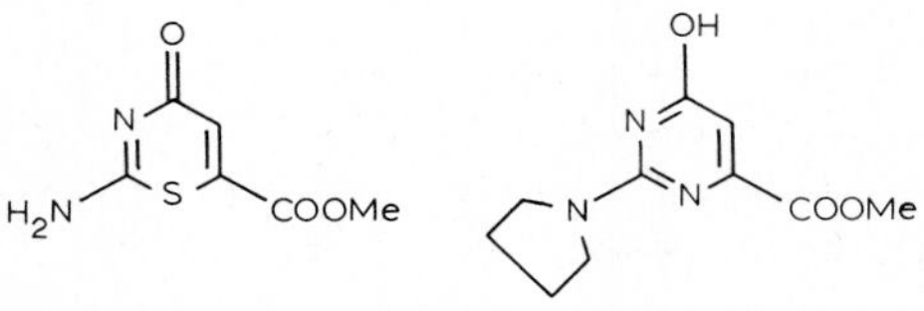

(CLXXXVI)   (CLXXXVII)

(iv) From purines and related heterocycles

Many purines when heated in aqueous alkali yield pyrimidines following fission of the imidazole ring. However, the reaction is of little preparative value as the 5,6-diaminopyrimidines so formed are usually the starting material for purine synthesis. An example is the hydrolysis of 6,9-dimethyl-2-methylthiopurine (CLXXXVIII) to 5-amino-4-methyl-6-methylamino-2-methylthiopyrimidine (CLXXXIX).

(CLXXXVIII)   (CLXXXIX)

Quaternised purines are very easily hydrolysed whether quaternised in the imidazole or pyrimidine ring. Thus the purine (CXC) in mild alkali gives the pyrimidine (CXCI) and the purine (CXCII) gives the pyrimidine (CXCIII).

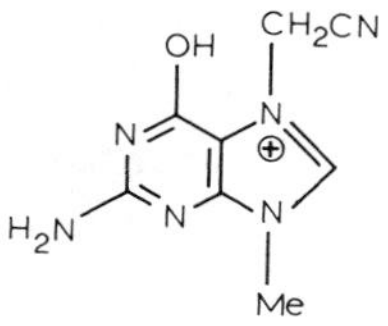

(CXC) 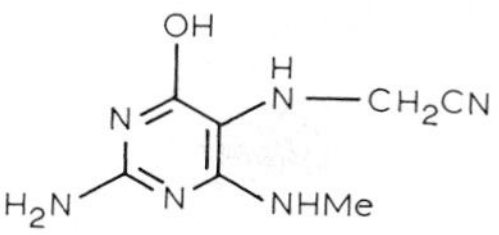 (CXCI)

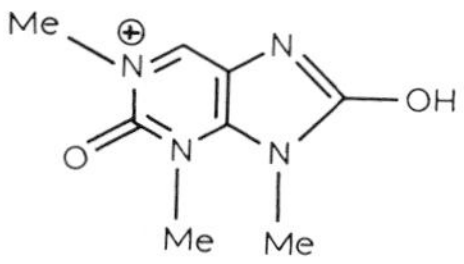

(CXCII) 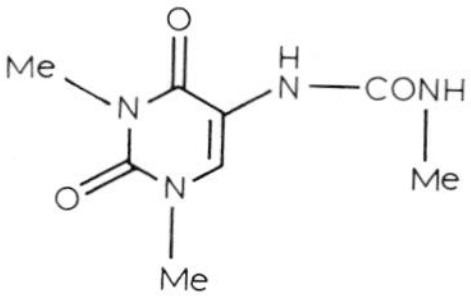 (CXCIII)

Triazolopyrimidines may be degraded to pyrimidines; for example, bromination of 5,7-dimethyl-*o*-triazolo[1,5-α]pyrimidine (CXCIV) gives 2-dibromomethyl-4,6-dimethylpyrimidine (CXCV). The imidazopyrimidine (CXCVI) on treatment with alkali gives 3-carboxymethyluracil (CXCVII).

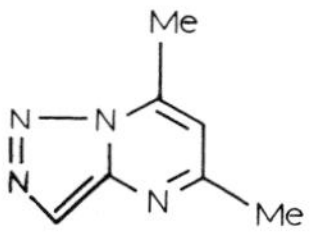

(CXCIV) 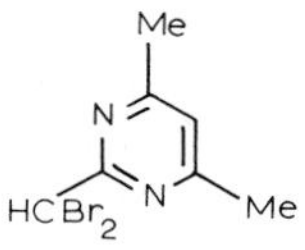 (CXCV)

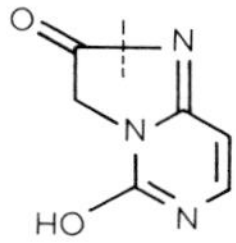

(CXCVI) 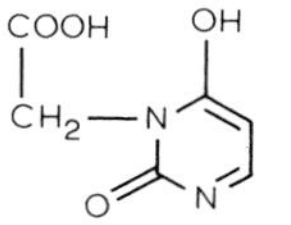 (CXCVII)

(v) Miscellaneous examples

There are many examples of pyrimidine synthesis from other heterocycles. Starting materials have been azoles, pteridines, azanaphthalenes, oxaza and thiaza bicyclic systems, azines, such as pyridazines, and isocyanates, isothiocyanates and active methylene compounds. These methods usually have a very limited, but nonetheless often important, specific use but are outside the range of this review and have been covered in considerable authoritative detail by *Brown* (*loc.cit.*).

(d) Some factors governing the synthesis of substituted pyrimidines

In a review of pyrimidine chemistry (*D.J. Brown*, in Comprehensive Heterocyclic Chemistry, Vol. 3, eds. *A. McKillop* and *R. Boulton*, Pergamon, Oxford, 1984, p. 57), the author has distilled the experience of a life-time working on pyrimidine chemistry into a summary of the preferred synthetic routes to pyrimidine derivatives. Anyone wanting to take advantage of this unique experience when planning a pyrimidine synthesis, should refer to the original text. Here, only a very brief summary of the general points made will be attempted and no reference will be made to specific examples.

It is recommended (*Brown, loc.cit.*, 1984) that for the synthesis of alkylpyrimidines, the necessary groups should be placed in position using an appropriate primary synthesis.

The synthesis chosen for an acyl derivative depends upon the position and nature of the functional groups. Thus pyrimidine-4-carbaldehydes can be made by the Principal Synthesis, but pyrimidine-2-carbaldehyde (CXCVIII) is best made by the reduction of pyrimidine-2-carboxylate (CXCIX). As simple pyrimidines cannot be formylated in the 5-position, suitably activated derivatives have to be used such that following formylation, the activating substituents can be removed. Selective oxidation of alkyl substituents at positions 4 and 6 (but not 5) can sometimes be achieved (*K.Y. Zee Chang* and *C.C. Cheng*, J. heterocyclic Chem., 1967, **4**, 163) and 5-formyluracil (CC; R = CHO) is usually prepared by the oxidation of 5-hydroxymethyluracil (CC; R = CH_2OH).

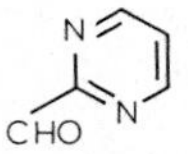

(CXCVIII) 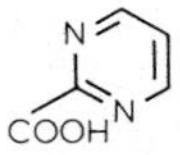 (CXCIX) 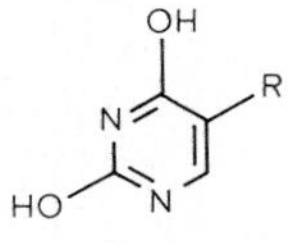 (CC)

Most 5-acetyl (or higher acyl) pyrimidines have to be made by primary synthesis and the Shaw synthesis is particularly valuable in this respect. Some 2- and 4-acylpyrimidines can be made from the corresponding nitrile or carboxylic acid *via* a lithium or magnesium intermediate.

Pyrimidinecarboxylic acids are usually best prepared by oxidation of alkyl- or hydroxymethyl-substituted pyrimidines (*J. Burckhalter, R.J. Siewald* and *W.C. Starborough*, J. Amer. chem. Soc., 1960, **52**, 991) as the pyrimidine ring itself is very resistant to attack. In practice, many pyrimidinecarboxylic acids are prepared by hydrolysis of esters, nitriles or amides as again the ring is relatively stable (*E.F. Godfroi*, J. org. Chem., 1962, **27**, 2264). Esters themselves are normally the products of a primary synthesis but, of course, can also be prepared by esterification of the carboxylic acid (*G.A. Archer et al.*, J. med. Chem., 1977, **20**, 1312). Carboxamides are produced in the normal way from esters or acid chlorides or by the controlled hydrolysis of nitriles which are often the products of a primary synthesis (*E.C. Taylor et al.*, J. Amer. chem. Soc., 1960, **82**, 5711). Once again the Shaw synthesis is particularly useful in the preparation of nitriles which can also be prepared by displacement reactions of halogen, amine or sulphur substituents. The usual organic reactions of dehydration of oximes or amides are also often used to prepare nitriles.

Pyrimidinamines can usually be made (*B. Roth, J.M. Smith* and *M.E. Hultquist*, J. Amer. chem Soc., 1951, **73**, 2864) by one of three methods: primary (usually Principal) synthesis, nucleophilic displacement of halogeno, mercapto, alkylthio and alkoxy groups or by reduction or modification of other groups. Considerable selectivity is possible in the nucleophilic displacement reactions because of the wide variation in activity of substituents at the 2-, 4/6-, and 5-positions. Also, the insertion of one amino group results in the deactivation of the remaining groups, hence increasing the selectivity (*W.F. Keir* and *N.C.S. Wood*, J. chem. Soc. Perkin I, 1976, 1847). Thus despite the fact that initially both halogen atoms in 4,6-dichloropyrimidine (CCI; R = Cl) have identical reactivity, when treated with ammonia, a good yield of 4-amino-6-chloropyrimidine (CCI; R = NH_2) may be obtained. Only a good leaving group in the 5-position can be displaced by ammonia because of the deactivating effect of both nitrogen atoms on substituents in this position; even then rather drastic conditions are required (*H.C. Van der Plas*, Rec. Trav. chim., 1965, **54**, 1101). However, 5-aminopyrimidines are easily prepared by the reduction of 5-nitro or 5-nitroso derivatives since the pyrimidine ring is stable.

(CCI)

Nitropyrimidines are only rarely made by primary synthesis but are usually made by nitration, particularly at the 5-position in a pyrimidine which contains at least one strongly electron-releasing group (*M.E. Bittum, D.J. Brown* and *T.L. Lee*, J. chem. Soc., 1967, 373). 5-Nitrosopyrimidines are also useful precursors for nitropyrimidine synthesis as they are easily oxidised. The nitrosopyrimidines themselves are usually easily formed under mild conditions by treatment of a pyrimidine with nitrous acid.

The preparation of a hydroxypyrimidine depends on whether the substituent is at the 2- or the 4/6-position (the pyrimidinones) or the 5-position (a true pyrimidinol). Practically all the pyrimidinones are prepared by primary synthesis, usually the Principal Synthesis, which provides a great variety of ways of preparing the large number of different isomers possible. In contrast, pyrimidin-5-ols cannot be prepared by primary synthesis and the preparation of precursor molecules bearing substituents capable of conversion into the hydroxyl group is necessary. Thus ethers (hydrogenolysis) or amino compounds (nitrous acid) are possible precursors (*J. Davoll* and *D.H. Laney*, J. chem. Soc., 1956, 2124).

The synthesis of alkoxypyrimidines again depends upon the position of the substituent. Normally the 2- and/or 4/6-alkoxypyrimidines are synthesised from the corresponding pyrimidinones *via* the chloropyrimidines whereas the 5-alkoxypyrimidines are made by the Principal Synthesis. *N*-Alkylated pyrimidinones can be made either by primary synthesis where the Principal Synthesis is normally used, although the Shaw synthesis is also applicable. However, the most widely used route is by alkylation of the corresponding hydroxy compound with an alkyl halide in DMF.

Mercaptopyrimidine synthesis provides a few problems. 2-Mercapto derivatives are easily prepared by the Principal Synthesis but the 4/6-derivatives have to be made by other ways which often involve either thiation of the corresponding pyrimidinones or thiolysis of halogenopyrimidines. Alkylthiopyrimidines can also be converted into thiones by treatment with phosphorus pentasulphide or ether cleavage with aluminium bromide or in the case of benzylthiopyrimidine, reductive cleavage. Very few pyrimidine-5-thiols are known and no general routes are available. 2-Alkylthiopyrimidines are made by primary synthesis but replacement of halogen by an alkylthio group even in the 5-position can often be used. The alkylthio substituent can also be introduced directly into the 5-position of activated pyrimidines such as uracil by using methanesulphenyl chloride which is generated *in situ*.

Halogeno derivatives are normally made from the corresponding hydroxypyrimidines using standard reagents. Direct halogenation will usually result in the formation of a 5-halogenopyrimidine and the normal route from an amino compound *via* diazotization and treatment with copper(I) halide can be used for the introduction of halogen at any pyrimidine position. Transhalogenation has also been used particularly for iodo- and fluoro-pyrimidines but not normally at the 5-position, although there are exceptions.

Finally, pyrimidine-*N*-oxides can be made directly or *via* their *N*-alkoxy analogues using the Principal Synthesis or by peroxide oxidation of the parent pyrimidine.

4. Chemical reactivity and properties of pyrimidine derivatives

(a) General properties

The pyrimidine ring is an aromatic system which has been deactivated towards electrophilic attack by the presence of the two electron-withdrawing nitrogen atoms, each of which reinforces the effect of the other. Thus positions 2,4 and 6 are particularly electron deficient and reactions at these positions or of substituents at these positions are somewhat similar to those occurring at the α- and γ-positions in pyridine and the 2-, 4- and 6-positions in 1,3-dinitrobenzene. Thus a halogen substituted at one of these positions is easily displaced in a nucleophilic substitution reaction by, for example, ammonia, amines, water (under acidic or basic conditions) or a sodium alkoxide. Carboxylic acid groups at these positions are labile and alkyl groups are easily oxidised.

The 5-position in pyrimidine is the least electron-deficient position and substituents here exhibit more normal aromatic properties. The properties of substituents here are more like those found in the β-position of pyridine and electrophilic reagents, if they react at all, react at this position.

However, the properties of pyrimidines are greatly modified by the introduction of electron-releasing substituents into positions 2, 4 or 6. The presence of such a substituent tends to offset the deactivating effect of the ring nitrogen atoms and the reactivity of the ring, and its substituents, takes on a more benzenoid character. Thus normal electrophilic substitution reactions can occur at position 5 and substituents at the position 2, 4 and 6 are no longer so easily replaced.

Electron-withdrawing substituents such as nitro have until recently only been introduced at the 5-position where their effect is to increase the electron deficiency of the remaining positions and make substituents there even more susceptible to displacement by nucleophiles. However, the synthesis of 2-nitropyrimidine has been reported (*E.C. Taylor, C.-P. Tseng* and *J.B. Rampal*, J. org. Chem. 1982, **47**, 552).

As mentioned previously, a pyrimidine having a hydroxy, mercapto or amino group in position 2, 4 or 6 is tautomeric with the oxo, thio or imino form, with the hydrogen located on a ring nitrogen atom. Thus, alkylation of a compound such as 2-hydroxypyrimidine can occur at nitrogen or oxygen. Hydroxypyrimidines prefer the oxo form and mercaptopyrimidines the thio form but aminopyrimidines exist in the amino form.

(b) Reactivity of the ring positions

(i) To electrophiles

As already noted, positions 2, 4 and 6 in unsubstituted pyrimidine are electron deficient and hence one could expect electrophilic attack to take place only at position 5, if at all, but firstly we need to consider attack at nitrogen to give quaternary compounds. As expected, alkyl halides will react with simple pyrimidines to give 1-alkylpyrimidinium halides (CCII), however if an amino, hydroxy or mercapto substituent is present at position

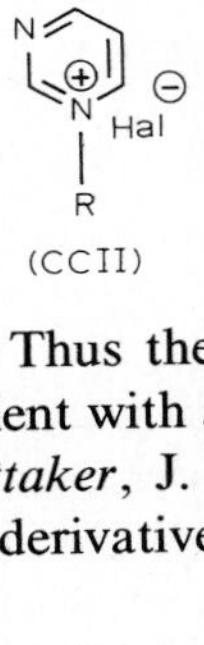

(CCII)

2 or 4, other reactions intervene. Thus the quaternary compound (CCIII) forms the imine (CCIV) on treatment with alkali and the imine undergoes a Dimroth rearrangement (*N. Whittaker*, J. chem. Soc., 1958, 1646) to give the more stable *N*-methylamino derivative (CCV). Pyrimidones are weak

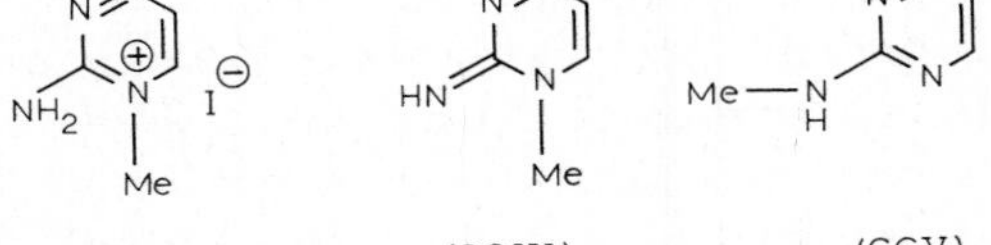

(CCIII)   (CCIV)   (CCV)

bases and do not quaternise, but the derived anion reacts easily with an alkyl halide to give (CCVI) which may then react with a further mole of alkyl halide to quaternise the other nitrogen atom to give (CCVII). Mercapto derivatives react at sulphur rather than at the ring nitrogen atom.

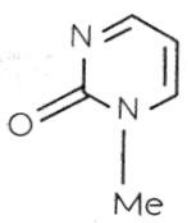

(CCVI) 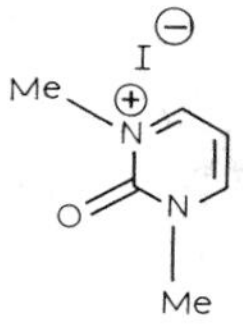 (CCVII)

Electrophilic aromatic substitution reactions of pyrimidine are difficult to achieve. The vigorous conditions required mean that the reaction involves the cation, which makes matters worse, and the cation eventually decomposes. Thus few electrophilic aromatic substitution reactions of pyrimidine itself are known but once even one electron-releasing substituent has been introduced into the molecule, nitration at the 5-position is possible using the drastic conditions of potassium nitrate in concentrated sulphuric acid at temperatures above 90° (*I. Wempen, H.U. Blank* and *J.J. Fox*, J. heterocyclic Chem., 1969, **6**, 593). Thus 2-hydroxypyrimidine (pyrimidin-2(1*H*)-one) (CCVIII; R = H) yields 2-hydroxy-5-nitropyrimidine (CCVIII; R = NO_2) under these conditions. The addition of further

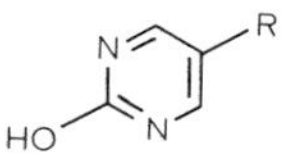

(CCVIII) 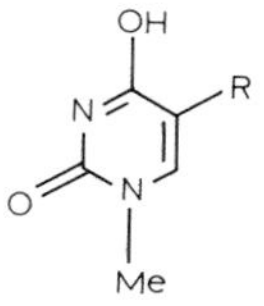 (CCIX)

electron-donating substituents increases the ease of electrophilic substitutions so that *N*-methyluracil (CCIX; R = H) will nitrate at room temperature to give the 5-nitro derivative (CCIX; R = NO_2) (*D.J. Brown, E. Hoerger* and *S.F. Mason*, J. chem. Soc., 1955, 211).

The reaction of unsubstituted pyrimidine with halogens can be achieved, although the conditions are rather drastic but the ring is stable to the conditions. Thus bromine will react with pyrimidine in the vapour phase at 230° to give 5-bromopyrimidine (CCX). Although the other 5-halopyrimi-

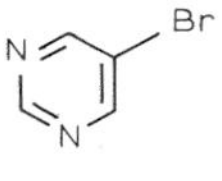

(CCX)

dines are known, they are not prepared by direct halogenation of the free base. Halogenation of alkyl-substituted pyrimidine is not satisfactory as the alkyl group is usually more susceptible to halogenation than is the pyrimidine ring. However, addition of electron-releasing substituents makes halogenation proceed very smoothly (*T. Nishiwaki*, Tetrahedron, 1966, **22**, 2401) although there are some anomalous reactions. Barbituric acid (CCXI; $R^1 = R^2 = H$) for instance gives initially the 5-bromo derivative (CCXI; $R^1 = H$, $R^2 = Br$) but then reacts further to give the 5,5-dibromo derivative (CCXI; $R^1 = R^2 = Br$). Uracil (CCXII; $R = OH$, $R' = H$) and cytosine

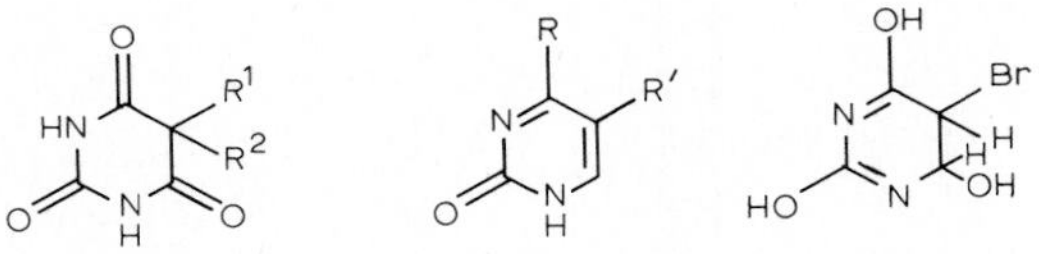

(CCXII; $R = NH_2$, $R' = H$) react with bromine water, the initial reaction being addition of hypobromous acid across the C-5,C-6 double bond, to give 5-bromo-5,6-dihydro-2,4,6-trihydroxypyrimidine (CCXIII). 5-Bromouracil is readily obtained from this but in the presence of an excess of bromine, further substitution can take place.

Other electrophilic substitution reactions which have been reported include sulphonation (as expected at the 5-position) but only on compounds bearing at least one electron-releasing group, formylation which requires compounds with 2 or 3 electron-releasing groups and hydroxymethylation (for example the reaction of uracil (CCXII; $R = OH$, $R' = H$) with paraformaldehyde to give the corresponding 5-hydroxymethyl derivative) (*R.E. Cline, R.M. Fink* and *K. Fink*, J. Amer. chem. Soc., 1959, **81**, 2521).

The latter reaction has been extended to the more general acid-catalysed hydroxyalkylation of uracil using an aromatic aldehyde containing an electron-deficient ring. Thus uracil (CCXII, $R = OH$, $R' = H$) reacts with 4-nitrobenzaldehyde in concentrated hydrochloric acid at 60° for 8 h to give a good yield of 5-(4-nitrophenylhydroxymethyl)uracil (CCXII; $R = OH$, $R' = NO_2C_6H_4CH_2OH$) (*B.L. Law* and *L.N. Pridgen*, J. org. Chem., 1986, **51**, 2592).

Unlike the case of pyridine where the *N*-oxides are much more susceptible to electrophilic attack, pyrimidine-*N*-oxides apparently show no such tendency.

(ii) To nucleophiles

Although the two nitrogen atoms in pyrimidine which are responsible for the deactivation of the ring towards electrophilic attack are meant to enhance the possibility of nucleophilic attack, there are in fact very few examples of this type of reaction with unsubstituted pyrimidines (*H. Bredereck, R. Gomper* and *H. Herlinger*, Ber., 1958, **91**, 2832). Some amination reactions using sodamide have been reported. The product of reaction with a Grignard or alkyllithium reagent is from addition across the 3,4-bond to give an adduct (CCXIV; R = MgBr or Li) which decomposes on work up to give a dihydropyrimidine (CCXIV; R = H).

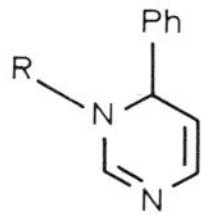

(CCXIV)

(iii) Photochemical reactions

In general there are few examples of photochemical reactions in the literature and the mechanisms of such reactions are not clear. The only comprehensive data available concerns the pyrimidines uracil, thymine and cytosine because of the DNA-damaging effects of u.v. irradiation. Thus thymine (CCXV; R = CH$_3$) has been studied most thoroughly. In aqueous solution, a series of oxidation products is formed (CCXV; R = CH$_2$OH,

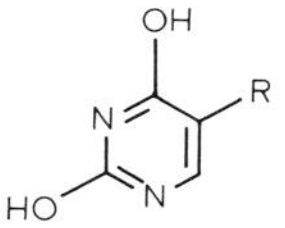

(CCXV)

CHO, COOH or H) but in frozen aqueous solution photodimers of the cyclobutane type are produced, a reaction which is more relevant to the DNA-damaging effect of radiation. The stereochemistry of the main product is (*cis-syn*).

Other related pyrimidines have been shown to undergo a similar reaction (*G.M. Blackburn* and *R.J.H. Davies*, J. chem. Soc., 1966, 2239).

(iv) Oxidation and reduction

In general, the pyrimidine ring is relatively stable towards oxidation and the oxidative degradation of the ring is of little preparative use. A hydroxyl group can be introduced into the 5-position of 2-hydroxy-4,6-dimethylpyrimidine (CCXVI; R = H) using the Elbs persulphate reaction (*R. Hull, J. chem. Soc.*, 1956, 2033) to give (CCXVI; R = OH). Another example is the production of 2,4-diamino-5,6-dihydroxypyrimidine (CCXVII; R = H) using the same reagent but in this case the intermediate (CCXVII; R = SO$_3$H) can be isolated. Substituted pyrimidine rings become more suscepti-

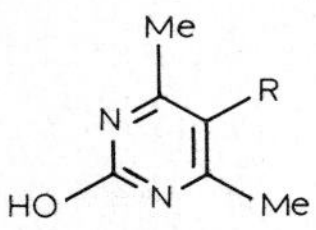

(CCXVI) 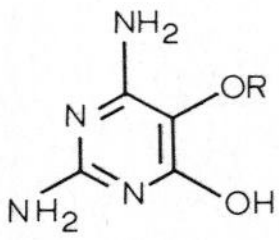　　　(CCXVII)

ble to oxidation and thus thymine, uracil and cytosine are easily oxidised by alkaline permanganate. The initial reaction is *cis*-hydroxylation of the 5,6-double bond followed by hydrolysis to give urea (and also biuret in the case of cytosine) (*M.H. Benn, B. Chatamra* and *A.S. Jones, J. chem. Soc.*, 1960, 1014; *Chatamra* and *Jones, ibid.*, 1963, 811). Reaction with a peracid, usually peracetic, gives pyrimidine -*N*-oxides.

Reduction of the pyrimidine ring is usually achieved by hydrogenation in the presence of a palladium or platinum catalyst. Any halogen substituent is removed and double bonds not substituted are usually reduced until the tetrahydropyrimidine is produced. Thus pyrimidine itself give the tetrahydro derivative (CCXVIII) whereas uracil gives the 5,6-dihydro derivative (CCXIX). Lithium aluminium hydride gives a di- or tetra-hydro derivative

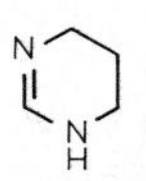

(CCXVIII) 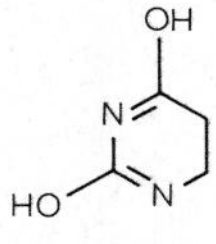　　　(CCXIX)

but sodium borohydride and many other classical reducing reagents have no effect.

(c) Properties of various pyrimidine types

(i) Pyrimidine, alkyl- and aryl-pyrimidines

Pyrimidine, m.p. 22.5°, b.p. 124°/758 mm, is best prepared by one of two following syntheses. 3-Aminoacrolein (CCXX) when heated with formamide and piperdinium acetate at

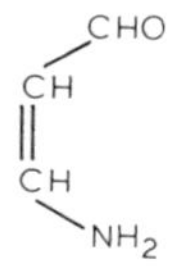

$$
\begin{array}{c}
\text{CH}-\text{CHO} \\
\parallel \\
\text{CH}-\text{NH}_2
\end{array}
$$

(CCXX)

140° gives pyrimidine in 60% yield. The alternative route starts from a 2-halopyrimidine which can be dehalogenated by heating with hydriodic acid for a few minutes to give pyrimidine in *ca.* 40% yield. Pyrimidine a water-soluble weak base (pK_a 1.3 and -6.92), gives a sparingly-soluble molecular complex with mercury(II) chloride (*N. Whittaker*, J. chem. Soc., 1953, 1646); *gold chloride* complex, m.p. 226°; *picrate*, m.p. 156°; *oxalate*, m.p. 160°; *methiodide*, m.p. 136°. It resists nitration but can be brominated and reactions with methylamine, cyclopropyllithium, hydrazine and pivalic acid have been reported.

Alkylpyrimidines are usually liquids or low melting solids with similar properties to pyrimidine and forming similar salts and complexes. They are more lipophilic than pyrimidine and the methylpyrimidines are slightly stronger bases (4-*methylpyrimidine*, pK_a 2.0).

Alkyl groups positioned at C-2, 4 or 6 are active in the same sense as the methyl group in 2,4-dinitrotoluene whereas alkyl groups in the 5-position show normal properties of such a group directly attached to an aromatic ring. Thus a pyrimidine having a methyl group in the 4/6 or 2- position, can be converted into a styrylpyrimidine (CCXXI) by reaction with benzaldehyde and a suitable catalyst such as acetic anhydride or zinc chloride. 5-Methylpyrimidines do not thus react (*C.E. Londer* and *C.J. Timmons*, J. chem. Soc. C, 1967, 1343).

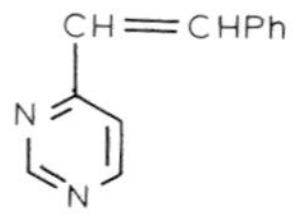

pyrimidine ring with CH=CHPh substituent

(CCXXI)

Oxidation of an alkyl- to a carboxy-pyrimidine can usually be achieved by permanganate, although sometimes it is preferable to proceed *via* the styryl derivative. Selenium dioxide can also be used as oxidant. It is also possible to control the oxidation and produce formylpyrimidines from pyrimidines having an active methyl group using selenium dioxide in acetic acid. Thus 2,4-dihydroxy-5,6-dimethylpyrimidine (CCXXII; R = CH$_3$) gives 6-formyl-2,4-dihydroxy-5-methylpyrimidine (CCXXII; R = CHO (*K.-Y. Zee Chang* and *C.C. Cheng*, J. heterocyclic chem., 1967, **4**, 163).

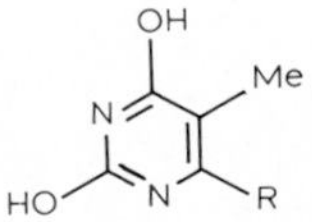

(CCXXII)

Halogenation of alkylpyrimidines with either elemental halogens or an *N*-halosuccinimide rarely gives one product in high yield. The reactions seem to be light-catalysed and clearly more work is required before a rational explanation of the formation of all the products can be made. Thus chlorination of 4-methylpyrimidine (CCXXIII; R = CH$_3$) gives the trichloro derivative (CCXXIII; R = CCl$_3$); 4,5-dimethylpyrimidine (CCXIV; R = CH$_3$) gives the corresponding 4-trichloromethyl derivative (CCXIV; R = CCl$_3$) whereas bromination of 4,6-dichloro-2,5-dimethylpyrimidine (CCXXV; R = CH$_3$) results in preferential monobromination at the C-5 substituent to give (CCXXV; R = CH$_2$Br) (*M. Hasegawa*, Chem. Pharm Bull. Japan, 1953, **1**, 387).

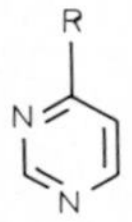

(CCXXIII) 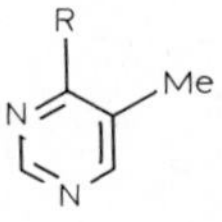   (CCXXIV) 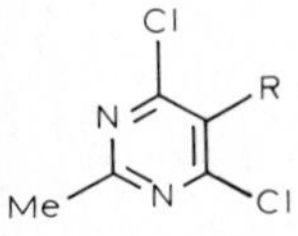   (CCXXV)

Nitric acid does not nitrate alkyl groups but they can be oxidised under these conditions. Nitrosation of activated 2 or 4/6 methyl groups can occur if the ring is not sufficiently activated so that aromatic electrophilic substitution at position 5 is prevented. If 4,6-dimethyl-2-hydroxypyrimidine (CCXXVI; R = CH$_3$) is treated with nitrous acid, then the C-nitroso compound (CCXXVI; R = CH$_2$NO) which is tautomeric with the aldoxime (CCXXVII) is formed.

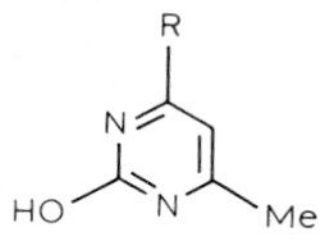

(CCXXVI) 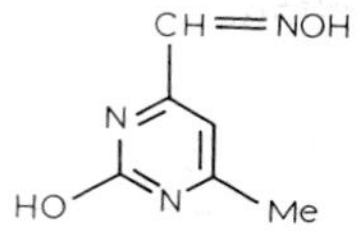   (CCXXVII)

Active methyl groups, 4/6 > 2, react in Claisen condensations provided that amino groups are not present to deactivate them. Thus diethyl oxalate with potassium ethoxide reacts with 4-methylpyrimidine (CCXXIII; R = CH$_3$) to give 4-ethoxyoxalylmethylpyrimidine (CCXXIII; R = CH$_2$CO-CO$_2$Et) (*W. Pfleiderer* and *H. Mosthaf*, Ber., 1957, **90**, 728). Similarly, 2- or 4/6-methylpyrimidines which do not contain more than two other electron-releasing groups will undergo a Mannich reaction at a methyl group. 4-Methylpyrimidine (CCXXIII; R = CH$_3$) reacts with formaldehyde and dimethylamine in acid solution giving a good yield of the Mannich base (CCXXIII; R = CH$_2$CH$_2$NMe$_2$) (*C.G. Overberger* and *I.C. Kogon*, J. Amer. chem. Soc., 1954, **76**, 1879). Methylpyrimidines will also react in the Vilsmeier reaction with dimethylformamide and phosphoryl chloride. Thus 4-methylpyrimidine (CCXXIII; R = CH$_3$) will react *via* the intermediate (CCXXIII; R = C(CHO)CHNMe$_2$) which decomposes on treatment with alkali to give the diformyl derivative (CCXXIII; R = CH(CHO)$_2$).

Arylpyrimidines have not been studied in any detail and even simple reactions like the nitration of 4-phenylpyrimidine (CCXXIII; R = Ph) give a complex series of products, the nature of which depends upon the exact conditions used (*B. Lynch* and *L. Poon*, Canad. J. Chem., 1967, **45**, 1431).

(ii) Halogenopyrimidines

Halogenopyrimidines are usually oils or low melting colourless solids with a characteristic odour and are very often skin irritants. Most of the work has been done on chloropyrimidines as they are most accessible and in cases where comparative reactivities of different halo-substituents can be assessed, there is unfortunately very little difference which can be exploited for preparative purposes. Because of the large variety of reactions which these compounds can undergo, the halogenopyrimidines are among the most versatile intermediates in pyrimidine synthesis.

One of the main uses of the halogeno group is that it can normally be removed reductively under mild conditions which often do not affect other substituents. More selective removal, however, can be achieved if necessary *via* reaction with hydrazine. For hydrogenolysis, zinc dust or hydrogen and palladium can be used. Pyrimidines containing more than one halogen substituent normally show considerable selectivity. Thus 2,4-dichloropyrimidine (CCXXVIII; R = Cl) on treatment with zinc dust in aqueous ammonium chloride gives 2-chloropyrimidine (CCXXVIII; R = H). When substituents are present in the ring which preclude the use of reducing conditions for halogen removal, the silver oxide oxidation of the hydrazino

(CCXXVIII)　　(CCXXIX)

derivative, formed by the reaction of the chloro compound with hydrazine, can be used. Thus 4,6-dichloro-5-nitropyrimidine (CCXXIX; R = Cl), can be converted into the dihydrazino derivative (CCXXIX; R = $NHNH_2$) which will give 5-nitropyrimidine (CCXXIX; R = H) is reasonable yield (*M.E.C. Biffen, D.J. Brown* and *T.C. Lee*, J. chem. Soc., 1967, 573).

Aminolysis of halogenopyrimidines must be one of the most widely used (and abused) reactions in pyrimidine chemistry. As *Brown* (*loc.cit.*) points out, very few kinetic or quantitative studies have been carried out so that most of the reaction conditions record an exercise in "overkill". There is a substantial effect of substituents on the reactivity of a halogen substituent as is shown by the following reactivity sequence: 2-chloro-4,6-dimethylpyrimidine, 1; 2-chloropyrimidine, 2; 5-bromo-2-chloropyrimidine, 200; 2-chloro-5-nitropyrimidine, 3×10^6. As is pointed out by *Brown*, all it requires to remove the guessing from the experiment is to do a single small-scale experiment and a chloride-ion titration to be able to obtain a reasonable estimate of the time and temperature required for a particular substrate (*D.J. Brown* and *P. Waring*, J. chem. Soc. Perkin II, 1974, 704).

Selective reaction is also possible. Thus 2,4-dichloro-6-methylpyrimidine (CCXXX; $R^1 = R^2 = Cl$) reacts with ethanolic ammonia at 90° to give a separable mixture of 4-amino-2-chloro-6-methylpyrimidine (CCXXX; $R^1 = NH_2$; $R^2 = Cl$) and 2-amino-4-chloro-6-methylpyrimidine (CCXXX; $R^1 = Cl$; $R^2 = NH_2$). The second chlorine in both these molecules is deactivated by the presence of the amino group and further reaction to the diamino compound (CCXXX; $R^1 = R^2 = NH_2$) requires treatment for 6 h at 105°.

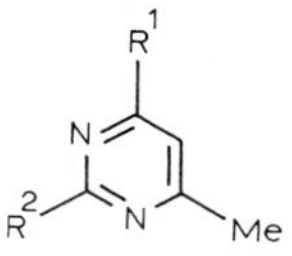

(CCXXX) 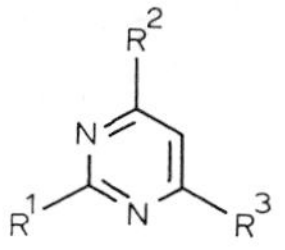   (CCXXXI)

Amination of 2,4,6-trichloropyrimidine (CCXXXI; $R^1 = R^2 = R^3 = Cl$) is remarkably specific. Alcoholic ammonia at 20–100° gives a separable mixture of 2-amino-4,6-dichloropyrimidine (CXXXI; $R^1 = NH_2$; $R^2 = R^3 = Cl$) and 4-amino-2,6-dichloropyrimidine (CCXXXI; $R^1 = R^3 = Cl$; $R^2 = NH_2$). At 160° each of these or the starting material gives 6-chloro-2,4-diaminopyrimidine (CCXXXI; $R^1 = R^2 = NH_2$, $R^3 = Cl$) and at 200°, all the compounds yield 2,4,6-triaminopyrimidine (CCXXXI; $R^1 = R^2 = R^3 = NH_2$) (S. Gabriel, Ber., 1901, **34**, 3362).

The presence of a nitro group increases the reactivity of any halogen substituent so that in the case of 2,4-dichloro-5-nitropyrimidine (CCXXXII; R = Cl), monoamination occurs in a few minutes at 0° with aqueous ammonia to give only one product, 4-amino-2-chloro-5-nitropyrimidine (CCXXXII; $R = NH_2$).

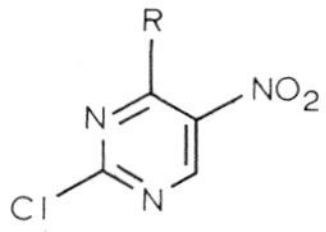

(CCXXXII)

Factors governing the effect of amine structure on the reaction rate are as follows: (i) the rate constant for an *n*-alkylamine is almost independent of chain length or chain branching beyond the γ-carbon atom. (ii) A β-branch has a small and an α-branch a large (one branch reduces the reactivity to 5%, two branches to 0.1%) effect. (iii) Di-*n*-alkylamines react at about the same rate as an α-branched alkylamine. All reactions show a slight preference for the 4/6-chloro substituent rather than for the 2-chloro substituent.

As previously mentioned, hydrazine reacts with chloropyrimidines to give the corresponding hydrazinopyrimidines (CCXXIX; $R = NHNH_2$). Substituted hydrazines react similarly. Sodium azide also reacts with halogenopyrimidines to give the corresponding azidopyrimidine which is usually in equilibrium with the tautomeric tetrazolopyrimidine. Thus 2-chloropyrimidine (CCXXXIII) gives 2-azopyrimidine (CCXXXIV) which exists predominantly as the tetrazolopyrimidine (CCXXXV).

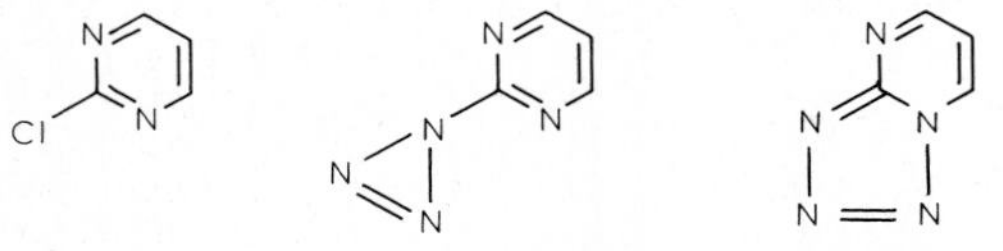

(CCXXXIII)   (CCXXXIV)   (CCXXXV)

Alkoxy groups can easily be substituted for chloro groups in any position in the ring but once again the rate is influenced by substituents already present in the ring and this also results in some selectivity of reaction being possible (*R.G. Shepherd, W.E. Taft* and *H.M. Crazinski*, J. org. Chem., 1961, **26**, 2674). When 2,4-dichloropyrimidine (CCXXXVI; $R^1 = R^2 = Cl$) is treated with one mole of sodium methoxide at room temperature, only 2-chloro-4-methoxypyrimidine (CCXXXVI; $R^1 = Cl$; $R^2 = OCH_3$) is formed whereas even at room temperature, the use of 2 moles gives 2,4-dimethoxypyrimidine (CCXXXVI; $R^1 = R^2 = OCH_3$). 2,4,6-Trifluoro-5-methylpyrimidine (CCXXXVII; R = F) when treated with two moles of sodium benzoxide at 25° gives 2,4-dibenzyloxy-6-fluoro-5-methylpyrimidine (CCXXXVII; $R = C_6H_5CH_2O$). Chloronitropyrimidines react extremely easily with alkoxides and 4,6-dichloro-5-nitropyrimidine (CCXXXVIII; $R^1 = R^2 = Cl$) reacts in the cold with one mole of sodium ethoxide to give exclusively 4-chloro-6-ethoxy-5-nitropyrimidine (CCXXXVIII; $R^1 = Cl$, $R^2 = OEt$). Addition of more ethoxide gives the 4,6-diethoxy derivative (CCXXXVIII; $R^1 = R^2 = OEt$).

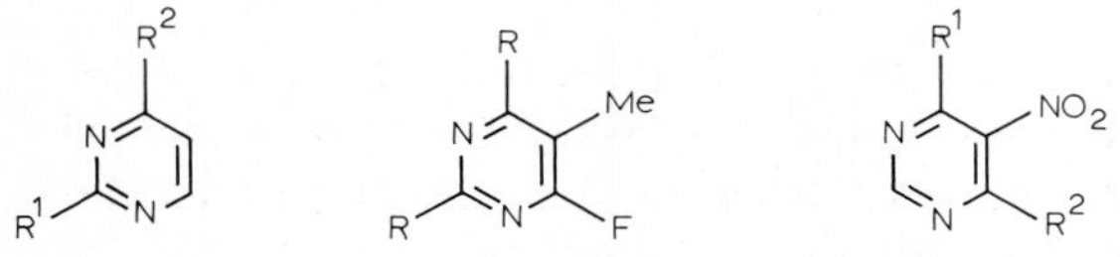

(CCXXXVI)   (CCXXXVII)   (CCXXXVIII)

The hydrolysis of halogenopyrimidines can be done directly either under acidic or basic conditions, although the usual (and often unnecessary) intermediary of an alkoxide is used. 4,6-Dichloropyrimidine (CCXXXIX;

R = Cl) on boiling with hydrochloric acid gives 4-chloro-6-hydroxypyrimidine (CCXXXIX; R = OH), and 4,6-dichloro-2-methylpyrimidine (CCXL; R = Cl) with boiling aqueous sodium hydroxide solution gives 4-chloro-6-hydroxy-2-methylpyrimidine (CCXL; R = OH). Alkaline hydrolyses can be catalysed by the addition of hydrogen peroxide. In the absence of peroxide, the hydrolysis of 2,4-dichloro-6-methylpyrimidine (CCXLI; R = Cl) to 2,4-dihydroxy-6-methylpyrimidine (CCXLI; R = OH) takes 27 h but the addition of peroxide reduces the reaction time to 5 min at 20° (*L.F. Ovechkina, V.I. Gunar* and *S.I. Zavyalov*, Izvest. Akad. Nauk. SSSR Ser. Khim., 1969, 2035).

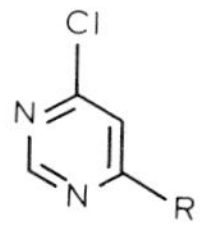

(CCXXXIX) 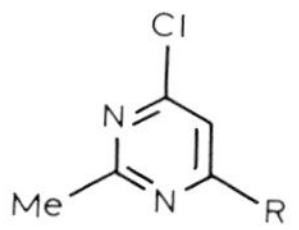   (CCXL) 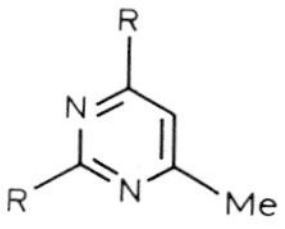   (CCXLI)

Halogeno substituents at any position of a pyrimidine ring can be substituted by an alkyl- or aryl-thio group although the value of the former reaction is not too great as the products are better made by *S*-alkylation of the corresponding thione. Reaction is easier at the 2 or the 4/6 position but a halogen at position 5 can also react. Thus 5-bromopyrimidine is converted into 5-methylthiopyrimidine (CCXLII) after 12 h in boiling ethanolic

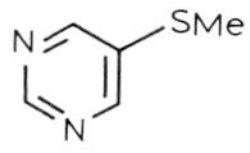

(CCXLII)

sodium methanethiolate. Thiolysis of halogen is possible in all but the 5-position. Either sodium hydrogen sulphide or treatment with thiourea and subsequent alkaline hydrolysis of the resulting thiouronium salt can be used. Thus 6-chloro-4-hydroxy-2-methylpyrimidine (CCXLIII; R = Cl) reacts with sodium hydrogen sulphide in ethanol under reflux to give a high yield of 4-hydroxy-6-mercapto-2-methylpyrimidine (CCXLIII; R = SH). The use of thiourea is illustrated by the conversion of the chloropyrimidine (CCXLIV; R = Cl) *via* the thiouronio intermediate (CCXLV) into the mercapto derivative (CCXLIV; R = SH)

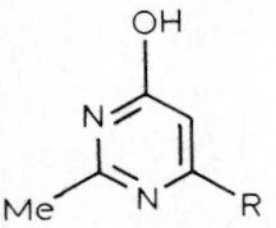

(CCXLIII) 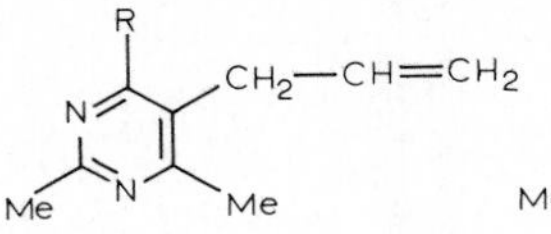   (CCXLIV) 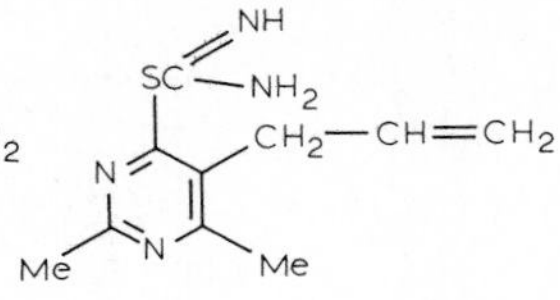   (CCXLV)

A halogen in the 5-position can be replaced by cyanide so that 5-bromo-4-hydroxy-6-phenylpyrimidine (CCXLVI; R = Br) can be converted into the 5-cyano derivative (CCXLVI; R = CN) in high yield by boiling it with copper(I) cyanide in quinoline (*D.J. Brown* and *M.N. Paddon-Row*, J. chem. Soc. C, 1967, 1928).

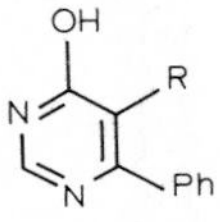

(CCXLVI)

Halogenopyrimidines react with active methylene compounds and other reagents which contain acidic hydrogen. Sodiodiethyl malonate condenses with 2-amino-4-chloro-6-methyl-5-nitropyrimidine (CCXLVII; R = Cl) to give the 4-diethoxycarbonylmethyl derivative (CCXLVII; R = CH(CO$_2$Et)$_2$) which can be hydrolysed and decarboxylated to give 2-amino-4,6-dimethyl-

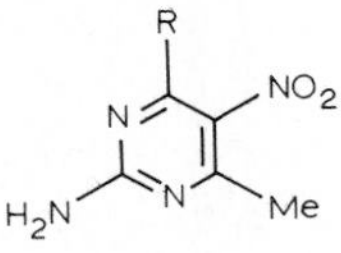

(CCXLVII)

5-nitropyrimidine (CCXLVII; R = CH$_3$). 18-Crown-6 ether has been used to increase the rate of the reaction of the potassium salt of diethyl malonate with 5-bromo-2-fluoropyrimidine (CCXLVIII; R = F) to give a poor yield of 5-bromo-2-diethoxycarbonylmethylpyrimidine (CCXLVIII; R = CH(CO$_2$Et)$_2$). Alkynyl and alkenyl residues also react with halogenopyrimidines. Prop-2-ynol reacts with 4-iodo-2,6-dimethylpyrimidine (CCXLIX; R = I) (in the presence of bistriphenylphosphine palladium dichloride, copper(I) iodide and triethylamine) to give compound (CCXLIX; R =

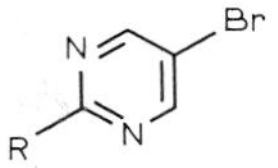

(CCXLVIII) 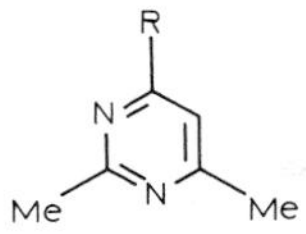   (CCXLIX)

C ≡ CCH$_2$OH). 1,4,5,6-Tetrafluoropyrimidine (CCL; R = F) reacts with hexafluoropropene in the presence of caesium fluoride to give (finally) 5-fluoro-2,4,6-trisheptafluoroisopropylpyrimidine (CCL; R = CF(CF$_3$)$_2$) by successive replacement of the fluorine atoms at C-6, C-4 and finally at C-2.

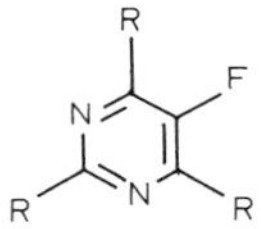

(CCL)

Bipyrimidines can be formed by a variety of reactions (*D.B. Bly* and *M.G. Mellon*, J. org. Chem., 1962, **27**, 2945). 2-Bromopyrimidine will react by the Ullmann reaction (copper in dimethylformamide) to give 2-pyrimidin-2'-ylpyrimidine (CCLI).

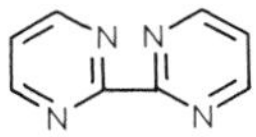

(CCLI)

The Busch reaction on 5-bromo-2-phenylpyrimidine (heating in methanolic potassium hydroxide, hydrazine hydrate and palladium/calcium carbonate) gives the diphenylbipyrimidine (CCLII) (*M.P.L. Caton et al.*, J. chem. Soc. C, 1967, 1204).

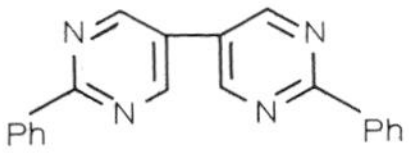

(CCLII)

5-Bromo-2,4-dimethoxypyrimidine (CCLIII) gives the bipyrimidine (CCLIV) upon reaction with butyllithium in tetrahydrofuran at −65°, followed by treatment with carbon dioxide. However, 5-bromopyrimidine on treatment with butyllithium and copper(I) chloride gives 5,5′-bipyrimidine (CCLV).

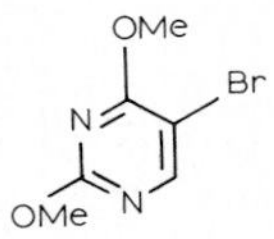

(CCLIII)

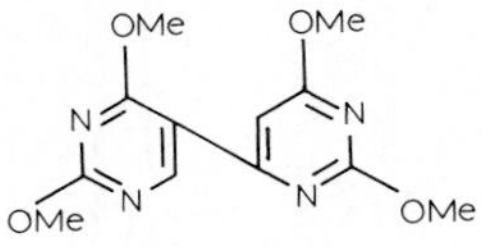

(CCLIV)

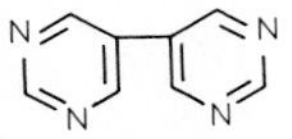

(CCLV)

(iii) Nitro- and nitroso-pyrimidines

Nowadays these derivatives are of little importance and are only normally used as a method of introducing an amino group at the same position, something which now can usually be done more satisfactorily in other ways. Thus the only significant reaction of the nitro and the nitroso group is its reduction to an amino group.

There are many methods described for the reduction of nitro to amino including hydrogenation over platinum (rarely used), palladium or Raney nickel of which the latter is the most popular. The classical method of using sodium hydrogen sulphide is still used, as is dithionite, tin(II) chloride, zinc dust or iron. Many other reactions of the nitro groups involve ring cleavage and subsequent rearrangement reactions and are not discussed here.

The nitroso derivatives, usually 5-nitrosopyrimidines, have mainly an historical interest only as they have long been recognised because of their colour. There is still doubt about their tautomeric structure and/or *cis–trans* configuration. The 5-nitroso group activates leaving groups in the other positions of the ring but its primary chemical reaction is that of reduction to give the amino compound. This again can be done by a variety of methods depending upon the other groups present and varies from the classical reagent ammonium sulphide to sodium dithionite, zinc, tin(II) chloride and the increasing use of catalytic hydrogenation. Catalysts used include Raney nickel, palladium and platinum.

5-Nitroso groups can be acylated as long as there is a proton available for the nitroso group to react in its isonitroso form. Thus 1,2,3,4-tetrahydro-

1,3-dimethyl-6-methylamino-5-nitroso-2,4-dioxopyrimidine (CCLVI) reacts in its isonitroso form (CCLVII; R = H) with acetic anhydride in tetrahydrofuran to give the 5-acetoxyimino derivative (CCLVII; R = Ac).

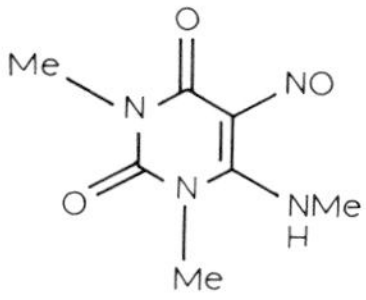

(CCLVI)

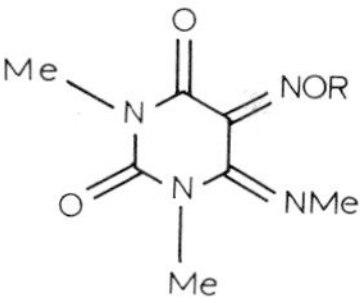

(CCLVII)

(iv) Pyrimidinamines

As previously stated, amino- and alkylamino-pyrimidines are found as such and not in the dihydroimino tautomeric form. Most simple amines are water soluble and have a high (> 100°) melting point. Several naturally occurring aminopyrimidines are of importance, amongst which are cytosine (CCLVIII) and its derivatives, gougerotin (CCLIX), and blasticidin S (CCLX). Aminopyrimidines with significant biological activity include trimethoprim (CCLXI; R = OMe), diaveridine (CCLXI; R = H) and FIAC (CCLXII).

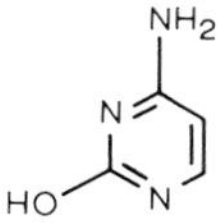

(CCLVIII)

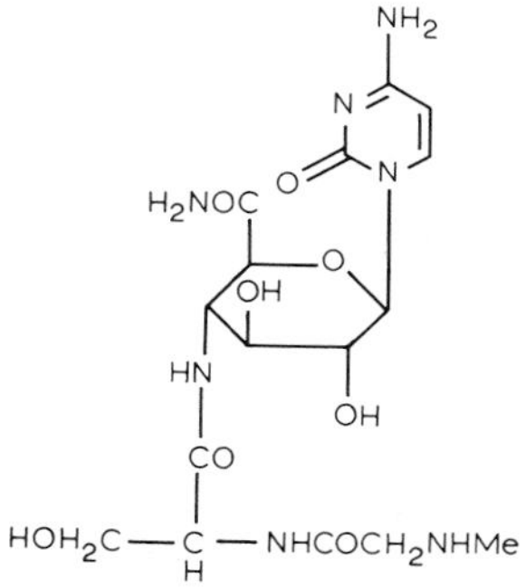

(CCLIX)

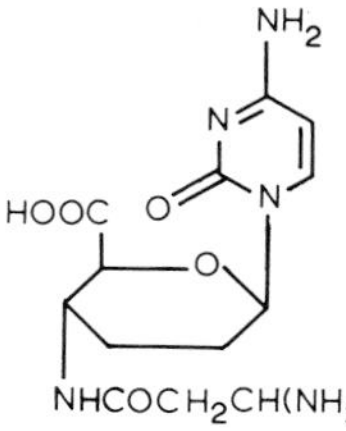

(CCLX)

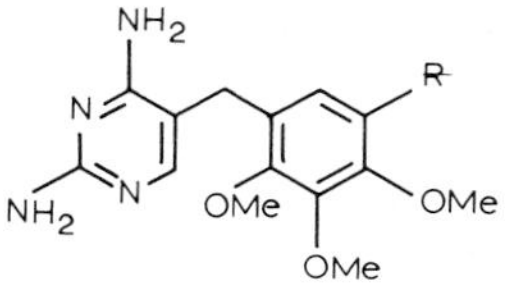

(CCLXI)

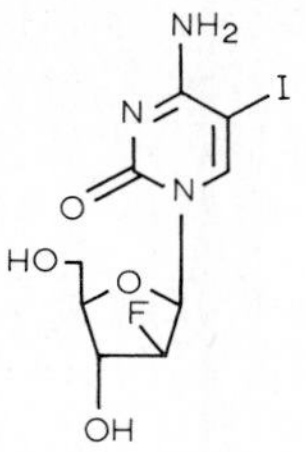

(CCLXII)

Aminopyrimidines can be hydrolysed under acidic conditions to the corresponding hydroxy compounds (*E.A. Oostveen* and *H.C. Van der Plas*, Rec. Trav. chim., 1974, **93**, 233). Thus 4-amino-2,6-dimethylpyrimidine (CCLXIII; R = NH$_2$) gives the corresponding 4-hydroxy compound (CCLXIII; R = OH) by heating it at 180° with hydrochloric acid. Alkaline hydrolysis can also be used so that 2-aminopyrimidine (CCLXIV; R = NH$_2$)

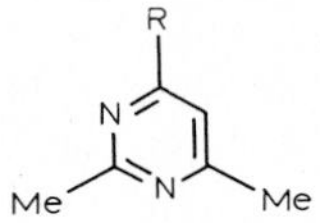

(CCLXIII) 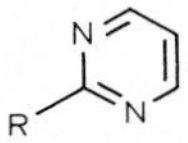   (CCLXIV)

gives 2-hydroxypyrimidine (CCLXIV; R = OH) by heating it in 10 *M* sodium hydroxide at 120°. Interestingly, this compound has two forms with differing melting points (160° and 178°) and once the latter form has been isolated it appears to be impossible to generate the former again. Primary amines can also be converted into the corresponding hydroxy compounds with nitrous acid (this also works for 5-amino derivatives) (*J.A. Bee* and *F.L. Rose*, J. chem. Soc. C, 1966, 2031) and a simple example is the conversion of 2-amino-4-hydroxy-5-(β-hydroxyethyl)pyrimidine (CCLXV; R = NH$_2$) into the corresponding 2,4-dihydroxy derivative (CCLV; R = OH) by treatment with aqueous acetic acid and sodium nitrite initially at 0° and subsequently at 15° for several hours.

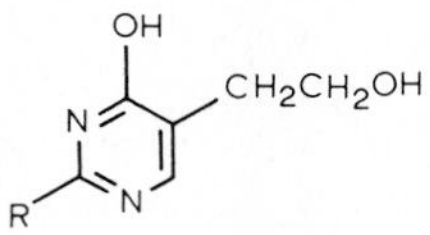

(CCLXV)

Aminopyrimidines can also be converted into chloropyrimidines by treatment of the former with nitrous acid in the presence of an excess of chloride ion (*D.J. Brown* and *M.N. Paddon-Row*, J. chem. Soc. C., 1967, 1928). Thus 4,6-diaminopyrimidine can be converted into 4,6-dichloropyrimidine by the action of nitrous acid and copper(I) chloride. An amino group at any position will react with an aldehyde or its acetals to form a Schiff base so that, 2,4,5-triaminopyrimidine will react with dimethylformamide dimethyl acetal to give the triple Schiff base (CCLXVI) (*F. Yoneda* and *M. Higuchi*, J. chem. Soc. Perkin I, 1972, 1819). Normally the Schiff

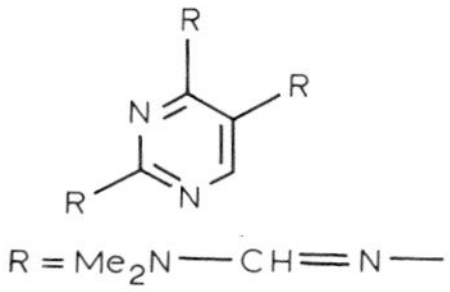

$R = Me_2N-CH=N-$
(CCLXVI)

base forms more easily with the 5-amino substituent and preferential reaction can be achieved with compounds such as 4,5-diaminopyrimidines; thus 4-amino-5-(1'-ethoxyethylidene)amino-2-mercaptopyrimidine will react selectively at the 5-amino group with triethylorthoacetate and acetic anhydride to give compound (CCLXVII). Even reaction at an *N*-amino group can occur to give compounds of the type shown (CCLXVIII).

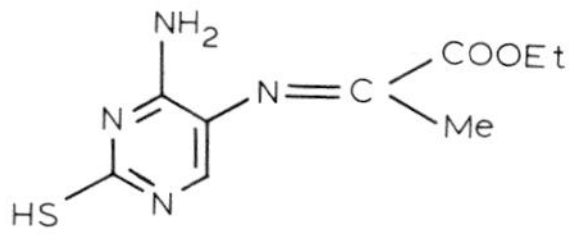

(CCLXVII) 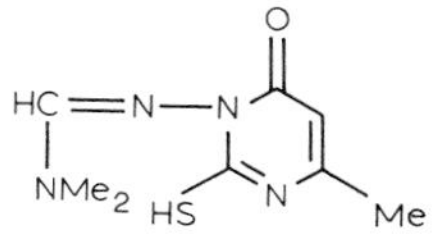   (CCLXVIII)

Aminopyrimidines can be acylated – a 5-amino group being more reactive than one at position 2, 4 or 6. Selective reaction is easily attained so that 2,5-diaminopyrimidine with acetic anhydride gives only the 5-acetamido-2-aminopyrimidine (CCLXIX). Formylation is normally achieved

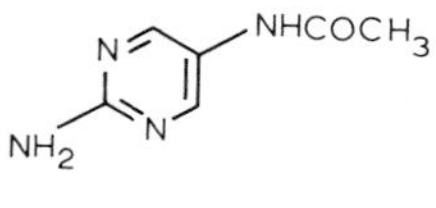

(CCLXIX)

with boiling 90–98% formic acid (*D.J. Brown et al.*, J. chem. Soc. Perkin I, 1972, 1819) and is invariably selective for the 5-amino group. Benzoylation can be achieved with benzoyl chloride either in pyridine or under Schotten-Baumann conditions and like acetylation, this can be selective but eventually all amino groups can be acylated (*D.J. Brown* and *N.W. Jacobsen*, J. chem. Soc., 1965, 3770).

Diazotization of aminopyrimidines is possible with an amino group in any position, but the intermediate diazonium compound has only been isolated for the 5-amino derivative. Otherwise, the intermediate is decomposed *in situ* giving a method of converting an aminopyrimidine into either a hydroxy- or a halo-pyrimidine. 2,4,5,6-Tetraaminopyrimidine (CCLXX) when treated with sodium nitrite at 0° in aqueous tetrafluoroboric acid gives an isolable solid (CCLXXI) which on heating at 225° gives 2,4,6-triamino-5-fluoropyrimidine (CCLXXII).

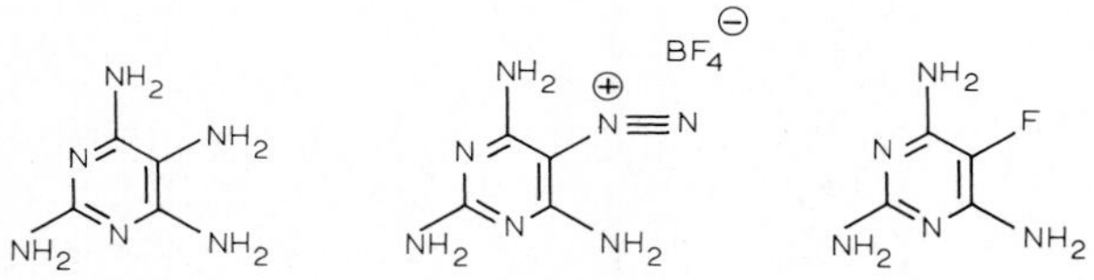

5-Amino-3-methyluracil (CCLXXIII) on treatment with aqueous sodium nitrite in hydrochloric acid followed by alkali gives 5-diazo-3-methyluracil (CCLXXIV).

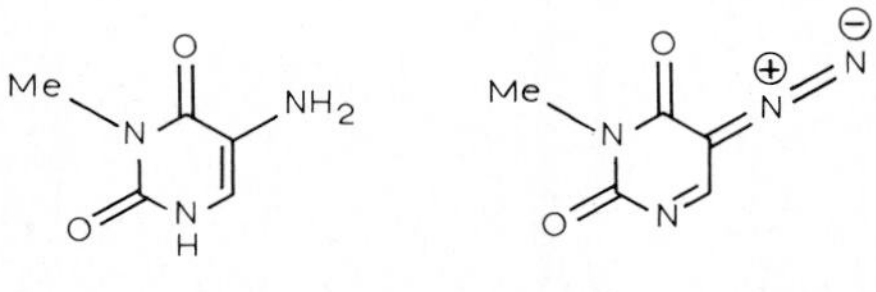

Alkylation of a pyrimidinamine with an alkyl halide will usually give selective alkylation at one of the ring nitrogens to give an *N*-alkylated pyrimidinamine. However, these compounds are often unstable, particularly if electron-withdrawing groups are present and in alkali, the pyrimidine ring opens and then recloses to involve the imino group leaving an alkylaminopyrimidine. This sequence of events is known as a Dimroth rearrangement (*G. L'Abbé*, Ind. chim. Belg., 1971, **3**, 36) and occurs in many π-deficient heteroaromatic systems. For example 2-aminopyrimidine (CCLXXV) can be alkylated with methyl iodide to give the imine (CCLXXVI). In alkaline solution the imine ring opens to give the intermediate (CCLXXVII) which ring closes to give the methylamino compound (CCLXXVIII).

(CCLXXV)   (CCLXXVI)

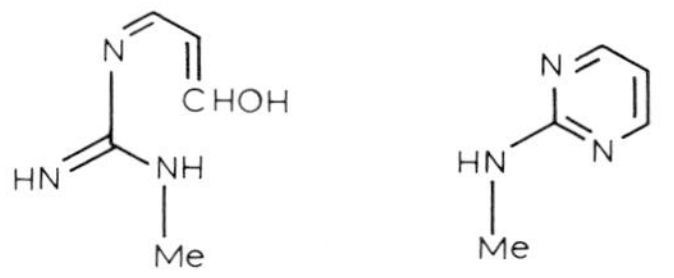

(CCLXXVII)   (CCLXXVIII)

An alkyl chloroformate will however react directly with an amino group in any position to give the urethane. Thus 6-amino-3-methyl-5-methylaminouracil (CCLXXIX; R = H) reacts with ethyl chloroformate in alkali to give directly the urethane (CCLXXIX; R = CO_2Et).

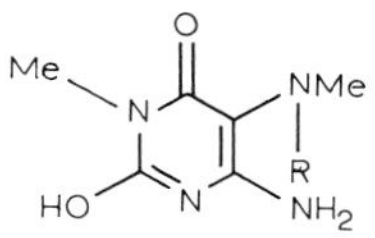

(CCLXXIX)

(v) Hydroxy- and alkoxy-pyrimidines

As has previously been explained, 2-, 4- and 6-hydroxypyrimidine are in tautomeric equilibrium with the oxo form and all the evidence suggests that the equilibrium is almost completely in favour of the latter in practically all compounds. However, for simplicity we shall continue to refer to the compounds in their hydroxy tautomeric form as the oxo-nomenclature is rather cumbersome and sometimes misleading. The monohydroxypyrimidines are soluble in water but this decreases as the number of hydroxyl substituents increases. The methoxypyrimidines are very water soluble and increase in lipophilicity as the size of the alkoxy group increases.

The most important reaction of hydroxypyrimidines is their conversion into the corresponding halopyrimidine. This reaction is not possible for the 5-hydroxy derivative. The most common reagent used is phosphoryl chloride (*D.J. Brown* and *B.T. England*, J. chem. Soc. C, 1967, 1922) often in the presence of a tertiary organic base to speed up the reaction. Phosphorus pentachloride can be used (*S. Yanagida et al.*, J. org. Chem., 1969, **34**, 2972) but often leads to side reactions and phosphoryl bromide can be used for the production of the relevant bromo derivatives. Phosphorus halides also halogenate extranuclear hydroxyl groups so that 5-hydroxymethyluracil (CCLXXX; R = OH) gives the trichloro derivative (CCLXXX; R = Cl). Thionyl chloride, however, only chlorinates the extranuclear hydroxyl group

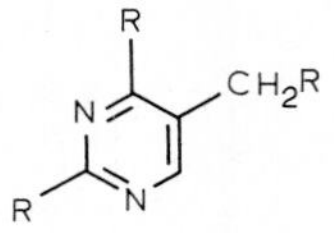

(CCLXXX)

(*J.D. Farkas* and *F. Sorm*, Coll. Czech. Chem. Comm., 1961, **26**, 893). The latter compound can also be made by stirring 5-hydroxymethyluracil in concentrated hydrochloric acid at room temperature for 30 min. The corresponding bromo or iodo derivatives can be made in a similar fashion.

Conversion into a mercaptopyrimidine requires that the hydroxypyrimidine is heated with phosphorus pentasulphide in pyridine (*W.L.F. Aramego*, J. chem. Soc., 1965, 2778). The reaction of phosphorus pentasulphide with 2-hydroxypyrimidines require severe conditions and occasionally reaction can also occur at a 5-hydroxyl group. Some examples are: the conversion of 3-amino-6-methyl-2-thiouracil (CCLXXXI; X = O) into 3-amino-6-methyl-dithiouracil (CCLXXXI; X = S); 4-acetoxymethyl-2,6-dihydroxypyrimidine (CCLXXXII; R = OH) into 4-acetoxymethyl-2-hydroxy-6-mercaptopyrimidine (CCLXXXII; R = SH), and 6-amino-1-benzyluracil (CCLXXXIII; R = OH) into 6-amino-1-benzyl-4-thiouracil (CCLXXXIII; R = SH).

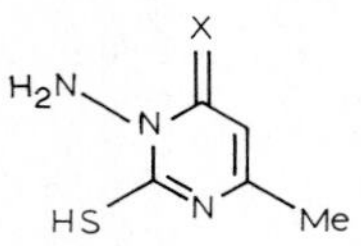

(CCLXXXI)   (CCLXXXII)   (CCLXXXIII)

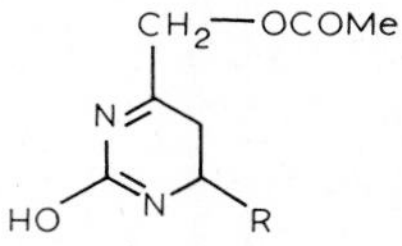

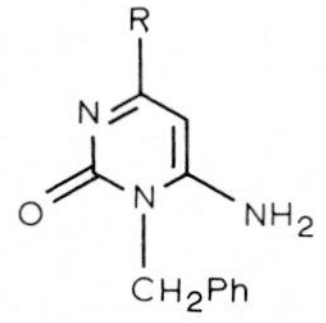

There are few preparations of 2-thiopyrimidines recorded but typical reaction conditions are used for the conversion of 1-methyl-4-thiouracil (CCLXXXIV; X = O) into 1-methyldithiouracil (CCLXXXIV; X = S) in low yield by treatment with phosphorus pentasulphide in tetralin at 180°.

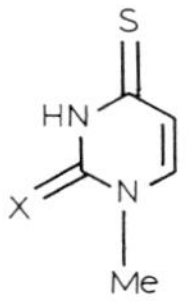

(CCLXXXIV)

Acetylation of any hydroxypyrimidine is possible usually using an acid anhydride or chloride but normally 5- or extranuclear hydroxyl groups react preferentially. However, 5-butyl-2-dimethylamino-4-hydroxy-6-methylpyrimidine (CCLXXXV; R = H) reacts with acetic anhydride under reflux to give the acetyl derivative (CCLXXXV; R = Ac). More normal is the reaction of 5-hydroxy-6-methyluracil (CCLXXXVI; R = H) which gives the corresponding monoacetoxy compound (CCLXXXVI; R = Ac) with acetic anhydride under reflux.

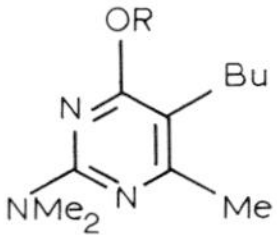

(CCLXXXV) 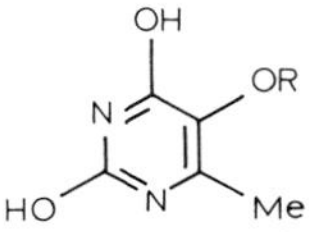   (CCLXXXVI)

Alkylation of pyrimidinones (*D.J. Brown, E. Hoerger* and *S.F. Mason*, J. chem. Soc., 1955, 211) usually results in alkylation at N-1 or N-3 even if diazomethane is used. Treatment of 2-thiouracil (CCLXXXVII) with dimethyl sulphate gives an immediate product of the *S*-methyl derivative (CCLXXXVIII) which is followed by the appearance of two *N*-methyl derivatives (CCLXXXIX and CCXC) but no *O*-methyl derivative is formed.

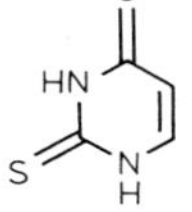

(CCLXXXVII) 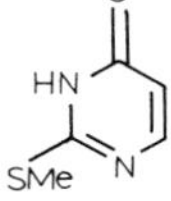   (CCLXXXVIII)

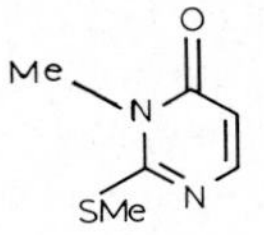

(CCLXXIX) 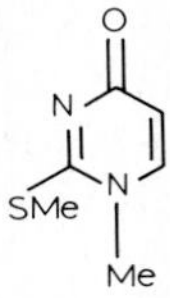   (CCXC)

However *O*-trimethylsilyl derivatives are easily formed by treatment of a hydroxypyrimidine with hexamethyldisilazane and/or chlorotrimethylsilane (*E. Wittenburg*, Coll. Czech. Chem. Comm., 1971, **36**, 246). The resulting products are volatile and hence suitable for g.l.c. separation and also soluble in non-polar solvents. The hydroxypyrimidine is easily regenerated by addition of water. Thus uracil when dissolved in hexamethyldisilazane and heated under reflux until solution is complete will give a quantitative yield of 2,4-bistrimethylsilyloxypyrimidine (CCXCI).

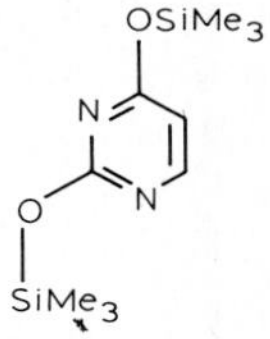

(CCXCI)

Usually the conversion of a 2- or 4/6-hydroxypyrimidine into the corresponding aminopyrimidine has had to go *via* the intermediate chloro compound or the alkylthio or alkoxy derivative. However, if the pyrimidine can stand up to the drastic conditions involved, it is possible to achieve the reaction directly. Thus if phosphoric trisdimethylamide [$O = P(NMe_2)_3$] is heated with uracil at 235° for 1 h, a good yield of 2,4-bisdimethylamino-pyrimidine (CCXCII) is obtained (*E.A. Arutyunyan et al.*, Izvest. Akad. Nauk. S.S.S.R. Ser. Khim., 1969, 655; *E.P. Gracheva et al.*, *ibid*., 1970, 420).

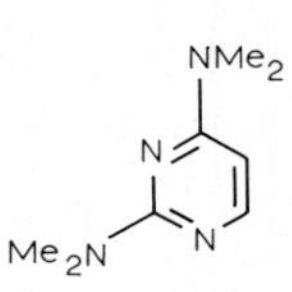

(CCXCII)

Alkoxypyrimidines having the substituent at position 2, 4/6 or extranuclear, can be hydrolysed to the corresponding hydroxypyrimidines (*M. Prystas* and *F. Sorm*, Coll. Czech. Chem. Comm., 1967, **32**, 1298). 2- or 4/6-Alkoxypyrimidines hydrolyse particularly easily in acidic conditions and selectivity using only one equivalent of acid is a practical proposition; thus 4,6-dibenzyloxypyrimidine (CCXCIII; R = OBz) can be hydrolysed to

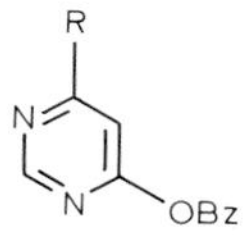

(CCXCIII)

6-benzyloxypyrimidine (CCXCIII; R = H) in good yield. Hydrolysis of an alkoxy group at the 5-position requires much more drastic conditions such as potassium hydroxide in boiling ethylene glycol or concentrated hydrochloric acid at 150°. As usual, benzyloxy groups can be removed by hydrogenation.

Transalkoxylation is also possible, particularly if the group is activated by the presence of electron-withdrawing groups at C-5. Silver oxide has been found to catalyse this process (*D.J. Brown* and *T.Sugimoto*, J. chem. Soc. C, 1970, 2661) and thus 2-methoxy-5-nitropyrimidine (CCXCIV; R = H) can be converted into the propoxy homologue (CCXCIV; R = Et) by

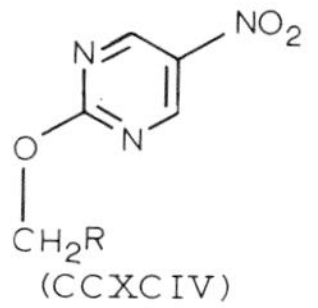

(CCXCIV)

boiling it in propanol containing silver oxide. The aminolysis of 2- and 4/6-alkoxypyrimidines is particularly easy and is accelerated by the presence of electron-withdrawing groups at C-5. Reaction conditions vary from ethanolic ammonia at 125° in a sealed tube to ethanolic hydrazine below 0°. 2- and 4/6-Alkoxypyrimidines can be rearranged into *N*-alkylpyrimidinones either thermally or by the Hilbert–Johnson reaction. Thus heating 4-methoxypyrimidine (CCXCV) at 190° gives the *N*-methyl derivative (CCXCVI). This reaction can be catalysed by tertiary bases with the

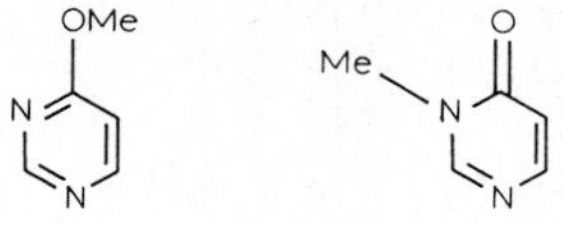

(CCXCV)   (CCXCVI)

more basic ones being more efficient. The reaction is accelerated by the presence of electron-withdrawing groups, the mechanism is intermolecular but not free radical.

The Hilbert–Johnson reaction is not a rearrangement but provides the basis for the synthesis of many pyrimidine nucleosides using a halogeno sugar (*A.R. Katritzky* and *A.J. Waring*, J. chem. Soc., 1962, 1540). A simpler case is the treatment of 2,4-diethoxypyrimidine (CCXCVII) with methyl iodide to give 2,4-diethoxy-1-methylpyridinium iodide (CCXCVIII) which loses ethyl iodide to give the *N*-methyl derivative (CCXCIX).

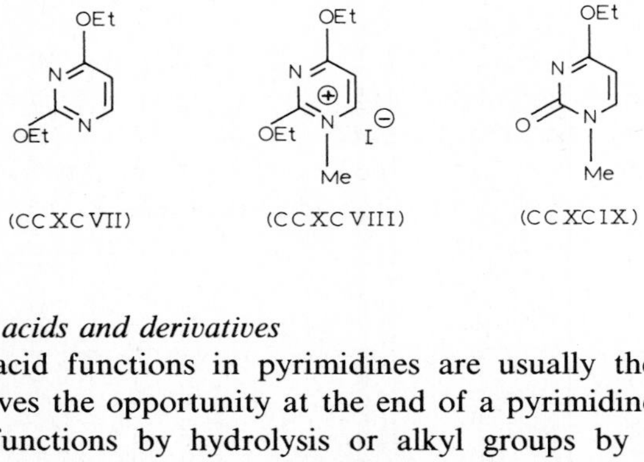

(CCXCVII)   (CCXCVIII)   (CCXCIX)

(vi) Carboxylic acids and derivatives

Carboxylic acid functions in pyrimidines are usually there to be removed! This gives the opportunity at the end of a pyrimidine synthesis to convert ester functions by hydrolysis or alkyl groups by oxidation, to carboxylic acid groups. The carboxypyrimidines are then easily decarboxylated by heating at 200–300°. Some selectivity is possible so that 4,5-dicarboxy-2,6-diphenylpyrimidine (CCC; $R^1 = R^2 = CO_2H$) can be de-

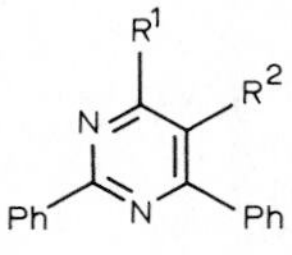

(CCC)

carboxylated to give 5-carboxy-2,4-diphenylpyrimidine (CCC; $R^1 = H$; $R^2 = CO_2H$) and by heating at a higher temperature to give 2,4-diphenylpyrimidine (CCC; $R^1 = R^2 = H$; note the numbering change in this compound once it becomes symmetrical). 5-Cyano-1-cyclohexyluracil (CCCI; $R = CN$) when heated in orthophosphoric acid at 170° gives a good yield of 1-cyclohexyluracil (CCCI; $R = H$) which presumably is formed *via* the

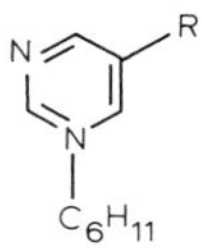

(CCCI)

intermediate carboxylic acid (CCCI; $R = CO_2H$). Esters of carboxypyrimidines are rarely required; indeed the most usual reaction of such esters is their hydrolysis. However, if required, esters can be made by the normal reactions used in organic synthesis. Acid chlorides and anhydrides are not usually made but syntheses of these derivatives have been recorded (*Brown, loc. cit.*).

As previously mentioned, the most widely used reaction of esters is their hydrolysis to carboxylic acids; alkali is usually used, sometimes in ethanolic solution; benzyl groups can be removed by hydrogenolysis. Aminolysis of esters proceeds normally to give amides or hydrazides. The amides can be hydrolysed to carboxypyrimidines under strong acidic conditions and can be dehydrated to nitriles, thus boiling with phosphoryl chloride 2-carbamoyl-4-hydroxy-6-methylpyrimidine (CCCII) is converted into 4-chloro-2-cyano-6-methylpyrimidine (CCCIII). The Hofmann reaction can also be used. Hydrazides can be converted into azides with nitrous acid and can be acylated. Azides give urethanes when treated with alcohols.

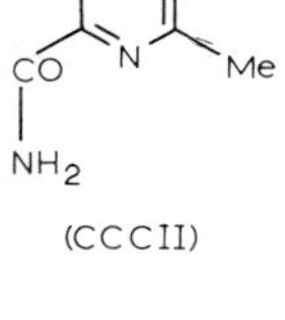

(CCCII) 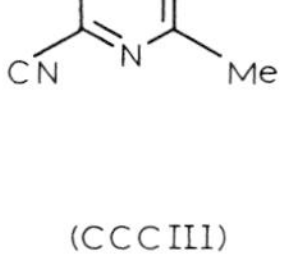   (CCCIII)

As already explained, cyano groups may be removed by hydrolysis without the isolation of the intermediate carboxylic acid. Most other reactions of nitriles are normal in that they can be hydrated, hydrogen sulphide can be added to them, the addition of alcohols and amines give iminoethers and amidines, respectively.

Formylpyrimidines can be oxidised to carboxypyrimidines using a variety of oxidising agents including hydrogen peroxide in aqueous alkali, alkaline permanganate and silver oxide in ethanol. Formylpyrimidine may also be reduced to the corresponding alcohol with lithium aluminium hydride and sodium borohydride being the reagents of choice.

Aldehydes have also been converted into acetals, hydrazones, aldoximes, semicarbazones and give Schiff bases with amines.

Pyrimidine ketones are a relatively neglected group of compounds, although the reactions which have been reported show that these derivatives are potentially as useful as other ketones as is illustrated by the reaction of 5-acetyl-2,4-dihydroxypyrimidine (CCCIV) with phosphoryl chloride and dimethylaniline to give 2,4-dichloro-5-(α-chlorovinyl)pyrimidine (CCCV) which on treatment with alkali gives 5-ethynyluracil (CCCVI) in reasonable yield.

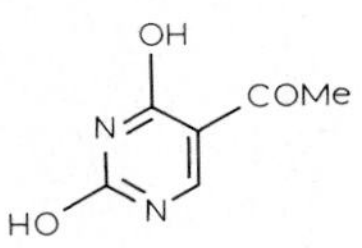

(CCCIV) 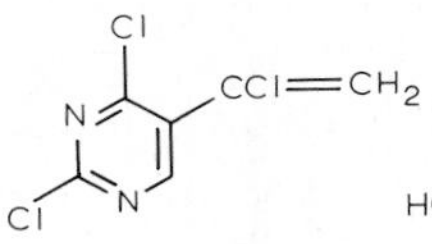   (CCCV) 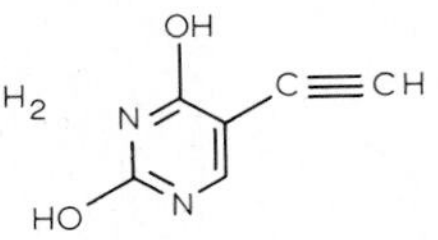   (CCCVI)

The most biologically important pyrimidine carboxylic acid is undoubtedly orotic acid, 4-carboxy-2,6-dihydroxypyrimidine (CCCVII). This compound plays a key part in pyrimidine biosynthesis but apart from the fact that it took 25 years from its original isolation in 1905 to correctly identify its structure, little interesting novel chemistry has been reported for this compound.

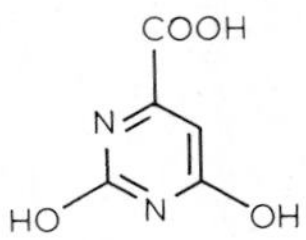

(CCCVII)

(vii) Sulphur-containing pyrimidines

The sulphur-containing pyrimidines form a closely related group of compounds of considerable importance because their many reactions mean that they are very useful synthetic intermediates. In addition, the use of thiourea in the Principal Synthesis means that many 2-thiopyrimidines are thus easily prepared and in addition, 2-thiouracil is of considerable biological importance. The parent mercapto derivatives (RSH) can be alkylated to give a thio ether (RSR') and this can be oxidised to a sulphoxide (RSOR') or sulphone (RSO_2R'). The original mercaptopyrimidine may be oxidised through a sulphide (RSR) or disulphide (RSSR) to a sulphinic acid (RSO_2H) and finally a sulphonic acid (RSO_3H). Very few reactions of 5-mercaptopyrimidines have been recorded and the focus will be on the 2- and 4/6-mercaptopyrimidines which exist as thiones but react as mercaptans or thioamides.

Mercapto groups can be removed from the 2- and 4/6-positions by boiling the compound in solution with Raney-nickel catalyst (*A.G. Beaman, J.F. Gerster* and *R.K. Robins*, J. org. Chem., 1962, **27**, 986). An example of this method is the conversion of 4,5-diamino-6-hydroxy-2-mercaptopyrimidine (CCCVIII; R = SH) into 4,5-diamino-6-hydroxy-pyrimidine (CCCVIII; R = H).

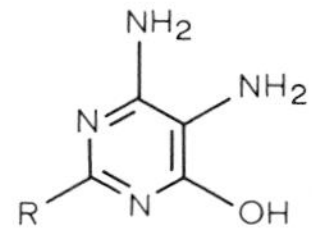

(CCCVIII)

S-Alkylation occurs easily and usually there is little problem in avoiding *O*- or *N*-alkylation as this requires much more drastic conditions. The usual conditions required are to make the anion with just over one equivalent of aqueous alkali and to add one equivalent of alkyl halide at room temperature (*H. Vorbruggen* and *P. Strehlke*, Ber., 1973, **106**, 3039). For compounds such as 2-thiouracil (CCCIX; R = SH) where reaction is sluggish the yield is increased by using dimethylformamide as the solvent to give, with methyl iodide, the 2-methylthio derivative (CCCIX; R = S–Me). Phase-transfer catalysis can also be used to increase the yield.

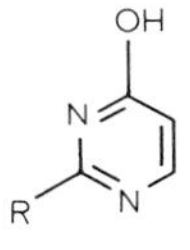

(CCCIX)

Mercaptopyrimidines can rarely be hydrolysed directly to the corresponding alcohol but the overall result can be achieved in three ways: (1) *S*-alkylation followed by acidic or alkaline hydrolysis; (2) oxidation to a sulphinic or sulphonic acid followed by hydrolysis; and (3) *S*-alkylation followed by oxidation to a sulphoxide or sulphone followed by hydrolysis.

Mercaptopyrimidines undergo preferential aminolysis at the 4-position so that 2,4-dimercaptopyrimidine (CCCX; R = SH) can be converted into 4-amino-2-mercaptopyrimidine (CCCX; R = NH_2) in good yield by treat-

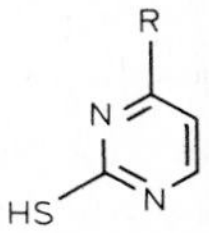

(CCCX)

ment with hot aqueous ammonia. In fact the aminolysis of 2-mercaptopyrimidine was thought to be extremely difficult but some examples of even this transformation have been reported. Thus 5-ethoxycarbonylmethyl-4-hydroxy-2-mercaptopyrimidine (CCCXI) is converted into 2-amino-5-carbamoylmethyl-4-hydroxypyrimidine (CCCXII) by reaction with ethanolic ammonia at 120° (*J.D. Fissekis* and *F. Sweet*, Biochemistry, 1970, **9**, 3136).

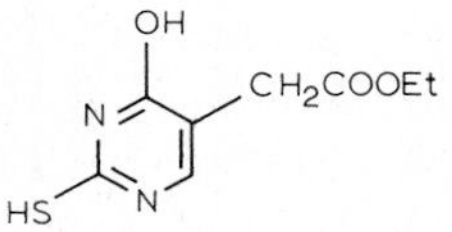

(CCCXI) 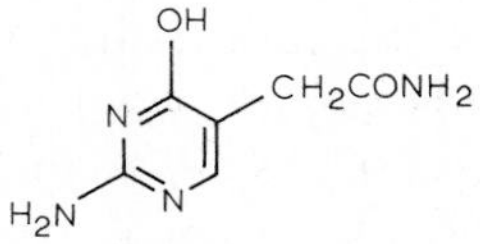 (CCCXII)

Acylation of mercaptopyrimidines is complicated by the fact that 5-acyl or *N*-acyl (nuclear or extranuclear) derivatives can be formed according to the conditions used (*G.I. Podzigun et al.*, Izvest. Akad. Nauk S.S.S.R., Ser. Khim., 1980, 2346). As an example, 4,6-diamino-2-mercaptopyrimidine (CCCXIII; R = H) and benzoyl chloride in dimethylformamide containing triethylamine at 20° gives 4,6-diamino-2-benzoylthiopyrimidine (CCCXIII; R = PhCO). At 80°, 4,6-diamino-1-benzoyl-1,2-dihydro-2-thiopyrimidine (CCCXIV) is formed and at 120° the product is 4-amino-6-benzamido-2-mercaptopyrimidine (CCCXV).

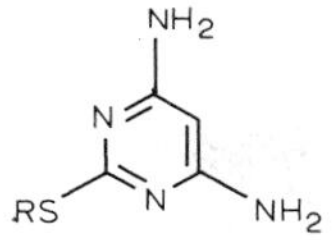

(CCCXIII) 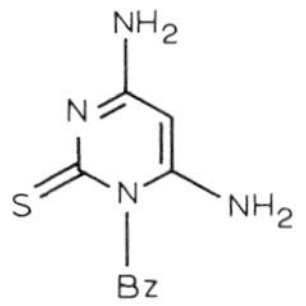   (CCCXIV) 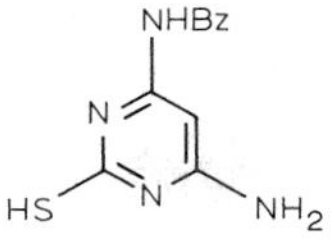   (CCCXV)

5-Mercaptopyrimidines oxidise on standing in air to the disulphide but 2- and 4/6-mercaptopyrimidines normally require more drastic conditions. Thus 4-mercapto-6-methylpyrimidine (CCCXVI) gives the corresponding disulphide (CCCXVIII), when a solution in aqueous alkali is treated with iodine in potassium iodide solution.

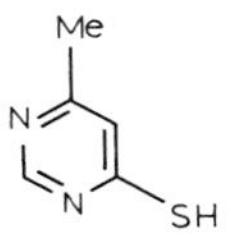

(CCCXVI) 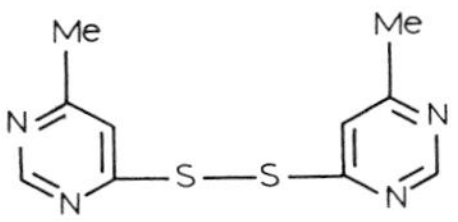   (CCCXVII)

There is usually little purpose in removing thioether groups as the mercapto groups from which they are invariably made are more easily removed. However, Raney nickel can be used to remove an alkylthio group if necessary. Methylthiopyrimidines are readily hydrolysed to the corresponding alcohol under acidic conditions, for example 5-ethoxycarbonyl-1,6-dimethyl-2-hydroxy-4-methylthiopyrimidine (CCCXVIII; R = SMe) in boiling 2 M hydrochloric acid for 10 min. gives the corresponding uracil derivative (CCCXVIII; R = OH) leaving the ester group intact.

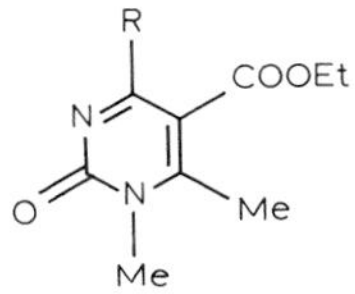

(CCCXVIII)

Alkylthiopyrimidines undergo aminolysis and like the corresponding chloro derivative (*D.J. Brown* and *R. Foster*, Austral. J. Chem., 1966, **19**, 2321), the 4/6-alkylthio derivatives react more easily than the 2-alkylthiopyrimidines. As would be expected, activation by nitro or nitroso groups is found to occur. Examples of this ammonolysis reaction are the conversion of 4-hydroxy-5-methyl-2-methylthiopyrimidine (CCCXIX; R = SCH_3) into the corresponding 2-isopropylamino derivative (CCCXIX; R = iPrNH) in isopropylamine at 180°; 4-amino-6-hydroxy-2-methylthio-5-nitrosopyrimidine (CCCXX; R = SCH_3) into the corresponding 2-dimethylamino derivative (CCCXX; R = NMe_2) in aqueous dimethylamine under reflux for 20 min., and the methylthio compound (CCCXXI; R = SCH_3) into the corresponding amino derivative (CCCXXI; R = NH_2) with methanolic ammonia at 170°.

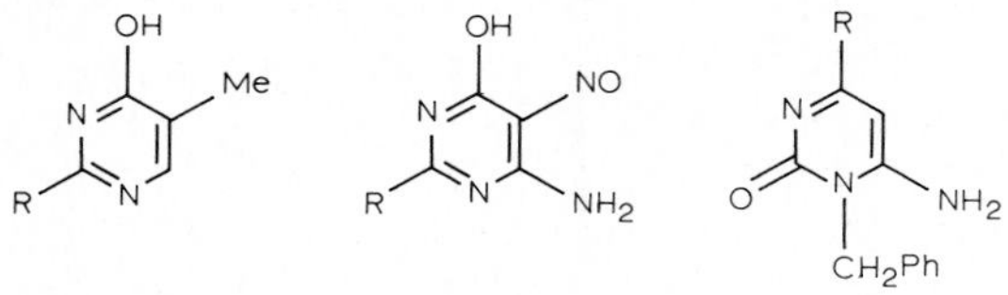

(CCCXIX)   (CCCXX)   (CCCXXI)

Dealkylation of alkylthiopyrimidines can be effected using one of the following four methods. (1) Acidic hydrolysis using hydrogen iodide or hydrochloric acid. (2) Phosphorus pentasulphide in xylene which used to be the reagent of choice but gives low yields and has been superceded by other methods. (3) Reduction of benzylthio groups with sodium in liquid ammonia. (4) Cleavage by a Lewis acid such as aluminium bromide in an aprotic solvent.

Thioethers can be readily oxidised to sulphones and sulphoxides. The practical problem with the production for the latter group of compounds is that it is sometimes difficult to find conditions under which all the starting material reacts and yet at the same time the sulphoxide is not over oxidised to give the sulphone (*D.J. Brown* and *P.W. Ford*, J. chem. Soc. C, 1967, 568). The main use of these derivatives is that the oxidised sulphur group is a much better leaving group in nucleophilic displacement reactions. However, at least the alkylthio group is more easily oxidised than the sulphoxide and thus as long as exact molar quantities of oxidising agent are used, good yields can be obtained. The usual oxidising agents are permanganate, *m*-chloroperbenzoic acid, and peracetic acid. Thus *m*-chloroperbenzoic acid will oxidize 2-methylthiopyrimidine (CCCXXII; R = SMe) into the corresponding sulphoxide (or 2-methylsulphinylpyrimidine) (CCCXII; R = SOMe), and 2-methylsulphonylpyrimidine (CCCXXII; R = SO_2Me) can be

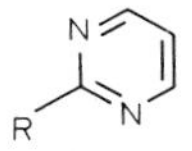

(CCCXXII)

made either from the alkylthiopyrimidine or the sulphinylpyrimidine by the action of the same oxidising agent. 2-, 4/6- or 5-Alkylthiopyrimidines are equally easily oxidised. The only useful reactions of sulphides and disulphides are their oxidation by permanganate or performic acid to sulphones and the reduction by sodium borohydride of the disulphide to the corresponding mercapto derivative (*S.B. Greenbaum* and *W.L. Holmes*, J. Amer. chem. Soc., 1954, **76**, 2899).

(viii) Metallopyrimidines

There is surprisingly little information available on the reactivity of metallopyrimidines despite the fact that in many other areas of synthetic organic chemistry, the use of metallo derivatives is common practice. Practically all the early work has been concerned with 5-lithiopyrimidines made from the corresponding bromo derivatives. One of the problems in dealing with organometallic intermediates in pyrimidine chemistry is to find a solvent which is unreactive and in which the reagent is soluble. More recently work has changed to pyrimidine nucleosides where the trimethylsilyl ether is made to protect the sugar hydroxyl groups and this derivative is soluble in solvents suitable for metallic intermediates to be formed and react (*E. De Clercq* and *R.T. Walker*, Pharmac. Ther., 1984, **26**, 1). Thus pyrimidin-5-yllithium(CCCXXIII) will react with carbon dioxide to give the corresponding carboxylic acid (*M.D. Mehta, D. Miller* and *E.F. Mooney*,

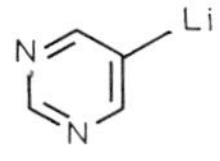

(CCCXXIII) 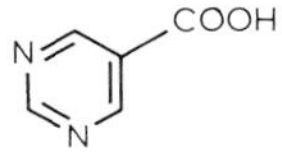   (CCCXXIV)

J. chem. Soc., 1965, 6695) and many other substituted pyrimidines react in a similar manner. 5-Lithio derivatives of uracil have also been reacted with trichloroacetaldehyde, 1,1-dichlorodifluoroethene and with perfluoropropene to give a regio-specific reaction. The final products of all these reactions were 5-(fluorovinyl)pyrimidines (*P.L. Coe et al.* J. med. Chem., 1982, **25**, 1329). These reactions have also been used at the nucleoside level.

The incentive to produce commercial quantities of the anti-herpes-virus agent *E*-5-(2-bromovinyl)-2′-deoxyuridine (CCCXXV; R = 2′-deoxyribosyl)

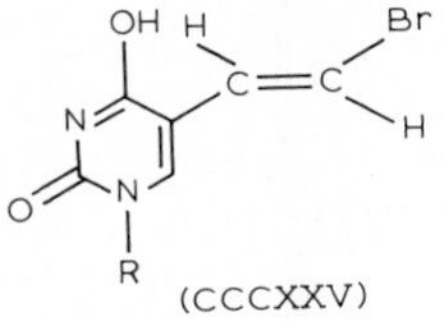

(CCCXXV)

resulted in the investigation of the use of organopalladium intermediates (*A.S. Jones et al.*, Tetrahedron Letters, 1979, 4415). This resulted in the replacement of a more than seven-stage synthesis with an overall yield of less than 10% by a three-stage preparation with a yield of more than 40%. The key reaction is the production of a pyrimidin-5-ylpalladium intermediate using work pioneered by *Heck* (*R.F. Heck*, Acc. chem. Res., 1979, **12**, 146) and first applied to pyrimidines by *Bergstrom* and coworkers (*D.E. Bergstrom et al.*, J. org. Chem., 1981, **46**, 1432). The pyrimidin-5-ylpalladium derivative can be made either from the 5-chloromercuri intermediate with lithium palladium chloride or *in situ* by the action of a catalytic amount of palladium acetate on a 5-iodopyrimidine in the presence of triethylamine and triphenylphosphine. These palladium compounds are then treated with an allyl chloride or halo-alkene. Apart from compound (CCCXXV) many hundreds of analogues have now been synthesised (*E. de Clercq* and *R.T. Walker*, Pharmac. Ther., 1984, **26**, 1) including 5-vinyl-2′-deoxyuridine (CCCXXVI; $R^1 = CH=CH_2$; $R^2 = $ 2′-deoxyribosyl) and 5-ethynyl-2′-deoxyuridine (CCCXXVI; $R^1 = C \equiv CH$; $R^2 = $ 2′-deoxyribosyl). The reactions of some pyrimidin-5-ylcopper derivatives with iodo-alkenes have been reported (*M. Bobek*, Nucleic Acids Res. Symp. Ser., 1984, **12**, 81).

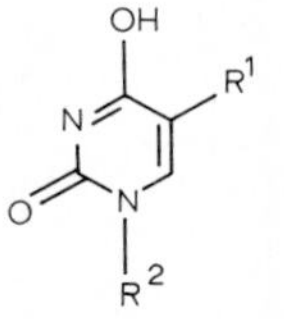

(CCCXXVI)

Recently a series of substitution reactions at C-5 or C-6 of pyrimidine nucleosides *via* the corresponding lithio derivatives has been reported (*H. Tanaka et al.*, Tetrahedron, 1982, **38**, 2635; *H. Hayakawa et al.*, Tetrahedron Letters, 1987, 87).

(ix) Pyrimidine-N-oxides

These derivatives are usually most easily prepared by the direct oxidation of the corresponding pyrimidine with a peroxy carboxylic acid. This reaction can be reversed by catalytic reduction but chloro substituents are usually removed first. *N*-Oxides can also be converted into the parent pyrimidine by the action of phosphorus trichloride or triethylphosphite (*E. Ochiai* and *H. Yamanaka*, Chem. Pharm. Bull. Japan, 1955, **3**, 175). Pyrimidine-*N*-oxides undergo electrophilic halogenation, nitration and nitrosation at C-5. They also undergo Reissert-like reactions (*H. Yamanaka*, *ibid*., 1958, **6**, 633) so that treatment of 4-ethoxy-6-methylpyrimidine-*N*-oxide (CCCXXVII) with aqueous potassium cyanide followed by the ad-

(CCCXXVII) (CCCXXVIII)

dition of benzoyl chloride and alkali gives 4-ethoxy-6-methylpyrimidine-2-carbonitrile (CCCXXVIII). Similar rearrangements occur with other acylating agents such that pyrimidine-*N*-oxide on treatment with acetic anhydride gives pyrimidin-4-yl acetate (CCCXXIX) (*H. Bredereck*, *R. Gomper* and *H. Herlinger*, Ber., 1958, **91**, 2832) but if the 4-position is blocked such as in 4,6-dimethylpyrimidine-1-oxide (CCCXXX) the product is 6-methylpyrimidin-4-ylmethyl acetate (CCCXXXI) (*R.R. Hunt*, *J.F.W. McOmie* and *E.R. Sayer*, J. chem. Soc., 1959, 525).

(CCCXXIX) (CCCXXX) (CCCXXXI)

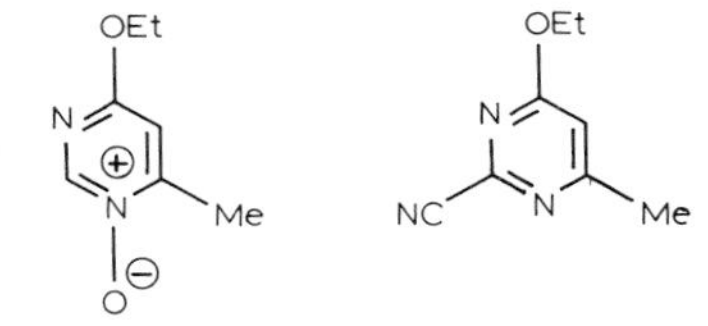

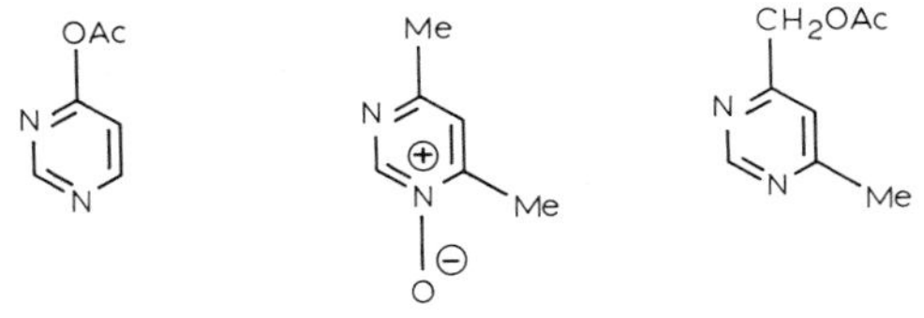

5. Quinazolines

In this review, quinazolines, the benzo-1,3-diazines (CCCXXXII) are regarded as derivatives of pyrimidine and not as a class of compounds in their own right. Thus the emphasis on the synthesis and reactions of these compounds will be to highlight the differences brought about because of the presence of the benzene nucleus substituted across the 5:6 bond of pyrimidine. The numbering system of the rings is as shown.

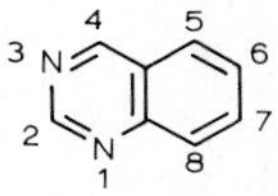

(CCCXXXII)

Several quinazolines have interesting or useful biological properties. Perhaps pride of place ought to go to tetrodotoxin (CCCXXXIII), the powerful neurotoxin found in the liver and ovaries of the puffer fish

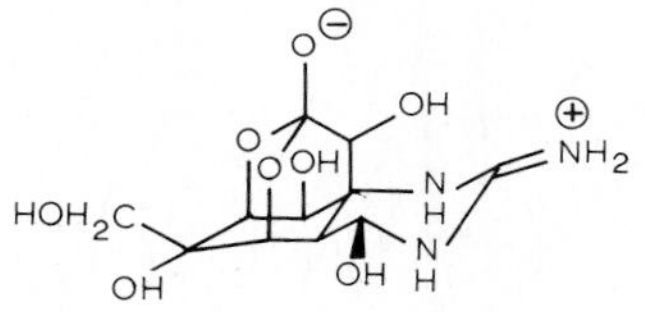

(CCCXXXIII)

Sphoerides rubripes and *S. phyreus* which is a delicacy for the Japanese and a gastronomic challenge to the foreigner (*Y. Kishi et al.*, J. Amer. chem. Soc., 1972, **94**, 9219). So far all attempts to increase the lethality of this compound have failed. Quinazoline alkaloids (*e.g.* CCCXXXIV) are also found in several plant families but by now the pyrimidine ring is being swamped by substituents and is of no relevance to this review. Another quinazoline alkaloid, febrifugine (CCCXXXV) has good anti-malarial properties but unfortunately is also toxic. The "pseudo-barbiturate" methaqualone (CCCXXXVI) shows a wide spectrum of biological activity. Prazosin (CCCXXXVII) is used as an anti-hypertensive while the quinazoline (CCCXXXVIII) has been used as an anti-histamine (*N.B. Chapman, K. Clarke* and *K. Wilson*, J. chem. Soc., 1963, 2256).

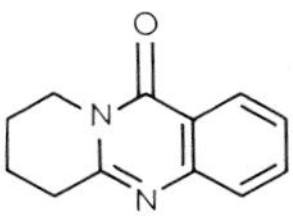

(CCCXXXIV) 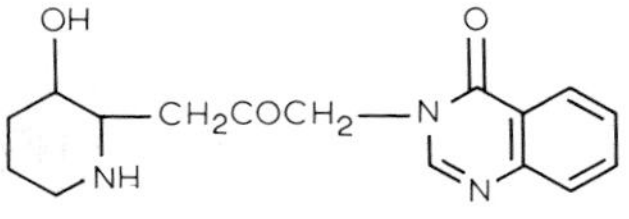   (CCCXXXV) 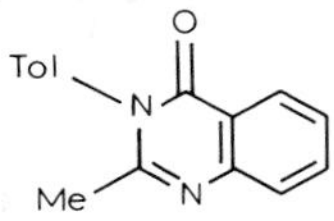   (CCCXXXVI)

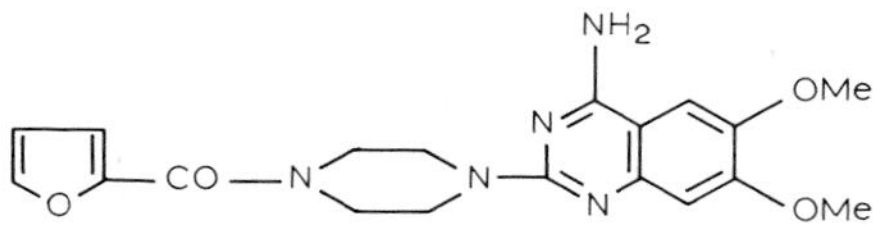

(CCCXXXVII)

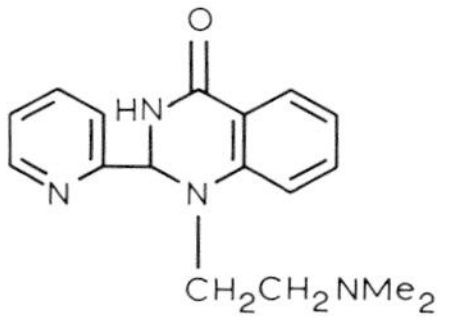

(CCCXXXVIII)

(a) General chemical and physical properties

Unlike pyrimidine, quinazoline has no symmetry. Electron-density calculations show marked variation depending on the method used but for comparative purposes the figures shown here are for the VESCF method (CCCXXXIX). These figures indicate that the 2- and the 4-position in

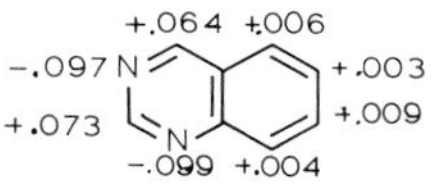

(CCCXXXIX)

quinazoline should be roughly comparable with the same positions in pyrimidine and that the benzenoid ring has been deactivated towards electrophilic attack by the electron-withdrawing effect of the pyrimidine ring. However substituents in position 5-8 should be similar in reactivity to those in the C-5 position in pyrimidine.

Somewhat surprisingly, quinazoline is a much stronger base (pK_a 3.51) than pyrimidine (pK_a 1.31) but this is because the cation is stabilised as a hydrate (CCCXL) (*W.L.F. Armarego*, Adv. heterocyclic Chem., 1979, **24**, 1).

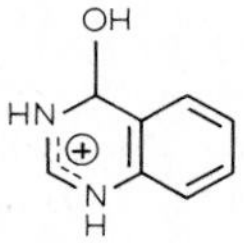

(CCCXL) 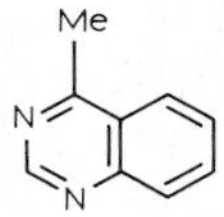   (CCCXLI)

Thus, the presence of the methyl group in 4-methylquinazoline (CCCXLI) interferes with hydration and the pK_a drops to 2.52 which is comparable with that of 4-methylpyrimidine (pK_a 2.0). 2-Methylquinazoline, however, which can form a hydrate and also has the benefit of the methyl substituent is really quite a strong base with a pK_a of 4.52. The same principle applies to most quinazoline derivatives, if the cation is capable of hydration then they are stronger bases, if no stabilisation is possible, then the quinazolines and corresponding pyrimidines have similar values.

The u.v. spectrum of quinazoline itself is quite complicated showing three absorption band at 220, 267 and 311 nm which are $\pi-\pi^*$ transitions. The $n-\pi^*$ transition is seen as an inflexion of low intensity at 330 nm. However, as already explained, the cation is hydrated and so its spectrum is very different with absorption bands at 208 and 260 nm. The mass spectrum of quinazolines is very similar in type to that of the pyrimidines. Hydrogen cyanide is lost twice with N-3 and C-4 being eliminated first. The remainder of the molecule behaves as ionised benzyne (*H. Culbertson, J.C. Decins* and *B.E. Christensen*, J. Amer. chem. Soc., 1952, **74**, 4834).

Quinazolin-2-(and -4-)one exist as oxo tautomers (*J.M. Hearn, R.A. Morton* and *J.C.E. Simpson*, J. chem. Soc., 1951, 3318) and the thiones exist in a similar tautomeric configuration. 4-Aminoquinazoline on the other hand exists entirely as the amino tautomer (CCCXLII).

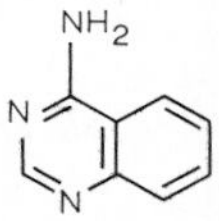

(CCCXLII)

(b) Synthesis of quinazolines

Much of the work on synthesis and properties of the quinazolines has been covered in detail elsewhere and the reader is referred to these reviews and the original references quoted therein (*D.J. Brown*, Pyrimidines and their Benzo Derivatives, in Comprehensive Heterocyclic Chemistry, Vol. 3, eds. *A. McKillop* and *R. Boulton*, Pergamon, Oxford, 1984, p. 57; *T. Amarego*, Quinazolines, Wiley, New York, 1967; *T.A. Williamson*, The Chemistry of Quinazoline, in Heterocyclic Compounds, Vol. 6, ed. *R.C. Elderfield*, Wiley, New York, 1957; *W.L.F. Armarego*, Adv. heterocyclic Chem., 1979, **24**, 1). In general, the synthesis of quinazolines is rather different from that of pyrimidines in that the equivalent of the Principal Synthesis is not used. Rather, most syntheses start from intermediates in which at least four atoms of the heterocyclic nucleus are linked together and the benzene ring is also present. Thus many quinazoline syntheses start with derivatives of *o*-aminoaryl-aldehydes, -ketones or -carboxylic acids and ring closure is achieved by the formation of the 2,3- or 3,4-bond.

A very early synthesis of the quinazoline nucleus involves the formation of only one bond using an example of the Rinkes synthesis where one group of a diamide (CCCXLIII) undergoes the first stages of a Hofmann reaction when treated with sodium hypochlorite and then cyclises at the isocyanate stage with the neighbouring amide group to give the quinazoline (CCCXLIV).

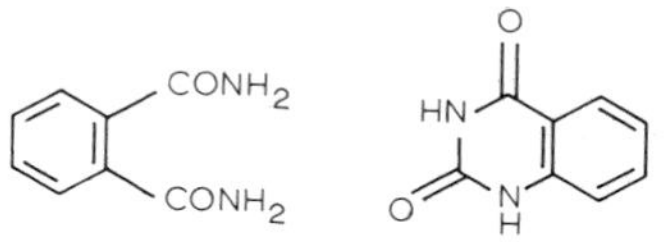

(CCCXLIII)   (CCCXLIV)

The Bischler synthesis (*A. Bischler*, Ber., 1891, **24**, 506; 1892, **25**, 3082; *K. Schofield et al.*, J. chem. Soc., 1952, 1924) involves the formation of two bonds with one nitrogen being supplied from ammonia. Thus *o*-acetylphenylurethane (CCCXLV) when heated with ammonia under pressure gives the quinazoline (CCCXLVI) (*W.L.F. Armarego* and *J.I.C. Smith*, J. chem. Soc. C, 1966, 234).

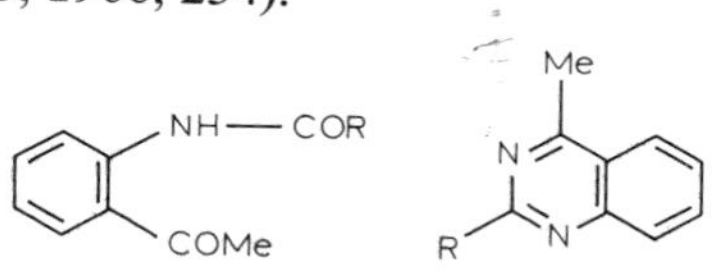

(CCCXLV)   (CCCXLVI)

The Grimmel synthesis (*H.W. Grimmel, A. Guenther* and *J.F. Morgan*, J. Amer. chem. Soc., 1946, **68**, 542) is similar and gives 2,3-disubstituted quinazolin-4-ones. When *o*-acetamidobenzoic acid (CCCXLVII) reacts with aniline and phosphorus trichloride, the product is the quinazolinone (CCCXLVIII).

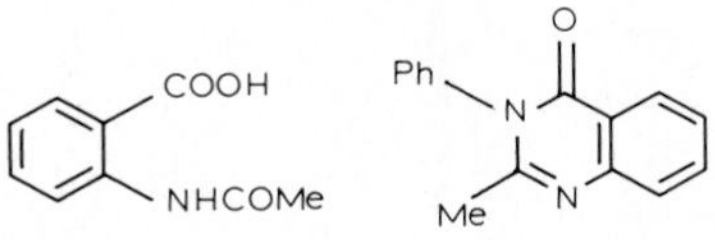

(CCCXLVII)   (CCCXLVIII)

There are also a few syntheses of quinazolines involving the formation of two bonds where the two-atom fragment supplies C-2 and N-3. If 4-nitroanthranilic acid (CCCXLIX) is heated with urea, the intermediate ureido acid (CCCL) is formed which cyclises in the acidic medium (*F.H.S. Curd, J.K. Landquist* and *F.L. Rose*, J. chem. Soc., 1948, 1759) to give 7-nitroquinazolin-2,4(1H,3H)-dione (CCCLI).

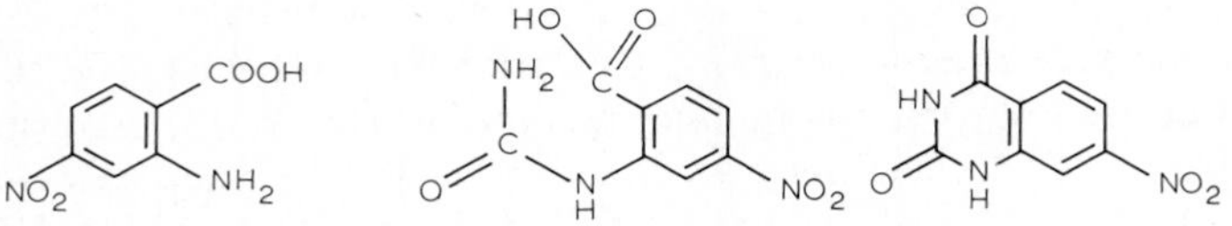

(CCCXLIX)   (CCCL)   (CCCLI)

Similarly *o*-aminobenzaldehyde reacts with methyl isocyanate to give 3-methylquinazolin-2(3H)-one (CCCLII) (*A. Albert* and *G.B. Baslin*, ibid., 1962, 3129).

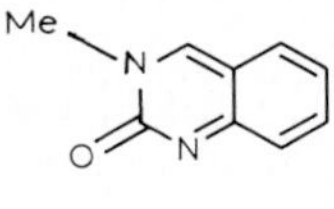

(CCCLII)

The Niementowski synthesis involves the condensation of an amide or thioamide with an aminomethylene derivative. An example is the reaction of anthranilic acid and formamide at 120° to give quinazolin-4(3H)-one (CCCLIII). Finally, the reaction of an iminoether or imidoyl chloride with an aminomethylene derivative can yield a quinazoline derivative. Thus ammonium anthranilate with N-phenylbenzimidoyl chloride gives 2,3-diphenylquinazolin-4(3H)-one (CCCLIV).

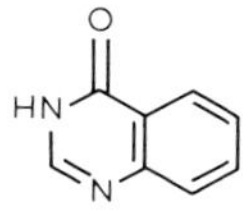

(CCCLIII) 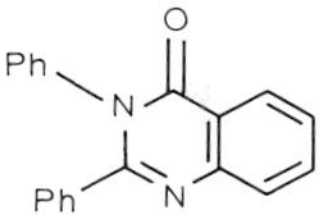   (CCCLIV)

The best synthesis of quinazoline (CCCXXXII) itself (*W. L. F. Armarego* and *J.I.C. Smith*, ibid., 1965, 5360) is to displace the halogen from 4-chloroquinazoline with toluene-*p*-sulphonic acid and to hydrolyse this in hot aqueous ethylene glycol. Alternatively, 4-chloroquinazoline undergoes hydrogenolysis in the presence of palladium.

Aryl- and alkyl-quinazolines can be made using Grignard reagents; a reaction which is seldom, if ever, used in the pyrimidine series. With quinazolines, the reaction only works with substituents at the 4-position and the best substituents are cyano or chloro (*T. Higashino*, Chem. Pharm. Bull. Japan, 1962, **10**, 1043). The required quinazoline-4-carbonitrile is made by addition of hydrogen cyanide across the 3,4-bond of quinazoline followed by oxidation of the adduct (CCCLV) with ferricyanide. Diazotization of the 5-aminoquinazolinone (CCCLVI; R = NH$_2$) in the usual way

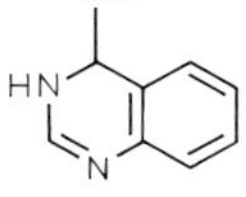

(CCCLV) 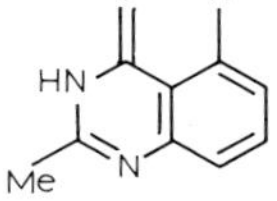   (CCCLVI)

and treatment with copper(I) cyanide gives the 5-cyano derivative (CCCLVI; R = CN) (*A. J. Tomisek* and *B. E. Christensen*, J. Amer. chem. Soc., 1945, **67**, 2122). Quinazalone-2(and -4)-thiones can be made by reactions similar to those used for the corresponding pyrimidines. However, one special method involves the reaction of *o*-aminobenzonitrile with thioacetic acid at 110° which gives the quinazolinthione (CCCLVII) (*H. L. Yale*,

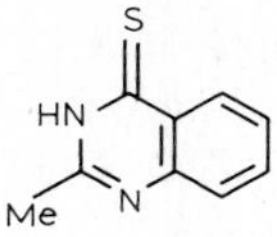

(CCCLVII)

ibid., 1953, **75**, 675). Quinazolin-2-(1*H*)-thione and methyl iodide react under alkaline conditions to give the thioether (CCCLVIII). Alternatively, thioethers can be made by the displacement of halide. Thus 2,4-dichloroquinazoline and sodium *p*-chlorothiophenate gives either the mono- or

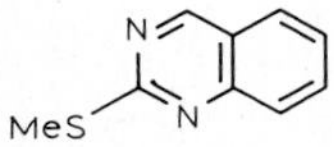

(CCCLVIII) 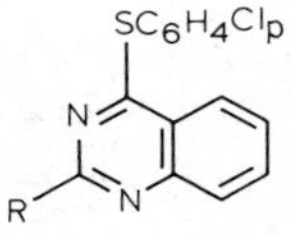 (CCCLIX)

di-chloroarylthio derivative (CCCLIX; R = H or R = $SC_6H_4Cl\,p$). The 4-substituent is displaced preferentially (*F.H.S. Curd* et al., J. chem. Soc., 1948, 1766). Although *N*-oxides are usually made by oxidation, quinazoline-3-oxides can be made by primary synthesis by condensing the oxime (CCCLX) with triethylorthoformate to give the *N*-3 oxide (CCCLXI).

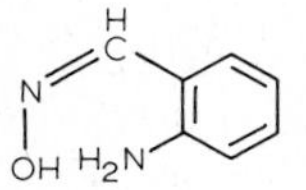

(CCCLX) 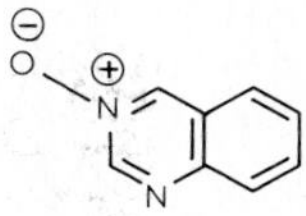 (CCCLXI)

(c) Chemical reactions of quinazolines

In general, one would expect electrophilic attack on quinazoline to occur in the benzene ring (*M.J.S. Dewar* and *P.M. Maitlis*, *ibid.*, 1957, 2521). Reaction conditions need to be quite severe and usually the 6-substituted compound is formed. Thus the 6-nitro, 6-chloro (and eventually, 6,8-dichloro) and 6-chlorosulphonyl derivatives are known (*E. Cohen*, *B. Klarbeg* and *J.R. Vaughan*, J. Amer. chem. Soc., 1960, **82**, 2731). Otherwise the

the reaction of electrophiles is to quaternise N-3 but no electrophilic substitution reactions in the pyrimidine ring are known.

As has already been mentioned, nucleophilic addition occurs readily at the 3,4-bond with even the unsubstituted quinazoline cation (CCCXL) existing as the hydrate. Upon oxidation, the hydrate gives quinazolin-4(3H)-one (CCCLIII). Hydrogen cyanide, ketones and sodium hydrogen sulphite all give similar adducts. As just mentioned, oxidation of quinazoline in acidic solution gives the quinazolinone (CCCLIII) because it is the hydrate being oxidised. In alkaline solution with, for example, permanganate the benzene ring is destroyed to give pyrimidine-4,5-dicarboxylic acid (CCCLXII).

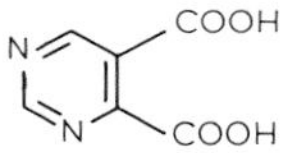

(CCCLXII)

Hydrogenation of quinazoline gives firstly the 3,4-dihydro derivative even if the 4-position is already substituted. Many quinazoline derivatives then react further with, for example, sodium in ethanol or lithium aluminium hydride to give tetrahydro derivatives.

(d) Chemical reactivity of substituted quinazolines

Substituents in the pyrimidine ring of quinazoline are activated in the same way as they are in pyrimidine except that there is much more of a differential in activity between substituents at the 2- and the 4-position with the latter being the more activated one. Substituents in the benzene ring show nearly normal aromatic properties. Thus, in general, the reactivities of substituents in quinazoline can be predicted from those of pyrimidine and benzene and only a few typical examples will be given here. Details on individual compounds, either their preparation or properties are to be found in the comprehensive reviews (*loc. cit.*). Thus 2- and 4-methylquinazolines show enhanced activity and condense with benzaldehyde to give styryl compounds (*W. Ried* and *H. Keller*, Ber., 1956, **89**, 2578). 2-Methylquinazoline reacts with chloral to give the intermediate (CCCLXIII) which can be dehydrated and hydrolysed under alkaline condi-

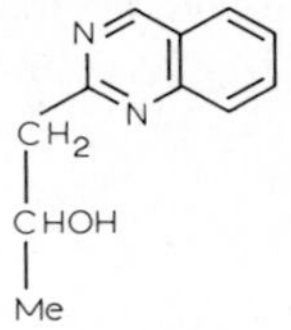

(CCCLXIII) 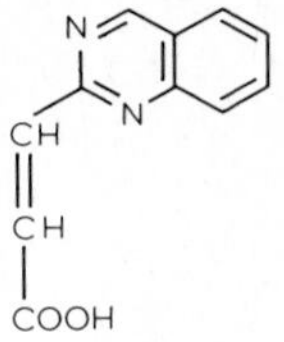 (CCCLXIV)

tions to give the acrylic acid derivative (CCCLXIV). Only the methyl group at the 4-position is sufficiently activated to give a Mannich base (*T. Siegle and B.E. Christensen*, J. Amer. chem. Soc., 1951, **73**, 5777). Thus 2,4-dimethylquinazoline reacts with dimethylamine and formaldehyde only at the 4-position to give the Mannich base (CCCLXV). Acyl derivatives of

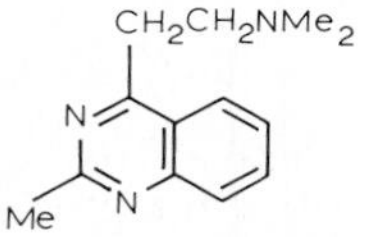

(CCCLXV)

quinazolines exhibit normal reactivities and form the usual derivatives such as Schiff bases, hydrazones and semicarbazones. Quinazoline carboxylic acids decarboxylate as expected if the carboxyl group is present at the 2- or 4-position but if in position 5 or 8, then very drastic and usually not realistic conditions have to be used. Acid chlorides can be formed normally using thionyl or phosphoryl chloride. There is a difference in the reactivity of the 2- and the 4-carbonitrile towards Grignard reagents. As already mentioned, the 4-carbonitrile undergoes displacement of the cyano group to give a 4-alkylquinazoline whereas the 2-carbonitrile reacts to give the corresponding ketone (*E. Hayashi* and *T. Higashino*, Chem. Pharm. Bull. Japan, 1964, **12**, 43; *T. Higashino*, ibid., 1962, **10**, 1043).

Quinazolinamines react with sulphonic acids to give sulphonamides (*H.J. Rodda*, J. chem. Soc., 1956, 3509) and react under alkylating conditions at N-3 to give a product which then undergoes a Dimroth rearrangement, as already described for the pyrimidines, to give a 2-alkyl derivative. Thus 2-aminoquinazolin-4(3H)-one (CCCLXVI; R = H) with methyl iodide gives the N-3 methyl derivative (CCCLXVI; R = CH_3) which undergoes a

Dimroth rearrangement to give 2-methylaminoquinazolin-4(3H)-one (CC-CLXVII; R = CH$_3$) (*R.J. Grout* and *M.W. Partridge, ibid.,* 1962, 3540).

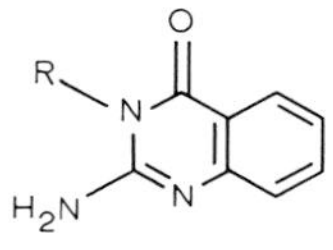

(CCCLXVI) 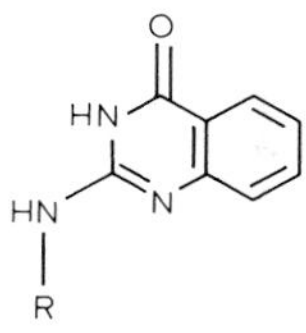   (CCCLXVII)

Phosphorus pentasulphide will convert the quinazolinone (CCCLXVIII; X = O) into the thione (CCCLXVIII; X = S) in a reaction similar to that found in the pyrimidine series but a hydroxyl group at position 5, 6, 7 or 8

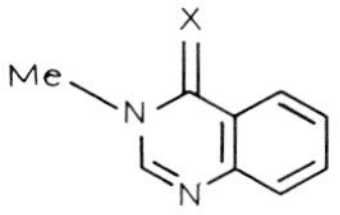

(CCCLXVIII)

will not react. The quinazolin-(5–8)-ols are phenolic in character but there are relatively few examples of their *O*-alkylation in the literature (*A. Albert* and *G.B. Berlin, ibid.,* 1962, 3129). One example is the *O*-alkylation of quinazolin-8-ol (CCCLXIX; R = H) to 8-allyloxyquinazoline (CCCLXIX;

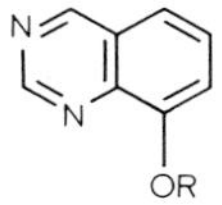

(CCCLXIX)

R = CH$_2$–CH = CH$_2$) by allyl bromide in methanolic sodium methoxide (*A. Albert* and *A. Hampton, ibid.,* 1954, 505). Trimethylsilyl ethers can be made from quinazolindiones so that the oxygen is protected and *N*-alkylation can be performed. Thus starting with quinazolin-2,4(1H,3H)-dione (CCCLXX; R = H), treatment with hexamethyl disilazane gives the bistri-methylsilyl derivative (CCCLXXI) which after methylation followed by acidic hydrolysis gives the *N*-methyl compound (CCCLXX; R = Me). Alkoxyquinazolines readily take part in thermal and Hilbert–Johnson reactions to give *N*-alkylated quinazolinones (*R. J. Grout* and *M.W. Partridge, ibid.*, 1960, 3546).

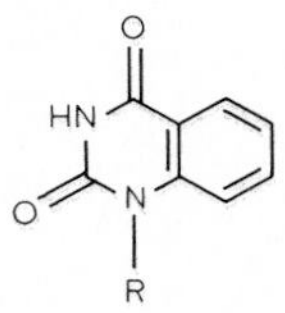

(CCCLXX) 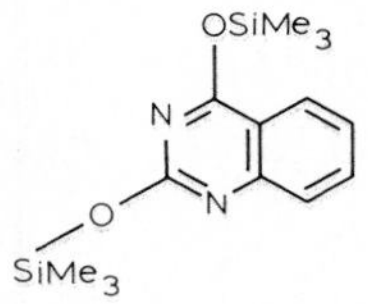 (CCCLXXI)

Quinazoline-4-thiones react readily with alkylamines but much less readily with arylamines to give *N*-substituted quinazolin-4-amines.

Haloquinazolines can be dehalogenated in the usual way by hydrogenolysis. Halogeno substituents in the benzene ring of quinazoline are completely resistant to ammonolysis but 2- and 4-chloroquinazoline are readily attacked by a variety of nucleophilic reagents. Preferential reactivity in this reaction is again shown by the 4-substituent so that the 4-substituted product can always be isolated before the final 2,4-disubstituted product is obtained from a 2,4-dichloroquinazoline. Thus halogenoquinazolines react in a predictable manner with hydroxide, alkoxides, phenoxides, sulphides, thiols and amines; they will also react with compounds containing active methylene groups, for example, 4-chloroquinazoline reacts with benzyl sodiocyanoacetate to give the cyanoester (CCCLXXII) (*A. Albert, W.L.F. Armarego* and *E. Spinner, ibid.*, 1961, 2689).

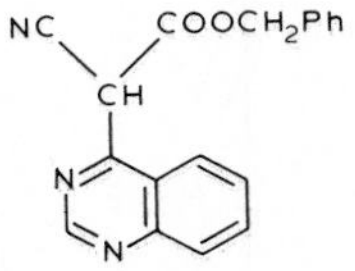

(CCCLXXII)

Quinazoline-*N*-oxides can have the *N*-oxide substituent removed in the usually way with phosphorus trichloride or by catalytic hydrogenation.

Sulphur dioxide has also been used for example with 4-ethoxyquinazoline-1-oxide (CCCLXXIII) to give the parent ethoxyquinazoline (*H. Yamanaka*, Chem. Pharm. Bull. Japan, 1959, 7, 152).

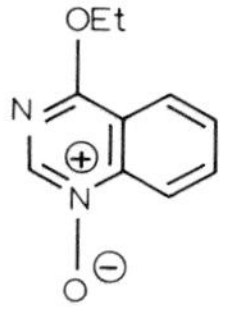

(CCCLXXIII)

Chapter 44

Pyrazines and Related Ring Structures

K.J. McCULLOUGH

1. Pyrazines, 1,4-diazines

(a) Introduction

Pyrazines occur extensively in nature in relatively small quantities and exhibit a range of biological properties. Fusel oils contain di- and tri-alkyl-pyrazines. In addition, pyrazine and simple mono- and poly-alkylated pyrazines have been identified as important flavour components in cooked foodstuffs (*J.A. Maga* and *C.E. Sizer*, J. agric. food Chem., 1973, **21**, 22). More complex naturally occurring pyrazines include the antibiotic aspergillic acid (I) and its structural analogues isolated from cultures of the aspergillus species (*E.C. White* and *J.H. Hill*, J. Bact., 1943, **45**, 433) and the red pigment pulcherrimin which has been shown to be a ferric ion complex of pulcherrimic acid (II) (*J.C. MacDonald*, Canad. J. Chem., 1963, **41**, 165). In certain coelenterates, enzyme-catalysed oxidations of pyrazine substrates, *e.g.* (III) to (IV), are responsible for the phenomenon of bioluminescence (*F. McCapra*, Acc. chem. Res., 1976, **9**, 201; *F. McCapra* and *R. Hart*, Nature, 1980, **286**, 660).

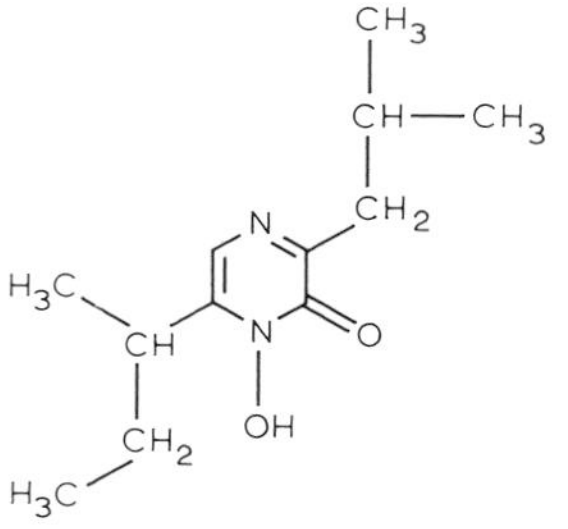

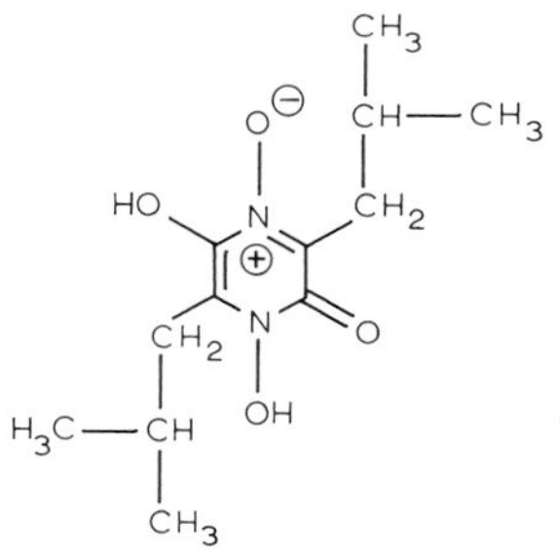

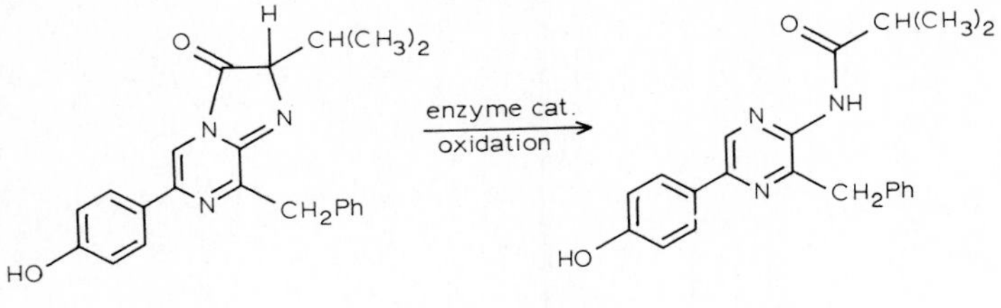

Synthetic pyrazines also possess a broad spectrum of physiological activity. Sulfalene (V) is a long acting sulphonamide (*M. Ghione et al.*, Chemotherapia, 1963, **6**, 344), pyrazinamide (VI) is a tuberculostatic drug (Merck Index, 9th Edn., 1976, 7740) and the di-*N*-oxide (VII) has herbicidal activity (U.S.P. 3,753,989; C.A., 1973, **79**, 92281). The biological properties of pyrazines have been reviewed (*L. Novacek*, Cesk. Farm., 1966, **15**, 323; C.A., 1967, **66**, 5712).

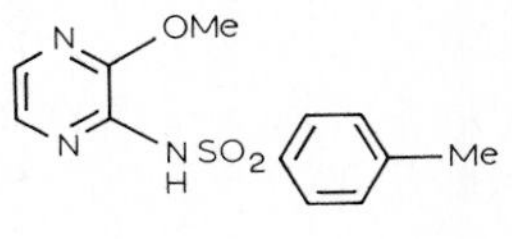

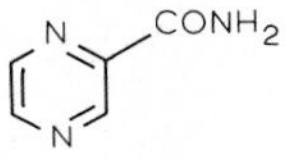

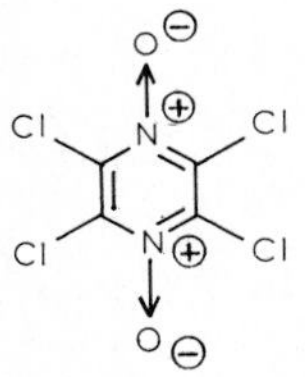

As a consequence of having two nitrogen atoms in the six-membered ring, pyrazines generally behave as weak bases. In 60% sulphuric acid, pyrazine is monoprotonated; complete protonation is achieved in concentrated sulphuric acid (*H. Kamei*, J. phys. Chem, 1965, **69**, 2791). Reaction with alkyl halides affords only the corresponding, thermally stable monoquaternary salts (*C.T. Bahner* and *L.L. Norton*, J. Amer. chem. Soc., 1950, **72**, 2881). The diquaternary salts, which can be prepared by using oxonium salts (*e.g.* triethyloxonium tetrafluoroborate) as alkylating agents, are unstable and undergo reduction in air and/or protic solvents to produce the radical cation (VIII) [Scheme 1] (*T.J. Curphey* and *K.S. Prasad*, J. org. Chem., 1972, **37**, 2259).

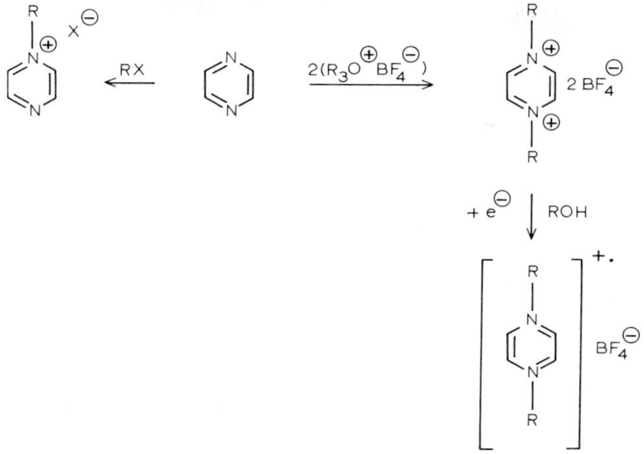

Scheme 1

The nitrogen atoms in pyrazine deactivate all the carbon positions to electrophiles and conversely activate them to nucleophiles. Electrophilic reagents often attack the ring preferentially at nitrogen, *via* the formally non-bonding lone pairs of electrons, forming an adduct which is not susceptible to further reaction with electron deficient species. Ring substituents such as halogen or methyl groups are activated as in the analogous α-substituted pyridines. Since pyrazine is less stable to permanganate oxidation than pyridine, pyrazine carboxylic acids are less readily obtained by oxidation of homologues.

Pyrazines, being potentially bidentate ligands, form 1 : 1 and 1 : 2 coordination complexes with a range of metal ions (*W. Kaim*, Angew. Chem. intern. Edn., 1983, **22**, 171 and references therein).

(b) Physical and spectroscopic properties

The pyrazine ring can be represented as a resonance hybrid of canonical forms as shown below.

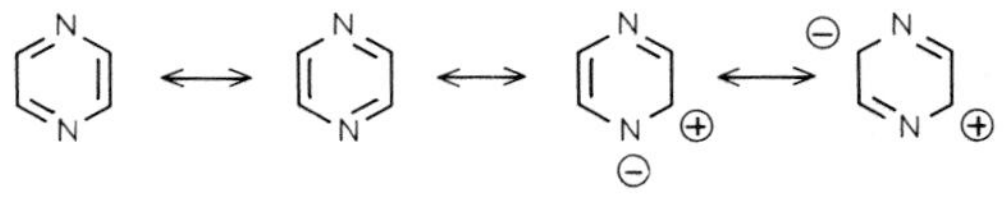

X-ray diffraction (*P.J. Wheatley*, Acta Cryst., 1957, **10**, 182; *G. deWith et al.*, *ibid.*, 1976, **B32**, 3178) and electron diffraction (*B.J.M. Bormans et al.*, J. mol. Struct., 1977, **42**, 121) studies show that pyrazine is a planar molecule of D_{2h} symmetry. The C–N and C–C bond distances are 1.339 and 1.403 Å, respectively, comparable with corresponding distances in pyridine and benzene. A number of theoretical studies have been carried out on the pyrazine ring system in order to predict a variety of electronic, structural and physical properties. Although there are differences in the ordering of the pyrazine molecular orbitals owing to inaccuracies and/or assumptions inherent in the respective methods, the results of SCF LCAO-MO (*M. Hachmeyer* and *J.L. Whitten*, J. chem. Phys., 1971, **54**, 3739) and extended Hückel theory (EHT) (*R. Gleiter et al.*, Helv., 1972, **55**, 255) calculations do, however, agree that the nitrogen lone pairs should interact and be significantly different in nature. The lower energy pair is essentially localised on nitrogen whereas the other higher energy pair is delocalised into the σ-skeleton of the ring. Moreover, re-examination of the photoelectron spectrum of pyrazine suggests that the highest occupied molecular orbital (HOMO) of pyrazine is the latter lone pair σ-obital in good agreement with e.s.r. studies of the radical cation generated by an alternative route (*T. Kato* and *T. Shida*, J. Amer. chem. Soc., 1979, **101**, 6869). The first ionisation potential of pyrazine (9.63 eV) is greater than that of benzene (9.25 eV) and pyridazine (9.31 eV) but slightly less than that of pyrimidine (9.73 eV). On the other hand, pyrazine has a low electron affinity, estimated to be *ca.* +0.40 eV (*I. Nenner* and *G.J. Schulz*, J. chem. Phys., 1975, **62**, 1747), and consequently affords a comparatively long-lived radical anion (*K.B. Wiberg* and *T.P. Lewis*, J. Amer. chem. Soc., 1970, **92**, 7154). The resonance energy of pyrazine is estimated to be *ca.* 18 kcal/mol (*M.H. Palmer*, "The Structure and Reactions of Heterocyclic Compounds", Arnold, London, 1967, p. 66). Calculations of the electronic distribution indicate that the ring carbons of pyrazine and the α-carbons of pyridine should be electron deficient to about the same extent (*R.D. Brown* and *B.A.W. Coller*, Theor. Chim. Acta, 1967, **7**, 259; and *R.J. Pugmire* and *D.M. Grant*, J. Amer. chem. Soc., 1968, **90**, 259).

The ionisation constants of pyrazine have been measured as pK_{a1} 0.65 and pK_{a2} −5.8 which makes pyrazine a weaker base than pyridine (pK_a 5.2) or the other diazines, pyridizine (pK_a 2.30) and pyrimidine (pK_a 1.30) (*A.S.-C. Chia* and *R.F. Trimble*, J. phys. Chem., 1961, **65**, 863; a fuller compilation of values for substituted pyrazines is given in *D.D. Perrin*, "Dissociation Constants for Organic Bases in Aqueous Solution", Butterworths, London, 1965, and Supplement, 1972). As in pyridine, methyl substituents increase the base strength by *ca.* 0.7 pK_a units. From theoretical studies, a similar order of relative proton affinities has been obtained (*J.E. Del Bene*, J. Amer. chem. Soc., 1977, **99**, 3617).

The infrared and Raman spectra of pyrazine have been studied in detail (*J.D. Simmons et al.*, J. mol. Spectrosc., 1964, **14**, 190 and references therein). The absorption bands at 1600–1575, 1550–1465, 1500–1465, and 1420–1370 cm^{-1}, which have their origins in ring skeleton vibrations and C–H in-plane bending modes, are the most diagnostic. The ultraviolet spectrum of pyrazine shows maxima at 260 nm (log ε 3.75) and 328 nm (log ε 3.02) associated with the $\pi-\pi^*$ and $n-\pi^*$ transitions, respectively (*S.F. Mason*, in "Physical Methods in Heterocyclic Chemistry", Vol. II, ed. *A.R. Katrizky*, Academic Press, New York, 1963). Auxochromes give rise to predictable hypsochromic and bathochromic shifts.

In deuterochloroform, the ^{1}H-n.m.r. spectrum of pyrazine consists of a singlet at δ 8.59 (*Uchimaru et al.*, Chem. pharm. Bull. Japan, 1972, **20**, 2204). The coupling constants between the ring protons ^3J_o, ^4J_m and ^5J_p, observed in the ranges 2.5–3, 1.1–1.4 and 0 Hz, respectively, show little variation with the nature of the substituent. The ^{1}H-n.m.r. spectra of some simple pyrazine derivatives have been tabulated (*T.J. Batterham*, "NMR Spectra of Simple Heterocycles", Wiley, New York, 1973). The proton-decoupled ^{13}C-n.m.r. spectrum of pyrazine contains one signal at 145.4 ppm with reference to TMS. From the ^{13}C-n.m.r. spectra of a series of mono-substituted pyrazines, values of the carbon–hydrogen coupling constants $^1J_{CH}$ 180–196, $^2J_{CH}$ 10–13, $^3J_{CH}$ 9–12, and $^4J_{CH}$ −1.5 Hz have been determined (*C.J. Turner* and *G.W.H. Cheeseman*, Org. magn. Resonance, 1974, **6**, 663).

Pyrazine and its derivatives generally produce mass spectra containing a strong molecular ion, often as the base peak. Subsequent fragmentation processes in pyrazine involve loss of one

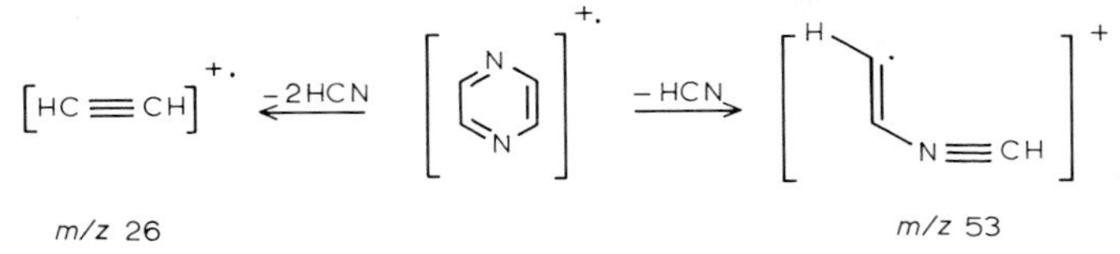

Scheme 2

or two molecules of hydrogen cyanide giving ions at m/z 53 and 26 [Scheme 2] (*Q.N. Porter*, "Mass Spectrometry of Heterocyclic Compounds", 2nd edn., Wiley, New York, 1985)

(c) General methods of synthesis

In the more general synthetic approaches, the pyrazine ring skeleton is conveniently constructed by the condensation of two appropriately functionalised aliphatic components, *e.g.* the dimerisation of α-amino-ketones or the reaction of 1,2-diamines with α,β-dicarbonyl compounds. Since the primary product from these reactions is usually a dihydropyrazine, a final aromatisation step may be required. Alternatively, pyrazines can be obtained from heterocyclic compounds which already contain the required C_4N_2 moiety by, for example, (i) dehydrogenation of piperazines, or (ii) ring cleavage of quinoxalines or pteridines. A variety of pyrazine derivatives can obtained from other pyrazines by ring substitution reactions, *e.g.* halopyrazines are prepared by halogenation, or from hydroxypyrazines.

(i) Self-condensation of α-aminocarbonyl compounds

α-Aminoketones undergo self-condensation with concommitant cyclisation to produce dihydropyrazines (IX) which are oxidised by mercury(II) chloride to the symmetrically substituted pyrazines (*J.W. Cornforth*, J.

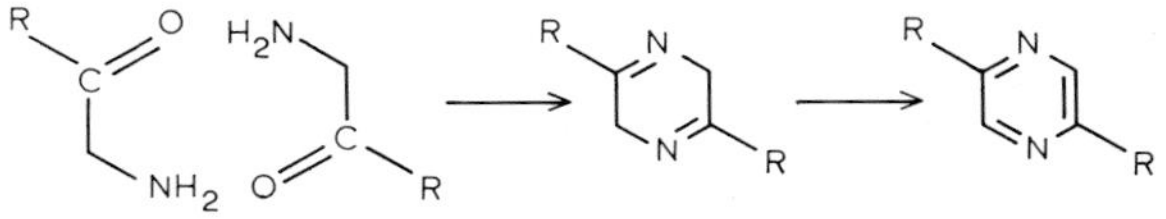

(IX)

chem. Soc., 1958, 1174). The required α-aminoketones can be prepared from α-amino-acids by the Dakin-West reaction (see G.H. Cleland, J. Amer. chem. Soc., 1949, **71**, 841; N.L. Allinger, J. org. Chem., 1974, **39**, 1730), by reduction of oximinoketones, or by rearrangement of an oxime p-toluene-sulphonate with potassium hydroxide in ethanol (P.W. Neber et al., Ann.,

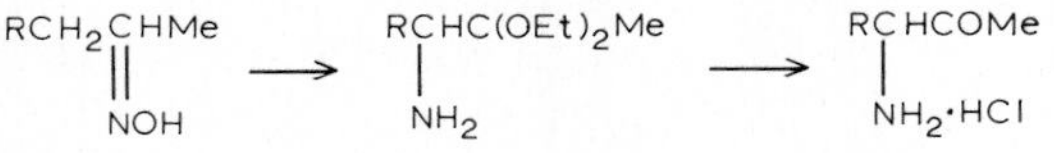

$$RCH_2CMe\text{=NOH} \longrightarrow RCHC(OEt)_2Me\text{, }NH_2 \longrightarrow RCHCOMe\text{, }NH_2\cdot HCl$$

1936, **526**, 277). Owing to their tendency to undergo self-condensation, the free bases are difficult to isolate, although the hydrochloride salts are usually stable.

As an alternative, aminoketones may be generated *in situ* from a variety of precursors and dimerised without prior isolation.

(*a*) Aminolysis of α-haloketones. Thus treatment of ω-bromoacetophenone with ammonia affords 2,5-diphenylpyrazine (X) together with some of the 2,6-isomer (F. Tutin, J. chem. Soc., 1910, **97**, 2495.).

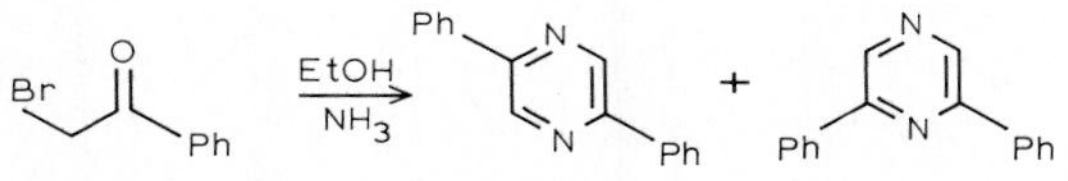

$$BrCH_2COPh \xrightarrow[NH_3]{EtOH} \text{2,5-Ph}_2\text{-pyrazine (X)} + \text{2,6-Ph}_2\text{-pyrazine}$$

(*b*) Heating α-hydroxyketones with ammonium acetate provides a convenient route to tetrasubstituted pyrazines (J. Wiemann et al., Bull Soc. chim. Fr., 1965, 3476; N. Vinot and J. Pinson, ibid., 1968, 4970).

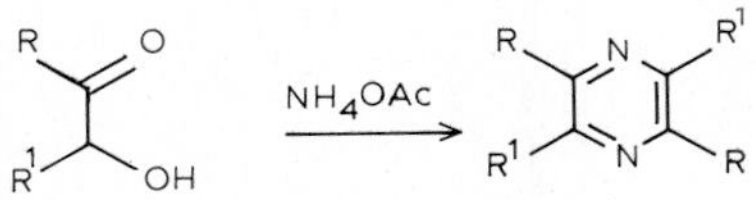

$$\underset{R^1\text{—OH}}{R\text{—CO}} \xrightarrow{NH_4OAc} \text{tetrasubstituted pyrazine}$$

(*c*) Reduction of α-diazoketones, prepared from acid chlorides and diazomethane, in the presence of a palladium catalyst leads to 2,5-disubstituted pyrazines (L. Birkhofer, Ber., 1947, **80**, 83). Under similar reaction conditions, α-azidoketones also produce di- and tetra-substituted pyrazines in high yield (> 80%) (J.-P. Anselme et al., Org. Prep. Proced. Int., 1980, **12**, 265).

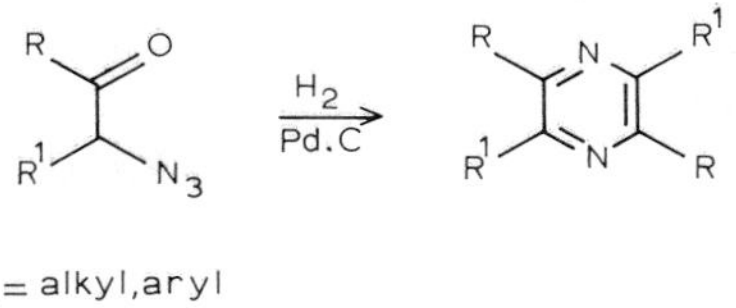

R = alkyl, aryl
R¹ = H, Et

(d) Treatment of an α-hydroxyiminoketone with an aqueous sodium bisulphite solution saturated with sulphur dioxide affords an adduct, which on reaction with potassium cyanide followed by hydrolysis, gives a 2,5-dicyanopyrazine (G. Gastaldi, Gazz., 1921, **51**, I, 233).

(ii) Condensation of α,β-diamines with α,β-carbonyl compounds

Although pyrazine cannot be prepared from the condensation of ethylenediamine and glyoxal (A. Mason, J. chem. Soc., 1893, **63**, 1284), this procedure represents one of the classical and general synthetic routes to substituted pyrazines. Condensation of ethylenediamine with 2,3-dioxoalkanes gives initially the corresponding dihydropyrazine (XI) which can be dehydrogenated at 300° over a copper chromite catalyst [Scheme 3] (*I.*

Scheme 3

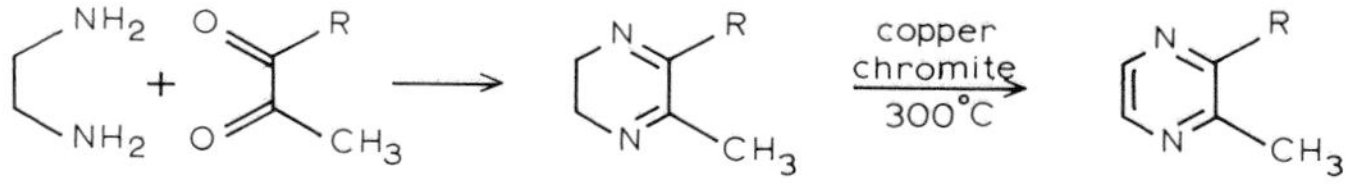

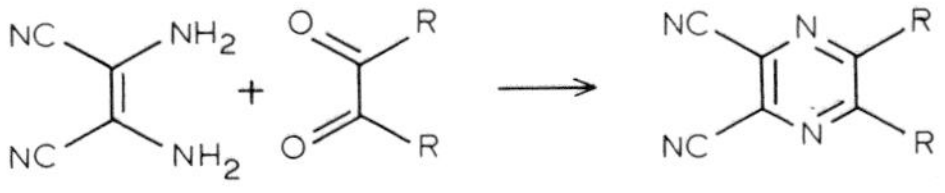

Flament and *M. Stoll*, Helv., 1967, **50**, 1754). Similarly, the reaction of 1,2-diaminopropane and diacetyl followed by air oxidation of the intermediate dihydropyrazine in the presence of potassium hydroxide pellets afforded 2,3,5-trimethylpyrazine in reasonable yield (*J.P. Marion*, Chimia, 1967, **21**, 510). Oxidation of 5,6-dihydropyrazines using manganese dioxide or copper oxide in ethanolic potassium hydroxide is reported to produce pyrazines of high purity in good yields (*T. Akiyama et al.*, J. agric. food Chem., 1978, **26**, 1176). From diaminomaleonitrile (DAMN), 2,3-dicyanopyrazines (XII) can be prepared directly in high yield without the requirement for an oxidation step (*H.W. Rothkopf et al.*, Ber., 1975, **108**, 875). The reaction of DAMN with β-ketosulphones affords a similar series of pyrazines (*S. Kano et al.*, Synthesis, 1978, 372).

In addition, the reaction of α,β-diketones with molecules containing the α,β-diamino, or equivalent, functionality provides entry to a further range of pyrazine derivatives.

(*a*) α-Amino-acid amides give hydroxypyrazines (*R.G. Jones*, J. Amer. chem. Soc., 1949, **71**, 78; *R.M. Siefert et al.*, J. agric. food Chem., 1970, **18**, 246).

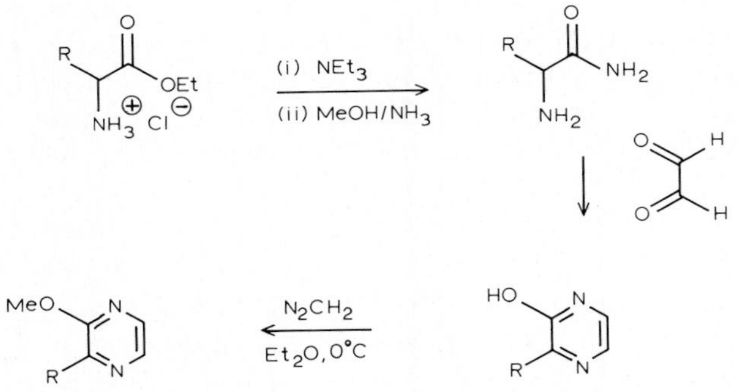

(*b*) α,β-Diaminocarboxylic acids give pyrazinecarboxylic acids (*E. Felder et al.*, Ber., 1967, **100**, 555).

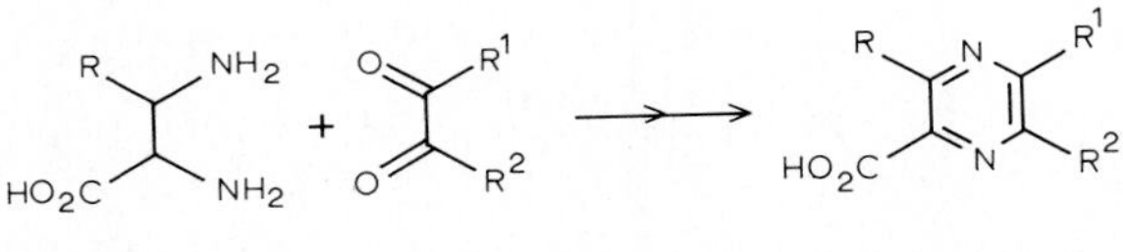

(*c*) Aminoacetamidine gives aminopyrazines (*D. Pitre* and *S. Boveri*, ibid., 1967, **100**, 560).

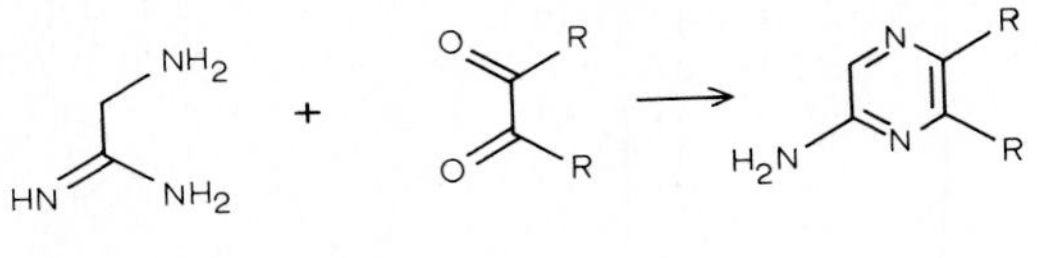

By analogy, α-amidino-α-amino-acetamides and acetate esters are useful precursors of aminopyrazine amides and esters, respectively (*O. Vogl* and

E.C. Taylor, J. Amer. chem. Soc., 1959, **81**, 2472; W.F. Weir, A.H. MacLennan and H.C.S. Wood, J. chem. Soc. Perkin I, 1978, 1002).

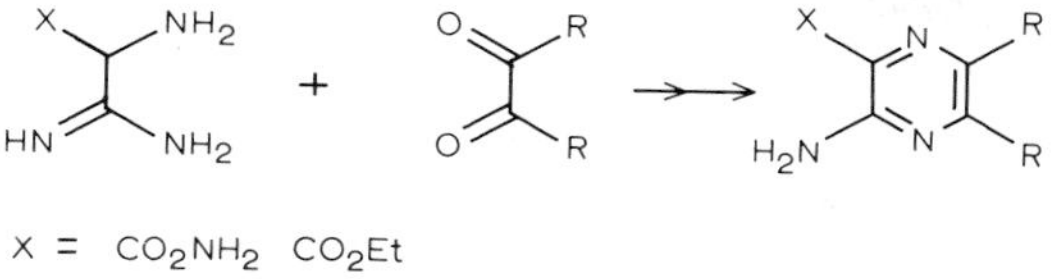

X = CO_2NH_2 CO_2Et

(iii) Miscellaneous condensation and dimerisation procedures

2-Hydroxypyrazines (XIII) can be prepared from α-aminoketones and α-halogenoacid halides followed by reaction of the product with ammonia (*Y.A. Tota* and *R.E. Elderfield*, J. org. Chem., 1942, **7**, 313).

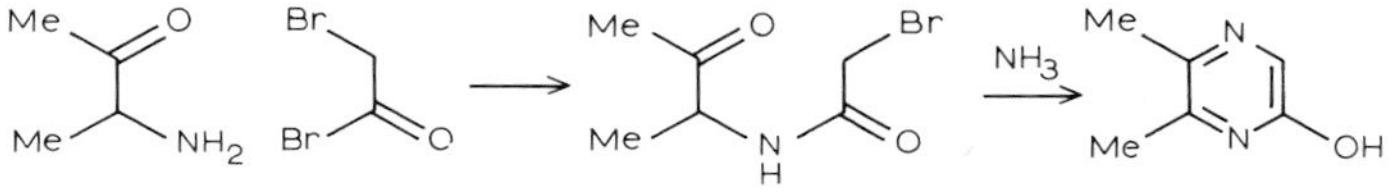

(XIII)

2,3-Disubstituted pyrazines (XIV) can be prepared by the sequential treatment of α-bromoketones with *N*-phenyltriflamide followed by ethylenediamine (*R.J. Bergerson* and *P. Hoffman*, ibid., 1980, **45**, 161).

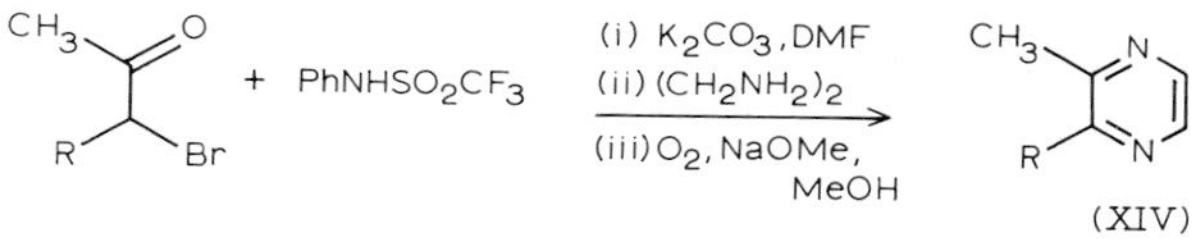

(XIV) 60-70%

DAMN reacts with acylnitriles to produce isolable mono-imines which can be selectively hydrolysed to amides prior to cyclisation to pyrazines (XV) (*Y. Ohtsuka*, J. org. Chem., 1979, **44**, 827). Similarly, when mono-im-

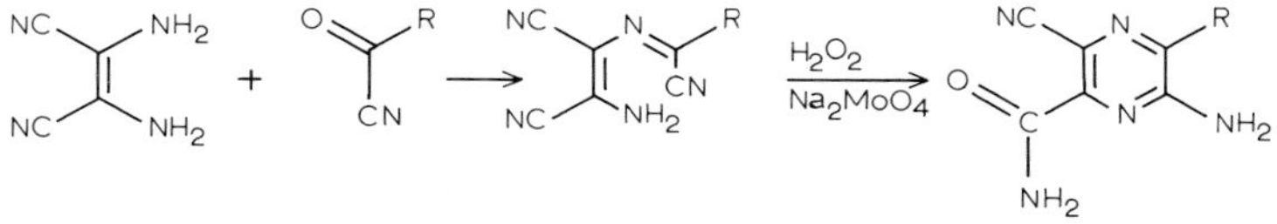

(XV)

ines, derived from DAMN and aromatic aldehydes, react with a second molecule of aldehyde, hydrolysis of one of the CN groups also occurs. The resulting di-imines cyclise readily to dihydropyrazines which are in turn oxidised to pyrazines (XVI) (Y. Ohtsuka et al., ibid, 1979, **44**, 4871).

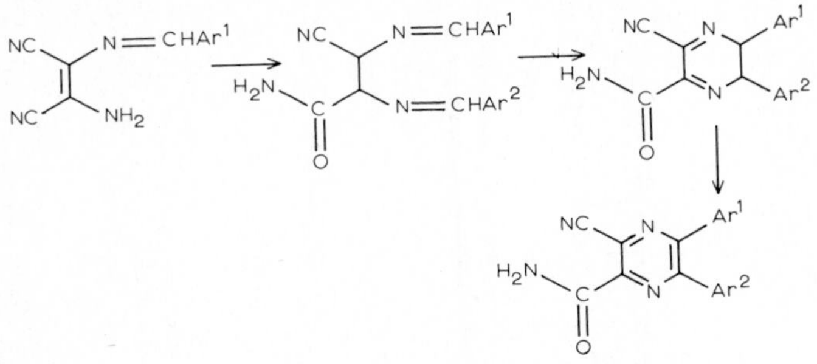

(XVI)

Glyceraldehyde condenses with DAMN to give a mono-imine which on treatment with mercuric chloride produces 2,3-dicyano-5-methylpyrazine (M. Sakaguchi et al., Chem. pharm. Bull. Japan, 1979, **27**, 1094).

The reaction of N-alkylaminoacetonitriles with oxalyl chloride in o-dichlorobenzene at 80–100° affords 2,6-dichloropyrazin-3-ones (XVII) in moderate yields (J. Vekemans et al., J. heterocyclic Chem., 1983, **20**, 919).

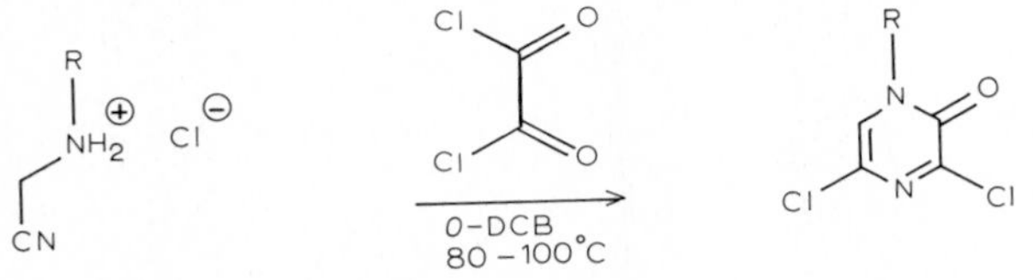

(XVII)

Treatment of α-azidoketones with triphenylphosphine generates the corresponding P–N ylids (XVIII) as intermediates which dimerise to give 2,5-disubstituted pyrazines (E. Zbiral and J. Stohl, Ann., 1969, **727**, 231).

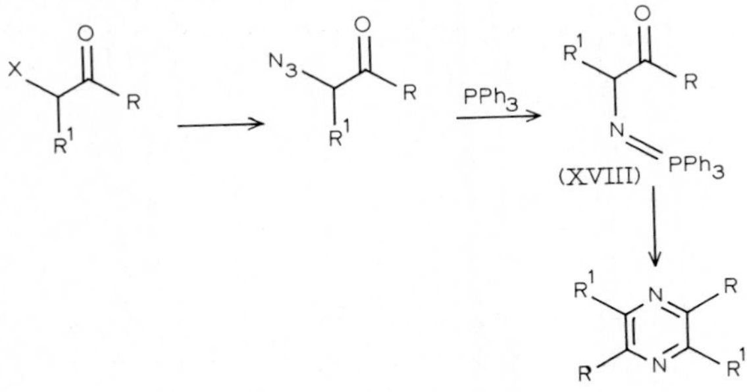

(XVIII)

Catalytic hydrogenolysis of cyclic nitro-enamines (XIX) results in the formation of 2,5-di(ω-aminoalkyl)pyrazines (XX); the product yields varying markedly with ring size (*S. Rajappa* and *R. Sreenivasan*, Tetrahedron Letters, 1978, 2217).

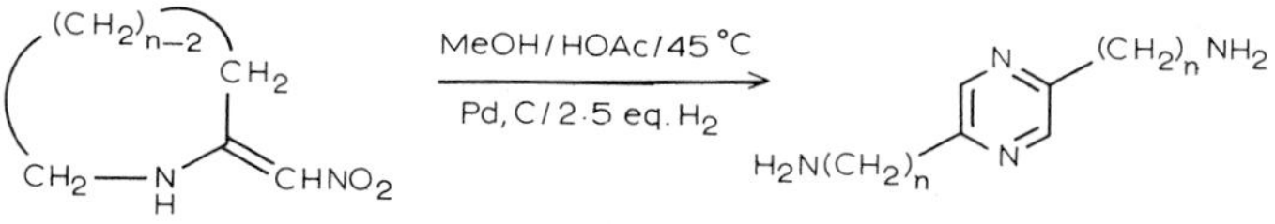

The bis-annelated pyrazine (XXI) has been obtained from the boron trifluoride catalysed ring opening and coupling of the corresponding 2H-azirine (*H. Bader* and *H.-J. Hansen*, Helv., 1978, **61**, 286).

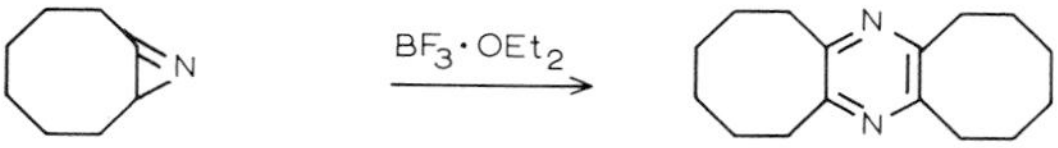

(iv) Dehydrogenation of piperazines

Owing to the comparatively severe conditions required, the preparation of pyrazine and its simple alkyl derivatives by dehydrogenation of the corresponding piperazines is essentially limited to a number of patented industrial processes. Thus the passage of an aqueous solution of piperazine over a reduced copper chromite catalyst at 350–370° (U.S.P. 3,005,820; C.A., 1962, **56**, 7335) or a copper oxide–zinc oxide catalyst at 400–420° (Jap. P. 76: 56479; C.A., 1977, **86**, 29872) is claimed to produce pyrazine in high yield.

The reaction of piperazine-2,5-diones with triethyloxonium tetrafluoroborate gives diethoxydihydropyrazines (XXII) which are readily oxidised to 2,5-diethyoxypyrazines (*K.W. Blake, A.E.A. Porter* and *P.G. Sammes*, J. chem. Soc. Perkin I, 1972, 2494).

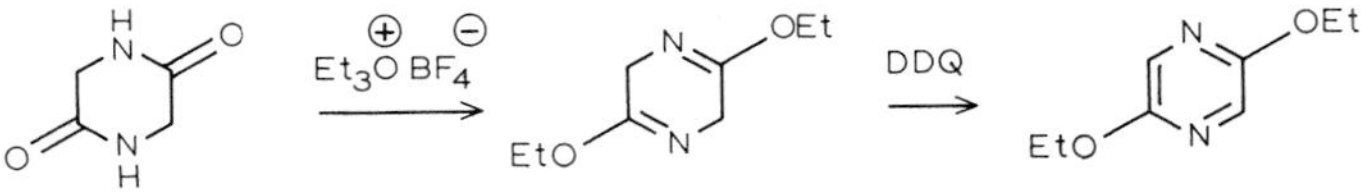

(v) Cleavage of quinoxalines and pteridines

Since the pyrazine ring is generally more resistant to oxidation than a benzene ring, pyrazinecarboxylic acids can be obtained by permanganate oxidation of either quinoxalines (S. Gabriel and A. Sonn, Ber., 1907, **40**, 4850; R.G. Jones and K.C. McLaughlin, Org. Synth., 1950, **30**, 86) or phenazines (B.P. 565,778). Moreover, the pyrazine ring is also less susceptible to acid- or base-catalysed hydrolysis than the other two diazine ring systems. On refluxing with dilute sulphuric acid, pteridine is converted to 2-amino-3-formylpyrazine (XXIII); under milder conditions and with sodium carbonate and hydroxylamine, the aldoxime (XXIV) is obtained (A. Albert, D.J. Brown and H.C.S. Wood, J. chem. Soc., 1956, 2066).

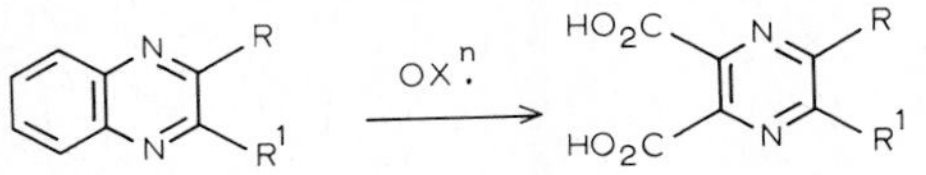

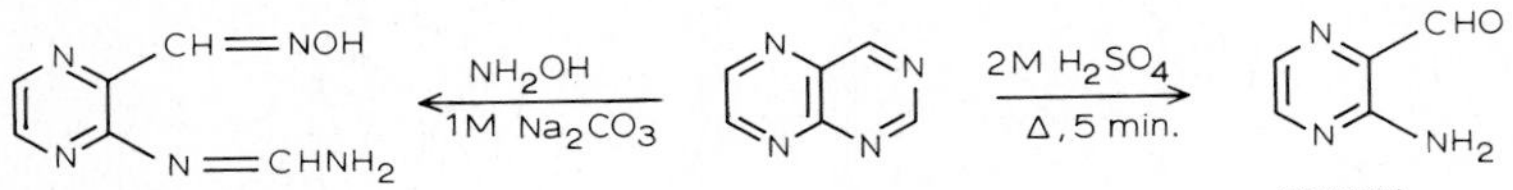

(d) Pyrazine, its homologues and derivatives

(i) Pyrazine, alkyl- and aryl-pyrazines

Pyrazine (p-diazine), m.p. 57°, b.p. 116° (*picrate*, m.p. 157°, *mercuric chloride salt*, m.p. 273°), sublimes at room temperature and has an odour of heliotrope. Although it is not readily synthesised, a number of industrial processes have been developed including (1) catalytic vapour phase dehydrogenation of precursors such as ethanolamine and diethylenetriamine at temperatures around 300–400° (see J.G. Aston et al., J. Amer. chem. Soc., 1934, **56**, 153; U.S.P. 2,414,522; C.A., 1947, **41**, 2756) and (2) vapour phase contact of ethylenediamine and ethylene glycol over a Cu–Cr catalyst at 400° (Jap. P. 74: 117,480; C.A., 1975, **82**, 171,053). In each case, piperazine is presumably formed as an intermediate. Catalytic dehydrogenation of piperazines has been discussed above (general method iv).

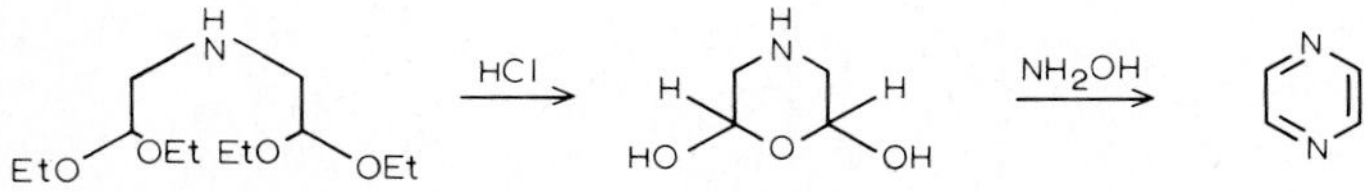

Other synthetic routes include (1) hydrochloric acid hydrolysis of diacetalylamine to give 2,6-dihydroxymorpholine which on treatment with hydroxylamine affords pyrazine in 78% yield (L. Wolff and R. Marburg, Ann., 1908, **363**, 169); (2) decarboxylation of pyrazinecarboxylic acid suspended in a high boiling solvent (*e.g.* dibutyl phthalate) with removal of the pyrazine by distillation (B.P. 560,965, 1944; C.A., 1946, **40**, 5074); (3) acid hydrolysis of aminoacetal followed by basification with alkali forms a product which is converted into pyrazine with mercury(II) chloride (*L. Wolff*, Ber., 1893, **26**, 1830; *S. Gabriel* and *G. Pinkus, ibid.*, 2197).

Methylpyrazine, b.p. 135° (*picrate*, m.p. 133°), has been prepared (1) by decarboxylation of 2-carboxy-5-methylpyrazine (*C. Stoehr*, J. pr. Chem. [2], 1895, **51**, 449); (2) as the major pyrazine derivative from the reaction of D-glucose with ammonia (*P. Brandes* and *C. Stoehr, ibid.*, 1896, **54**, 681); (3) by catalytic dehydrogenation of 2-methylpyrazine over copper chromite (*L.J. Kitchen* and *E.S. Hanson*, J. Amer. chem. Soc., 1951, **73**, 1838).

2,3-**Dimethylpyrazine**, b.p. 156° (*picrate* m.p. 150°), is obtained (1) by mild oxidation of the condensation product from diacetyl and ethylenediamine (see general method ii); (2) by decarboxylation of 2,3-dicarboxy-5,6-dimethylpyrazine in acetic acid at 180° (*R.A. Pages* and *P.E. Spoerri*, J. org. Chem., 1963, **28**, 1702); (3) from catalytic dehydrogenation of 2,3-dimethylpiperazine (*T. Ishiguro et al.*, J. pharm. Soc. Japan, 1960, **80**, 314).

2,5-**Dimethylpyrazine**, m.p. 15°, b.p. 153° (*picrate*, m.p. 157°), can be readily synthesised by a variety of procedures outlined in general method ii. It is also obtained (1) from the reaction of glycerol with ammonium salts (*Stoehr, loc. cit.*); and (2) from oximinoacetone in 40% yield by an improved laboratory procedure (*H.I.X. Mager* and *W. Berends*, Rec. Trav. chim., 1958, **77**, 827).

2,6-**Dimethylpyrazine**, m.p. 48°, b.p. 155° (*picrate*, m.p. 175–176°), (1) isolated, together with pyrazine and other homologues, from the reaction of ammonia and D-glucose (*Brandes* and *Stoehr, loc. cit.*); (2) prepared in high yield (91%) by dehydrogenation of 2,6-dimethylpiperazine over copper chromite (U.S.P. 3,005,820, 1962; C.A., 1962, **56**, 7335).

Trimethylpyrazine, b.p. 172° (*picrate*, m.p. 138–141°), is prepared from 2,5-dimethylpyrazine by (1) heating the methobromide salt at 270–280° (*P. Brandes* and *C. Stoehr*, J. pr. Chem. [2], 1896, **53**, 501); (2) direct methylation using methyllithium at 0° (*B. Klein* and *P.E. Spoerri*, J. Amer. chem. Soc., 1951, **73**, 2949).

Tetramethylpyrazine, m.p. 86° (*trihydrate*, m.p. 75°), b.p. 190° (*picrate*, m.p. 192°), formed from (1) the reaction of β-bromolaevulic acid and ammonia (*L. Wolff*, Ber., 1887, **20**, 425); (2) the reduction of diacetylmonoxime (*E. Braun*, Ber., 1889, **22**, 569; *J.G. Pritchard* and *I.A. Siddiqui*, J. chem. Soc. Perkin II, 1972, 1309); (3) nickel-catalysed hydrogenation of dimethylglyoxime (*C.F. Winnans H.D. Adkins*, J. Amer. chem. Soc., 1933, **55**, 4167).

2,5-**Diethyl**-, b.p. 185–186°, and 2,5-**diisopropyl-pyrazine**, b.p. 206°, are prepared by treatment of the products derived from reduction of the appropriate oximinomethyl ketones with mercury(II) chloride (*M. Conrad* and *H. Hock*, Ber., 1899, **32**, 1199).

Pyrazine and its lower alkyl homologues are soluble in water giving weakly basic solutions. With mercury(II) chloride, they give rise to characteristic, sparingly soluble salts which can be used to facilitate recovery from solution since the salts decompose in acid or alkali. Pyrazine forms an azeotrope with water, b.p. 95.5°, comprising 60% w/w pyrazine (*H.F. Pflann*, J. Amer. chem. Soc., 1944, **66**, 155).

Although strongly aromatic in character, pyrazines, like pyridines, do not undergo electrophilic substitution reactions, therefore sulphonation and nitration do not take place under the normal conditions. On the other hand, direct nucleophilic substition of hydrogen in pyrazine is comparatively uncommon.

Ammination of pyrazine with sodamide in liquid ammonia (U.S.P. 2,394,963, 1946; C.A., 1946, **40**, 3143) and other solvents (*R.N. Shreve* and *L. Berg*, J. Amer. chem. Soc., 1947, **69**, 2116) generally results in a low yield of 2-aminopyrazine. Similarly, alkylation of a pyrazine, with the exception of 2,5-dimethylpyrazine (*Klein* and *Spoerri, loc. cit.*), with alkyllithium reagents usually produces low (< 10%) yields of mono-substituted products (*W. Schwager* and *J.P. Ward*, Rec. Trav. chim., 1971, **90**, 513). Pyrazines are, however, alkalated in moderate to good yield by aldehydes and ketones in the presence of alkali or alkaline-earth metals dissolved in liquid ammonia; 2-alkylpyrazines afford predominently the 2,6-isomer (*A.F. Bramwell et al.*, J. chem. Soc. C, 1971, 1627). In strongly acidic media, pyrazine undergoes homolytic substitution reactions with alkyl and carbamoyl ($\dot{C}ONH_2$) radicals (*F. Menisci* and co-workers, Tetrahedron Letters, 1976, 1731; 1970, 25).

Halogenation of pyrazine occurs in the vapour phase at elevated temperatures giving mixtures of mono- and poly-halogenated pyrazines (*J.K. Dixon et al.*, U.S.P. 2,524,431; C.A., 1951, **45**, 2513).

Pyrazines are reduced to piperazines by sodium and ethanol, tin and hydrochloric acid, and catalytic hydrogenation. Although the pyrazine ring is comparatively resistent to oxidation, treatment of an alkylpyrazine with permanganate usually gives the corresponding carboxylic acid in poor yield. With peracids, *e.g.* peracetic acid, pyrazines are transformed to either mono- or di-*N*-oxides depending on the reaction conditions employed (*B. Klein* and *J. Berkowitz*, J. Amer. chem. Soc., 1959, **81**, 5166). For the selective preparation of di-*N*-oxides, trifluoroperacetic acid is a favoured reagent (*K.W. Blake* and *P.G. Sammes*, J. chem. Soc. C, 1970, 1070). *N*-Oxides are smoothly converted into chloropyrazines in high yield by phosphoryl chloride (*G.T. Newbold* and *F.S. Spring*, J. chem. Soc., 1947, 1183).

Alkylpyrazines can be halogenated in the side chain using conventional methods, *e.g.* 2,3-dimethylpyrazine reacts with either *N*-chlorosuccinimide or chlorine with irradiation to give 2,3-di(chloromethyl)pyrazine (XXV) (*R.A. Pages* and *P.E. Spoerri*, J. org. Chem., 1963, **28**, 1702). Analogous

(XXV)

side-chain bromination can be effected with *N*-bromosuccinimide (*B.D. Mookherjee* and *E.M. Klaiber*, *ibid*., 1972, **37**, 511). Ethylpyrazines can be oxidised to the corresponding acetylpyrazines using hot chromic acid (*J. Wolt*, *ibid*., 1975, **40**, 1178).

Methylpyrazine, on reaction with sodamide in liquid ammonia, forms pyrazylmethylsodium which reacts readily with alkyl halides, carbonyl compounds (*J.D. Behun* and *R. Levine*, J. org. Chem., 1961, **26**, 3379; J. Amer. chem. Soc., 1959, **81**, 5157; 1958, **81**, 5666) and epoxides [Scheme 4] (*M.R. Karmal* and *R. Levine*, J. org. Chem., 1962, **27**, 1360). Coupling of

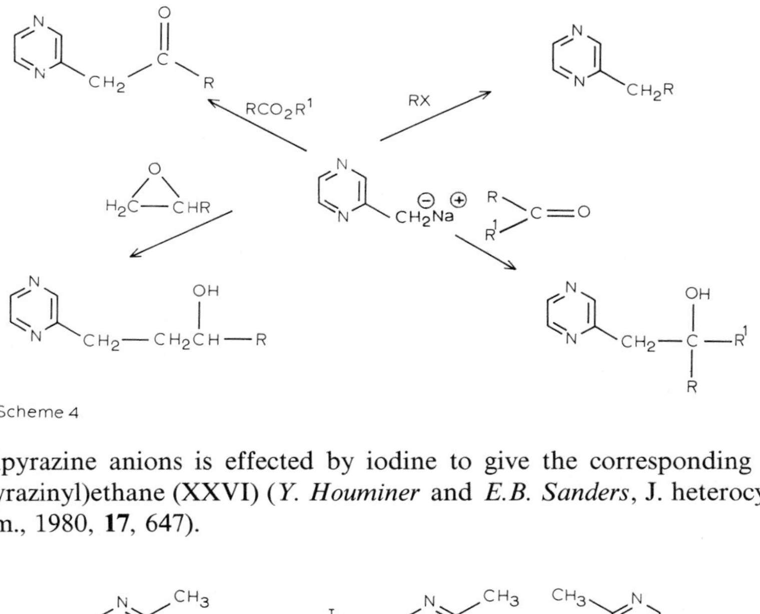

Scheme 4

alkylpyrazine anions is effected by iodine to give the corresponding 1,2-di(pyrazinyl)ethane (**XXVI**) (*Y. Houminer* and *E.B. Sanders*, J. heterocyclic Chem., 1980, **17**, 647).

(XXVI)

Methylpyrazines also react with aromatic aldehydes in the presence of zinc chloride to produce styrylpyrazines. Thus methylpyrazine and benzaldehyde is reported to give styrylpyrazine in high yield (*A.S. Elma* and *I.S. Musatova*, C.A., 1969, **70**, 87747) and 2,5-dimethylpyrazine and benzaldehyde gives a mixture of 5-methyl-2-styryl- and 2,5-distyryl-pyrazine (*R. Franke*, Ber., 1905, **38**, 3724).

Simple alkylpyrazines do not readily undergo cycloaddition reactions such as the Diels-Alder reaction. Prolonged reaction of pyrazine or methyl-pyrazine with 1-diethylaminopropyne produces small quantities of the pyridine derivatives (XXVII) whose formation is rationalised in terms of a

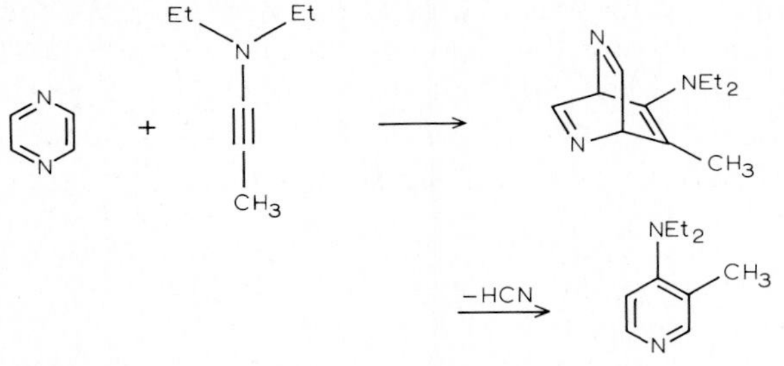

Diels-Alder reaction with inverse electronic demand followed by an elimination of hydrogen cyanide (*H. Nuenhoeffer* and *G. Werner*, Ann., 1972, **761**, 39). Methyl- and 2,6-dimethyl-pyrazine also react with dimethyl acetylenedicarboxylate (DMAD) to produce pyrrolopyrazines (XXVIII) in low yield (*R.M. Acheson* and *M.W. Foxton*, J. chem. Soc. C, 1966, 2218). 2,5-Dimethylpyrazine and DMAD gives initially a 1 : 2 adduct which undergoes a further cycloaddition with a third molecule of DMAD followed by elimination of acetonitrile to yield a pyridoazepine (XXIX) [Scheme 5] (*R.M. Acheson et al., ibid.*, 1968, 926).

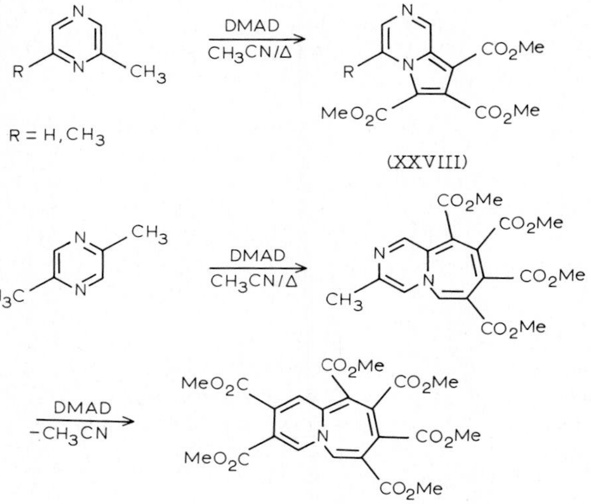

Scheme 5

Phenylpyrazine, m.p. 72–73°, is prepared by decarboxylation of the corresponding 2,3-dicarboxylic acid (*A. Ohta et al.*, Heterocycles, 1977, **6**, 1881) or by direct phenylation of pyrazine, in THF with phenyllithium at −75° with air oxidation [60% yield] (*R.E. van der Stoel* and *H.C. van der Plas*, Rec. Trav. chim., 1978, **97**, 116).

Benzylpyrazine, b.p. 107–108°/1.3 mm, is obtained by phenylation of methylpyrazine (*J.D. Behun* and *R. Levine*, J. org. Chem., 1961, **26**, 3379).

Reduction of the appropriate diazomethyl ketones affords the centrosymmetric 2,5-**dibenzyl**-, m.p. 76°, and 2,5-**diphenyl-pyrazine**, m.p. 196°, (*L. Birkhofer*, Ber., 1947, **80**, 83). The 2,5-diphenyl compound has been prepared by many of the general self-condensation procedures outlined above.

2,6-**Diphenylpyrazine**, m.p. 90°, together with the 2,5-isomer is obtained from reaction of ω-haloacetophenone and ammonia (*F. Tutin*, J. chem. Soc., 1910, **97**, 2495).

2,3-**Diphenylpyrazine**, m.p. 118–121°, is prepared from ethylenediamine and benzil (*A.T. Mason*, J. chem. Soc., 1889, **55**, 97). The diphenylpyrazines are soluble in hot concentrated hydrochloric acid but precipitate on dilution with water. 2-*Methyl*-5,6-*diphenyl*-, m.p. 86–87° (*picrate* m.p. 137–138°), and 2,5-*diphenyl*-3,6-*dimethyl-pyrazine*, m.p. 126° (*picrate*, 153–154°), the latter from α-bromopropiophenone and alcoholic ammonia, behave similarly.

Tetraphenylpyrazine, m.p. 246°, is obtained by heating benzoin and ammonia or ammonium salts (*R. Leuckart*, J. pr. Chem. [2], 1890, **41**, 333).

(ii) *Halopyrazines*

Chloropyrazine, b.p. 62.5°/29 mm, is prepared by a number of procedures, generally applicable to other substituted pyrazines, including (1) from hydroxpyrazine and phosphoryl chloride (*G.W.H. Cheeseman*, J. chem. Soc., 1960, 242); (2) from pyrazine-1-oxide and phosphoryl chloride (*B. Klein et al.*, J. org. Chem., 1963, **28**, 1682); (3) the vapour phase chlorination of pyrazine either in the presence of water at 400° (*J.M. Sayward*, U.S.P. 2,391,745, 1945; C.A., 1946, **40**, 1888) or with nitrogen as diluent (*P.S. Winnek*, U.S.P. 2,396,066, 1946; C.A., 1946, **40**, 3143).

2,6-**dichloropyrazine**, m.p. 51–58°, b.p. 120–122°/40 mm, is obtained by treatment of pyrazine-1,4-dioxide with phosphoryl chloride (*Klein et al., loc. cit.; Bernardi et al.*, Gazz., 1961, **91**, 1431). Reaction of 2,3-dihydroxy- and 5-chloro-2-hydroxy-pyrazine with phosphoryl chloride gives 2,3-**dichloro**- (m.p. 23–24°) and 2,5-**dichloro-pyrazine** (b.p. 72°/12 mm), respectively (*G. Palamidessi et al.*, C.A., 1967, **66**, 37884: J. org. Chem., 1964, **29**, 2491).

Tetrachloropyrazine, m.p. 97–100°, b.p. 100°/0.1 mm, is prepared from chloropyrazine or hydroxypyrazine or piperazine-2,5-dione and phosphorus pentachloride (*W.K.R. Musgrave et al.*, Chem. Ind. (London), 1966, 1721).

Reaction of (±)-alanine anhydride (3,6-dimethylpiperazine-2,5-dione) with phosphoryl chloride gives a mixture of 2-**chloro**-, b.p. 78°/15 mm, and 2,5-**dichloro**-3,6-**dimethylpyrazine**, m.p. 73°; the latter is non-basic whereas the former dissolves in dilute hydrochloric acid. In the presence of dimethylaniline only the monochloro compound is obtained (*R.A. Baxter* and *F.S. Spring*, J. chem. Soc., 1947, 1149).

Phosphoryl chloride and pyrazine mono-*N*-oxides give 2-chloropyrazines in good yield however conversion of di-*N*-oxides to the corresponding 2,5-dichloro compounds is generally poorer. 2-**Chloro**-, b.p. 116°/12 mm, and 2,5-**dichloro**-, m.p. 59–61°, 3.6-**di**(*sec*-**butyl**)**pyrazine** have been prepared by the above procedure (*G.T. Newbolt* and *F.S. Spring*, ibid., 1947, 1183).

Bromopyrazine, b.p. 58°/9 mm, is prepared by heating pyrazine hydrobromide (or hydrochloride) with bromine in carbon tetrachloride at 150–250° (*J. K. Dixon* and *J.M.*

Sayward, U.S.P. 2,403,710, 1946; C.A., 1946, **40**, 6103). Reaction of hydroxypyrazine with (1) phosphorus tribromide alone or together with phosphoryl bromide yields bromopyrazine (*G. Karmas* and *P.E. Spoerri*, J. Amer. chem. Soc., 1956, **78**, 2141); (2) phosphoryl bromide and phosphorus pentabromide gives a mixture 2-**bromo**- and 2,6-**dibromo-pyrazine**, m.p. 52°, b.p. 97°/10 mm (*P.E. Spoerri* and co-workers, *ibid.*, 1946, **68**, 400; 1949, **71**, 2043); (3) bromine, anhydrous iron(II) bromide and a trace of phosphorus tribromide gives 2,3-**dibromopyrazine**, m.p. 57–58° (*G. Karmas* and *P.E. Spoerri, ibid.*, 1957, **79**, 680). 2,5-**Dibromopyrazine**, m.p. 47–48°, is prepared *via* the bromoamino compound (*R.C. Ellingson* and *R.L. Henry, ibid.*, 1949, **71**, 2798).

Although not satisfactory for alkylhydroxypyrazines, phosphorus tribromide transforms phenyl-substituted 2-hydroxypyrazines to the corresponding 2-bromo derivatives. The reaction of phosphorus tribromide with (±)-α-phenylglycine anhydride gives 2-**bromo**-3,6-**diphenyl-pyrazine**, m.p. 119–120°, in 60% yield rather than the expected 2,5-dibromo compound (*idem, ibid.*, 1956, **78**, 2141).

Dihydroxyphenylpyrazines are converted to dibromopyrazines with phosphorus tribromide at 180–200° (*N. Sato* and *J. Adichi*, J. org. Chem., 1978, **43**, 340).

Other dichloro- and dibromo-pyrazines include 2,3-*dichloro*-, m.p. 80–81°, 2,3-*dibromo*-5,6-*dimethylpyrazine*, m.p. 87–88°, the corresponding *diphenyl* derivatives, m.p. 182–183° and 212–214°, respectively; and 2,5-*dichloro*-3,6-*diphenylpyrazine*, m.p. 159–160° (*G. Karmas* and *P.E. Spoerri*, J. Amer. chem. Soc., 1957, **79**, 4071).

Unlike 2-aminopyridine, diazotisation of aminopyrazine followed by treatment with copper(I) bromide does not produce the bromo derivative.

Fluoropyrazine, b.p. 108–110°/atm., is obtained (1) by diazotisation of aminopyrazine in fluoroboric acid containing copper powder (*H. Rutner* and *P.E. Spoerri*, J. heterocyclic Chem., 1965, **2**, 492); (2) from chloropyrazine and anhydrous potassium fluoride in either refluxing dimethylsulphoxide (*idem, ibid.*, 1966, **3**, 4359), or *N*-methyl-2-pyrrolidone at 185° (*G.W.H. Cheeseman* and *R.A. Godwin*, J. chem. Soc. C, 1971, 2973).

Tetrafluoropyrazine, m.p. 53–54°, is prepared from the tetrachloro compound and anhydrous potassium fluoride (*W.K.R. Musgrave et al., ibid.*, 1970, 1023).

Iodopyrazine, b.p. 109–110°/34 mm, and other simple alkyliodo derivatives are accessible from the corresponding chloropyrazine by treatment with a saturated solution of potassium iodide in methyl ethyl ketone containing hydriodic acid (*A. Hirschberg* and *P.E. Spoerri*, J. org. Chem., 1961, **26**, 1907).

Simple halogenopyrazines readily undergo displacement reactions with a range of nucleophiles to give mono-substituted pyrazines (*G.W.H. Cheeseman*, J. chem. Soc., 1960, 242). Alkyl substituents tend to reduce the reactivity of the halogen, *e.g.* 2-chloro-3,6-dimethylpyrazine, although converted into the hydroxy and ethoxy derivatives, fails to produce the 2-amino compound, and 2-chloro-3,6-di(*sec*-butyl)pyrazine is unchanged by 20% aqueous hydroxide (*R.A. Baxter* and *F.S. Spring, ibid.*, 1947, 1179).

Bromopyrazines are converted into the corresponding cyano derivatives by refluxing in γ-picoline with copper(I) cyanide (*G. Karmas* and *P.E. Spoerri*, J. Amer. chem. Soc., 1956, **78**, 2141).

Chloropyrazines are readily dechlorinated using tetrakis(triphenylphosphine)palladium in sodium formate (*Y. Akita* and *A. Ohta*, Heterocycles, 1981, **16**, 1325); the palladium reagent also promotes cyanodechlorination with potassium cyanide in dimethylformamide (*Y. Akita et al.*, Synthesis, 1981, 974).

Fluoropyrazines are significantly more susceptible to nucleophilc substitution reactions than chloropyrazines, *e.g.* reaction with anhydrous sodium sulphite affords the sodium salt of pyrazinesulphonic acid (*Rutner* and *Spoerri, loc. cit.*).

Treatment of iodopyrazines with *n*-butyllithium results in the formation of lithio derivatives which react with carbon dioxide and carbonyl compounds to give pyrazinecarboxylic acids and pyrazinylcarbinols, respectively (*P.E. Spoerri* and co-workers, J. heterocyclic Chem., 1965, **2**, 209; 1969, **6**, 239).

(iii) Aminopyrazines

The preparation of aminopyrazines can be accomplished by a variety of routes including the following.

(1) Direct synthesis by the condensation of aminoacetamidines and α,β-diketones gives 5,6-disubstituted aminopyrazines (see general method iii).

(2) The reaction of α-aminonitriles and α-oximinoketones to give 3,5-disubstituted 2-aminopyrazine-1-oxides (not successful with the 3,5-diphenyl compound) which are reduced using sodium dithionite (Table 1, Ref. 2; *E.C. Taylor* and *K. Lennard*, J. Amer. chem. Soc., 1968, **90**, 2424).

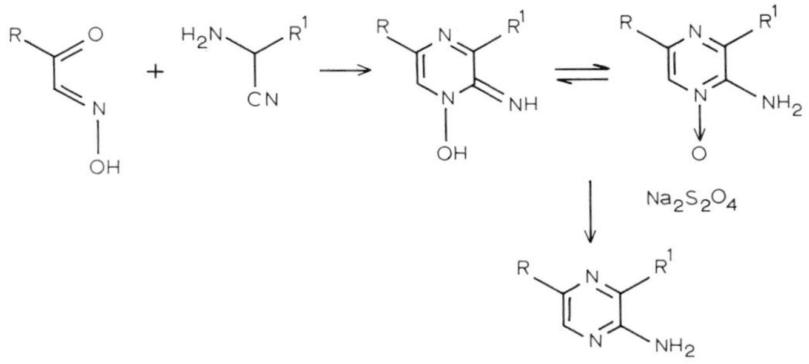

(3) Hofmann degradation of pyrazine-2-carboxamides, *e.g.* pyrazinamide affords aminopyrazine in 80% yield (*B. Camerino* and *G. Palamidessi*, Gazz., 1960, **90**, 1807); pyrazine-2,3-dicarboxamides in this reaction yield 2-amino-3-carboxylic acids which are readily decarboxylated in an inert solvent (*S. Gabriel* and *A. Sonn*, Ber., 1907, **40**, 4850).

TABLE 1
MELTING POINTS OF 2-AMINOPYRAZINES AND THEIR DERIVATIVES

Compounds	M.p. (°C)	Derivative	m.p. (°C)	Ref.
2-Aminopyrazine	118–120	Picrate	214–215 (dec.)	1
		Acetyl	133	1
5-Methyl-	116–118			1
6-Methyl-	124–125			1
3,5-Dimethyl-	96	Picrate	230–240 (dec.)	2
5,6-Dimethyl-	140–144			1
3,6-Dimethyl-	111–112	Phenylacetyl	130	
		Picrate	206	3,4,5
3,6-Di-sec-butyl-	(b.p. 112–114/0.5 mm)	Chloroplatinate	183	4
3,6-Diphenyl-	186	Phenylacetyl	194	5
5,6-Diphenyl-	227–228			1

1 J. Weijlard, M. Tishler and A.E. Erickson, J. Amer. chem. Soc., 1945, **67**, 802.
2 W. Sharp and F.S. Spring, J. chem. Soc., 1951, 932, 2679.
3 R.R. Joiner and P.E. Spoerri, J. Amer. chem. Soc., 1941, **63**, 1929.
4 R.A. Baxter, G.T. Newbold and F.S. Spring, J. chem. Soc., 1947, 370.
5 G.T. Newbold, F.S. Spring, and W. Sweeny, ibid., 1949, 300.

(4) Direct amination of pyrazines. For pyrazine, this is most successfully carried out with sodamide in liquid ammonia at room temperature in an autoclave (*M.L. Crossley* and *J.P. English*, U.S.P. 2,394,963; C.A., 1946, **40**, 3143); 3-amino-2,5-dimethylpyrazine is obtained in 33% yield by heating 2,5-dimethylpyrazine with sodamide in dimethylaniline at 165° (Table 1, Ref. 3).

(5) Aminolysis of halopyrazines; this reaction is most convenient with fluoropyrazines, e.g. fluoropyrazine on treatment with aqueous ammonia at ambient temperature for three days gives aminopyrazine in 70% yield (*H. Rutner* and *P.E. Spoerri*, J. heterocyclic Chem., 1966, **3**, 435) whereas the analogous transformation with chloropyrazine is carried out in a sealed tube at 180° (*G.W.H. Cheeseman*, J. chem. Soc., 1960, 242).

(6) Hydrolysis of lumazines (a) by acid to 2-aminopyrazines or (b) by alkali to the corresponding 2-amino-3-carboxylic acids which are then decarboxylated (Table 1, Ref. 1; *D.M. Sharefkin*, J. org. Chem., 1959, **24**, 345).

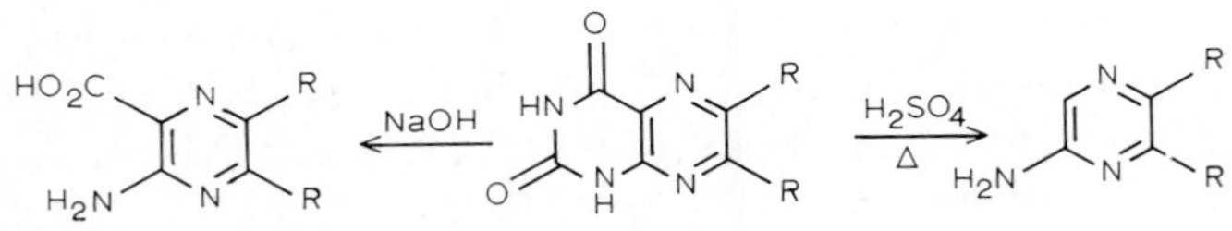

(7) 2-Aminopyrazine-5-carboxylic acids are conveniently prepared from 2-(*N*-acylamino)quinoxalines (*E. Felder et al.*, Helv., 1964, **47**, 873).

Aminopyrazine, m.p. 118–120°, is prepared from pyrazine by shaking it with sodamide in liquid ammonia at room temperature (*Crossley* and *English, loc. cit.*); by ammonolysis of (i) 2-bromo- or 2-chloro-pyrazine (*P.S. Winnek* and *Q.P. Cole*, U.S.P. 2,396,067; C.A., 1946, **40**, 4089) or (ii) 2-fluoropyrazine at room temperature (*Rutner* and *Spoerri, loc. cit.*)

2-Alkoxy-6-aminopyrazines are obtained from *N*-nitroso-di(cyanomethyl)amines by treatment with sodium methoxide or ethoxide; the nitroso group is essential to the success of the reaction (*D. Swern et al.*, J. org. Chem., 1979, **44**, 1129).

$(R = CH_3, CH_2CH_3)$

Other 2-aminopyrazines are listed in Table 1.

2,6-Diaminopyrazine, m.p. 136°, is prepared by heating 2,6-dibromopyrazine with ammonia at 195–200° (*A.E. Erickson* and *P.E. Spoerri*, J. Amer. chem. Soc., 1946, **68**, 400).

Comparison of the ionisation constants and the u.v.-absorption spectra of the series aminopyrazine, methylaminopyrazine and dimethylaminopyrazine indicates that aminopyrazine does not exist to any appreciable extent in the tautomeric imino form (*Cheeseman, loc. cit.*). Aminopyrazines, on treatment with nitrous acid, form diazonium salts which decompose to pyrazinones (hydroxypyrazines) (Table 1, Ref. 4). The amino substituent actives the pyrazine ring to electrophilic attack at the *ortho-* and the *para-*position. Thus, bromination of aminopyrazine in glacial acetic acid affords 2-**amino**-3,5-**dibromopyrazine**, m.p. 114–116° (*B. Camerino* and *G. Palamidessi*, B.P. 928,152; C.A., 1964, **60**, 2971).

(iv) Hydroxypyrazines

Although formally hydroxypyrazines, they are often found as the tautomeric keto form and hence referred to as pyrazinones. The position of

the tautomeric equilibrium can, however, be influenced by ring substituents such as halogens, and solvent (*G.W.H. Cheeseman* and *E.S.G. Westeruik*, Adv. heterocyclic Chem., 1972, **42**, 99; *W.K.R. Musgrave et al.*, J. chem. Soc., 1970, 1023). The i.r. spectra of hydroxypyrazines generally exhibit strong carbonyl absorptions consistent with the amide structure, *e.g.* for hydroxypyrazine in $CHCl_3$ ν_{CO} 1730 cm^{-1} (*S.F. Mason, ibid.*, 1957, 4874). Contrary, however, to structural preferences, hydroxypyrazines undergo a number of reactions which are typical of phenols (*vide infra*).

TABLE 2
MELTING POINTS OF 2-HYDROXYPYRAZINES

Compound	m.p. (°C)	Ref.
2-Hydroxypyrazine	187–188	1–4
3-Methyl-	140–142	4
3,5-Dimethyl-	146–147	4,5
3,6-Dimethyl-	210–211	3
5,6-Dimethyl-	199–200	4,6
3,6-Di-isobutyl-	144.5–147	7
3,6-Di-sec-butyl-	122–124	8
3,5,6-Trimethyl-	197–199	4,5,8
3-Phenyl-	172–173	4
3-Methyl-5-phenyl-	222–223	4,5
3-Phenyl-5,6-dimethyl-	235–238	4,5
3,5-Diphenyl-	270–272	5
5,6-Diphenyl-	225–227	4
3-Methyl-5,6-diphenyl-	212–214	9

1 J. Weijlard, M. Tishler and A.E. Erickson, J. Amer. chem. Soc., 1945, **67**, 802.
2 A.E. Erickson and P.E. Spoerri, ibid., 1946, **68**, 400.
3 R.A. Baxter, G.T. Newbold and F.S. Spring, J. chem. Soc., 1947, 370.
4 R.G. Jones, J. Amer. chem. Soc., 1949, **71**, 78.
5 G. Dunn et al., J. chem. Soc., 1949, 2707.
6 Y.A. Tota and R.C. Elderfield, J. org. Chem. 1942, **7**, 313.
7 G. Dunn, G.T. Newbold and F.S. Spring, J. chem. Soc., 1949, 2586.
8 G.T. Newbold and F.S. Spring, ibid., 1947, 373.
9 M.E. Hultquist, U.S.P. 2,805,223; C.A., 1958, **52**, 2935.

Hydroxypyrazines can be prepared as follows by the following methods.

(1) Condensation of α,β-dicarbonyl compounds with α-amino-acid amides (Table 2, Ref. 4). Similarly, α,β-dicarbonyl compounds condense

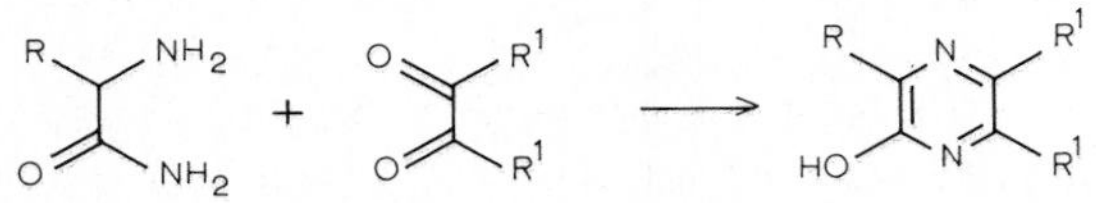

with aminomalonamides to give the corresponding 3-carbamoyl-2-hydroxypyrazines (*F.L. Muelchmann* and *A.R. Day*, J. Amer. chem. Soc., 1956, **78**, 242).

(2) Haloacylation of α-aminoketones followed by treatment with ammonia and oxidation provides a general entry to 5,6-disubstituted and 3,5,6-trisubstituted 2-hydroxypyrazines (XXX) (Table 2, Ref. 6).

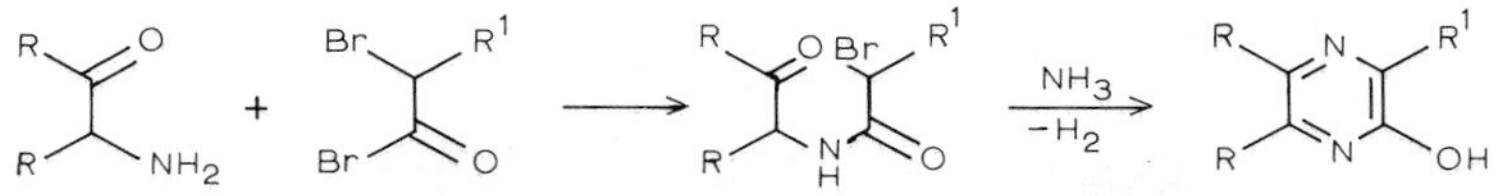

[Scheme leading to (XXX)]

(3) Reaction of amino-acid nitriles with α,β-diketones in 50% NaOH gives, after acidification, hydroxypyrazines (Table 2, Ref. 9).

(4) Treatment of aminopyrazines with nitrous acid (Table 2, Ref. 8) or nitrosylsulphuric acid (Table 2, Ref. 2).

(5) By hydrolysis of halopyrazines (discussed previously); this route is more synthetically useful than (4) since aminopyrazines are less readily available.

(6) Hydroxypyrazinecarboxylic acids are decarboxylated by heating alone or in a solvent (Table 2, Ref. 1).

(7) Alkaline hydrolysis of dinitriles followed by decarboxylation proceeds by the following sequence (*G. Gastaldi*, Gazz., 1921, **51**, I, 233).

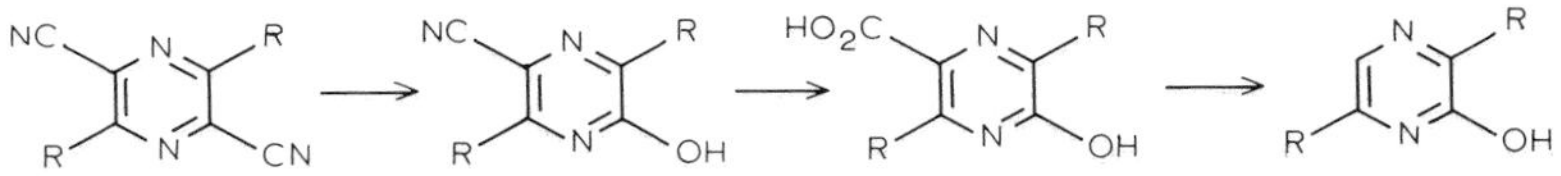

[Scheme showing sequence of pyrazine intermediates]

(R = CH$_3$, Ph)

A number of important hydroxypyrazines are included in Table 2.

Hydroxypyrazines are significantly more activated towards electrophilic substitution reactions, *e.g.* halogenation and nitration, than other pyrazine derivatives. Bromination of 2-hydroxy-2-phenylpyrazine with one equivalent of bromine and two equivalents of pyridine (as a hydrogen bromide acceptor) in acetic acid gives 5-**bromo**-2-**hydroxy**-3-**phenylpyrazine**, m.p. 192–193°. 2-Hydroxy-3,6- and 5,6-disubstituted pyrazines behave similarly. Displacement of the bromine in 3-bromo-2-hydroxypyrazines by hydroxyl ion gives 2,3-hydroxypyrazines (XXXI) in good yield. Treatment of bromohydroxypyrazines with phosphoryl chloride yields the corresponding dichlorides (XXXII) (*G. Karmas* and *P.E. Spoerri*, J. Amer. chem. Soc., 1956, **78**, 4071).

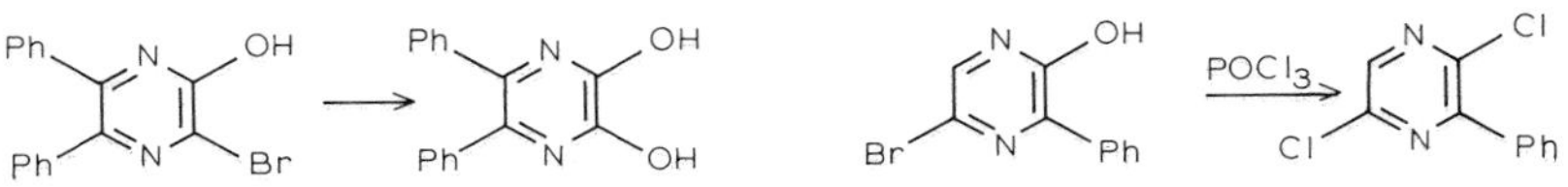

[Structures (XXXI) and (XXXII)]

2-Hydroxy-3-phenylpyrazine with one equivalent of nitric acid initially forms an nitrate, m.p. 123–125°, which in refluxing acetic acid is converted into 2-**hydroxy**-5-**nitro**-3-**phenylpyrazine**, m.p. 254–256°.

Alkylhydroxypyrazines cannot be nitrated under analogous reaction conditions. 5,6-Diphenyl-2-hydroxypyrazine is nitrated smoothly under mild conditions with one equivalent of nitric acid in acetic acid at room temperature to give the corresponding 3-nitro derivative, m.p. 210–213°. With 3,6-diphenyl-2-hydroxypyrazine, the reaction mixture is heated to reflux to produce the 5-nitro derivative, m.p. 274–276° (dec) (*Karmas* and *Spoerri, loc. cit.*).

The nitro group is labile and readily undergoes dispacement reactions as illustrated in Scheme 6.

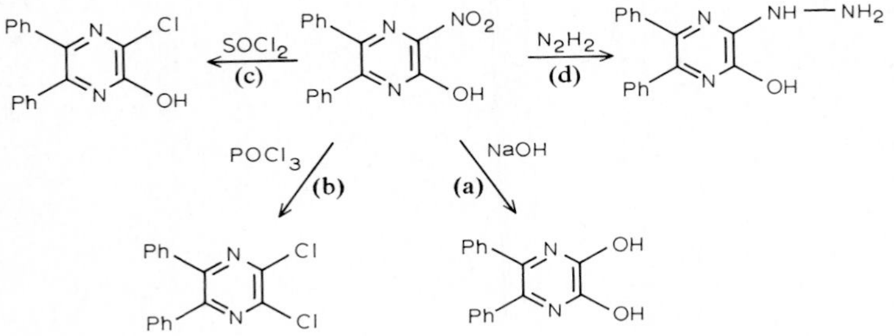

Scheme 6

(a) G. Karmas and P.E. Spoerri, J. Amer. Chem. Soc., 1953, **75**, 5517.
(b) idem, ibid., 1956, **78**, 4071.
(c) A. Hirchberg and P.E. Spoerri, J. heterocyclic Chem., 1969, **6**, 975.
(d) G.W.H. Cheeseman and M. Raffiq, J. Chem. Soc. C, 1971, 452.

Hydroxypyrazines couple with phenyl- (or aryl-) diazonium salts in neutral or weakly alkaline solution to produce the corresponding 5-azo compounds (*C. Gastaldi* and *E. Princivalle*, Gazz., 1928, **58**, 412; 1929, **59**, 751). In 1 M sodium hydroxide, decomposition of the phenyldiazonium chloride occurs with concommitant phenylation of the pyrazine, *e.g.* hydroxypyrazine gives mainly 2-**hydroxy**-3-**phenylpyrazine** and a small amount of 2-**hydroxy**-3,6-**diphenylpyrazine**, m.p. 292–293° (*G. Karmas* and *P.E. Spoerri*, J. Amer. chem. Soc., 1956, **78**, 4071).

Alkylation of hydroxypyrazines can result in *O*- or *N*-alkylated products, or, more commonly, a mixture of both. 2-Hydroxypyrazine with either diazomethane or dimethyl sulphate and anhydrous potassium carbonate in acetone is reported to yield 1-**methyl**-2-**oxo**-1,2-**dihydropyrazine**, m.p. 84–85°

(*J.D. Dutcher*, J. biol. Chem., 1947, **171**, 321; *G.W.H. Cheeseman*, J. chem. Soc., 1960, 242). The reaction of 2-hydroxy-3-isobutylpyrazine with diazomethane, however, affords a mixture of the *O*- and the *N*-methyl derivatives in a ratio of almost 1:2 (*R.M. Seifert et al.*, J. agric. food Chem., 1970, **18**, 246). *O*-silylation of 2-hydroxypyrazine can be effected with trimethylsilylchloride and bis(trimethylsilyl)amine (*M. Robek* and *A. Bloch*, J. med. Chem., 1972, **15**, 164).

In general, alkoxypyrazines are more conveniently prepared from the corresponding halopyrazines by nucleophilic displacement of the halogen atom by alkoxide anion, *e.g.* ethoxypyrazines are formed from chloro- and bromo-pyrazines with sodium ethoxide in ethanol; **ethoxypyrazine**, b.p. 72–73°/30 mm (Table 2, Ref. 2); and 2-**ethoxy**-3,6-**dimethylpyrazine**, b.p. 81°/15 mm (*R.A. Baxter* and *F.S. Spring*, J. chem. Soc., 1947, 1182).

(A) N-Oxides

The *N*-oxides of 2-hydroxypyrazines are tautomeric with cyclic hydroxamic acids, give deep red solutions with iron(III) chloride, liberate carbon

dioxide from sodium hydrogen carbonate solution, titrate as monobasic acids and exhibit characteristic absorption spectra. On treatment with hydrazine, they are converted into the corresponding 2-hydroxypyrazines.

2-Hydroxypyrazine-1-oxides cannot be prepared by reaction of 2-hydroxy- or 2-alkoxy-pyrazines with peroxides since peroxidation invariably takes place at the distal nitrogen atom (N-4) (*F.S. Spring et al.*, J. chem. Soc., 1948, 519; 1859). Fluoropyrazine, on treatment with potassium persulphate in concentrated sulphuric acid, is reported to yield 2-fluoropyrazine-1-oxide which undergoes nucleophilic displacement with hydroxide ion (*M.V. Jovanovic*, Heterocycles, 1984, **22**, 1105).

Alternative approaches include the reaction of α-aminohydroxamic acids (XXXIII, R = CH_3, Ph) with α,β-dicarbonyl compounds. With diacetyl, 2-**hydroxy**-3,5,6-**trimethyl**- (XXXIV, R = CH_3), m.p. 176–177°, and 2-**hydroxy**-5,6-**dimethyl**-3-**phenyl-pyrazine**-1-**oxide** (XXXIV, R = Ph), m.p. 154–156°, are obtained. With unsymmetrical ketoaldehydes, the 3,5-disubstituted pyrazines are formed selectively; the 3,6-isomer is not observed.

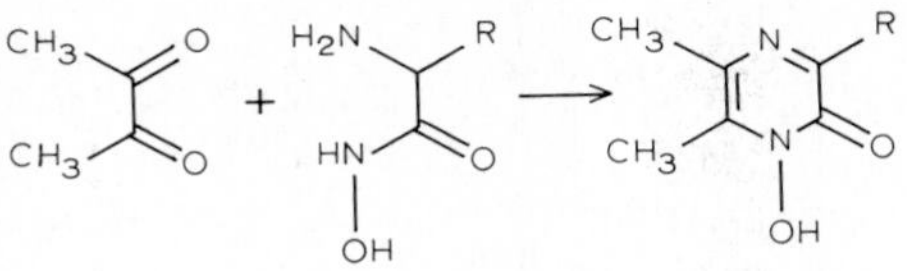

Thus, methylglyoxal gives exclusively **2-hydroxy-3,5-dimethylpyrazine-1-oxide**, m.p. 135°. **2-Hydroxy-3-methyl-5-phenyl-**, m.p. 185°, and **2-hydroxy-3,5-diphenyl-pyrazine-1-oxide**, m.p. 165–166°, are obtained in a similar fashion (*F.S. Spring et al.*, J. chem. Soc., 1949, 2707). The 3,6-disubstituted isomer (XXXV) can be prepared by condensation of an α-aminoketone with a 2-oxohydroxamic acid, *e.g.* pyruvohydroxamic acid and aminoacetone gives **2-hydroxy-3,6-dimethylpyrazine**-1-oxide, m.p. 194–195° (*D.W.C. Ramsay* and *F.S. Spring, ibid.*, 1950, 3409).

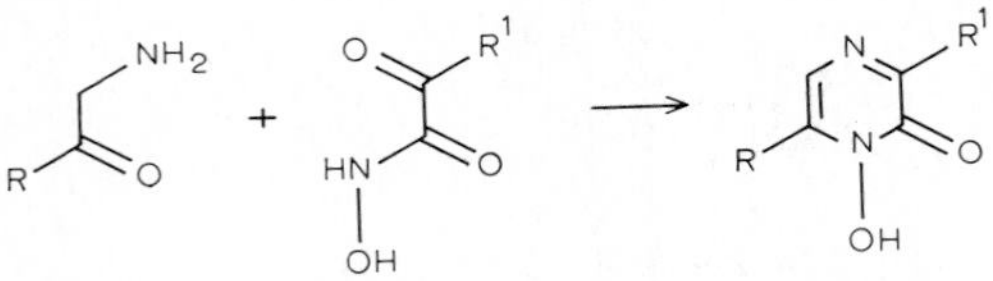

Hydroxypyrazines and their 1-oxides have attracted considerable interest on account of the naturally occurring antibiotic aspergillic acid and related compounds. **Flavacol**, **2-hydroxy-3,6-diisobutylpyrazine**, m.p. 144.5–147°, is a by-product in the isolation of **aspergillic acid**, **2-hydroxy-3-isobutyl-6-*sec*-butylpyrazine-1-oxide**, m.p. 93°, $[\alpha]_D + 14°$. On treatment with hydrazine, aspergillic acid is converted to **deoxyaspergillic acid**, **2-hydroxy-3-isobutyl-6-*sec*-butylpyrazine**, m.p. 102°. The structural characterisation and syntheses of the aforementioned compounds has been extensively investigated (*J.D. Dutcher et al.*, J. biol. Chem., 1944, **155**, 359; 1947, **171**, 321; *F.S. Spring et al.*, J. chem. Soc., 1949, S126; 1949, 2586; 1951, 2679; 1952, 4870). Several synthetic routes to compounds containing the aspergillic acid skeleton have been reported (*M. Ohta* and co-workers, J. org. Chem., 1964, **29**, 3165; 1966, **31**, 4143; Tetrahedron Letters, 1967, 845; *R.J. Bergeron* and *P.G. Hoffman*, J. org. Chem., 1980, **45**, 163).

(B) Dialkoxy- and dihydroxy-pyrazines

2,3- and 2,5-Dimethoxypyrazines are prepared from dibromopyrazines and sodium methoxide (Table 3, Ref. 1); dichloropyrazines react similarly (*G.W.H. Cheeseman* and *R.A. Godwin*, J. chem. Soc. C, 1971, 2973). The

TABLE 3
MELTING POINTS OF DIALKOXY- AND DIHYDROXY-PYRAZINES

Compound	m.p. (°C)	Ref.
2,3-Dimethoxy-	(b.p. 108–110/50 mm)	1
2,6-Dimethoxy-	32–31.5	1
2,6-Diethoxy-	(b.p. 112–114/26 mm)	3
2,3-Dimethoxy-5,6-dimethyl-	62–63	1
2,3-Dimethoxy-5,6-diphenyl-	140–141	1
2,5-Dimethoxy-3,6-dimethyl-	63–65	1
	69–70	3
2,5-Dimethoxy-3,6-diphenyl-	146–147	1
2,3-Dihydroxy-	> 320	1
2,3-Dihydroxy-5-methyl-	301–303	2
2,3-Dihydroxy-5-phenyl-	288–290	2
2,3-Dihydroxy-5-methyl-6-phenyl-	327–328	2
2,5-Dihydroxy-3,6-dimethyl-	320	1
2,5-Dihydroxy-3,6-diphenyl-	295–300 (dec)	1
2,6-Dihydroxy-3,5-diphenyl-	258–259	4

1 G. Karmas and P.E. Spoerri, J. Amer. chem. Soc., 1957, **79**, 680.
2 J. Adachi and N. Sato, J. org. Chem., 1972, **37**, 221.
3 G.W.H. Cheeseman and R.A. Godwin, J. chem. Soc. C, 1971, 2973.
4 idem, ibid., 1971, 2977.

diethers can be cleaved by the action of methanolic sodium methoxide or hydrogen bromide to give the corresponding hydroxypyrazyl ethers or dihydroxypyrazines (Table 3, Ref. 1.).

Several pyrazine ring syntheses producing 2,3-dihydroxypyrazines directly are given below.

(1) Condensation of α-amino-ketals with ethyl oxamate affords oxamylamino ketones (XXXVI) which are cyclised in acetic acid to 2,3-dihydroxypyrazines (Table 3, Ref. 2).

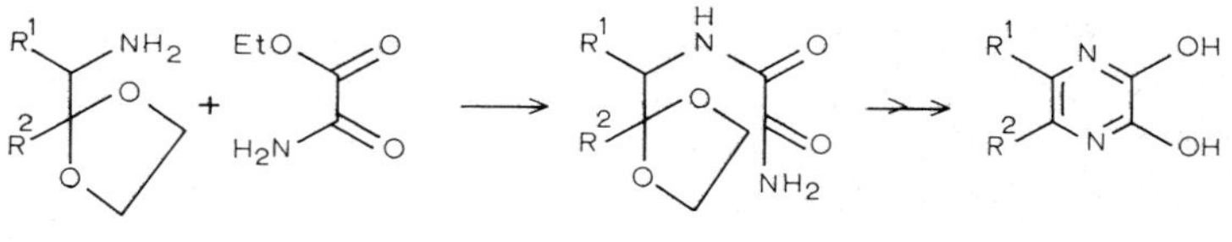

(XXXVI)

(2) The condensation product from α-aminoacetaldehyde acetal and diethyl oxalate, on treatment with ammonia, affords a linear intermediate which cyclises in acetic acid containing a trace of hydrochloric acid to

2,3-dihydroxypyrazine (*G. Palamidessi* and *L. Panizzi*, C.A., 1965, **62**, 1674; *G. Palamidessi* and *M. Bonanomi*, C.A., 1967, **66**, 37884).

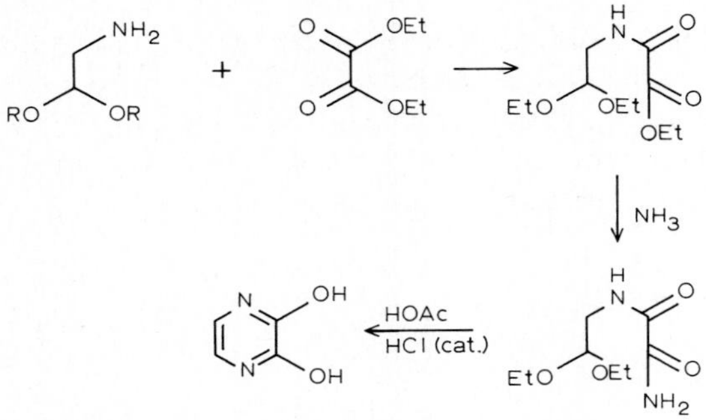

(3) Reaction of DAMN with oxalyl chloride affords 5,6-**dicyano**-2,3-**dihydroxypyrazine** (XXXVII), m.p. > 270° (*H. Bredereck* and *G. Schmotzer*, Ann., 1956, **600**, 95).

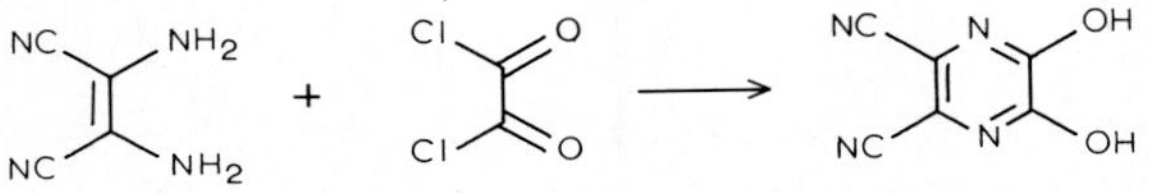

(4) Permanganate oxidation of 2,3-dichloroquinoxaline gives 2,3-dihydroxypyrazine-5,6-dicarboxylic acid (*H.I.X. Mager* and *W. Berends*, Rec. Trav. chim., 1958, **77**, 842).

2,6-Dihydroxy-3,5-diphenylpyrazine (XXXVIII) is prepared from the corresponding hydroxamic acid by the sequence shown below (Table 3, Ref. 4):

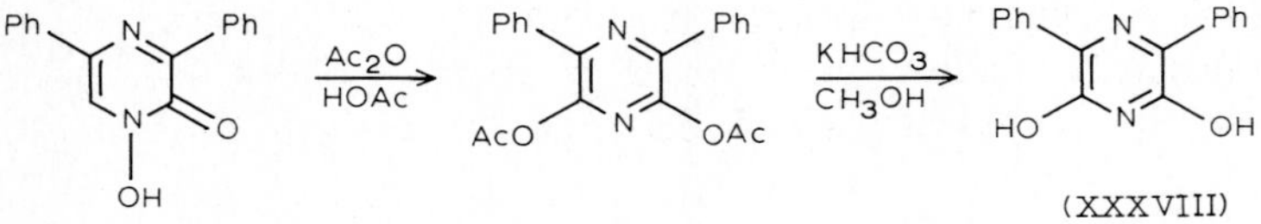

Dihydroxypyrazines are decomposed by acids and the 2,5-dihydroxy isomers require substituents at the 3- and 6-positions in order to be isolable. Alkylation of 2,3-dihydroxypyrazines with dimethyl sulphate in alkali gives the *N,N*-dimethyl derivatives, whereas with excess diazomethane, a mixture of *N,N*-, *O,O*- and *N,O*-dimethyl derivatives is obtained (*G.W.H. Cheeseman* and *E.S.G. Torzs*, J. chem. Soc., 1965, 6681). 2,5-Diethoxypyrazines can be obtained from 2,5-piperazinediones using excess triethyloxonium fluoroborate (*P.G. Sammes* and co-workers, J. chem. Soc. Perkin I, 1972, 2494).

2,5-Dihydroxypyrazines readily participate in cycloaddition reactions with dienophiles like DMAD and norbornene, *e.g.* 3-benzyl-2,5-dihydroxy-6-methylpyrazine forms a cycloadduct (XXXIX) with DMAD which, on thermolysis, eliminates a molecule of isocyanic acid producing the pair of isomeric pyridones [Scheme 7] (*idem, ibid.*, 1973, 404).

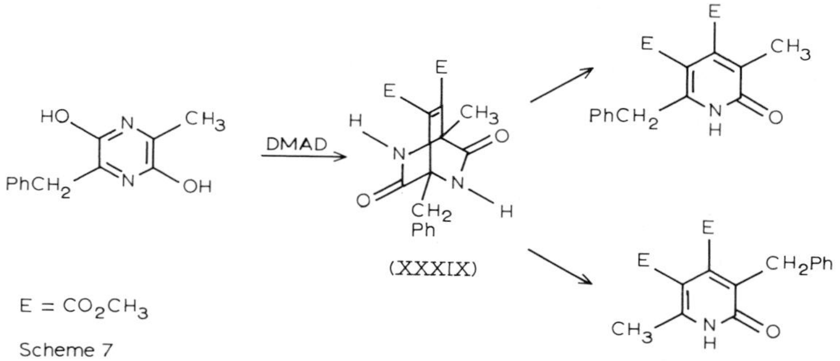

$E = CO_2CH_3$

Scheme 7

By analogy, photo-oxygenation of 2,5-dihydroxy- and 2,5-dialkoxy-pyrazines gives rise to a series of bicyclic endoperoxides, *e.g.* 2,5-diethoxy-3,6-dimethylpyrazine yields quantitatively the corresponding endoperoxide (XL) (*idem, ibid.*, 1979, 1885).

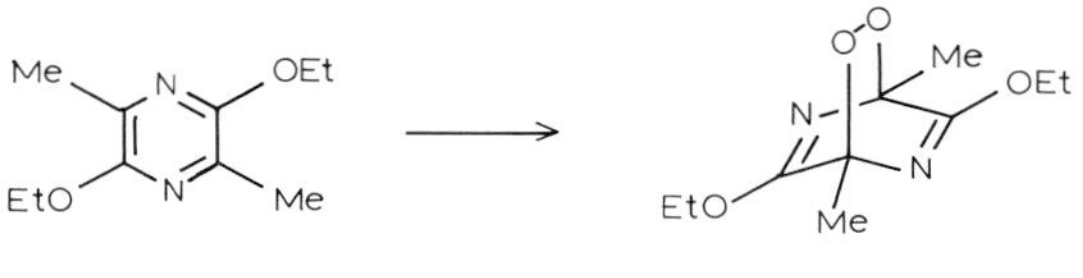

2,6-Dihydroxypyrazines have been shown to undergo an unusual cycloaddition with electron-deficient dienophiles to yield diazobicyclo[3.2.1]octanes (XLI) (*G.W.H. Cheeseman et al., ibid.*, 1980, 1603).

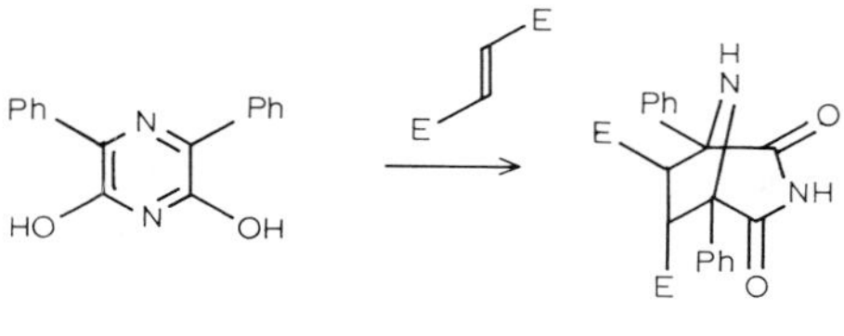

$E = CO_2Et$

(v) Pyrazinecarboxylic acids

Pyrazinecarboxylic acids are prepared by the following methods.

(1) Simple mono-, di-, tri-, and tetra-carboxylic acids are obtained in widely varying yields by oxidation of alkyl- and styryl-pyrazines, and quinoxalines. Typical oxidising agents include potassium permanganate, sodium dichromate and selenious acid. Electrolytic oxidation of quinoxaline in an alkaline permanganate electrolyte affords the 2,3-dicarboxylic acid in high yield (*T. Kimura et al.*, J. pharm. Soc. Japan, 1957, **77**, 891).

(2) Reduction of ethyl oximinoacetoacetate to the amino compound which on self-condensation followed by oxidation produces **diethyl 3,6-dimethylpyrazine-2,5-dicarboxylate** (XLII), m.p. 87° (*H. Adkins* and *E.W. Reeve*, J. Amer. chem. Soc., 1938, **60**, 1328).

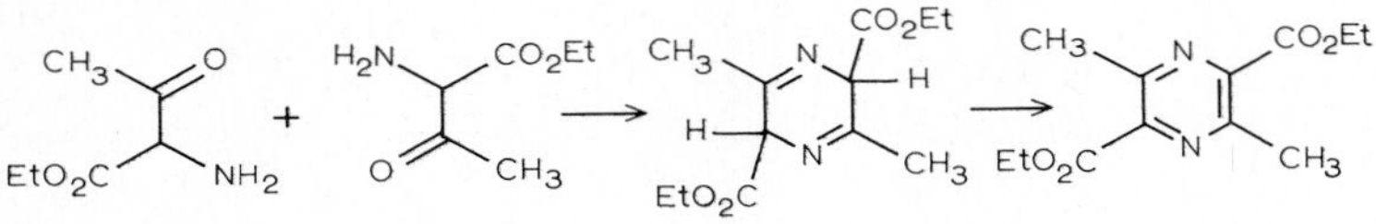

(3) Condensation of α,β-diaminopropionic acid with α,β-dicarbonyl compounds, *e.g.* with phenylglyoxal, a mixture of the separable isomeric **5-phenyl-**, m.p. 190°, and **6-phenyl-pyrazinoic acid**, m.p. 205°, is obtained (*E. Felder et al.*, Ber., 1967, **100**, 555).

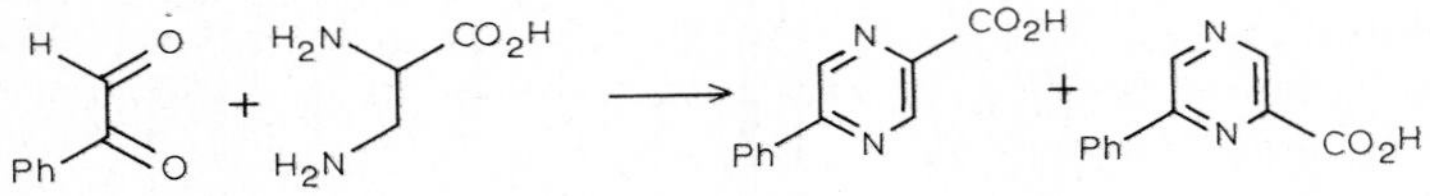

(4) Condensation of DAMN, the hydrogen cyanide tetramer, and α,β-dicarbonyl compounds gives rise to 2,3-dicyanopyrazines which can be subsequently hydrolysed (*R.P. Linstead et al.*, J. chem. Soc., 1937, 911; *L.E. Hinkel et al.*, ibid., 1937, 1432; *P.D. Popp*, J. heterocyclic Chem., 1974, **11**, 79).

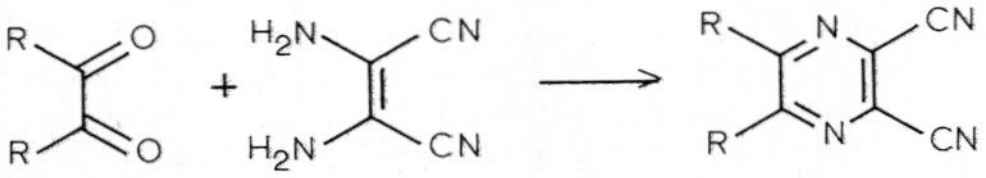

(5) Conversion of bromopyrazines into cyanopyrazines followed by hydrolysis.

Pyrazinoic acid (*methyl ester*, m.p. 59°) is prepared in good yield by permanganate oxidation of methylpyrazine at elevated temperatures (*L.H. Beck*, U.S.P. 3,154,549; *C.A.*, 1965, **62**, 1673). Selenious acid is reported to be a convenient oxidising agent for large scale preparations (*H. Gainer*, J. org. Chem., 1959, **24**, 691).

5-Methylpyrazinoic acid is obtained in superior yield by permanganate oxidation of 2-(hydroxymethyl)-5-methylpyrazine rather than by oxidation of the corresponding 2,5-dimethyl derivative (*D. Pitre et al.*, Ber., 1966, **99**, 364).

Pyrazine-2,3-dicarboxylic acid (*dimethyl ester*, m.p. 50°) is obtained from quinoxaline by either permanganate oxidation (75% yield) (*R.G. Jones* and *K.C. McLaughlin*, Org. Synth., 1950, **30**, 86), or electrolytic oxidation (92% yield) (*T. Kimura, loc. cit.*). Oxidation of 5-methylquinoxaline with potassium permanganate affords 5-methylpyrazine-2,3-dicarboxylic acid which on decarboxylation gives 6-methylpyrazinoic acid (Table 4, Ref. 8).

Pyrazinetricarboxylic acid (*trimethyl ester*, m.p. 80.5°) is obtained in excellent yield (87%) by oxidation of 2-(D-arabo)-tetrahydroxybutylquinoxaline (XLIII). This yield is substantially

(XLIII)

better than that obtained by oxidation of ethoxymethylquinoxaline (Table 4, Ref. 13). Decarboxylation of the tricarboxylic acid under reduced pressure at 210° gives pyrazine-2,6-dicarboxylic acid, whereas in refluxing dimethylformamide the 2,5-isomer is formed (*H.I.X. Mager* and *W. Berends*, Rec. Trav. chim., 1958, **77**, 827).

Pyrazinetetracarboxylic acid (*tetra-ethyl ester*, m.p. 104°) is prepared by oxidation of tetramethylpyrazine, or quinoxaline-2,3-dicarboxylic acid (Table 4, Ref. 14), or phenazine (*R.J. Light* and *C.R. Hauser*, J. org. Chem., 1961, **26**, 1296). It is isolated as a crystalline dipotassium salt which on heating at 200° decarboxylates to the 2,5-di-acid. With sulphur tetrafluoride, the tetra-acid is transformed to tetra(trifluoromethyl)pyrazine (*W.R. Hasek et al.*, J. Amer. chem. Soc., 1960, **82**, 543).

Pyrazine carboxylic acids are generally transformed to acid chlorides, esters and amides under standard conditions, *e.g.* treatment with thionyl chloride affords the corresponding acid chloride (*W.L. MacKenzie* and *W.O. Foye*, J. med. Chem., 1972, **15**, 570). Decarboxylation of the acids occurs at the melting point to give corresponding pyrazines. With iron(II) sulphate, the acids give red-violet solutions which are decolourised by the addition of mineral acid. Other pertinent data on pyrazine carboxylic acids are compiled in Table 4.

Pyrazinamide, m.p. 189°, is conveniently prepared from the methyl ester and ammonia. Owing to the antitubercular activity of pyrazinamide, a number of other amide derivatives have been prepared (*S. Kushner et al.*, J. Amer. chem. Soc., 1952, **74**, 3617). Dehydration of pyrazinamide with phophoryl chloride gives the cyanopyrazine in high yield (*R. Delaby et al.*,

TABLE 4
PYRAZINECARBOXYLIC ACIDS AND HOMOLOGUES

Compound	m.p. anhydrous acid (°C)	Ref.
Monocarboxylic (pyrazinoic) acid	229–230 (dec.)	1–4
3-Methyl-2-carboxylic acid	177	5
5-Methyl-2-carboxylic acid	166–167	6
6-Methyl-2-carboxylic acid	200–202	6
3,6-Dimethyl-2-carboxylic acid	117	7
5,6-Dimethyl-2-carboxylic acid	182	2
2,3-Dicarboxylic acid (dihydrate)	193	2,3
2,5-Dicarboxylic acid (dihydrate)	255–256 (dec.)	8
2,6-Dicarboxylic acid	217–218	8,9
5-Methyl-2,3-dicarboxylic acid	175	10
5,6-Dimethyl-2,3-dicarboxylic acid	200 (dec.)	2,11
3,6-Dimethyl-2,5-dicarboxylic acid	200–201	12
Tricarboxylic acid (dihydrate)	180, 191	9,13
Tetracarboxylic acid (monohydrate)	204–205	14

1 C. Stoehr, J. pr. Chem., 1895, **51**, 468.
2 S. Gabriel and A. Sonn, Ber., 1907, **40**, 4850.
3 S.A. Hall and P.E. Spoerri, J. Amer. chem. Soc., 1940, **62**, 664.
4 T.C. Daniels and H. Iwamoto, ibid., 1941, **63**, 257.
5 T. Ishiguro, M. Matsumura and H. Murai, J. pharm. Soc. Japan, 1960, **80**, 349.
6 D. Pitre, S. Boveri and E.B. Grabitz, Ber., 1966, **99**, 364.
7 C. Stoehr, J. pr. Chem., 1893, **47**, 482.
8 C. Stoehr and W. Detert, ibid., 1897, **55**, 248.
9 K.H. Schaaf and P.E. Spoerri, J. Amer. chem. Soc., 1949, **71**, 2043.
10 F. Leonard and P.E. Spoerri, ibid., 1946, **68**, 526.
11 R.A. Pages and P.E. Spoerri, J. org. Chem., 1963, **28**, 1702.
12 S. Wleügel, Ber., 1882, **15**, 1051; H. Adkins and E.W. Reeve, J. Amer. chem. Soc., 1938, **60**, 1328.
13 J. Bradshaw et al., J. chem. Soc., 1915, 107, 813.
14 L. Wolff, Ber., 1893, **26**, 721; F.D. Chattaway and W.G. Humphrey, J. Chem. Soc., 1929, 645.

Compt. rend., 1958, **247**, 822). In the Hofmann reaction with pyrazinamide, the intermediate sodium carbamate is isolable and requires treatment with acid to produce the amine (S.F. Hall and P.E. Spoerri, J. Amer. chem. Soc., 1940, **62**, 664).

The methyl ester of pyrazine-2,5-dicarboxylic acid yields a *diamide*, m.p. > 270°, and a *dihydrazide*, m.p. 270°; these are converted into the *diisocyanate*, m.p. 205°, and *diurethane*, m.p. > 270°, which do not hydrolyse to the diamine (P.E. Spoerri and A. Erickson, ibid., 1938, **60**, 400).

Pyrazine-2,3-**dicarboxamide**, m.p. 240° (dec), is converted into the *o*-dinitrile (XLIV) with thionyl chloride (M.G. Gal'pern and E.A. Luk'yanets, C.A., 1968, **68**, 2789). With one and two equivalents of potassium hypobromite, the 3-aminocarboxylic acid (XLV) and the lumazine (XLVI) are obtained, respectively (Table 4, Ref. 2) [Scheme 8].

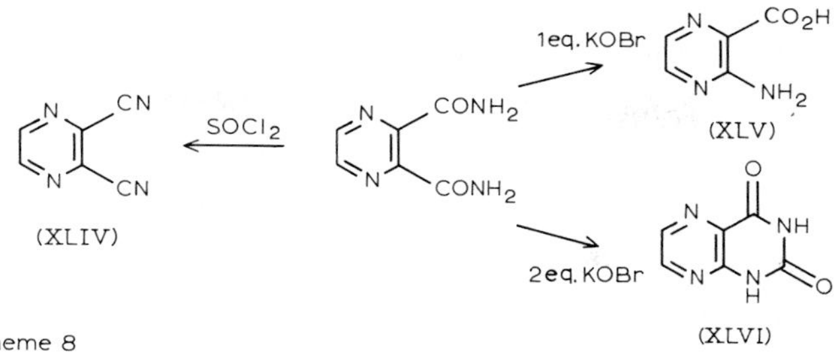

Scheme 8

Several amides are derived from nitriles by heating them with concentrated sulphuric acid including *pyrazine-2,6-dicarboxamide*, m.p. > 355° (*K.H. Schaaf* and *P.E. Spoerri*, J. Amer. chem. Soc., 1949, **71**, 2043) and 3-*methyl*-, m.p. 164–165°, 3-*phenyl*-, m.p. 171–172°, and 3,5,6-*trimethylpyrazine-2-carboxamide*, m.p. 165–166° (*G. Karmas* and *P.E. Spoerri, ibid.*, 1956, **78**, 2141).

Cyanopyrazines can be prepared either by the reaction of bromopyrazines with copper(I) cyanide in refluxing pyridine or γ-picoline (*G. Karmas* and *P.E. Spoerri, loc. cit.*), or by dehydration of the appropriate amide as discussed in the previous section.

Cyanopyrazine, b.p. 116–117°/50 mm, is obtained from bromopyrazine in 30% yield; 3-*methyl*-, 3-*ethyl*-, 3,5-*dimethyl*-, 5,6-*dimethyl*-, 3,5,6-*trimethyl*-, 3-*phenyl*-, 3,5,6-*triphenyl-2-cyanopyrazine*, b.ps. 125–126°/50 mm, 102–103°/15 mm, 113–115°/20 mm, m.ps. 29–30°, 68–69°, 77–78°, 255–256° respectively, are prepared similarly.

2,6-**Dicyanopyrazine**, m.p. 162–163°, is also obtained from the 2,6-dibromo derivative. (*G. Karmas* and *P.E. Spoerri, loc. cit.*).

Condensation of DAMN with the appropriate α,β-dicarbonyl compound provides a convenient entry to a series of 2,3-dicyano compounds including 2,3-*dicyanopyrazine*, m.p. 113° (*E. Gryszkiewicz-Trochimowski*, C.A., 1928, **22**, 4675), 5,6-*dimethyl*-, m.p. 171°, and 5,6-*diphenyl-2,3-dicyanopyrazine*, m.p. 246° (see Refs. on p. 270). 2,5-**Dicyanopyrazine**, m.p. 207°, is obtained from oximinoacetone *via* the Gastaldi procedure (see general method i; *E. Golombok* and *F.S. Spring*, J. chem. Soc., 1949, 1364).

Cyanopyrazines are hydrolysed to the corresponding amides and carboxylic acids by refluxing with concentrated sulphuric acid and aqueous alkali, respectively. Reaction with methylmagnesium bromide affords 2-acetylpyrazines; 2-*acetyl*-, m.p. 76–78°, 2-*acetyl*-3,5,6-*trimethyl*-, m.p. 61–62°, and 2-*acetyl*-5,6-*diphenylpyrazine*, m.p. 152–153° (*Kushner et al., loc. cit.*).

Cyanopyrazine reacts with hydrazine to give the corresponding amidrazone (XLVII) (*H. Foks et al.*, C.A., 1971, **74**, 125632). The amidino derivative (XLVIII) is obtained by the reaction of the nitrile with methanol/hydrogen chloride to form the imino-ether which is then treated with ammonia (*G. Karmas* and *P.E. Spoerri*, J. Amer. chem. Soc., 1957, **79**, 680).

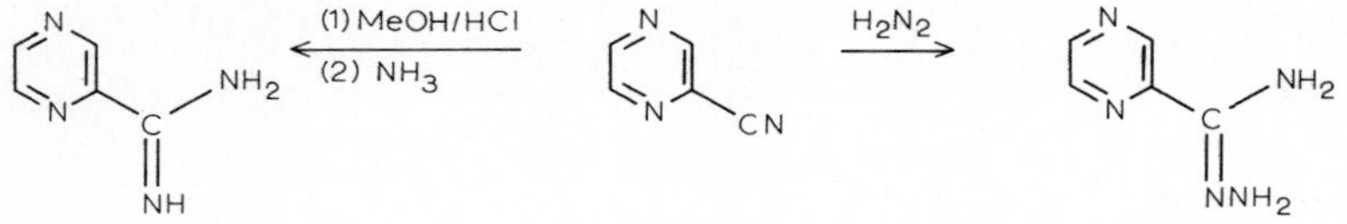

Treatment of 2,3-dicyanopyrazines with an alcohol and triethylamine in DMF results in displacement of one of the cyano groups to give alkoxy cyanopyrazines (*T. Kojima et al.*, J. heterocyclic Chem., 1980, **17**, 455).

Heating 2,3-dicyanopyrazine with copper(I) chloride yields the blue copper tetrapyrzinotetra-azaporphorin tetrahydrate, an analogue of copper phthalocyanine. From the analogous magnesium complex, free tetrapyrazinotetra-azaporphorin has been obtained (*R.P. Linstead et al.*, J. chem. Soc., 1937, 921).

The hydrolysis of lumazines by alkali at 170° provides a route to 2-**aminopyrazine**-3-**carboxylic acid**, m.p. 201° (dec), and the related 6-*methyl*, m.p. 211–212°, 5,6-*dimethyl*, m.p. 208–209° (dec), and 5,6-*diphenyl*, m.p. 189° (dec), derivatives (*J. Weijland et al.*, J. Amer. chem. Soc., 1945, **67**, 802).

2-**Amino**-5-**pyrazinoic acid**, m.p. 82–83°, is prepared from (*N*-acetylamino)quinoxaline (*E. Felder et al.*, Helv. 1964, **47**, 873). Permanganate oxidation of 2-(*N*-acetylamino)-6-methylquinoxaline (XLIX) followed by removal of the acyl protecting group yields 2-**amino**-6-**pyrazinoic acid**, m.p. 120–121° (*M. Sharefkin*, J. org. Chem., 1959, **24**, 345).

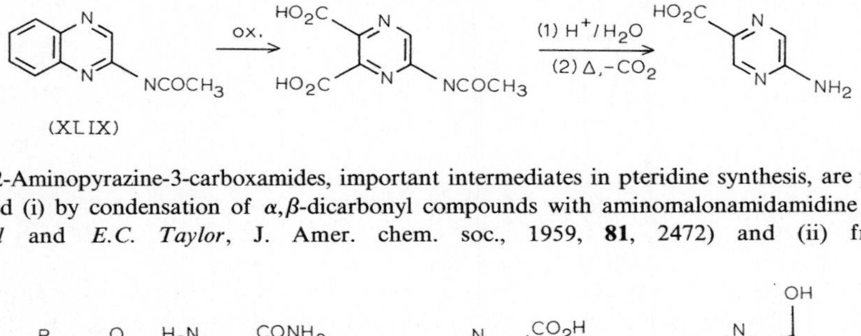

2-Aminopyrazine-3-carboxamides, important intermediates in pteridine synthesis, are prepared (i) by condensation of α,β-dicarbonyl compounds with aminomalonamidamidine (*O. Vogl* and *E.C. Taylor*, J. Amer. chem. soc., 1959, **81**, 2472) and (ii) from

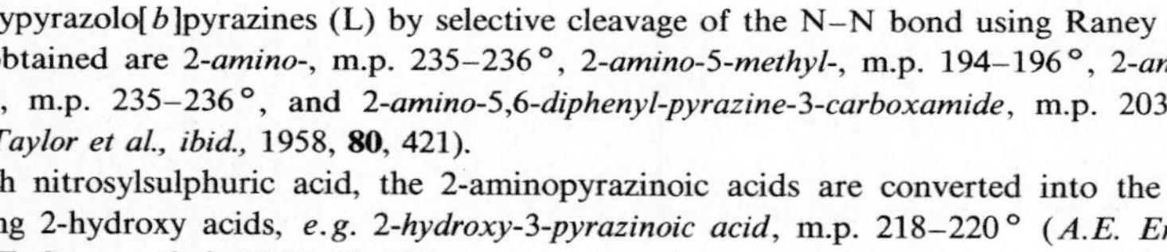

hydroxypyrazolo[*b*]pyrazines (L) by selective cleavage of the N–N bond using Raney nickel. Thus obtained are 2-*amino*-, m.p. 235–236°, 2-*amino*-5-*methyl*-, m.p. 194–196°, 2-*amino*-6-*methyl*-, m.p. 235–236°, and 2-*amino*-5,6-*diphenyl-pyrazine*-3-*carboxamide*, m.p. 203–205° (*E.C. Taylor et al., ibid.*, 1958, **80**, 421).

With nitrosylsulphuric acid, the 2-aminopyrazinoic acids are converted into the corresponding 2-hydroxy acids, *e.g.* 2-*hydroxy*-3-*pyrazinoic acid*, m.p. 218–220° (*A.E. Erickson* and *P.E. Spoerri, ibid.*, 1946, **68**, 400).

Condensation of aminomalonamides with α,β-dicarbonyl compounds yields 2-hydroxy-3-carboxamides (LI), e.g. 2-*hydroxypyrazine-3-carboxamide*, m.p. 265° (dec), and the 5,6-*dimethyl*, m.p. 231–232°, and the 5,6-*diphenyl*, m.p. 174–175°, derivatives which can be

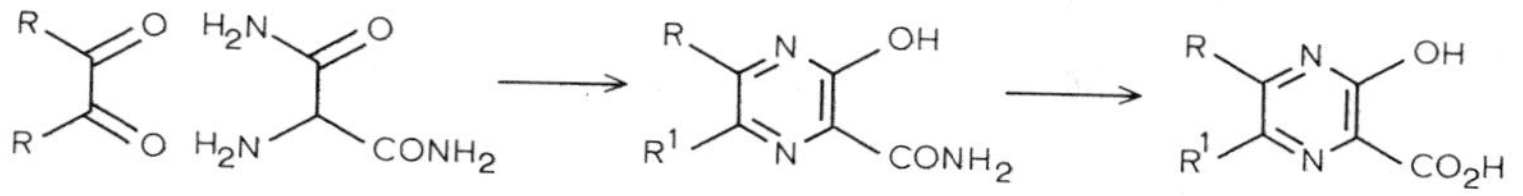

hydrolysed to the corresponding acids. From the unsymmetrical methylglyoxal, 2-**hydroxy-5-methyl-3-pyrazinoic acid**, m.p. 155–157°, is obtained selectively (*R.G. Jones*, J. Amer. chem. Soc., 1949, **71**, 78).

2-Hydroxy-5-pyrazinoic acids are available by alkaline hydrolysis of the corresponding 2,5-dicyano derivatives, e.g. 2-*hydroxy-3,6-dimethyl-5-pyrazinoic acid*, m.p. 265°; the 3,6-*diethyl* compound, m.p. 164°, is similarly obtained (*E. Golombok* and *F.S. Spring*, J. chem. Soc., 1949, 1364).

(e) Reduced pyrazines

(i) Dihydropyrazines

Dihydropyrazines can be classified acording to the four structural types illustrated below.

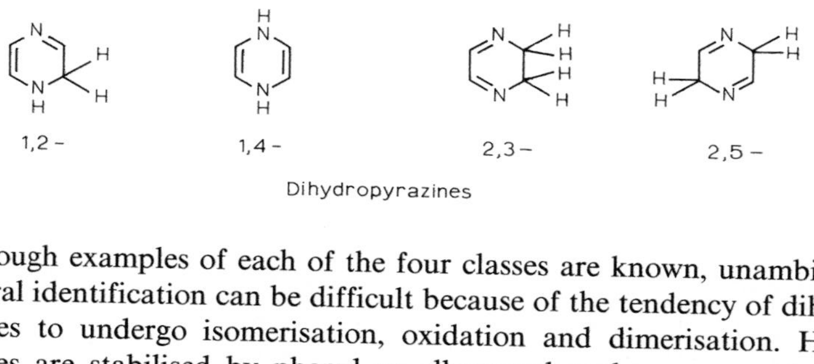

Although examples of each of the four classes are known, unambiguous structural identification can be difficult because of the tendency of dihydropyrazines to undergo isomerisation, oxidation and dimerisation. Hydropyrazines are stabilised by phenyl or alkoxycarbonyl groups, or by substituents on nitrogen, or by a geminally disubstituted ring carbon atom.

A. 1,2-Dihydropyrazines

The addition of organolithium reagents to pyrazine followed by hydrolysis produces 1,2-dihydropyrazines (LII) which rapidly oxidise in air to pyrazines (*B. Klein* and *P.E. Spoerri*, J. Amer. chem. Soc., 1950, **72**, 1844; 1951, **73**, 2949; *R.E. van der Stoel* and *H.C. van der Plas*, Rec. Trav. chim., 1978, **97**, 116).

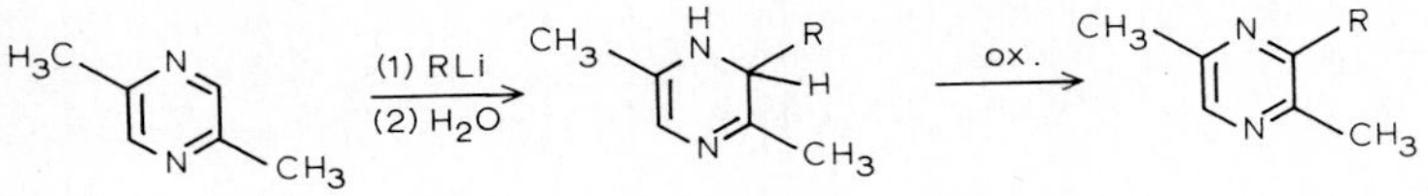

The base-catalysed self-condensation of *N*-phenacylbenzylamine hydrobromide has been shown to produce a mixture of two isomeric 1,2-dihydropyrazines (LIII), m.p. 153–157°, and (LIV), m.p. 101–102.5°, wrongly identified as 1,4-dihydropyrazine (LV) in earlier reports (*S.-J. Chen* and *F.W. Fowler*, J. org. Chem., 1970, **35**, 3987; *J.W. Lown* and *M.H. Akhtar*, J. chem. Soc. Perkin I, 1973, 683).

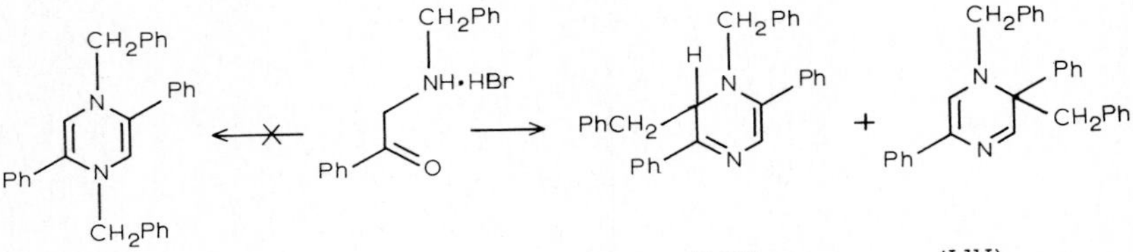

In general, *N*-alkylphenacylamines are found to furnish 1,2-dialkyl-2,5-diphenyl-1,2-dihydropyrazines in good yield. By analogy, the reaction of *N*-benzyl-*N*,*N*-diphenacylamine with benzylamine at 120–130° also gives a rearranged 1,2-dihydropyrazine (LVI), m.p. 94–98°, rather than the expected 1,4-isomer (LVII) (*Chen* and *Fowler*, *loc. cit.*).

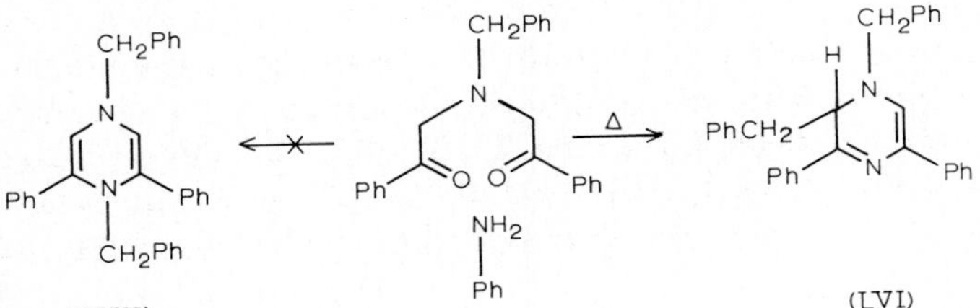

Re-examination of the catalytic hydrogenation of di-, tri-, and tetra-alkoxycarbonylpyrazines indicates that 1,2- rather than 1,4-dihydropyrazines are obtained, *e.g.*, from 2,6-dimethoxycarbonylpyrazine, the reduced ester (LVIII), m.p. 202–204°, is obtained (*J.R. Williams et al.*, J. org. Chem., 1972, **37**, 2963).

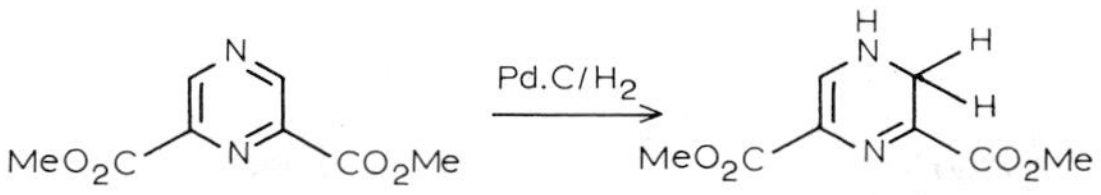

DIHYDROPYRAZINES

(LVIII)

When 2-hydroxypyrazines are *N*-alkylated, the resulting products are usually stabilised in the amide form, *e.g.* 2-hydroxypyrazine with diazomethane or dimethyl sulphate yields 2-**oxo-1-methyl**-1,2-**dihydropyrazine**, m.p. 84–85° (see Refs. p. 264). 3,6-Dimethyl- and 3,6-diphenyl-2-hydroxypyrazine with methyl iodide form the corresponding methiodides which on heating give 2-oxo-1-methyl-1,2-dihydropyrazines (LIX, R = CH_3) (*hydrochloride*, m.p. 227°) and (LIX, R = Ph), m.p. 168°, respectively (*C. Gastaldi* and *E. Princivalle*, Gazz., 1928, **58**, 412).

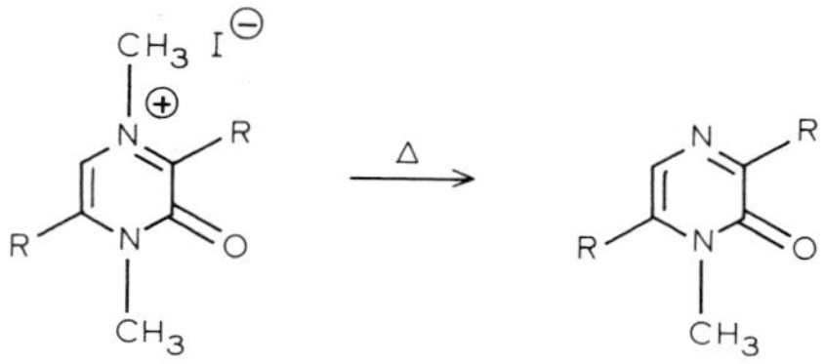

(LIX)

B. *1,4-Dihydropyrazines*

This system has attracted considerable theoretical interest since it formally possesses eight π-electrons and hence is potentially anti-aromatic. Earlier claims for the isolation of 1,4-dihydropyrazines, prior to the availability of n.m.r. spectroscopy, have been shown to be incorrect (*vide infra*). 1,4-Dihydropyrazines are stabilised by electron-withdrawing groups such as the acyl group, and, to avoid unfavourable ground states, also undergo ring distortion or rearrangement (*R.R. Schmidt*, Angew. Chem. intern. Edn., 1975, **14**, 581; *W. Kaim*, ibid., 1983, **22**, 171). Air-oxidation of 1,4-dihydropyrazines does, however, result in the generation of comparatively persistent radical cations (*J.W. Lown* and *M.H. Ahktar*, Chem. Comm., 1974, 829).

Reductive silylation of pyrazine produces the air-sensitive 1,4-**di(trimethylsilyl)**-1,4-**dihydropyrazine** (LX), m.p. 65° (*R.A. Sulzback* and *A.F.M. Iqbal*, Angew. Chem. intern. Edn., 1971, **10**, 127). A series of air-stable 1,4-diacyl derivatives have also been obtained from pyrazine under reducing conditions (cathodic reduction or with zinc metal), *e.g.* 1,4-**diacetyl**-1,4-**dihydropyrazine** (LXI, R = CH_3), m.p. 188–191° (*R. Gottlieb* and *W.*

Pfleiderer, Ann., 1981, 1451). Electrochemical reduction of pyrazine with simultaneous saturation of the catholyte with carbon dioxide, followed by the addition of ethyl bromide, gives 1,4-di(ethoxycarbonyl)-1,4-dihydropyrazine (LXI, R = OEt) in 76% yield (*D. Michelet*, Fr.P. 2444030; C.A., 1981, **94**, 164821).

Treatment of 2,3-diphenyl-5,6-dihydropyrazine with acetyl chloride in the presence of two equivalents of pyridine gives 1,4-**diacetyl**-2,3-**diphenylpyrazine**, m.p. 194–195°, in 30% yield together with 2,3-diphenylpyrazine and 1,4-**diacetyl**-2,3-**diphenyl**-1,4,5,6-**tetrahydropyrazine**, m.p. 131–132°; in the absence of pyridine, the latter compound is obtained but none of the 1,4-dihydro derivative (*S.-J. Chen* and *F.W. Fowler*, J. org. Chem., 1971, **36**, 4025; *cf. A.T. Mason* and *G.R. Winder*, J. chem. Soc., 1893, **65**, 1355).

As discussed above, neither the self-condensation of *N*-alkylphenacylamines nor the reaction of *N*-alkyl *N,N*-diphenacylamines with benzylamine under the conditions described by *Mason* and *Winder* (*loc. cit.*) produces the expected 1,4-dihydropyrazine. These reactions do proceed *via* 1,4-dihydropyrazines which undergo rearrangement by 1,3-sigmatropic migration of an *N*-alkyl group (*J.W. Lown* and *M.H. Ahktar*, Tetrahedron Letters, 1974, 179). Condensation of *N*-benzyl-*N,N*-diphenacylamine with benzylamine at 40° affords 1,4-**dibenzyl**-2,6-**diphenyl**-1,4-**dihydropyrazine**, m.p. 107–108.5°, which rearranges at higher temperatures with first order kinetics (*J.W. Lown* and co-workers, Chem. Comm., 1973, 511; J. org. Chem., 1974, **39**, 1998). This reaction is quite general and applies to products derived from other primary amines (*idem*, Chem. Comm., 1972, 829; J. chem. Soc. Perkin I, 1973, 683).

Dimerisation of *N*-aryl-*N*-phenacylamines does not give the expected centro-symmetric 1,4-dihydropyrazines (LXII); the products are the isomeric 1,4-dihydropyazines, *e.g.* (LXIII; R^1 = Ph, R^2 = *p*-Cl-C_6H_4, m.p. 202–206°), probably arising from *in situ* formation of a diphenacylamine and an amine which subsequently condense together (*R.R. Schmidt et al.*, Ber., 1976, **109**, 2395).

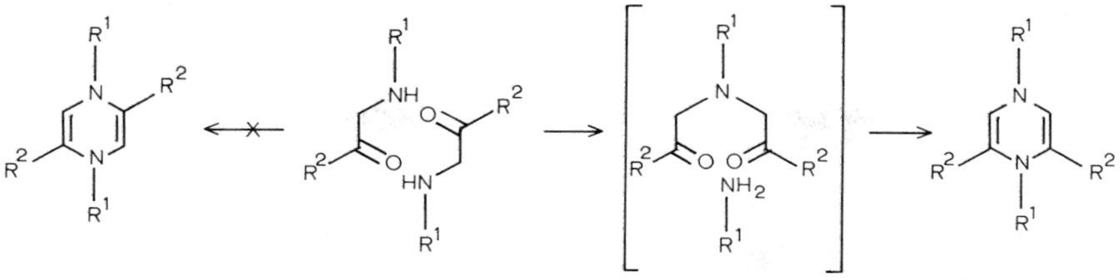

C. 2,3-Dihydropyrazines

Condensation of α,β-dicarbonyl compounds with α,β-diamines gives 2,3-dihydropyrazines which are readily dehydrogenated to pyrazines (see general method ii, p. 247), e.g. 5,6-**diphenyl**-2,3-**dihydropyrazine** (LXIV), m.p. 181°, one of the more stable derivatives, is obtained from benzil and ethylenediamine (*L.H. Amundsen*, J. chem. Educ., 1939, **16**, 566). Heating this with acetic anhydride results in the formation of the tetrahydropyrazine (LXV), not the 1,4-dihydropyrazine derivative previously reported, and 2,3-diphenylpyrazine (*S.-J. Chen* and *F.W. Fowler*, J. org. Chem., 1971, **36**, 4025).

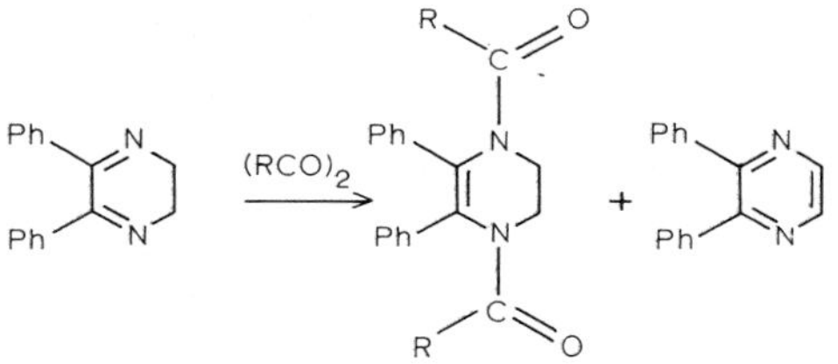

Addition of hydrogen peroxide to (LXIV) gives a hydroperoxide which on thermolysis rearranges to N,N'-dibenzoyl ethylenediamine (*H.I.X. Mager* and *W. Berends*, Rec. Trav. chim., 1965, **84**, 314). Dihydropyrazine (LXIV) with two equivalents of hydrogen cyanide forms 2,3-dicyano-2,3-diphenylpyrazine (*P. Beak* and *J. Yamamoto*, J. heterocyclic Chem., 1972, **9**, 155). With potassium cyanide in alcohol, 5,6-**diphenyl**-2-**carboxamide**, m.p. 197–198°, is obtained (*A.T. Mason* and *L.A. Dryfoos*, J. chem. Soc., 1893, **63**, 1293).

The reaction of dihydropyrazine (LXIV) with malononitrile gives a 1:2 adduct (LXVI) [Scheme 9], which results from a complex rearrangement (*P. Beak et al.*, J. org. Chem., 1976, **41**, 3389). Compound (LXIV) enters into cycloadditions with diphenylketene and with diethyl fumarate to give the

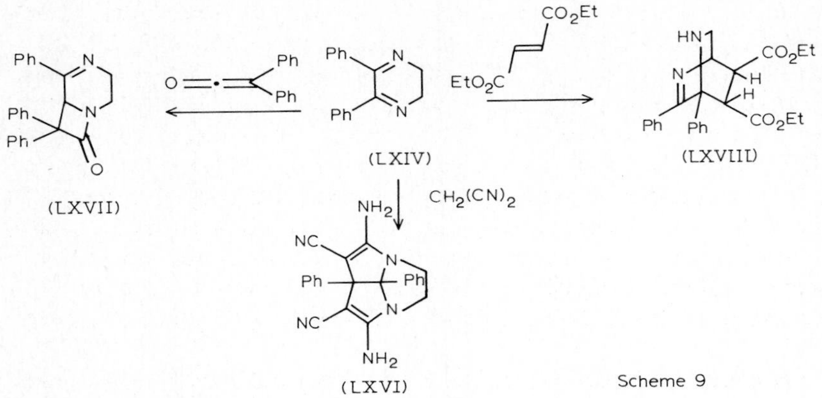

[2 + 2] cycloadduct (LXVII) and the unusual 2,5-adduct (LXVIII), respectively (*M. Sakamoto* and *Y. Tomimatsu*, J. pharm. Soc. Japan, 1970, **90**, 1386; 544).

2,3-Disubstituted, 2,3-dihydropyrazines can exist in *cis* and *trans* forms. Thus *meso-* and *rac.*-stilbenediamine with glyoxal yield respectively *cis-*, m.p. 166.5–167.5°, and *trans-*, m.p. 202–203°, 2,3-**diphenyl**-2,3-**dihydropyrazine** (*T. Hayashi*, C.A., 1947, **41**, 5886; 6258).

D. 2,5-Dihydropyrazines

Self-condensation of α-aminoketones (general method i, p. 245) gives 2,5-dihydropyrazines which can be characterised if air is excluded. Only symmetrically substituted derivatives can be prepared in this way, *e.g.*

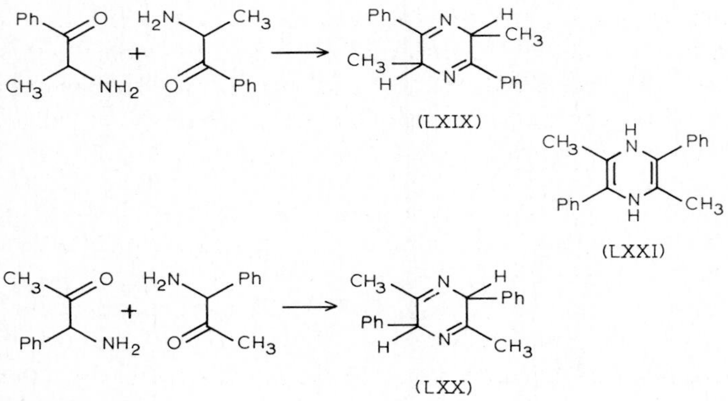

Scheme 10

3,6-*dimethyl*-, b.p. 100°/50 mm, 2,2,3,5,5,6-*hexamethyl*-, m.p. (anhyd.) 65–68°, (hexahydrate) 87–89°, 3,6-*diphenyl*-, m.p. 166–167°, 2,5-*dimethyl*-3,6-*diphenyl*- (LXIX), m.p. 140°, and 3,6-*dimethyl*-2,5-*diphenyl*-2,5-*dihydropyrazine* (LXX), m.p. 99–100° (dec.) (*hydrochloride*, m.p. 167–168°) (*S. Gabriel* and *J. Colman*, Ber., 1902, **35**, 3807; *S. Gabriel*, Ber., 1908, **41**, 1127; *P.W. Neber* and *A. von Friedelsheim*, Ann., 1926, **449**, 109).

The independent formation of the isomeric 2,5-dihydropyrazines (LXIX) and (LXX) from phenyl α-aminoethyl ketone and α-amino α-phenylacetone, respectively, suggests that the reaction does not proceed *via* the isomeric 1,4-dihydropyrazine (LXXI) and that the rate of tautomeric interconversion is slow [Scheme 10] (*S. Gabriel*, Ber., 1908, **41**, 1127; 1911, **44**, 57).

^{1}H-n.m.r. spectroscopic studies of the self-condensation of α-aminoacetone show conclusively that the expected product 3,6-dimethyl-2,5-dihydropyrazine (LXXII) is formed together with small quantities of either the 1,4- or the 1,2-isomer. At room temperature, (LXXII) dimerises on standing (*S. Wilen*, Chem. Comm., 1970, 25).

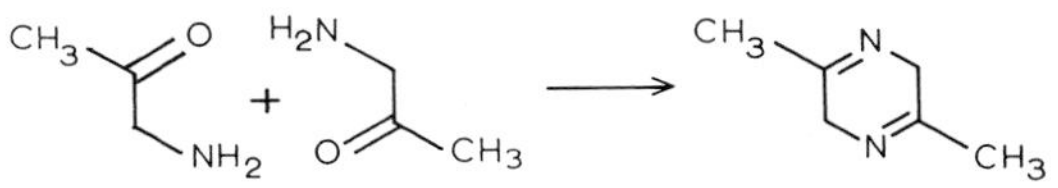

(LXXII)

Treatment of isobutyraldehyde with warm alkaline potassium ferricyanide gives 2,2,5,5-**tetramethyl**-2,5-**dihydropyrazine**, m.p. 83–84°, in moderate yield; with methyl isopropyl ketone, the hexamethyl compound mentioned above is obtained. The ferricyanide ion is the sole source of nitrogen in the reaction mixture (*J.B. Conant* and *J.G. Aston*, J. Amer. chem. Soc., 1928, **50**, 2783). Modification of this procedure allows isolation of the dihydropyrazine or α-aminoketones depending on the method of work-up (*D.G. Farnum* and *G.R. Carlson*, Synthesis, 1972, 191).

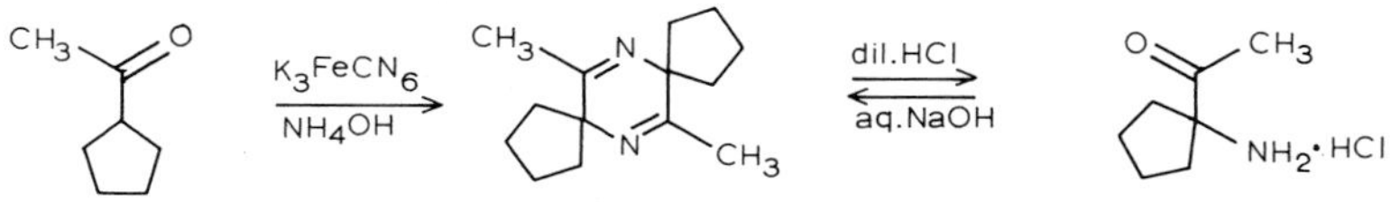

Dimerisation of 3-phenylazirine affords the comparatively stable, 3,6-diphenyl-2,5-dihydropyrazine (LXXIII) (*L. Horner et al.*, Ber., 1963, **96**, 399).

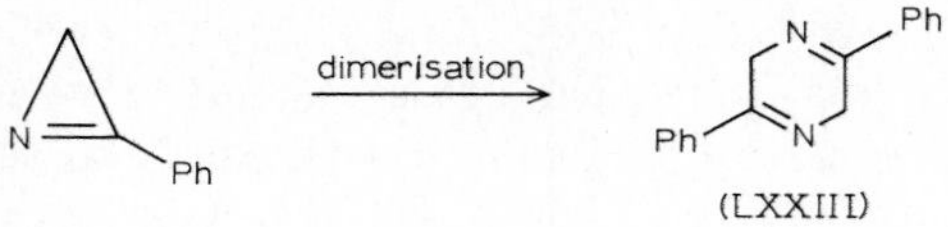

2,5-Diketopiperazine can be alkylated, *via* the silver salt, with benzyl chloride to give 2,5-**dibenzyloxy**-2,5-**dihydropyrazine** (LXXIV), m.p. 163–164° (*P. Karrer* and *C. Granacher*, Helv., 1923, **6**, 1108; 1924, **7**, 763). The corresponding *ethoxy* derivative, m.p. 84°, is prepared by treatment of 2,5-diketopiperazine with triethyloxonium fluoroborate (*P.G. Sammes* and co-workers, J. chem. Soc. C, 1970, 1070; 1972, 2494). The latter compound is metallated with lithium diisopropylamide (LDA) at −78°, and may be subsequently either alkylated with alkyl or allyl halides, or condensed with aldehydes and ketones (*P.G. Sammes* and *J.L. Markham*, J. chem. Soc. Perkin I, 1979, 1889).

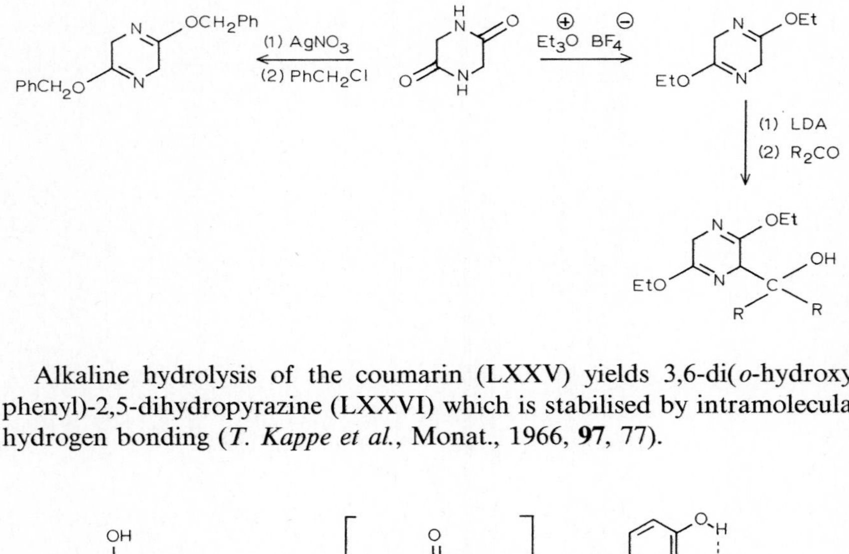

Alkaline hydrolysis of the coumarin (LXXV) yields 3,6-di(*o*-hydroxyphenyl)-2,5-dihydropyrazine (LXXVI) which is stabilised by intramolecular hydrogen bonding (*T. Kappe et al.*, Monat., 1966, **97**, 77).

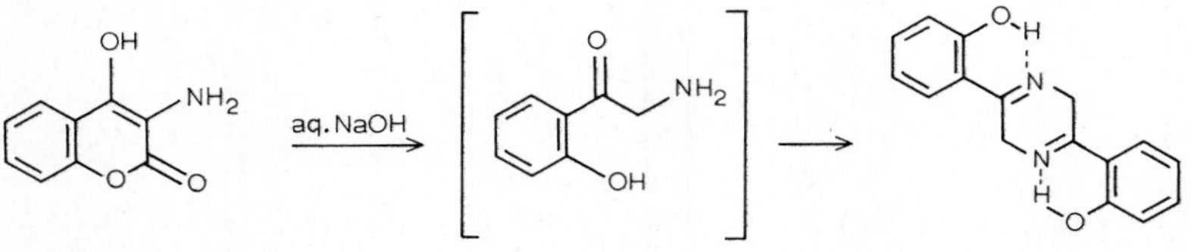

(ii) Tetrahydropyrazines

In the synthesis of tetrahydropyrazines, there is often uncertainty, particularly in earlier work, regarding the precise location of the double bond. Certain compounds are readily converted into piperazines with the necessary hydrogen coming from the solvent or by intermolecular disproportionation with concomitant formation of pyrazines.

Tetrahydropyrazines have been prepared by the following methods.

(1) Condensations between ethylenediamine and *N*-substituted derivatives. Reaction of *N*-phenylethylenediamine with benzoin gives the triphenyltetrahydropyrazine (LXXVII) (*S. Gabriel* and *G. Eschenbach*, Ber., 1898, **31**, 1581). By analogy, *N,N*-diphenylethylenediamine and phenacyl bromide yield 1,2,4-**triphenyl**-1,4,5,6-**tetrahydropyrazine** (LXXVIII), m.p. 130–131°

(*L. Garzino*, Gazz., 1893, **23**, 9) which can be hydrogenated to 1,2,3-triphenylpiperazine. The tetrahydropyrazine (LXXIX), b.p. 140°/0.01 mm, is obtained in 40% yield from *N,N*-dimethylethylenediamine and ethyl α-benzoyl-α-chloroacetate (*F. Korte* and co-workers, Tetrahedron, 1970, **26**, 3993).

The reaction of the epoxyether (LXXX) with ethylenediamine affords, in high yield, the tetrahydropyrazine (LXXXI), b.p. 85°/0.05 mm, which may be acetylated with acetic anhydride in pyridine (*C.L. Stevens*, J. org. Chem., 1971, **29**, 3574).

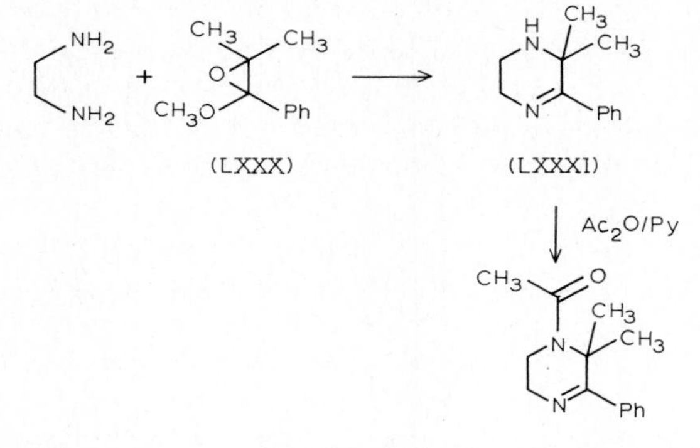

(2) Cycloaddition reactions. Dimethylketene added to dianils derived from α-diketones gives tetrahydropyrazines, *e.g.* compound (LXXXII), m.p. 160–161°, is obtained in 55% yield (*R.D. Burpitt et al.*, J. org. Chem., 1971, **36**, 2222). Phenyl- and diphenyl-ketene are reported to give similar adducts (*R. Pfleger* and *A. Jager*, Ber., 1957, **90**, 2460), however, a later investigation suggests that the adducts are β-lactams (*M. Sakamoto* and *Y. Taminatsu*, J. pharm. Soc. Japan, 1970, **90**, 1368).

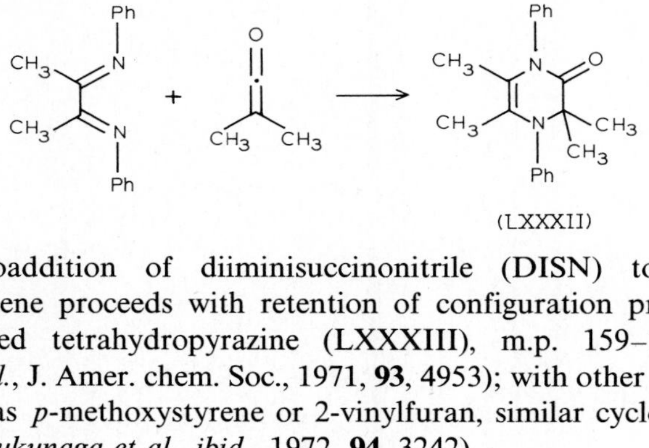

The cycloaddition of diiminisuccinonitrile (DISN) to *cis*-1,2-dimethoxyethylene proceeds with retention of configuration producing the tetrasubstituted tetrahydropyrazine (LXXXIII), m.p. 159–162° (*W.A. Sheppard et al.*, J. Amer. chem. Soc., 1971, **93**, 4953); with other electron-rich olefins such as *p*-methoxystyrene or 2-vinylfuran, similar cycloadducts are formed (*T. Fukunaga et al.*, *ibid.*, 1972, **94**, 3242).

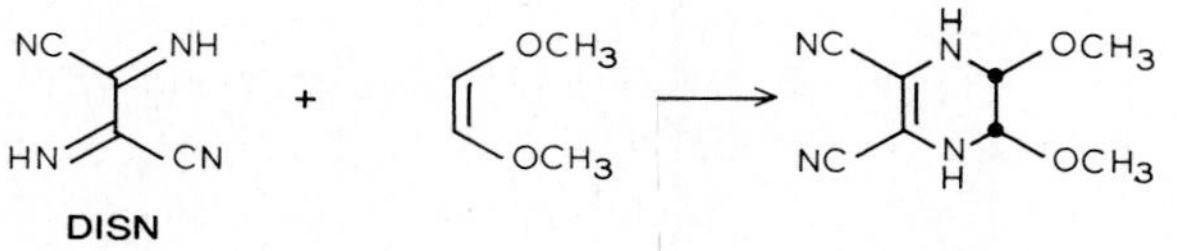

(3) Cyclisation reactions. 5-Amino-2,2,5,5-tetramethyl-3-aza-1-hexanol is cyclised to the hygroscopic 2,2,5,5-**tetramethyl**-2,3,4,5-**tetrahydropyrazine** (LXXXIV), m.p. 83–84° [*hydrochloride*, m.p. 167–169° (dec)], on distillation from Raney nickel at atmospheric pressure. The hydrogenation of (LXXXIV) to tetramethylpiperazine only requires 30% of the theoretical

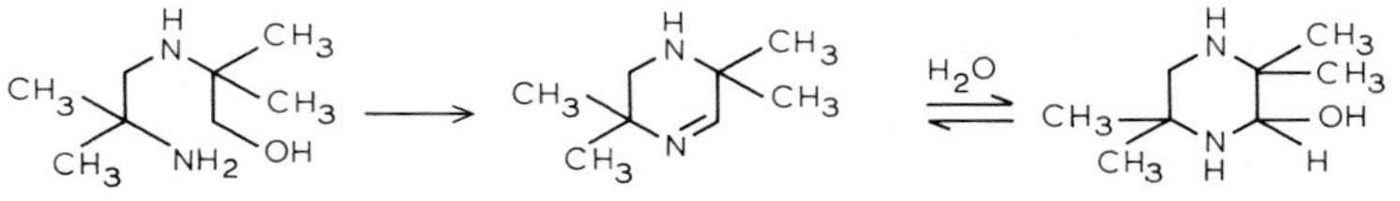

(LXXXIV)

quantity of molecular hydrogen; the remainder is apparently supplied by the solvent (ethanol). Nitrosation or acylation of (LXXXIV) also results in formation of some of the corresponding piperazine derivatives (*L.B. Clapp et al.*, J. org. Chem., 1956, **21**, 82).

Although condensation of α-(β-chloroethylamino)-α-phenylacetophenone or its *N*-benzyl derivative with benzylamine or ammonia, respectively, almost certainly gives the expected tetrahydropyrazine (LXXXV), the isolated product is a peroxide (LXXXVI), m.p. 117–118°, arising from autoxidation of (LXXXV). Thermolysis of (LXXXVI) gives the acyclic diamide

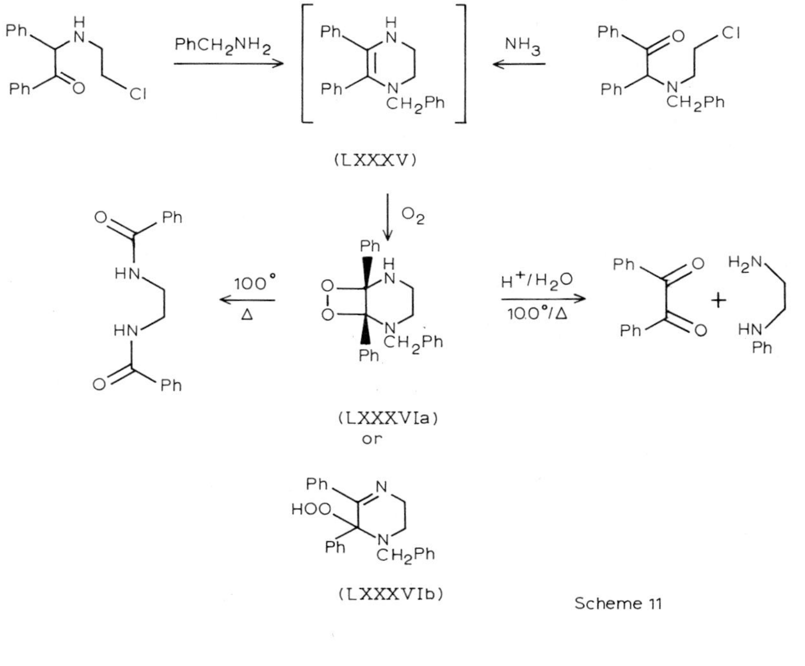

(LXXXV)

(LXXXVIa)
or

(LXXXVIb)

Scheme 11

whereas in aqueous acid at 100°, benzil and *N*-benzylethylenediamine are obtained (*C.D. Lunsford et al.*, J. org. chem., 1955, **20**, 1513) [Scheme 11]. The peroxide (LXXXVI) has been assigned a bicyclo[4.2.0] structure (LXXVIa) on the basis of the decomposition products. Since similar products are obtained from thermolysis of the hydroperoxide derived from peroxidation of 5,6-diphenyl-2,3-dihydropyrazine, structure (LXXXVIb) is more probable (*H.I.X. Mager* and *W. Berends*, Rec. Trav. chim., 1965, **84**, 314).

(4) From dihydropyrazines. Treatment of certain dihydropyrazine methiodides, *e.g.* (LXXXVII, Y = I), or the corresponding pseudobase (LXXXVII, Y = OH), with respectively one or two equivalents of a Grignard reagent affords tetrahydropyrazine derivatives, *e.g.* 1,2,5,5,6-*hexamethyltetrahydropyrazine* (LXXXVIII, R = CH$_3$), b.p. 72–73°/13 mm, and the analogous 6-*phenyl* (LXXXVIII, R = Ph), b.p. 152–153°/10 mm and 6-*benzyl* derivative (LXXXVIII, R = CH$_2$Ph), m.p. 62–63° (*J.G. Aston et al.*, J. Amer. chem. Soc., 1934, **56**, 1163).

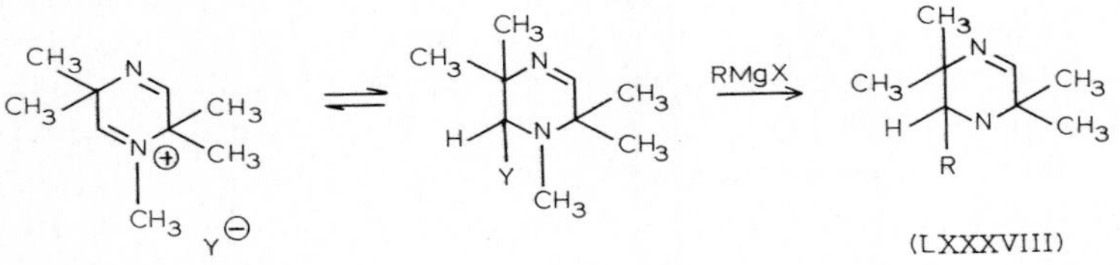

Acetylation of 5,6-diphenyl-2,3-dihydropyrazine affords the tetrahydropyrazine (LXXXIX), m.p. 131–132°, which on reduction with excess lithium aluminium hydride gives the corresponding 1,4-diethyl derivative (XC), m.p. 100–102° (*S.-J. Chen* and *F.W. Fowler*, J. org. Chem., 1971, **36**, 4025).

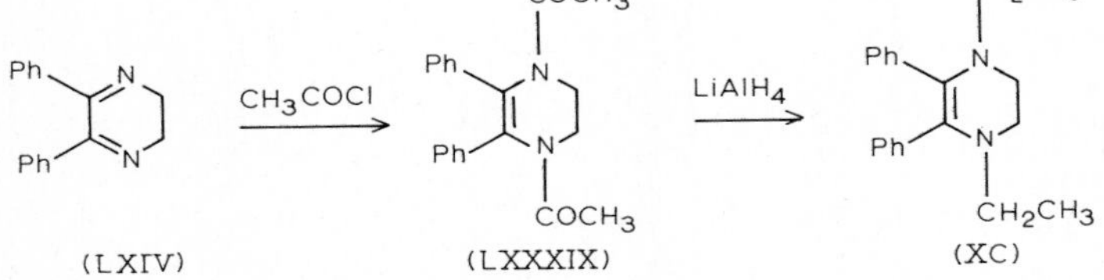

(5) From azirines. Reaction of 2,2-dimethyl-3-phenyl-2*H*-azirine (a) with ammonia gives the amino-tetrahydropyrazine (XCI), and (b) with α-amino-acid methyl esters affords the 1,2,3,6-tetrahydropyrazin-2-ones (XCII) (*A.V.*

Eremeev et al., C.A., 1982, **92**, 94346; Chem. heterocyclic Comp., 1975, **15**, 810).

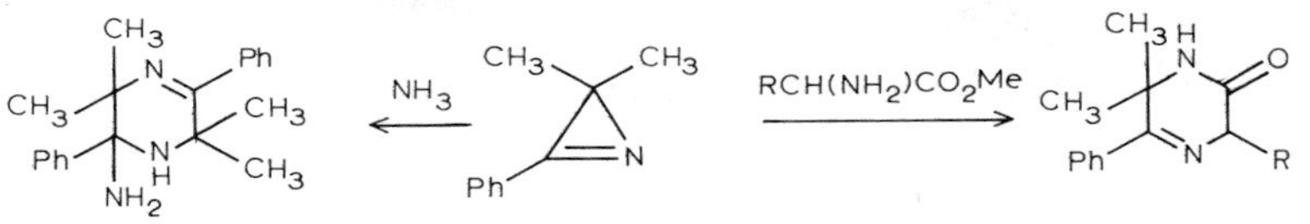

(XCI)   (XCII)

(iii) Piperazines (hexahydropyrazines)

The fully reduced pyrazines constitute an important class of 1,4-diazines. The parent compound, piperazine, was first prepared from 1,2-dibromoethane and aqueous ammonia (*S. Cloez*, Jahresber. Fortschr. Chem., 1853, 468) and the N,N'-diphenyl derivative, in better yield, from the dibromide and aniline (*A.W. Hofmann et al., ibid.*, 1858, 353).

Piperazines are conformationally flexible molecules, capable of undergoing ring inversion and pyramidal inversion at each nitrogen centre (*J.B. Lambert*, Top. Stereochem., 1971, **6**, 19). Electron diffraction studies of piperazine in the gas phase (*A. Yokozeki* and *K. Kuchitsu*, Bull. chem. Soc. Japan, 1971, **44**, 2352) and X-ray crystallographic studies of *trans*-2,5-dimethylpiperazine (*K. Okamoto et al., ibid.*, 1982, **55**, 945) both indicate that the piperazine ring has a chair conformation analogous to cyclohexane. Within this conformation, piperazines may also adopt three possible invertomeric forms (A–C). For $R = CH_3$, the relative populations A : B : C

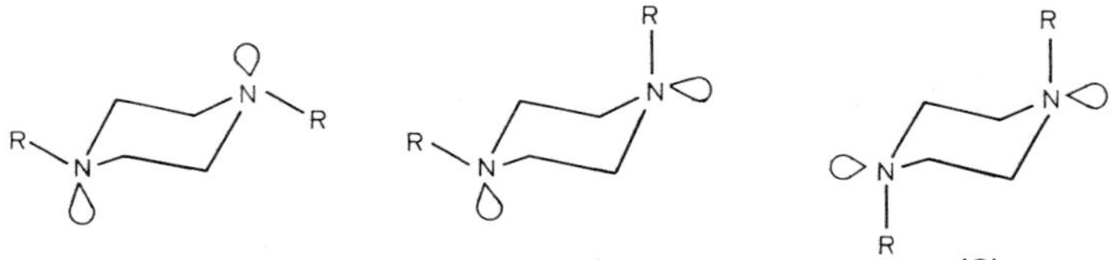

(A)   (B)   (C)

are found to be 0.880 : 0.116 : 0.004 in solution (*N.L. Allinger et al.*, J. Amer. chem. Soc., 1965, **87**, 1232; *I. Horikoshi et al.*, Chem. pharm. Bull. Japan, 1975, **23**, 754). Like cyclohexanes, piperazines favour conformations in which substituents on nitrogen and carbon are predominantly in equatorial positions (see *W.L.F. Armarego*, "Stereochemistry of Heterocyclic Compounds, Part 1", Wiley, New York, 1977).

Piperazine and its *N*-substitution derivatives have pharmacological applications as anthelmintics, sedatives and local anaesthetics.

The chemistry of piperazines is dominated by reactions at the ring nitrogen; C-substituted derivatives are obtained by reduction of the corresponding pyrazines or by ring synthesis.

A. *Methods of synthesis*

(1) Piperazine can be prepared from the 1,4-phenyl derivative by nitrosation followed by hydrolysis of the dinitroso compound in alkali or ammonia (*G. Sanna*, C.A., 1943, **37**, 1718).

(2) Cyclodehydration occurs on heating salts of 2-hydroxyethylamine at 230–240° in the presence of metallic halides (B.P. 595,430; C.A., 1948, **42**, 3438) or *N*-(2-hydroxyethyl)ethylenediamine (XCIII) with Raney nickel (*L.J. Kitchen* and *C.B. Pollard*, J. Amer. chem. Soc., 1947, **69**, 854) to give piperazine. *N*-Substituted ethanolamines undergo self-condensation with a copper chromite catalyst at 250–270° (*J.P. Bain* and *C.B. Pollard, ibid.*, 1939, **61**, 532; 2704).

$$\underset{(XCIII)}{\text{H-N-CH}_2\text{-CH}_2\text{-NH}_2\text{-CH}_2\text{-CH}_2\text{-OH}} \xrightarrow[\Delta]{\text{Raney Nickel}} \text{piperazine}$$

(3) Catalytic deamination of diethylenetriamine over Raney nickel or alumina (*W.B. Martin* and *A.E. Martell, ibid.*, 1948, **70**, 1817; *A.A. Anderson et al.*, Chem. heterocyclic Comp., 1967, **3**, 271).

(4) α-Halo-amines or sulphanamides undergo self-condensation to piperazines (*L. Knorr*, Ber., 1904, **37**, 3507).

(5) Reaction of *N*-alkyl bis(2-haloethyl)amines with primary amines affords piperazines with different 1- and 4-substituents (*V. Prelog* and *V. Stepan*, Coll. Czech. Chem. Comm., 1935, 7, 93).

(6) Pyrazines and di- and tetra-hydro derivatives are reduced to piperazines with sodium in alcohol or by catalytic hydrogenation. Although widely used for the preparation of alkylpiperazines, mixtures of stereoisomers are obtained for systems with two or more substituents.

(7) Reduction of *N*-benzylpyrazinium salts with sodium borohydride affords the corresponding piperazines (*R.E. Lyle* and *T.J. Thomas*, J. org. Chem., 1965, **30**, 1907).

Piperazine (hexahydropyrazine), m.p. 104°, b.p. 145–146°, is a hygroscopic solid, less basic than piperidine, which on steam distillation from alkaline solution forms a *hexahydrate*, m.p. 44°, b.p. 125–130°. It is conveniently prepared by either the cyclodehydration of *N*-(2-hydroxyethyl)ethylenediamine or the deamination of diethylenetriamine, both reactions being catalysed by Raney nickel (*Kitchen* and *Pollard*, and *Martin* and *Martell, loc. cit.*).

Derivatives include 1,4-*diacetyl*-, m.p. 138.5°, -*dibenzoyl*-, m.p. 194°, -*dinitroso*-, m.p. 158°, and *mono-tosyl-pyrazine*, m.p. 110°.

Piperazine, estimated annual demand of 900 tonnes for 1975, is manufactured primarily for use as an anthelmintic in human and veterinary medicine. It is usually formulated as the citrate, phosphate, hydrochloride or sulphate salt (see *K. Mjos*, "Kirth-Othmer Encycl. Chem. Technol.", 3rd Edn., 1978, Vol. 2, p. 295, and Refs. therein). The anthelmintic activity of several mono-quaternary piperazinium salts has been examined (*M. Harfenist*, J. Amer. chem. Soc., 1957, **79**, 2211).

Quantitative analysis of piperazine can be effected gravimetrically as the insoluble complex $C_4H_8N_2 \cdot Cr_2O_7$, formed on reaction with chromic acid. In paper chromatography, piperazine is visualised as a red-brown spot using iodine vapour (*A. Castigloni* and *M. Nivoli*, C.A., 1953, **47**, 5848).

(XCIV)

Piperazine forms a 1:2 adduct with maleic acid and fumaric acid, giving the acid (XCIV), *methyl ester* m.p. 159°, *ethyl ester*, m.p. 96°; methyl maleate is rapidly isomerised to methyl fumarate prior to the addition reaction (*C.B. Pollard* and co-workers, J. Amer. chem. Soc., 1935, **37**, 199). Similar Michael-type adducts are formed with α,β-unsaturated esters and ketones. These compounds are more stable than the corresponding piperidine adducts though, on treatment with dilute hydrochloric acid, they revert to starting materials (*idem, ibid.*, 1936, **58**, 1980; 1937, **59**, 1719).

B. N-Substituted piperazines

Standard methods of alkylation and acylation of piperazine give mixtures of *N*-mono- and *N,N′*-di-substituted products. By maintaining the acidity at around pH 3, the mono-acetyl, -benzoyl and -ethoxycarbonyl derivatives are prepared in good yield.

1-*Acetyl*-, m.p. 52°, 1-*benzoyl*-, m.p. 75°, and 1-*ethoxycarbonyl-piperazine*, b.p. 116–117°/12 mm, serve as useful intermediates for the synthesis of piperazines having different substituents on the nitrogen centres (*T.S. Moore et al.*, J. chem. Soc., 1929, 39; *K.R. Jacobi*, Ber., 1933, **66**, 113). Since the protecting group is easily removed, the ethoxycarbonyl compound (XCV) (*Stewart et al.*, J. org. Chem., 1948, **13**, 134) is preferred for the preparation of the antihistamines, *cyclizine* (XCVI, R = H), m.p. 258–260°, and *chlorocyclizine* (XCVI, R = Cl), m.p. 223–224° (*K.E. Hamlin et al.*, J.

(XCV) → (XCVI)

Amer. chem. Soc., 1949, **71**, 2731). The convenient removal of benzyl groups by hydrogenolysis makes 1-*benzylpiperazine*, b.p. 145–147°/12 mm, useful for preparing mono- and di-substituted piperazines (*R. Baltzly et al., ibid.*, 1944, **66**, 263).

Reductive cyclisation of the dinitriles (XCVII) affords a series of *N*-mono-alkylated and -acylated piperazines (*H.S. Mosher et al.*, J. Amer. chem. Soc., 1953, **75**, 4949). Cyclisation of the substituted ethylenediamine (XCVIII) with elimination of an ester group affords the 1,4-disubstituted piperazine which is hydrolysed to the mono-substituted compound (*K. Nakajima*, Bull. chem. Soc. Japan, 1961, **34**, 651; 655).

(XCVII) (XCVIII)

Derivatives of *N*-monosubstituted piperazine include: 1-*methyl*-, b.p. 134–136° (*dihydrochloride monohydrate*, m.p. 83°; *benzoyl* deriv., m.p. 240°) (*Stewart et al., loc. cit.; Prelog* and *Stepan, loc. cit.*); 1-*ethyl*-, b.p. 155–158° (*dihydrochloride*, m.p. 203–205°) (*Moore et al., loc. cit.*); 1-*butyl*-, b.p. 186–192°/747 mm (*Hamlin et al., loc. cit.*); 1-*phenyl*-, b.p. 156–157°/10 mm, prepared by heating aniline and ethanolamine hydrochlorides (*C.B. Pollard* and *L.G. McDowell*, J. Amer. chem. Soc., 1934, **56**, 2199); 1-*β-hydroxyethyl*-, b.p. 125–128°/12 mm, from piperazine and ethylene oxide (*O. Hromatka* and *E. Engel*, Ber., 1943, **76**, 712) and 1-*methoxycarbonyl-piperazine*, b.p. 112–116°/7 mm (*Stewart et al., loc. cit.*); *piperazine-1-acetic* acid, m.p. 279°, and *β-1-piperazinopropionic acid*, m.p. 215° (*Moore et al., loc. cit.*).

Dialkylaminoalkyl piperazine-1-carboxylates are prepared from the *N*-ethoxycarbonyl derivative by ester exchange at 60° under reduced pressure, *e.g.* 2-*dimethylaminoethyl piperazine-1-carboxylate*, b.p. 114–117°/0.5–1 mm (*R.J. Turner*, U.S.P. 2617803, C.A., 1954, **48**, 2124).

Piperazines with heterocyclic substituents on nitrogen have been prepared (*K.E. Hamlin et al.*, J. Amer. chem. Soc., 1949, **71**, 2734; *K.L Howard et al.*, J. org. Chem., 1953, **18**, 1484). The physiological properties of 4-acyl and 4-aroyl derivatives of ethyl piperazine-1-carboxylate have been examined; some of the former are active as sedatives (*L. Goldman* and *J.H.*

Williams, *J. Amer. chem. Soc.*, 1954, **76**, 6078; see also D. Lednicer and L.A. Mitscher in "The Organic Chemistry of Drug Synthesis", Wiley, New York, 1977, p. 277).

Hetrazan, 1-diethylcarbamyl-4-methylpiperazine, m.p. 47–49°, b.p. 108.5–111°/3 mm, a useful drug for the treatment of filarias, is prepared either by methylation of 1-*diethylcarbamylpiperazine*, b.p. 113.5–115.5°/3 mm, with formaldehyde in formic acid, or from 1-methylpiperazine and diethylcarbamyl chloride in chloroform, a reaction which initially gives the *hydrochloride*, m.p. 156.5–157° (S. Kushner et al., *J. org. Chem.*, 1948, **13**, 144).

Benzyl 4-carbamyl-1-piperazinecarboxylates (CI), which have anti-convulsant properties, are prepared in a straightforward fashion from 1-*benzyloxycarbonylpiperazine* (XCIX), b.p. 144–146°/1 mm, *via* the *acid chloride* (C), b.p. 203–208°/1.3 mm (L. Goldman and J.H. Williams, *ibid.*, 1953, **18**, 815).

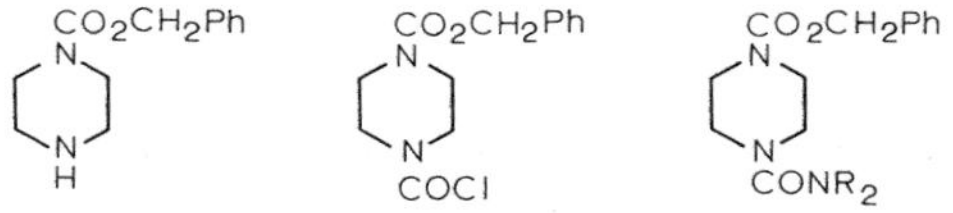

(XCIX)   (C)   (CI)

N,N'-Disubstituted piperazines are prepared from the corresponding N,N'-disubstituted ethylenediamines and 1,2-dibromides, *e.g.* with ethyl α,β-dibromopropionate gives 2-ethoxycarbonylpiperazines (CII); hydrogenolysis of (CII, R = CH₂Ph) gives **ethyl piperazinoate**, m.p. 59–61° (E. Jucker and E. Rizzi, *Helv.*, 1962, **45**, 2383). With the disodium salt of N,N'-ethyl-

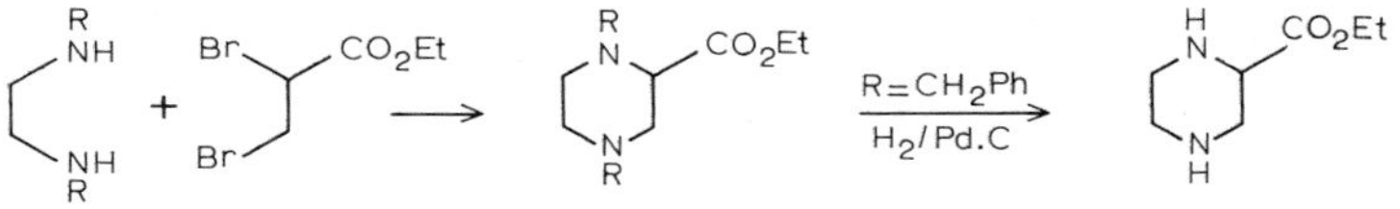

(CII)   (CIII)

enebis-*p*-toluenesulphonamide, the corresponding esters (CII, R = *p*-CH₃-C₆H₄-SO₂-) are obtained (J.R. Piper et al., *J. org. Chem.*, 1972, **37**, 4476). Heating of N,N'-dimethylethylenediamine with bis(diethylamino)ethylene at 120° affords the diethylaminopiperazine (CIV) which, on further heating at 170°, eliminates diethylamine to give 1,4-dimethyltetrahydropyrazine (CV) (A. Halleux and H.G. Viehe, *J. chem. Soc. C*, 1970, 881).

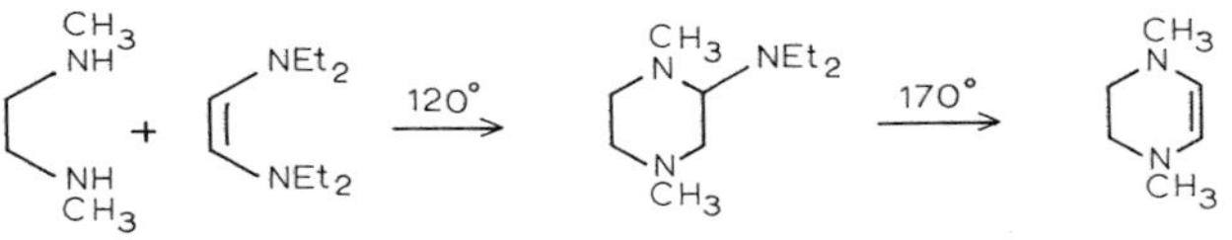

(CIV)   (CV)

N-Allyl-anilines are dimerised using mercury(II) acetate to give the organomercurial (CVI) which on reduction gives 1,4-diaryl-2,5-dimethylpiperazine (CVII) (J. Barluenga et al., *J. heterocyclic Chem.*, 1979, **16**, 1017).

1,4-Diphenylpiperazine has been prepared by the reaction of *N*-sulphinylaniline with ethylene carbonate in the presence of lithium bromide (*O. Tsuge et al.*, Bull. chem. Soc. Japan, 1967, **40**, 2709).

1,4-**Diazobicyclo**[2.2.2]**octane** (DABCO, triethylenediamine) (CVIII), m.p. 158–160°, an important industrial catalyst, was initially prepared in poor yield by intramolecular cyclisation of 1-β-hydroxyethylpiperazine and isolated as a 1:2 *p-nitrophenol* complex, m.p. 183° (*O. Hromatka* and *E. Engel*, Ber., 1943, **76**, 712). An improved laboratory procedure gives DABCO (*dihydrochloride*, m.p. 290° (dec); *dipicrate*, m.p. 298° (dec)) and the dimeric compound (CIX) (*tetrahydrochloride*, m.p. 291° (dec); *tetrapicrate*, m.p. 277°) (*F.G. Mann* and *D.P. Mukherjee*, J. chem. Soc., 1949, **22**, 98). (CVIII) is obtained commerically as a product of the reaction of diethanolamine and ethylenediamine (1:1) over an alumina–silica catalyst at 300–500° (*J.H. Kinse*, U.S.P. 2,977,363; C.A., 1961, **55**, 17664). DABCO forms stable complexes with lithium alkyls, catalyses the metallation of hydrocarbons such as toluene (*C.G. Screttas* and *J.F. Eastham*, J. Amer. chem. Soc., 1965, **87**, 3276), and is a useful reagent for cleavage of δ-keto-α,β-unsaturated esters and β-ketoesters (*D.H. Miles* and co-workers, J. org. Chem., 1974, **39**, 1592; 2647).

C. Stereochemistry of N,N'-disubstituted piperazines

Oxidation of 1,4-diphenylpiperazine with hydrogen peroxide affords a mixture of the *cis* and the *trans* 1,4-di-*N*-oxide, in a ratio of 1:10, reflecting a preference for oxidation at the axial position. The *trans-dioxide*, m.p. 281° (*picrate*, m.p. 221°) is obtained as the octahydrate whereas the *cis*-isomer forms a complex with hydrogen peroxide, $C_{16}H_{18}O_2N_2 \cdot H_2O_2 \cdot H_2O$, m.p. 210° (*G.M. Bennett* and *E. Glynn*, J. chem. Soc., 1950, 211).

Reaction of 1,4-dimethylpiperazine with ethylenechlorohydrin gives mainly the monoquaternary chloride (CX) together with two isomeric diquaternary salts (CXI) and (CXII), the *cis*- and the *trans*-isomers, respectively. Although the *cis* compound (CXI), m.p. 241° (*picrate*, m.p. 220°) can be converted into the corresponding diacetyl and dichloro derivatives by treatment with acetyl chloride and thionyl chloride, respectively, the *trans* compound is unreactive under similar conditions (*W.E. Hanby* and *H.N. Rydon, ibid.*, 1945, 833).

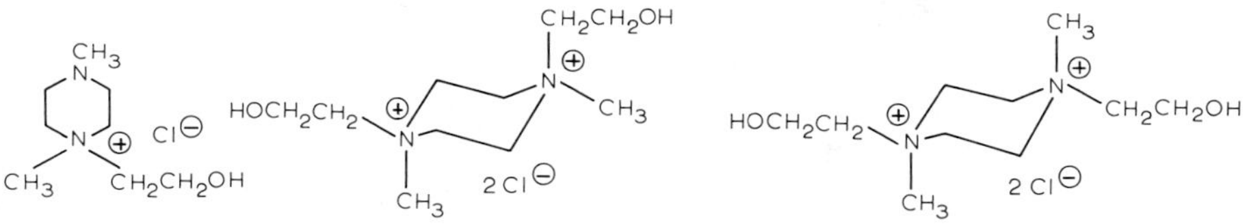

(CX)   (CXI)   (CXII)

D. Piperazines with substituents on carbon

Piperazine homologues are prepared by reduction of the corresponding pyrazines, or pyrazinium salts. With increasing substitution, however, the homologues can exist in an increasing number of stereoisomeric forms. Since the reduction procedure usually give rise to mixtures of stereoisomers, specific forms are best obtained by direct synthesis, *e.g.*, for optically active forms of 2-methyl-1,4-diphenylpiperazine (*F.B. Kipping* and *W.J. Pope*, J. chem. Soc., 1926, 2396).

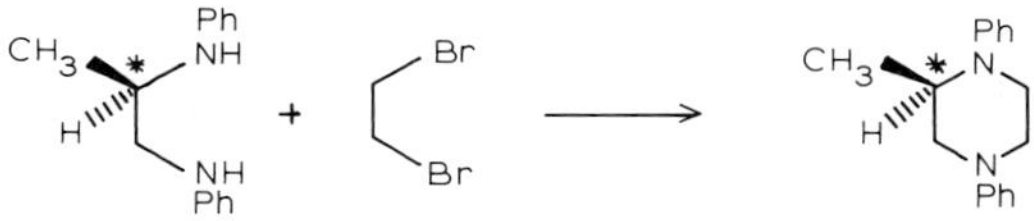

2-Methyl- and 2-phenyl-piperazine are also readily prepared from the diamine (CXIII, R = CH_3, Ph) (*L.J. Kitchen* and *C.B. Pollard*, J. Amer. chem. Soc., 1947, **69**, 854).

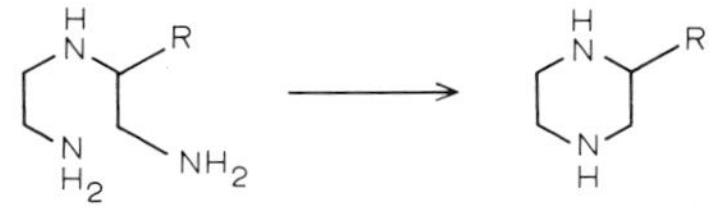

(CXIII)

2-*Methylpiperazine*, m.p. 62° (*dihydrochloride*, m.p. 248–249°), has been resolved (*C. Stoehr*, J. pr. Chem., 1895, **51**, 4499) while 2,5-*dimethylpiperazine* exists in a *cis*(*rac.*)-form, m.p. 114–115° (1,4-*dibenzoyl*, m.p. 147–148°, 1,4-*dinitroso* deriv., m.p. 205°), and a *trans*(*meso*)-form, m.p. 118–119° (1,4-*dibenzoyl*-, m.p. 224–225°, 1,4-*dinitroso* deriv., m.p. 205°), the latter isomer having a centre of symmetry (*C. Stoehr, ibid.*, 1897, **55**, 49; *F.B. Kipping* and *W.J. Pope*, J. chem. Soc., 1926, 1076). Reduction of tetramethylpyrazine by a variety of standard methods affords 2,3,5,6-**tetramethylpiperazine** as a mixture of isomers, of which five have been characterised; the α-, β-, γ-, δ- and ε-forms, m.p. 45°, 183° (b.p.), 67–68°, 53–55°, and 60°, respectively (*F.B. Kipping, ibid.*, 1929, 2889; 1937, 368).

2,2,5,5-**Tetramethylpiperazine**, m.p. 85–87°, b.p. 168–171°, is prepared by reduction of the corresponding dihydro compound with sodium in alcohol (*J.B. Connant* and *J.G. Aston*, J. Amer. chem. Soc., 1928, **50**, 2783), or by catalytic cyclodehydraton of 2-amino-2-methyl-1-propanol (*S.M. McElvain* and *L.W. Bannister, ibid.*, 1954, **76**, 1126); derivatives include 1,4-*dinitroso*, m.p. 210–212°, 1,4-*diethoxycarboxy*, m.p. 38–39°, 1,4-*dibenzoyl*, m.p. 270–272°. The latter compound is reduced with excess lithium aluminium hydride to the 1,4-dibenzyl derivative (*Clapp et al., loc. cit.*).

2-**Phenylpiperazine**, m.p. 88°, b.p. 138°/10 mm (*dihydrochloride*, m.p. 335° (dec), *picrate*, 276°) (*Kitchen* and *Pollard, loc. cit.*). Five of the possible isomers of 2,3,5,6-*tetraphenylpiperazine* have been separated by fractional crystallisation and characterised; α-, m.p. 161–162°, β-, m.p. 209.5–210.5°, γ-, m.p. 266–268°, δ-, m.p. 291–292°, and ε-form, m.p. 300–302° (*T. Hayashi*, C.A., 1955, **49**, 1050).

2,3-*Diphenylpiperazine* occurs in two forms, α, m.p. 122–123°, and β, m.p. 108–109°. Potassium cyanide with 2,3-diphenyldihydropyrazine gives 2,3-**dicyano**-2,3-**diphenylpiperazine**, m.p. 203–204° (*P. Beak* and *J. Yamamoto*, J. heterocyclic Chem., 1972, **9**, 155). 2,2,5,5-Tetramethyldihydropyrazine adds two equivalents of hydrogen cyanide affording 3,6-**dicyano**-2,2,5,5-**tetramethylpiperazine**, m.p. 193–194.5° (*dinitroso* deriv., m.p. 178°).

(iv) Piperazinones (ketopiperazines)

Piperazinone, m.p. 136° (*picrate*, m.p. 180°; *hydrochloride*, m.p. 208°), is prepared from ethyl chloroacetate and excess ethylenediamine followed by cyclisation of the resulting intermediate (CXIV) by heating under reduced pressure.

(CXIV)

The derivatives, 3-*phenyl*-, m.p. 139–139.5° (*W.R. Rodderick et al.*, J. med. Chem., 1966, **9**, 181), 3,3-*dimethyl*-, m.p. 134° (*S.R. Aspinall*, J. Amer. chem. Soc., 1940, **62**, 1202), 1,4-*diphenyl-piperazinone*, m.p. 148° (*C. Bischoff et al.*, Ber., 1892, **25**, 2931; 2942) are similarly prepared. The preparation of a series of 1,4-dialkyl-2-piperazinones has been reported (*W.B. Martin* and *A.E. Martell*, J. Amer. chem. Soc., 1950, **72**, 4301).

Condensation of the unsymmetrical 1-substituted ethylenediamines and ethyl α-bromopropionates affords 3,5-disubstituted piperazinones (CXV) as the major isomer, *e.g.* (CXV, $R^1 = R^2 = Ph$), m.p. 170–171.5° and (CXV, $R^1 = CH_3$, $R^2 = H$), b.p. 129–132°/2 mm (*K. Masuzawa et al.*, Bull. chem.

(CXV)

Soc. Japan, 1965, **38**, 2078). A similar condensation between *N*-benzyl-1-methylethylenediamine and α-bromopropionic acid gives 1-benzyl-3,5-dimethylpiperazinone as a mixture of the *cis*- and the *trans*-isomer; lithium aluminium hydride reduction yields the corresponding piperazine (*G. Cignarella* and *G.G. Gallo*, J. heterocyclic Chem., 1974, **11**, 985).

Reductive cyclisation of the acyclic adduct from the reaction of an α-chloro oxime with an α-amino-acid ester affords a 3,6-disubstituted piperazinone, *e.g.* (CXVI), m.p. 106° (*M. Masaki* and *M. Ohta*, Bull. chem. Soc. Japan, 1963, **36**, 922).

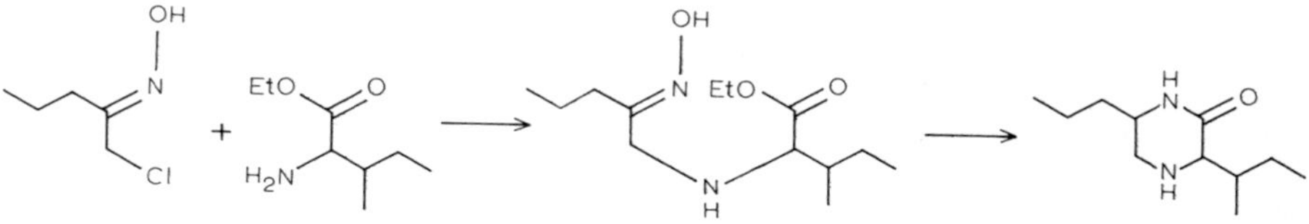

(CXVI)

The reaction of a trisubstituted ethylenediamine, a ketone (or the corresponding cyanohydrin), and chloroform with sodium hydroxide under phase transfer catalysis produces a separable mixture of the isomeric

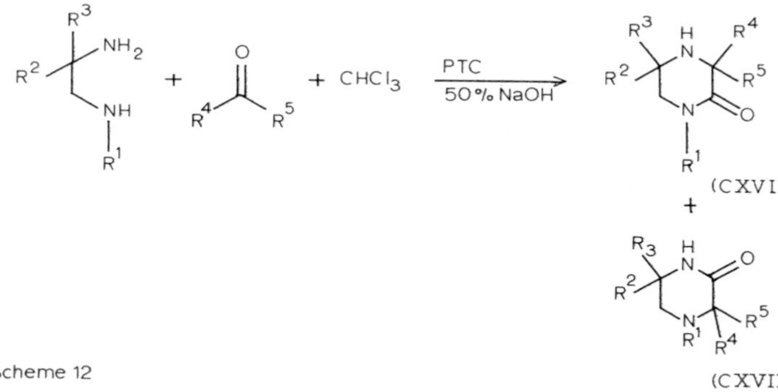

Scheme 12

(CXVII)
+
(CXVIII)

piperazinones (CXVII) and (CXVIII) [Scheme 12]. Compounds (CXVII) are oxidised using *m*-chloroperbenzoic acid to nitrosyl radicals which are useful polymer stabilisers and radical spin-traps; reduction of a piperazinone with lithium aluminium hydride gives a piperazines (*J.T. Lai*, J. org. Chem., 1980, **45**, 754; Synthesis, 1981, 40).

Piperazinones behave as simple amides, being hydrolysable to amino acids which cyclise readily. Derivatives bearing *N*-substituents can be oxidised first to the 2,3-di- and, on further treatment, to the 2,3,5,6-tetra-piperazinone.

3,3-Dimethylpiperazin-2-one undergoes *N*-alkylation only with aldehydes in formic acid by the Leuckhart procedure; 3,3,4-**trimethyl**-, m.p. 131–132°, and **4-ethyl**-3,3-**dimethylpiperazinone**, m.p. 164–165°, are obtained, respectively (*P. Ruby* and *P.L. de Benneville*, J. Amer. chem. Soc., 1953, **75**, 3027).

Piperazine-2,3-**dione**, m.p. 258°, is prepared from ethylenediamine and oxalic acid (*idem, ibid.*), or diamide (*C.E. Goulding* and *C.B. Pollard, ibid.*, 1948, **70**, 1967), or diester (*J.L. Riebsomer*, J. org. Chem., 1950, **15**, 68). The 1,4-**diphenyl**, m.p. 148°, and 1,4-**dimethoxycarbonyl**, m.p. 167°, derivatives are prepared either as above or by chromic acid oxidation of the appropriate pyrazinone. The 1,4-diphenyl compound has also been obtained by catalytic hydrogenation of 1,4-diphenyl-1,2,3,4-tetrahydropyrazine-2,3-dione (*J. Honzl*, Coll. Czech. Chem. Comm., 1960, **25**, 2651). Reaction of DISN with oxalyl chloride gives 5,6-dichloro-5,6-dicyanopiperazine-2,3-dione (CXIX) which is dechlorinated on treatment with ethanethiol to yield 5,6-**dicyanopiperazine**-2,3-**dione** (CXX), m.p. 268° (*R.W. Begland* and *D.R. Hartter*, J. org. Chem., 1972, **37**, 4136).

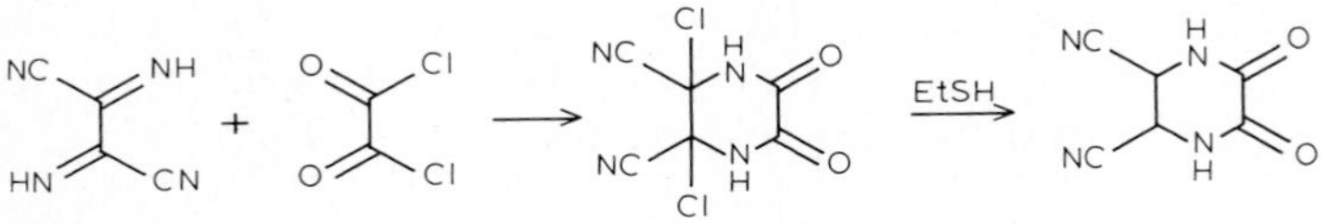

Piperazine-2,5-diones may be formally regarded as the cyclic dimeric anhydrides of α-amino-acids. The piperazine-2,5-dione moiety is found as a substructural unit of a range of antibiotics, *e.g.* bicyclomycin (CXXI) which is active against Gram-negative bacteria. An X-ray structure determination of (CXXI) (*Y. Tokuma et al.*, Bull. chem. Soc. Japan, 1974, **47**, 18), syntheses of the bicyclic unit (*T. Fukuyama et al.*, Tetrahedron Letters, 1981, **22**, 4155; *R.M. Williams, ibid.*, 1981, **22**, 2341) and a trimethyl derivative of (CXXI) (*S.-I. Nakatsuka et al., ibid.*, 1981, **22**, 4973) have been reported. For more detailed discussion of the naturally occurring piperazine-2,5-diones see *P.G. Sammes* (Fortschr. Chem. Org. Naturst., 1975, **32**, 51).

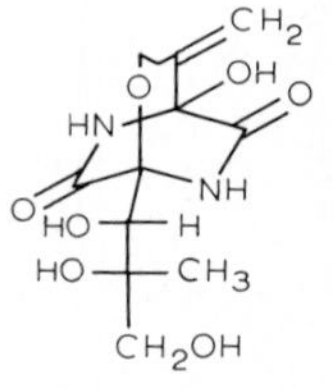

More generally, symmetrically substituted piperazine-2,5-diones are prepared (i) by dehydration of α-amino-acids in refluxing solvents such as ethylene- and diethylene-glycol, *e.g.* α-amino isobutyric acid in diethyleneglycol gives 3,3,6,6-**tetramethylpiperazine**-2,5-**dione** (CXXII), m.p. > 400°, which is reduced using sodium in ethanol or over a copper–chromium catalyst to the corresponding piperazine (*C. Sannie*, Bull. Soc. chim. Fr., 1942, **9**, 487; *S.M. McElvain* and *E.H. Pryde*, J. Amer. chem. Soc., 1949,

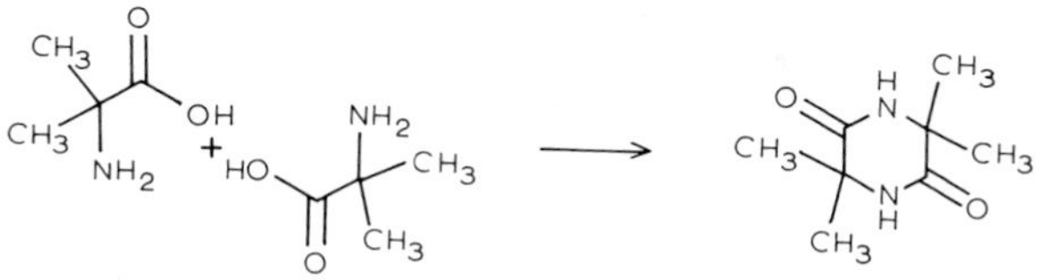

(CXXII)

71, 326); (ii) by stirring α-halocarboxamides with an ion-exchange resin in a two-phase system, a process which produces piperazinediones in good yields, *e.g.* 1,4-**dibenzylpiperazine**-2,5-**dione** (CXXIII), m.p. 174–175° (*T.*

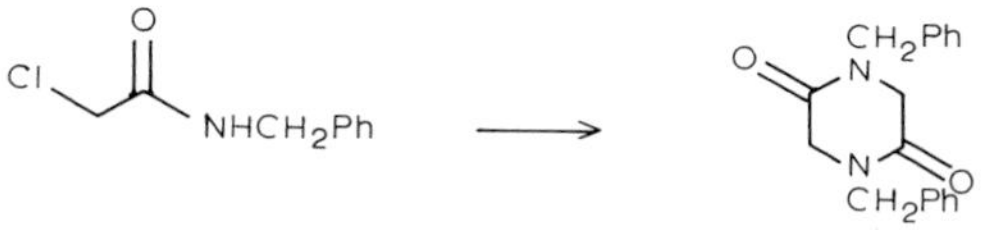

(CXXIII)

Okawara et al., Chem. Letters, 1981, 185); (iii) by reduction of a 2-bromo-5-hydroxypyrazine with zinc dust in acetic acid to produce a piperazine-2,5-dione directly (*K. Tatsuta et al.*, J. Antibiot., 1972, **25**, 674; 1973, **26**, 606).

N-Protected dipeptide esters can be cyclised without racemisation to give optically pure piperazine-2,5-diones by an initial deprotection in formic acid or 4 *M* hydrochloric acid followed by *in situ* cyclisation (*D.E. Nitecki*

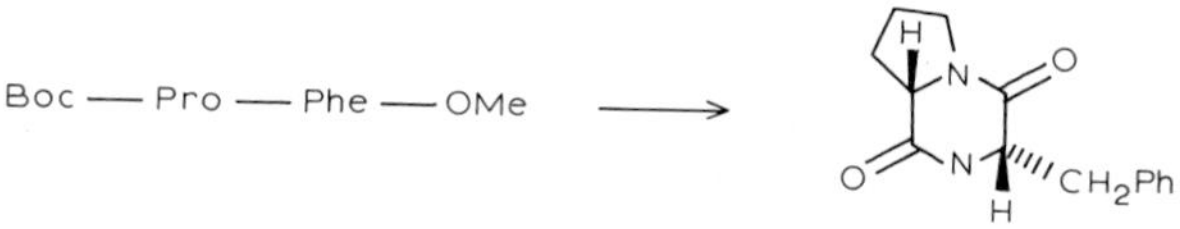

Boc — Pro — Phe — OMe ⟶

et al., J. org. Chem., 1968, **33**, 864; *K. Suzuki et al.*, Chem. pharm. Bull. Japan, 1981, **29**, 233).

With phosphoryl chloride, the piperazine-2,5-diones yield mixtures of the 2-chloro- and the 2,5-dichloro-pyrazine, however in the presence of dimethylaniline, only the monochloro compound is obtained (*F.S. Spring et al.*, J.

chem. Soc., 1947, 1179; 1183). The reaction of (±)-phenylglycine amhydride with phosphorus tribromide gives 2-bromo-3,6-diphenylpyrazine (*G. Karmas and P.E. Spoerri*, J. Amer. chem. Soc., 1956, **78**, 2141). Bromination of 1,4-diphenylpiperazine-2,5-dione with phosphorus pentabromide in *o*-dichlorobenzene at 140–150° affords the corresponding 3,6-dibromo derivative (*J. Honzl*, Coll. Czech. Chem. Comm., 1960, **25**, 2651). With *N*-bromosuccinimide, the tetrasubstituted piperazinedione (CXXIV) is brominated in both the ring and the side-chain to give the tetrabromide (CXXV) which has been elaborated to the *epi*-dithiopiperazine-2,5-dione (CXXVI), a substructural unit in several classes of antibiotic (*K. Matsunari et al.*, Bull. chem. Soc. Japan, 1975, **42**, 605).

Alkylation of either the silver or the sodium salt of a piperazine-2,5-dione usually gives the N,N'-dialkyl derivative, except that treatment of the silver salt of the parent compound with benzyl chloride produces the O,O'-dibenzyl compound (*P. Karrer et al.*, Helv., 1927, **5**, 140; *C. Granacher et al., ibid.*, 1928, **11**, 1228). The use of triethyloxonium fluoroborate affords the corresponding O,O'-dialkylated products (*P.G. Sammes* and co-workers, J. chem. Soc. Perkin I, 1972, 2494).

Owing to the presence of the carbonyl groups, the adjacent methylene groups are activated and piperazine-2,5-diones derived from optically active α-amino-acids can be readily racemised. Although there is no evidence for the formation of the enol form from piperazine-2,5-dione in 0.1 *M* sodium hydroxide solution (*H. Lenormant*, Bull. Soc. chim. Fr., 1948, 33), enolic forms are reported to have been obtained by heating piperazine-2,5-diones in aniline at 200° or glycerol at 180° (*E. Aberhalden et al.*, Z. physiol. Chem., 1925, **149**, 100, 298; 1926, **152**, 88; **153**, 83; **157**, 140). I.r. studies are generally consistent with the presence of a *cis*-amide bond in piperazine-2,5-dione (*K. Blaha et al.*, Coll. Czech. Chem. Comm., 1966, **31**, 4296).

Piperazine-2,5-dione condenses with benzaldehyde in acetic acid/acetic anhydride to yield the dibenzylidene derivative (CXXVII), m.p. 298–300° (*T. Sasaki*, Ber., 1921, **54B**, 163). Similarly, the methylpiperazinedione (CXXVIII) condenses with benzaldehyde in acetic anhydride giving the

(CXXVII)

benzylidene derivative (CXXIX) which on irradiation isomerises to (CXXX). The benzylidene compound (CXXXI) is prepared directly by the con-

(CXXVIII) (CXXIX) (CXXX)

(CXXXI)

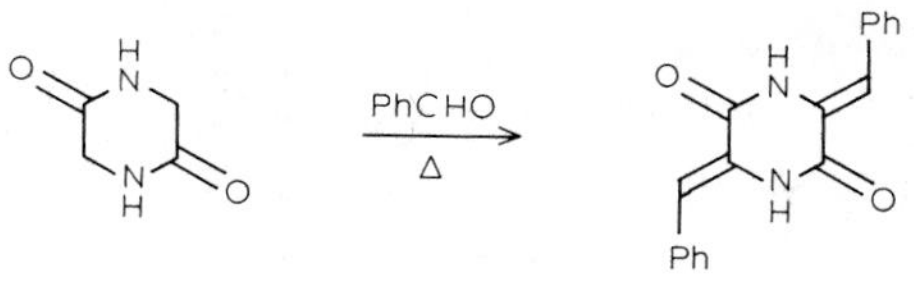

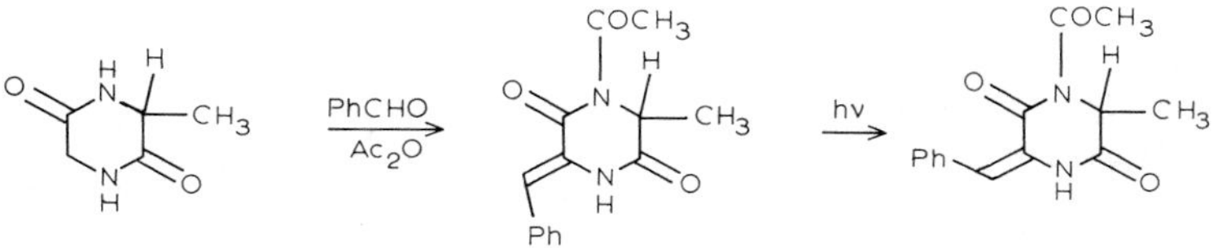

densation of *N*-dichloroacetyl phenylalanine with methylamine (*P.G. Sammes et al.*, J. chem. Soc. C, 1970, 980; 2530).

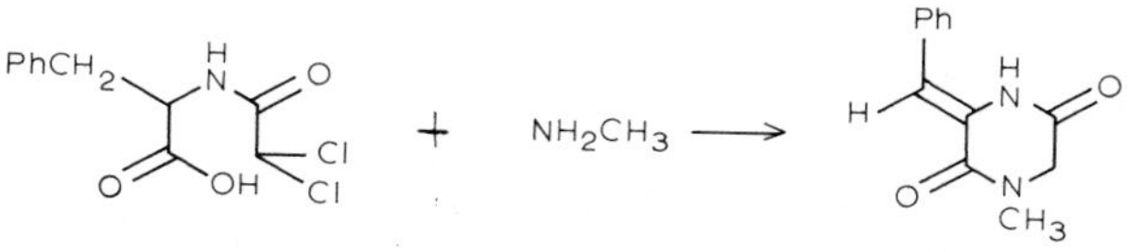

Treatment of 1,4-diacetyl-3,6-dibenzylpiperazine-2,5-dione with sulphur in dimethylformamide followed by hydrolysis affords the monobenzylidene derivative (CXXXII) (*P.J. Machin* and *P.G. Sammes*, J. chem. Soc. Perkin I, 1976, 624).

(CXXXII)

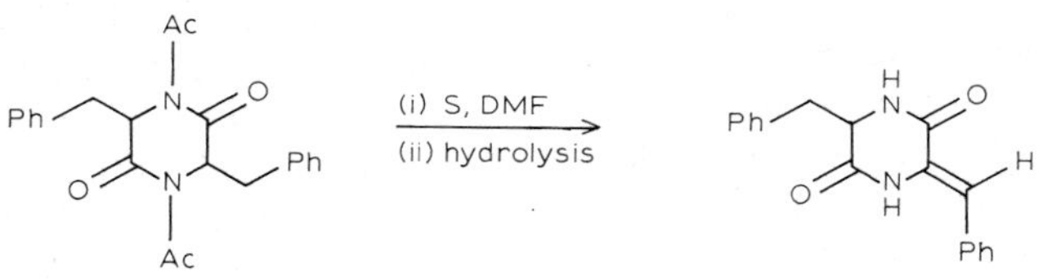

Piperazine-2,5-dithione, m.p. > 270 ° (dec), is obtained from aminoacetonitrile and hydrogen sulphide in ammoniacal solution (*E.S. Gatewood* and *T.B. Johnson*, J. Amer. chem. Soc., 1926, **48**, 2900).

Piperazine-2,6-dione, decomp. *ca.* 260°, is prepared from iminodiacetic acid by heating the ammonium salt or the mono-amide under reduced pressure, or by heating the diacid with urea (*J.V. Dubsky*, Ber., 1916, **49**, 1039). Treatment of *N*-alkyliminodiacetic acid mono-amides with acetic anhydride produces the corresponding piperazine-2,6-diones (*D.W. Henry*, J. heterocyclic Chem., 1966, **3**, 503; *M. Sorm* and *J. Honzl*, Tetrahedron, 1972, **28**, 203). Heating *N*-methyliminodiacetic acid and urea at 160–170° yields **4-methylpiperazine-2,6-dione**, m.p. 103–104°, which is benzylated with benzyl chloride and potassium carbonate in acetone to give 1-**benzyl**-4-**methylpiperazine**-2,6-**dione**, m.p. 165.5–166.5° (*I.O. Hromatta* and *H. Schramek*, Monatsh., 1961, **92**, 1242).

Cylisation of the glycine ester derivative (CXXXIII) using sodium hydride in xylene gives 3-**phenylpiperazine**-2,6-**dione**, m.p. 121–124° (*W.R. Rodderick et al.*, J. med. Chem., 1966, **9**, 181).

(CXXXIII)

The aminonitrile (CXXXIV) with potassium hydroxide in refluxing methanol is converted into the tetramethyl compound (CXXXV), m.p. 236–238° (*F. Yoshioka et al.*, Bull. chem. Soc. Japan, 1972, **42**, 1855).

(CXXXIV)   (CXXXV)

Piperazine-2,6-diones are reduced by excess lithium aluminium hydride to piperazines (*B.H. Chase* and *A.M. Downes*. J. Amer. chem. Soc., 1953, **54**, 3874). Oxidation of 1,4-diphenylpiperazine-2,6-dione with selenium dioxide in dioxane affords the corresponding tetraone, m.p. 290–295° (*T. Tanaka et al.*, Bull. chem. Soc. Japan, 1977, **50**, 1821).

Piperazinetrione, m.p. 240–424°, is prepared from the condensation of aminoacetamide and diethyl oxalate in methanolic sodium methoxide (*G. Palamidessi et al.*, C.A., 1967, **66**, 37885).

Piperazinetetraone, decomp. > 250°, is obtained by the self-condensation of two molecules of ethyl oxamate and isolated as the monosodium salt (*A.T. de Mouilpied* and *A. Rule*, J. chem. Soc., 1909, **95**, 549). *N*-Substituted derivatives are obtained by the reaction of substituted oxamides with oxalyl chloride (*J.V. Dubsky* and *F. Blumer*, Ber., 1919, **52**, 215). Treatment of oxanilic acid with thionyl chloride gives the 1,4-diphenyl compound (CXXXVI) which can also be obtained by oxidation of the corresponding 2,5- or 2,6-dione (*D. Buckley* and *H.B. Henbest*, J. chem. Soc., 1956, 1888; *T. Tanaka et al., loc. cit.*).

(CXXXVI)

2. Quinoxalines (benzopyrazines)

(a) Introduction

The one possible benzopyrazine is known as quinoxaline (I), a name which suggests a formal derivation from quinoline by substitution of a nitrogen atom for a methine group at the 4-position.

(I) (II)

There are comparatively few natural products which contain the quinoxaline ring system. The 1,4-di-N-oxide of quinoxaline-2-carboxylic acid (II), isolated from *Streptomyces ambrofaciens*, and its esters exhibit antibiotic activity (*A.S. Elma et al.*, Chem. heterocyclic Comp., 1969, **5**, 540; *G.W.H. Cheeseman* and *E.S.G. Westeriuk*, Adv. heterocyclic Chem., 1978, **22**, 367). There is also a family of quinoxaline-peptide antibiotics, including echinomycin (*A. Dell et al.*, J. Amer. chem. Soc., 1975, **97**, 2497), levomycin and actinoleutin (*K. Sato, O. Shiratori* and *K. Katagiri*, J. Antibiot., 1967, **A20**, 270) which contain one or more pendant quinoxaline substituents. Synthetic quinoxalines find a range of uses as fungicides, insecticides and anthelmintics (see *G.W.H. Cheeseman* and *R.F. Cookson*, in "Condensed Pyrazines", Chemistry of Heterocyclic Compounds, Vol. 35, Wiley, New York, 1979).

(b) Physical and spectroscopic properties

MO calculations on quinoxaline indicate that the 2(3)-positions adjacent to the nitrogen atoms are slightly electron deficient and the 5(8)-positions in the carbocyclic ring are slightly electron rich, consistent with the normal position for electrophillic substitution reactions (*R.J. Pugmire et al.*, J. Amer. chem. Soc., 1969, **91**, 6381). Like pyrazines, quinoxalines are weak bases (pK_{a1} 0.56; pK_{a2} −5.52) (*A.R. Katritzky et al.*, J. chem. Soc., 1967, 1233; *A. Albert*, in "Physical Methods in Physical Organic Chemistry", ed. *A.R. Katritzky*, Academic Press, New York, 1963). The basicities of a number of 5- and 6-substituted 2,3-dimethylquinoxalines have been measured spectroscopically (*P. Vetesnik* and co-workers, C.A., 1967, **67**, 63617; Coll. Czech. Chem. Comm., 1968, **33**, 556; 566).

The u.v. spectrum of quinoxaline in cyclohexane exhibits three principal absorption bands at 340 (log ϵ 2.84) [$n-\pi^*$], 312 (log ϵ 3.81) [$\pi-\pi^*$] and 232 (log ϵ 4.51) [$\pi-\pi^*$] nm; in protic solvents the weaker $n-\pi^*$ transition becomes obscured by a long wavelength $\pi-\pi^*$ band (*Cheeseman* and *Cookson*, loc. cit.).

Infrared spectra of quinoxaline and its substituted derivatives contain eight ring stretching bands in the region 1620–1350 cm^{-1} which vary little with the nature of the substituents. Absorptions in the region 1300–1000 cm^{-1} are assigned as C–H bending modes (*G.W.H. Cheeseman et al.*, J. chem. Soc., 1963, 3764).

In the ^{1}H-n.m.r. spectrum of quinoxaline (CCl$_4$ solution), the α-hydrogens H-2 and H-3 appear as a low field singlet (δ 8.74) and the hydrogens of the benzenoid ring appear as an AA′BB′ system, H-6(H-7) at δ 7.68 and H-5(H-8) at δ 8.07. The signals for H-6(H-7) are broadened owing to ^6J coupling with H-2(H-3) (*P.J. Black* and *M.L. Heffernan*, Austral. J. Chem., 1965, **18**, 707). The ^{1}H-n.m.r. spectra of a series of quinoxaline derivatives have been compiled (see *T.J. Batterham*, "NMR Spectra of Simple Heterocycles", Wiley, New York, 1973). Consistent with the symmetry of the quinoxaline molecule, the ^{13}C-n.m.r. spectrum in CDCl$_3$ consists of four signals at 129.4 (C-6, 7), 129.6 (C-5, 8), 142.8 (C-9, 10) and 144.8 (C-2, 3) ppm (*Pugmire et al.*, loc. cit.).

In the mass spectrum of a quinoxaline, the molecular ion is usually the base peak. Fragmentation of quinoxaline arises by sequential loss of hydrogen cyanide to give ions at m/z 103 and 76 [Scheme 1] (*A. Karjalamen* and *H. Krieger*, Suom. Kemistil. B, 1970, **43**, 273). Details of mass spectra of several quinoxaline derivatives have been discussed (*Q.N. Porter*, "Mass Spectrometry of Heterocyclic Compounds", 2nd Edn., Wiley, New York, 1985).

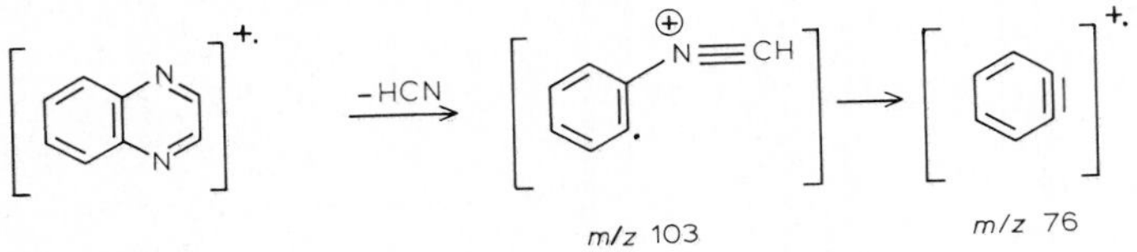

Scheme 1

(c) Methods of synthesis

Compounds containing the quinoxaline ring system may be prepared by one or more of the following general methods.

(1) The condensation of an *o*-phenylenediamine with α,β-dicarbonyl compound is known as the Hinsberg reaction (*O. Hinsberg*, Ann., 1887, **237**, 327), e.g. *o*-phenylenediamine with glyoxal, benzil and 2,3-dihydroxy-

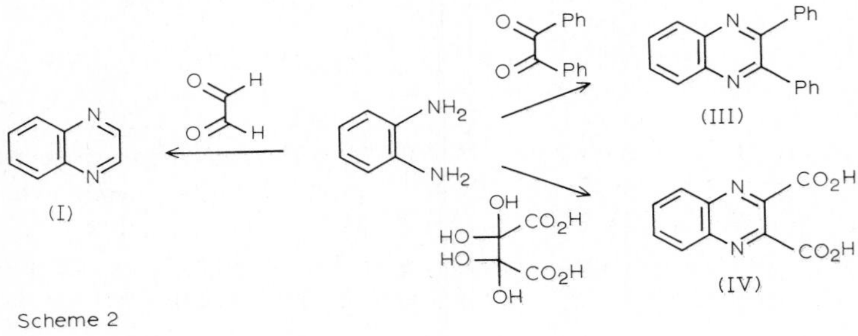

Scheme 2

tartaric acid gives quinoxaline (I), 2,3-diphenylquinoxaline (III) and quinoxaline-2,3-dicarboxylic acid (IV), respectively [Scheme 2] (*J.H. Billman* and *J.L. Rendall*, J. Amer. chem. Soc., 1944, **66**, 540; *J.H. Boyer* and *D. Straw, ibid.*, 1953, **75**, 1642; *R.A. Baxter* and *F.S. Spring*, J. chem. Soc., 1945, 229). 2,3-Dihydroxydioxane is reported to be a convenient non-aqueous substitute for glyoxal (*M.C. Venuti*, Synthesis, 1982, 61). When both precursors are unsymmetrical, mixtures of the corresponding isomeric quinoxalines are invariably obtained.

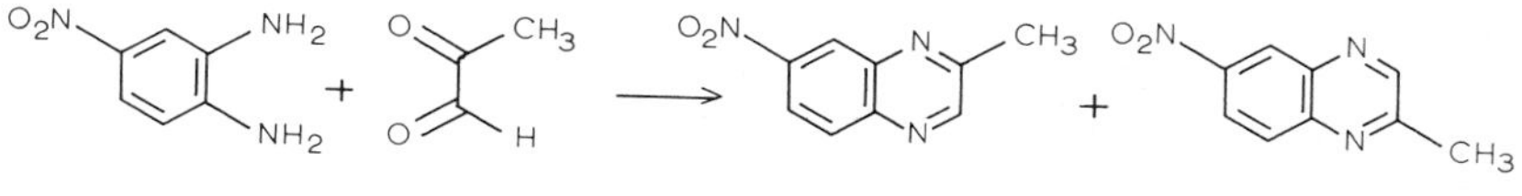

Reaction of *o*-phenylenediamines with oxalic acid (or esters) and α-keto-acids such as pyruvic acid affords 2,3-dihydroxy- and monohydroxy-quinoxalines (VII) and (VIII), respectively (*O. Hinsberg* and *J. Pollak*, Ber., 1896, **29**, 784; *D.C. Morrison*, J. Amer. chem. Soc., 1954, **76**, 4483).

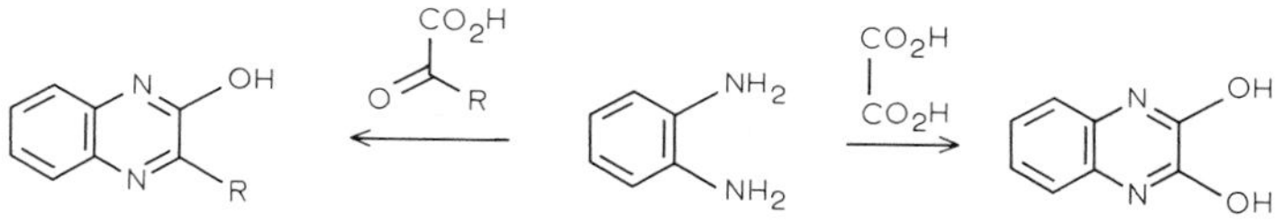

Although mixtures of isomers are obtained when unsymmetrically substituted reactants are used, as mentioned above (see *C.M. Atkinson et al.*, J. chem. Soc., 1956, 26), the condensation of 2-(methylamino)-5-nitroaniline with butyl glyoxylate gives the quinoxalinone (IX) as a single isomer (*G.W.H. Cheeseman, ibid.*, 1961, 1246).

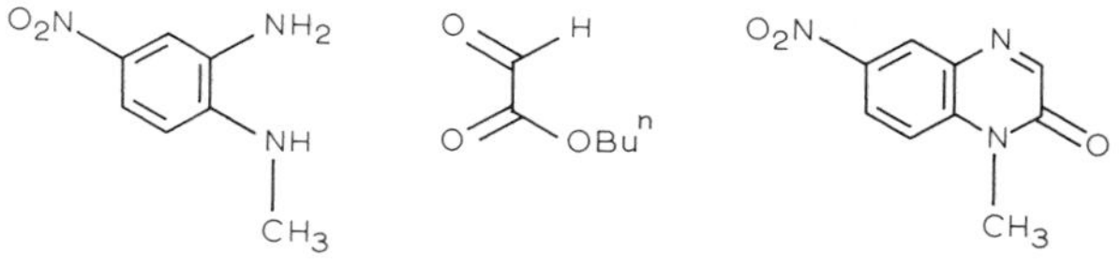

Treatment of *N,N*-dimethyl-*o*-phenylenediamines with alloxan affords the unexpected tetrahydroquinoxaline spiran (X); the mechanism of this transformation has been the subject of some controversy (*J.W. Clark-Lewis* and co-workers, Austral. J. Chem., 1964, **17**, 877; 1965, **18**, 907; 1970, **23**, 1249; *O. Meth-Cohn et al.*, J. Chem. Soc. C, 1969, 1438).

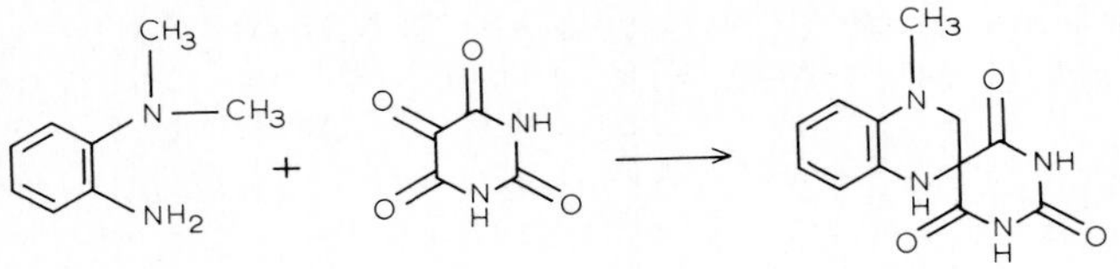

(X)

(2) o-Phenylenediamines react with a range of α,β-dicarbonyl compound equivalents including those discussed below.

(i) α-Halocarbonyl compounds give reduced quinoxalines which are oxidised to the corresponding quinoxalines. The reaction of phenacyl bromide with *o*-phenylenediamine has been shown to be a two-step process (*D.M. Aleksandrova et al.*, Zhur. org. Khim., 1973, **9**, 2107). By analogy,

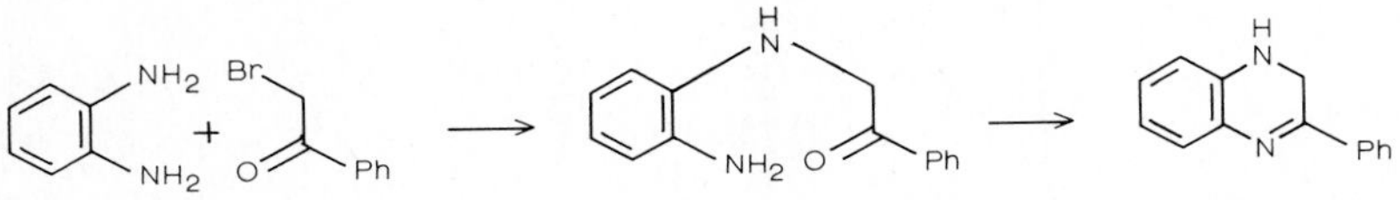

ethyl α-chlorophenylacetate in the presence of triethylamine affords a tetrahydroquinoxalinone which is oxidised using potassium permanganate or selenious acid to the corresponding quinoxalinone (*H. Zellner*, C.A., 1969, **71**, 70642).

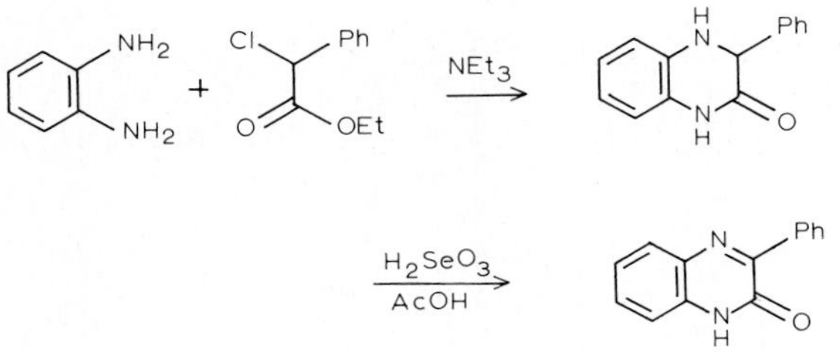

Other compounds including α-formyl and α-keto-alcohols, *e.g.* sugars such as glucose (*H. Ohle* and *M. Hielscher*, Ber., 1941, **74**, 13), condense with *o*-phenylenediamine in an similar fashion.

With mono-*N*-substituted *o*-phenylenediamines, the resulting dihydroquinoxalines are isolable and can be subsequently oxidised by iron(III) chloride to the corresponding quaternary salts (*O.N. Witt*, Ber., 1887, **20**, 1183; *K. Brand* and *O.N. Wild*, ibid., 1923, **56**, 105).

(ii) ω-Nitroacetophenones in the presence of sodium dithionite give

2-phenylquinoxaline derivatives, *e.g.* R = H, yield 48% (*A. Dornow* and *W. Sassenberg*, Ann., 1955, **594**, 185).

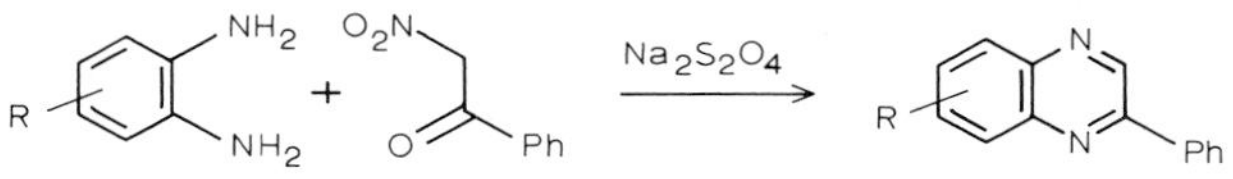

(iii) β-Ketosulphoxides produce mono- and 2,3-di-substituted quinoxalines; yields of the former compounds are generally compartively higher (*S. Kano et al.*, Synthesis, 1978, 372).

(iv) Oxazolinones derived from hexafluoroacetone and α-amino-acids yield quinoxalinones (*K. Burger* and *M. Eggersdorfer*, Ann., 1979, 1547).

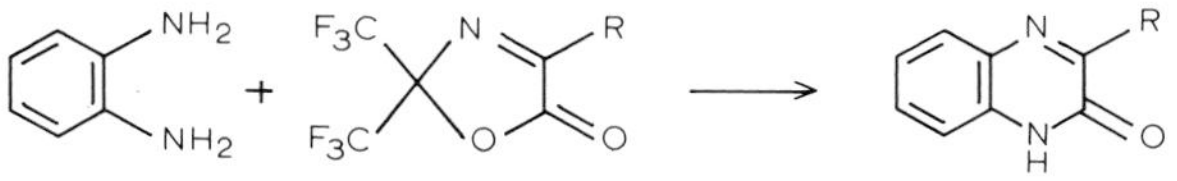

(v) Imidoyl chloride under acidic conditions gives 3-methylamino-quinoxalinones (*D. Bartholomew* and *I.T. Kay*, Tetrahedron Letters, 1979, 2827).

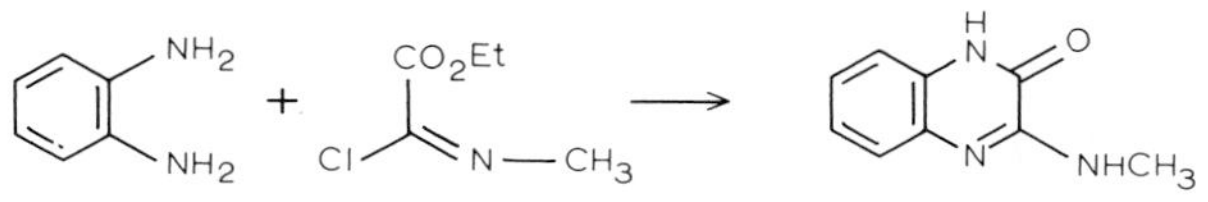

(vi) Styrene oxides in hot dimethylformamide give either phenylquinoxalines when X = *p*-tosyl, or cyanotetrahydroquinoxalines when X = CN [Scheme 3] (*E.C. Taylor et al.*, J. org. Chem., 1980, **45**, 2512).

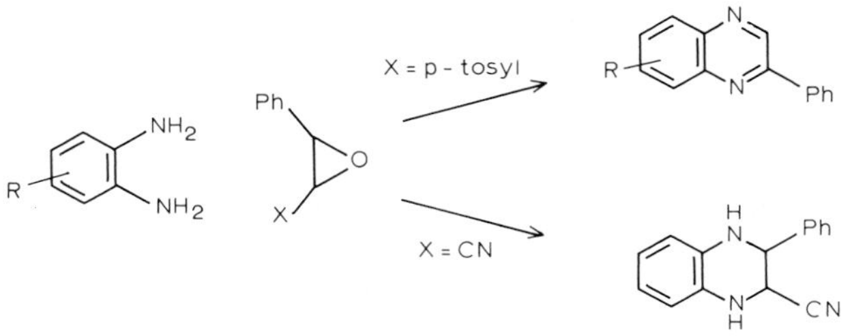

Scheme 3

(vii) Ketones and chloroform in 50% aqueous alkali under phase transfer conditions produce 3,3-dialkyltetrahydroquinoxalinones (*J.T. Lai*, Synthesis, 1982, 71).

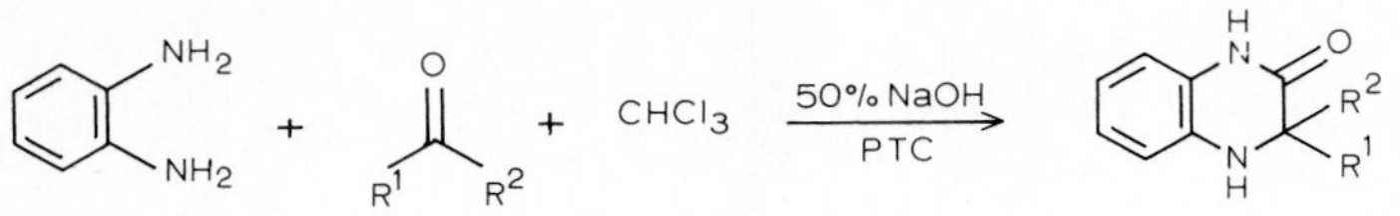

(viii) Propargyl alcohols and mercury(II) acetate give a direct but limited route to 3-alkyl-2-methylquinoxalines (*J. Barluenga et al.*, Synthesis, 1985, 313).

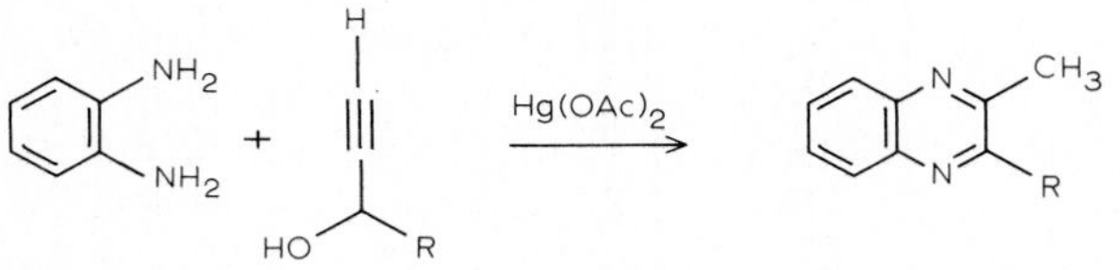

(3) Intramolecular cyclisation reactions provide entry to a range of quinoxalines and, with substituents on the carbocyclic ring, can ensure the unambiguous synthesis of a specific quinoxaline isomer.

Reductive cyclisation of α-(*o*-nitroanilino)-carboxylic acids (XI) affords the quinoxalinones (XII) after air oxidation (*B.C. Platt*, J. chem. Soc., 1948, 1310). Catalytic hydrogenation over Raney nickel may also be used in the reduction step (*R. van Dusen* and *H.P. Schultz*, J. org. Chem., 1956, **21**, 1326). Similarly, reduction of nitroaniline derivative (XIII) followed by reflux in ethanol produces 6-chlorotetrahydroquinoxaline (XIV) (*P. Clarke* and *A. Moorehouse*, J. chem. Soc., 1963, 4763).

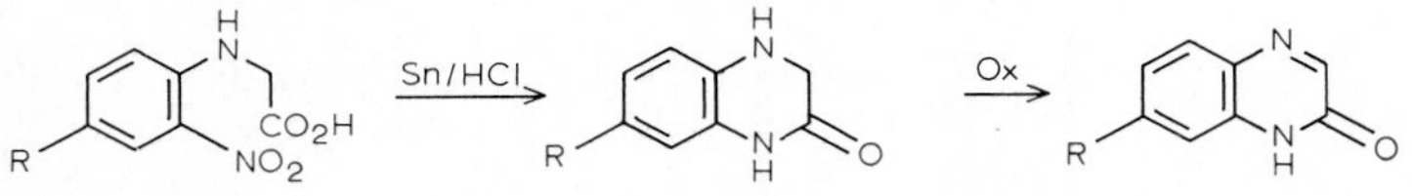

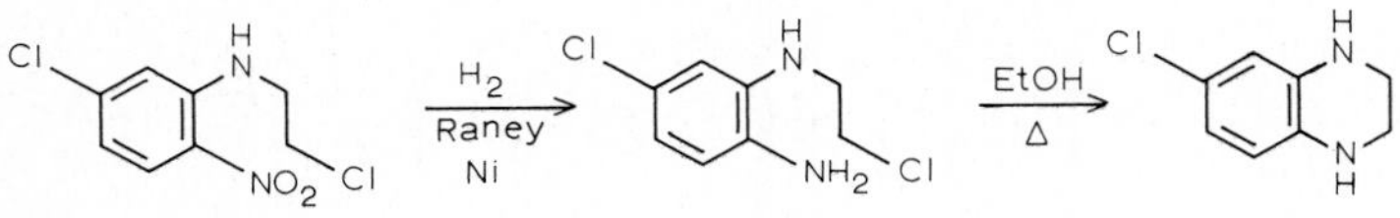

Treatment of *N*-cyanomethyl-*o*-phenylenediamine with hydroxylamine produces the intermediate amidoxime which cyclises to the oxime (XV) (*K. Harsanyi et al.*, Ber., 1972, **105**, 805). Reaction of the *N*-nitroso compounds

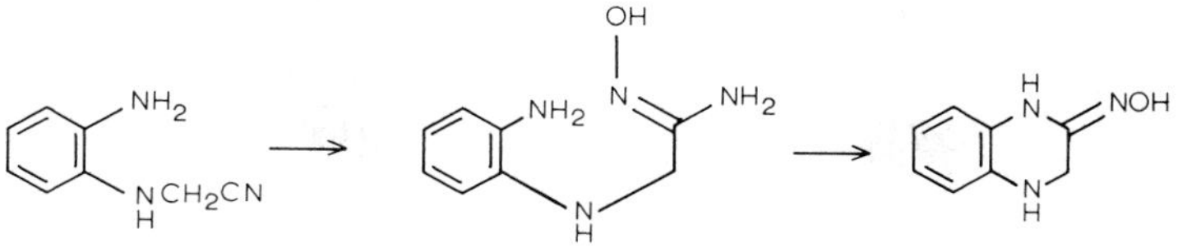

(XV)

(XVI) with acetic anhydride at 100–120° produces quinoxalines (XVII) which are hydrolysed in base to the corresponding mono-amides (*S.I. Burmistrov et al.*, Zhur. org. Khim., 1972, **8**, 1095).

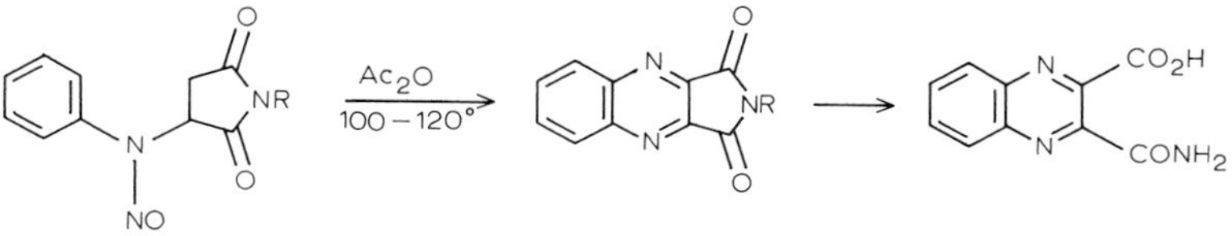

(XVI) (XVII)

Cyclisation of *o*-nitro-acetanilides (XVIII) to quinoxaline-*N*-oxides (XIX) can be effected by treatment with base; the resulting *N*-oxides are de-

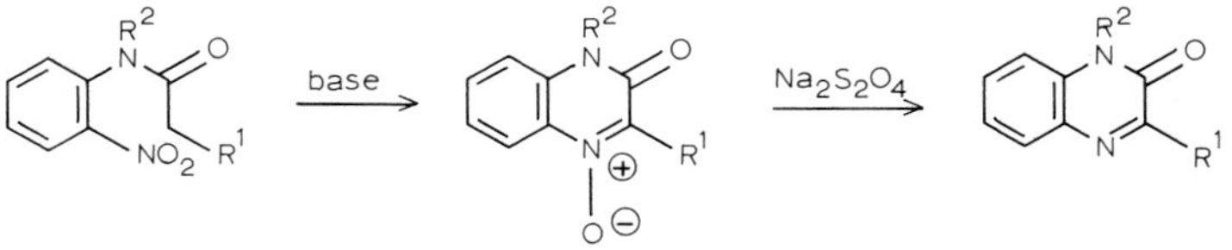

(XVIII) (XIX)

oxygenated using sodium dithionite (*G. Tennant*, J. chem. Soc., 1963, 2428; 1964, 2666; J. chem. Soc. C, 1966, 2285; *M.S. Habib* and co-workers, Tetrahedron, 1964, **20**, 1107; 1965, **21**, 861; *R. Furco* and *S. Rossi*, Gazz., 1964, **94**, 3). Treatment of the enaminones (XX), obtained from cyclohexan-1,3-dione and *o*-nitro-anilines, with base gives rise to the quinoxaline-*N*-oxides (XXI) in good yield, *e.g.* compound (XXI), R = CH_3 and X = H, is obtained in 92% yield (*S. Miyano et al.*, Synthesis, 1981, 60).

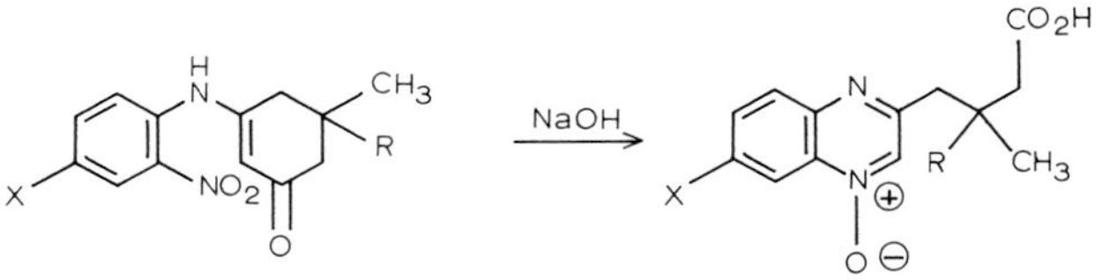

(XX) (XXI)

(4) Transformations of benzofuroxan to quinoxaline-*N*-oxides (the Beirut reaction). Reaction of benzofuroxan with an enamine or a compound containing an active methylene group, *e.g.* a β-diketone or a β-keto-ester, in the presence of base affords quinoxaline-1,4-dioxides (XXIII) and (XXIV) bearing a wide range of substituents including alkyl, aryl, amino and cyano groups [Scheme 4] (*C.H. Issidorides* and *M.J. Haddadin*, Tetrahedron

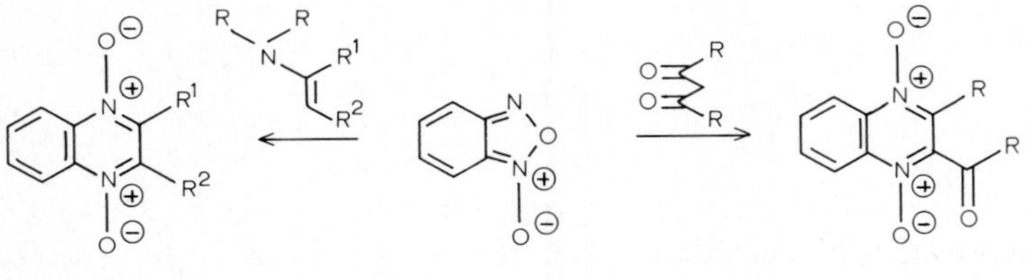

Scheme 4

Letters, 1965, 3253; J. org. Chem., 1966, **31**, 4067). Simple ketones also condense with benzofuroxan, *e.g.* ethyl methyl ketone gives 2,3-dimethyl-quinoxaline-1,4-dioxide (XXV) in good yield (*K. Lev et al.*, Angew. Chem. intern. Edn., 1968, **8**, 596).

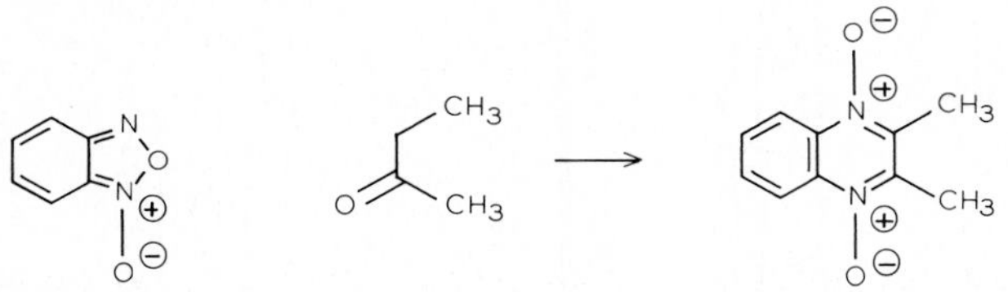

The condensation of *o*-benzoquinone dioximes with either α,β-dicarbonyl or α-hydroxycarbonyl compounds produces quinoxaline-1,4-dioxides like (XXVI) (*E. Abishanab*, J. org. Chem., 1970, **35**, 4279).

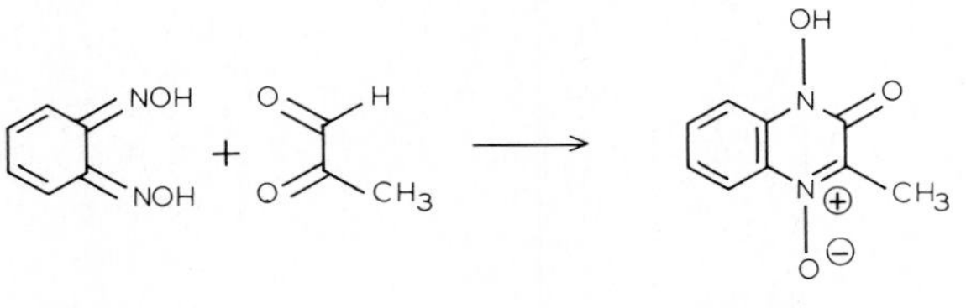

(5) Miscellaneous methods of synthesis.

(i) o-Phenylenediamine also condenses with cyanogen affording 2,3-diaminoquinoxalines which are hydrolysed by dilute acid to the corresponding dihydroxy compounds (XXVII) (*J.A. Bladin*, Ber., 1885, **18**, 666; *O. Hinsberg* and *E. Schwantes, ibid.*, 1903, **36**, 4039).

(XXVII)

(ii) The reaction of catechol and ethylenediamine at 210–220° gives 1,2,3,4-tetrahydroquinoxaline which can be dehydrogenated to quinoxaline (*F.G. Mann* and *B.B. Smith*, J. chem. Soc., 1951, 1906).

(iii) The condensation between an o-nitrosophenol, acetaldehyde and ammonia or a primary amine produces a hydroxydihydroquinoxaline or the related *N*-substituted derivative (XXVIII) (*M. Lange*, Ber., 1909, **42**, 574).

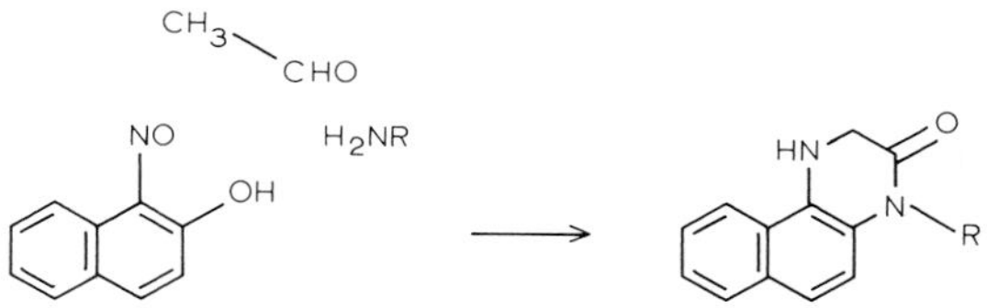

(XXVIII)

(iv) 2-Aminoquinoxalines containing substituents in the benzo ring, *e.g.* 2,7-diaminoquinoxaline (XXIX), can be unambiguously prepared by a base-catalysed condensation between o-nitrosoanilines and acetonitrile derivatives [Scheme 5] (*T.S. Osdene* and *G.M. Timmis*, J. chem. Soc., 1955, 2027; 4349).

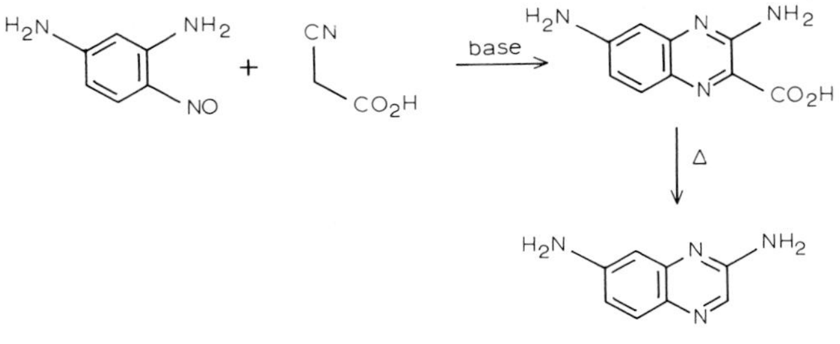

(XXIX)

Scheme 5

(v) *o*-Quinone diimides undergo cycloaddition reactions with electron-rich dienophiles to afford a range of highly substituted quinoxaline derivatives, *e.g.* (XXXI) [Scheme 6] (*W. Friedrichsen*, Angew. Chem. intern. Edn., 1974, **13**, 348; Ann., 1978, 1129).

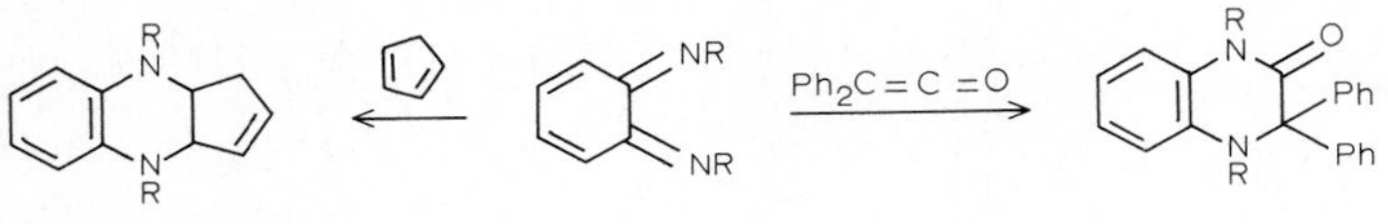

(XXX) (XXXI)

Scheme 6

(d) Quinoxaline, its homologues and derivatives

(i) Quinoxaline and its alkyl and aryl derivatives

Quinoxaline, m.p. 27°, b.p. 112–115°/17 mm, is obtained from the condensation of *o*-phenylenediamine and glyoxal, or glyoxal bisulphite (*J.H. Billman* and *J.L. Rendall*, J. Amer. chem. Soc., 1944, **65**, 1210; *J.C. Cavagnol* and *F.Y. Wyselogle*, ibid., 1947, **69**, 795; *A. Zmojdzinand* and *B. Hoffman*, C.A., 1974, **80**, 59962), or 2,3-dihydroxydioxane (*M.C. Venuti*, Synthesis, 1982, 61); yields in all cases are > 90%. Quinoxaline, although a weak base, forms a *hydrochloride*, m.p. 184° (dec), and a *sulphate*, m.p. 186–187° (*Hinsberg, loc. cit.*). Quaternary salts are formed, often with difficulty, by treatment of quinoxaline with either an alkyl halide or a dialkyl sulphate; *methiodide*, m.p. 175° (*O. Hinsberg*, Ann., 1896, **292**, 245); *ethiodide*, m.p. 141–143° (*R.F. Smith et al.*, J. org. Chem., 1959, **24**, 205; see also *W.K. Easley* and *C.T. Bahner*, J. Amer. chem. Soc., 1950, **72**, 3803). Urea and quinoxaline form a 1:1 molecular complex (*E.S. Lane* and *C. Williams*, Chem. Ind., 1953, 1230).

Like pyrazine, quinoxaline does not readily undergo electrophilic aromatic substitution reactions; nitration under forcing conditions (oleum/nitric acid/90°/24 h) gives however the 5-*nitro*- (1.5%), m.p. 95–96°, and 5,6-*dinitro*- (24%), m.p. 172–173°, quinoxaline (*M.J.S. Dewar* and *P.M. Maitlas*, J. chem. Soc., 1957, 2518).

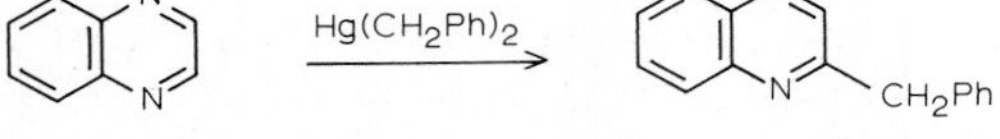

Quinoxaline is susceptible to radical substitution reactions, *e.g.* with dibenzyl mercury, 2-benzylquinoxaline (XXXII) is obtained as the sole product (*K.C. Bass* and *P. Nababsing*, Org. Prep. Proced. Int., 1971, **3**, 45). Other homolytic reactions which produce a 2-substituted product include alkylation (*F. Minisci* and co-workers, Ann. Chim (Rome), 1970, **60**, 746; J. chem. Soc. Perkin I, 1977, 865), acylation (*idem*, Chem. Comm., 1969, 201), α-oxyalkylation (*idem*, Tetrahedron, 1971, **27**, 3655), carboxylation (*idem*, Tetrahedron Letters, 1973, 645), *o*-aminoalkylation (*idem, ibid.*, 1976, 203), and carbamylation (*idem, ibid.*, 1970, 15). Irradiation of quinoxaline with light in acidic methanol affords 2-methylquinoxaline *via* a radical process involving a single electron transfer to an electronically excited protonated quinoxaline molecule (*S. Wake et al.*, Bull. chem. Soc. Japan, 1974, **47**, 1257).

Addition reactions take place readily in the heterocyclic ring, *e.g.* with bisulphite, hydrogen cyanide and Grignard reagents to give the appropriate 1,2,3,4-tetrahydroquinoxaline derivatives. With potassium amide, quinoxaline forms the dipotassium salt of fluorubin (*F.W. Bergstrom* and *R.A. Ogg*, J. Amer. chem. Soc., 1931, **53**, 245). The heterocyclic ring is also reduced to the tetrahydroquinoxaline under a variety of conditions including sodium in ethanol, catalytic hydrogenation over platinum (*Cavagnol and Wiselogle, loc. cit.*), with lithium aluminimum hydride in ether (*R.F. Smith et al., loc. cit.*), and sodium borohydride and trifluoroacetic acid (*R.C. Bugle* and *R.A. Osteryoung*, J. org. Chem., 1979, **44**, 1719). Treatment of quinoxaline with sodium in tetrahydrofuran at 20° yields the 1,4-*dihydroquinoxaline* (26%), m.p. 158–159° (*J. Hamer* and *R.E. Holliday, ibid.*, 1963, **28**, 2488). Total reduction of quinoxaline to the *cis-decahydro* compound, m.p. 56–58°, requires high pressure catalytic reduction over a rhodium–alumina catalyst at 100° (*H. Smith Broadbent et al.*, J. Amer. chem. Soc., 1960, **82**, 189). Quinoxaline undergoes reductive formylation with formic acid in formamide to give the 1,4-diformyl-tetrahydro compound (XXXIII) and 2,2'-biquinoxalinyl (*I. Baxter* and *D.W. Cameron*, J. chem. Soc. C, 1968, 2471) and with potassium borohydride in acetic acid to produce the 1,4-*diethyl* derivative (XXXIV), b.p. 101°/0.3 mm (*J.-M. Cosmao et al.*, J. heterocyclic Chem., 1979, **16**, 973).

Treatment of quinoxaline with one equivalent of peracetic acid in acetic acid gives the 1-*N*-oxide; excess reagent gives the 1,4-di-*N*-oxide (*J.K. Landquist*, J. chem. Soc., 1953, 2816). With 30% aqueous hydrogen peroxide in acetic acid, 2,3-dihydroxyquinoxaline is obtained (*M. Asai*, J. pharm. Soc. Japan, 1959, **79**, 260). Selective oxidative cleavage of the benzo ring in quinoxaline by either reaction with potassium permanganate or electrolysis provides pyrazine-2,3-dicarboxylic acid (*H.I.X. Mager* and *W. Berends*, Rec. Trav. chim., 1959, **78**, 5; *T. Kimura et al.*, J. pharm. Soc. Japan, 1957, **77**, 891).

2-Methylquinoxaline is conveniently prepared from *o*-phenylenediamine and either oximinoacetone (*K.A. Bottcher*, Ber., 1913, **46**, 3084) or methyl glyoxal bisulphite (Table 5, Ref. 2). The α-methyl group displays expected reactivity in condensing with formaldehyde and aromatic aldehydes to give 2-(β-hydroxyethyl), m.p. 78–79.5°, and 2-styryl (XXXV) deriva-

TABLE 5
ALKYL- AND ARYL-QUINOXALINES

Compound	m.p. (b.p.) (°C)	Compound	m.p. (b.p.) (°C)
Quinoxaline	27 [1] (225–226)	2-Methyl-	(125-7/11 mm) [2]
2-Ethyl-	(97–100/3 mm) [3]	2-Isopropyl-	(270) [4]
2-Hydroxymethyl-	78–79 [5]	2-Phenyl-	78 [7]
2-Benzyl-	38.5 [6] (80/0.1 mm)	2-4′-diphenylyl	128 [9]
2-m-4-Xylyl-	56–57 [8]	2-β-Naphthyl-	137 [10]
2-α-Naphthyl-	114 [8,10]	2-Methyl-3-phenyl-	64–65 [9], 57–58 [12]
2,3-Dimethyl-	106 [11]	(monohydrate)	
2,3-Diethyl-	50.5 [13]	2-Benzyl-3-phenyl-	97 [14,15]
2-Benzyl-3-p-tolyl-	112–113 [16]	2,3-Diphenyl-	126 (corr.) [7,11]
2-Phenyl-3-4′-diphenylyl-	163 [17]	2-Phenyl-3-α-naphthyl-	137–139 [18]
2-Phenyl-3-β-naphthyl-	108–110 [18]	2,3-Di-β-naphthyl-	192–193 [19]
		2,3,6-Triphenyl-	148 [20]

1 F.D. Chattaway and W.G. Humphrey, J. chem. Soc., 1929, 649.
2 R.G. Jones and K.C. McLaughlin, Org. Synth., 1950, 30, 86; R.G. Jones et al., J. Amer. chem. Soc., 1950, 72, 3539.
3 J.K. Landquist and G.J. Stacey, J. chem. Soc., 1953, 2822.
4 M. Conrad and K. Hock, Ber., 1899, 32, 1209.
5 A.S. Elina, C.A., 1962, 32, 2967.
6 H. Dahn and G. Rotzler, Helv., 1963, 43, 1555.
7 J.H. Boyer and D. Straw, J. Amer. chem. Soc., 1953, 75, 1642.
8 R.C. Fuson, W.S. Emerson and H.W. Gray, ibid., 1939, 61, 480.
9 F. Kröhnke and E. Borner, Ber., 1936, 69, 2006.
10 L.N. Goldyrev and I.Ya. Postovskii, C.A., 1940, 34, 4732.
11 R.W. Bost and E.E. Towell, J. Amer. chem. Soc., 1948, 70, 903.
12 K. von Auwers, Ber., 1917, 50, 1177.
13 E. Urion, C.A., 1934, 28, 2677.
14 T. Malkin and R. Robinson, J. chem. Soc., 1925, 369.
15 H. Burton and C.W. Shoppee, ibid., 1937, 546.
16 H. Jörlander, Ber., 1917, 50, 406.
17 F.D. Chattaway and E.A. Coulson, J. chem. Soc., 1928, 1080; 1361.
18 R. Ruggli and M. Reinert, Helv., 1926, 9, 67.
19 J.D. Fulton and R. Robinson, J. chem. Soc., 1939, 200.
20 F. Bell and J. Kenyon, ibid., 1926, 2705.

tives, respectively (A.S. Elina and L.G. Tsrul'nikova, J. gen. Chem. U.S.S.R., 1964, 34, 2089; G.M. Bennet and G.H. Willis, J. chem. Soc., 1928, 1960). 2-Methylquinoxaline also undergoes the Mannich reaction forming 2-(β-dialkylaminoethyl)quinoxalines (P.F. Wiley, J. Amer. chem. Soc., 1954, 76, 4924; 1950, 70, 2893) and the Claisen condensation with, for example, ethyl oxalate (W. Borsche and W. Doeller, Ann., 1939, 537, 39). Similarly, 2-methylquinoxaline and chloral affords the adduct (XXXVI), m.p. 106°, which on hydrolysis followed by hydrogenation over Raney nickel gives β-*quinoxalinylpropionic acid* (XXXVII), m.p. 115–115.5°, amide 152–152.5° (*R.G. Jones et al.*, J. Amer. chem. Soc., 1950, **72**, 3539).

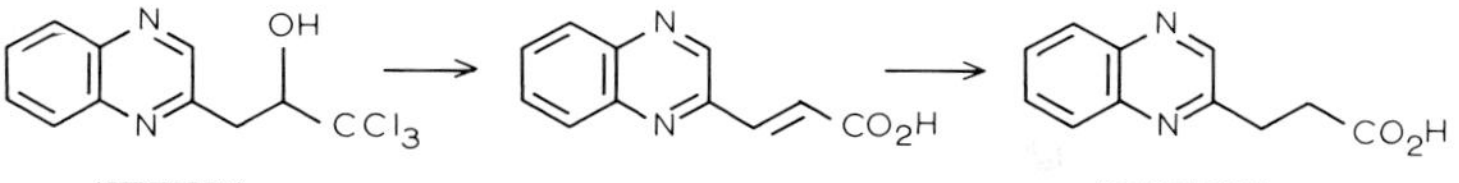

(XXXVI) (XXXVII)

Bromination of 2-methylquinoxaline in acetic acid/sulphuric acid at 70° gives a mixture of 2-*bromomethyl*- (37%), m.p. 67–68°, and 2-(*dibromomethyl*)- (27%), m.p. 120–121°, *quinoxaline* (*A.S. Elina*, J. gen. Chem. U.S.S.R., 1959, **29**, 2728); in the presence of sodium acetate at 100°, the 2-(*tribromomethyl*) derivative, m.p. 109°, is obtained (*Bennett* and *Willis, loc. cit.*) and this can be partially debrominated to the dibromomethyl compound on treatment with silver nitrate in aqueous alcohol (*B.R. Brown*, J. chem. Soc., 1949, 2577). Oxidation of 2-methylquinoxaline with selenium dioxide yields the 2-formyl derivative, whereas with alkaline potassium permanganate produces 5-methlpyrazine-2,3-dicarboxylic acid (*F. Leonard* and *P.E. Spoerri*, J. Amer. chem. Soc., 1946, **68**, 526; *H.I.X. Mager* and *W. Berends, loc. cit.*).

On treatment with hot dilute hydrochloric acid or heating at 200° with palladium on charcoal, 2-methylquinoxaline is transformed to a high melting, orange crystalline solid, ultimately shown to have the fused polycyclic structure (XXXVIII) (*G.W.H. Cheeseman* and

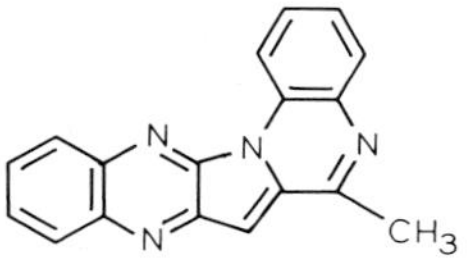

(XXXVIII)

B. Tuck, Tetrahedron Letters, 1968, 4851). With two equivalents of tetrachloro-*o*-benzoquinone, 2-methylquinoxaline affords a mixture of the unusual adduct (XXXIX) and the 1:1 adduct (XL) (*J.W. Lown et al.*, Canad. J. Chem., 1970, **48**, 327). Cycloadditions with DMAD

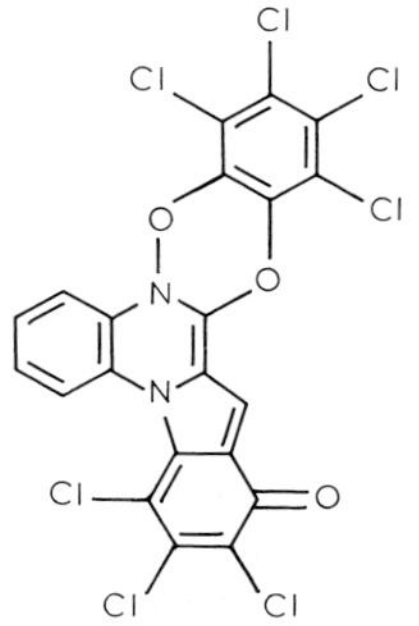

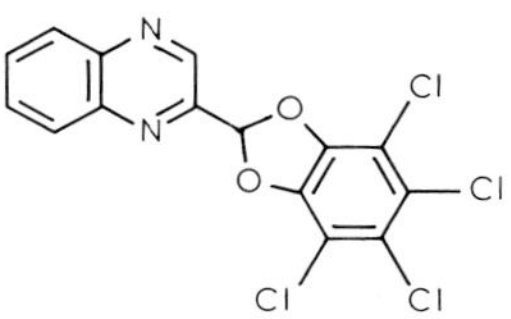

(XXXIX) (XL)

give mixtures of the isomeric azepino[1,2-*a*]quinoxalines (XLI) and (XLII) (*R.M. Acheson* and *M.W. Foxton*, J. chem. Soc. C, 1968, 378).

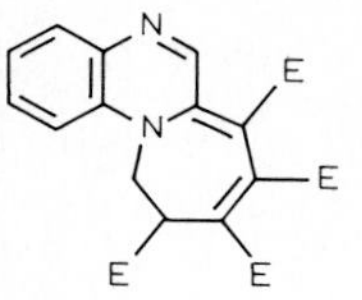

$E = CO_2Me$

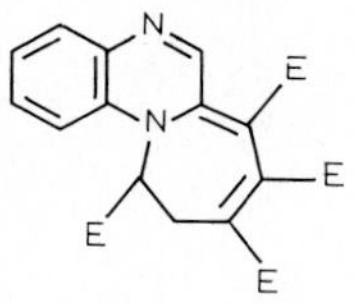

Treatment of 2-methylquinoxaline with dimethyl sulphate affords the quaternary salt (XLIII) [Scheme 7] which can be hydrolysed to 1,3-dimethyl-1,2-dihydroquinoxalinone (XLIV) (*H. Wahl et al.*, Bull. chim. Soc. Fr., 1973, 1285) or oxidised with concomitant dimerisation to the salt (XLV) (*idem, ibid.*, 1973, 1289).

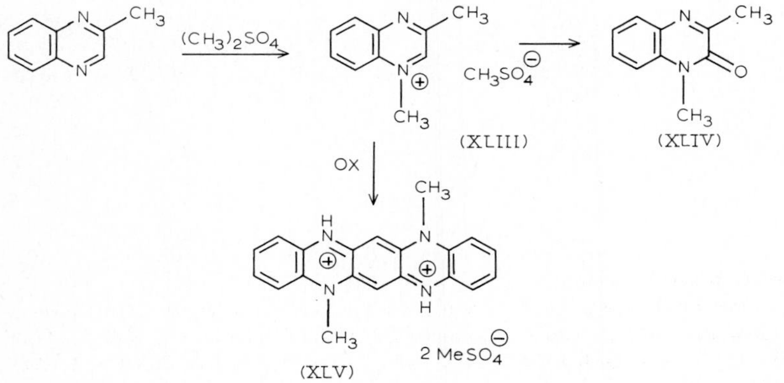

Scheme 7

2-Methylquinoxaline and diphenylcyclopropenone form the adduct (XLVI) (*J.W. Lown* and *K. Matsumoto*, Canad. J. Chem., 1971, **49**, 1165).

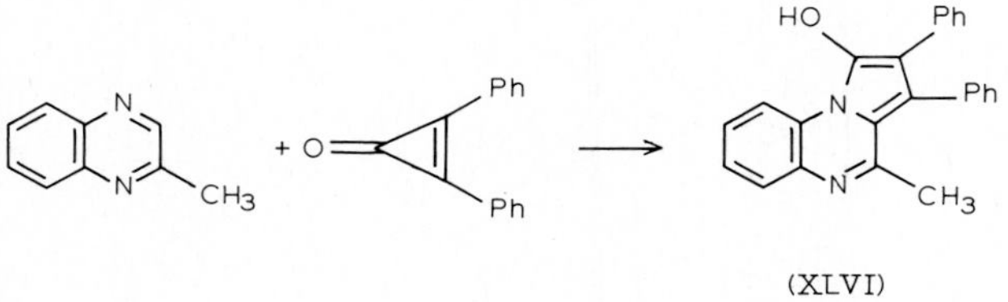

2,3-Dimethylquinoxaline, m.p. 106°, *picrate*, m.p. 189°, is obtained from the condensation of *o*-phenylenediamine and diacetyl (*W. Bost* and *E.E. Towell*, J. Amer. chem. Soc., 1948, **70**,

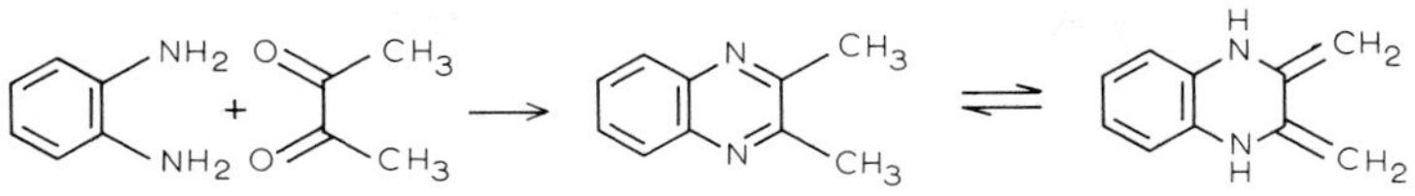

(XLVII)

903), or by dithionite reduction of the corresponding 1,4-di-*N*-oxide, prepared by the Beirut procedure (*K. Heynes et al.*, Ber., 1981, **114**, 240), or by treatment of 2,3-dichloroquinoxaline with methylmagnesium iodide (*R.A. Ogg* and *F.W. Bergstrom*, J. Amer. chem. Soc., 1931, **53**, 1846).

2,3-Dimethylquinoxaline reacts with maleic anhydride and *N*-phenylmaleimide to form the adducts (XLVIII) and (XLIX), respectively, which had been wrongly identified as Diels-Alder adducts derived from the tautomeric form (XLVII). Since similar adducts are also formed from a variety of heterocycles with only one α-methyl group, the intermediacy of the

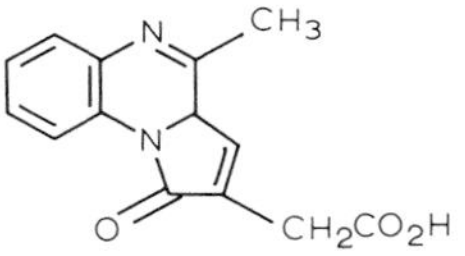

(XLVIII) 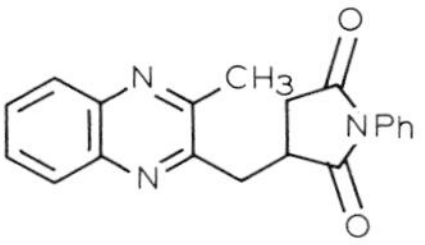   (XLIX)

tautomeric form (XLVII) is not essential. Moreover, the "benzoquinone adduct" reported earlier, turns out to be a 2:1 molecular complex between 2,3-dimethylquinoxaline and quinol (*C.W. Bird* and *G.W.H. Cheeseman*, J. chem. Soc., 1962, 3037; *E.C. Taylor* and *E.S. Hand*, J. org. Chem., 1962, **27**, 3734; J. Amer. chem. Soc., 1963, **85**, 770). The reaction of 2,3-dimethylquinoxaline with DMAD yields two exocyclic addition products (L) [*cf.* 2-methylquinoxaline] and (LI). An alternative structure, in which the carbonyl and ester groups indicated are interchanged, has been suggested for (LI) (*R.M. Acheson* and *M.W. Foxton*, J. chem. Soc. C, 1968, 378).

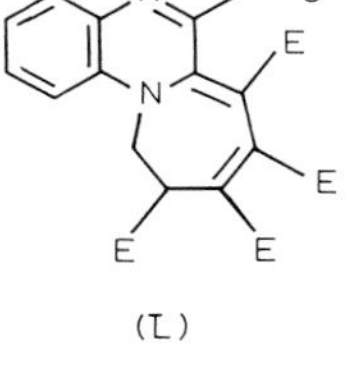

(L) 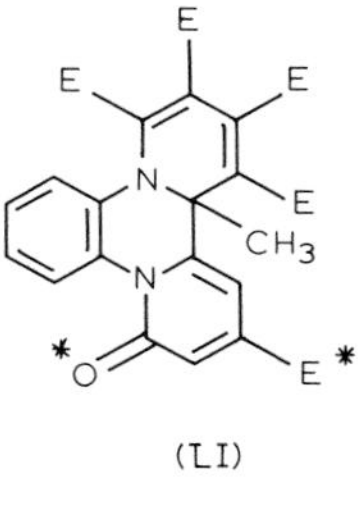   (LI)

Because of steric overcrowding, 2,3-dimenthylquinoxaline forms only monostyryl derivatives with certain aromatic aldehydes and reacts with only one molecule of ethyl oxalate (*G.M. Bennet* and *G.H. Willis*, J. chem. Soc., 1928, 1960).

Treatment of 2,3-dimethylquinoxaline with potassium amide in liquid ammonia affords the dipotassium salt which with ethyl iodide gives 2,3-*di-n-propylquinoxaline*, m.p. 43° (*Ogg* and *Bergstrom*, loc. cit.). The combination of lithium diisopropylamide in tetrahydrofuran/hexamethylphosphonamide is reported to promote mono- or vicinal di-alkylation of 2,3-dimethylquinoxaline (*E.M. Kaiser* and *J.D. Petty*, J. organometal. Chem., 1976, **108**, 139). 2,3-Dimethylquinoxaline undergoes dehydrodimerisation on treatment with phenyl lithium and copper(I) chloride to yield (LII) (*H. Schrechen et al.*, Angew. Chem. intern. Edn., 1968, **7**, 541).

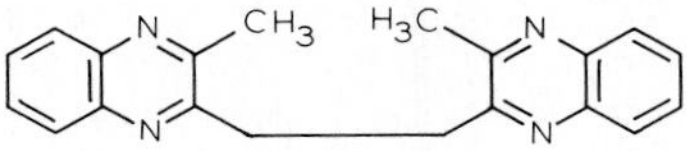

(LII)

Heating 2,3-dimethylquinoxaline with methyl iodide at 90° for 6 h yields the *monomethiodide* derivative, m.p. 192°. Quaternisation with dimethyl sulphate affords the corresponding quinoxalinium salt (LIII) which dimerises to (LIV, major) and (LV) at pH 7.5–8.0 (*M.T. LeBris*, Bull. chim. Soc. Fr., 1970, 563).

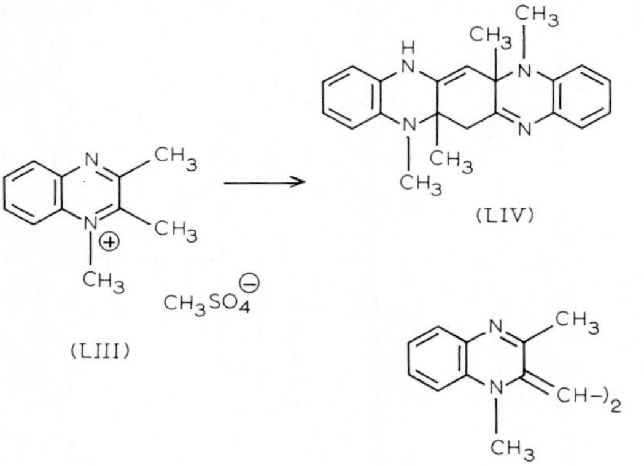

(LIII)   (LIV)

(LV)

1-Alkyl- or 1-aryl-2,3-dimethylquinoxalinium salts are conveniently prepared by direct synthesis in strong acid, e.g. (LVI) (*D. Schelz* and *M. Priester*, Helv. 1975, **58**, 317).

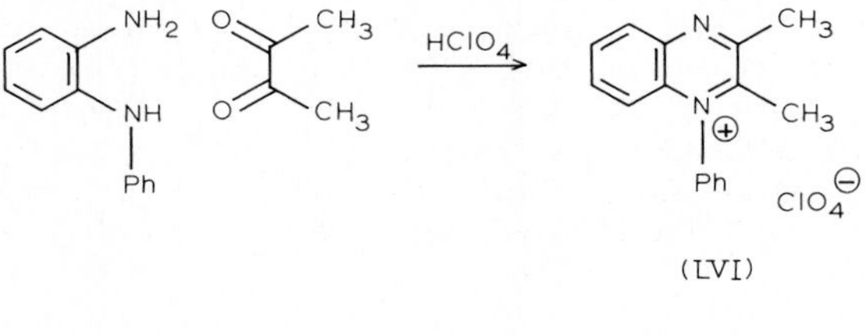

(LVI)

Bromination with six moles of bromine in acetic acid/sodium acetate gives 2,3-*bis*(*dibromomethyl*)*quinoxaline*, m.p. 228° (*Bennet* and *Willis, loc. cit.*) whereas with excess bromine in alcoholic solution, the product is tentatively identified as 5,6,7,8-*tetrabromo*-2,3-*dimethylquinoxaline*, m.p. 234° (dec.) (*S.T. Henderson*, J. chem. Soc., 1929, 466). The dibrominated compound, 2,3-*bis*(*bromomethyl*)*quinoxaline* (LVII), m.p. 150–151°, is more conveniently prepared by direct synthesis than by side-chain bromination of the parent compound (*O. Westphal* and *K. Jann*, Ann., 1957, **605**, 8; *W.E. Hahn* and *J.Z. Lesiak*, C.A., 1973, **78**, 58357; *E.J. Moriconi* and *A.J. Fritsch*, J. org. Chem., 1965, **30**, 1542).

(LVII) (LVIII)

Reduction of 2,3-dimethylquinoxaline by lithium aluminium hydride or catalytic hydrogenation over platinum affords *cis*-2,3-*dimethyl*-1,2,3,4-*tetrahydroquinoxaline* (LVIII), m.p. 114–115° (*R.F. Smith* and co-workers, J. org. Chem., 1959, **24**, 205; *R.C. DeSelma* and *H.S. Mosher*, J. Amer. chem. Soc., 1960, **82**, 3762).

Treatment of 2,3-dimethylquinoxaline with 20% nitric acid at 90° for 15 h gives a mixture of 6-nitro- and 6,7-dinitro-quinoxaline-2,3-dione- (LIX) and (LX); the reaction is postulated to proceed through the 2,3-dione (*A.S. Elina et al.*, Zhur. org. Khim., 1965, **1**, 147).

(LIX) (LX)

Irradiation of 2,3-dimethylquinoxaline in ether solution results in the formation of 1,2-adducts in low yield (< 20%), *e.g.* (LXI) in tetrahydrofuran (*T.T. Chen et al.*, Helv., 1968, **51**, 632).

(LXI)

Quinoxalines substituted in the carbocyclic ring are generally prepared by direct synthesis as outlined above in the the general methods. Thus, diaminotoluene with glyoxal gives 6-**methylquinoxaline**, b.p. 245° (*oxalate*, m.p. 135–135°), and with chloroacetone (*O. Hinsberg*, Ann., 1887, **237**, 327), or methylglyoxal, or oximinoacetone (*H. v. Pechman*, Ber., 1887, **20**, 2539), yields a mixture of 2,6- and 2,7-*dimethylquinoxaline*, m.p. 54°, b.p. 266–286°. Similarly prepared are 5-*methyl*-, m.p. 20–21°, b.p. 120°/15 mm, and 6,7-*dimethyl*-, m.p. 100–101° (*J.K. Landquist*, J. chem. Soc., 1953, 2819), 2,3,6-*trimethyl*-, m.p. 72–73°, and 2,3,6,7-*tetramethyl*-, m.p. 189–190°, *quinoxaline* (*F. Bell* and *J. Kenyon*, ibid., 1936, 2705).

6-Methylquinoxaline is brominated in the side-chain using N-bromosuccinimide to give 6-*bromomethylquinoxaline*, m.p. 100–105° (dec.) (*R.C. DeSelms et al.*, J. heterocyclic Chem., 1974, **11**, 595).

2-Phenylquinoxaline, m.p. 78°, is prepared by condensation of *o*-phenylenediamine with (i) phenylglyoxal (Table 5, Ref. 7; *G.Y. Sarkis* and *S. Al-Azawe*, J. chem. Eng. Data, 1973, **18**, 102); (ii) ω-nitroacetophenone in the presence of sodium dithionite (*A. Dornow* and *W. Sassenberg*, Ann., 1955, **594**, 185); (iii) ω-bromoacetophenone to give the corresponding dihydroquinoxaline which is aromatised in air or by a mild oxidising agent (*J. Figueras*, J. org. Chem., 1966, **31**, 803; *Sarkis and Al-Azawe, loc. cit.*); and (iv) β-tosyl styrene oxides (*E.C. Taylor et al.*, J. org. Chem., 1980, **45**, 2512).

Nitration of 2-phenylquinoxaline with a mixture of nitric and sulphuric acid affords a mixture of 2-(*p-nitrophenyl*)-, m.p. 186–188°, and 2-(*o-nitrophenyl*)-, mp. 110°, *quinoxaline*; the former compound is reduced to the 2-(*p-aminophenyl*) derivative, m.p. 167–168° (*Ng.Ph. Buu-Hoï*, and *Ng.H. Khôi*, Bull. Soc. chim. Fr., 1950, 733).

2-Aryl derivatives are conveniently prepared by appropriate modification of the condensation methods mentioned above, *e.g.* 2-*hydroxyphenylquinoxaline*, m.p. 204° (*O. Kovacs* and co-workers, C.A., 1952, **46**, 3514; 7573).

Reduction of 2-phenylquinoxaline with sodium in tetrahydrofuran affords the 1,4-derivative which rearranges to the thermodynamically more stable 1,2-isomer (LXII) (*M. Schellenberg*, Helv., 1970, **53**, 1151).

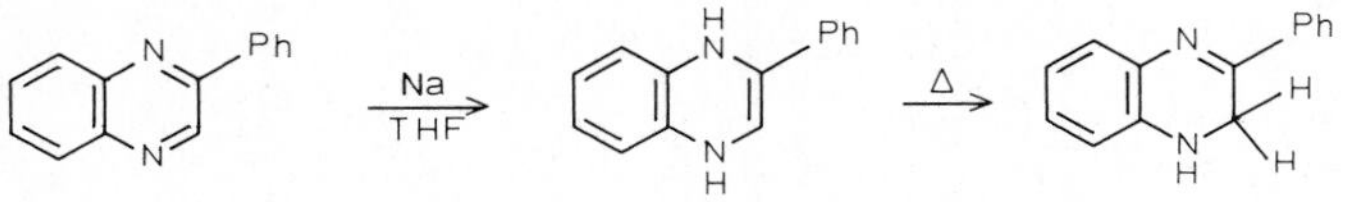

(LXII)

2,3-Diphenylquinoxaline, readily prepared from *o*-phenylenediamine and benzil, on treatment with potassium amide in liquid ammonia undergoes ring contraction with elimination of benzylidenimide to 2-*phenylbenzimidazole*, m.p. 289–290° (*E.C. Taylor* and *A. McKillop*, J. org. Chem., 1965, **30**, 2858). Reduction with sodium gives a dianion which can be trapped by suitable electrophiles to give the corresponding 1,4-dihydroquinoxaline, *e.g.* with ethyl chloroformate (XLIII) is obtained; analogous reactions with lithium results in cyclodehydrogenation to produce (XLIV) [Scheme 8] (*J.G. Smith* and *E.M. Levi*, J. organomet. Chem., 1972, **36**, 215).

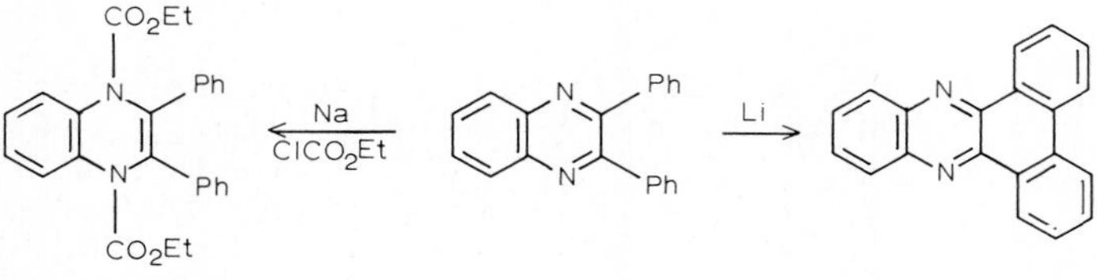

(LXIII)   (LXIV)

Scheme 8

(ii) Halogenated quinoxalines

2-Chloroquinoxaline and derivatives are prepared by chlorination of 2-hydroxyquinoxalines (quinoxalin-2(1*H*)-ones) with (a) phosphoryl chloride (91% yield) (Table 6, Ref. 2); or (b) a mixture of phosphoryl chloride and phosphorus pentachloride (*F.J. Wolf*, U.S.P. 2,537,870; C.A., 1951, **45**, 4274); or (c) phosphoryl chloride in pyridine (*J. Klicnar et al.*, Coll. Czech. Chem. Comm., 1971, **36**, 262). Phosphoryl chloride also reacts with quinoxaline-1-oxides to give 2-chloroquinoxalines (Table 6, Ref. 1). 2-Chloro-3-phenylquinoxaline is obtained from the reaction of acetyl chloride with the corresponding 1-oxide (Table 6, Ref. 7).

2-Bromoquinoxaline is prepared from quinoxalin-2-one and phosphoryl bromide at 120° (Table 6, Ref. 12). Treatment of 2-chloroquinoxaline with sodium iodide in hydroiodic acid yields the 2-iodo derivative (Table 6, Ref. 13). 2-Fluoroquinoxaline is obtained from the 2-amino compound *via* the Balz-Schiemann reaction (Table 6, Ref. 14).

Quinoxalines with halogen substituents in the benzo ring are usually synthesised directly from the appropriate *o*-phenylenediamine; a number of other halogenated quinoxaline derivatives are listed in Table 6.

The halogen atoms in 2-haloquinoxalines are readily displaced by a range of nitrogen, oxygen, and sulphur centred nucleophiles. A 2-fluoro atom is more labile than a 2-chloro atom in a given quinoxaline substrate (Table 6, Ref. 14; *G.B. Bress et al.*, J. chem. Soc. B, 1971, 2259). Reaction of 2-chloroquinoxaline with ammonia affords 2-aminoquinoxaline (*F.J. Wolf, loc. cit.*). The displacement process is not always straightforward, *e.g.* 2-chloroquinoxaline with aniline and sodium phenoxide gives the fused-ring systems (LXV) and (LXVI), respectively, in addition to the expected displacement products (*S.D. Carter* and *G.W.H. Cheeseman*, Tetrahedron,

(LXV)   (LXVI)

1978, **34**, 981; *R.K. Anderson* and *G.W.H. Cheeseman*, J. Chem. Soc. Perkin I, 1974, 129). 2-Haloquinoxalines react with potassium amide in liquid ammonia to give benzimidazole in addition to the displacement products; with X = Cl, Br, I, benzimidazole is the major or exclusive product whereas with X = F, only trace amounts of the ring contracted product are formed (*I.M. Ismail et al.*, Z. Chem., 1977, **17**, 15; *P.J. Pont et al.*, Rec. Trav. chim., 1972, **91**, 949).

TABLE 6
HALOGENATED QUINOXALINES

Compound	m.p. (°C)	Compound	m.p. (°C)
2-Chloro-	46–47 [1], 46–48 [2]	2-Chloro-3-phenyl-	127–128 [7,12]
5-Chloro-	60–62 [3]	2-Chloro-3-methyl-	86–87 [6]
6-Chloro-	64 [4], 60 [5]	2-Chloro-5-methyl-	95 [9]
2-Chloro-6-ethyl-	38–40 [8]	2-Chloro-6-methyl-	105–107 [10]
5-Chloro-2,3-dimethyl-	78–80 [11]	2-Chloro-7-methyl-	76 [10]
6-Chloro-2,3-dimethyl-	91–92 [11]		
2-Bromo-	58–59 [12]	2-Bromo-3-ethyl-	67–68 [13]
6-Bromo-	56 [3,5]	6-Bromo-2,3-dimethyl-	84–85 [11]
2-Iodo-	104–105 [12]	6-Iodo-2,3-dimethyl-	91–92 [11]
6-Iodo-	114–115 [3]		
2-Fluoro-	(b.p. 96–97.5/11 mm) [12,14]		
2,3-Dichloro-	149–150 [3,15–17]	2,6-Dichloro-	159–160 [18,19]
2,7-Dichloro-	151 [10,18]	6,7-Dichloro-	210 [3]
5,8-Dichloro-	205–207 [3]		
2,3-Dibromo-	171–164 [20] 165–170 [21]	2,3-Difluoro-	94–95 [22]
Hexachloro-	200–201 [22]	Hexafluoro-	142–144 [22]

1 A.H. Gowenlock, G.T. Newbold and F.S. Spring, J. chem. Soc., 1954, 622.
2 G.W.H. Cheeseman, ibid., 1957, 3236.
3 J.K. Landquist, ibid., 1953, 2816.
4 J.C. Cavagnol and F.Y. Wiselogle, J. Amer. chem. Soc., 1947, **69**, 795.
5 F.D. Chattaway and W.G. Humprey, J. chem. Soc., 1929, 645.
6 G.T. Newbold and F.S. Spring, ibid., 1948, 519; F.J. Wolf et al., J. Amer. chem. Soc., 1948, **70**, 2572.
7 Y. Ahmed et al., J. org. Chem., 1966, **31**, 2613.
8 Y.J. L'Italien and C.K. Banks, J. Amer. chem. Soc., 1951, **73**, 3246.
9 B.C. Platt and T.M. Sharp, J. chem. Soc., 1948, 2129.
10 B.C. Platt, ibid., 1949, 1310.
11 J.K. Landquist and G.J. Stacey, ibid., 1953, 2822.
12 P.J. Lont and H.C. Van der Plas, Rec. Trav. chim., 1972, **91**, 850.
13 H. Reinheckel, Monatsh., 1968, **99**, 2215.
14 H. Rutner and P.E. Spoerri, J. heterocyclic Chem., 1966, **3**, 435.
15 R.D. Haworth and S. Robinson, J. chem. Soc., 1948, 777.
16 J.R. Stevens, K. Pfister and F.J. Wolf, J. Amer. chem. Soc., 1946, **68**, 1035.
17 G.W.H. Cheeseman, J. chem. Soc., 1955, 1804.
18 A.F. Crowther, F.H.S. Curd, D.G. Davey and G.T. Stacey, ibid., 1949, 1260.
19 H.G. Petering and J. van Geissen, J. org. Chem., 1961, **26**, 2818.
20 E.H. Usherwood and M.A. Whitely, J. chem. Soc., 1923, 1069.
21 C. Sasnowski and L. Wojciechowski, C.A., 1971, **24**, 125728.
22 C.G. Allison et al., J. fluorine Chem., 1971, **1**, 59.

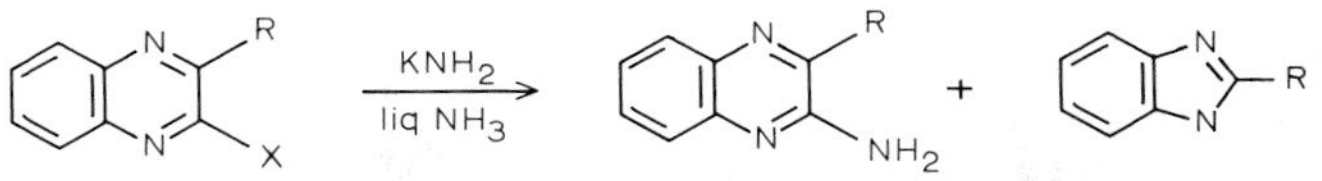

The acid-catalysed condensation of 2-chloroquinoxaline with 4-methylquinazoline affords the novel fused pyrrol (LXVII) (*R.K. Anderson, S.D. Carter* and *G.W.H. Cheeseman*, Tetrahedron, 1979, **35**, 2463). With α-carbanions derived from phenylacetonitrile (*R.N. Castle et al.*, J. org. Chem., 1956, **21**, 139) and ketones (*C. Iijima* and *E. Hayashi*, J. pharm. Soc. Japan, 1972, **92**, 729), 2-chloroquinoxaline gives the substitution products (LXVIII) and (LXIX), respectively. The analogous reaction with the potassium enolate of pinacolone affords the *ipso*-substitution product (LXX, 70%) and the furoquinoxaline (LXXI, 15%); (LXX) is postulated to arise from a radical substitution process [Scheme 9] (*D.R. Carver et al.*, J. org. Chem., 1982, **47**, 1036).

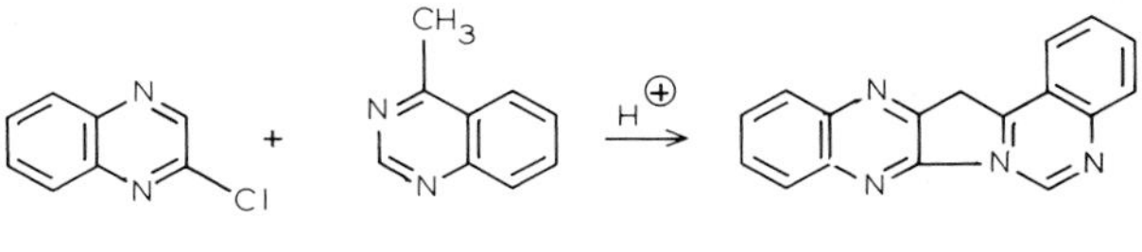

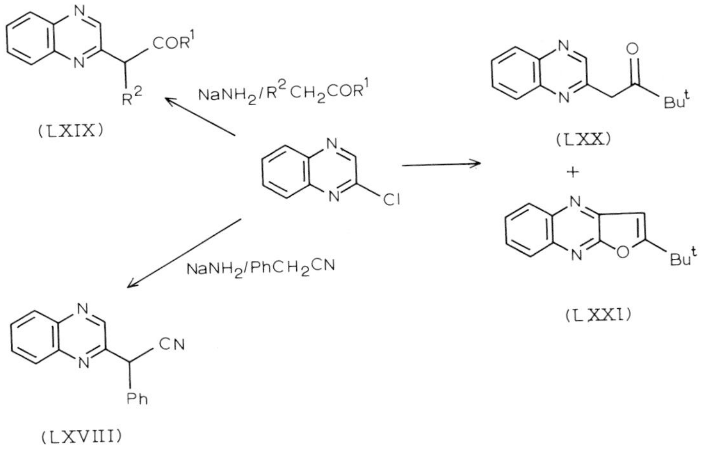

Scheme 9

2,3-Dichloroquinoxalines are generally prepared from the corresponding 2,3-dihydroxyquinoxalines (quinoxaline-2,3(1H, 4H)-diones) (Table 6, Refs. 3, 17), or 1,4-di-N-oxides (Table 6, Ref. 3). The parent compound is obtained in high yield by treatment of the 2,3-dione with (1) phosphoryl chloride using aromatic amines as solvent (Table 6, Refs. 3, 17), or (ii) thionyl chloride (*A.P. Komin* and *M. Carmack*, J. heterocyclic Chem., 1976, **13**, 13). The use of phosphorus pentachloride and phosgene has also been reported in the patent literature (*K. Sasse et al.*, G.P. 1,194,631; C.A., 1965, **63**, 8381; *H. Weidinger* and *G. Wellenreuther*, B.P. 927,974; C.A., 1964, **60**, 2987). From the 2,3-dione with phosphorus pentachloride under more vigorous conditions, the hexachloro derivative is obtained (Table 6, Ref. 22). Treatment of quinoxaline-1,4-dioxide with phosphoryl chloride also produces 2,3-dichloroquinoxaline (Table 6, Ref. 3).

2,3-Dibromoquinoxaline is prepared from the 2,3-dione by reaction with phosphoryl bromide in N,N-dimethylaniline (Table 6, Ref. 21) and 2,3-difluoroquinoxaline is obtained from the 2,3-dichloro compound by halogen exchange (Table 6, Ref. 22).

One or both of the chlorine atoms of 2,3-dichloroquinoxaline may be displaced by appropriate nucleophiles, *e.g.* partial replacement has been achieved with ammonia (*H. Saikachi* and *S. Tagami*, Chem. pharm. Bull. Tokyo, 1961, **9**, 941), amines (*W.R. Vaughan* and *M.S. Habib*, J. org. Chem., 1962, **27**, 324), and carbanions derived from malononitrile and ethyl cyanoacetate (*E.F. Pratt* and *J.C. Keresztesy, ibid.*, 1967, **32**, 49). Similarly, 2,3-dichloroquinoxaline with β-(diethylamino)ethylamine gives a monosubstitution product, b.p. 160–170°/0.01 mm, which on heating cyclises with loss of ethyl chloride to (LXXII), m.p. 155–156° (*R.D. Haworth* and *S. Robinson*, J. chem. Soc., 1948, 777).

(LXXII)

Complete displacement of the chlorines in 2,3-dichloroquinoxaline is effected in ammonia at 90° to give 2,3-diaminoquinoxaline (*W. Deuschel* and *W. Vilsmeier*, Belg. Pat., 612,092; C.A., 1961, **57**, 16634), or by alkoxide and aryloxide ions to produce 2,3-diethers (*H.I.X. Mager* and *W. Berends*, Rec. Trav. chim., 1959, **78**, 5). 2,3-Dialkynylquinoxalines (LXXIII) are prepared in good yield from a Pd–Cu catalysed coupling of 2,3-dichloroquinoxaline with terminal alkynes (*D.E. Ames* and *M.I. Brohi*, J. chem. Soc. Perkin I, 1980, 1384).

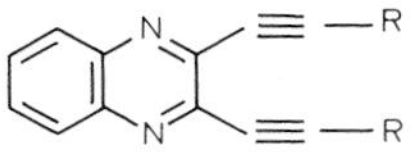

(LXXIII)

Condensation of 2,3-dichloroquinoxaline with *o*-disubstituted aromatic compounds gives rise to a number of fused polycyclic systems. Thus 2,3-dichloroquinoxaline reacts with (i) *o*-phenylenediamine to give fluoflavin (LXXIV) (*O. Hinsberg* and *J. Pollak*, Ber., 1896, **29**, 784; *K. Asano* and *S. Asia*, J. pharm. Soc. Japan, 1959, **79**, 661); (ii) with 2,3-diaminoquinoxaline to give fluorubin (LXXV) (*S. Noguchi*, C.A., 1961, **55**, 4516; *Deuschel* and *Vilsmeier, loc. cit.*); (iii) with catechol to give (LXXVI) (*F. Kehrmann* and *C. Bener*, Helv., 1925, **8**, 16; *D.E. Ames* and *R.J. Ward*, J. chem. Soc. Perkin, I, 1975, 534); (iv) *o*-amino-phenol and -thiophenol to give (LXXVII) and (LXXVIII), respectively (*G. Walter et al.*, Monatsh., 1933, **63**, 186; *Carter* and *Cheeseman, loc. cit.*).

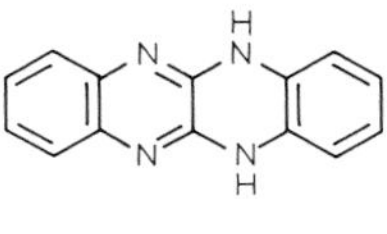

(LXXIV) 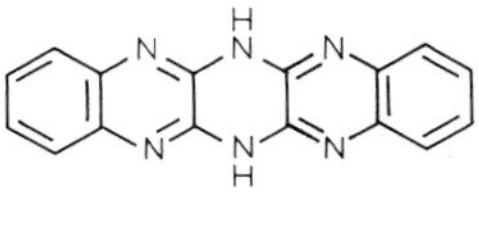   (LXXXV)

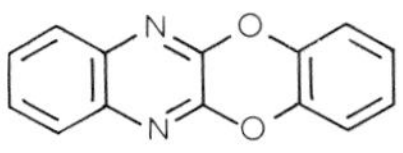

(LXXVI) 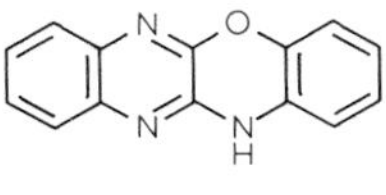   (LXXVII) 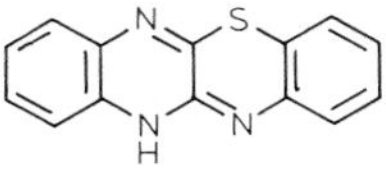   (LXXVIII)

(iii) Nitroquinoxalines

Direct nitration of quinoxaline using oleum and nitric acid affords a mixture of 5-*nitro*-, m.p. 96–97°, and 5,6-*dinitro*-, m.p. 172–173°, quinoxa-

TABLE 7
NITROQUINOXALINES

Compound	m.p. (°C)	Ref.
5-Nitro-	96–97	1
6-Nitro-	177	2
2,3-Dimethyl-5-nitro-	131–131.5	3
2,3-Dimethyl-6-nitro-	133–134	3
2,3-Diphenyl-6-nitro-	188	4
6,7-Dinitro-	193–194	5

1 H.P. Shultz, J. Amer. chem. Soc., 1950, **72**, 3824.
2 O. Hinsberg, Ann., 1896, **292**, 245.
3 J.K. Landquist and G.J. Stacey, J. chem. Soc., 1953, 2822.
4 A. Mangini and C. Deliddo, Gazz., 1933, **63**, 612.
5 Belg. Pat. 631,044, C.A., 1964, **60**, 15891.

line in 1.5% and 24% yield, respectively (*M.J.S. Dewar* and *P.M. Maitlas*, J. chem. Soc., 1957, 2518; *F.H. Case* and *J.A. Brennan*, J. Amer. chem. Soc., 1959, **81**, 6297); 4-nitrobenzotriazole has also been reported as a product of the nitration of quinoxaline (*R. Nasielski-Hinkens* and *M. Benedek-Vamos*, J. chem. Soc. Perkin I, 1975, 1229). Methoxy groups activate the benzo ring to nitration. Thus, 5-methoxy- and 6-methoxy-quinoxalines give 5-*methoxy-6,8-dinitro-*, m.p. 204–206°, and 6-*methoxy-5-nitro-*, m.p. 203°, *quinoxaline*. Catalytic hydrogenation (10% palladium on charcoal) of the nitro compounds affords the corresponding amino compounds (*H. Otomasu* and *S. Nakajima*, Chem. pharm. Bull. Tokyo, 1958, **6**, 566). Other nitro derivatives are prepared by standard condensation methods from nitro-*o*-phenylenediamines (Table 7).

(iv) Aminoquinoxalines

2-Aminoquinoxalines are most often prepared from (i) the reaction of 2-chloroquinoxalines with ammonia (Table 8, Refs. 2–5), or (ii) the hydrolysis of alloxazines (LXXIX) by sulphuric acid at 200° (Table 8, Ref. 1).

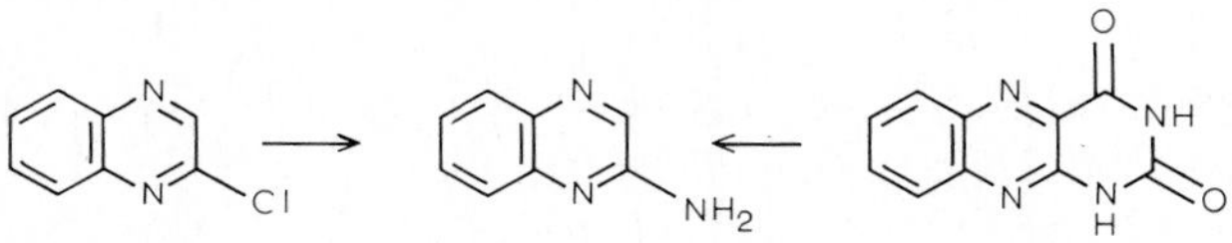

(LXXIX)

The Hofmann degradation of 2-quinoxalinecarboxamides in 5% sodium hypochlorite solution has been used to synthesis unambiguously the 6- and

TABLE 8
AMINOQUINOXALINES

	m.p. (°C)		m.p. (°C)
2-Aminoquinoxalines			
Parent	155–156 [1–3]	3-Methyl-	163–165 [3,4]
5-Methyl-	201–202 [4]	6-Methyl-	181–182 [5]
7-Methyl-	178–180 [5]	8-Methyl-	129 [4]
6,7-Dimethyl-	275–278 [3]	6-Chloro-	220–221 [6]
2-Phenyl-	163–164 [7]	7-Chloro-	219–220.5 [6]
5-Aminoquinoxalines			
Parent	92 [3,8]	7-Methyl-	103 [4]
2-Methyl-	83–84 [9]	2,3-Dimethyl-	162–163 [9]
6-Aminoquinoxalines			
Parent	158–159 [4,8,10]	7-Methyl-	194–195 [4]
2-Methyl-	162–163 [9]	8-Methyl-	158–160 [4]
2,3-Dimethyl-	186–187 [9]	2,3-Diphenyl-	175 [8,10]
Diaminoquinoxalines			
2,3-Diamino-	331 [2,8,13], > 360 [11]	3,6-Diamino-	204 [12]
6-Methyl-2,3-diamino-	249 [13]		
6-Chloro-2,3-diamino-	280–282 [13]		

1 J. Weijlard, M. Tishler and A.E. Erickson, J. Amer. chem. Soc., 1944, **66**, 1957; C.W. Atkinson, C.W. Brown and J.C.E. Simpson, J. chem. Soc., 1956, 26.
2 A.H. Gowenlock, G.T. Newbold and F.S. Spring, ibid., 1945, 622; G.W.H. Cheeseman, ibid., 1947, 3236; F.J. Wolf, U.S.P. 2,537,870; C.A., 1951, **45**, 4274.
3 F.J. Wolf, R.H. Bentel and J.R. Stevens, J. Amer. chem. Soc., 1948, **70**, 2572; 1949, **71**, 6.
4 B.C. Platt and T.M. Sharp, J. chem. Soc., 1948, 2129.
5 B.C. Platt, ibid., 1948, 1310.
6 H.R. Moreno and H.P. Schultz, J. org. Chem., 1971, **36**, 1158.
7 F. Krönhke and H. Leister, Ber., 1958, **91**, 1479.
8 K.A. Jensen, Acta Chem. Scand., 1948, **2**, 91.
9 P. Vestesnik, Coll. Czech. Chem. Comm., 1968, **33**, 556.
10 O. Hinsberg, Ann, 1887, **237**, 327.
11 H. Weidinger and J. Kranz, Ber., 1964, **97**, 1599.
12 T.S. Osdene and G.M. Timmis, J. chem. Soc., 1955, 2027.
13 E. Schipper and A.R. Day, J. Amer. chem. Soc., 1951, **73**, 5672.

7-chloro-2-aminoquinoxalines (Table 8, Ref. 6); alternative approaches result in mixtures of isomers.

Base-catalysed cyclisation of 2-amino-anilinoacetonitrile affords the 2-amino-3,4-dihydroquinoxaline (LXXX) which is subsequently aromatised using hydrogen peroxide (K. Pfister et al., J. Amer. chem. Soc., 1951, **73**, 4955).

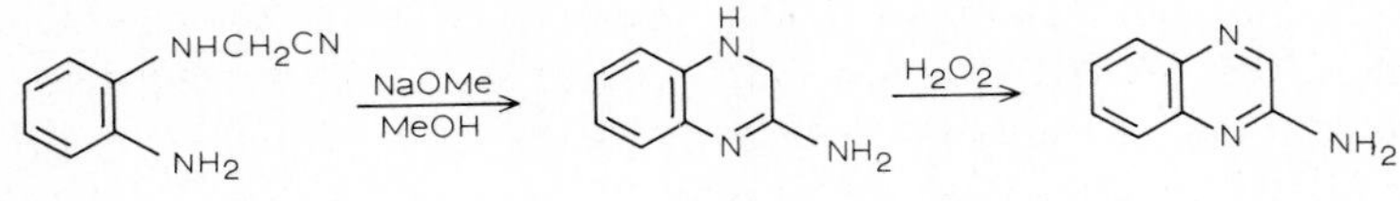

Condensation of *o*-phenylenediamine with *p*-(dimethylamino)anils of aroylnitriles in acetic acid gives 3-aryl-2-aminoquinoxalines (Table 8, Ref. 7; *F. Kröhke* and *K.F. Gross*, Ber., 1959, **92**, 22).

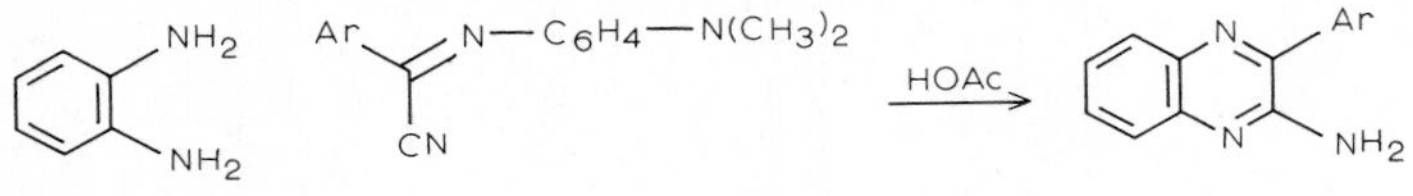

Substituted 2-aminoquinoxalines can be unambiguously prepared by the base-catalysed reaction of *o*-amino-nitrosobenzenes and phenylacetonitrile (*T.S. Osdene* and *G.M. Timmis*, J. chem. Soc., 1955, 2027).

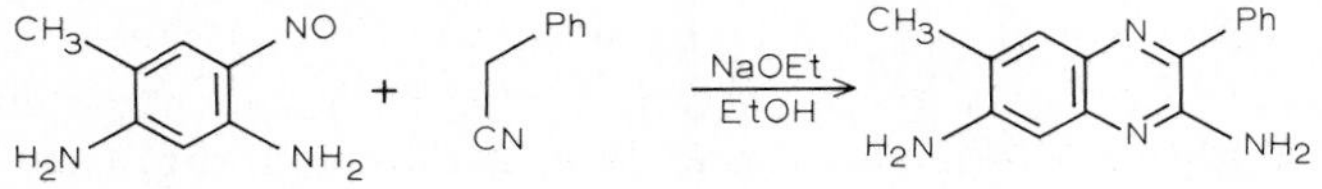

2-**Amino**-3-**chloroquinoxaline**, m.p. 139°, is obtained from 2,3-dichloroquinoxaline by nucleophilic displacement of one chlorine atom by ammonia (*R.D. Haworth* and *S. Robinson, ibid.*, 1948, 781; 1351).

2-Aminoquinoxaline is a relatively weak base (pK_a 3.96 in water at 25°) which exists predominantly in the amino rather than the imino form (*G.W.H. Cheeseman, ibid.*, 1958, 108). Hydrolysis of 2-aminoquinoxalines in either aqueous acid or alkali results in the formation of 2-hydroxyquinoxalines (quinoxalin-2(1*H*)-ones) (*J.A. Barltrop et al., ibid.*, 1959, 1132); the same transformation is effectively accomplished using sodium nitrite in conc. sulphuric acid (Table 8, Ref. 7).

2,3-Diaminoquinoxaline has been prepared by heating the 2,3-dichloro compound with anhydrous ammonia at 90° (*W. Deuschel* and *G. Riedel*, G.P. 1,135,471; C.A., 1963, **58**, 537; *W. Deuschel, W. Vilsmeier* and *G. Riedel*, Belg. Pat. 612,092; C.A., 1962, **57**, 16634; *A.P. Komin* and *M. Carmack*, J. heterocyclic Chem., 1976, **13**, 13). Primary and secondary aliphatic amines (*L.G.S. Brooker* and *E.J. VanLare*, U.S.P. 3,431,111; C.A., 1970, **72**, 68222; *I.N. Gorcharova* and *I.Ya. Postovskii*, J. gen. Chem. U.S.S.R., 1962, **32**, 3271; *W.R. Vaughan* and *M.S. Habib*, J. org. Chem.,

1962, **27**, 324) and anilines (*S.D. Carver* and *G.W.H. Cheeseman*, Tetrahedron, 1978, **34**, 981) participate in similar substitution reactions to give the corresponding 2,3-diamino derivatives.

2,3-Diaminoquinoxalines are prepared directly by base-catalysed condensation of *o*-phenylenediamines with cyanogen (*O. Hinsberg* and *E. Schwantes*, Ber., 1903, **36**, 4039; Table 8, Ref. 13) or the oxaldiimido ester derived from cyanogen (LXXXII, X = OMe) (Table 8, Ref. 11; *D.J. Sam* and *M.A. Wounola*, Ger. Offen., 2,339,012; C.A., 1974, **80**, 108578).

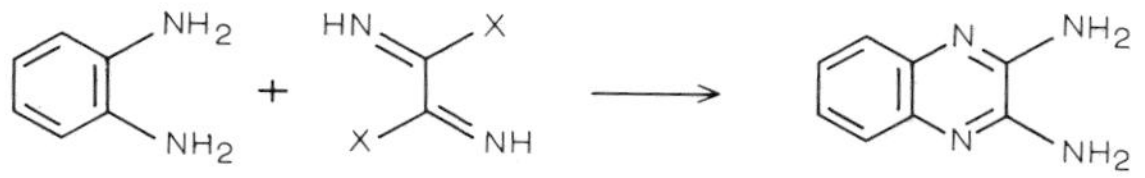

An analogous reaction between *o*-phenylenediamine and DISN (LXXXII, X = CN) under neutral conditions gives 2,3-diaminoquinoxaline (*R.W. Begland* and *D.R. Hartter*, J. org. Chem., 1972, **37**, 4136).

The fused ring system may be further extended linearly by condensation of 2,3-diaminoquinoxaline with α,β-dicarbonyl compounds (*E.B. Nyquist* and *M.M. Joullie*, J. chem. Soc. C, 1968, 947), or *o*-aromatic diamines [Scheme 10] (*H.M. Woodburn* and *W.E. Hoffman*, J. org. Chem., 1958, **23**, 262; *Deuschel, Vilsmeier* and *Riedel, loc. cit.*).

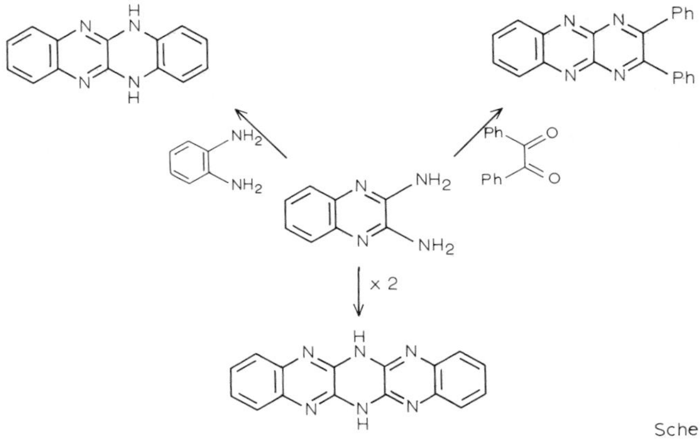

Scheme 10

(v) *Hydroxyquinoxalines (quinoxalin-2(1H)-ones)*

Spectroscopic (ν_{CO} 1660–1690 cm^{-1}) and X-ray crystallographic data indicate that 2-hydroxyquinoxaline exists predominantly in the amide tautomeric form (*G.W.H. Cheeseman, A.R. Katritzky* and *J. Okone*, J. chem. Soc., 1961, 3983; *A. Stephen et al.*, Acta Cryst., 1976, **B32**, 2048).

2-Hydroxyquinoxaline is prepared directly by condensation of butyl glyoxylate and *o*-phenylenediamine (*C.M. Atkinson et al.*, J. chem. Soc., 1956, 26; *G.W.H. Cheeseman, ibid.*, 1957, 3236; *K. Pfister et al.*, J. Amer. Chem. Soc., 1949, **71**, 6). 3-Substituted derivatives are obtained similarly by use of arylglyoxylic acids (*J. Druey* and *A. Huni*, Helv., 1952, **35**, 2301) and α-keto-acids and esters (*D.C. Morrison*, J. Amer. chem. Soc., 1954, **46**,

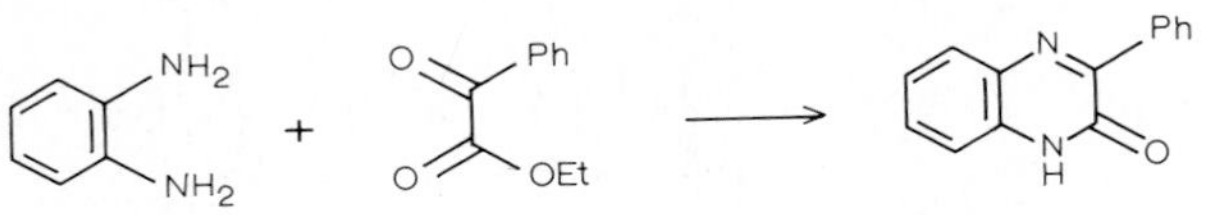

4483). Although the reactions of *o*-phenylenediamines with polyfunctional carbonyl compounds could give rise to larger ring heterocycles, 2-hydroxy-quinoxalines are normally obtained, *e.g.* ethyl mesoxalate gives (LXXXIII) (*R.M. Acheson*, J. chem. Soc., 1956, 4731) and ethyl ethoxalylacetate (*D.D. Chapman*, J. chem. Soc. C, 1966, 806) or diethyl acetylenedicarboxylate (*Y. Iwanami*, J. chem. Soc. Japan, 1961, **82**, 778; see also *H. Sushitzky et al.*, J. chem. Soc. Perkin I, 1975, 401) gives (LXXXIV) [Scheme 11]. Hydrolysis and decarboxylation of (LXXXIII) and (LXXXIV) affords the parent and 3-methyl-hydroxy compounds, respectively. The latter compound is also

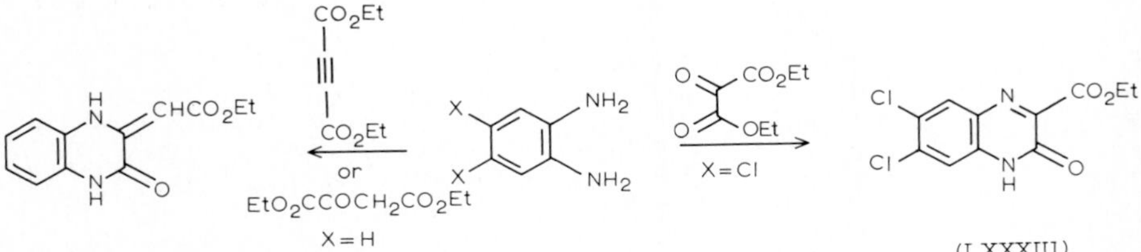

Scheme 11

directly prepared from the reaction of *o*-phenylenediamine and ethyl α-oximinoacetoacetate (*H. Dahn, J.P. Leresche* and *H.P. Schlunke*, Helv., 1966, **49**, 26).

TABLE 9
2-HYDROXYQUINOXALINES (QUINOXALINE-2(1H)-ONES)

Compound	m.p. (°C)	Compound	m.p. (°C)
Parent	269 [1], 267–269 [2]	3-Methyl-	250 [3], 251–252 [4]
5-Methyl-	282–283 [5]	6-Methyl-	274 [6]
7-Methyl-	270–272 [6]	3-Ethyl-	210 [3]
3,5-Dimethyl-	256.5–257.5 [7]	3,6-Dimethyl-	354 [8]
3,7-Dimethyl-	243–244 [9]	3,8-Dimethyl-	256.5–257.5 [7]
3-Benzyl-	196 [10]	3-Phenyl-	247 [11]
6-Chloro-	312–313 [12]	7-Chloro-	269–270 [12]
6-Nitro-	300–302 [13], 294 [14]	7-Nitro-	272–274 [13], 273 [14]
6,7-Dichloro-	343 (dec.) [15]		

1 O. Hinsberg, Ann., 1896, **292**, 245.
2 G.W.H. Cheeseman, J. chem. Soc., 1957, 3236.
3 Y.J. L'Italien and C.K. Banks, J. Amer. chem. Soc., 1951, **73**, 3246.
4 H. Suschitzky, B.J. Wakefield and R.A. Whittaker, J. chem. Soc. Perkin I, 1975, 401.
5 J.K. Landquist, J. chem. Soc., 1953, 2816.
6 B.C. Platt, ibid., 1948, 1310.
7 G. Kyryacos and H.P. Schultz, J. Amer. chem. Soc., 1953, **75**, 3597.
8 M. Munk and H.P. Schultz, ibid., 1952, **74**, 3433.
9 B. Marks and H.P. Schultz, ibid., 1951, **73**, 1368.
10 A.H. Cook and C.A. Perry, J. chem. Soc., 1943, 394.
11 H. Burton and C.W. Shoppee, ibid., 1937, 546.
12 W. Morano and H.P. Schultz, J. org. Chem., 1971, **36**, 1158.
13 G.W.H. Cheeseman, J. chem. Soc., 1961, 1246.
14 L. Horner, U. Schwenk and E. Junghauns, Ann., 1953, **579**, 212.
15 R.M. Acheson, J. chem. Soc., 1956, 4731.

Hydrolysis of carboxyureides, the condensation products of o-phenylenediamines and alloxan, in neutral or weakly acidic solution produces

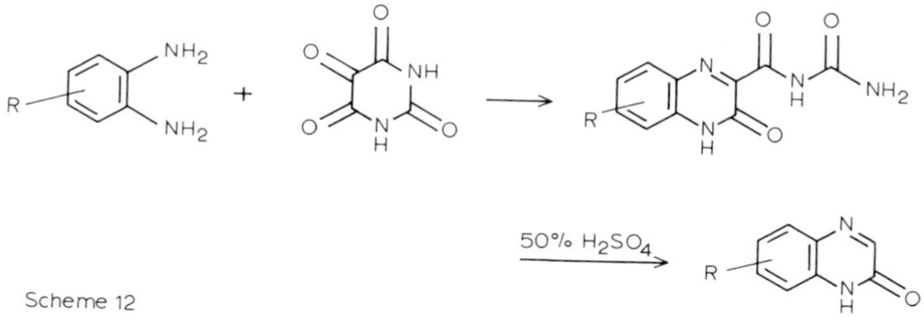

Scheme 12

2-hydroxyquinoxaline unsubstituted in the 3-position [Scheme 12] (*H.G. Petering* and *G.J. van Giessen*, J. org. Chem., 1961, **26**, 2818). Reaction of o-phenylenediamines with α-azido acids (*J.H. Boyer* and *D. Straw*, J. Amer. chem. Soc., 1953, **75**, 1642), and oxazolones (LXXXV) and (LXXXVI) (*F. Weygard et al.*, Ann., 1962, **658**, 128; *K. Burger* and *M. Eggersdrofer, ibid.*, 1979, 1547) gives rise to 3-substituted 2-hydroxyquinoxa-

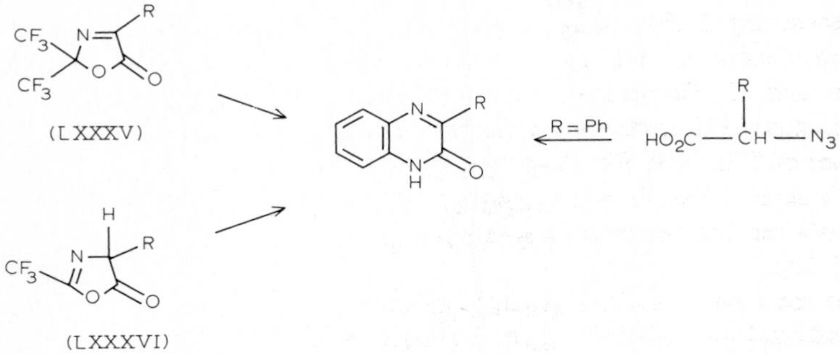

Scheme 13

lines [Scheme 13]. Condensation of imidoyl chloride with o-phenylenediamine gives 3-amino derivatives (*D. Bartholomew* and *I.T. Kay*, Tetrahedron Letters, 1979, 2827).

Unambiguous synthesis of 2-hydroxyquinoxaline derivatives can be accomplished by reductive cyclisation of α-(o-nitroanilino)carboxylic acids, derived from the appropriate anilines and α-halo acids, followed by oxidation of the resulting dihydro-2-hydroxyquinoxalines (*B.C. Platt*, J. chem. Soc., 1948, 1310; *R.C. van Dusen* and *H.P. Schultz*, J. org. Chem., 1956, **21**, 1326). Ring closure of the o-nitroanilides (LXXXVII) affords the 2-hydroxyquinoxaline-4-oxides which are subsequently reduced using sodium dithionite (*G. Tennant*, J. chem. Soc. C, 1966, 2285).

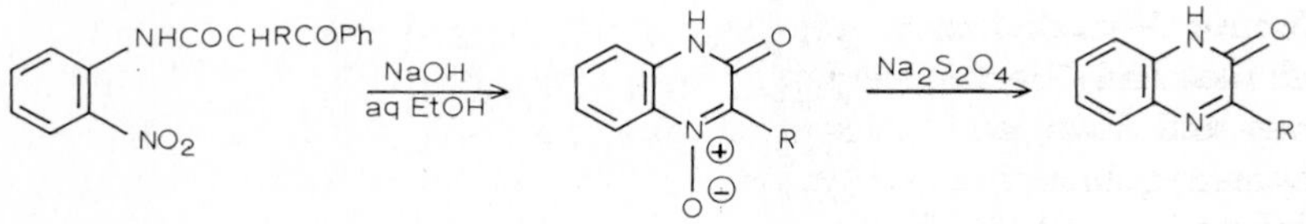

(LXXXVII)

2-Hydroxyquinoxalines are also prepared by (i) hydrolysis of 2,3-diaminoquinoxaline in dilute hydrochloric acid to give 3-amino-2-hydroxy compounds (*E. Schipper* and *A.R. Day*, J. Amer. chem. Soc., 1951, **73**, 5672); (ii) diazotisation of 2-aminoquinoxalines using nitrous acid (*F.*

Kronke and *H. Leister*, Ber., 1958, **91**, 1479); (iii) treatment of 3-substitued quinoxaline-1-oxides with acetic anhydride (*E. Hayashi* and *C. Iijima*, J. pharm. Soc. Japan, 1962, **82**, 1093).

2-Hydroxyquinoxalines, though sparingly soluble in organic solvents, dissolve readily in aqueous ammonia or alkali. They are transformed to the corresponding 2-chloroquinoxalines on treatment with either phosphoryl chloride (Table 9, Ref. 2), or thionyl chloride in dimethylformamide (*F. Eiden* and *G. Bachmann*, Arch. Pharm., 1973, **306**, 401). Nitration or chlorination of 2-hydroxyquinoxaline in acetic acid gives the 7-substituted derivatives (Table 9, Ref. 13; *H. Otomasu* and *K. Yoshida*, Chem. pharm. Bull. Tokyo, 1960, **8**, 475). In concentrated sulphuric acid, the nitration proceeds *via* the protonated substrate to give the 6-nitro compound (Table 9, Ref. 13).

The reaction of Grignard reagents with 2-hydroxyquinoxalines followed by mild oxidation gives 3-substituted products (*A. Maxer et al.*, Helv. 1971,

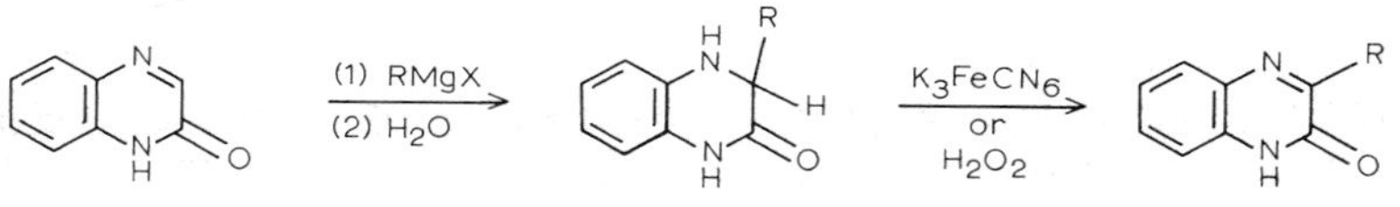

54, 2507). With diazomethane, 2-hydroxyquinoxalines afford a mixture of the *O*-methyl (LXXXVIII) and the *C,N*-dimethyl (LXXXIX) derivative.

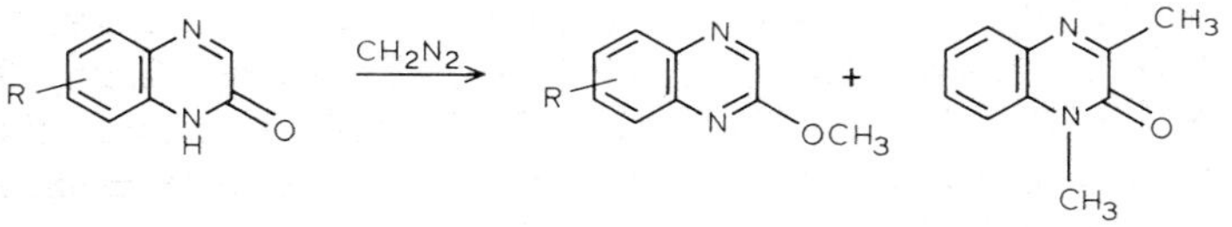

Selective *N*-methylation is effected by treating the 2-hydroxyquinoxaline with base and dimethyl sulphate or methyl iodide (Table 9, Ref. 13: see also *Druey* and *Huni, loc. cit.* for more specific alkylation procedures). 3-Amino-2-hydroxyquinoxaline is selectively methylated at the 1-position by dimethyl sulphate in alkali (*G.W.H. Cheeseman*, J. chem. Soc., 1955, 1804). *O*-Trimethylsilyl derivatives are prepared using hexamethyldisilazane in ammonium sulphate or bistrimethylsilyltrifluoroacetamide (*U. Langenbeck et al.*, J. Chromatogr., 1975, **115**, 65; *F. Reisser* and *W. Pfleiderer*, Ber., 1966, **99**, 547).

N-Substituted quinoxalinones can also be prepared directly from *N*-substituted *o*-phenylenediamines by condensation with α-keto-esters (Table 9,

TABLE 10
1-SUBSTITUTED-QUINOXALIN-2(1*H*)-ONES

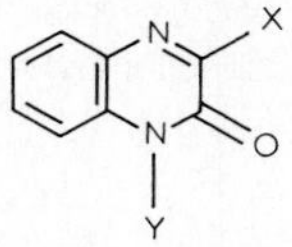

Substituents		m.p. (°C)
X	Y	
H	CH_3	122 [1]
CH_3	CH_3	87 [2,3]
CH_3	CH_2Ph	99–100 [2]
CH_3	Ph	195 [2,3]
CH_3	NH_2	273–275 [4]
Ph	CH_3	134–136 [5]

1 O. Kühling and O. Kaselitz, Ber., 1906, **39**, 1314.
2 F. Kehrmann and J. Messinger, Ber., 1892, **25**, 1627.
3 A.H. Cook and C.A. Perry, J. chem. Soc., 1943, 394.
4 G.W.H. Cheeseman, ibid., 1955, 1804.
5 J. Druey and A. Huni, Helv., 1952, **35**, 2301.

Ref. 13) or *via* alloxazines which are in turn hydrolysed and decarboxylated (Table 10, Ref. 1) [Scheme 14].

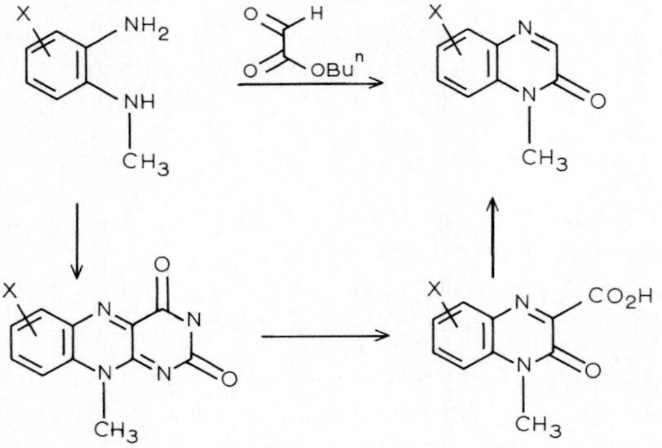

Scheme 14

Reduction of *N*-substituted quinoxalinones with lithium aluminium hydride affords the corresponding 1,2-disubstituted tetrahydroquinoxalines which can be oxidised to quaternary salts (*Druey* and *Huni, loc. cit.*).

A. *2,3-Dihydroxyquinoxalines (quinoxaline-2,3(1H,4H)-diones)*

I.r.- and u.v.-spectroscopic studies generally favour the diketo form for 2,3-dihydroxyquinoxalines (*Cheeseman, Katritzky* and *Okone, loc. cit.*). Condensation of *o*-phenylenediamines with excess of either diethyl oxalate or oxalic acid affords 2,3-dihydroxyquinoxalines in good yield (Table 11, Refs. 1–3). Alternative synthetic procedures include (i) the hydrolysis of 2,3-difluoroquinoxalines (*J. Hamer*, J. heterocyclic Chem., 1966, **3**, 244; *W.R.K. Musgrave et al.*, J. fluorine Chem., 1971, **1**, 59); (ii) the hydroxylation of 2-hydroxyquinoxaline with hydrogen peroxide (*A.H. Gowenlock, G.T. Newbold* and *F.S. Spring*, J. chem. Soc., 1945, 622); (iii) the diazotisation of 3-amino-2-hydroxyquinoxaline by aqueous nitrous acid (*D. Shiho* and *S. Tagami*, Chem. pharm. Bull. Tokyo, 1957, **5**, 45); (iv) oxidative decarboxylation of 2-hydroxyquinoxaline-3-carboxylic acid and quinoxaline-2,3-dicarboxylic acid (*G.T. Newbold* and *F.S. Spring*, J. chem.

TABLE 11
2,3-DIHYDROXYQUINOXALINES (QUINOXALINE-2,3(1*H*,4*H*)-DIONES)

Compound	*m.p. (°C)*	*Compound*	*m.p. (°C)*
Parent	360 [1,2], > 360 [3]	6-Methyl-	> 300 [4]
6-Chloro-	380 [5]	6-Bromo-	339–340 [6]
		1-Methyl deriv.	214–216 [6]
		1,4-Dimethyl deriv.	205–206 [7]
6,7-Dichloro-	> 360 [6]	5-Nitro-	284 [8], 296 (dec.) [9]
6,7-Dinitro-	> 360 [6]	6-Nitro-	344–346 [7,8], 355 [10]
5-Quino-	344 [8]	5-Methoxy-	377–274 [11]
	339–340 [9]		
6-Amino-	350 [8]	6-Methoxy-	
		1,4-Dimethyl deriv.	182–183 [7]

1 O. Hinsberg and J. Pollak, Ber., 1896, **29**, 784.
2 G.T. Newbold and F.S. Spring, J. chem. Soc., 1948, 519.
3 H.I.X. Mager and W. Berends, Rec. Trav. chim., 1958, **77**, 842.
4 O. Hinsberg, Ann., 1887, **237**, 327.
5 F. Kehrmann and C. Bener, Helv., 1925, **8**, 20; R.D. Haworth and S. Robinson, J. chem. Soc., 1948, 777.
6 G.W.H. Cheeseman, ibid., 1962, 1170.
7 F.H.S. Curd, D.G. Darg and G.J. Stacey, ibid., 1949, 1271.
8 E.H. Huntress and J.V.K. Gladding, J. Amer. chem. Soc., 1942, **64**, 2644.
9 H.I.X. Mager and W. Berends, Rec. Trav. chim., 1959, **78**, 5.
10 L. Horner, U. Schrenk and E. Junghanns, Ann., 1953, **579**, 212.
11 E.S. Lane and C. Williams, J. chem. Soc., 1956, 2983.

Soc., 1948, 519); and (v) oxidation of quinoxaline with aqueous nitric acid (*M. Asia*, J. pharm. Soc. Japan 1959, **79**, 269).

2,3-Dihydroxyquinoxaline is insoluble in water and organic solvents and behaves as a monobasic acid. It has been used as a specific reagent for barium (*A. Steigmann*, J. Soc. chem. Ind., 1943, **62**, 42).

The **mono-*N*-methyl**, m.p. 278–280° (*G.W.H. Cheeseman*, J. chem. Soc., 1955, 1804) and the *N,N*-**dimethyl**, m.p. 252–253° (*idem, ibid.*, 1961, 1246) derivative have been obtained by methylation of 2,3-dihydroxyquinoxaline using dimethyl sulphate in aqueous alkali. The latter compound is clearly distinguished from the isomeric 2,3-*dimethoxy* compound, m.p. 92–94°, prepared from 2,3-dichloroquinoxaline (*G.T. Newbold* and *F.S. Spring, ibid.*, 1948, 519). Other 2,3-dialkoxy derivatives are prepared by nucleophilic displacement reaction of alkoxide ion with 2,3-dichloroquinoxaline (see *R. Patton* and *H.P. Schultz*, J. Amer. chem. Soc., 1951, **73**, 5899; Table 11, Ref. 7; and *A.F. Crowther et al.*, J. chem. Soc., 1949, 1260).

2,3-Dihydroxyquinoxalines are converted into the corresponding 2,3-dichloro compounds by a mixture of phosphoryl chloride and dimethylaniline (*G.W.H. Cheeseman, ibid.*, 1955, 1804), or phosphorus pentachloride (*J.K. Landquist, ibid.*, 1953, 2816), or thionyl chloride in dimethylformamide (*A.F. Komin* and *M. Carmack*, J. heterocyclic Chem., 1976, **13**, 13). Similarly, 2,3-dibromoquinoxaline is obtained from 2,3-dihydroxyquinoxaline with phosphoryl bromide in dimethylaniline (*C. Sosnowski* and *L. Wojciechowski*, C.A., 1971, **74**, 125728).

Electrophilic substitution of 2,3-dihydroxyquinoxaline takes place at the 6- and 7-position, *e.g.* nitration with one or two equivalents of potassium nitrate in concentrated sulphuric acid yields the 6-nitro and the 6,7-dinitro compound, respectively (Table 11, Ref. 6).

Reaction of 2,3-dihydroxyquinoxaline with *o*-phenylenediamine affords 2,2′-dibenzimidazoyl (XC) (*E.S. Lane*, J. chem. Soc., 1955, 1079).

(XC)

Quinoxalines which are hydroxylated in the carbocyclic ring are normally prepared by condensation of the appropriate methoxy *o*-phenylenediamine with glyoxal, followed by demethylation of the adduct using aluminium chloride. Several simple benzohydroxylated quinoxalines together with their respective derivatives are listed in Table 12.

(vi) Quinoxaline-N-oxides

A large number quinoxaline-*N*-oxides have been prepared in the search for biologically active compounds. The 1,4-dioxide of 6-chloroquinoxaline-2-carboxylic acid (XCI), isolated from culture of *Streptomyces ambofaciens*

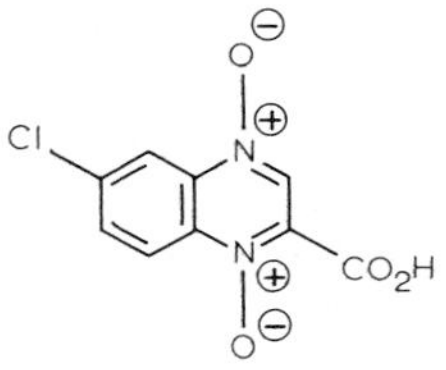

(XCI)

NRRL 3455, and its ester derivatives exhibit anti-biotic activity (*E.P. Stapley et al.*, U.S.P. 3,692,633; C.A., 1972, **77**, 163041). The 2-carboxylic acid derivatives of several quinoxaline-1,4-dioxides are anti-bacterial agents (*A.S. Elina et al.*, Chem. heterocyclic Compd., 1969, **5**, 540; *G.W.H. Cheeseman* and *E.S.G. Westeriuk*, Adv. heterocyclic Chem., 1978, **22**, 367). *In vivo* hydroxylation of the side-chain results in higher biological activity (*J. Francis* and co-workers, Biochem. J., 1956, **63**, 455). Various other quinoxaline-1,4-dioxide derivatives are reported as bactericides (*M.L. Edwards et al.*, J. med. Chem., 1975, 8, 637; *A. Monge et al.*, An. Quim., 1975, **71**, 248), fungicides (*R.A. Burrell et al.*, J. chem. Soc. Perkin I, 1973, 2707; *R.R. Shaffer*, U.S.P. 3,560,616; C.A., 1971, **75**, 47839), and growth promoters (*R.H.B. Galt*, U.S.P. 3,479,354; C.A., 1970, **72**, 79095; *J.D. Johnston*, U.S.P. 3,444,022; C.A., 1967, **67**, 111452).

Quinoxalines are conventionally transformed to the corresponding mono- and di-*N*-oxides by oxidising agents such as 30% hydrogen peroxide or peracids. With one equivalent of peracetic acid in acetic acid, quinoxaline affords the mono-*N*-oxide; with an excess of reagent, the di-*N*-oxide is

TABLE 12
HYDROXYQUINOXALINES WITH HYDROXYL GROUP IN BENZO RING

	m.p. (°C)		
	Hydroxy-	Acetoxy-	Methoxy-quinoxaline
5-	100–101 [1]	103–104 [1]	72–73 [1]
6-	242 [1]	80–81 [1]	60 [2]
5,6-di	190 [1]	112 [1]	69–70 [1]
5,7-di	250 (dec.) [1]	113 [1]	110 [1]
5,8-di	230 [1]	209 [1]	146 [1]
6,7-di	260 (dec.) [1]	112 [1]	150–151 [3]

1 *F.E. King, N.G. Clark* and *P.M.H. Davis*, J. chem. Soc., 1949, 3012.
2 *J.G. Cavagnol* and *F.Y. Wiselogle*, J. Amer. chem. Soc., 1947, **69**, 795.
3 *J. Ehrlich* and *M.T. Bogert*, J. org. Chem., 1947, **12**, 522.

TABLE 13
QUINOXALINE-N-OXIDES

	m.p. (°C)		
	1-Oxide	4-Oxide	1,4-Dioxide
Parent	122–123 [1]	–	241–243 [1]
2-Methyl-	93–94 [2]	118 [3]	180–181 [4]
2,3-Dimethyl-	92–93 [4]	–	189 [9]
2-Phenyl-	154 [5]	137–138 [4]	202–203 [4,5]
2,3-Diphenyl-	192–193 [4,6] 197 [5]	–	209–211 [4]
2-Chloro-	114 [6] 114–115 [7]	150–152 [1,7]	–
2-Amino	191–192 [8]	276 (dec.) [8]	242 (dec.) [8]
2-Cyano-	130–130.5 [3]	163 [3]	–
2-Formyl-	129 [3]	169 [3]	189–190 [10]
2-Carboxy-	180–181 [10]	180–182 [10]	208–209 [10]

1 J.K. Landquist, J. chem. Soc., 1953, 2816.
2 A.S. Elina, J. gen. Chem. U.S.S.R., 1962, **32**, 2919.
3 E. Hayashi and C. Iijima, J. pharm. Soc. Japan, 1964, **84**, 163.
4 J.K. Landquist and G.J. Stacey, J. chem. Soc., 1953, 2822.
5 E. Hayashi and C. Iijima, J. pharm. Soc. Japan, 1962, **82**, 1093.
6 J.K. Landquist, J. chem. Soc., 1953, 1830.
7 G.W.H. Cheeseman and E.S.G. Torz, J. chem. Soc. C, 1966, 157.
8 A.S. Elina and I.G. Tsyrul'nikova, J. gen. Chem. U.S.S.R., 1963, **33**, 1507.
9 K. Ley and F. Seng, Synthesis, 1975, 415.
10 A.S. Elina and O.Yu. Magidson, J. gen. Chem. U.S.S.R., 1955, **25**, 145.

formed (*J.K. Landquist*, J. chem. Soc., 1953, 2816). Other peracids such as monoperphthalic acid (*E. Hayashi* and *C. Iijima*, J. pharm. Soc. Japan, 1964, **84**, 163), *m*-chloroperbenzoic acid (*Galt, loc. cit.*), and permaleic acid (*R.A. Burrell et al., loc. cit.*) have also been used. When the heterocyclic ring is unsubstituted, 2,3-dihydroxy compounds may also be obtained in addition to N-oxides (*M. Asia*, J. pharm. Soc. Japan, 1959, **79**, 260; *J.K. Landquist* and *G.J. Stacey*, J. chem. Soc., 1953, 2822).

Peroxidation of 5-substituted quinoxalines generally stops at the 1-oxide, except for the 5-methoxy derivative which readily gives the 1,4-dioxide. Quinoxalines with electron donating substituents at the 6-position generally produce the 1,4-dioxide. Conversely, quinoxalines with electron withdrawing substituents at the 6-position afford significant quantities of the corresponding 2,3-dihydroxy compounds (*Landquist, loc. cit., J.A. Silk, ibid.*, 1956, 2058). With 2- and 3-alkyl-quinoxalines, mixtures of the 1- and 4-N-oxides are often obtained (*Hayashi* and *Iijima, loc. cit.*). Oxidation of

2-aminoquinoxaline with permaleic acid gives the 1-oxide in high yield (*Burrell et al., loc. cit.*). In contrast, electron withdrawing groups (chloro, carboxylic acid, carboxy derivatives) or a phenyl group at the 2-position direct oxidation to the more remote 4-position (*G.T. Newbold* and *F.S. Spring*, J. chem. Soc., 1948, 519; *G.W.H. Cheeseman, ibid.*, 1957, 3236).

Quinoxaline-*N*-oxides are also available by direct synthesis from *o*-nitroaniline derivatives. Treatment of 2-nitro-aniline with α-substituted acetyl chlorides, followed by base-catalysed cyclisation of the resulting adduct produces mono-*N*-oxides (XCII) in good yield (*G. Tennant, ibid.*, 1963, 2428; 1964, 2666; J. chem. Soc. C, 1966, 2285; *Y. Ahmad et al.*, Tetrahedron, 1964, **20**, 1107; 1965, **21**, 861; *R. Fusco* and *S. Rossi*, Gazz.,

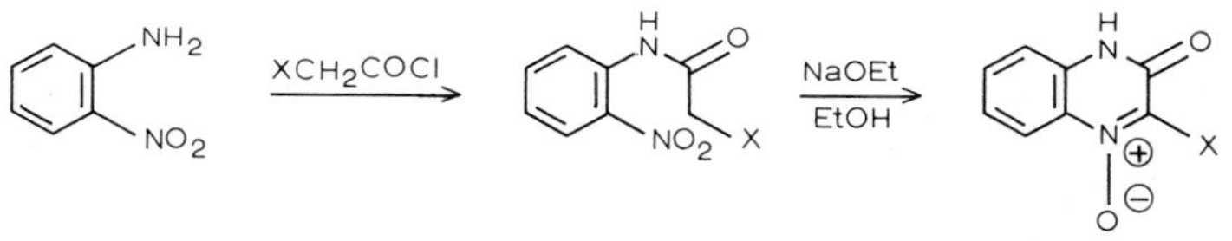

(XCII)

1964, **94**, 3). By analogy, reductive cyclisation of the cyanoformyl derivative of *o*-nitroaniline affords the mono-*N*-oxide (XCIII) (*C.W. Jefford* and *E.C. Taylor*, Chem. Ind. (London), 1963, 1559). The reaction of 2,4-di-

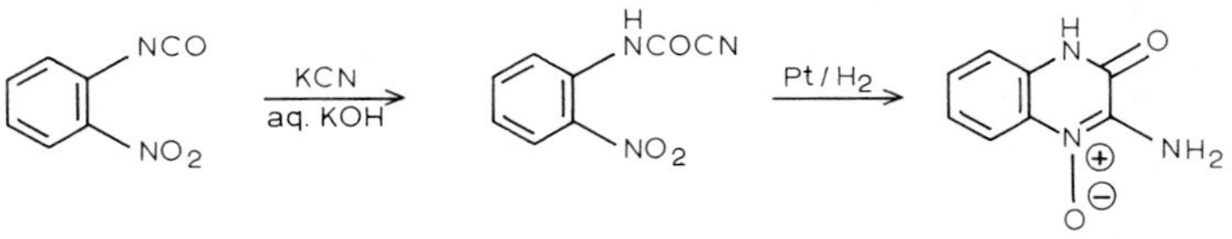

(XCIII)

nitrofluorobenzene (Sangor's reagent) and α-phenylacetamidines provides a useful one-step synthesis of quinoxaline mono-*N*-oxides (XCIV) (*M.J. Strauss et al.*, J. org. Chem., 1978, **43**, 2041).

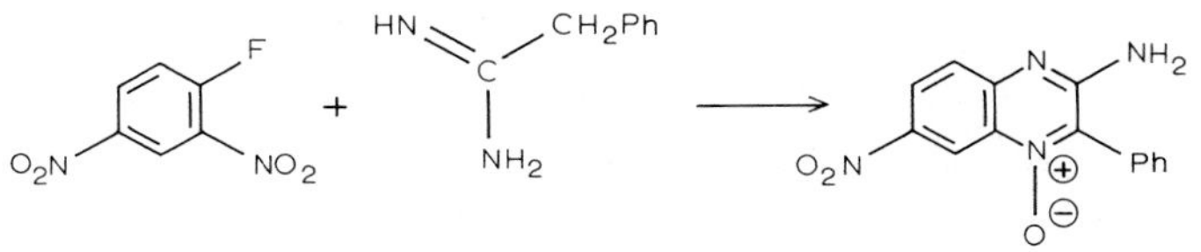

(XCIV)

More recently, many quinoxaline-1,4-dioxides have been prepared from benzofuroxans *via* the Beirut procedure which overcomes many of the difficulties, indicated above, associated with the direct oxidation methods (reviews: *M.J. Haddadin* and *C.H. Issidorides*, Heterocycles, 1976, **4**, 767; *K. Ley* and *F. Seng*, Synthesis, 1975, 415; *A. Gasco* and *A.J. Boulton*, Adv. heterocyclic Chem., 1981, **29**, 251). Thus, the reaction of benzofuroxan and an enamine affords the corresponding quinoxaline-1,4-dioxide in one step (*M.J. Haddadin* and *C.H. Issidorides*, Tetrahedron Letters, 1965, 3253;

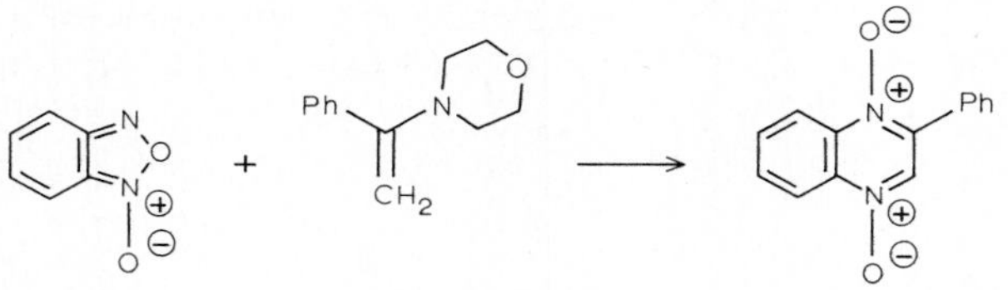

M.J. Haddadin and co-workers, J. chem. Soc. Perkin I, 1972, 965). The enamine may be conveniently generated *in situ, e.g.* benzofuroxan and ethyl methyl ketone in methanolic ammonia affords 2,3-dimethylquinoxaline-1,4-dioxide in 91% yield (*Ley* and *Seng, loc. cit.*). Similarly, enolate anions of ketones, β-diketones, β-keto-esters, β-keto-amides, β-cyanoketones and other related species, being electronic equivalents of enamines, all react with bezofuroxan to give the corresponding 1,4-dioxides (*J.C. Mason* and *G. Tennant*, Chem. Comm., 1971, 586; *E. Abushanab*, J. org. Chem., 1973, **38**, 3105; *D.P. Claypool et al., ibid.*, 1972, **37**, 2372; *F. Seng et al.*, Angew. Chem. intern. Edn., 1969, **8**, 596); *e.g.* dibenzoylmethane and benzofuroxan in alcoholic potassium hydroxide gives 2-**phenyl**-3-**benzoylquinoxaline** (XCV), m.p. 234° (*C.H. Issidorides* and *M.J. Haddadin*, J. org. Chem., 1966, **31**, 4067). Malononitrile, ethyl cyanoacetate and ynamines also react

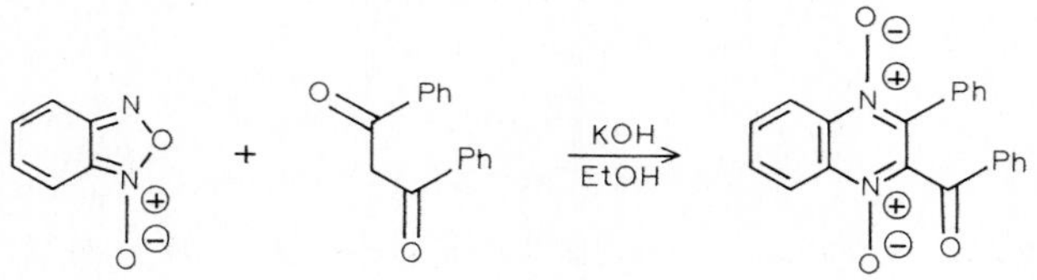

(XCV)

readily with benzofuroxan to produce a series of 2,3-disubstituted quinoxaline derivatives [Scheme 15] (*Ley* and *Seng, loc. cit.*).

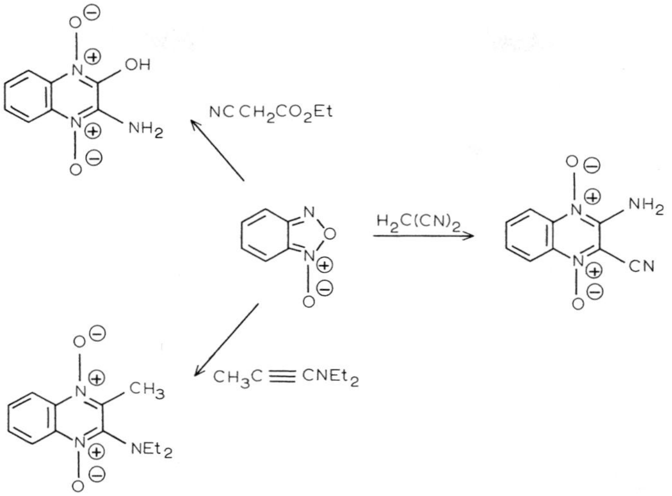

Scheme 15

With unsymmetrically substituted benzofuroxans and/or unsymmetrical co-reactants, mixtures of products could be, in principle, formed. In some instances, however, a degree of selectivity is observed, *e.g.* only products of structural formula (XCVI) are obtained from the reaction of 5-substituted benzofuroxan and benzoylacetonitrile (*Mason* and *Tennant, loc. cit.*).

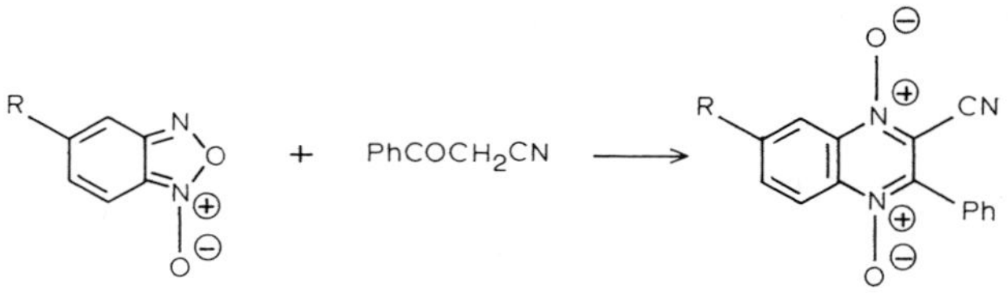

(XCVI)

Although the scope is limited, mono-*N*-oxides can also be obtained from benzofuroxans by reaction with:

(i) benzofuranones to give 3-(*o*-hydroxyaryl)quinoxaline-1-oxides (XCVII) which on treatment with acetic anhydride cyclise to the fused polycyclic system (XCVIII) (*C.H. Issidorides* and co-workers, J. chem. Soc. Perkin I, 1974, 1687);

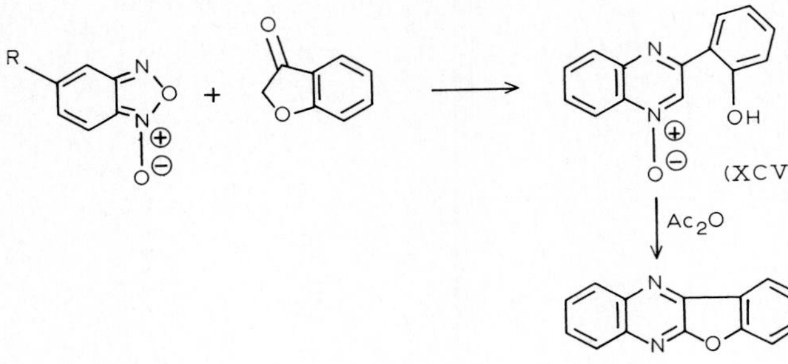

(ii) the sodium salt of ethyl oxaloacetate to give 2,3-bis(ethoxycarbonyl)quinoxaline-1-oxide (XCIX) (*M.L. Edwards et al.*, J. med. Chem., 1976, **19**, 330);

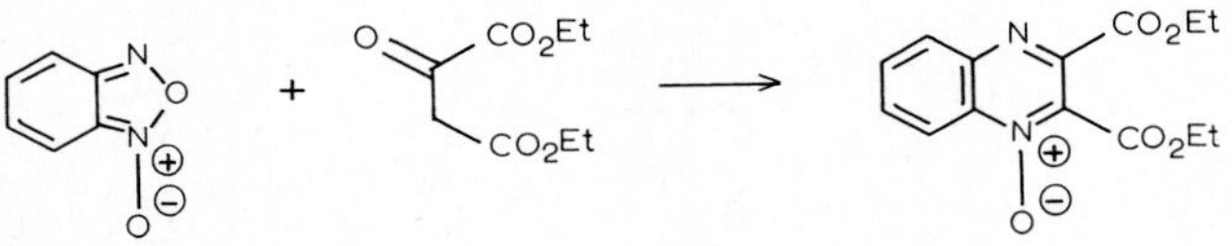

(iii) enones in the presence of a primary or secondary amine; the course of the reaction depends on the nature of the base (*M.L. Maddox et al.*, J. org. Chem., 1980, **45**, 1909).

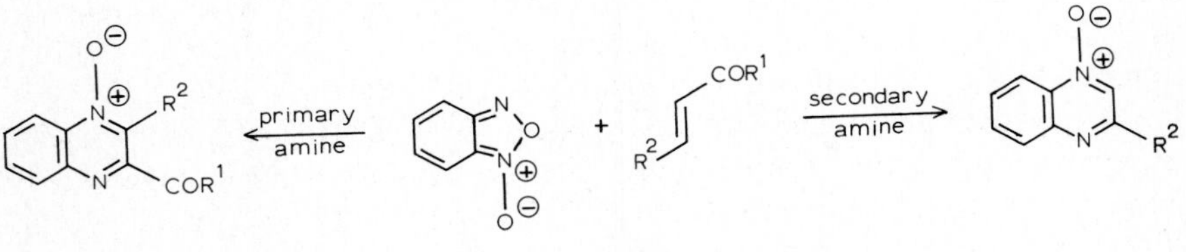

Reaction of *o*-benzoquinone dioxime with α-diketones or α-hydroxyketones, and α,β-unsaturated ketones yields the quinoxaline-1,4-dioxides (C) and (CI), respectively (*E. Abushanab*, J. org. Chem., 1970, **35**, 4279; B.P. 1,271,194; C.A., 1972, **77**, 34575). With glyoxal or pyruvaldehyde, the hydroxamic acids (CII, R = H or CH$_3$) are obtained [Scheme 16] (*Abushanab, loc. cit.; E. Abushanab* and *N.A. Alterni*, J. org. Chem., 1975, **40**, 157).

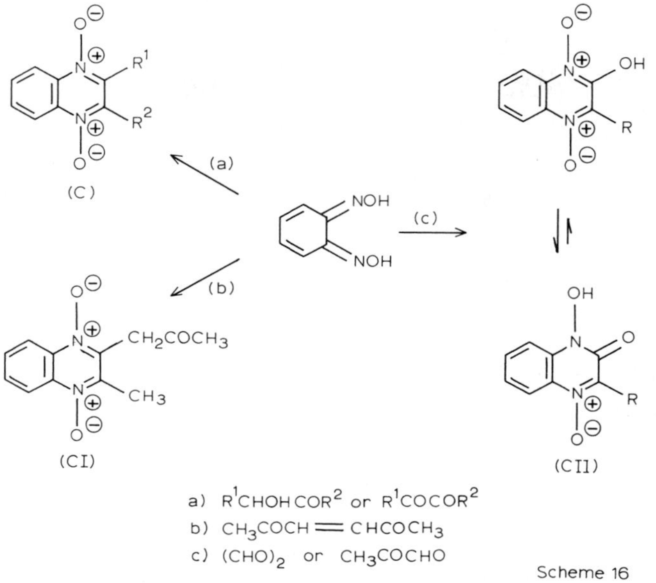

a) $R^1CHOHCOR^2$ or R^1COCOR^2
b) $CH_3COCH = CHCOCH_3$
c) $(CHO)_2$ or CH_3COCHO

Scheme 16

Phosphoryl chloride converts mono- and di-oxides to the 2-chloro and 2,3-dichloro derivatives, respectively (*J.K. Landquist*, J. chem. Soc., 1953, 2816). Mixtures of 2- and 6-chloroquinoxalines are reported from the reaction of the 1-oxide with phosphoryl chloride, or sulphuryl chloride, or acetyl chloride (*C. Iijima*, J. pharm. Soc. Japan, 1967, **87**, 942). With acetyl chloride at room temperature, quinoxaline-1,4-dioxide gives the 6-chloroquinoxaline-1-oxide (*Y. Ahmed et al.*, J. org. Chem., 1973, **38**, 2176) whereas with benzenesulphonyl chloride the isomeric 2-chloroquinoxaline-1-oxide is obtained (*A.S. Elina*, Chem. heterocyclic Compd., 1967, **3**, 576).

Deoxygenation of quinoxaline-*N*-oxides is achieved conventionally by treatment with either sodium dithionite or phophorus tribromide, or by catalytic hydrogenation over Raney nickel. Several newer reagents including titanous chloride in methanol or tetrahydrofuran (*B.W. Cue et al.*, Org. Prep. Proced. Int., 1977, **9**, 263), hexachlorosilane, iodomethylsilane, trifluoroacetic acid/sodium iodide, and titanium tetrachloride/zinc (*F.R. Homaidan* and *C.H. Issidorides*, Heterocycles, 1981, **16**, 411) have been developed to facilitate efficient bis-deoxygenation of 1,4-dioxides. Trimethyl phosphite in refluxing *n*-propanol is reported to effect mono-deoxygenation of certain 1,4-dioxides (*J.P. Dirlam* and *J.W. McFarlane*, J. org. Chem., 1977, **42**, 1360).

Quinoxaline-*N*-oxides undergo rapid photochemical deomposition; in water, the mono- and di-oxides give 2-hydroxyquinoxaline and 2-hydroxyquinoxaline-4-oxide, respectively (*J.K. Landquist*, J. chem. Soc., 1953, 2830). Irradiation of 1-phenyl-, or 3-phenyl-, or 2,3-diphenyl-quinoxaline-1-oxide affords benzoxadiazepines, *e.g.* (CIII) from 2,3-diphenylquinoxaline-1-oxide (*S. Kameko et al.*, Tetrahedron Letters, 1967, 1873; *O. Buchardt et al.*, Acta Chem. Scand., 1967, **21**, 1399; 1968, **22**, 877; see also *A. Albini et al.*, J. chem. Soc. Perkin I, 1978, 924).

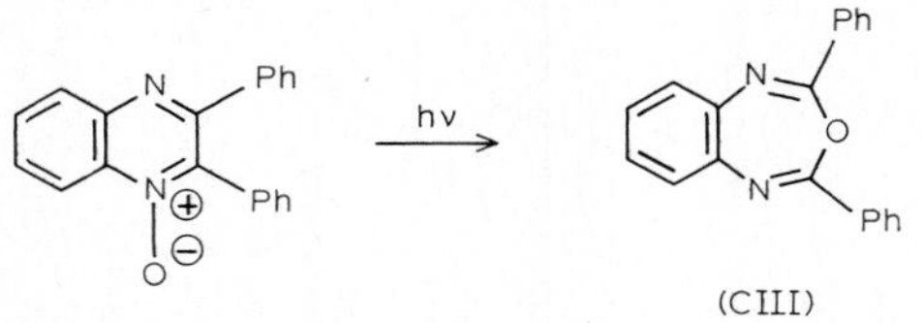

Treatment of 2-substituted quinoxaline-4-oxides with alkaline hydrogen peroxide in methanol results in contraction of the heterocyclic ring to give benzimidazole oxides (CIV) (*E. Hayashi* and *Y. Miura*, J. pharm. Soc. Japan, 1967, **87**, 648).

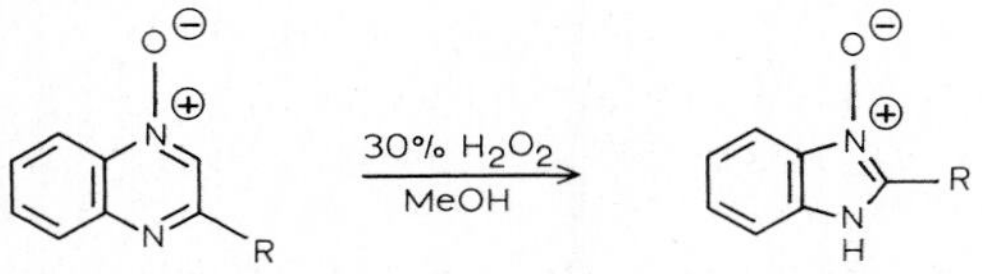

(vii) Polyhydroxyalkylquinoxalines

The condensation of *o*-phenylenediamines with aldoses can give rise to either quinoxalines or benzimidazoles (*P. Greiss* and *G. Harrow*, Ber., 1887, **20**, 2205; 3111). The formation of 2-polyhydroxyalkylquinoxaline derivatives is favoured by the addition of hydrazine and boric acid. Thus, glucose and fructose both give rise to 2-(D-*arabotetrahydroxybutyl*)*quinoxaline* (CV),

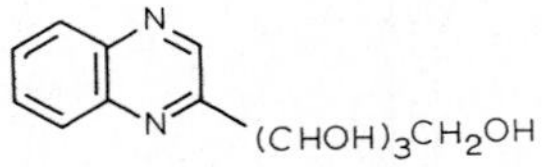

m.p. 152° (dec.), *monohydrate*, m.p. 189–190°, in 35% and 62%, respectively (*H. Ohle* and *M. Hielscher*, Ber., 1941, **74**, 13; *F. Weygard* and *A. Bergmann*, Ber., 1947, **80**, 255). Pentoses such as L-arabinose and L-sorbose condense with *o*-phenylenediamines in the presence of pyridine, hydrazine and acetic acid yielding the analogous quinoxaline derivatives (*H. Ohle* and *J.J. Kruyff*, Ber., 1944, **77**, 507; *W.S. Chilton* and *R.C. Krahn*, J. Amer. chem. Soc., 1968, **90**, 1318).

Compound (CV) has also been prepared by the reaction of *o*-phenylenediamine with (i) *p*-tolyl-D-isoglucosamine and hydrazine in acetic acid at 100° (90% yield) (*H.I.X. Mager* and *W. Berends*, Rec. Trav. chim., 1958, **77**, 827), or (ii) sucrose in dilute acetic acid at reflux with simultaneous aeration of the reaction mixture (20% yield) (*H.P. Schultz et al.*, J. med. Chem., 1966, **9**, 266).

2-Tetrahydroxybutylquinoxalines have been prepared by treatment of osone hydrazones from D-galactose, L-sorbose and D-fructose with *o*-phenylenediamines (*G. Hensche* and *C. Bauer*, Ber., 1959, **92**, 501; *H. Horton* and *M.J. Miller*, J. org. Chem., 1965, **30**, 2457). 3-Hydroxy-2-polyhydroxyalkylquinoxalines are obtained by condensation of *o*-phenylenediamine with the sodium or potassium salt of α-ketogluconic acids in dilute hydrochloric acid (*H. Ohle*, Ber., 1934, **67**, 155).

Oxidative cleavage of the carbohydrate side-chain occurs readily with (i) alkaline hydrogen peroxide or sodium peroxide to give quinoxaline-2-carboxylic acids (*H.R. Moreno* and *H.P. Schultz*, J. med. Chem., 1970, **13**, 119), and (ii) sodium periodate to give quinoxaline-2-carboxyaldehydes (*C. Leese* and *H.N. Rydon*, J. chem. Soc., 1955, 303). With alkaline potassium permanganate, more extensive oxidative degradation occurs and pyrazine tricarboxylic acid is produced (87% yield from (CV)) (*Mager* and *Berends*, loc. cit.).

Treatment of (CV) with sulphuric acid at 100° results in cyclodehydration of the side-chain to give 2-(2-furyl)quinoxaline (CVI) rather than the glucazidone (CVII) originally reported. The furyl derivative (CVI) has been independently synthesised from 2-furylglyoxylic acid [Scheme 17]. In (CVI), electrophilic substitution reactions take place preferentially in the furan ring, *e.g.* nitration of (CVI) gives the 4-*nitrofuryl* derivative (CVIII), m.p. 215° (*A. Gomez-Sanchez* and co-workers, An. Quim., 1954, **50B**, 431; 1955, **51B**, 423).

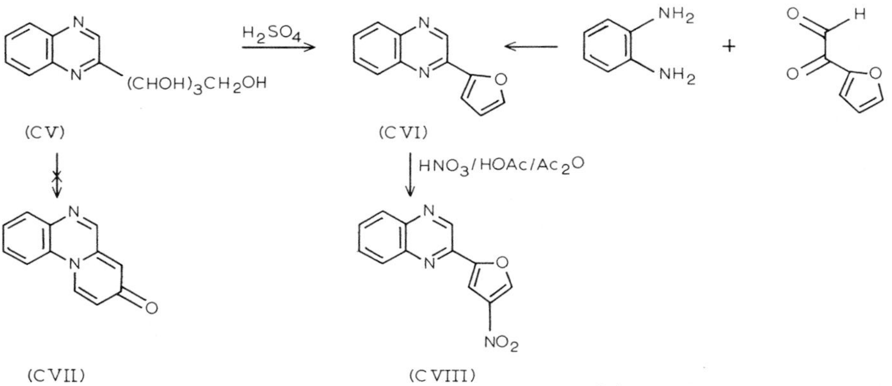

Scheme 17

2-Polyhydroxyalkylquinoxalines react with hydrazine to yield flavazoles such as (CIX) and (CX) (*W.S. Chilton* and *R.C. Krahn*, J. Amer. chem. Soc., 1967, **89**, 4129).

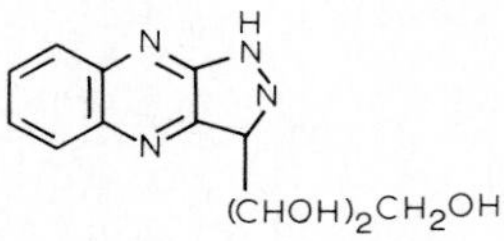

(CIX) 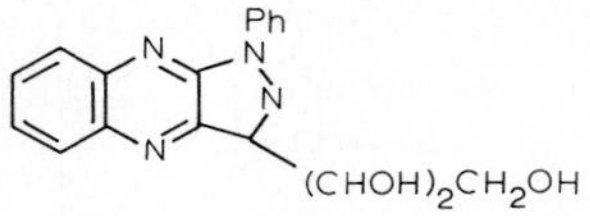   (CX)

(viii) Quinoxalinecarboxaldehydes

Quinoxaline-2-carboxaldehyde is obtained by oxidation of 2-methylquinoxaline with selenium dioxide (*J.K. Landquist* and *J.A. Silk*, J. chem. Soc., 1956, 2025), or by oxidative cleavage of 2-polyhydroxyalkylquinoxalines using either lead tetraacetate (*A. Muller* and *I. Varga*, Ber., 1939, **72**, 1993) or sodium periodate (*C.L. Leese* and *H.N. Rydon*, J. chem. Soc., 1955, 303). Oxidation of 2-(bromomethyl)quinoxaline by dimethylsulphoxide affords the corresponding aldehyde (*E.J. Moriconi* and *A. Fitsch*, J. org. Chem., 1965, **30**, 1542). Treatment of quinoxaline with a mixture of trioxane, hydrogen peroxide and iron(II) sulphate under acidic conditions yields 2-trioxanylquinoxaline (CXI) *via* a free-radical substitution process. Subsequent hydrolysis in acid gives the aldehyde in 17% overall yield (*G.P.*

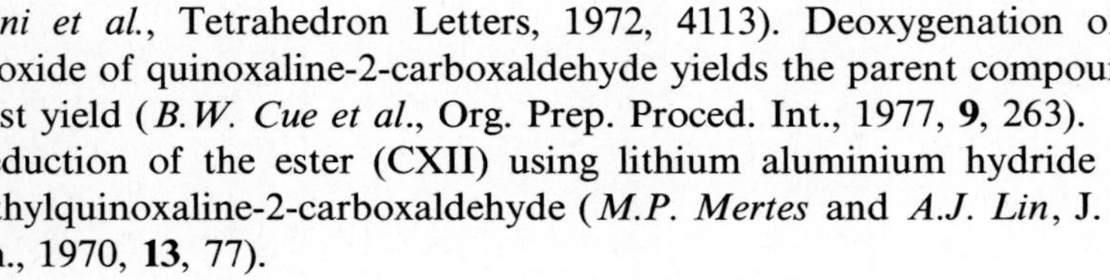

(CXI)

Gardini et al., Tetrahedron Letters, 1972, 4113). Deoxygenation of the di-*N*-oxide of quinoxaline-2-carboxaldehyde yields the parent compound in modest yield (*B.W. Cue et al.*, Org. Prep. Proced. Int., 1977, **9**, 263).

Reduction of the ester (CXII) using lithium aluminium hydride gives 3-methylquinoxaline-2-carboxaldehyde (*M.P. Mertes* and *A.J. Lin*, J. med. Chem., 1970, **13**, 77).

(CXII)

Quinoxaline-2,3-dicarboxaldehyde is prepared by (i) the oxidation of 2,3-bis(bromomethyl)quinoxaline in dimethylsulphoxide, or (ii) treatment of the 1,4-dioxide (CXIII) with alkali (*A.S. Elina et al.*, Chem. heterocyclic Compd., 1969, **5**, 115); both reactions proceed through the cyclic monohydrate (CXIV) [Scheme 18].

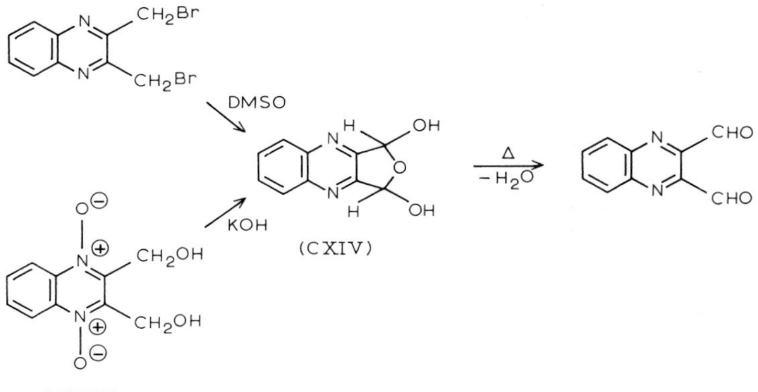

Scheme 18

(ix) Quinoxaline-2-carboxylic acids

Many of the general methods of synthesis of quinoxaline carboxylic acids, usually substituted in the 3-position, are derived from the condensation of a *o*-phenylenediamine with a tricarbonyl compound. With unsymmetrically substituted *o*-phenylenediamines, isomeric mixtures are obtained. Quinoxaline-2-carboxylic acids are decarboxylated at their melting point or on attempted distillation under reduced pressure.

Some general methods of synthesis are described below.

(1) Selenium dioxide oxidation of 2-alkyl groups, *e.g.* methyl, gives the carboxylic acid together with the aldehyde (*W. Borsche* and *W. Doeller*, Ann., 1939, **357**, 39). Condensation of 2-methylquinoxaline with aromatic aldehydes affords 2-styryl derivatives which are oxidatively cleaved with potassium permanganate to the corresponding carboxylic acids (*A.S. Elina* and *O.Yu. Magidson*, J. gen. Chem. U.S.S.R, 1955, **25**, 145). Oxidative cleavage of tetrahydroxybutylquinoxalines with dilute alkaline hydrogen peroxide or sodium peroxide also gives quinoxaline carboxylic acids (*H.R. Moreno* and *H.P. Schultz*, J. med. Chem., 1970, **13**, 119; 1005).

(2) Mesoxalic acid and its esters condense with *o*-phenylenediamines to give 3-hydroxyquinoxaline carboxylic acids and esters, respectively. The

latter are hydrolysed to the acids by treatment with mild alkali (*A.B. Sen* and *O.P. Madan*, J. Indian chem. Soc., 1961, **38**, 225).

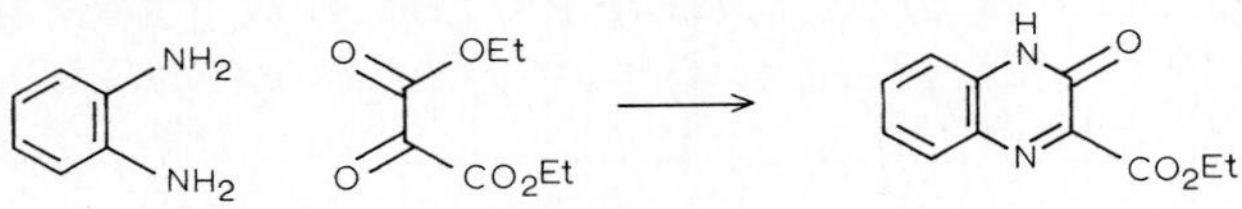

o-Phenylenediamines will condense with a variety of other tricarbonyl compounds or their equivalents to yield 3-substituted quinoxaline carboxylic acid derivatives including the amide (CXV) (*I.J. Pachter* and *P.E. Nemeth*, J. org. Chem., 1963, **28**, 1203; *H. Dahn* and *G. Rotzler*, Helv., 1960, **43**, 1555); lactone (CXVI) which can be ring opened with ammonia to the corresponding amide (*H. Dahn* and *H. Moll, ibid.*, 1964, **47**, 1860); and ester (CXVII) (*H. Taki* and *T. Mukaiyama*, Bull. chem. Soc. Japan, 1970, **43**, 3607) [Scheme 19].

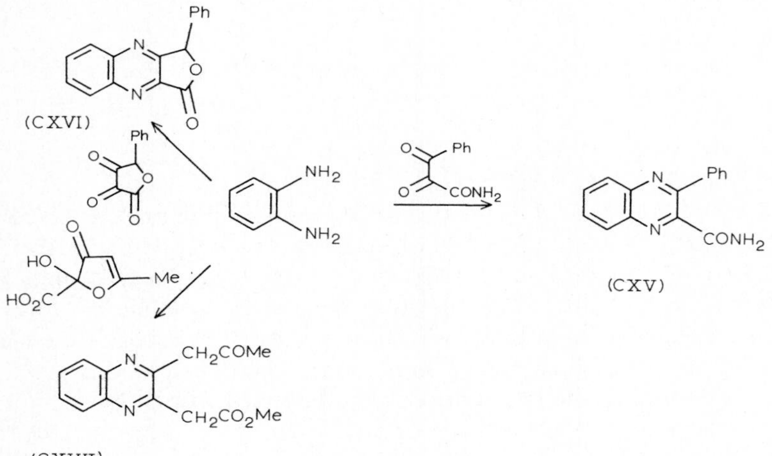

Scheme 19

(3) The product obtained from the reaction of an *o*-phenylenediamine with alloxan depends on pH; in dilute acid, alloxazine (CXVIII) is formed (*H. Bredereck* and *W. Pfleiderer*, Ber., 1954, **87**, 1119; *H.G. Petering* and *G.J. van Giessen*, J. org. Chem., 1961, **26**, 2818), but at pH > 7, cleavage of the alloxazine ring occurs and the quinoxaline ureide (CXIX) is obtained (*F.E. King et al.*, J. chem. Soc., 1951, 3329; *R.B. Barlow et al., ibid.*, 1951, 3242). Cleavage of the pyrimidine ring of (CXVIII) in aqueous ammonia affords the amino carboxylic acid (CXX) (*F.J. Wolf et al.*, J. Amer. chem.

TABLE 14
QUINOXALINE-2-CARBOXYLIC ACID AND ETHYL ESTERS

Compound	m.p. (°C)	
	Acid	Ester
Parent	210–213 [1-3]	85 [2]
3-Phenyl-	153 [1,4]	65–66 [5]
3-Amino-	209–211 [6], 212–213 [7,8] (dec.)	165–166
3-Methyl-	154.5 [9]	73–74 [7]
3-Chloro-	146–147 (dec.) [10]	42.5 [10]
3-Hydroxy-	265 [10,12], 268 [11]	175.5–176.5

1 K. Maurer, B. Schiedt and H. Schroeter, Ber., 1935, **68**, 1716.
2 K. Maurer and B. Boettger, Ber., 1938, **71**, 1383.
3 H.R. Moreno and H.P. Schultz, J. med. Chem., 1970, **13**, 119; 1005.
4 A.T. Blomquist and E.A. LaLancette, J. Amer. chem. Soc., 1962, **84**, 220.
5 H. Bohme and H. Schneider, Ber., 1958, **91**, 988.
6 C.D. Hurd and V.C. Bethune, J. org. Chem., 1970, **35**, 1471.
7 R.A. Baxter, F.S. Spring, J. chem. Soc., 1945, 229; A.H. Gowenlock, G.T. Newbold and F.S. Spring, ibid., 1948, 517.
8 J. Weijlard, M. Tishler and A.E. Erickson, J. Amer. chem. Soc., 1944, **66**, 1957.
9 A.S. Elina, J. gen. Chem. U.S.S.R., 1964, **34**, 2089.
10 A.H. Gowenlock, G.T. Newbold and F.S. Spring, J. chem. Soc., 1945, 622.
11 M.S. Habib and C.W. Rees, ibid., 1960, 3384.
12 H. Ohle and W. Gross, Ber., 1935, **68**, 2262.

Soc., 1948, **70**, 2572; J. Weijlard et al., ibid., 1944, **66**, 1957). Hydrolysis of (CXIX) or (CXX) in aqueous alkali gives the 3-hydroxy acid [Scheme 20] (Table 14, Ref. 10; R. Kuhn and F. Bar, Ber., 1934, **67**, 898).

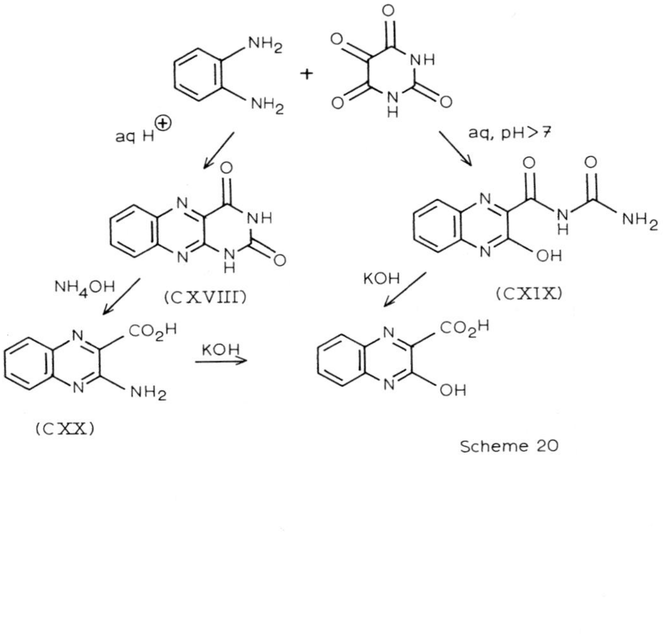

Scheme 20

Quinoxaline-2,3-dicarboxylic acids are prepared directly by condensation of the appropriate *o*-phenylenediamine with dihydroxytartaric acid (*R.A. Baxter* and *F.S. Spring*, J. chem. Soc., 1945, 229); diester derivatives are obtained by the analogous reaction using dioxosuccinate esters (*M.G. Gal'pern* and *E.A. Luk'yanets*, Chem. heterocyclic Compd., 1971, **7**, 257). Controlled oxidative cleavage of phenazine by alkaline potassium permanganate affords quinoxaline dicarboxylic acid in high yield (*F.D. Chattaway* and *W.G. Humphrey*, J. chem. Soc., 1929, 645; *I. Yoshioka* and *Y. Otomasu*, Chem. pharm. Bull. Tokyo, 1957, **5**, 277); more vigorous oxidation produces pyrazinetricarboxylic acid (*H.I.X. Mager* and *W. Berends*, Rec. Trav. chim., 1959, **78**, 5).

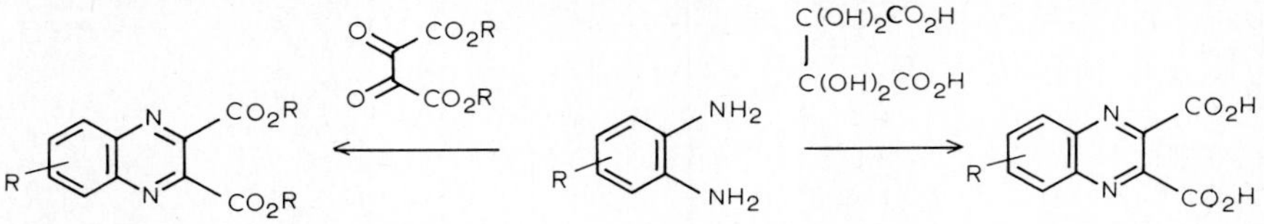

Reaction of the di-*N*-oxide (CXIII) with dimethylsulphoxide affords the mono-*N*-oxide (CXXI) which on treatment with alkaline hydrogen peroxide gives the quinoxaline-2,3-dicarboxylic acid in good yield (*A.S. Elina et al.*, Chem. heterocyclic Compd., 1969, **5**, 115).

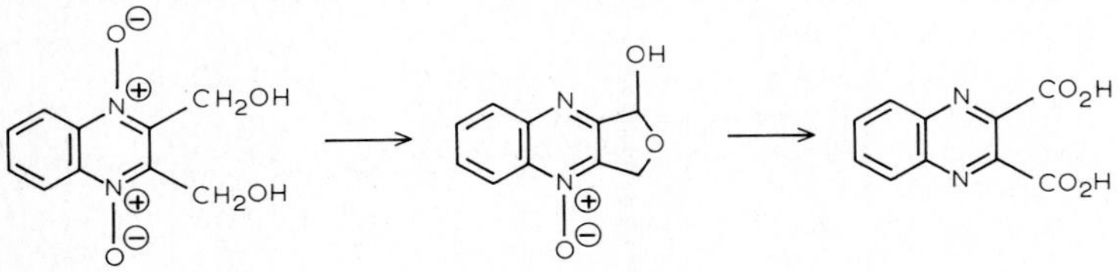

(CXIII)   (CXXI)

TABLE 15
QUINOXALINE-2,3-DICARBOXYLIC ACIDS AND ETHYL ESTERS

Compound	m.p. (°C)	
	Acid	Ester
Parent	190 (dec.) [1,2]	81.5–83 [2]
6-Chloro-	175 (dec.) [2]	60 [2]
6-Bromo-	172 (dec.) [2]	69 [2]
6-Methyl-	145 (dec.) [3]	–
6-Nitro-	298–300 [4]	81–82 [5]
6,7-Dimethyl-	180–182 [6]	–

1 R.A. Baxter and F.S. Spring, J. chem. Soc., 1945, 229.
2 F.D. Chattaway and W.G. Humphrey, ibid., 1929, 645.
3 O. Hinsberg, Ber., 1885, **18**, 1228; Ann, 1887, **237**, 327.
4 J. Klienar, F. Kosek and S. Panusova, Coll. Czech. Chem. Comm., 1964, **29**, 206.
5 M.G. Gal'pern and E.A. Luk'yanets, Chem. heterocyclic Compd., 1971, **7**, 257.
6 T.G. Koksharova et al., C.A., 1972, **76**, 153715.

Quinoxaline-2,3-diacids or their corresponding ammonium salts are readily decarboxylated on heating to quinoxalines. The diacids are converted into anhydrides with acetic anhydride (*Chattaway* and *Hymphreys, loc. cit.*). **Quinoxaline-2,3-dicarboxylic anhydride**, m.p. 250–260° (dec.), undergoes normal transformations to (i) acid esters by treatment with alcohols alone, (ii) diesters with alcohols and hydrogen chloride, *e.g.* from methanol, the *dimethyl ester*, m.p. 130°, (iii) the *diamide*, m.p. 328° (dec.), with ammonia in methanol (*R.A. Baxter* and *F.S. Spring*, J. chem. Soc., 1945, 22).

(e) Reduced quinoxalines

(i) Dihydroquinoxalines

The heterocyclic ring of quinoxaline is selectively reduced either by sodium in tetrahydrofuran (*J. Hamer* and *R.E. Holliday*, J. org. Chem., 1963, **28**, 2488), or electrochemically in alkaline solution (*J. Pinson* and *J. Armand*, Coll. Czech. Chem. Comm., 1971, **36**, 585) to give **1,4-dihydroquinoxaline** (CXXII), m.p. 158–159°. Electrochemical reduction of 2-methyl-

(CXXII)

quinoxaline in neutral or alkaline media affords 1,2-dihydro-3-methylquinoxaline (CXXIII), which on exposure to air is rapidly oxidised to starting material (*Pinson* and *Armand, loc. cit.*).

(CXXIII)

Similarly, condensation of chloroacetone with *o*-phenylenediamine affords only the 2-methylquinoxaline since the 1,2-dihydro compound undergoes spontaneous *in situ* oxidation (*O. Hinsberg*, Ber., 1886, **19**, 483; Ann., 1887, **237**, 327). Reduction of 2-phenylquinoxaline with zinc amalgam yields initially the 1,4-dihydro derivative which isomerises to the thermodynamically more stable 1,2-**dihydro-3-phenylquinoxaline**, m.p. 166° (*Pinson* and *Armand, loc. cit.; M. Schellenberg*, Helv., 1970, **53**, 1151). A series of related 3-aryl-1,2-dihydroquinoxalines has been prepared by condensation of *o*-phenylenediamine with 2-haloacetophenones (*J. Figueras*, J. org. Chem., 1966, **31**, 803).

2,3-Diphenyl-1,2-dihydroquinoxaline (CXXIV), m.p. 146°, is obtained either by reduction of the corresponding quinoxaline with tin(II) chloride in hydrochloric acid (*O. Hinsberg* and *F. Konig*, Ber., 1894, **27**, 2181), or by direct synthesis from *o*-phenylenediamine and benzoin (*O. Fischer*, Ber., 1891, **24**, 719; 1893, **26**, 192). The isomeric 1,4-*dihydro* compound (CXXV), m.p. 101°, is obtained by electrochemical reduction of the diphenylquinoxaline (*Pinson* and *Armand, loc. cit.*).

(CXXIV)   (CXXV)

(ii) Tetrahydroquinoxalines

1,2,3,4-Tetrahydroquinoxalines can be prepared by reduction of quinoxaline derivatives. Thus, quinoxaline is reduced to 1,2,3,4-tetrahydroquinoxaline by (i) sodium in ethanol (*V. Merz* and *C. Riz*, Ber., 1887, **20**, 1190); (ii) catalytic hydrogenation over transition metal catalysts (*J.C. Cavagnol* and *F.Y. Wiselogle*, J. Amer. chem. Soc., 1947, **69**, 795; *R.C. DeSelms, R.J. Greaves* and *W.R. Schleigh*, J. heterocyclic Chem., 1974, **11**, 595), and (iii) lithium aluminium hydride (*F. Bohlmann*, Ber., 1952, **85**, 390). The 5- and 6-substituted quinoxalines are similarly reduced by sodium borohydride in glacial acetic acid to the corresponding tetrahydroquinoxalines (*K.V. Rao* and *D. Jackman*, J. heterocyclic Chem., 1973, **10**, 213). An analogous reduction of quinoxaline in acetic acid using potassium borohydride affords 1,4-diethylquinoxaline in high yield (*J.-M. Cosmao, N. Collignon* and *G. Quequiner, ibid.*, 1979, **16**, 973).

With 2,3-disubstituted 1,2,3,4-tetrahydroquinoxalines, there is the possibility of the formation of *cis*- and *trans*-isomers. Catalytic hydrogenation (*R. Aguilera, J.C. Duplan* and *C. Nofre*, Bull. Soc. chim. Fr., 1968, 4491; *R.A. Archer* and *H.S. Mosher*, J. org. Chem., 1967, **23**, 1378), and hydride reduction (*R.C. DeSelms* and *H.S. Mosher*, J. Amer. chem. Soc., 1960, **82**, 3762) of 2,3-disubstituted quinoxalines give predominantly the *cis*-isomer; with sodium in ethanol, significant quantities of the corresponding *trans*-isomer are formed (*C.S. Gibson*, J. chem. Soc., 1927, 342; *M.J. Haddadin, N.N. Alkaysi* and *S.E. Saheb*, Tetrahedron, 1970, **26**, 1115). Reduction of 2,3-disubstituted quinoxaline-1,4-dioxides with sodium borohydride also affords the *cis*-2,3-disubstituted compounds (*Haddadin, Alkaysi* and *Saheb, loc. cit.*).

Nucleophilic addition of organo-lithium and Grignard reagents to quinoxaline affords 2,3-disubstituted 1,2,3,4-tetrahydroquinoxalines (*E.S. Lane* and *C. Williams*, J. chem. Soc., 1954, 4106; *H. Stetter*, Ber., 1953, **86**, 197; *H. Gilman, J. Eisch* and *T. Soddy*, J. Amer. chem. Soc., 1957, **79**, 1245).

Cyclisation of *o*-disubstituted benzene derivatives provides useful synthetic routes to tetrahydroquinoxalines. The reaction of catechol with

TABLE 16
1,2,3,4-TETRAHYDROQUINOXALINES

Compound	m.p. (°C)	Derivatives	m.p. (°C)
Parent	91–95 [1], 97–98 [2]	1-Acetyl-	109–110 [2]
	(b.p. 153–154 °/14 mm)	1,4-Diacetyl-	147–47.5 [3]
		1-Benzoyl-	152–153 [2]
		1,4-Dibenzoyl-	206–207 [2,3]
		Picrate	128–129.5 [2]
		1,4-Ditosyl-	162 [4]
2-Methyl-			
R-form	90.5–91 [5] ($[\alpha]_D^{25}$ −60.3°)	1,4-Diacetyl-	143–144 [5]
S-form	90–90.5 [5] ($[\alpha]_D^{25}$ +60.2°)		
RS-form	70–71 [6]	1,4-Diacetyl-	138–139 [5]
6-Methyl-	104.5–105.5 [3]	1,4-Diacetyl-	105–106 [3]
2-Phenyl-	77–78 [13]		
5-Nitro-	127–128 [7]		
6-Nitro-	114–115 [7], 116 [8]		
6-Chloro-	114 [3]	1,4-Dibenzoyl-	105–106 [3]
2,3-Dimethyl-			
cis	112–113 [9]	1,4-Diacetyl-	147 [10]
trans dl	101–102 [11]		
d	94.5 [11] ($[\alpha]_D^{20}$ +112.3°)		
l	94.5 [11] ($[\alpha]_D^{20}$ −112.0°)		
2,3-Diphenyl-			
cis	142–143 [12]		
trans dl	106 [12]		
d	135–135.5 [12] ($[\alpha]_{546}^{20}$ +155.7°)		
l	135–135.5 [12] ($[\alpha]_{546}^{20}$ −156.2°)		

1 F. Bohlman, Ber., 1952, **85**, 390.
2 J.S. Morley, J. chem. Soc., 1952, 4002.
3 J.C. Cavagnol and F.Y. Wiselogle, J. Amer. chem. Soc., 1947, **69**, 795.
4 H. Stetter, Ber., 1953, **86**, 197.
5 G.H. Fisher, P.J. Whitman and H.P. Schultz, J. org. chem., 1970, **35**, 2240.
6 M. Munk and H.P. Schultz, J. Amer. chem. Soc., 1952, **72**, 3433.
7 K.V. Rao and D. Jackman, J. heterocyclic Chem., 1973, **10**, 213.
8 G.R. Ramage and R. Trappe, J. chem. Soc., 1952, 4406.
9 R.C. DeSelms and H.S. Mosher, J. Amer. chem. Soc., 1960, **82**, 3762.
10 S. Maffei and S. Pietra, Gazz., 1958, **88**, 556.
11 C.S. Gibson, J. chem. Soc., 1927, 342.
12 G.M. Bennett and C.S. Gibson, ibid., 1923, **123**, 1570.
13 J. Figueras, J. org. Chem., 1966, **31**, 803.

ethylenediamine and 1,2-propylenediamine at high temperature affords 1,2,3,4-tetrahydroquinoxaline and the 2-methyl derivative, respectively (*Merz* and *Riz, loc. cit.; Cavagnol* and *Wiselogle, loc. cit.; F.G. Mann* and

B.B. Smith, J. chem. Soc., 1951, 1906). *N,N'*-Ditosyl-*o*-phenylenediamines react with 1,2-dibromoethane to give, after hydrolysis, the corresponding 1,2,3,4-tetrahydroquinoxalines (*Stetter, loc. cit.; R.M. Acheson*, J. chem. Soc., 1956, 4731; *E. Negishi* and *A.R. Day*, J. org. Chem., 1965, **30**, 43).

Tetrahydroquinoxalines have also been prepared by acid-catalysed cyclisation of *o*-amino-*N*-(β-hydroxyethyl)anilines (CXXVI) (*G.R. Ramage* and *G. Trappe*, J. chem. Soc., 1952, 4406; *W. Knobloch* and *G. Lietz*, J. pr. Chem., 1967, **36**, 113), and reductive cyclisation of *N*-(β-chloroethyl)-*o*-nitroanilines (CXXVII) (*P. Clarke* and *A. Moorehouse*, J. chem. Soc, 1963, 4763; *V. Beksha* and *Yu. Deguts*, J. org. Chem. U.S.S.R., 1965, **1**, 1873).

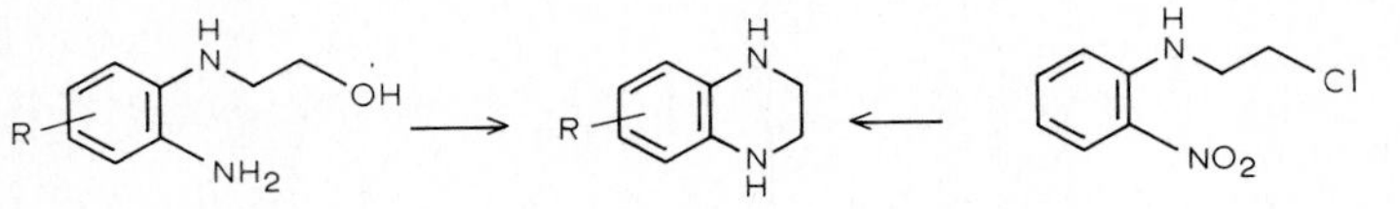

Sequential elaboration of the methyl groups of 2,3-dimethylpyrazine gives rise to the acyclic intermediate (CXXVIII) which on base-catalysed cyclisation affords the uncommon 5,6,7,8-tetrahydroquinoxaline (CXXIX) [Scheme 21].

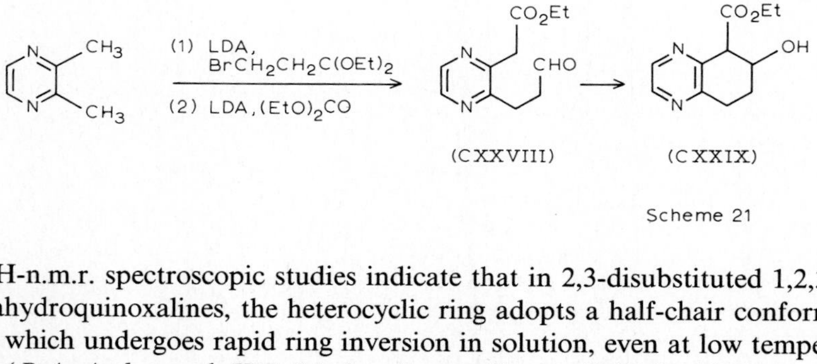

Scheme 21

^{1}H-n.m.r. spectroscopic studies indicate that in 2,3-disubstituted 1,2,3,4-tetrahydroquinoxalines, the heterocyclic ring adopts a half-chair conformation which undergoes rapid ring inversion in solution, even at low temperature (*R.A. Archer* and *H.S. Mosher*, J. org. Chem., 1967, **32**, 1378).

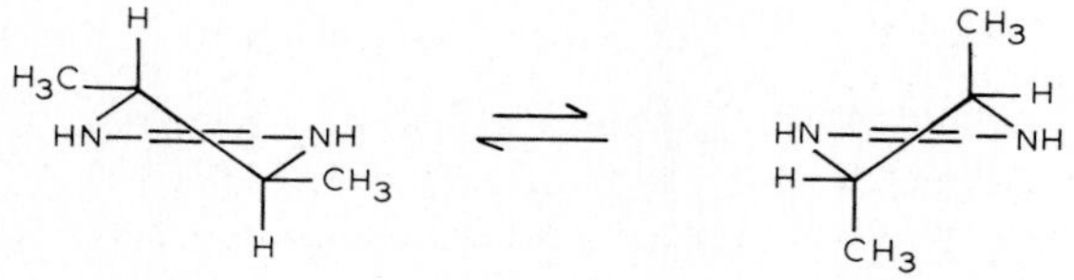

2-Methyl-1,2,3,4-tetrahydroquinoxaline has been resolved *via* the corresponding dibenzoyl-*d*-tartrate salts and the absolute configuration of the *S*-enantiomer confirmed by synthesis from 2,4-dibromonitrobenzene and *S*-alanine (*G.H. Fisher, P.J. Whitman* and *H.P. Schultz*, *J. org. Chem.*, 1970, **35**, 2240). The *trans*-isomers of 2,3-dimethyl- and 2,3-diphenyl-1,2,3,4-tetra-hydroquinoxaline have also been resolved into their respective optically active forms (Table 16, Refs. 11 and 12).

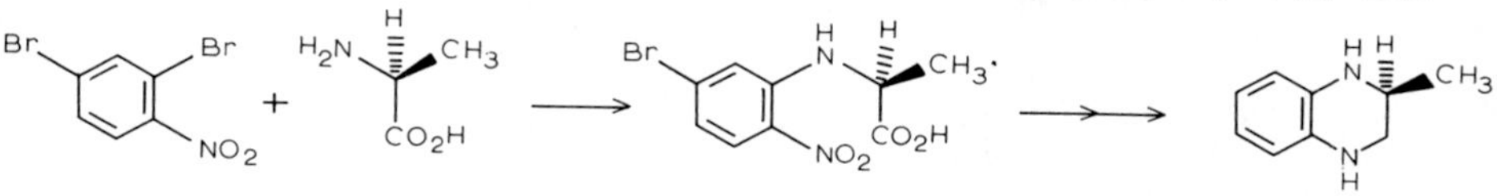

Tetrahydroquinoxaline reacts with acetic anhydride in alcohol at room temperature to give the 1-acetyl derivative; the 1,4-diacetyl compound results from the use of acetic anhydride alone. The 1-*ethoxycarbonyl*, m.p. 151–152°, and 1,4-*diethoxycarbonyl*, m.p. 40–41°, derivatives are prepared by the addition of ethyl chloroformate to a suspension of the tetrahydroquinoxaline in alcohol.

The 1,4-*ethano-bridged tetrahydroquinoxaline* (CXXX), m.p. 138–141°, has been unambiguously synthesised from the 1-acetyl derivative (CXXXI) [Scheme 22] (*G.V. Shishkin* and *A.A.*

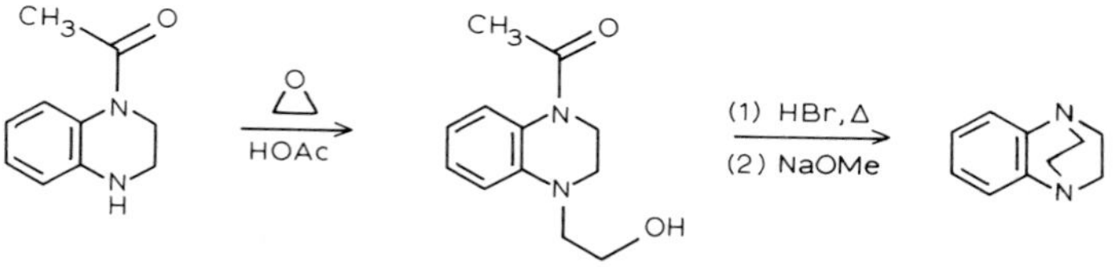

Scheme 22

Gall', *Chem. heterocyclic Compd.*, 1980, **16**, 645; 648). The isomeric 1,3-*ethano-bridged tetrahydroquinoxaline* (CXXXII), m.p. 131–134°, has also been synthesised (*H.C. Cunningham* and *A.R. Day*, *J. org. Chem.*, 1973, **38**, 1225).

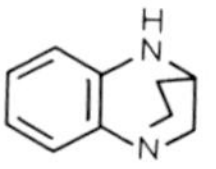

(CXXXIII)

(iii) Decahydroquinoxalines

This totally saturated ring system can exist in either the *cis* or the *trans* form; both forms are known.

trans-**Decahydroquinoxaline**, *dl*-form, m.p. 150–151°, has been synthesised (i) by cyclisation of *cis*-1-chloro-2-(β-chloroethylamino)cyclohexane in aqueous ammonia at 130° for 5 h (*M. Mousseron* and *G. Combes*, *Bull. chem. Soc. Fr.*, 1947, 82); (ii) by reduction of the product from *trans*-1,2-diaminocyclohexane and glyoxal (*H.S. Broadbent et al.*, *J. Amer. chem. Soc.*, 1960, **82**, 189); (iii) by lithium aluminium hydride reduction of 2-oxo- and 2,3-dioxo-decahydroquinoxaline, prepared from *trans*-1,2-diaminocyclohexane with chloracetic acid or diethyl oxalate, respectively (*E. Brill* and *H.P. Schultz*, J. org. Chem., 1963, **28**, 1135); (iv) catalytic hydrogenation of quinoxaline with palladium on charcoal at elevated temperatures and pressures (*S. Maffei* and *S. Pietra*, Gazz., 1958, **88**, 556); (v) reduction of 2-oxo-1,2,5,6,7,8-hexahydroquinoxaline with sodium in amyl alcohol (yield 58%) (*E. Brill* and *H.P. Schultz*, J. org. Chem., 1964, **29**, 579).

cis-**Decahydroquinoxaline**, m.p. 56–57°, b.p. 85–87°/0.25 mm, is prepared (i) by catalytic hydrogenation of either quinoxaline over rhodium on alumina or Raney nickel (*Maffei* and *Pietra, loc. cit.*), or 1,2,3,4-tetrahydroquinoxaline with platinum in acidic medium (*Brill* and *Schultz*, 1963, *loc. cit.*); (ii) from *cis*-1,2-diaminocyclohexane as for the *trans*-isomer (*Brill* and *Schultz*, 1963, *loc. cit.*); and (iii) from cyclohexane-1,2-dione and glycine amide [Scheme 23] (*Brill* and *Schultz*, 1963, *loc. cit.*).

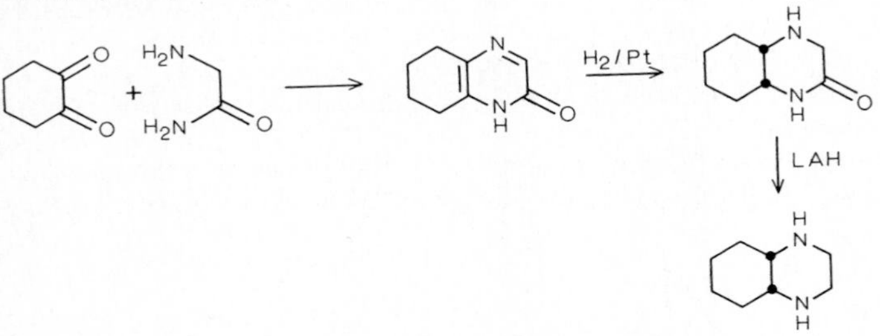

Scheme 23

cis-Decahydroquinoxaline is a symmetrical molecule and consequently *meso*. The *trans*-isomer, being racemic, has been resolved into its enantiomers by fractional crystallisation of the dibenzoyl tartrate salts (*Brill* and *Schultz*, 1963, *loc. cit.*). The absolute configuration has been unambiguously established by synthesis (*D. Gracian* and *H.P. Schultz*, J. org. Chem., 1971, **36**, 3989).

3. Phenazines (dibenzopyrazines)

(a) Introduction

The formal substitution of the 9- and 10-methine groups in anthracene by nitrogen atoms would give rise to the dibenzo-1,4-diazine, phenazine (I). The commonly accepted numbering system is depicted on structural formula (I). This differs from an earlier system, directly related to that of anthracene, in which the nitrogen atoms were labelled as the 9- and 10-positions.

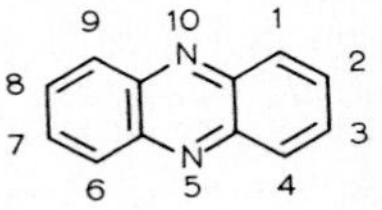

(I)

INTRODUCTION

Some thirty phenazine derivatives are known to be produced by microorganisms including iodinin (II), the violet pigment isolated from cultures of *Chromobacterium iodinium*, and pyocyanine (III), the blue pigment produced by the organism *Pseudomonas aeruginosa*. Although many of the

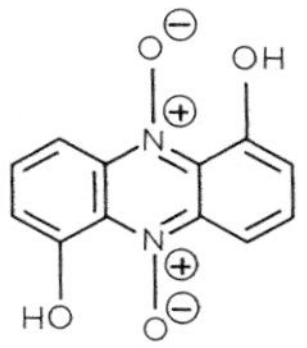

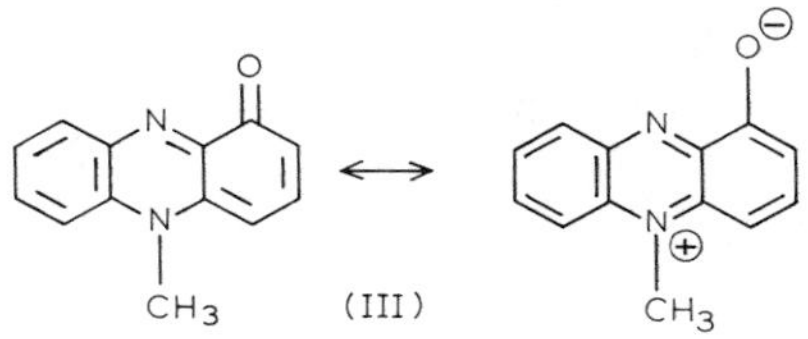

naturally occurring phenazines possess antibiotic activity, their relatively high toxicity limits clinical use. Myxin, stabilised as the copper complex (IV), is used as a topical formulation in verterinary medicine (*J. Berger*, "Kirk-Othmer Encycl. Chem. Tech.", 3rd Edn., Wiley, New York, 1978,

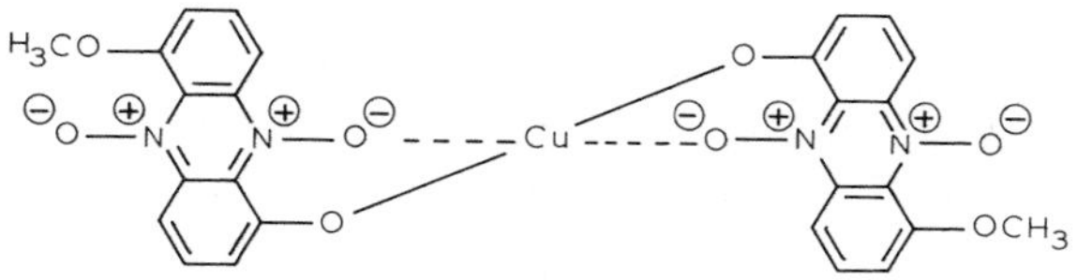

Vol. 3, pp. 1–21). Synthetic phenazines, structurally related to the safranines, also show antitubercular activity (*V.C. Barry*, Sci. Proc. Roy. Dublin Soc., 1969, **3A**, 153). Simple alkyl- and halo-phenazines are reported to be effective herbicides (*B. Cross* and co-workers, J. Sci. Food Agric., 1969, **20**, 8; 340).

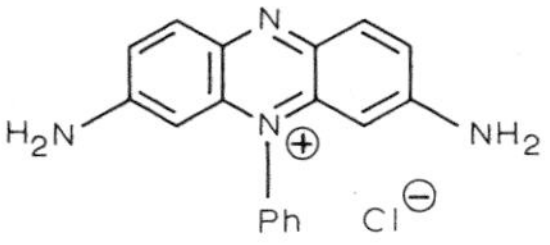

Historically, phenazine dyes and pigments, e.g. phenosafranine (V), prepared by the reaction of *p*-phenylenediamine with a primary amine under oxidising conditions, were of considerable importance. On account of their poor light feastness, they have been largely superceded by azo- and anthraquinone dyes (see *G.A. Swan* and *D.G.I. Felton*, "Phenazines",

Chemistry of Heterocyclic Compounds, Vol. 11, Wiley, New York, 1957; C.C.C., 1st Edn., Vol IVc; *J.F. Corbett*, J. Soc. Dyers Colour., 1972, **88**, 438).

More recent interest in the chemistry of phenazines and their related 5,10-dihydro derivatives stems from their inherent ability to form molecular and charge-transfer complexes which exhibit a range of desirable solid-state electrical properties including conduction and paramagnetism. 5-Alkylphenazines form persistent, solid-state free-radicals, *e.g.* 5-ethylphenazyl (VI). Moreover, phenazines readily undergo one-electron redox reactions

(VI)

with associated proton transfer (*H.J. Keller* and *Z.G. Soos*, Top. curr. Chem., 1985, **127**, 169–210 and references therein).

As a consequences of the fused ring structure in phenazines, the carbon centres immediately adjacent to the nitrogen atoms cannot participate in substitution reactions which are common to both the pyrazines and quinoxalines. The carbocyclic rings are nonetheless deactivated towards electrophilic attack on account of the electron-withdrawing effects of the hetero atoms. Thus, nitration and chlorination require vigorous reaction conditions and give rise to mixtures of products with substitution taking place primarily at the 1- and symmetry related positions. Phenazines with electron-donating substituents undergo electrophilic substitution more readily. Conversely, nucleophilic substitution reactions are facilitated in mono- and di-*N*-oxides, quaternary salts, and phenazines possessing electron-withdrawing substituents such as nitro groups.

(b) Physical and spectroscopic properties

X-ray crystal structure analysis indicates that phenazine (α-form) is a planar molecule of D_{2h} symmetry with C–N bond lengths which are similar to those in pyrazine and C–C bond lengths comparable to those in anthracene (*F.L. Hirschfeld* and *G.M. Schmidt*, J. chem. Phys., 1957, **26**, 923).

From molecular orbital calculations, the electronic distributions at the α- and β-positions are similar to those at the corresponding positions in the carbocyclic ring of quinoxaline (*R.J. Pugmire et al.*, J. Amer. chem. Soc., 1969, **91**, 6381). The observed first and second ionisation potentials of phenazine, 8.4 and 9.9 eV, are assigned to loss of an electron from a π-type orbital (HOMO) and a non-bonding orbital, respectively, in reasonable agreement with theoretical calculations (*M. Sundborn*, Acta Chem. Scand., 1971, **25**, 487), and spectroscopic studies on the phenazine radical cation (*S. Kato* and *T. Shida*, J. Amer. chem. Soc., 1979, **101**, 6869). Polarographic reduction of phenazine in solution ($E_{1/2}$ +0.94 V versus the standard calomel electrode) results in generation of the radical cation which has a lifetime of *ca.* 3 min. (*K.H. Hausser, A. Habich* and *V. Franzen*, Z. Naturforsch., 1961, **16A**, 836).

The ionisation constants of phenazine (pK_{1a} 1.23 and pK_{2a} −4.9) indicate that it is a weak base, similar to the other 1,4-diazines (*A. Albert et al.*, J. chem. Soc., 1948, 2240; *M. Iwaizumi* and *H. Azumi*, J. chem. Soc. Japan, 1963, **84**, 694).

The solution phase infrared spectra of phenazine and several derivatives have been studied in detail (*C. Stammer* and *A. Taurins*, Spectrochim. Acta, 1963, **19**, 1625; *T.J. Durnick* and *S.C. Wait*, J. mol. Spectrosc, 1971, **39**, 536). Major infrared bands appear at 3065, 1515, 1437, 1362, 1112, 1029, 958, 905, 820, 752 and 745 cm^{-1}. The u.v.-visible absorption spectrum of phenazine in 95% ethanol consists of two broad bands with maxima at 250 nm (log ϵ 4.1) and 370 nm (log ϵ 3.15); spectral data on several simple derivatives have also been reported (*A. Gray* and *F.G. Holliman*, Tetrahedron, 1962, **18**, 1095).

The ring protons of phenazine give rise to an AA'BB' splitting pattern in its ^{1}H-n.m.r. spectrum with multiplets centred at δ 8.26 and 7.97 (DMSO-d$_6$). The magnitudes of the H–H coupling constants J^{12}, J^{23}, J^{13} and J^{14} have been estimated to be 8.9, 6.37, 1.41 and 0.36 Hz, respectively (*N.S. Angerman* and *S.S. Danyluk*, Org. mag. Resonance, 1972, **4**, 895). ^{1}H-n.m.r. spectral data for some substituted phenazines together with data on the effects of substituents on chemical shifts and coupling constants have been reported (*A. Roemer, ibid.*, 1982, **19**, 66). The proton-decoupled ^{13}C-n.m.r. spectrum of phenazine consists of three signals at 143.5, 130.4 and 129.7 ppm which have been assigned to the quaternary carbons (4a, 5a, 9a, 10a), the α-carbons (1, 4, 6, 9) and the β-carbons (2, 3, 7, 8), respectively (*idem, ibid.*, 1983, **21**, 130).

The mass spectrum of phenazine is dominated by an intense molecular ion at m/z 180 with significant contributions from the doubly charged molecular ion. Fragmentation occurs by loss of CN and C_2H_2 (M−26), and HCN and C_2H_3 (M−27) (*S. Eguchi*, Bull. chem. Soc. Japan, 1978, **51**, 1128). Other phenazine derivatives also give rise to strong molecular ions in their respective mass spectra (*Q.N. Porter*, "Mass Spectrometry of Heterocyclic Compounds", 2nd Edn., Wiley, New York, 1985).

(c) General methods of synthesis

In the more widely used approaches to the synthesis of the phenazine ring system, construction of the central heterocyclic ring represents the crucial step of the sequence. Because phenazines do not readily participate in electrophilic substitution reactions, derivatives are usually prepared from appropriately substituted aromatic precursors. Phenazines may be prepared by the following methods.

(1) Condensation of *o*-phenylenediamine with *o*-benzoquinones affords phenazines directly (*F. Kehrmannn et al.*, Helv., 1924, **7**, 973; 1927, **10**, 62); polynuclear *o*-quinones, *e.g.* 1,2-naphthaquinone, condense with *o*-phenylenediamine in a similar fashion (*C. Liebermann* and *O.N. Witt*, Ber., 1887,

20, 2442; *T.G.H. Jones* and *R. Robinson*, J. chem. Soc., 1917, **111**, 927). The requisite *o*-benzoquinones are prepared by oxidation of catechols using *o*-chloranil, or cerium(IV) sulphate (*Y. Omote et al.*, Bull. chem. Soc. Japan, 1974, **47**, 1957), or manganese dioxide (*B. Cross et al.*, J. Sci. Food Agric., 1969, **20**, 340) and, to minimise the effect of self-condensation, are best trapped *in situ* by *o*-phenylenediamine to give the corresponding phenazines.

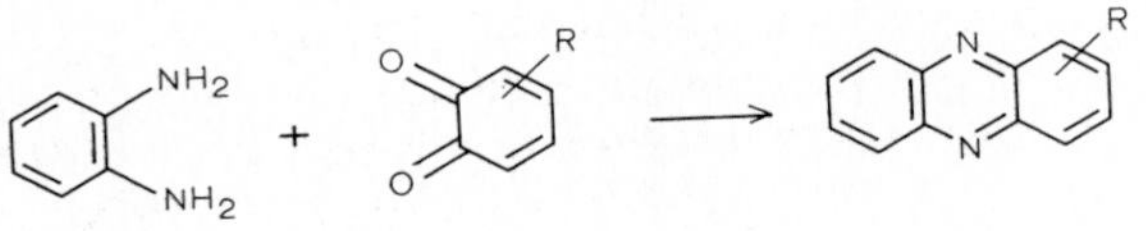

The problems associated with *o*-benzoquinones can be overcome by the Ris procedure which utilises catechols directly. Thus, heating a mixture of *o*-phenylenediamine and a catechol at 220–240° in a sealed tube gives initially a 5,10-dihydrophenazine which is readily oxidised to the phenazine (*C. Ris*, Ber., 1886, **19**, 2206). Phenazine and several of its simple alkyl derivatives have been prepared by this procedure (*J.S. Morley*, J. chem. Soc., 1952, 4008; *B. Cross et al., loc. cit.*).

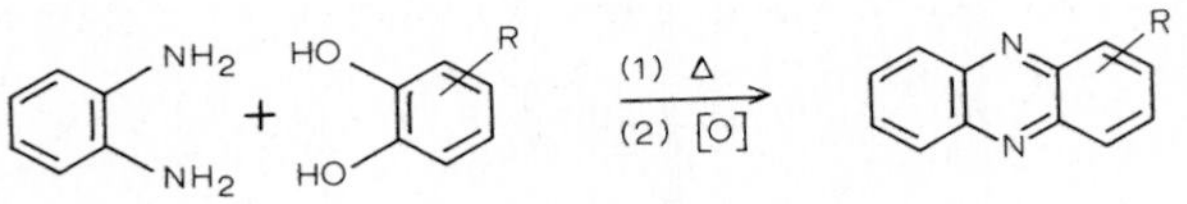

(2) Substituted 2-nitrodiphenylamines (VII), obtained from *ipso* substitution of the reactive halogen atom in *o*-nitrohalobenzenes by an aromatic amine, undergo cyclisation to phenazines on heating with iron(II) oxalate and granulated lead for a short time at 200° (the Vivian method). The reduction process is considered to be catalysed by iron(II) oxide formed *in situ* under the reaction conditions.

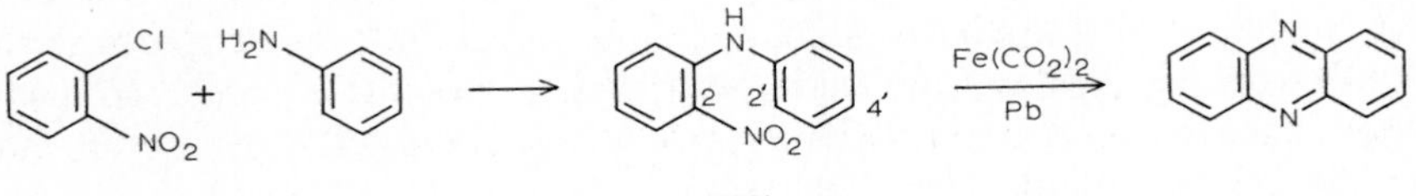

Amino and methoxy substituents at the 2'-position, and chlorine and bromine atoms at the 2'- and 4'-positions are frequently eliminated during the cyclisation process. Similar substituents in the same ring as the nitro group are unaffected (*D.L. Vivian* and co-workers, J. org. Chem., 1949, **14**, 289; 1953, **18**, 1065; 1955, **20**, 797; 1956, **21**, 565; 824).

o-Nitrodiphenylamines can also be cyclised (i) on treatment with sodium borohydride in sodium ethoxide/ethanol to produce the corresponding phenazines (*S.R. Challand, R.B. Herbert* and *F.G. Holliman*, Chem. Comm., 1970, 1423) and (ii) in oleum to yield phenazine-*N*-oxides (*B. Cross et al.*, J. Sci. Food Agric., 1968, **20**, 8).

2,2'-Dinitrodiphenylamines in alcoholic potassium hydroxide undergo reductive cyclisation with loss of one nitro group on treatment with hydrazine in the presence of ruthenium and charcoal, *e.g.* 4-chloro-2,2'-dinitrodiphenylamine (VIII) gives 2-chlorophenazine in 74% yield (*B. Cross, P.J. Williams* and *R.E. Woodall*, J. chem. Soc. C, 1971, 2085).

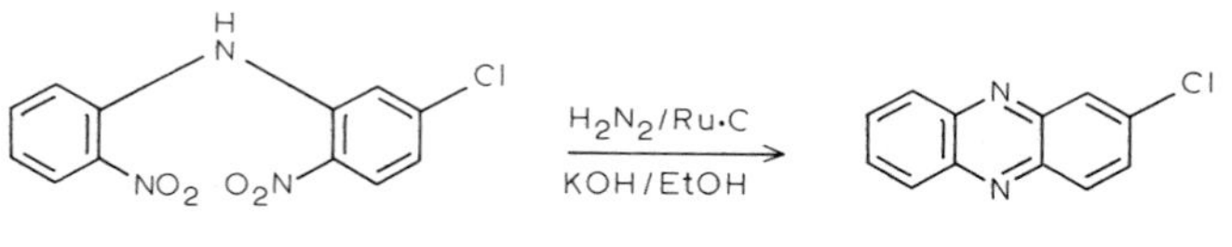

(VIII)

(3) In the Wohl–Aue method, an aniline derivative reacts with a nitrobenzene in the presence of powdered potassium hydroxide in refluxing toluene to give rise to a phenazine mono-*N*-oxide; at higher temperatures (140°) the parent phenazine is obtained (*A. Wohl* and *W. Aue*, Ber., 1901, **34**, 2442; *B. Cross et al.*, J. Sci. Food Agric., 1969, **20**, 8). Although the

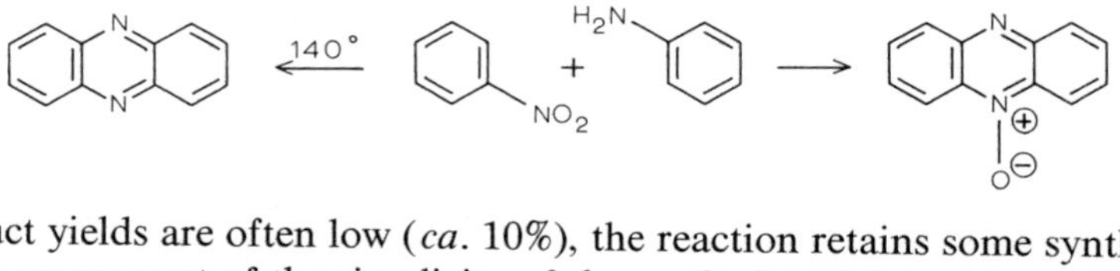

product yields are often low (*ca.* 10%), the reaction retains some synthetical utility on account of the simplicity of the method and the ready availability of the reactants. Moreover, in the preparation of halophenazines, improved product yields have been reported, *e.g.* 2-chlorophenazine (30–35%) (*I.J. Pachter* and *M.C. Kloetzel*, J. Amer. chem. Soc., 1951, **73**, 4958; 1952, **74**, 971) and 1,4-dichlorophenazine (40%) (*B. Cross et al., loc. cit.*). The potassium hydroxide may be replaced by sodamide (*E.I. Abramova* and *I.Ya. Postovskii*, C.A., 1953, **47**, 2182).

(4) The Beirut reaction offers an alternative route to a range of phenazine-*N*-oxide derivatives. As outlined in Scheme 1, benzofuroxan will condense with a variety of substrates including:

(a) cyclohexane-1,2-dione in the presence of a secondary or tertiary amine to give a mixture of 1-hydroxyphenazine mono- and di-oxides which can be oxidised by peracid to yield the di-N-oxide (IX) (*C.H. Issidorides et al.*, Tetrahedron, 1978, **34**, 217);

(b) phenolate anions to give the 2-hydroxy-5,10-dioxide compound (X) (*K. Ley* and co-workers, Angew. Chem. intern. Edn., 1969, **8**, 596; *A. Roemer* and *M. Sammet*, Z. Naturforsch., 1983, **38B**, 866);

(c) β-naphthol in sodium methoxide/methanol to yield benzo[*a*]phenazine dioxide (XI) (*M.J. Abu El-Haj et al.*, J. org. Chem., 1972, **37**, 589);

(d) enamines derived from cyclohexanone to produce the 1,2,3,4-tetrahydrophenazine (XII) (*M.J. Haddadin* and *C.H. Issidorides*, Heterocycles,

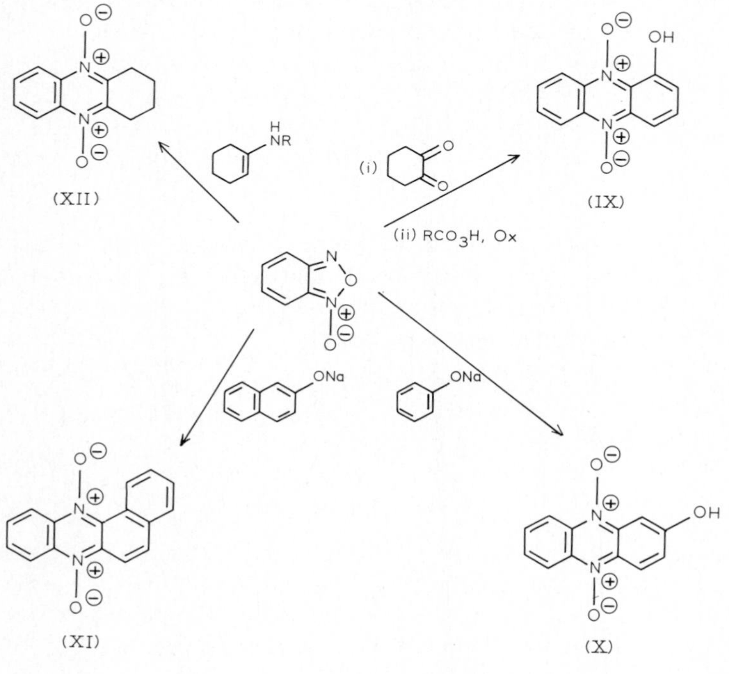

Scheme 1

1976, **4**, 767). Cyclohexanone azines or anils also react with benzofuroxan in triethylamine to give (XII) in high yield (> 90%) (*L. Marchetti* and *G. Tosi*, Ann. Chim. (Rome), 1967, **57**, 1414).

The N-oxide products are usually obtained in high yield from this reaction and can be conveniently deoxygenated by a variety of methods, e.g. sodium dithionite, or phosphoryl chloride, or catalytic hydrogenation, or heating with aniline.

(5) Treatment of 2,2′-diaminodiphenylamines with iron(III) chloride in hydrochloric acid results in cyclisation with concomitant elimination of ammonia to give initially dihydrophenazines which undergo rapid air oxidation to the corresponding phenazines (*M.L. Tomlinson*, J. chem. Soc., 1939, 158). Although of general application, this procedure is reported to be particularly useful for the preparation of 1-aminophenazines and related compounds (*R.C. Elderfield, W.J. Gensler* and *O. Birstein*, J. org. Chem., 1946, **11**, 812).

(6) The condensation of cyclohexane-1,2-diones with *o*-phenylenediamine affords 1,2,3,4-tetrahydrophenazines which can be subsequently dehydrogenated to the fully aromatic compounds using either iodine and acetic acid (*G.R. Clemo* and *H. McIlwain*, J. chem. Soc., 1934, 1991; 1935, 738) or palladium on charcoal (*idem, ibid.*, 1936, 258; 1698).

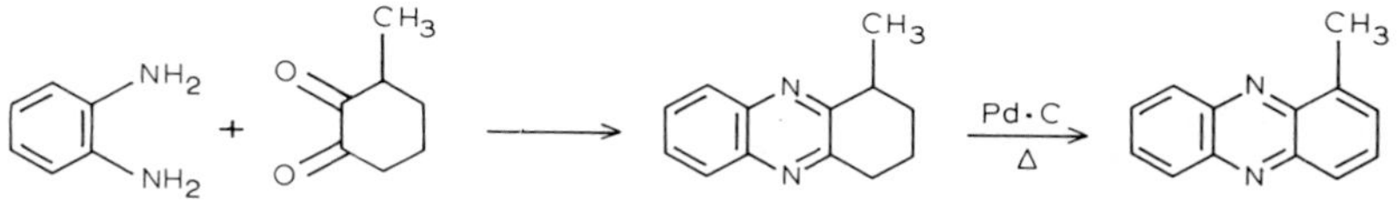

(7) The elaboration of quinoxalines to phenazines has been reported in only a limited number of cases. 2,3-Dimethylquinoxaline-1,4-dioxide (XIII) condenses with diketones, *e.g.* benzil, in alcoholic potassium hydroxide to give a mixture of the mono- and di-*N*-oxides which are deoxygenated using sodium dithionate (*C.H. Issidorides et al.*, Tetrahedron, 1978, **34**, 217).

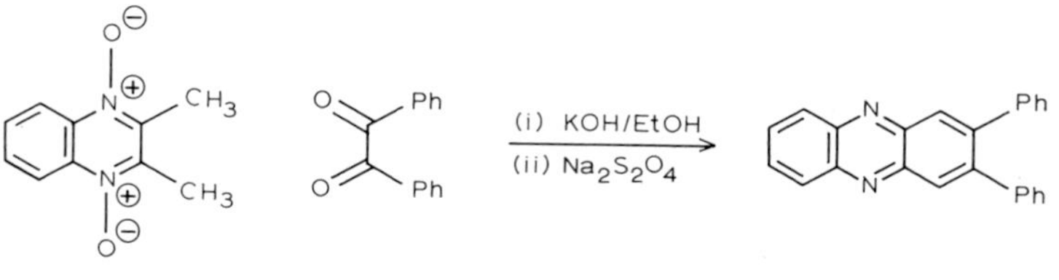

(XIII)

2-Benzoyl-3-methylquinoxaline undergoes a base catalysed double condensation with acetophenone to give 1,3-diphenylphenazine (XIV) in 15% yield (*M.J. Haddadin* and *M.A. Aftah*, J. org. Chem., 1982, **47**, 1772).

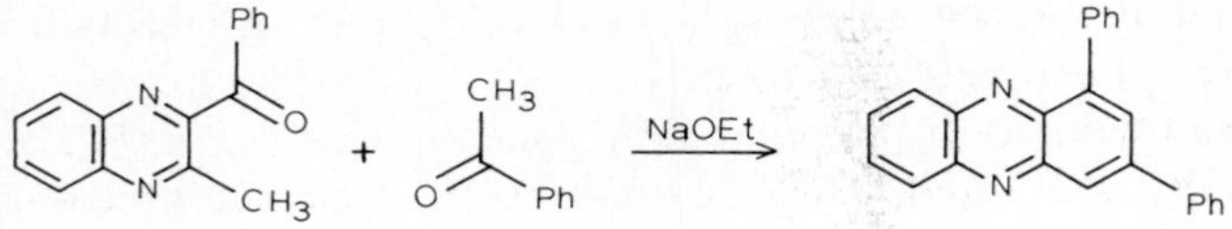

(XIV)

(8) Phenazines are often isolated as significant components of the mixtures resulting from pyrolysis or high temperature oxidation of aromatic nitrogen compounds. In many cases, the phenazines arise from radical coupling reactions. Such methods include:

(a) pyrolysis of aniline by passage through a red hot tube (*E. Bernthsen*, Ber., 1886, **19**, 3257);

(b) heating a mixture of *o*-amino- and *o*-nitro-diphenylamine with anhydrous sodium acetate (*F. Kehrmann* and *E. Havas*, Ber., 1913, **46**, 341);

(c) oxidation of *o*-aminodiphenylamine with lead oxide (*H. McCombie et al.*, J. chem. Soc., 1928, 353; *I.M.G. Campbell et al., ibid.*, 1938, 404);

(d) the iron(II) oxalate catalysed rearrangement of azobenzenes (*R.A. Abramovitch* and *B.A. Davis*, J. heterocyclic Chem., 1968, **5**, 793) [Scheme 2, a];

(e) oxidation of benzenesulphenanilides (XV) with lead dioxide to give 2,7-disubstituted phenazines (*H. Sayo, K. Mori* and *T. Michida*, Chem. pharm. Bull. Tokyo, 1979, **27**, 351) [Scheme 2, b];

(f) oxidative dimerisation of 2-amino-4-*tert*-butylanisole with simultaneous loss of the methoxy group affords 2,7-di-*tert*-butylphenazine (*F.N. Mazitova et al.*, Zhur. org. Khim., 1967, **3**, 878) [Scheme 2, c];

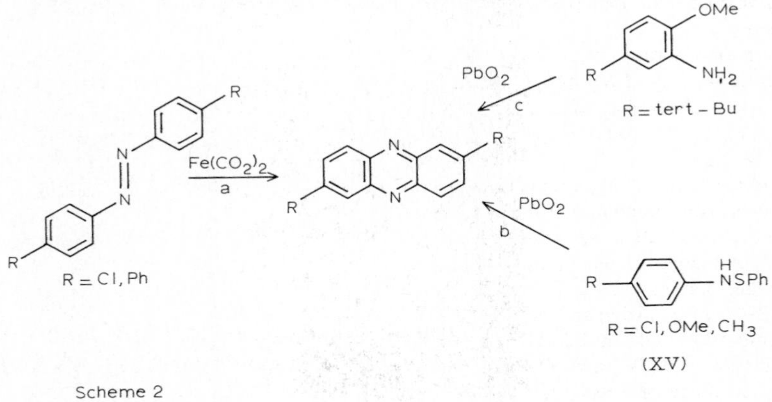

Scheme 2

(g) oxidation of certain aromatic amine to give symmetrically substituted phenazines, *e.g.* 2,4-dimethylaniline with silver carbonate on celite gives 1,3,6,8-tetramethylphenazine (*M. Hedayatullah et al.*, Tetrahedron Letters, 1975, 2039). β-Naphthylamine with *tert*-butylpotassium in oxygen gives dibenzo[*a,h*]phenazine (*L. Horner* and *J. Dehnert*, Ber., 1963, **96**, 786).

(d) Phenazine, its homologues and derivatives

(i) Phenazine, alkyl- and aryl-phenazines

The parent compound, phenazine, has been prepared by several general methods including (a) from *o*-phenylenediamine and catechol followed by oxidation of the resulting dihydrophenazine (*J.S. Morley*, J. chem. Soc., 1952, 4008); (b) by ring closure of 2-nitrodiphenylamine using the Vivian procedure (*H.C. Waterman* and *D.L. Vivian*, J. org. Chem., 1949, **14**, 289; *R.H. Smith* and *H. Suschitzky*, Tetrahedron, 1961, **16**, 80), or sodium borohydride in sodium ethoxide/ethanol (*S.R. Challand, R.B. Herbert* and *F.G. Holliman*, Chem. Comm., 1970, 1423), or by the action of alkali metal alkoxides in DMSO at 110–120° (Jap. Pat. 75, 95281; C.A., 1976, **84**, 105648); (c) from cyclisation of 2,2′-diaminodiphenylamine with iron(III) chloride (*M.L. Tomlinson*, J. chem. Soc., 1939, 158); (d) by catalytic dehydrogenation of reduced phenazines, *e.g.* 1,2,3,4-tetrahydro- or 1,2,3,4,6,7,8,9-octahydro-phenazine, using palladium on charcoal (*G.R. Clemo* and *H. McIlwain*, J. chem. Soc., 1936, 258; C.A., 1982, **96**, 162743; 1983, **98**, 198279).

Phenazine, yellow needles, m.p. 175–176° (given as 171° in earlier literature), b.p. 360° (*picrate*, m.p. 181°), forms a monohydrochloride and a diperchlorate, sublimes readily, is steam volatile, and, being only weakly basic, may be extracted by ether from dilute acid solution.

Phenazine forms crystalline molecular complexes with: 5,10-dihydrophenazine (1:1), *phenazhydrin*, m.p. 209° (dec.); diphenylamine (2:1), m.p. 57–58° (*G.R. Clemo* and *H. McIlwain*, J. chem. Soc., 1935, 728); trinitrobenzene (1:1), m.p. 151–153°, quinol (2:1), m.p. 232° (dec.), and catechol (2:1), m.p. 184° (*Th. Zerewitinoff* and *Iw. Ostromisslensky*, Ber., 1911, **44**, 2402).

Photoreduction of phenazine with iron(III) chloride or tin(II) chloride in isopropanol affords the corresponding 1:1 chelates (*H. Inoue et al.*, Bull. chem. Soc. Japan, 1966, **39**, 555).

Reduction of phenazine with hydrogen sulphide (*C. Ris*, Ber., 1886, **19**, 2206), or sodium dithionite in aqueous alkaline ethanol (*R. Scholl* and *W. Neuberger*, Monatsh., 1918, **39**, 238), or lithium aluminium hydride (*L. Birkofer* and *A. Birkofer*, Ber., 1952, **85**, 286; *F. Bohlmann, ibid.*, 390), or hydrazine and palladium on charcoal (*G.F. Bettinetti, S. Maffei* and *S. Pietra*, Synthesis, 1976, 748) affords 5,10-dihydrophenazine which is highly susceptible to air oxidation. Reductive silylation with lithium metal and trimethylsilyl chloride gives 5,10-*di*(*trimethylsilyl*)-*5,10-dihydrophenazine*, m.p. 92° (*L. Birkhofer* and *N. Ramadan*, Ber., 1975, **108**, 3105).

Electrochemical reduction of phenazine in the presence of methyl chloroformate gives rise to a mixture of 5-*methoxycarbonyl*-, (48%) m.p. 171–172°, and 5,10-*di*(*methoxycarbonyl*)-, (32%) m.p. 161–162°, 5,10-*dihydrophenazine*; with lithium perchlorate as co-electrolyte, the former compound is obtained exclusively (*J. Armand et al.*, J. org. Chem., 1983, **48**, 2847). Irradiation ($\lambda > 330$ nm) of phenazine in the presence of simple aliphatic aldehydes affords *N*-acylated 5,10-dihydrophenazines (*M. Takagi, S. Goto* and *T. Matsuda*, Bull. chem. Soc. Japan, 1980, **53**, 1777).

Oxidation of phenazine by heating with hydrogen peroxide in glacial acetic acid at 50° gives the 5,10-*di-N-oxide*, orange needles m.p. 204°, which on refluxing in acetic anhydride gives the 5-*N-oxide*, m.p. 223° (*G.R. Clemo* and *H. McIlwain*, J. chem. Soc., 1938, 479). The mono-*N*-oxide can be prepared directly from phenazine (53% yield) by reation with cerium(IV) ammonium nitrate in methanol (*B. Rindone* and *C. Scolastico*, J. chem. Soc. Perkin I, 1975, 1398). Treatment with alkaline potassium permanganate results in cleavage of the phenazine ring system to give either quinoxaline-2,3-dicarboxylic acid (*I. Yoshioka* and *Y. Otomasu*, Chem. pharm. Bull. Tokyo, 1957, **5**, 277), or, under more vigorous conditions, pyrazinetetra-carboxylic acid (*H.I.X. Mager* and *W. Berends*, Rec. Trav. chim., 1959, **78**, 5). Ozonolysis also results in selective cleavage of the carbocyclic rings to give a tetraozonide (*N.F. Tyupalo et al.*, C.A., 1976, **85**, 46577).

Phenazine is strongly resistant to electrophilic substitution. Treatment with a mixture of concentrated nitric and sulphuric acid produces, in low yield, a mixture of 1-nitro- and dinitro-phenazine (*A. Albert* and *H.Duewell*, J. Soc. chem. Ind., 1947, **66**, 11). Similarly, prolonged heating of phenazine with mercury(II) chloride in oleum gives *phenazine-2-sulphonic acid*, m.p. > 380° (dec.), together with polysulphonated products (*S. Maffei*, Gazz., 1950, **80**, 651). Direct chlorination of phenazine in refluxing caron tetrachloride containing suspended sodium acetate affords a mixture of 1-chloro- and 1,4-dichloro-phenazine (*V.P. Chernetskii* and *A.I. Kiprianov*, C.A., 1955, **49**, 14774). Phenazine does not undergo Friedel-Crafts acylation (*H. Gilman* and *J.J. Dietrich*, J. Amer. chem. Soc., 1957, **79**, 6178).

Phenazine, particularly as its protonated form, is susceptible to radical substitution reactions. For example, (a) irradiation of an ethanolic solution of phenazine produces a mixture of the 1- and the 2-(1-hydroxyethyl) derivative (*A. Albini et al.*, Gazz., 1970, **100**, 700), (b) iron(II) ion catalysed reaction with a chloramine in sulphuric acid gives the corresponding 2-alkylamino compound (*A. Citterio* and *T. Crolla*, Org. Prep. Proced. Int., 1978, **10**, 63), and (c) irradiation of aqueous or alcoholic solutions of phenazine containing either phosphoric acid or *p*-toluenesulphonic acid gives 1-hydroxy- and 1-alkoxy-phenazines, respectively (*S. Wake* and co-workers, Tetrahedron Letters, 1970, 2418; Bull. chem. Soc. Japan, 1974, **47**, 1251).

With appropriate alkylating agents, phenazine undergoes *N*-alkylation. Treatment of phenazine with excess ethyl iodide gives ethylphenazinium iodide (*O. Fischer* and *E. Hepp*, Ber, 1897, **30**, 391).

Methylphenazinium methylsulphate (phenazine methosulphate) (XVI), m.p. 155–157°, is prepared by heating a solution of phenazine and dimethyl sulphate in nitrobenzene at 110°

(XVI)

(*F. Kehrmann* and *E. Havas*, Ber., 1913, **46**, 341); stronger heating or omitting the reaction solvent results in nuclear methylation (*H. Hilleman*, Ber., 1938, **71**, 34). The methylphenazinium salt undergoes a series of transformations including (a) demethylation in aqueous alkali, (b) decomposition to a mixture of phenazine, 5-methyl-5,10-dihydrophenazine and formaldehyde, *via* the unstable hydroxide salt, in aqueous solution, (c) oxidative decomposition to phenazine and 10-*methyl*-2(10*H*)-*phenazinone* (XVII), m.p. 200°, and (d) photochemical oxidation to give predominantly 5-methyl-1(5*H*)-phenazinone (pyocyanine) (XVIII) together with some (XVII) (*H. McIlwain*, J. chem. Soc., 1937, 1704). Phenazinone (XVIII) is readily demethylated in aqueous sodium hydroxide solution to give 1-hydroxyphenazine (*D.L. Vivian*, Nature, 1956, **178**, 753).

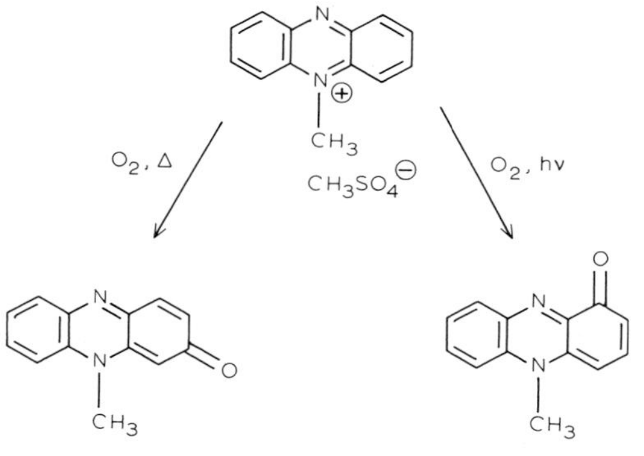

(XVII)   (XVIII)

Methylphenazinium methylsulphate reacts readily with nucleophiles: (a) with ammonia to give the corresponding 2-amino derivative (XIX) (*F. Kehrmann* and *E. Havas*, Ber., 1913, **46**, 341); (b) with an aqueous, concentrated potassium cyanide solution, 3-*cyano*-5-*methyl*-5,10-*dihydrophenazine* (XX), m.p. 155°, is formed in 60% yield together with the oxidised form (XXI) which precipitates as a blue crystalline solid, m.p. 145° (*McIlwain, loc. cit.*); and (c) with methylmagnesium iodide to give 5,10-dimethyldihydrophenazine (*McIlwain, loc. cit.; cf. H. Hilleman*, Ber., 1938, **71**, 42).

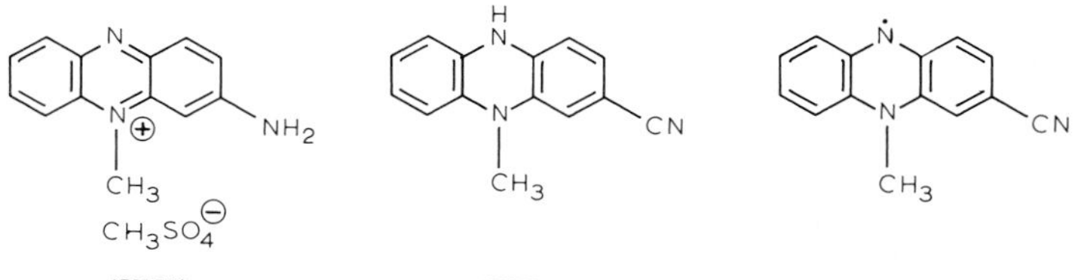

(XIX)   (XX)   (XXI)

Ethylphenazinium ethylsulphate, m.p. 190°, is prepared by heating a mixture of phenazine and diethyl sulphate at 150° without solvent. Although more resistant to dealkylation than (XVI), it undergoes a similar range of oxidation processes. Thus, reaction with aqueous potassium ferricyanide gives 10-*ethyl*-2(10*H*)-*phenazinone* (dark red), m.p. 174°, whereas photochemical oxidation produces the isomeric 10-*ethyl*-1(10*H*)-*phenazinone* (dark blue), m.p. 187°. The ethylphenazinium salt is reduced with zinc dust in sulphuric acid under nitrogen to 5-*ethyl*-5,10-*dihydrophenazine*, m.p. 99°, which can be oxidised to the stable radical 5-*ethylphenazyl* (VI), m.p. 102°, by lead peroxide (*McIlwain, loc. cit.; K.L. Hausser*, Naturwiss., 1956, **43**, 14).

Deamination of phenosafranine (V) affords 5-*phenylphenazinium chloride* (XXII) (*F. Kehrmann*, Ber., 1896, **29**, 2316) which reacts with nucleophilic reagents in a similar manner to the alkylphenazinium salts mentioned above, *e.g.* with sodium hydroxide, aposafranone (XXIII) is obtained in 50% yield (*F. Kehrmann* and *W. Schappschinkoff*, Ber., 1897, **30**, 2620; *O. Fischer* and *E. Hepp*, Ber., 1900, **33**, 1485).

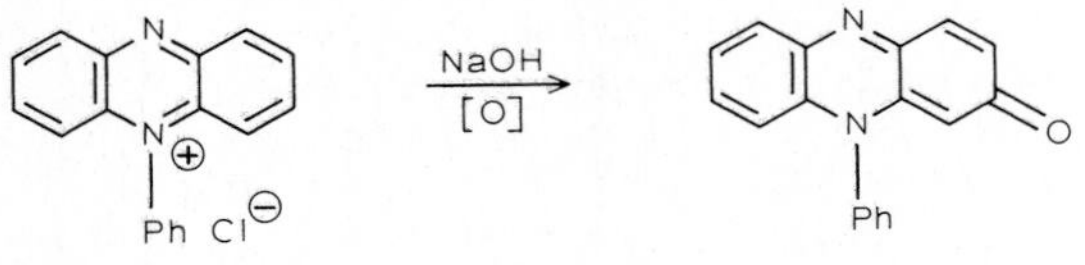

Carbanions derived from activated methylene compounds usually attack methylphenazinium salts at the 3-position, even when the 1-position is substituted. When the 3-position is substituted, the nucleophilic attack is directed to the 7-position in the unsubstituted carbocyclic ring [Scheme 3] (*V.N. Rudenko, A.Ya. Il'Chenko* and *Yu.S. Rosum*, C.A., 1970, **73**, 45468).

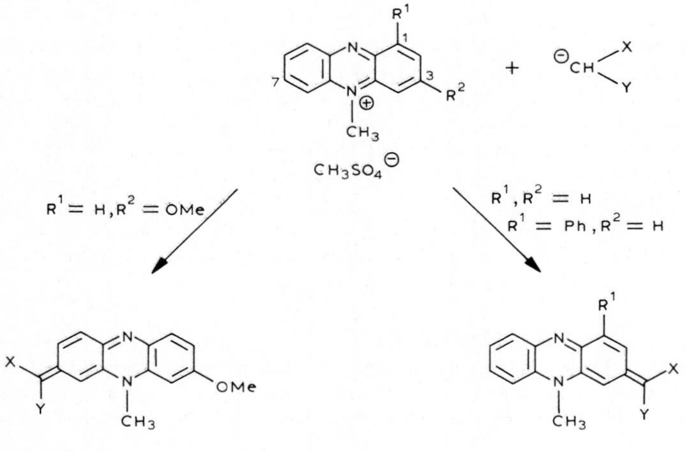

Scheme 3

Although alkylphenazines can be prepared by a number of general procedures, the condensation of *o*-phenylenediamine with an appropriately substituted catechol followed by oxidation of the resulting dihydrophenazine (Ris procedure) is reported to be one of the more synthetically useful methods. Yields of 2-alkylphenazines are generally higher than those of the corresponding 1-isomers. The more direct condensation between *o*-phenylenediamine and *o*-benzoquinones is found to be less satisfactory for the synthesis of simple 1- and 2-alkylphenazines because of the required *o*-benzoquinones tend to undergo self-condensation (*B. Cross et al.*, J. Sci. Food Agric., 1968, **20**, 340).

1- and 2-Methylphenazines have also been prepared by catalytic dehydrogenation of the appropriately substituted tetrahydrophenazines using palladium on charcoal (*G.R. Clemo* and *H. McIlwain*, J. chem. Soc., 1934, 1991; 1935, 738). 1,7- and 1,8-Dimethylphenazine have been prepared unambiguously *via* the Vivian procedure (*A. Turins* and *A.D. Long*, Canad. J. Chem., 1974, **52**, 1307).

Centrosymmetrically substituted alkylphenazine derivatives have been obtained in variable yield from a variety of oxidation–dimerisation reactions. Treatment of (a) 2-amino-4-*tert*-butylanisole with lead dioxide gives 2,7-di-*tert*-butylphenazine (*F.N. Mazitova et al.*, Zhur. org. Khim., 1967, **3**, 878); (b) 2,4-dimethylaniline with silver carbonate on celite gives 1,3,6,8-tetramethylphenazine (*M. Hedayatullah et al.*, Tetrahedron Letters, 1975, 2039); and (c) benzenesulphen-*p*-toluidide with lead dioxide gives 2,7-dimethylphenazine (*A. Sayo et al.*, Chem. pharm. Bull. Tokyo, 1979, **27**, 351).

Oxidation of 1- and 2-methylphenazine with chromic acid affords the corresponding carboxylic acids (*S. Maffei*, Gazz., 1950, **80**, 651). The methylphenazines also undergo side-chain bromination with *N*-bromosuccinimide (NBS) giving respectively 1-*bromomethyl*-, m.p. 164°, and 2-*bromomethyl*-, m.p. 105–160°, *phenazine*. With two equivalents of NBS, 1-methylphenazine affords the 1-*dibromomethyl* derivative, m.p. 153°, which is hydrolysed in dilute sulphuric acid to the 1-*carboxaldehyde*, m.p. 179° (*I. Yoshioka* and *K. Ueda*, Chem. pharm. Bull. Tokyo, 1964, **12**, 1247).

1- and 2-Phenylphenazine have been prepared in low yield (9–13%) from the base-catalysed reaction of aniline with nitrobiphenyls (Wohl–Aue procedure) (*Yu.S. Rozum*, Zhur. obshchei Khim., 1959, **29**, 1299; *D.B. Paul*, Austral. J. Chem., 1972, **25**, 2283). Condensation of 2,3-dimethylquinoxaline-1,4-dioxide with benzil affords, after reduction, 2,3-diphenylphenazine (*C.H. Isaidorides et al.*, Tetrahedron, 1978, **34**, 217). *p*-Phenylazobenzene undergoes rearrangement on heating with either iron(II) oxalate or a mixture of aluminium chloride and sodium chloride to give 2,7-diphenylphenazine (15% yield) (*R.A. Abramovich* and *B.A. Davis*, J. heterocyclic Chem., 1968, **5**, 793). Base-catalysed condensation of acetophenone with 2-benzoyl-3-methylquinoxaline affords the 1,3-diphenyl derivative (*M.J. Haddadin* and *M.A. Aftah*, J. org. Chem., 1982, **47**, 1772).

A selection of alkyl- and aryl-phenazines are listed in Table 17. The syntheses of phenazines bearing higher alkyl side-chains have been reported (*Cross et al.* and *Omote et al.*, loc. cit.).

(ii) Benzo- and dibenzo-phenazines

Benzo[*a*]phenazine (XXIV), 1,2-benzophenazine, m.p. 142°, is obtained (88% yield) by the condensation of β-naphthaquinone and *o*-phenylenediamine in ether containing anhydrous sodium sulphate (*F. Kehrmann* and *C. Mermod*, Helv., 1927, **10**, 62). Acid-catalysed rearrange-

(XXIV)

TABLE 17
ALKYL- AND ARYL-PHENAZINES

Compound	m.p. (°C)
1-Methyl-	107–109 [1]
2-Methyl-	116–116.5 [2], 117 [3]
1-Isopropyl-	141–142 [4]
2-Isopropyl-	90–91 [2], 91–92 [4]
2-Ethyl-	59–60 [4]
2-tert-Butyl-	84 [4]
1,3-Dimethyl-	123 [3]
2,3-Dimethyl-	174–175 [3]
1,6-Dimethyl-	235 [4a]
1,7-Dimethyl-	114–117 [5]
1,8-Dimethyl-	146.5–147 [5]
1,9-Dimethyl-	142 [4a]
2,7-Dimethyl-	162.5–163 [6]
2,7-Di-tert-butyl-	215.5–216 [7]
1-Benzyl-	120–12 [8a]
2-Benzyl-	106–107.5 [8a]
1,4-Dibenzyl-	158 [8]
1,3,6,8-Tetramethyl-	224 [9]
1-Phenyl-	156–157 [10,11]
2-Phenyl-	147 [10,12]
1,3-Diphenyl-	167–168 [13]
2,3-Diphenyl-	170–172 [14]
2,7-Diphenyl-	271–272 [15]

1 U. Hollstein, J. heterocyclic Chem., 1968, **5**, 299.
2 Y. Omote, T. Hirama and T. Komatsu, Bull. chem. Soc. Japan, 1974, **47**, 1957.
3 H.J. Teuber and G. Staiger, Ber., 1955, **88**, 802.
4 Neth. Appl. 6,511,395; C.A., 1966, **65**, 2279.
4a E. Breitmaier and U. Hollstein, J. org. Chem., 1976, **41**, 2104.
5 A. Turins and A.D. Long, Canad. J. Chem., 1974, **52**, 1307.
6 H. Sayo, K. Mori and T. Michida, Chem. pharm. Bull. Tokyo, 1979, **27**, 351.
7 F.N. Mazitova, R.R. Shagidulun and V.V. Abushaeva, Zhur. org. Khim., 1967, **3**, 878.
8 H. McIlwain, J. chem. Soc., 1937, 1701.
8a W.A. Waters and D.H. Watson, ibid., 1959, 2085.
9 M. Hedayatullah, J.P. Dechatre and L. Denivelle, Tetrahedron Letters, 1975, 2039.
10 D.B. Paul, Austral. J. Chem., 1972, **25**, 2283.
11 A.E.J. Herbert and M. Tomlinson, J. chem. Soc., 1958, 4492.
12 Yu.S. Rozum, Zhur. obshchei Khim., 1959, **29**, 1299.
13 M.J. Haddadin and M.A. Atfah, J. org. Chem., 1982, **47**, 1772.
14 C.H. Issidorides et al., Tetrahedron, 1978, **34**, 217.
15 R.A. Abramovitch and B.A. Davis, J. heterocyclic Chem., 1968, **5**, 793.

ment of 2-anilino-1-benzeneazonaphthalene affords (XXIV) quantitatively (*G. Schroeder* and *W. Luttke*, Ber., 1972, **105**, 2175).

The condensation of 2,3-dihydroxynaphthalene and *o*-phenylenediamine gives the 5,10-dihydro derivative which on oxidation with potassium dichromate in acetic acid yields **benzo[*b*]phenazine** (XXV), m.p. 233° (*O. Hinsberg*, Ann., 1901, **319**, 257).

(XXV)

The centrosymmetric **dibenzo[*a,h*]phenazine** (XXVI), m.p. reported variously as 283–284° and 286–287°, has been prepared by (a) reaction of 1-nitroso-2-naphthylamine and 1-naphthylamine (*O. Fischer* and *A. Junk*, Ber., 1893, **26**, 183), (b) dimerisation of 2-naphthylamine in the presence of *tert*-butylpotassium and molecular oxygen (*L. Horner* and *J. Dehnert*, Ber., 1963, **96**, 786), (c) from the thermal decomposition of 1- or 2-azonaphthalene (*B. Nay et al.*, J. chem. Soc. Perkin I, 1980, 611).

(XXVI)   (XXVII)   (XXVIII)

Reaction of 1-nitroso-2-naphthylamine with 2-naphthylamine hydrochloride gives a mixture of two isomeric adducts, **dibenzo[*a,j*]phenazine** (XXVII), m.p. 243°, and **dibenzo[*a,i*]phenazine** (XXVIII), m.p. 295–296°. Compound (XXVII) has also been isolated from (a) the cyclisation of *N*-nitroso-2,2′-dinaphthylamine in acetic acid (*Fischer* and *Junk*, *loc. cit.*), (b) the molecular rearrangement of benzeneazo-2,2′-dinaphthylamine with concomitant elimination of a molecule of aniline, and (c) the reaction of 1-benzeneazo-2-naphthylamine with 2-naphthol (*F. Ullmann* and *J.S. Ankersmitt*, Ber., 1905, **38**, 1811).

(iii) Phenazine-N-oxides

The isolation of the antibiotics *iodinin* (*G.R. Clemo* and *H. McIlwain*, J. chem. Soc., 1938, 479; *H. McIlwain*, *ibid.*, 1943, 322) and the structurally related *myxin* (*H.P. Sigg* and *A. Toth*, Helv., 1967, **50**, 716) has stimulated the synthesis of a range of phenazine-*N*-oxides.

Phenazines are conventionally transformed into the corresponding mono- and di-oxides by reaction with either hydrogen peroxide in acetic acid (*Clemo* and *McIlwain*, *loc. cit.*), or a peracid, *e.g.* *m*-chloroperbenzoic acid (*M. Weigele* and *W. Leimgruber*, Tetrahedron Letters, 1967, 715) in the appropriate ratio. The mixture of the *N*-oxides, which is often obtained, can be separated by chromatographic methods (*Yu.S. Rozum* and *N.N.*

Lisovskaya, Zhur. obshchei Khim., 1959, **29**, 228). With substituted phenazines, oxidation occurs initially to the less sterically hindered nitrogen centre. Phenazine has also been oxidized to the mono-*N*-oxide using cerium(IV) ammonium nitrate (*B. Rindone* and *C. Scolastico*, J. chem. Soc. Perkin I, 1975, 1398).

Where appropriate, the 5- and the 10-oxide can be prepared unambiguously by application of the Wohl–Aue procedure, although the yield is generally low (*Yu.S. Rozum*, Zhur. obschchei Khim., 1960, **30**, 1661).

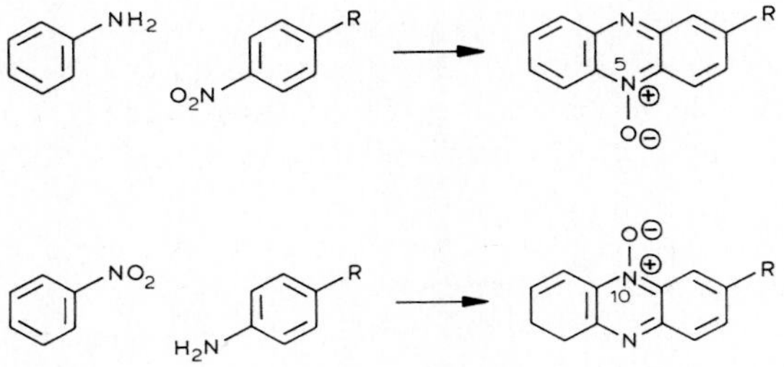

Intramolecular cyclisation of 2-nitrodiphenylamines in acid or alkali (*B. Cross, P.J. Williams* and *R.E. Woodall*, J. chem. Soc. C, 1971, 2085), or amines (Jap. Pat. 75 70,376; 75 95,282, C.A., 1976, **84**, 17425; 105649) also gives rise to the corresponding mono-*N*-oxides in moderate yield. Mono-*N*-oxides are also obtained by dimerisation of nitrosobenzenes (*E. Bamberger* and *W. Ham*, Ann., 1911, **382**, 82).

Although the scope of the Beirut reaction for the preparation of phenazine di-*N*-oxides is more limited than for quinoxaline di-*N*-oxides (*vide supra*), it does find application in the synthesis of hydroxy derivatives which are of pharmacological interest. The condensation of benzofuroxan and a phenolate anion provides a convenient, direct synthetic route to di-*N*-oxides derived from 2-hydroxyphenazine. Thus, 2-hydroxyphenazine 5,10-dioxide (**XXIX**) is obtained as the major product from the base-cata-

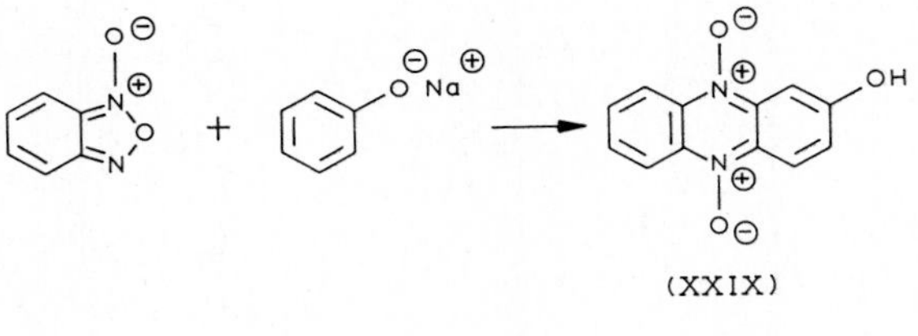

(XXIX)

lysed reaction of benzofuroxan with either phenol, or resorcinol, or hydroquinone, or benzoquinone (*El-Haj et al.*, J. org. Chem., 1972, **34**, 589; *K. Ley et al.*, Angew. Chem. intern. Edn., 1969, **8**, 596; *K. Ley* and *F. Seng*, Synthesis, 1975, 415; *M.L. Edwards et al.*, J. heterocyclic Chem., 1976, **13**, 653). Substituted benzofuroxans and/or phenols have been used to produce a range of 2-hydroxyphenazine derivatives (F.P. 2,003,273; C.A., 1970, **72**, 111500; Ger. Offen 1,927,473; C.A., 1970, **72**, 79100). The observed product ratios from the reaction of benzofuroxan and substituted hydroquinones can be rationalised in terms of substituent stabilisation of certain preferred resonance forms of the intermediate phenoxide anion. Thus, with methoxy and alkoxycarbonyl groups, the reaction is highly stereoselective, producing the 3- and the 1-substituted 2-hydroxyphenazine dioxide respectively whereas the presence of a chloro substituent leads to equal quantities of the three possible isomers (*A. Roemer* and *M. Sammet*, Z. Naturforsch., 1983, **38B**, 866).

1-Phenazinol dioxides (XXX) are prepared by a modified procedure involving condensation of a benzofuroxan with cyclohexane-1,2-dione in diethylamine followed by oxidation of the resulting mixture of mono- and di-oxides with *m*-chloroperbenzoic acid (*C.H. Issidorides et al.*, Tetrahedron, 1978, **34**, 217).

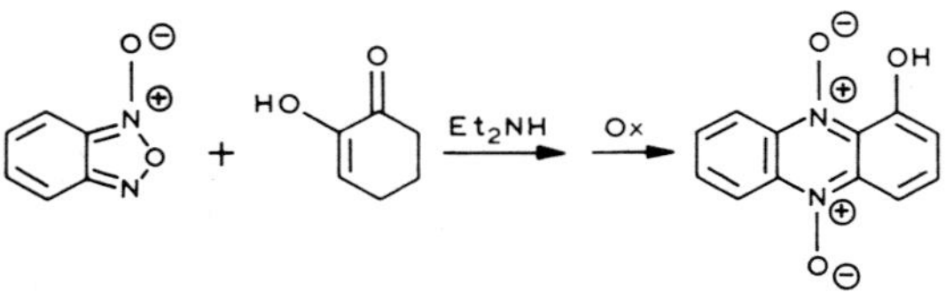

When *p*-aminophenol is used in the Beirut reaction, 2-aminophenazine dioxides are obtained, *e.g.* reaction with the benzofuroxan (XXXI) affords a mixture (70:30) of the separable isomeric dioxides (XXXIIa) and (XXXIIb) (*G.W. Ludwig* and *H. Baumgaertel*, Ber., 1982, **115**, 2380).

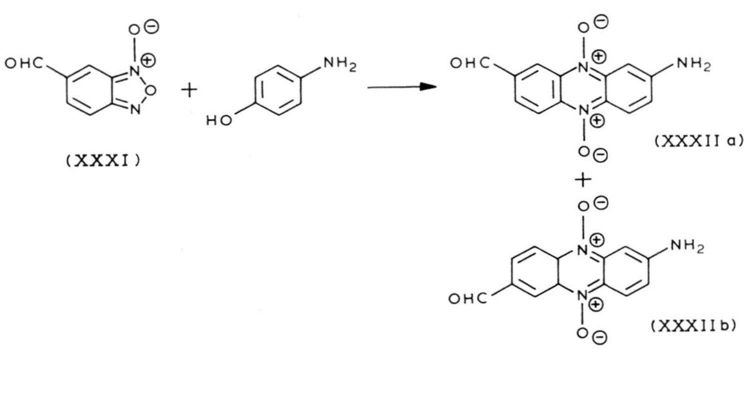

The base-catalysed condensation of 2,3-dimethylquinoxaline-1,4-dioxide with α-diketones has been exploited to a limited extent for the preparation of phenazine dioxides because mixtures of the mono- and di-oxides are invariably obtained and overall yields are low (*Issidorides et al., loc. cit.*).

The *N*-oxide functionality in phenazines activates the carbocyclic rings to nucleophilic substitution reactions. Phenazine mono- and di-oxides are converted into the 2-chloro- and 2,6-dichloro-phenazine, respectively (*I.Ya. Postovskii* and *E.I. Abramova*, Zhur. obshchei Khim., 1954, **24**, 485). By analogy, the reaction of the mono-oxide with tosyl chloride yields the 1-tosyloxy derivative (XXXIII) (*E. Matsumura* and *H. Takeda*, J. chem. Soc. Japan, 1960, **81**, 515). This effect is enhanced by electron withdrawing groups, *e.g.* the reaction of 2-nitrophenazine-10-oxide with carbanions (*S. Pietra* and *S. Casiraghi*, Gazz., 1966, **96**, 1630) or cyanide ion (*idem, ibid.*, 1967, **97**, 1817) affords the corresponding 1-substituted-2-nitro compounds. In contrast, Grignard reagents attack at the remote nitrogen atom to give phenazinium salts (XXXIV) [Scheme 4] (*A.I. Kiprianov* and *G.M. Prilutskii*, Zhur. obshchei Khim., 1959, **29**, 1020).

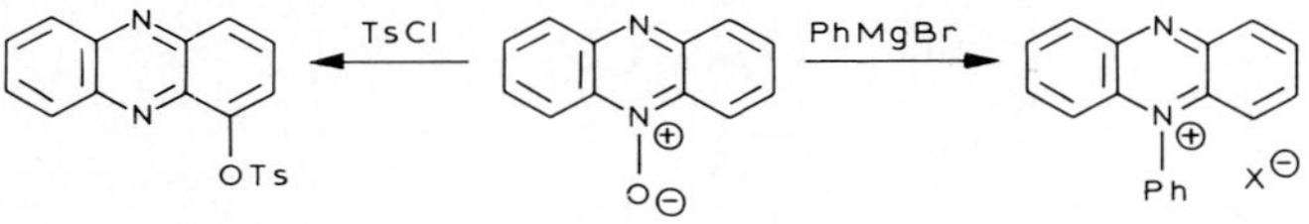

Scheme 4

Halogenated phenazine-*N*-oxides exhibit greater susceptibility to *ipso* nucleophilic substitution than the parent compounds (*H. Endo, M. Tada* and *K. Katagiri*, Bull. chem. Soc. Japan, 1969, **42**, 502). Halogen atoms at the 1- and the 2-position with respect to the *N*-oxide group are most affected, *e.g.*, in 2,7-dichlorophenazine-5-oxide, the adjacent 7-chloro atom is selectively displaced by ethoxide ion (*I.J. Pachter* and *M.C. Kloetzel*, J. Amer. chem. Soc., 1952, **74**, 971).

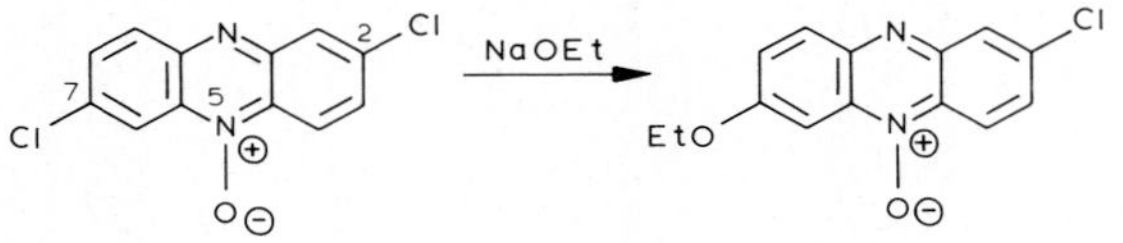

In general, phenazine-*N*-oxides are nitrated under milder conditions than the parent compounds. Phenazine-5-oxide gives the 3-nitro-5-oxide as the major product with some of the 1-nitro isomer (*H. Otomasu*, Chem. pharm.

TABLE 18
PHENAZINE-N-OXIDES

Compound	m.p. (°C)		
	5-N-*oxide*	10-N-*oxide*	5,10-di-N-*oxide*
Parent	223 [1a]	–	204 [1a]
			189–191 [1b]
			202–203 [1b]
1-Methyl-	142 [2]	158 [2]	141–142 [3]
	143 [4]		
2-Methyl-	122 [5a]	155–156 [5b]	171–172 [6]
1-Phenyl-	200 [7]	172 [7]	–
2-Phenyl-	168 [7]	135 [7]	–
2,3-Dimethyl-	204–205 [8]	–	–
2,3-Diphenyl-	177–179 [9]	–	196–198 [9]
1-Chloro-	158–158.5 [10]	–	190–191 [12]
	159–160 [11]		
2-Chloro-	176–177 [10]	174–176 [10,11]	182 [10]
	178 [11]	178–179 [7]	190–191 [12]
2,7-Dichloro-	238 [8], 236 [11]	–	201–203 [13]
1-Bromo-	188–189 [14]	–	–
2-Bromo-	163–164 [14]	165–167 [14]	–
1-Hydroxy-	190 [11,15]	165–167 [4,11]	189–190 [1]
			184–185 [9]
2-Hydroxy-	–	258 [4,11]	234 [6], 232 [16]
1-Methoxy-	204 [17]	175–176 [17]	195–196 [18]
2-Methoxy-	176–177 [17]	179 [7]	184 [16]
	170 [19]		188 [19]
1-Carboxy-	223 [4]	–	–
2-Carboxy-	–	263 [20]	*ca.* 280 [20]
2-Formyl-	–	–	195–196 [6]
1-Amino-	226–227 [21]	–	–
2-Amino-	260–261 [21]	253–254 [21]	219 [16]

1a G.R. *Clemo* and H. *McIlwain*, J. chem. Soc., 1938, 479.
1b Z.V. *Pushareva, L.V. Varyukhina* and *Z.Yu. Kokoshko*, Dokl. Akad. nauk SSSR, 1956, **108**, 1098.
2   Z.H. *Pushareva* and G.I. *Agibalova*, J. gen. Chem. U.S.S.R., 1938, **8**, 151.
3   B. *Cross* and P.J. *Williams*, B.P. 1,182, 617; C.A., 1970, **72**, 90513.
4   Ger. Offen., 1,935,705; C.A., 1970, **72**, 134173.
5a K. *Yoshinori* and H. *Otomasu*, Chem. pharm. Bull. Tokyo, 1956, **4**, 117.
5b A. *Albini, et al.*, J. chem. Soc. Perkin II, 1978, 238.
6   M.L. *Edwards, R.E. Ramsbury* and H.K. *Kim*, J. heterocyclic Chem., 1976, **13**, 653.
7   Yu.S. *Rozum*, Zhur. obshchei Khim., 1960, **30**, 1661.
8   E. *Bamberger* and W. *Ham*, Ann., 1911, **382**, 82.
9   C.H. *Issidorides, et al.*, Tetrahedron, 1978, **34**, 217.
10 B. *Cross*, J. Sci. Food Agric., 1969, **20**, 8.
11 J. *Pachter* and M.C. *Kloetzel*, J. Amer. chem. Soc., 1952, **74**, 971.
12 D.L. *Vivian, ibid.*, 1949, **71**, 1139; 1951, **73**, 457.

TABLE 18 (continued)

13 *M.F. Grundon, B.T. Johnston* and *W.L. Matier*, J. chem. Soc. B, 1966, 260.
14 *V.P. Chernetskii* and *A.I. Kiprianov*, Zhur. obshchei Khim., 1956, **26**, 3032.
15 *N.N. Gerber*, Biochemistry, 1966, **5**, 3824.
16 *D.J. Johnson* and *M. Abuel-Haj*, Ger. Offen., 1,927,473; C.A., 1970, **72**, 79100.
17 *I. Yoshioka*, J. pharm. Soc. Japan, 1953, **73**, 23.
18 *H. Endo, M. Tada* and *K. Katagiri*, Sci. Rep. res. Inst., Tohoku Univ., 1967, **14C**, 164.
19 *A. Roemer* and *M. Sammet*, Z. Naturforsch., 1983, **38B**, 866.
20 *S. Maffei, S. Pietra* and *A.M. Rivolta*, Ann. Chim. (Rome), 1952, **42**, 519.
21 *M. Tada*, Bull. chem. Soc. Japan, 1975, **48**, 3405.

Bull. Tokyo, 1954, **2**, 283; *S. Maffei* and *G.F. Bettinetti*, Ann. Chim. (Rome), 1955, **45**, 1031; *A.D. Grabenko* and *S.B. Serebryanyi*, Ukrain. khim. Zhur., 1955, **21**, 249). Nitration of the 1- and the 2-substituted phenazine oxides has also been reported (*H. Otomasu* and *K. Yoshida*, J. pharm. Soc. Japan, 1961, **81**, 861; *Otomasu, loc. cit.*).

Irradiation of phenazine mono-oxide with u.v. light affords a mixture of products including the parent phenazine and the novel rearrangement products (XXXV) and (XXXVI) (*A. Albini* and co-workers, Tetrahedron Letters, 1972, 3657; J. chem. Soc. Perkin II, 1978, 238).

Heating phenazine-5,10-dioxide in acetic anhydride under reflux gives, after 10 min., the mono-oxide; over an extended period, phenazine is obtained (*G.R. Clemo* and *H. McIlwain*, J. chem. Soc., 1938, 479). In a similar fashion, phenazine-*N*-oxides are de-oxygenated on heating with aniline (*Pachter* and *Kloetzel, loc. cit., Cross et al., loc. cit.*) or phosphorus trichloride (*C.W. Rees* and *T.R. Emerson*, J. chem. Soc., 1964, 2319). The deoxygenation of the *N*-oxides can also be effected conveniently by (a) alkaline sodium dithionite (*D.L. Vivian*, J. Amer. chem. Soc., 1951, **73**, 457; *Issidorides et al., loc. cit.*), (b) iron in acetic acid (*V.P. Chernetskii*, Ukrain. obshchei Khim., 1960, **26**, 36; *Cross et al., loc. cit.*), (c) zinc in aqueous sodium hydroxide (*M.F. Grundon et al.*, J. chem. Soc. B, 1966, 206), (d) tin(II) chloride in hydrochloric acid (*Bamberger* and *Ham, loc. cit.*), and (e) catalytic hydrogenolysis over palladium on charcoal (*S. Maffei* and *S. Pietra*, Gazz., 1958, **88**, 556; *Roemer* and *Sammet*, loc. cit.). Under dissolving metal conditions, e.g. sodium amalgam in ethanol, or zinc in hydrochloric acid, or iron in acetic acid, the phenazine produced may be reduced *in situ* to the corresponding dihydrophenazine (*A. Wohl* and *W. Aue*, Ber., 1901, **34**, 2442; *Chernetskii*, loc. cit.).

(iv) Halogenated phenazines

Simple halogenated phenazines can be prepared directly *via* either the Wohl–Aue or the Vivian procedure described above. In the latter reaction, halogen substituents at the 2'- or 4'-positions are usually eliminated on cyclisation; halogen atoms in the same ring as the nitro group are unaffected (*D.L. Vivian et al.*, J. org. Chem., 1953, **18**, 1065).

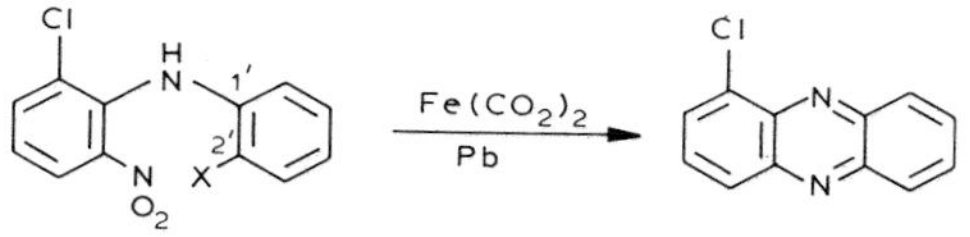

Intramolecular cyclisation of 2,2'-dinitrodiphenylamines in alcoholic potassium hydroxide by hydrazine in the presence of either Raney nickel or ruthenium on charcoal provides an alternative synthetic route to chlorophenazines, e.g. 2-chlorophenazine from 4-chloro-2,2'-dinitrodiphenylamine in 74% yield (*B. Cross, P.J. Williams* and *R.E. Woodall*, J. chem. Soc. C, 1971, 2085).

Treatment of benzenesulphenanilides with lead peroxide affords the 2,7-dibromo- or dichloro-phenazines directly (*C. Balboni et al.*, J. chem. Soc. Perkin I, 1983, 2111).

Several procedures including (a) the Wohl–Aue reaction at lower temperatures (*ca.* 110°), (b) the cyclisation of halo-2-nitrodiphenylamines in oleum (*B. Cross et al.*, J. Sci. Food Agric., 1969, **20**, 8) and (c) the cyclodimerisation of *p*-halonitrosobenzenes (*E. Bamberger* and *W. Ham*, Ann., 1911, **382**, 82; *S.N. Sawhney* and *D.W. Boykin*, J. heterocyclic Chem., 1979, **16**, 397; *I.J. Rinkes*, C.A., 1919, **13**, 1579) give rise to phenazine mono-*N*-oxides which can be subsequently deoxygenated to the corresponding phenazines.

TABLE 19
HALOGENATED PHENAZINES

Compound	m.p. (°C)
1-Fluoro-	208–209 [1]
1-Chloro-	121–122 [2]
1-Bromo-	132–134 [2,3]
1-Iodo-	142–144 [3]
2-Fluoro-	181 [2]
2-Chloro-	138–139 [4–6]
2-Bromo-	149–150 [2,6]
2-Iodo-	169–170 [2]
1,2-Dichloro-	172–173 [7]
1,3-Dichloro-	189–190 [7]
1,4-Dichloro-	191–192 [7], 195 [8]
1,6-Dichloro-	265–266 [7], 266–267 [9]
1,9-Dichloro-	206.5–207.5 [7]
2,3-Dichloro-	246–247 [7]
2,7-Dichloro-	265.5–268 [7], 265–266 [10]
2,8-Dichloro-	230–231 [7], 232–234 [11]
2-Chloro-8-iodo	157–159 [11]
1,4-Dibromo-	190.5 [12]
2,7-Dibromo-	251–252 [13–15]
2,8-Dibromo-	226–228 [11]
2,7-Diiodo-	235 [13]
1,2,3-Trichloro-	202 [7]
1,2,4-Trichloro-	185–186 [7]
1,2,9-Trichloro-	204.5–205.5 [7]
1,4,6-Trichloro-	215–216 [7], 213 [8]
1,4,7-Trichloro-	220–221 [7]
1,2,3,4-Tetrachloro-	235 [7,16]
1,4,6,8-Tetrachloro-	210 [7,17]
1,4,6,9-Tetrachloro-	322–323 [8]
2,3,7,8-Tetracloro-	223–224 [17]
1,4,6,9-Tetrabromo-	340–341 [18]
1,2,3,4-Tetrafluoro-	233–234 [19], 262–263 [20]
Octafluoro-	238.5–239.5 [19], 259–260 [20]

1 V.P. Chernetskii, L.M. Yagupol'skii and S.B. Serebryanyi, Zhur. obshchei Khim., 1955, **25**, 2166.
2 D.L. Vivian and J.L. Hartwell, J. org. Chem., 1953, **18**, 1065.
3 D.L. Vivian, ibid., 1956, **21**, 1188.
4 H.C. Waterman and D.L. Vivian, ibid., 1949, **14**, 289.
5 I.J. Pachter and M.C. Kloetzel, J. Amer. chem. Soc., 1952, **74**, 971.
6 H. McCombie, H.A. Scarborough and W.A. Waters, J. chem. Soc., 1928, 353.
7 B. Cross, et al., J. Sci. Food Agric., 1969, **20**, 8.
8 S. Maffei, S. Pietra and A. Cattaneo, Gazz., 1953, **83**, 327.
9 I.J. Pachter and M.C. Kloetzel, J. Amer. chem. Soc., 1951, **73**, 4958; S. Maffei, Gazz., 1946, **76**, 239.

TABLE 19 (continued)

10 D.L. Vivian, J. Amer. chem. Soc., 1951, **73**, 457.
11 D.L. Vivian, R.M. Hogart and M. Belkin, J. org. Chem., 1961, **26**, 112.
12 Neth. Appl. 6,511,395; C.A., 1966, **65**, 2279.
13 E. Bamberger and W. Ham, Ann., 1911, **382**, 82.
14 S.N. Sawhney and D.W. Boykin, J. heterocyclic Chem., 1979, **16**, 397.
15 D.L. Vivian, J. org. Chem., 1956, **21**, 824.
16 L. Horner and H. Metz, Ann., 1950, **570**, 89.
17 S. Maffei, S. Pietra and A. Catanneo, Gazz., 1953, **83**, 812.
18 H. Gilman and J.J. Dietrich, J. Amer. chem. Soc., 1957, **79**, 6178.
19 V.D. Shteingarts and A.G. Budnik, C.A., 1968, **68**, 114218.
20 A.G. Hudson, A.E. Pedler and J.C. Tatlow, Tetrahedron, 1970, **26**, 3791.

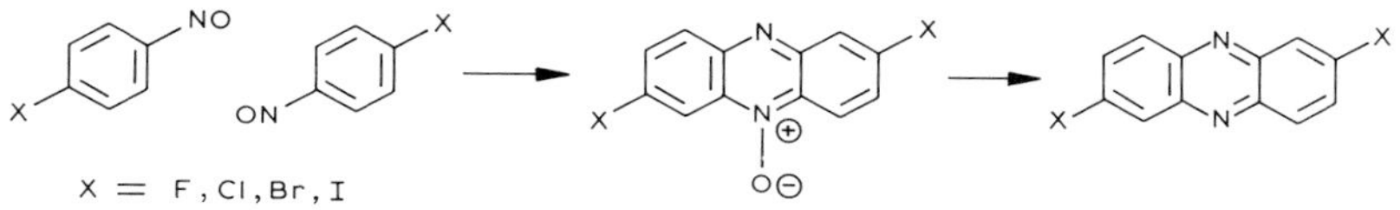

Although stable to chlorine in concentrated sulphuric acid, water and carbon tetrachloride at room temperature, phenazine undergoes direct chlorination in hydrochloric acid to give a mixture of separable mono- and poly-chlorinated compounds (S. Maffei, S. Pietra and A. Cattaneo, Gazz., 1953, **83**, 327). In refluxing carbon tetrachloride containing suspended sodium acetate, 1-chloro- (32%) and 1,4-dichloro- (25%) phenazine are obtained (V.P. Chernetskii and A.I. Kiprianov, C.A., 1955, **49**, 14774). Treatment of phenazine with bromine in acetic acid affords 1,4,6,9-tetrabromophenazine (H. Gilman and J.J. Dietrich, J. Amer. chem. Soc., 1957, **79**, 6178).

Phenazines bearing electron-donating groups such as alkoxy or amino are more susceptible to direct halogenation than the parent compound. Thus, the reaction of 2-methoxyphenazine with a mixture of hydrochloric acid and potassium chlorate introduces a chlorine atom at the 1-position (S. Pietra and G. Casiraghi, Gazz., 1970, **100**, 149). 1- and 2-Methoxyphenazine react with bromine in acetic acid to give 4-*bromo*-1-*methoxy*-, m.p. 153–154°, and 4-*bromo*-2-*methoxy*- m.p. 179–180°, *phenazine*, respectively (I. Yoshioka and S. Arafune, Chem. pharm. Bull. Tokyo, 1959, **7**, 581; K. Ueda and I. Yoshioka, ibid., 1968, **16**, 1521).

Halogen atoms, particularly at the 2-position, can be displaced by suitable nucleophiles, e.g. hydroxide ion (D.L. Vivian, J. Amer. chem. Soc.,

1951, **73**, 457), alkoxide and mercaptide ions (*B. Cross et al.*, J. Sci. Food Agric., 1968, **20**, 340), ammonia (*I.J. Pachter* and *M.C. Kloetzel*, J. Amer. chem. Soc., 1951, **73**, 4958; 1952, **74**, 971) and amines (*D.L. Vivian et al.*, J. org. Chem., 1961, **26**, 112), thus providing entry to a range of other phenazine derivatives. In phenazine-*N*-oxides, halogen substituents show significantly increased activation towards nucleophilic substitution; the observed ease of displacement is in the order dioxide > mono-oxide > parent.

(v) Nitrophenazines and phenazinesulphonic acid

Phenazine is nitrated under forcing conditions to give mixtures of mono- and di-nitro compounds. The usual nitrating conditions are concentrated nitric acid in concentrated sulphuric acid, acetic acid or acetic anydride, or potassium nitrate in concentrated sulphuric acid.

On the basis of theoretical calculations, phenazine would be expected to undergo nitration preferentially at the α-positions in a similar fashion to other electrophilic reactions. Consistent with this prediction, direct nitration of phenazine affords 1,6- and 1,9-dinitrophenazine as the major products together with lesser quantities of the 1-nitro derivative (*S. Maffei* and *M. Aymon*, Gazz., 1954, **84**, 677; *H. Otomasu*, Chem. pharm. Bull.

Tokyo, 1958, **6**, 77; *cf. A. Albert* and *H. Duewell*, J. Soc. chem. Ind., 1947, **66**, 11). The structural assignment for each dinitro compound has been confirmed by ^{13}C-n.m.r. spectroscopy (*E. Breitmaier* and *U. Höllstein*, J. org. Chem., 1976, **41**, 2104). An earlier claim for the formation of the 1,3-dinitro compound by nitration of phenazine is consequently incorrect (*F. Kehrmann* and *E. Havas*, Ber., 1913, **46**, 341). Phenazines containing electron-donating substituents, *e.g.* amino or methoxy groups, or phenazine-*N*-oxides (*vide supra* and *H. Otomasu*, Chem. pharm. Bull. Tokyo, 1954, **2**, 283; 1956, **4**, 117) are activated towards nitration.

Nitration of 1-aminophenazine gives 1-amino-4-nitrophenazine which, on subsequent treatment with sodium hydroxide, is tranformed to 4-*nitro*-1(5*H*)-*phenazinone* (XXXVII), m.p. 200° (dec.), the more stable tautomeric form of the resulting hydroxy compound (*C. Stammer* and *A. Taurins*,

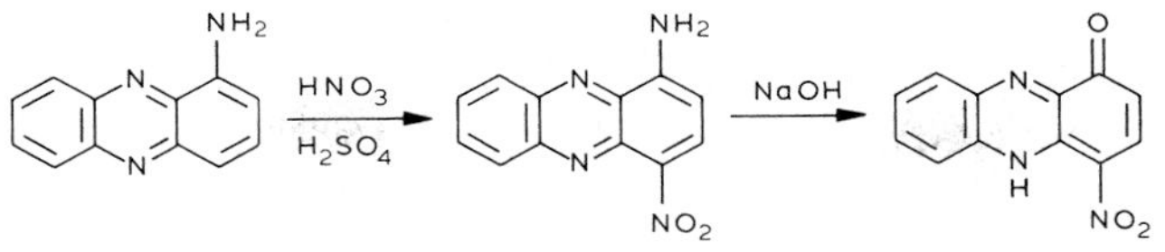

(XXXVII)

Canad. J. Chem., 1963, **41**, 228). By analogy, 1-methoxy- and 1-acetamidophenazine would also be expected to undergo nitration at the 4-position. Although the structure of each nitration product was assigned as the 4-nitro derivative, the product obtained in each case has physical and chemical properties more in keeping with those of the 2-nitro compound (*A. Gray* and *F.G. Holliman*, Tetrahedron, 1962, **18**, 1095; *H. Otomasu*, 1954, *loc. cit.; Stammer* and *Taurins, loc. cit.*). The 2-hydroxy-, 2-methoxy- (*B. Hegedus*, C.A., 1947, **41**, 6262), and 2-amino- (*Stammer* and *Taurins, loc. cit.*) phenazines are specifically nitrated at the 1-position. The nitration of 2-aminophenazine has been shown to proceed *via* the intermediate nitramine (XXXVIII).

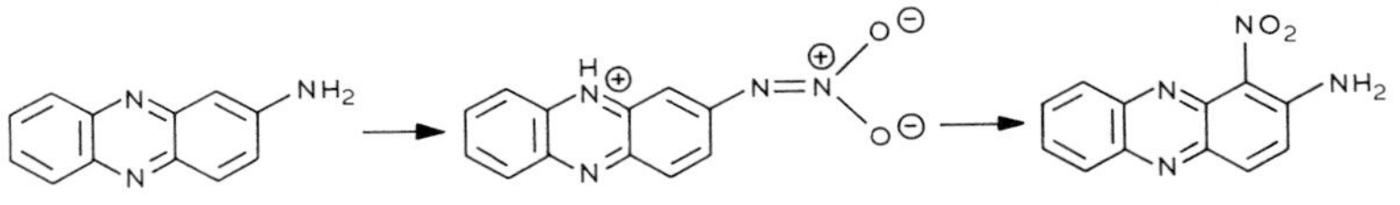

(XXXVIII)

2-Nitrophenazine is obtained from the nitration of 5,10-diacetyl-5,10-dihydrophenazine (*Kehrmann* and *Havas, loc. cit.*) and by the deoxygenation of 2-nitrophenazine-10-oxide using phosphorus trichloride (*A. Albini et al.*, J. chem. Soc. Perkin 1, 1978, 299).

The direct synthesis of nitrophenazines is comparatively uncommon, *e.g.* from *N,N*-diarylamines, because the nitro groups are often reduced under the reaction conditions required for the cyclisation process (*vide infra*). Nitro groups are, however, found to survive the conditions of the Vivian method. Thus, 2,6-dinitro-4′-methoxydiphenylamine yields 7-methoxy-1-nitrophenazine though, in the absence of the methoxy group, no useful products are isolated (*D.L. Vivian*, J. org. Chem., 1956, **21**, 824).

Reduction of nitrophenazines by either catalytic hydrogenation over palladium on charcoal, or zinc in acetic acid affords the corresponding amino compounds.

Nitrophenazines and their related *N*-oxides show enhanced reactivity towards nucleophiles. The reactions of 2-nitrophenazine have been studied

TABLE 20
NITROPHENAZINES

Compound	m.p. (°C)
1-Nitro-	192–195 [1], 195 [2]
2-Nitro-	214 [3], 225–226 [4]
1,6-Dinitro-	343 [5]
1,9-Dinitro-	273 [5]
2,8-Dinitro-	230–232 [6]
2-Methoxy-1-nitro-	233 [7]
2-Hydroxy-1-nitro	223 [7]
1-Amino-4-nitro-	281–282 [8]
1-Amino-2-nitro-	260 [9]
2-Amino-1-nitro-	264–265 [8]
6-Amino-2-nitro-	264–266 [4]
1-Cyano-2-nitro-	260 [10]

1 R.W.G. Preston, S.H. Tucker and J.M.L. Cameron, J. chem. Soc., 1942, 500.
2 S. Pietra and G. Casiraghi, Gazz., 1970, **100**, 138.
3 F. Kehrman and E. Havas, Ber., 1913, **46**, 341.
4 A. Albini et al., J. chem. Soc. Perkin I, 1978, 299.
5 S. Maffei and M. Aymon, Gazz., 1954, **84**, 667.
6 H. Endo, M. Tada and K. Katagiri, Sci. Rep. res. Inst., Tohoku Univ., 1967, **14C**, 171.
7 B. Hegedus, C.A., 1947, **41**, 6262.
8 C. Stammer and A. Taurins, Canad. J. Chem., 1963, **41**, 228.
9 S. Pietra and G. Casiraghi, Gazz., 1967, **97**, 1826.
10 idem, ibid., 1967, **97**, 1817.

in detail, e.g. treatment with hydroxylamine hydrochloride at room temperature gives 1-amino-2-nitrophenazine. The 10-oxide with potassium cyanide yields 1-cyano-2-nitrophenazine (S. Pietra and G. Casiraghi, Gazz., 1966, **96**, 1630; 1967, **97**, 1817; 1826; S. Pietra, C.A., 1970, **72**, 43505).

Irradiation of 1- and 2-nitrophenazine in the presence of primary amines gives mixtures of the corresponding alkylamino substituted products (A. Albini, et al., loc. cit.). Although no adducts are formed with triethylamine under similar photochemical conditions, 2-nitrophenazine is, however, photoreduced to the corresponding phenazinyl radical which apparently exists in solution as a dimer (A. Albini, G.F. Bettinetti and G. Minoli, J. chem. Soc., Perkin I, 1980, 191).

Sulphonation of phenazine also requires vigorous reaction conditions, oleum and mercury(II) chloride at 160–170° for seven hours, and affords *phenazine-2-sulphonic acid*, m.p. > 380° (dec.), together with some polysulphonic acids. The sodium salt gives (a) phenazine on heating alone; (b)

2-cyanophenazine on treatment with potassium cyanide at 150°; and (c) 1,2-dihydroxyphenazine on heating with sodium hydroxide at 280° (*S. Maffei*, Gazz., 1950, **80**, 651).

Several aminophenazine sulphonic acids and sulphonamides are structurally related to microbial pigments, *e.g. aeruginosin B* (*R. Herbert* and *F.G. Holliman*, Tetrahedron, 1965, **21**, 663; *R.K. Bentley* and *F.G. Holliman*, J. chem. Soc. C, 1970, 2530).

(vi) Aminophenazines

Since amino groups exert a strong bathochromic effect on the phenazine chromaphore, many dyes and pigments, trivially called *eurhodines*, have been derived from mono- and di-aminophenazines and their quaternary salts. The chemistry relating to such dyes, which are now of little commercial importance, is described elsewhere (see C.C.C., 1st Edn., Vol. IVC; *Swan* and *Felton, loc. cit.*).

Simple mono- and di-aminophenazines have been prepared by several of the general procedures mentioned above. Some of the more useful methods are discussed below.

(a) Condensation of amino-1,2-benzoquinones, derived from the *in situ* oxidation of the appropriate amino-pyrocatechols by silver oxide or lead peroxide, with *o*-phenylenediamine yields aminophenazines (*F. Kehrmann* and *C. Mermod*, Helv., 1927, **10**, 62; *J.H. Boyer* and *L.R. Morgan*, J. org. Chem., 1961, **26**, 1654).

(b) The Vivian procedure has been used to prepare 1-aminophenazine from 6-amino-2-nitrodiphenylamine. This reaction is, however, not widely applicable to the synthesis of aminophenazines because 2′-amino groups are eliminated during the cyclisation process (*D.L. Vivian, ibid.*, 1956, **21**, 565).

(c) Polynitrodiphenylamines are converted into aminophenazines with simultaneous loss of one nitrogen centre by either (i) treatment with tin(II) chloride (*F. Ullmann* and *G. Engi*, Ann., 1909, **366**, 82; *A. Albert* and *H. Duewell*, J. Soc. chem. Ind., 1947, **66**, 11), or (ii) catalytic reduction to the corresponding amino compound (not isolated) followed by oxidative cyclisation with iron(II) chloride (*R.C. Elderfield, W.J. Gensler* and *O. Birstein*, J. org. Chem., 1946, **11**, 812; *B. Hegedus*, Helv., 1950, **33**, 766).

(d) Heating 2,2'-diaminodiphenylamine in a large excess of nitrobenzene results in cyclisation without deamination to give 1-aminophenazine in good yield. Under similar conditions, 2-aminophenazine and a series of 8-substituted derivatives have been obtained from the appropriate 2,4- and 2,4'-diaminodiphenylamines [Scheme 5] (*C. Stammer* and *A. Taurins*, Canad. J. Chem., 1963, **41**, 228; *G. Gaertner, A. Gray* and *F.G. Holliman*, Tetrahedron, 1962, **18**, 1105). Similarly, 2,4,4'-triaminodiphenylamine yields the 2,8-diamino compound (*J. Fernando, W.S. Morgan* and *J.W. Hausser*, J. org. Chem., 1967, **32**, 1120). This reaction can been catalysed by palladium on charcoal which is thought to act as a hydrogen transfer agent (*F.G. Holliman, B.A. Jeffrey* and *D.J.H. Brock*, Tetrahedron, 1963, **19**, 1841).

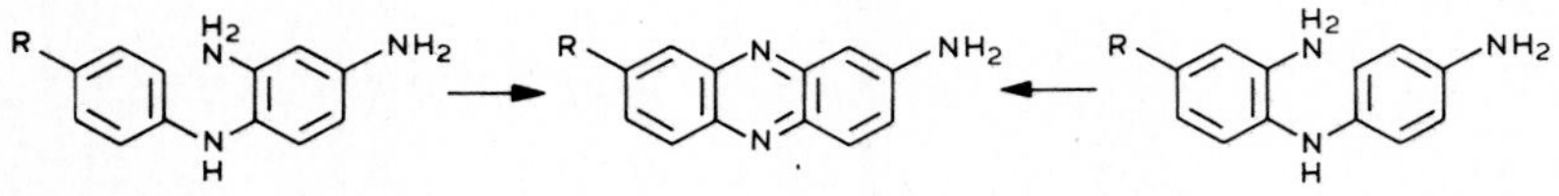

Scheme 5

(e) Oxidation of *o*-phenylenediamine using either iron(III) chloride (*F. Ullmann* and *F. Mauthner*, Ber., 1902, **35**, 4302), or iodine in ethanol (*E. Knoevenagel*, J. pr. Chem., 1914, **89**, 1; *M.M. Richter*, Ber., 1911, **44**, 3466), or potassium persulphate in acetic acid (*R.C. Gupta* and *S.P. Srivastava*, Indian J. Chem., 1971, **9**, 1303) affords 2,3-diaminophenazine in good yield together with 3-amino-2-hydroxyphenazine as a minor product. With 4-halo-*o*-phenylenediamines, 7-halo-2,3-diaminophenazines are obtained (*F. Ullmann* and *F. Mauthner*, Ber., 1903, **36**, 4026). In a similar fashion, autoxidation of 1,2,4-triaminobenzene in isopropanol yields 2,3,6-triaminophenazine (*A.A. Kiryushkin et al.*, Zhur. org. Khim., 1984, **20**, 1052; C.A., 1984, **101**, 110874).

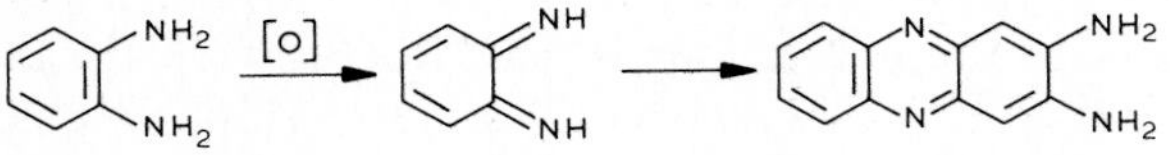

The condensation of *o*-phenylenediamine with 3,3-dimethylpentane-2,4-dione produces unexpectedly the readily isolable Schiffs base (XXXIX) derived from 2,3-diaminophenazine (*S.E. Drewes* and *P.C. Coleman*, Tetrahedron Letters, 1975, 91; Chem. Ind., 1976, 995).

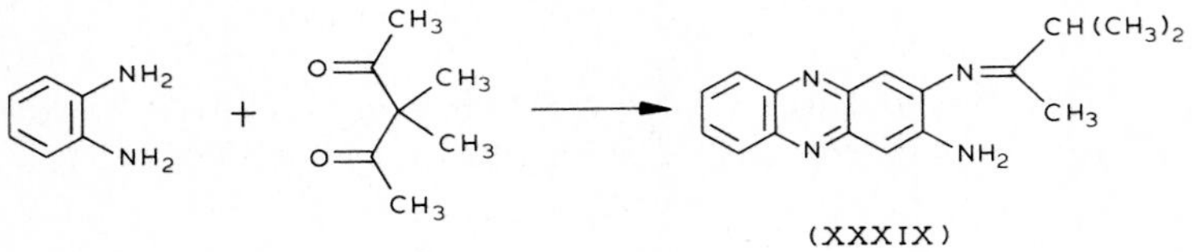

(XXXIX)

TABLE 21
AMINOPHENAZINES

Compound	m.p. (°C)
1-Amino-	179–181 [1], 183–184 [2]
	181–182 [3]
2-Amino-	279 [4], 267–268 [5]
1,2-Diamino-	180 [6], 189–191 [3]
1,3-Diamino-	284–285 [4,7]
1,4-Diamino-	> 300 [3]
1,6-Diamino-	245 [8], 250 [9]
1,7-Diamino-	221 [10]
1,9-Diamino-	264–265 [8]
2,3-Diamino-	264 [11]
2,7-Diamino-	> 320 [12]
2,8-Diamino-	280 [13], 284 [14]

1 B. *Hegedus*, Helv., 1950, **33**, 766.
2 J.H. *Boyer* and L.R. *Morgan*, J. org. Chem., 1961, **26**, 1654.
3 C. *Stammer* and A. *Taurins*, Canad. J. Chem., 1963, **41**, 228.
4 A. *Albert*, R. *Goldacre* and J. *Phillips*, J. chem. Soc., 1948, 2240.
5 M. *Tada*, Bull. chem. Soc. Japan, 1975, **48**, 3405.
6 S. *Pietra* and G. *Casiraghi*, Gazz., 1967, **97**, 1826.
7 F. *Kehrmann* and P. *Prunier*, Helv., 1924, **7**, 472; 984.
8 S. *Maffei* and M. *Aymon*, Gazz., 1954, **84**, 667.
9 E. *Breitmaier* and U. *Hollstein*, J. org. Chem., 1976, **41**, 2104.
10 A. *Albini et al.*, J. chem. Soc. Perkin I, 1978, 299.
11 F. *Ullmann* and F. *Mauthner*, Ber., 1902, **35**, 4302.
12 G.P. 148,113, 1904.
13 R. *Neitzki* and O. *Ernst*, Ber., 1890, **23**, 1854.
14 J. *Fernando*, W.S. *Morgan* and J.W. *Hausser*, J. org. Chem., 1967, 32; 1120.

(f) Ammonolysis of 2-chlorophenazine at 190–200° in a pressure vessel with copper(I) chloride affords 2-aminophenazine in essentially quantitative yield (*Gray* and *Holliman, loc. cit.; cf. I.J. Pachter* and *M.C. Kloetzel*, J. Amer. chem. Soc., 1951, **73**, 4958). Diaminophenazines related to Neutral Red (XL), a biological stain, have been prepared in a similar manner (*D.L. Vivian*, J. org. Chem., 1956, **21**, 565).

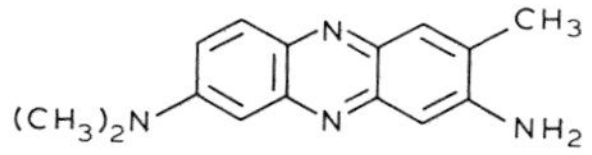

(XL)

(g) Treatment of 1- and 2-chlorophenazine with sodium azide in DMSO or DMF yields the corresponding aminophenazines in 77% and 38% yield,

respectively; the intermediate azido derivatives must decompose *in situ* under the reaction conditions (*M. Tada*, Bull. chem. Soc. Japan, 1975, **48**, 3405). 2-Aminophenazine is isolated from the thermal or photolytic decomposition of the 2-azido derivative in hydrocarbon solvents; in solutions containing alcohols or amines, 2-alkoxy or 2-alkylamino-1-aminophenazines are formed (*G.F. Bettinetti et al.*, Gazz., 1979, **109**, 175; 1980, **110**, 135; 1982, **112**, 13).

(h) Reduction of nitrophenazines by either catalytic hydrogenation, or with zinc and acetic acid, provides a route to aminophenazine derivatives (*Gray* and *Holliman, loc. cit.; H. Otomasu*, Chem. pharm. Bull. Tokyo, 1954, **2**, 283; *Stammer* and *Taurins, loc. cit.*). Reaction of 2-nitrophenazine with hydroxylamine hydrochloride at 65° gives 1,2-diaminophenazine (*S. Pietra* and *G. Casiraghi*, Gazz., 1967, **97**, 1826).

References to specific aminophenazines are given in Table 21.

Diazotisation of 1-aminophenazine by nitrous acid in the presence of β-napththol does not give an azo compound because the intermediate diazonium salt reacts faster with water to form instead 1-diazo-2(1*H*)-phenazinone (XLI) which has also been obtained from the diazotisation of 1-amino-2-ethoxyphenazine (*E.S. Olson*, J. heterocyclic Chem., 1977, **14**, 1255).

(XLI)

(vii) Hydroxyphenazines (phenazinols)

Interest in the chemistry of hydroxyphenazines has been closely related to the antibiotics iodinin and myxin (referred to earlier p. 355), and the bacterial pigment pyocyanine (III). The syntheses of several derivatives of iodinin and myxin have been reported in the patent literature (C.A., 1969, **71**, 61423; 1971, **74**, 22884; 1972, **76**, 59565; 1976, **84**, 164846). Structurally simpler compounds including 1- and 2-hydroxy-, 6-methoxy-1-hydroxy-, 2,3-dihydroxy-, and 2,3,7-trimethoxy-phenazine, which have been isolated from several bacterial sources, exhibit some biological activity (*N.N. Gerber*, J. org. Chem., 1967, **32**, 4055; *A. Roemer* and co-workers, Tetrahedron Letters, 1979, 509; *S. Yamanaka*, C.A., 1972, **77**, 162986; see review *J. Berger*, Kirk-Othmer Encycl. Chem. Tech., Vol. 3, Wiley, New York, 1978, p. 1).

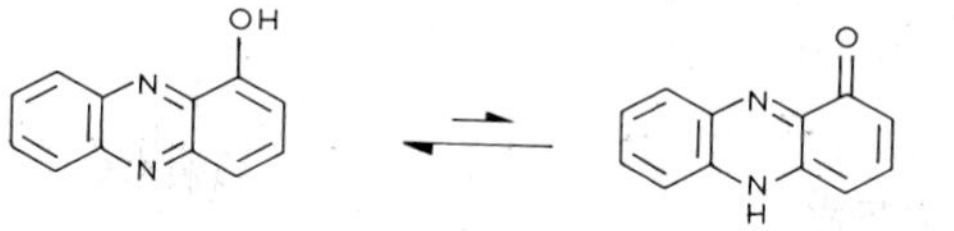

The hydroxyphenazines are potentially tautomeric substances; 1-hydroxyphenazine is tautomeric with phenazin-1(5H)-one and the 2-hydroxy compound is tautomeric with phenazin-2(10H)-one. I.r.- and u.v.-spectroscopic studies indicate that the parent compounds exist predominantly in the hydroxy form both in solution (methanol and carbon tetrachloride) and in the solid state. Substituted 2-hydroxyphenazines such as (XLII) and (XLIII) are, however, known to exist in the oxo-form. In the case of the trimethyl derivative, both forms have been isolated (*J.F. Corbett*, Spectrochim. Acta, 1964, **20**, 1665).

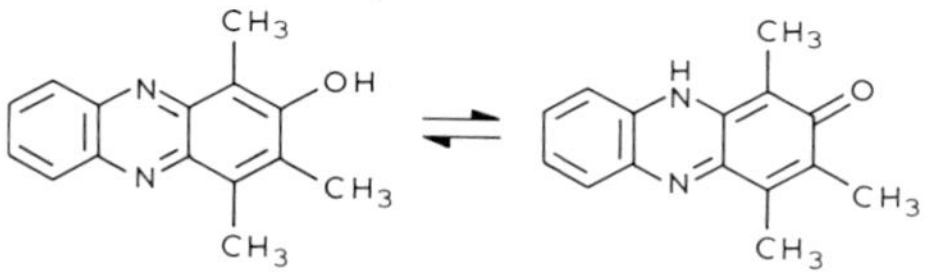

(XLII) 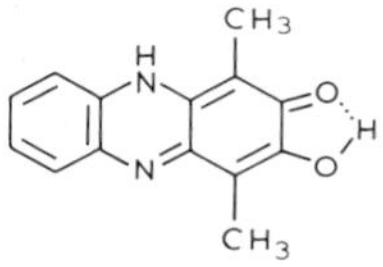 (XLIII)

The simpler mono- and di-hydroxyphenazines are usually prepared by adaptation of the general synthetic procedures outlined previously. Moreover, alkoxyphenazines, which are conveniently obtained by either direct synthesis, or nucleophilic displacement reactions, are readily dealkylated by hydrobromic acid in acetic acid (*M.L. Tomlinson*, J. chem. Soc., 1939, 158) or aluminium trihalides (*F.E. King, N.G. Clark* and *P.M.H. Davis*, J. chem. Soc., 1949, 3012; *M. Weigle* and *W. Leimgruber*, Tetrahedron Letters, 1967, 715).

Hydroxyphenazines may undergo both *N*- and *O*-methylation; with dimethyl sulphate *N*-methylation results (*A.R. Surrey*, Org. Synth. Coll. Vol. III, 1955, 785; *H. Hilleman*, Ber., 1938, **71**, 34) whereas with diazomethane, methoxy derivatives are obtained (*S. Maffei*, Gazz., 1950, **80**, 651).

1-**Hydroxyphenazine** (phenazin-1-ol) is prepared by a variety of synthetic routes including (i) demethylation of 1-methoxyphenazine which is obtained from either the condensation of 3-methoxy-*o*-benzoquinone and *o*-phenylenediamine (*L. Michaelis et al.*, Z. physiol. Chem.,

1932, **255**, 70; *A.R. Surrey, loc. cit.*), or the cyclisation of 6-methoxy-2-nitrodiphenylamine (*D.L. Vivian* and co-workers, J. org. Chem., 1953, **18**, 1065; 1954, **19**, 1136); (ii) from pyrogallol and *o*-phenylenediamine (*V. Melo*, C.A., 1950, **44**, 9450); (iii) by hydrolysis of 1-aminophenazine in dilute phosphoric acid at 150° for 36 hours (*B. Hegedus*, Helv., 1950, **33**, 766); (iv) from the photochemical oxidation of phenazine methosulphate to pyocyanine which is subsequently demethylated in alkaline solution (*D.L. Vivian*, Nature, 1956, **178**, 753); and (v) by hydrolysis of 1-(*p*-tosyloxy)phenazine, obtained from the reaction of phenazine-5-oxide and tosyl chloride (Jap. Pat., 25,682, 1963; C.A., 1964, **60**, 5523).

A series of chloro derivatives of 1-methoxyphenazine have been obtained (i) by the Wohl–Aue procedure to give 6-*chloro*-, m.p. 222–223°, 8-*chloro*-, m.p. 206–207°, 7-*chloro*-, m.p. 164–165°, and 9-*chloro*-, m.p. 184–185°, 1-*methoxyphenazine* (*I. Yoshioka* and *R. Ashikawa*, C.A., 1960, **54**, 549); (ii) from 2-nitrophenazine which on reaction with sodium hypochlorite in methanolic potassium hydroxide gives 2-chloro-1-methoxyphenazine; and (iii) by chlorination of 1-methoxyphenazine using potassium hypochlorite in hydrochloric acid which yields a mixture of the 4-*chloro*-, m.p. 151–152°, and 2,4-*dichoro*-, m.p. 190°, compound (B.P. 1,206,866; C.A., 1971, **74**, 3659).

Both 1-hydroxy- and 1-methoxy-phenazine undergo *N*-methylation on treatment with dimethyl sulphate; prolonged reaction results in nuclear methylation (*H. Hilleman, loc. cit.*).

Pyocyanine, 5-methylphenazin-1(5*H*)-one (III), *hemihydrate*, m.p. 133°, the unstable blue pigment produced by *Ps. aeruginosa* (*H. McCombie* and *H.A. Scarborough*, J. chem. Soc., 1923, 3279), is prepared from 1-hydroxy-5-methylphenazinium methylsulphate by either treatment with alkali (*Surrey, loc. cit.*), or irradiation of an aqueous solution (*H. McIlwain*, J. chem. Soc., 1937, 1708). Pyocyanine is demethylated to give 1-hydroxyphenazine on treatment with hydroiodic acid (*F. Werde* and *E. Strack*, Z. physiol. Chem., 1924, **140**, 1; Ber., 1929, **62**, 2051), or in alkaline solution on exposure to air, or heating at 80° in alkaline hydrogen peroxide (*H. McIlwain*, J. chem. Soc, 1937, 1704). Reduction of pyocyanine with zinc and acetic acid gives dihydropyocyanine, which reacts with oxalyl chloride to form the adduct (XLIV), thereby confirming that the methyl group is located at the 5-position (*Hilleman, loc. cit.*).

(III)   (XLIV)

2-Hydroxyphenazine (phenazin-2-ol) is obtained (i) by reductive cyclisation of 2-nitro-2′,5′-dihydroxydiphenylamine in alkaline sodium hydrosulphide. Similarly prepared in high yield are 6-*chloro*-, m.p. 217° (dec.), 6-*methyl*-, m.p. 245° (dec.), and 6-*methoxy*-, m.p. 200° (dec.), 2-*hydroxyphenazine* (G.P. 1,197,462; C.A., 1965, **63**, 18120); (ii) from the corresponding 2-alkoxyphenazines, 2-ethoxy- is found to be more readily dealkylated than 2-methoxy-phenazine (*H. McCombie et al.*, J. chem. Soc., 1928, 353; *D.L. Vivian et al.*, J. org. Chem., 1949, **14**, 289; 1953, **18**, 1065; *I.J. Pachter* and *M.C. Koetzel*, J. Amer. chem. Soc., 1952, **74**, 971; *Y. Fellion*, Ann. Chim. (Paris), 1957, **2**, 426); (iii) by heating 2-chlorophenazine-5,10-dioxide in aqueous alcoholic sodium hydroxide followed by reduction of the resulting 2-hydroxy di-*N*-oxide by alkaline sodium dithionite (*D.L. Vivian*, J. Amer. chem. Soc., 1951, **73**, 457). The latter di-*N*-oxide is also readily prepared *via* the Beirut reaction (see p. 370).

The reaction of 2-hydroxyphenazine with dimethyl sulphate yields 10-*methylphenazin-2(10H)-one* (parapyocyanine), m.p. 197–198°, (*Fellion, loc. cit.*), an isomer of (III), which is also obtained by oxidation of methylphenazinium methylsulphate (*McIlwain, loc. cit.*). Formaledehyde condenses with 2-hydroxyphenazine to give the 1:2 adduct (XLV) which, as a consequence of intramolecular hydrogen bonding, exhibits thermochromic properties (*A.G. Cairns-Smith*, J. chem. Soc., 1961, 182).

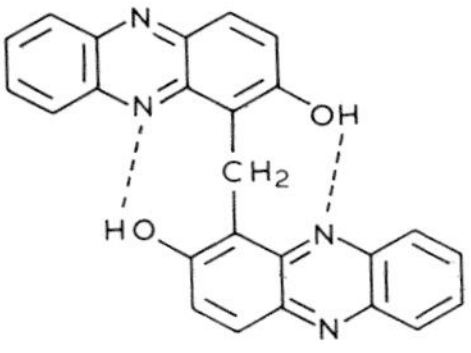

(XLV)

All of the ten possible dihydroxyphenazines have been synthesised by extensions of the general methods either as the free hydroxy compounds or *via* the corresponding dimethyl ethers. Thus, 1,2-dimethoxyphenazine is prepared by dehydrogenation of the condensation product from 3,4-dimethoxy-*o*-phenylenediamine and 1,2-cyclohexanedione (*F.E. King, N.G. Clark* and *P.M.H. Davis*, J. chem. Soc., 1949, 3012) or from (XLVI) by heating with iron at 250° (*P.Z. Slack* and *R. Slack*, Nature, 1947, **160**, 437) [Scheme 6]. Cyclisation of 2,2′-dinitro-3,6′-dimethoxy-diphenylamine affords 1,6-dimethoxyphenazine (*G.R. Clemo* and *A.F.*

Scheme 6

(XLVI)

Daglish, J. chem. Soc., 1950, 1481) from which can be derived the antibiotics iodinin and myxin (*Weigele* and *Leimgruber, loc. cit.*). Condensation of 2,5-dihydroxy-*p*-benzoquinone with *o*-phenylenediamine affords directly the 2,3-dihydroxy compound (*R. Nietzki* and *G. Hasterlik*, Ber., 1891, **24**, 1337) which may also be obtained from the corresponding dimethyl ether (*Slack* and *Slack, loc. cit.*). 2-Aminoindophenols (XLVII) undergo intramolecular cyclisation at pH > 8 to give 2,8-dihydroxyphenazines (*K.C. Brown* and *J.F. Corbett*, J. chem. Soc. Perkin II, 1981, 886).

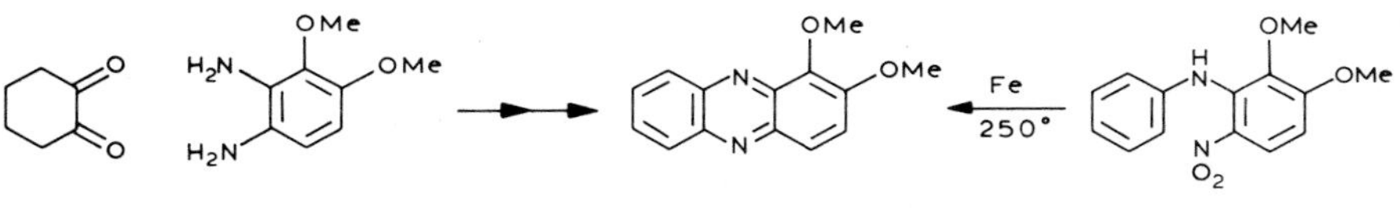

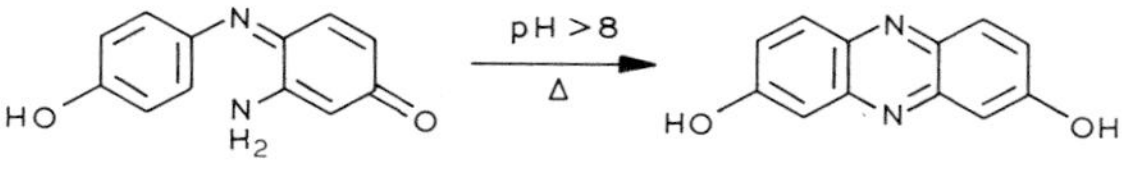

Additional hydroxyphenazines together with their respective derivatives are listed in Table 22.

Condensation of tetrahydroxy-*p*-benzoquinone with an *o*-phenylenediamine yields the corresponding 1,2,3,4-tetrahydroxyphenazine (*S. Skujins* and *G.A. Webb*, Tetrahedron, 1969, **25**, 3935).

Nitrosation of phenols, *e.g. m*-methoxyphenol, in the presence of copper salts yields the coordination complex (XLVIII) which on treatment with triphenylphosphine gives 1,6-dihy-

TABLE 22
HYDROXY- AND METHOXY-PHENAZINES

Compound	m.p. (°C)	Derivative	m.p. (°C)
1-Hydroxy-	153–155 [1], 155–157 [2,3], 157–158 [4,5]	Acetyl	122–123 [6]
1-Methoxy-	167–169 [1], 169 [5]		
2-Hydroxy-	253–254 [7,8]	Acetyl	160–161 [6], 152 [7]
2-Methoxy-	112 [9], 123 [10], 123–124 [6]		
1,2-Dihydroxy-	261 [11], 270–275 [9]	Diacetyl	168 [6]
1,2-Dimethoxy-	138–139 [11,12], 145–146 [9]		
1,3-Dihydroxy-	275 [13,14], 280 [15]	Diacetyl	163 [13]
1,3-Dimethoxy-	228 [15], 145 [16]		
1,4-Dihydroxy-	230 [17], 232–234 [15]	Diacetyl	195 [15], 193–194 [17,18a]
1,4-Dimethoxy-	185 [12,17]		
1,6-Dihydroxy-	274 [19,20], 274–275 [21]	Diacetyl	233 [21], 228–230 [22], 233–235 [23]
1,6-Dimethoxy-	246–247 [23], 247 [19], 251 [20], 249–253 [3]		
1,7-Dihydroxy-	305–306 [22]	Diacetyl	148–149 [22]
1,7-Dimethoxy-	174–175 [24]		
1,8-Dihydroxy-	247–248 [25], 230 [26], 216 [27]	Diacetyl	181 [25], 178–181 [26], 166 [27]
1,8-Dimethoxy-	154–155 [25], 150–152 [26]		
1,9-Dihydroxy-	295 [19,28], 296–297 [22]	Diacetyl	257–258 [22]
1,9-Dimethoxy-	253–254 [19], 259–260 [24]		
2,3-Dihydroxy-	> 340 [15]	Diacetyl	238–240 [29]
2,3-Dimethoxy-	226 [15], 230–231 [30]		
2,7-Dihydroxy-	> 300 [25]	Diacetyl	266–267 [18b,25]
2,7-Dimethoxy-	246 [25,31], 250–251 [32]		
2,8-Dihydroxy-	> 260 [33]	Diacetyl	224 [34]
2,8-Dimethoxy-	163 [10]		

1 A.R. Surrey, Org. Synth. Coll. Vol. III, 1955, 785.
2 B. Hegedus, Helv., 1950, **33**, 766.
3 N.N. Gerber, J. org. Chem., 1967, **32**, 4055.
4 A. Albert and H. Duewell, J. chem. Soc. Ind., 1947, **66**, 11.
5 F. Werde and F. Strack, Z. physiol. Chem., 1928, **177**, 177.
6 I. Yoshioka and Y. Kidani, J. pharm. Soc. Japan, 1952, 72, 1301.
7 F. Kehrmann and F. Cherpillod, Helv., 1924, **7**, 973.
8 D.L. Vivian, J. Amer. chem. Soc., 1949, **71**, 1139.
9 B. Hegedus, C.A., 1947, **41**, 6262.
10 M.L. Tomlinson, J. chem. Soc., 1939, 158.
11 S. Maffei, Gazz., 1950, **80**, 651.
12 P.Z. Slack and R. Slack, Nature, 1947, **160**, 437.
13 G.R. Clemo and A.F. Daglish, J. chem. Soc., 1948, 2318.
14 M.J. Haddadin, H.E. Bittmar and C.H. Issidorides, Heterocycles, 1979, **12**, 323.

TABLE 22 (continued)

15 I. *Yoshioka* and H. *Otomasu*, Chem. pharm. Bull. Tokyo, 1954, **2**, 53.
16 Yu.S. *Rozum*, C.A., 1955, **49**, 1065.
17 F.E. *King, N.G. Clark* and P.M.H. *Davis*, J. chem. Soc., 1949, 3012.
18 (a) H. *Endo, M. Tada* and K. *Katagiri*, Sci. Rep. res. Inst., Tohoku Univ., 1967, **14C**, 164;
    (b) *idem, ibid.*, 1969, **16C**, 18.
19 C.R. *Clemo* and A.F. *Daglish*, J. chem. Soc., 1950, 1481.
20 I. *Yoshioka* and Y. *Kidani*, J. pharm. Soc. Japan, 1952, **72**, 847.
21 Jap. Pat. 25,682, 1963; C.A., 1964, **60**, 5523.
22 S.B. *Serebryanyi*, Zhur. obshchei Khim., 1952, **22**, 702.
23 A. *Albini et al.*, J. chem. Soc. Perkin I, 1978, 299.
24 S.B. *Serebryanyi* and V.P. *Chernetskii*, Zhur. obshchei Khim., 1951, **21**, 2033; A.I.
    Kiprianov, S.B. Serebryanyi and V.P. *Chernetskii, ibid.*, 1949, **69**, 651.
25 I. *Yoshioka* and H. *Otomasu*, Chem. pharm. Bull. Tokyo, 1953, **1**, 66.
26 N.N. *Gerber*, J. heterocyclic Chem., 1969, **6**, 297.
27 Yu.S. *Rozum*, Zhur. obshchei Khim., 1955, **25**, 583.
28 I. *Yoshioka*, Chem. pharm. Bull. Tokyo, 1954, **2**, 25.
29 W.M. *Horspool, P.I. Smith* and J.M. *Tedder*, J. chem. Soc. C, 1971, 138.
30 D.L. *Vivian*, J. org. Chem., 1956, **21**, 824.
31 A. *Sugimoto et al.*, Bull. Univ. Osaka Prefect., 1971, **20A**, 181.
32 P. *Walker* and W.A. *Waters*, J. chem. Soc., 1962, 1632.
33 K.C. *Brown* and J.F. *Corbett*, J. chem. Soc. Perkin II, 1981, 886.
34 F. *Kehrmann* and E. *Haenny*, Helv., 1925, **8**, 676.

droxy-3,8-dimethoxyphenazine (XLIX) in 50% yield [Scheme 7]. Deoxygenation of 5-methoxy-2-nitrosophenol with triphenylphosphine in pyridine or chloroform gives the 5,10-dihydro derivative of (XLIX) (*J. Charalambous, M.J. Kensett* and *J.M. Jenkins*, Chem. Comm., 1977, 400).

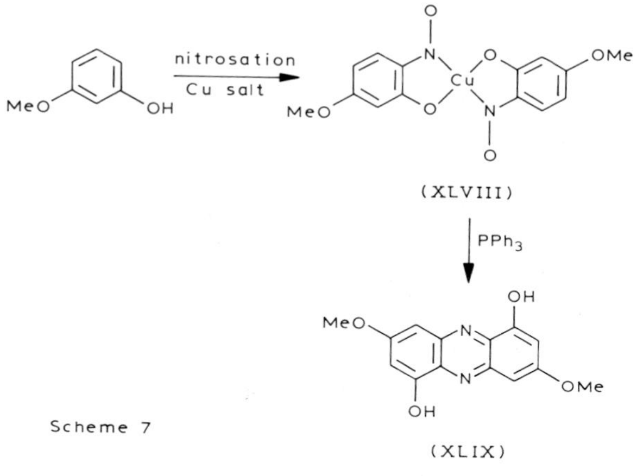

Scheme 7

(viii) Phenazinecarboxylic acids

Phenazine-1-carboxylic acid together with several alkyl, hydroxy and alkoxy derivatives (*I.K. Isono, K. Anzai* and *S. Suzuki*, J. Antibiot., 1958, **11A**, 264; *J. Tax et al.*, Coll. Czech. Chem. Comm., 1983, **48**, 527; *E.A. Kiprianova* and *A.S. Rabinovich*, Mikrobiologiya, 1969, **38**, 224; Jap. Pat. 76 32,790; C.A., 1976, **85**, 107461) and 2,3,7-trihydroxyphenazine-1,6-dicarboxylic acid (*A. Roemer et al.*, Tetrahedron Letters, 1979, 509), which have been isolated from a range of microbial sources, exhibit antibiotic activity. The 5-methyl quaternary salt derived from phenazine-1-carboxylic acid has been shown to be a key intermediate in the biosynthesis of the bacterial pigment, pyocyanin (III) (p. 355) (*F.G. Holliman, M.E. Flood* and *R.B. Herbert*, Chem. Comm., 1970, 1514; J. chem. Soc. Perkin I, 1972, 622).

Structural studies on the air-sensitive green pigment *chlororaphine*, m.p. 228–229°, isolated from *Ps. chlororaphis*, suggest that it is a 3:1 molecular complex between phenazine-1-carboxamide and the corresponding 5,10-dihydro derivative; the former component is obtained on air-oxidation of chlororaphine (*C. Dufraise, A. Etiene* and *E. Toromanoff*, Compt. rend., 1952, **235**, 920).

Phenazinecarboxylic acids are prepared by many of the general methods described previously, *e.g.* condensation of 3-carboxy-1,2-benzoquinone with *o*-phenylenediamine gives the 1-carboxylic acid (*L.R. Morgan*, J. org. Chem, 1962, **27**, 2634). The diethyl ester of phenazine-1,4-dicarboxylic acid has been obtained from the reaction of *o*-phenylenediamine and diethyl 1,2-cyclohexanedione-3,6-dicarboxylate followed by dehydrogenation of the resulting adduct (U.S.P. 3,379,728; C.A., 1968, **69**, 19216).

Heating the *o*-aminodiphenylamine (L) in nitrobenzene at reflux temperature affords 2-methoxyphenazine-1-carboxylic acid (*R.B. Herbert, F.G. Holliman* and *J.D. Kynnersley*, Tetrahedron Letters, 1968, 1907).

Reductive cyclisation of a series of appropriately substituted *o*-nitrodiphenylamines using sodium borohydride under basic conditions gives rise to the corresponding methoxyphenazine-1-carboxylic acids and esters, *e.g.* (LI) affords 6-methoxyphenazine-1-carboxylic acid in 34% yield (*K.P. Brooke et al.*, J. chem. Soc. Perkin I, 1976, 2248).

(LI) → [structure with CO₂Me, NaBH₄/NaOEt/HOEt, giving phenazine with CO₂Me and OMe groups]

Reaction of anthranilic acid with nitrobenzene and with *o*-nitrobenzoic acid by the Wohl–Aue procedure affords the 1-carboxylic acid (*F. Kogl* and *J.J. Postowsky*, Ann., 1930, **480**, 280; *J. Ou* and *Y. Lu*, C.A., 1982, **96**, 217798) and the 1,6-dicarboxylic acid (*L. Birkofer* and *W. Widmann*, Ber., 1953, **86**, 1295) respectively in low yield (*ca.* 5%). Oxidative dimerisation of anthranilic acid by manganese dioxide also produces the 1,6-dicarboxylic acid, which on heating at 270° in the presence of copper gives a mixture of phenazine and the mono-acid (*L.R. Morgan* and *C.C. Aubert*, Proc. chem. Soc., 1962, 73; J. org. Chem., 1962, **27**, 4092). The 2,7-dicarboxylic acid is synthesised from methyl *p*-nitrosobenzoate *via* the Bamberger–Ham procedure followed by reduction of the resulting *N*-oxide with zinc in hydrochloric acid (*M. Schubert*, Ann., 1947, **558**, 10; *H. Hopff* and *Ch. Ziegler*, Chimia, 1958, **12**, 112).

Hydrolysis of 1- and 2-*cyanophenazine*, m.p. 170–172° and 236° respectively, prepared by the Vivian method, gives the corresponding carboxylic acid (*D.L. Vivian, J.L. Hartwell* and *H.C. Waterman*, J. org. Chem., 1954, **19**, 1641).

Side-chain oxidation of a methyl-substituted phenazine produces the corresponding carboxylic acid in moderate yields (*S. Maffei*, Gazz., 1950, **80**, 651; *Y. Kidani*, Chem. pharm. Bull. Tokyo, 1959, **7**, 83). With selenium dioxide as the oxidising agent, the 1- and 2-*phenazinecarboxaldehydes*, m.p. 176–177° and 187° respectively, have been obtained from the respective methylphenazines (*W.A. Waters* and *D.H. Watson*, J. chem. Soc., 1959, 2085). Hydrolysis of 1-(dibromomethyl)phenazine also gives the 1-carboxaldehyde (*I. Yoshioka*, and *K. Ueda*, Chem. pharm. Bull. Tokyo, 1964, **13**, 1247).

Specific mono- and di-phenazinecarboxylic acids together with their simple derivatives are listed in Table 23.

(e) Reduced phenazines

The central heterocyclic ring in phenazine is highly susceptible to reduction, resulting in the formation of the corresponding 5,10-dihydrophenazine which may also be obtained directly from the reaction of *o*-phenylenedi-

TABLE 23
PHENAZINE MONO- AND DI-CARBOXYLIC ACIDS

Compound	m.p. (°C)	Derivative	m.p. (°C)
1-Carboxylic acid	237–239 [1,2]	Methyl ester	124–125 [1]
	242–243 [3]	Amide	245–246 [1]
2-Carboxylic acid	296–297 [1]	Methyl ester	154–155 [1]
	292 [4]	Amide	316–317 [1]
1,4-Dicarboxylic acid	> 310 [5]	Diethyl ester	103–104 [5]
1,6-Dicarboxylic acid	300–320 (char) [6]	Diethyl ester	143 [6]
		Dimethyl ester	229–230 [7]
			228–230 [8]
1,8-Dicarboxylic acid	–	Dimethyl ester	159.5–160 [9]
1,9-Dicarboxylic acid	–	Dimethyl ester	156–157 [8]
2,7-Dicarboxylic acid	–	Dichloride	214 [10]

1 E. Toromanoff, Ann. Chim. (Paris), 1956, **1**, 115.
2 G.R. Clemo and H. McIlwain, J. Chem. Soc., 1934, 1991.
3 L.R. Morgan, J. org. Chem., 1962, **27**, 2634.
4 Y. Kidani, Chem. pharm. Bull. Tokyo, 1959, **7**, 88.
5 H.R. Schweizer, Helv., 1969, **52**, 322.
6 L.R. Morgan and C.C. Aubert, J. org. Chem., 1962, **27**, 4092.
7 N.N. Gerber, J. heterocyclic Chem., 1969, **6**, 297.
8 E. Breitmaier and U. Hollstein, J. org. Chem., 1976, **41**, 2104.
9 M.E. Flood, R.B. Herbert and F.G. Holliman, J. Chem. Soc. Perkin I, 1972, 622.
10 M. Schubert, Ann., 1947, **558**, 10.

amine with catechol (Ris reaction). 1,2,3,4-Tetrahydrophenazines, the condensation products from *o*-phenylenediamine and cyclohexane-1,2-diones, may be further reduced by catalytic hydrogenation or in dissolving metal reactions to give octa- and tetradeca-hydrophenazine.

(i) 5,10-Dihydrophenazines

The air-sensitive parent compound is obtained (a) from the direct condensation of *o*-phenylenediamine and catechol (*J.S. Morley*, J. chem. Soc., 1952, 4008); (b) from the pyrolysis of aniline at 400° in the presence of hydrogen sulphide (*M.G. Voronkov, E.N. Deryagina* and *E.N. Sukhomazova*, Khim. Geterot. Soedin., 1977, 270); and (c) by the reduction of phenazine using a variety of reagents including hydrogen sulphide in alcoholic ammonia solution (*A. Claus*, Ann., 1873, **168**, 1; *C. Ris*, Ber., 1886, **19**, 2206), sodium dithionite in alcoholic sodium hydroxide solution (*R. Scholl*, Monatsh., 1918, **39**, 231; *E. Toromanoff*, Ann. Chim. (Paris), 1956, **1**, 115), lithium aluminium hydride (*L. Birkofer* and *A. Birkofer*, Ber., 1952, **85**, 286; *F. Bohlmann, ibid.*, 1952, **85**, 390), and catalytic hydrogenation (*Y. Fellion*, Ann. Chim. (Paris), 1957, **2**, 426).

The simple mono-*N*-alkyl derivative, particularly methyl and ethyl, are prepared by reduction of the corresponding 5-alkylphenazinium salts with zinc in hydrochloric or sulphuric acid (*Morley, loc. cit.; A. Hantzch*, Ber., 1916, **49**, 511; *H. McIlwain*, J. chem. Soc., 1937, 1704).

The reaction of phenazine in an ether solvent with an alkali metal followed by the addition of an alkyl halide affords a 5,10-dialkyl-5,10-dihydrophenazine, *e.g.* with sodium metal and methyl iodide, the 5,10-dimethyl derivative (LII) is obtained (*B.M. Mikhailov* and *A.N. Blokhina*, C.A., 1950, **44**, 9452). (LII) has also been synthesised from (a) 5-methyl-phenazinium methylsulphate and methylmagnesium chloride (*H. Gilman* and *J.J. Dietrich*, J. Amer. chem. Soc., 1957, **79**, 6178) and (b) the condensation of *N,N'*-dimethyl-*o*-phenylenediamine and cyclohexane-1,2-dione followed by oxidation of the resulting adduct (*G.R. Clemo* and *H. McIlwain*, J. chem. Soc., 1935, 738).

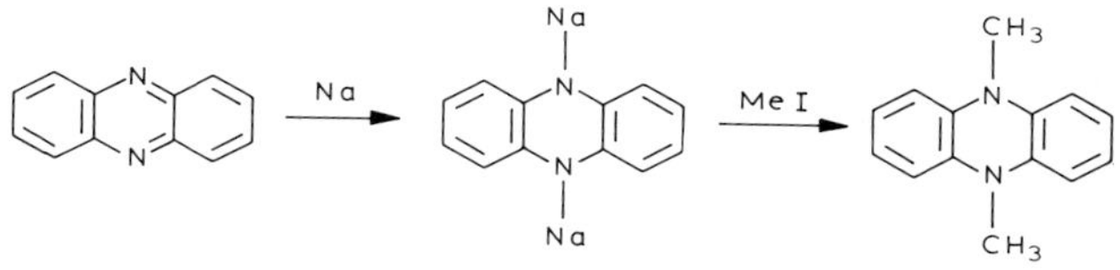

A series of symmetrically substituted 5,10-dihydrophenazines has been conveniently prepared by the sequence outlined in Scheme 8. Thus, 5,10-dihydrophenazine, obtained from the reduction of phenazine by hydrazine in the presence of palladium on charcoal, is converted *in situ* into the 5,10-di(methoxymethyl) compound (LIII) which is subsequently treated with the appropriate Grignard reagent (*G.F. Bettinetti, S. Maffei* and *S. Pietra*, Synthesis, 1976, 748).

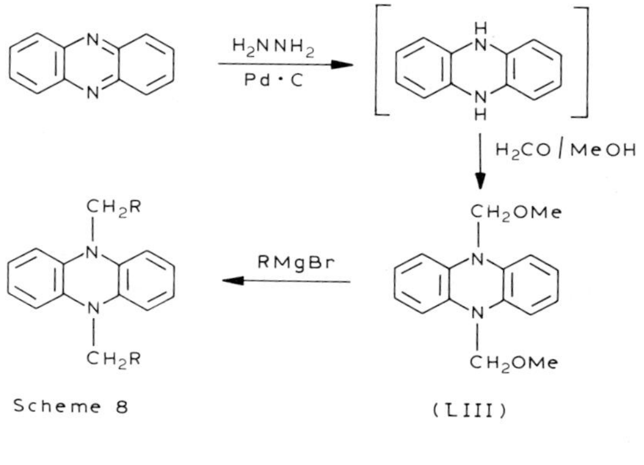

Scheme 8

Treatment of phenazine in dimethoxyethane with phenyl lithium followed by methyl iodide yields 5-methyl-10-phenyl-5,10-dihydrophenazine (*Mikhailov* and *Blokhina, loc. cit.*). A similar reaction in tetrahydrofuran, with the addition of iodobenzene and copper powder instead of methyl iodide, gives the 5,10-diphenyl derivative (*Gilman* and *Dietrich, loc. cit.*).

The 5,10-dimethyl compound (LII) has a small dipole moment of 0.4D which has been interpreted as evidence for the molecule existing as a mixture of isomeric forms, each existing in a folded conformation (*I.M.G. Campbell et al.*, J. chem. Soc., 1938, 404). In the 1:1 molecular complex between (LII) and tetracyanoquinodimethane, a crystallographic study reveals that the dihydrophenazine molecule has a folded geometry, though the deviation from planarity is not as great as earlier postulated, with each methyl group in a quasi-equatorial position (*I. Goldberg* and *U. Shmueli*, Acta Cryst., 1973, **29B**, 421).

Acylation of 5,10-dihydrophenazine at room temperature, with an excess of reagent, *e.g.* acetyl or benzoyl chloride, in the absence of air usually affords the mono-acyl derivative; heating the appropriate reaction mixture at reflux produces the 5,10-diacetyl compound (*Morley, loc. cit.*). The photochemical addition of acetaldehyde to phenazine also produces the 5-acetyl compound (*M. Takagi, S. Goto* and *T. Matsuda*, Bull. chem. Soc. Japan, 1980, **53**, 1777). Reductive silylation of phenazine gives the 5,10-di(trimethylsilyl) adduct (LIV) which on treatment with either acetyl chloride or acetic anhydride, is transformed into 5,10-diacetyl-5,10-dihydrophenazine (*L. Birkofer* and *N. Ramadan*, Ber., 1975, **108**, 3105). Electro-

chemical reduction of phenazine in the presence of methyl chloroformate affords a mixture of the 5-mono- and the 5,10-di-methoxycarbonyl-5,10-dihydro compound; with lithium perchlorate as co-electrolyte, the mono adduct is obtained selectively (*J. Armand et al.*, J. org. Chem., 1983, **48**, 2847).

5,10-Dihydrophenazines substituted in the carbocyclic rings appear to be intrinsically more stable than the parent compound. Such derivatives may be prepared by (a) reduction of the corresponding phenazine (*Toromanoff, loc. cit.*) or dehydrogenation of a 1,2,3,4-tetrahydrophenazine (*H.R.*

TABLE 24
5,10-DIHYDROPHENAZINES

Compound	m.p. (°C)
Parent	280 [1], 317 [2]
5-Acetyl-	253–254 [1], 258–260 [3]
5-Benzoyl-	224–225 [1]
5,10-Diacetyl-	179–180 [1,4]
5-Methoxycarbonyl-	171–172 [5]
5-Ethoxycarbonyl-	109–110 [1]
5,10-Dimethoxycarbonyl-	161–162 [5]
5-Methyl-	163–164 [1]
5-Ethyl-	99 [6]
5,10-Dimethyl-	153 [7], 153–155 [8]
5,10-Diethyl-	127–128 [9]
5,10-Dibenzyl-	214–215 [9]
5-Phenyl-	143 [10]
5,10-Diphenyl-	283–285 [8]
5-Methyl-10-phenyl-	116–117 [8,11]
1-Methyl-	176–178 [12]
2-Methyl-	269–270 [12]
1-Carboxy-	275–276 [12]
1-Methoxycarbonyl-	132–133 [12]
1-Carboxamide	204–205 [12]
2-Carboxy-	261–262 [12]
1,4-Diethoxycarbonyl-	156–157 [13]
1-Nitro-	158 [14]
2-Nitro-	154 [14]
1-Hydroxy-	174–175 [15]
1-Methoxy-	140–141 [15]
2-Hydroxy-	265–266 [15]
2-Methoxy-	230–231 [15]

1 J.S. Morley, J. chem. Soc., 1952, 4008.
2 L. Birkofer and A. Birkofer, Ber., 1952, **85**, 286.
3 M. Takagi, S. Goto and T. Masuda, Bull. chem. Soc. Japan, 1980, **53**, 1777.
4 L. Birkofer and N. Ramadan, Ber., 1975, **108**, 3105.
5 J. Armand et al., J. org. Chem., 1983, **48**, 2847.
6 H. McIlwain, J. chem. Soc., 1937, 1704.
7 G.R. Clemo and H. McIlwain, ibid., 1935, 738.
8 H. Gilman and J.J. Dietrich, J. Amer. chem. Soc., 1957, **79**, 6178.
9 G.F. Bettinetti, S. Maffei and S. Pietra, Synthesis, 1976, 748.
10 F. Kehrmann, Ber., 1896, **29**, 2316; Ann., 1902, **322**, 67.
11 B.M. Mikhailov and A.N. Blokhina, Izvest. Akad. Nauk SSSR, 1950, 304.
12 E. Toromanoff, Ann. Chim. (Paris), 1956, **1**, 115.
13 H.R. Schweizer, Helv., 1969, **52**, 322.
14 S.P. Gupta and D.M.L. Garg, J. Indian chem. Soc., 1964, **41**, 52.
15 Y. Fellion, Ann. Chim. (Paris), 1957, **2**, 426.

Schweizer, Helv., 1969, **52**, 322) and (b) reaction of *o*-halonitrobenzenes with *o*-phenylenediamines (*S.P. Gupta* and *D.M.L. Garg*, J. Indian chem. Soc., 1964, **41**, 52).

5,10-Dihydrophenazine and phenazine form molecular complexes known as phenazhydrins. Heating approximately equimolar quantities of phenazine and 5,10-dihydrophenazine in ethanol yields the blue, 1 : 1 complex, **phenazhydrin**, m.p. 255–256°; with a large excess of phenazine, a violet, 3 : 1 complex, m.p. 216–217°, is obtained (*C. Dufraise et al.*, Compt. rend., 1951, **232**, 2399; *Toromanoff, loc. cit.*). The melting points reported for phenazhydrin vary considerably suggesting that it may have several polymorphic forms (see *G.R. Clemo* and *H. McIlwain*, J. chem. Soc., 1934, 1991; *Morley, loc. cit.; W. Schlenk* and *E. Bergmann*, Ann., 1928, **463**, 306). Although these molecular complexes have been known for some time, their solid state structures have not been reported. In phenazhydrin, the two phenazine components are thought to be held together by bridging hydrogen atoms as indicated in structural formula (LV). This would be consistent with the observation that **methylphenazhydrin**, purple prisms, m.p. 157–160°, is obtained from either phenazine and 1-methyl-5,10-dihydrophenazine, or 1-methylphenazine and 5,10-dihydrophenazine. 1-Methyl-

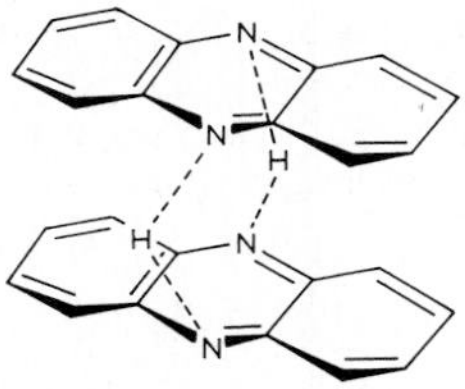

(LV)

phenazine and its dihydro derivative similarly afford *dimethylphenazhydrin*, m.p. 116° (*Clemo* and *McIlwain, loc. cit.*). Phenazhydrins on air-oxidation or heating at 100° revert to phenazines.

Although phenazine does not form phenazhydrins with either 5-methyl- or 5,10-dimethyl-5,10-dihdyrophenazine, 1 : 1 complexes of unknown structure have been isolated as the hydrochloride salt in each case (*G.R. Clemo* and *H. McIlwain*, J. chem. Soc., 1935, 738). Molecular charge-transfer complexes between either 5-methyl- or 5,10-dimethyl-dihydrophenazine and tetracyano-*p*-quinodimethane have attracted considerable interest as one-dimensional organic electrical conductors (*H.J. Keller et al.*, Canad. J. Chem., 1979, **57**, 1033; *D.J. Sandman*, J. Amer. chem. Soc., 1978, **100**, 5230).

(ii) 1,2,3,4-Tetrahydrophenazines

The condensation of *o*-phenylenediamine and cyclohexane-1,2-dione yields, 1,2,3,4-**tetrahydrophenazine**, m.p. 92.5° (*monomethiodide*, m.p. 207°) (*G.R. Clemo* and *H. McIlwain*, J. chem. Soc., 1934, 1991; *J.S. Morley, ibid.*, 1952, 4008). Similarly prepared are 1-*methyl*-, m.p. 37°, b.p. 160–165°/20 mm, from 3-methylcyclohexane-1,2-dione (*Clemo* and *McIlwain, loc. cit.*), 2-*methyl*-, b.p. 144–148°/0.1. mm, from 4-methylcyclohexane-1,2-dione (*idem*, J. chem. Soc., 1935, 738), and 1,4-*di(ethoxycarbonyl)*-, m.p. 99°, 1,2,3,4-*tetrahydrophenazine*, from diethyl cyclohexane-1,2-dione-3,6-dicarboxylate (U.S.P. 3,385,855; C.A., 1968, **69**, 27456). Appropriately substituted *o*-phenylenediamines with cyclohexane-1,2-dione give rise to the following derivatives, 7-*methyl*-, m.p. 78° (*Clemo* and *McIlwain, loc. cit.*), 6-*methoxy*-, m.p. 118–119° (*A.I. Kiprianov* and *E.A. Ponomareva*, Ukrain. khim. Zhur., 1960, **26**, 633), 6,7-*dimethoxy*-, 82–82°, 6,8-*dimethoxy*-, m.p. 119°, 6,9-*dimethoxy*-, m.p. 152°, and 7,8-*dimethoxy*-, m.p. 119–120°, 1,2,3,4-*tetrahydrophenazine*, (*F.E. King et al.*, J. chem. Soc., 1949, 3012).

Most 1,2,3,4-tetrahydrophenazines are readily dehydrogenated to the corresponding phenazine either by iodine in acetic acid or by heating with palladium on charcoal (*G.R. Clemo* and *H. McIlwain*, J. chem. Soc., 1936, 258); the 6,7- and 6,9-dimethoxy compounds are reported to require more forcing conditions (*King et al., loc. cit.; Clemo* and *Daglish*, J. chem. Soc., 1950, 1481).

The reaction of *o*-phenylenediamine with α-methylsulphinylcyclohexanone also gives 1,2,3,4-tetrahydrophenazine. Since the reported yields are low (*ca.* 16%), this procedure offers no advantage over those described above (*S. Kano, S. Shibuya* and *Y. Yuasa*, J. heterocyclic Chem., 1980, **17**, 1559). In a variation on the Beirut reaction (p. 359), benzofuroxan reacts with either the pyrrolidino enamine of cyclohexanone, or cyclohexanone in triethylamine at room temperture to produce 1,2,3,4-**tetrahydrophenazine**-9,10-**dioxide** (LVI), m.p. 185° (dec.) (*M.J. Haddadin* and *C.H. Issidorides*, Tetrahedron Letters, 1965, 3253; *L. Marchetti* and *G. Tosi*, Ann. Chim. (Rome), 1967, **57**, 1414). (LVI) undergoes reduction with sodium dithionite to 1,2,3,4-tetrahydrophenazine and with excess sodium borohydride to *cis*-1,2,3,4,4a,5,10,10a-octahydrophenazine [Scheme 9].

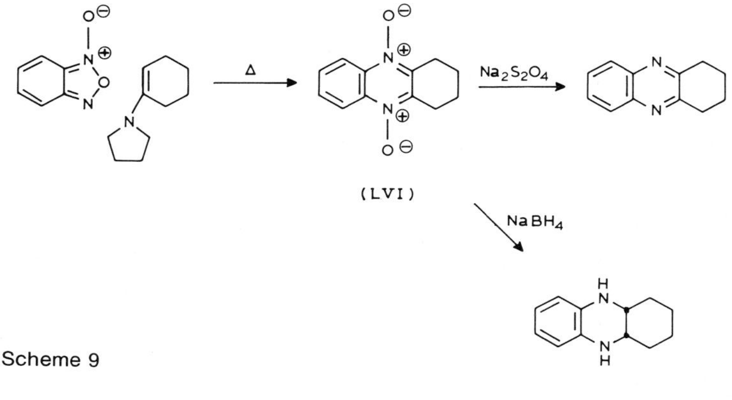

Scheme 9

The reaction of *o*-phenylenediaminie with 2,6-bisoximinocyclohexanone yields 1-**oximino**-1,2,3,4-**tetrahydrophenazine**, m.p. 213° (dec.), which can react further with *o*-phenylenediamine to give **quinoxalo[2,3-*a*]phenazine** (LVII), m.p. 304° (*G.H. Cookson*, J. chem. Soc., 1953, 1328).

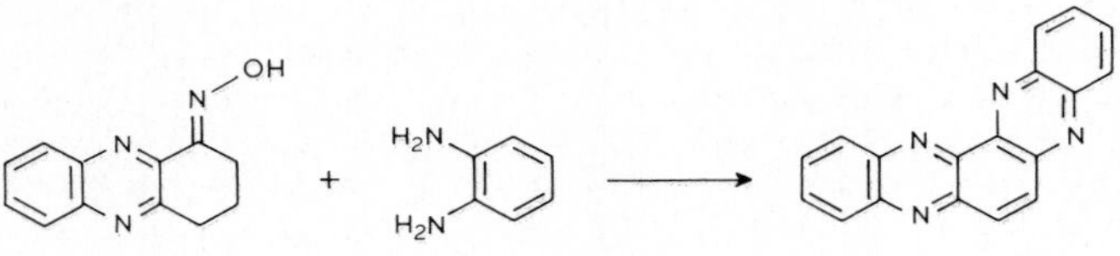

(LVII)

1,2,3,4-Tetrahydrophenazine, being the equivalent of a 2,3-dialkylquinoxaline, has reactive methylene groups at the 1- and 4-positions, and with benzaldehyde and its *p*-nitro and *p*-dimethylamino derivatives affords a series of bis-adducts, m.ps. 158°, 250° and 207°, respectively. On the basis of their reactions these compounds were initially regarded as dibenzylphenazines (*H. McIlwain*, J. chem. Soc., 1937, 1701; 1704). However on reinvestigation of these reactions, the mono- (LVIII) and the bis- (LIX) benzylidene compound, m.p. 125–126° and 155–156° respectively, have been obtained and each subsequently transformed into the corresponding epoxide on treatment with perbenzoic acid (*W.E. Hahn* and *J.Z. Lesiak*, Roczniki Chem., 1973, **47**, 1175; *W.E. Hahn, J.Z. Lesiak* and *B. Muszkiet*, *ibid.*, 1974, **48**, 177).

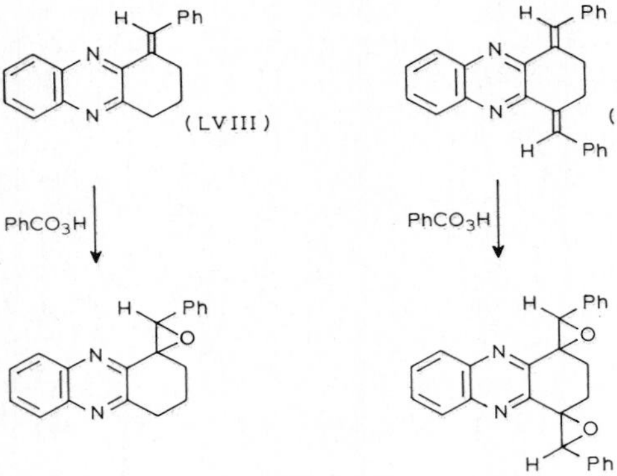

(LVIII)   (LIX)

(iii) Octahydrophenazines

1,2,3,4,4a,5,10,10a-**Octahydrophenazine** can exist in the *cis*- and the *trans*-form. The *trans*-isomer (LX), m.p. 156° (*monohydrochloride*, m.p. 315°; *dinitroso* deriv., m.p. 126°), is obtained from the 1,2,3,4-tetrahydro compound either by reduction with sodium in ethanol (*G.R. Clemo* and *H. McIlwain*, J. chem. Soc., 1936, 258) or by gas-phase hydrogenation over a nickel catalyst (*H. Rupe*, Helv., 1918, **1**, 453). Although (LX) can exist in optically active forms, these have not been obtained.

At around neutral pH, (LX) undergoes only monoacylation to give, for example, the 5-*acetyl*, m.p. 292–294°, and the 5-*benzenesulphonyl* compound, m.p. 150–151°, whereas at pH 3, the diacyl compounds are obtained, *e.g.* 5,10-*diacetyl*, m.p. 175–176°, and 5,10-*dibenzenesulphonyl*, m.p. 169–170° or 185–186°, derivatives (*Morley, loc. cit.*).

The *cis*-isomer (LXI), m.p. 147° (*dinitroso* deriv., m.p. 109°), is obtained (a) from tetrahydrophenazine either by liquid-phase catalytic hydrogenation in acetic acid solution or by reduction with sodium and ethanol in the presence of acetic acid (*Clemo* and *McIlwain*, J. chem. Soc., 1936, 258; 1698); (b) by reduction of 1,2,3,4-tetrahydrophenazine-5,10-dioxide with excess sodium borohydride (*M.J. Haddadin* and *C.S. Issidorides*, Tetrahedron Letters, 1965, 3253) and (c) as the major product of the reaction of *o*-phenylenediamine and cyclohexane-1,2-diol, irrespective of the configuration of the latter (*S. Miyano* and *M. Nakao*, Chem. pharm. Bull. Tokyo, 1972, **20**, 1328). Instead of the expected hexahydrophenazine, the reaction of *o*-phenylenediamine and 2-hydroxycyclohexanone affords an equimolar mixture of the *cis*-octahydro- and the tetrahydro-phenazine. (*P.J. Earle* and *M.L. Tomlinson*, J. chem. Soc., 1956, 794).

(LX)   (LXI)

The *cis*-isomer (LXI) forms the 5-*acetyl*, m.p. 155°, the 5,10-*diacetyl*, m.p. 148°, and the 5,10-*dibenzoyl*, m.p. 203°, derivative (*Earle* and *Tomlinson, loc. cit.*).

Both the (LX) and (LXI) give a purple colour, characteristic of *o*-phenylenediamines, with iron(III) chloride and are smoothly dehydrogenated to phenazine on heating with palladium at 200°. Moreover (LX) and (LXI) undergo isomerisation on passage over a catalyst derived from nickel nitrate (see *L. Gatterman*, Laboratory Methods of Organic Chemistry, 24th Edn., English translation, MacMillan, 1960, p. 379) to give the symmetrical 1,2,3,4,6,7,8,9-**octahydrophenazine** (LXII), m.p. 107–108° (*methiodide*, m.p. 175°) which is a bis-annelated pyrazine (*G.R. Clemo* and *H. McIlwain*, J. chem. Soc., 1936, 258). This octahydrophenazine has also been prepared directly by several methods including: (a) the reaction of cyclohexanone in dry ammonia (*M. Godchot*, Compt. rend., 1925, **180**, 444); (b) the reaction of

(LXII)

2-hydroxycyclohexanone and formamide (*A. Novelli*, An. Asoc. Quim. Argent., 1964, **52**, 95); (c) ammonolysis of 1-chlorocyclohexene oxide (*C. Herzig* and *J. Gasteiger*, Ber., 1981, **114**, 2348); and (d) the treatment of 1-nitroso-2-chlorocyclohexane with ammonia in methanol (F.P. 1,438,222; C.A., 1967, **65**, 20150).

Treatment of 1-morpholino-6-oximinocyclohexene with hydrogen chloride in chloroform affords the dichloro-*N*-oxide (LXIII) as a mixture of the *cis*- and the *trans*-isomer which are separable by chromatography, m.ps. 170° and 166° respectively (*R.H. Fischer* and *H.M. Weitz*, Synthesis, 1975, 791).

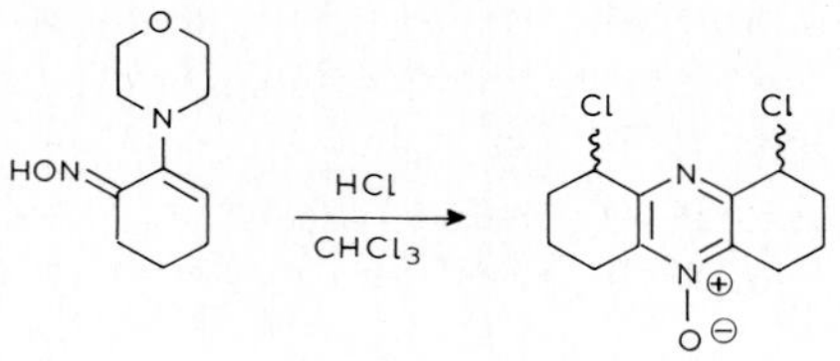

2-Aminocyclohexane-1,3-dione undergoes an oxidative self-dimerisation mediated by potassium nitrososulphonate (Fremy's salt) to yield 1,6-**dioxo**-1,2,3,4,6,7,8,9-**octahydrophenazine** (LXIV), m.p. 160° (dec.) (*H.J. Teuber, J. Hohn* and *A. Gholami*, Ber., 1983, **116**, 1309).

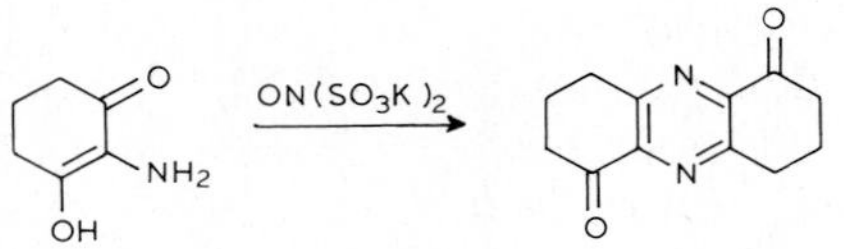

The α-methylene groups in (LXII) are reactive in a similar fashion to those in tetrahydrophenazine. Thus, (LXII) condenses with benzaldehyde to give the di- and the tetra-benzylidene compound, each giving the corresponding epoxide with perbenzoic acid [Scheme 10] (*W.E. Hahn* and *B. Muszkiet*, Roczniki Chem., 1973, **47**, 943; *W.E. Hahn, J.Z. Lesiak* and *B. Muszkiet*, ibid., 1974, **48**, 177).

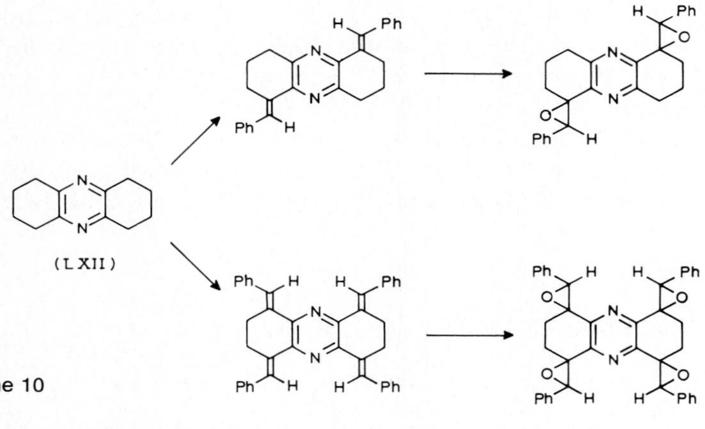

Scheme 10

(iv) Decahydrophenazine

The parent compound, although formed directly from the self-dimerisation of 2-aminocyclohexanone, is not readily isolable because it undergoes rapid *in situ* oxidation to give the symmetrical octahydrophenazine (LXII). Oxidative dimerisation of 2-methylcyclohexanone by potassium ferricyanide in ammonium hydroxide solution, however, affords the the dimethyl derivative (LXV) as a mixture of geometrical isomers. In (LXV), the angular methyl groups prevent further oxidation of the system.

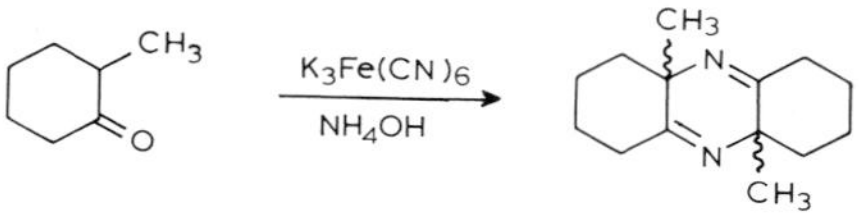

(LXV)

(v) Tetradecahydrophenazines

The fully saturated tricyclic ring system (LXVI) can in principle exist in five isomeric forms. Of these three (designated as α, β and γ) have been isolated, although their respective structures have not been unambiguously determined. Some derivatives of the three isomers are listed in Table 25.

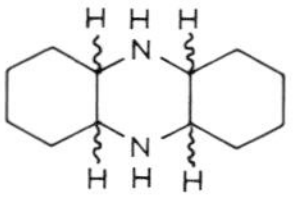

(LXVI)

The reduction of the symmetrical octahydrophenazine (LXII) under a variety of conditions yields the α-form. With hydrogen and platinum in acetic acid, the β- and γ-forms are also obtained (*M. Godchot* and *M. Mousseron*, Bull. Soc. chim. Fr., 932, **51**, 360; 528; *G.R. Clemo* and *H. McIlwain*, J. chem. Soc., 1936, 1698). The tetradecahydrophenazines may also be obtained by reduction of 1,2,3,4-tetrahydrophenazine and the *cis*- and the *trans*-octahydrophenazine (LXI) and (LX) (*Clemo* and *McIlwain*,

TABLE 25
TETRADECAHYDROPHENAZINES

Isomer	m.p. (°C)	Dinitroso deriv., m.p. (°C)	Dipicrate deriv., m.p. (°C)
α	135	168	278
β	95	183	252
γ	62	107	242

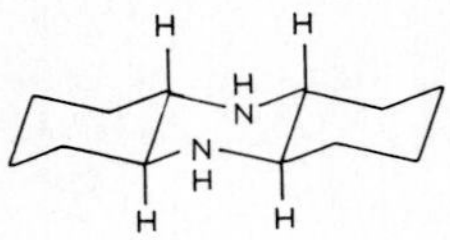

trans-trans

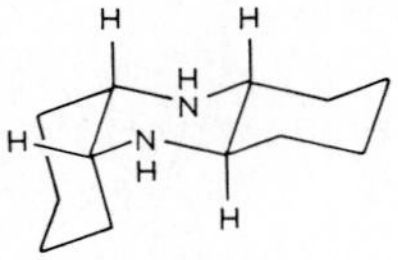

cis-trans

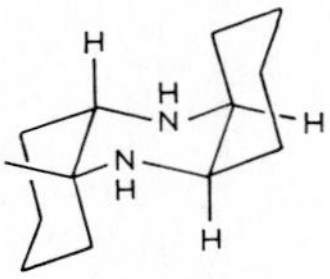

cis-cis

loc. cit.), and by reductive deamination of 1,2-diaminocyclohexane over a nickel–alumina catalyst (Canad. Pat. 989,405; C.A., 1976, **85**, 14314).

Since both the β- and γ-forms can be isomerised to the α-form with hydrogen in the presence of a nickel catalyst, it seems likely that the α-form is the most stable. In addition, if the carbo- and hetero-cyclic rings each adopt a chair conformation, the three forms obtained probably correspond (not necessarily respectively) to the *trans–trans-*, the *cis–trans-* and the *cis–cis-*isomer.

Chapter 45

Phenazine, Oxazine, Thiazine and Sulphur Dyes

N. HUGHES

1. Introduction

Dyes of the azine, oxazine and thiazine series, (I, X = NR, O and S, respectively), described in this chapter, are historically important since many of the early synthetic textile dyes were of these types. Their dyeing and fastness properties are, by present day standards, generally only moderate and they have, to a large extent, been superseded by more modern dyes. They do, however, retain some commercial importance in the fields of paper and leather dyeing and, in some cases, have been successfully adapted to the requirements of modern synthetic textile materials, for example, polyacrylonitrile fibres. Blue fibre-reactive derivatives of triphendioxazine have gained considerable commercial importance for the dyeing of cellulosic textiles as a result of their brightness of shade and high fastness against photochemical degradation (light fastness) which is combined with excellent economics. In addition to the numerous basic and acid dyes of these types there is a large class of colouring matters, the sulphur dyes, many of which are structurally phenothiazine derivatives, which are still of major importance for the dyeing of cellulosic fibres.

In azine, oxazine, and thiazine dyes the basic onium chromophore is normally substituted in position 3 or in positions 3 and 7 (numbering as shown) by auxochromic groups capable of donating electrons *via* mesomerism, for example amino, substituted amino and hydroxyl. The positive ions of dyes so substituted exist as resonance hybrids as depicted below, and their deep and intense colours are due to electronic transitions within the charged form.

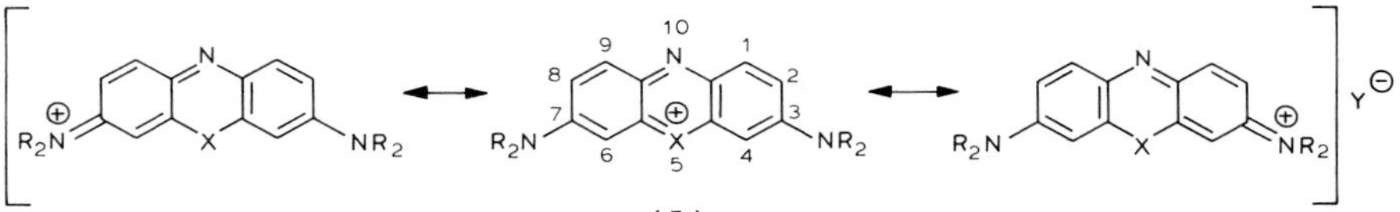

(I)

From the point of view of colour and constitution they have been compared with indamine dyes, for example Bindschedlers Green (II) with the bridging atom X regarded as a mesomeric electron donating group attached simultaneously to two unstarred positions (*J. Griffiths*, Colour and Constitution of Organic Molecules, Academic Press, New York, 1976).

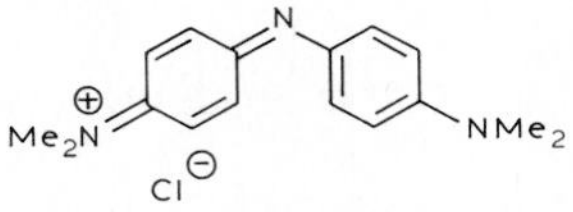

$$Me_2\overset{\oplus}{N}\text{=}\!\!\!\bigcirc\!\!\!\text{=}N\text{—}\bigcirc\text{—}NMe_2 \quad Cl^{\ominus}$$

(II)

General references for this chapter are: "The Colour Index", 3rd Edn., The Society of Dyers and Colourists, Bradford, and the American Association of Textile Chemists and Colorists, U.S.A., 1971; *K. Venkataraman*, "The Chemistry of Synthetic Dyes", Vol II, Academic Press, New York, 1952; *P.F. Gordon* and *P. Gregory*, "Organic Chemistry in Colour", Springer Verlag, Berlin, 1983.

2. Phenazine dyes*

The chromophore of the phenazine dyes is the phenazonium ion which is usually substituted at position 3 or at positions 3 and 7 by amino, substituted amino or hydroxyl groups. The dyes are classified (see Table 1) according to the mode of substitution and are best considered under these headings.

Following these and fluorindines, and in order of increasing complexity, come the commercially important indules and nigrosines and, finally, the aniline blacks.

Phenazonium compounds which do not carry auxochromic groups in either the 3 or the 7 position may also be effective dyestuffs, for example Flavinduline O (C.I. 50000, *C. Schraube*, B.P. 18374/1893) a brownish-yellow dyestuff which has been used in the colouration of leather and in calico printing. Such compounds are synthesised by condensing an *o*-diamine (I) with an *o*-quinone (II) (see *W. Schepss* and *O. Bayer*, U.S.P. 2,215,859/1941; C.A., 1941, **35**, 911).

---

* *G.A. Swan* and *D.G.I. Felton*, in "Chemistry of Heterocyclic Compounds, Vol XI, Phenazines, ed. *A. Weissberger*, Interscience, London, 1957.

TABLE 1
PHENAZINE DYES

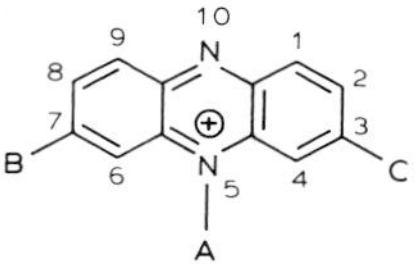

[Phenazine core structure with positions 1-10, substituent A at N5, B at position 7, C at position 3]

Class	A	B	C	Other substituents
(a) Eurhodines and Eurhodols	H	–N<	H	—
	H	–OH	–OH	—
	H	–N<	–N<	—
	H	–OH	–N<	—
(b) Aposafranines				
Phenylaposafranines	Ph	H	–N<	
Rosindulines	Ph	H	–N<	benzo[a]
Rosindones	Ph	H	–OH	benzo[a]
Isorosindulines	Ph	–N<	H	benzo[a]
(c) Safranines				
Phenosafranines	Ph	–N<	–N<	—
Benzophenosafranines	Ph	–N<	–N<	benzo[a]
Dibenzophenosafranines	Ph	–N<	–N<	dibenzo[a,j]

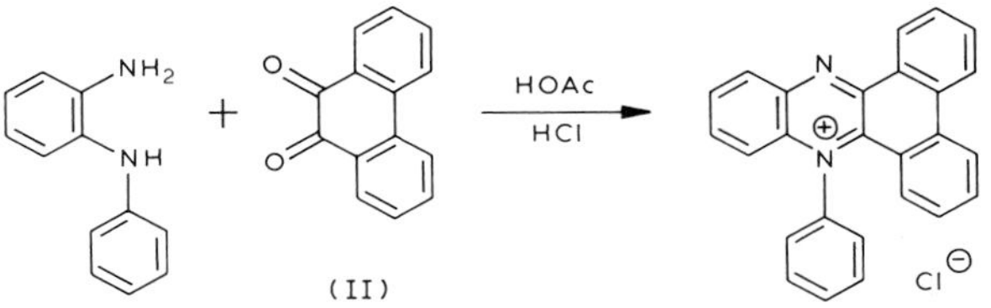

[Reaction scheme: (I) 2-aminodiphenylamine + (II) phenanthrene-9,10-quinone → (HOAc, HCl) → Flavinduline O (with Cl⁻ counterion)]

(a) Eurhodines and eurhodols

A general method of synthesis is illustrated by that of Neutral Red (C.I. Basic Red 5; 50040, O. Witt, B.P. 4846/80). Condensation of 4-nitrosodimethylaniline with m-toluylene diamine forms first the indamine (III) which undergoes intramolecular 1,4-quinonoid addition to form the phenazine (IV) which is finally oxidised to the phenazonium derivative.

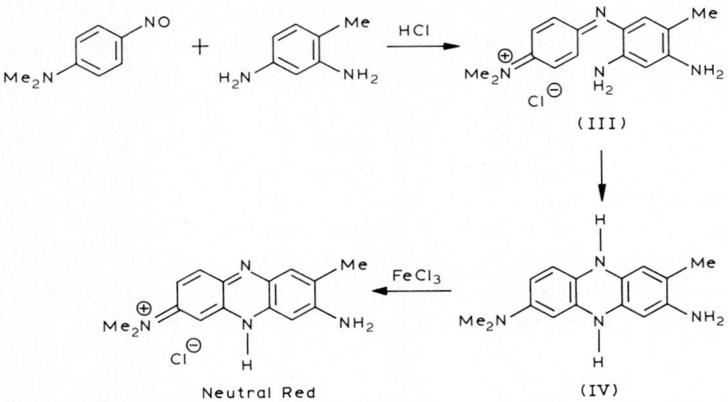

Neutral Red (III), (IV)

By modification of the Flavinduline O synthesis hydroxy- or aminophenanthraquinones may be used to prepare eurhodols and eurhodines (*A.C. Sircar* and *S. Dutt*, J. chem. Soc., 1922, **121**, 1944). Less important syntheses are the condensation of 1,4-benzoquinone dichloroimines with *para*-substituted aromatic amines and condensations of 2,4-diaminoazo compounds such as chrysoidine (V) with, for example, 2-naphthol (*F. Ullman* and *J.S. Ankersmit*, Ber., 1905, **38**, 1811).

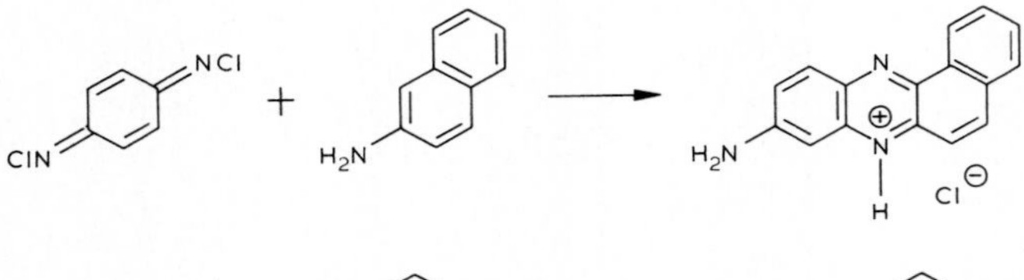

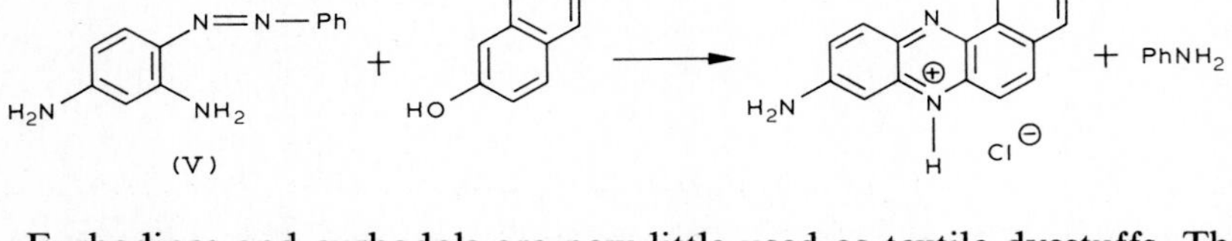

Eurhodines and eurhodols are now little used as textile dyestuffs. They are used as dyes for polypropylene fibres (*Ube Nitto*, Jap. P. 12,951/1966) and as non-toxic stains for animal tissues (*D.L. Vivian, R.M. Hogart* and *M. Balking*, J. org. Chem., 1961, **26**, 112–115).

(b) Aposafranines

Aposafranine (3-amino-5-phenylphenazonium chloride) is obtained as the major product when a mixture of 4-aminoazobenzene hydrochloride, aniline and aniline hydrochloride is heated in aqueous media, under pressure, at 160–170° (*P. Barbier* and *P. Sisley*, Bull. Soc. chim. Fr., 1907 [iv], **1**, 470). It is deaminated by boiling an alcoholic solution of the derived diazonium salt (*F. Kehrmann*, Ber., 1896, **29**, 2316) and the product of deamination, 5-phenylphenazonium chloride (VI), reacts with ammonia in the presence of oxygen to re-introduce the 3-amino group thereby regenerating aposafranine (*Kehrmann, loc.cit.; W. Schaposchnikow*, Ber., 1897, **30**, 2624). Treatment of the iron(III) chloride double salt of 5-phenylphenazonium chloride with aqueous alcoholic sodium hydroxide affords aposafranone (*Kehrmann* and *H. Bürgin*, Ber., 1896, **29**, 1819).

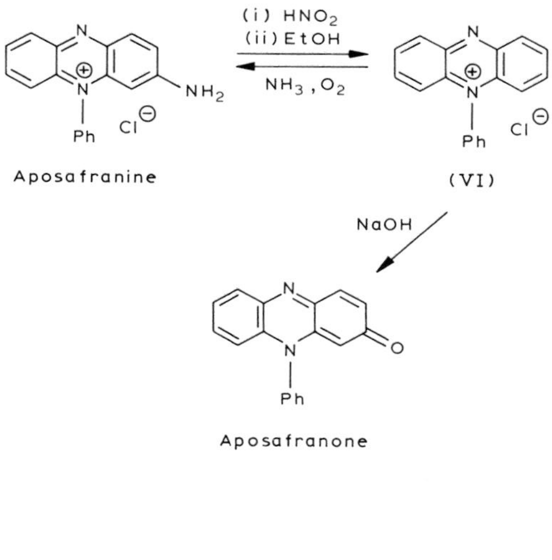

Two syntheses of aposafranines have been used commercially; the "azocarmine synthesis" (*O. Fischer* and *E. Hepp*, Ann., 1890, **256**, 233; *C. Schraube*, U.S.P. 430975/1890), leading to, for example, Azo Carmine G (C.I. Acid Red 101, 50085) and the "oxidative synthesis" (F.I.A.T. 1313, II, 382) leading, for example, to the isorosinduline dye intermediate, Blue I.

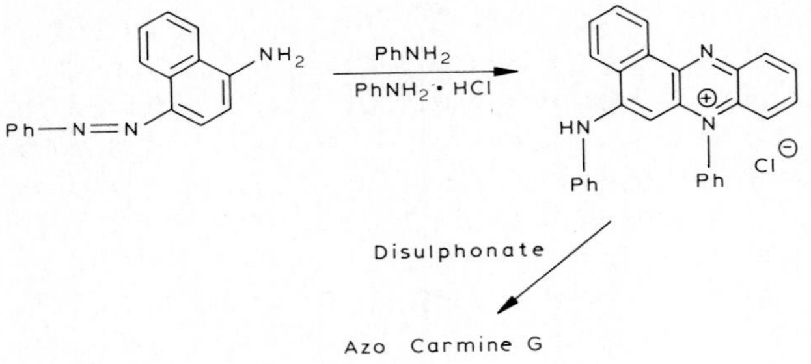

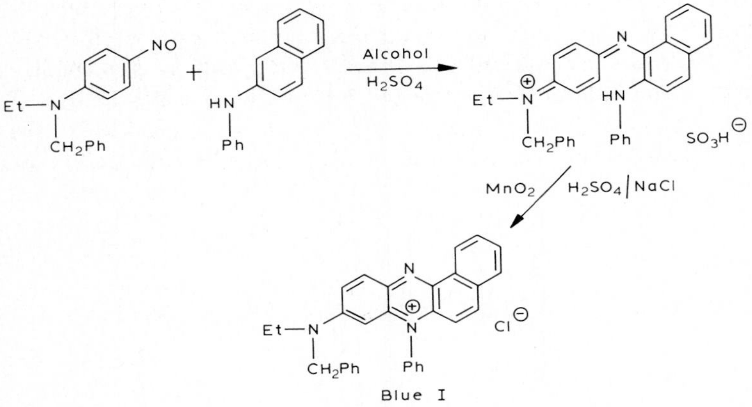

5-Alkyl-rosindulines, for example Induline Scarlet (C.I. Basic Dye 50080, C. *Schraube*, B.P. 10138/1892), may be obtained by interaction of *o*-aminoazo compounds and amines containing a free reactive *para*-position.

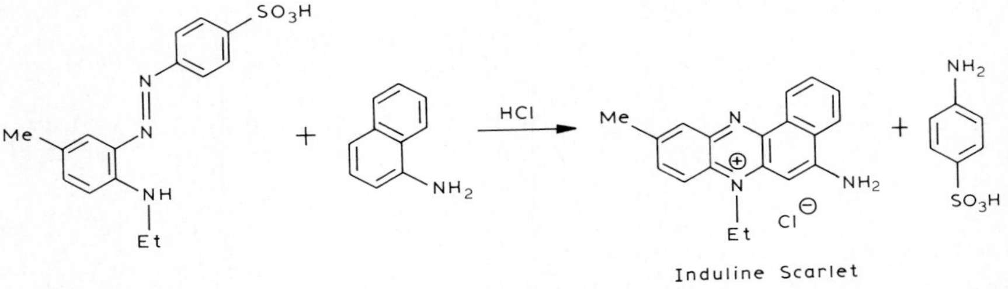

The compound without the 5-ethyl group (a eurhodine) can be obtained in a similar manner (*O. Witt*, Ber., 1885, **18**, 1119; 1886, **19**, 441) and

Fischer and *Hepp* (*ibid*., 1890, **23**, 846; 2788) used *o*-phenylene diamine and 4-phenylazo-1-naphthylamine in an analogous synthesis.

Aposafranines are no longer of commercial importance. They have been used for the dyeing of wool and silk and as biological stains. Induline Scarlet has been used as a catalyst in the preparation of the indigo vat and for printing vat colours by discharge methods.

(c) Safranines

The first synthetic dye to be manufactured, Mauveine (C.I. 50245), belongs to the safranine class. It was discovered by *W.H. Perkin* who, in 1856, oxidised a solution of impure aniline (containing toluidines) sulphate with chromic acid in an attempt to synthesise the anti-malarial drug quinine. He isolated instead a black, insoluble material from which he extracted a low yield of a bluish-red colouring material which, he found, could be used to dye silk a brilliant mauve shade. (*Perkin*, B.P. 1984/1856).

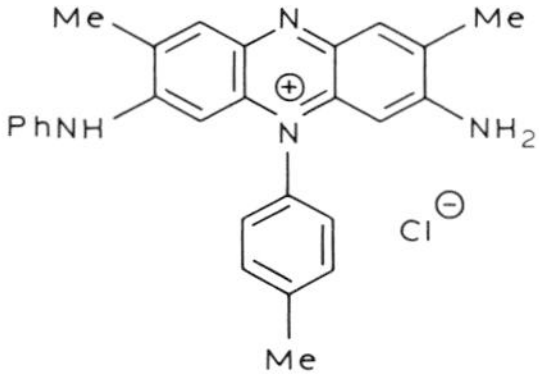

Mauveine

Perkin's discovery laid the foundation of a rapidly expanding dyestuffs industry and helped to establish organic chemistry as a major branch of science.

The simplest safranine, Safranine B or Phenosafranine (C.I. Basic Dye 50200) of commerce, can be obtained by oxidation of a mixture of *p*-phenylenediamine and aniline.

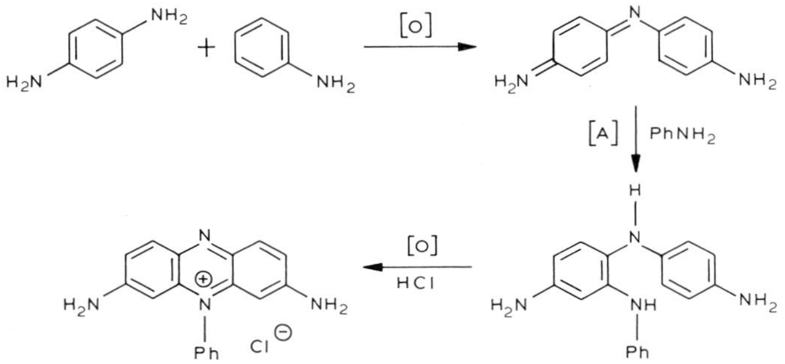

Phenosafranine

In general, one mole of a *para*-diamine with a free *ortho*-position and two moles of the same or different monoamine(s) with both free *meta*- and *para*-positions, (the amine reacting at [A] being primary), will give rise to a safranine on oxidation.

An ingenious method of obtaining the requisite amine mixture is shown in the synthesis of Safranine T (C.I. Basic Red 2, 50240; F.I.A.T. 1313, II, 376).

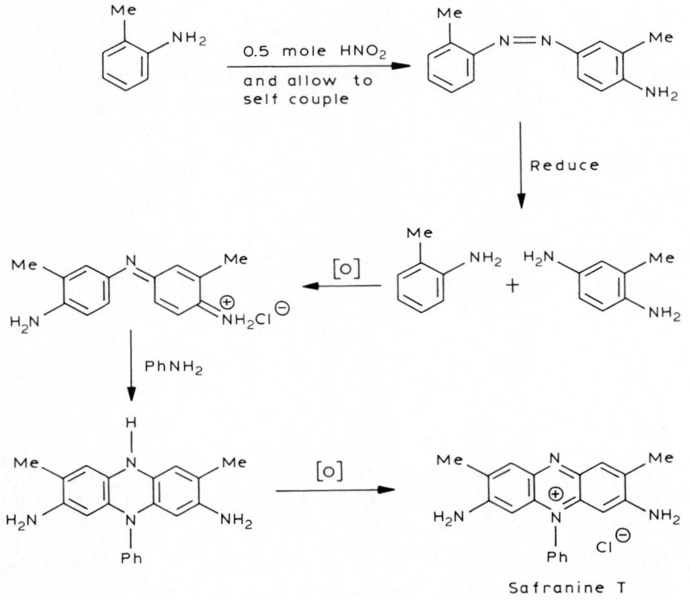

Safranine MN (C.I. Basic Violet 8, 50210) and Rhoduline Heliotrope 3B (C.I. Basic Violet 6, 50055) may be prepared in an analogous manner.

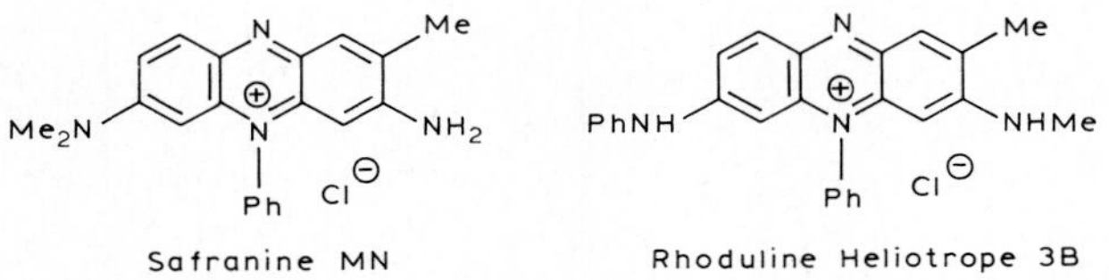

These dyestuffs were used extensively, particularly for paper colouration, but are no longer of commercial significance.

Sulphonated safranines, especially benzophenosafranines, have been used as wool dyes but, because of their poor fastness properties, in particular light fastness, their use has declined. They are, however, relatively inexpensive dyestuffs and may be used where economy rather than fastness is the main consideration. Their preparation from sulphonated intermediates is illustrated by that of Wool Fast Blue BL (C.I. Acid Blue 59, 50315; F.I.A.T. 1313, II, 380).

Wool Fast Blue BL

The use of this type of dyestuff in a colour photographic process has been claimed (W.A. Schmidt et al., Ind. Eng. Chem., 1953, **45**, 1726; O. Wohl, Angew. Chem., 1952, **64**, 259).

Similarly constituted dye intermediates such as Blue II, are formed by oxidising a mixture of an intermediate such as Blue I and an N,N-dialkyl-p-phenylenediamine (F.I.A.T. 1313, II, 383) and sulphonating the product.

Fibre-reactive phenosafranines are used for dyeing cotton (*A. Crabtree*, B.P. 1,374,852/1974). PROCION Blue MX-7RX (C.I. Reactive Blue 161; I.C.I. plc) has gained considerable commercial importance as a result of the uniquely brilliant shades it produces on cellulosic textiles.

Azo dyes with basic properties prepared by diazotisation of safranines having primary amino groups and coupling the resultant diazonium salt with various aromatic amines and phenols have also found some use, for example, for colouration of polyacrylonitrile textile materials (*Du Pont*, B.P. 942,844/1963; *D.G. Coe*, U.S.P. 3,121,711/1960).

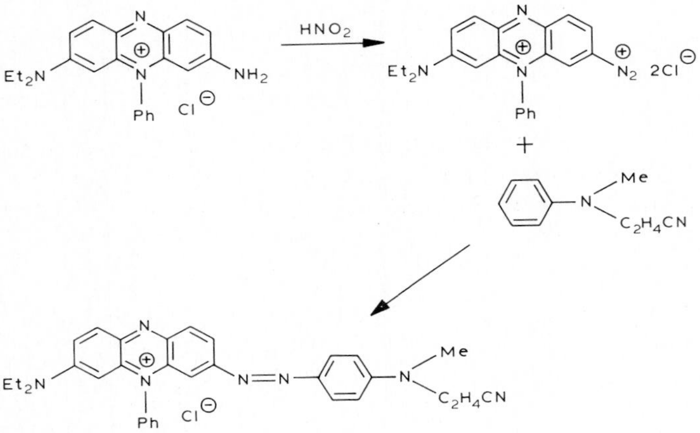

Many of the safranine dyes are oxidation–reduction indicators (*R.D. Stiehler, T-T. Chen* and *W.M. Clark*, J. Amer. chem. Soc., 1933, **55**, 891; 4097) and are capable of forming semiquinones. They are also claimed to act as polymerisation inhibitors for vinyl-substituted nitrogen compounds, for example vinyl pyridines (*R.E. Reussen* and *A.M. Schitzer*, U.S.P. 2,761,864/1956).

(d) *Fluorindines*

Fluorindines (5,12-dihydroquinoxalo[2,3-b]phenazines) are structurally related to the triphendioxazines. They do not appear to have been used commercially and, as a result, little is known of them from the colour point of view. The parent compound, fluorindine, is a deep bluish-purple compound with red fluorescence in solution in organic solvents.

It is the stable member of the series, and is obtained by heating 2,3-diaminophenazine with *o*-phenylenediamine in benzyl alcohol.

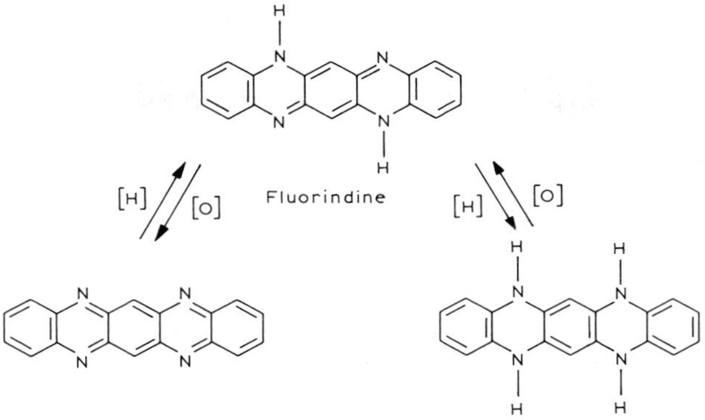

A process for the preparation of fluorindines in high yield and in a high state of purity is claimed to proceed from *o*-phenylenediamine and 1,4-benzoquinone (*C.E. Osborne* and *E.R. Shelton*, B.P. 1,159,130/1969).

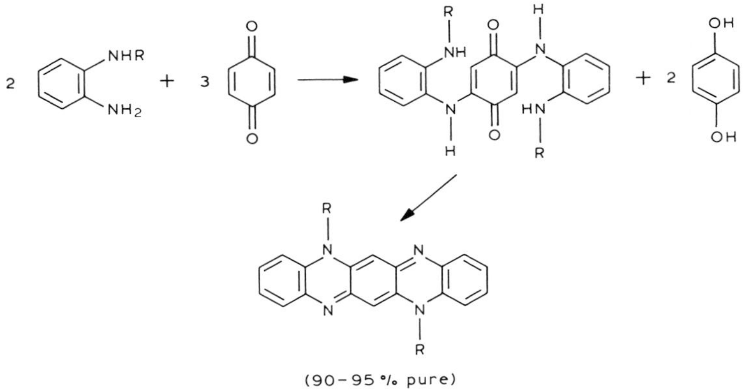

(90–95 % pure)

The products thus obtained may be quaternised with, for example, dimethyl sulphate and thereby afford soluble dyes useful for the dyeing of polyacrylonitrile fibres (*J.M. Stanley* and *R.C. Harris*, U.S.P. 3,390,948/1968). Derivatives of fluorindine may be present in nigrosines (see below).

(e) Indulines and nigrosines

Complex azines of the induline series were first obtained by *J. Dale* and *H. Caro* (B.P. 3307/1863) by fusing aniline hydrochloride with sodium nitrite, the aminoazobenzene so produced reacting with more aniline hydrochloride to form poly(phenylamino)-substituted safranines, *e.g.* Indulines 3B and 6B.

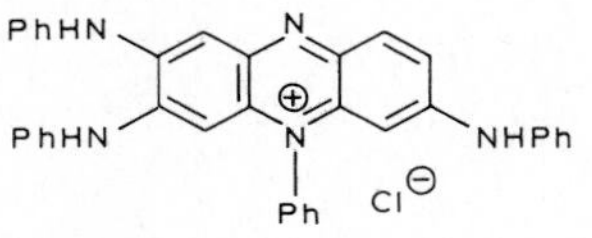

Induline 3B

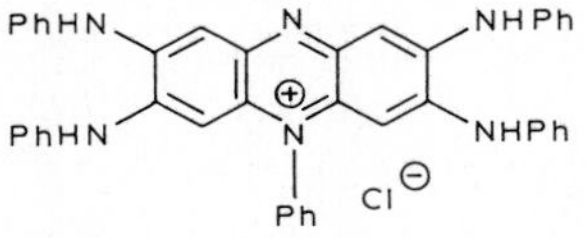

Induline 6B

Most manufacturing processes are of the "induline melt" type which consists of heating *p*-aminoazobenzene with aniline and aniline hydrochloride whereupon a complex mixture is produced depending on the experimental conditions. Besides Induline 3B and Induline 6B, *O. Fischer* and *E. Hepp* (Z. Farben Textilchem., 1901, **I**, 457; Chem. Ztbl., 1902, **II**, 902) were able to isolate and characterise other phenazines such as 2,8-diamino-3-phenylamino-5-phenylphenazonium chloride (VII) and 8-amino-2,3-bis(phenylamino)-5-phenylphenazonium chloride (VIII) as well as *p*-phenylenediamine, 4,4′-diaminodiphenylamine and 2,5-bis(phenylamino)-1,4-bis(phenylimino)benzene ("azophenine") (IX).

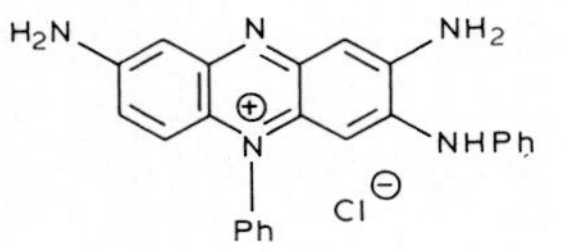

(VII)

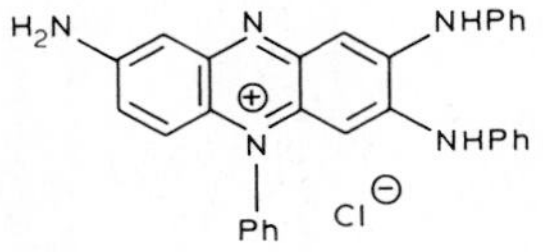

(VIII)

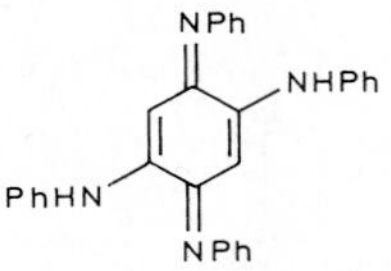

(IX)

For a more detailed summary of the indulines, including postulated reaction mechanisms, see *G.A. Swan* and *D.G.I. Felton, op. cit.* Indulines 3B and 6B (the mixture is Induline Spirit Soluble; C.I. Solvent Blue 7, 50400) have been synthesised separately (*F. Kehrmann* and *W. Klopfenstein*, Ber., 1923, **56**, 2394; *Kehrmann* and *L. Stanoyévitch*, Helv., 1925, **8**, 661).

The oxidation of aniline with nitrobenzene and iron(III) chloride to form the blue-black to black nigrosines, *e.g.* Nigrosine Spirit Soluble (C.I. Solvent Black 5, 50415) was first studied by *J.T. Coupier* (B.P. 3657/1867). The shade of the product, a complex mixture, depends on the reaction conditions, in particular fusion temperature. The precise structure of many of the components is unknown but the presence of triphenazineoxazines (X) and (XI) and fluorindines (XII) and (XIII) is postulated.

Nigrosine C New (Indamine Blue R, C.I. 50204) has been assigned the structure (XIV) (B.I.O.S. 1433 p. 82).

The properties of nigrosines may be further modified by alkylation with an alkyl or alkenyl halide in an organic solvent (*Orient Kagaku Kogyo*, Jap. P. J5-9068-374-A/82).

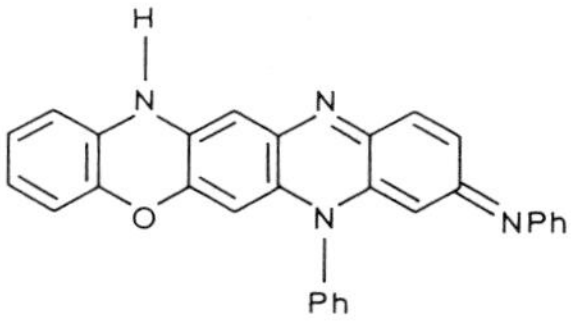

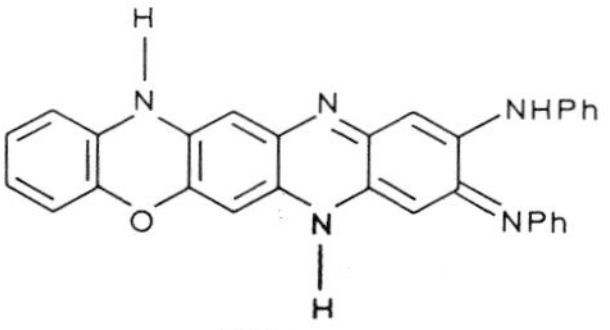

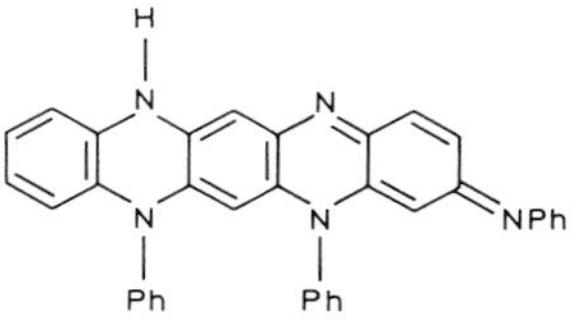

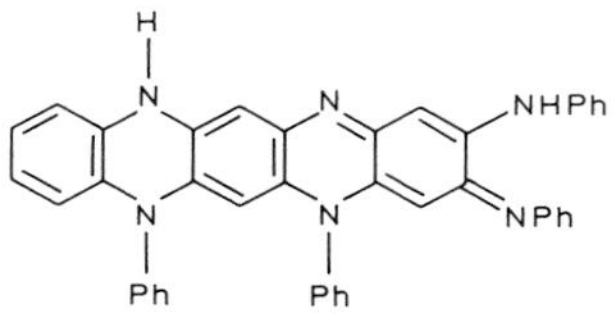

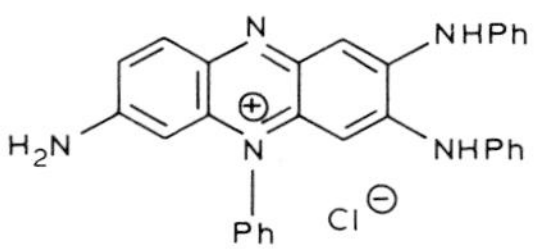

Indulines and nigrosines are widely used as colours for oil, spirit and waxes, *e.g.* boot polishes, printing inks, typewriter ribbons, wood stains and lacquer colours. As sulphonated water-soluble derivatives they find use in the dyeing of leather and paper. More recently they have been employed as charge-control agents in resin-based toner pigments for electrophotography, e.g. Xerography (*Hitachi*, Jap. P. 8-4007-382-B/79; *Agfa-Gevaert*, E.P. 122-650-A/83).

(f) Aniline black (C.I. Pigment Black 1, 50440)

The production of green-black and blue-black colours from aniline salts by oxidation has been known since 1840 (*J. Fritzsche*, J. pr. Chem., 1840, **20**, 453; 1843, **28**, 198). Numerous oxidising agents have been used such as chromic acid, chromates, persulphates and chlorates often in the presence of catalytic amounts of transition metal salts, for example copper salts. It was from one such experiment that Perkin extracted a small yield of mauveine. The major product, aniline black, is an insoluble black powder which is very stable to the action of acids, alkalis, oxidising and reducing agents. Its formation *in situ* by oxidising water-soluble aniline salts with, for example, chromic acid in the presence of vanadium salts as catalyst has been used as a method for the colouration of textile materials (*K.A. Hofmann, F. Quoos* and *O. Schneider*, Ber., 1914, **47**, 1991).

A considerable amount of research has been carried out to determine the structure of aniline black. *N*-Phenyl-1,4-benzoquinodiimine is regarded as a precursor (as in induline formation), itself produced possibly by a free radical mechanism from aniline. The phenazine structure (XV) due to *H.Th. Bucherer* (Ber., 1907, **40**, 3412; 1909, **42**, 2931), also owes much to the researchers of *R. Wilstäter* and *S. Dorogi* (*ibid.*, 1909, **42**, 4118), *A.G. Green* and *A.E. Woodhead* (J. chem. Soc., 1910, **97**, 2388; 1912, **101**, 1117) and *Green* and *S. Wolff* (Ber., 1911, **44**, 2570).

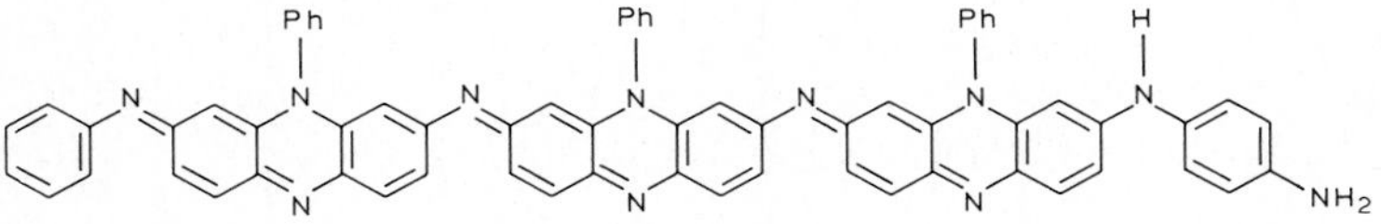

(XV)

The presence of the phenazine ring system has, however, never been proved nor has the size of the molecule been accurately determined. For a review of the structural evidence see *Swan* and *Felton, op. cit.*

Instead of aniline salt as starting material, the use of other materials has been patented, for example 4-aminodiphenylamine as a water-soluble derivative such as the sulphamate (*R. Lanz*, B.P. 498,755/1939) and 4-phenylaminomaleanilic acid (XVI) (*G. Fuortes, E. Graktani* and *E. Greppi*, B.P. 787,357/1957).

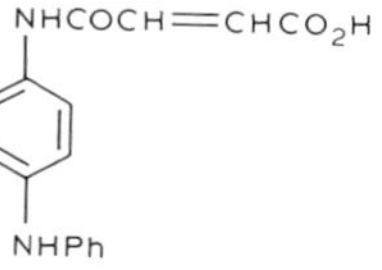

$$\underset{\text{(XVI)}}{\underset{\text{NHPh}}{\text{NHCOCH}=\text{CHCO}_2\text{H}}}$$

The oxidation of 2-aminodiphenylamine with iron(III) chloride in acid solution yields phenylamino phenazines which are less complex than aniline black and said to be useful as pigments (*V.C. Barry et al.*, J. chem. Soc., 1956, 888; 893; 896).

3. Phenoxazine dyes*

(a) Basic type

The chromophore of the phenoxazine dyes is the phenoxonium ion (formula I on page 403, $X = O$), which, like the phenazine dyes of the previous section, is normally substituted in either position 3 or in positions 3 and 7 by electron-donating substituents.

The most general method of preparation involves reaction of the *p*-nitroso derivative of an aromatic amine at the *ortho*-position of a phenol or naphthol followed by oxidative cyclisation of the resultant indophenol and is illustrated by the preparation of Meldola's Blue (C.I. Basic Blue 6, 51175) from *p*-nitrosodimethylaniline and 2-naphthol.

---

* K. Venkataraman, *op. cit.*, p. 781.

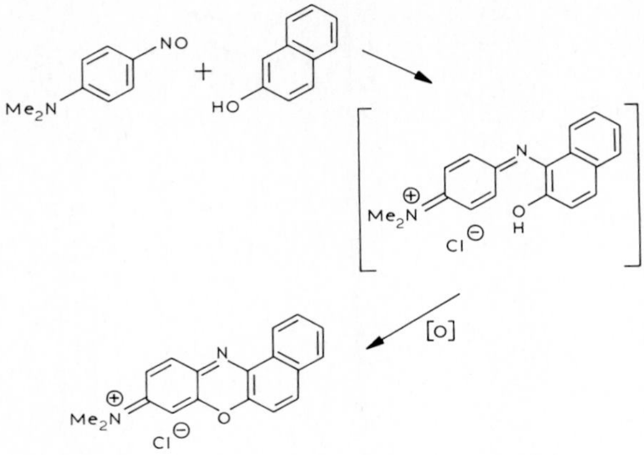

[Scheme showing synthesis of Meldola's Blue from p-nitrosodimethylaniline and 2-naphthol via indophenol intermediate, followed by oxidative cyclisation [O].]

Meldola's Blue

In practice three moles of *p*-nitrosodimethylaniline hydrochloride are used with two of 2-naphthol, the third mole of the former being the oxidising agent necessary to convert the intermediate indophenol into the dye; 4-aminodimethylaniline is recovered as a by-product.

By substituting 3-diethylamino-4-cresol for 2-naphthol in the above synthesis, the dye known as Capri Blue GON (C.I. Basic Dye 51015) is obtained (*F. Bender*, B.P. 13565 and 18623/1890).

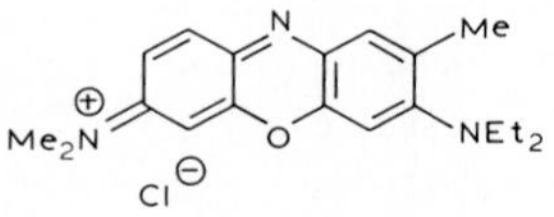

Capri Blue GON

The same principle is applied when the nitroso derivative of an *N*-substituted 3-aminophenol is condensed with an amine and the resulting indamine is oxidatively cyclised. In this way Nile Blue A (C.I. Basic Blue 12, 51180) is obtained from 3-diethylamino-6-nitrosophenol and 1-naphthylamine.

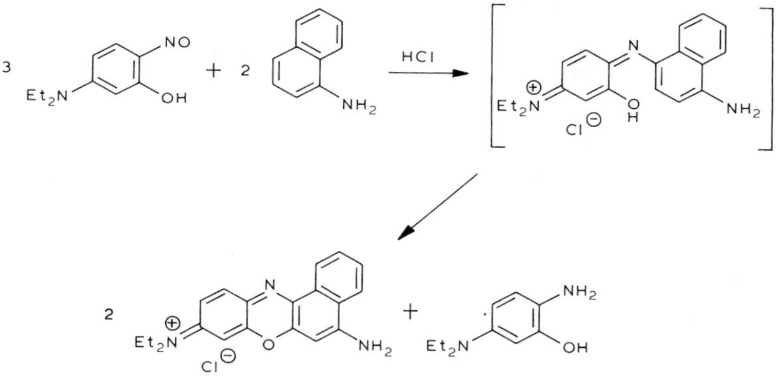

[Scheme: synthesis of Nile Blue A from 3-diethylamino-nitrosophenol and 1-naphthylamine with HCl, via intermediate, yielding Nile Blue A plus byproduct]

Nile Blue A

The phenoxazine dyes are readily reduced with reagents such as sodium dithionite (Hydros) to their leuco compounds from which the dyestuff can be quantitatively regenerated by careful oxidation. Strong oxidising agents such as sodium chlorate or ferricyanide destroy the colour permanently.

The 7-position in compounds such as Meldola's Blue is reactive. Thus condensation, under oxidising conditions, with N,N-dimethyl-p-phenylenediamine, with dimethylamine and with Michler's Hydrol gives New Blue B (C.I. Basic Blue 10, 51190), New Methylene Blue GG (C.I. Basic Dye, 51195) and New Fast Blue F (C.I. Basic Dye, 51200), respectively.

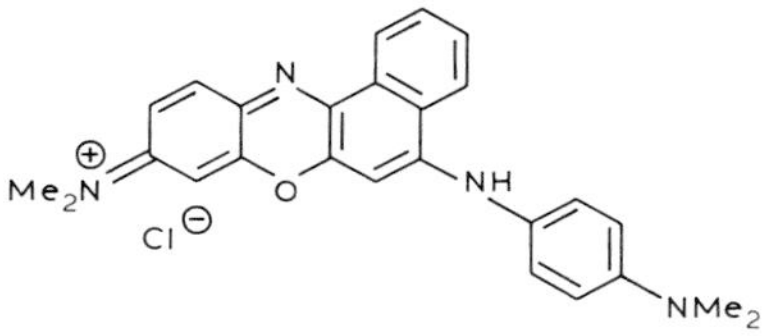

New Blue B

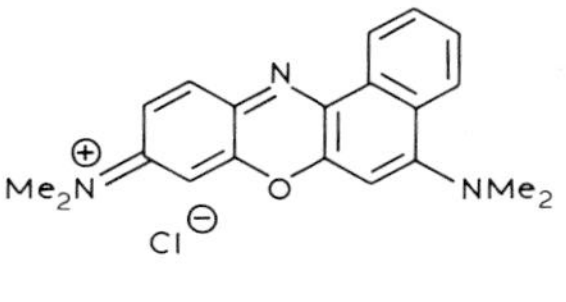

New Methylene Blue GG

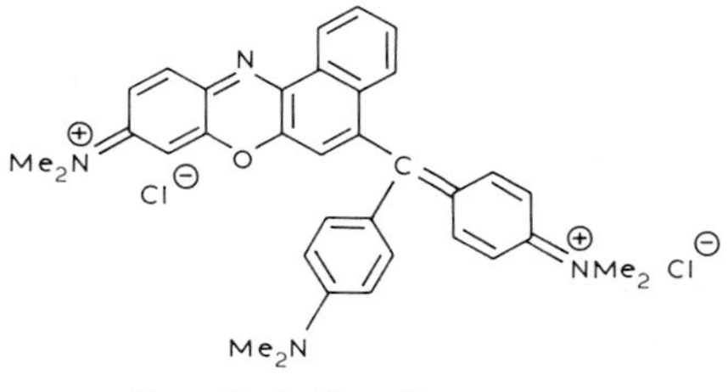

New Fast Blue F

Similarly, the 7-amino group in Nile Blue A is labile and can be removed in dilute acid to yield the phenoxazone (I) (*J.F. Thorpe*, J. chem. Soc., 1907, **91**, 324).

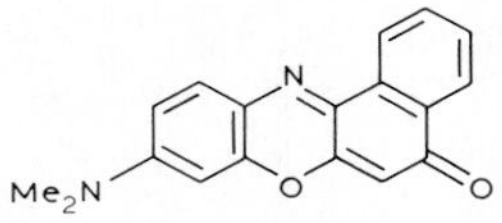

(I)

The oxazine dyes have been used for such purposes as calico printing and the dyeing of mordanted cotton but, because of their ready photochemical degradation, they are now little used. They did find some use for the colouration of polyacrylonitrile fibres (*Ciba*, B.P. 751,150/1956) and specific dyes, in particular those carrying bis-(β-cyanoethyl)-amino groups in the 3-position, are claimed to have improved light fastness on this substrate (*F. Zwilgmeyer*, U.S.P. 2,741,605/1956).

(b) Gallocyanines

Gallocyanines were discovered during attempts to fix nitrosophenol on to textile fibres in the presence of tannic acid as mordant (*H. Köchlin*, B.P. 4899/1881; Monit. Sci., 1883, III, **13**, 292) and their formulation as oxazines was proposed some years later by *R. Nietski* and *R. Otto* (Ber., 1888, **21**, 1736) who, at the same time, recognised Meldola's Blue as an oxazine. Gallocyanine can be obtained in excellent yield, by a modification of Meldola's synthesis, from 4-nitrosodimethylaniline and gallic acid.

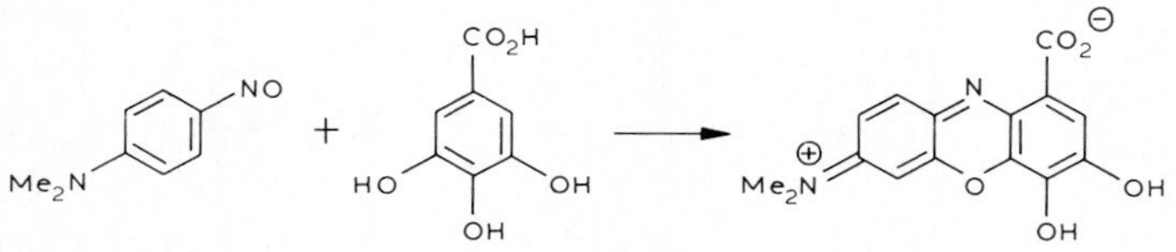

Gallocyanine

The reaction takes the same course as the Meldola's Blue synthesis and the same proportions of reactants are used. Similar syntheses starting from methyl gallate, gallamide and gallanilide give Gallo Blue E (C.I. Mordant Violet 54, 51040), Gallamine Blue (C.I. Mordant Dye, 51060) and Gallanil

Violet (C.I. Mordant Blue 45, 51065), respectively (*A. Kern*, B.P. 5953/1887, *C. de la Harpe* and *R. Burckhardt*, B.P. 12067/1908; *J. Mohler* and *C. Meyer*, B.P. 11848/1889).

Chemically the interesting features of the gallocyanine molecule are the ready decarboxylation and the substitution reactions at positions 1 and 2. Gallocyanines are readily reduced to their leuco derivatives and these are decarboxylated by heating with steam. Air, blown through a hot solution of gallocyanine hydrochloride (II) in aniline, results in replacement of the carboxyl group by a phenylamino group, as in (III) and the same treatment carried out cold results in the latter group entering the 2-position to yield (IV).

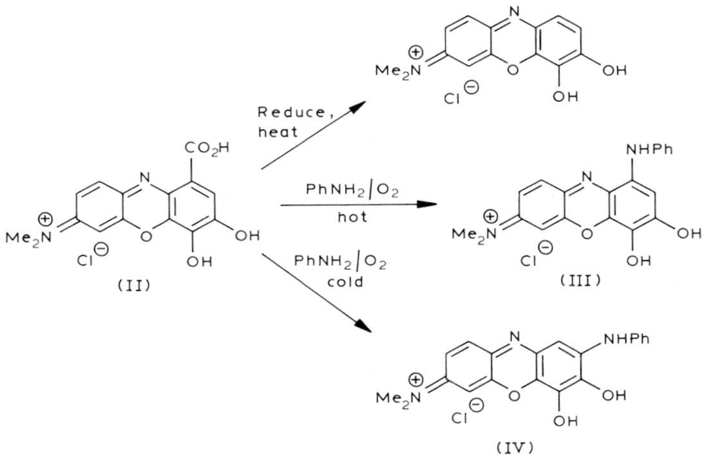

Phenols can replace aniline, although the reactions appear to be less specific and mixed 1- and 2-aryloxy derivatives are obtained. Thus, reaction of gallocyanine and resorcinol affords the 1- and 2-(3-hydroxyphenoxy) derivatives Phenocyanine VS (C.I. Mordant Dye, 51140; *de la Harpe*, B.P. 24802/1893) and Ultracyanine B (C.I. Mordant Dye, 51130; *A. Steiner*, B.P. 6270/1909). Alizarin Green G (C.I. Mordant Dye, 51405; *Elsässer*, B.P. 5153/1895) is prepared by condensation of 1,2-naphthoquinone-4-sulphonic acid with 1-amino-2-naphthol-6-sulphonic acid.

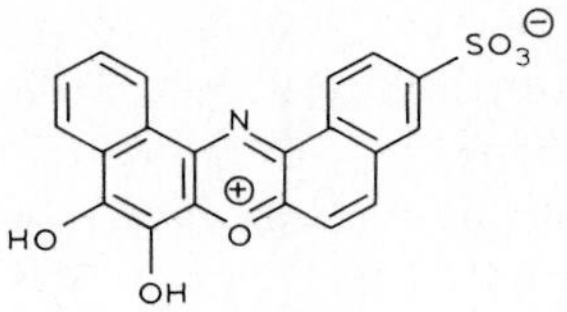

Alizarin Green G

The gallocyanines were at one time the most important of the commercial oxazine dyes. Their 1,2-dihydroxy structure makes them ideal for such dyeing processes where metal coordination is involved, particularly where the fibre may act as one of the ligands in the metal complex so formed, for example, the dyeing of chrome mordanted wool. The introduction of sulphonic acid groups converts them into chromable acid wool dyes. Delphine Blue (C.I. Mordant Blue 56, 51120) and Chromazurin Blue E (C.I. Mordant Blue 59, 51125) which are sulphonated derivatives of (III) and (IV), respectively, are examples of such dyes.

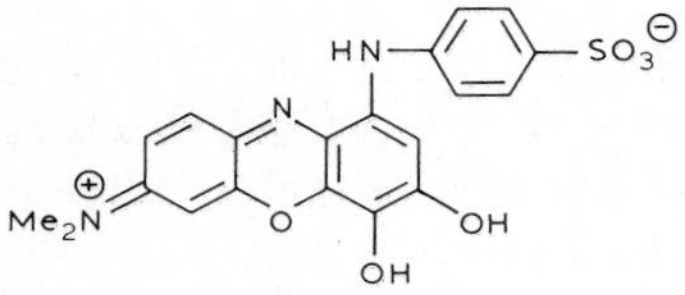

Delphine Blue 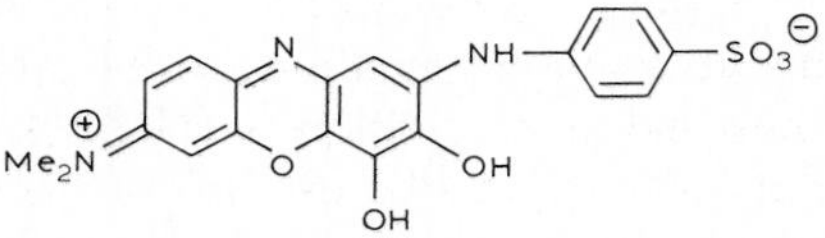   Chromazurin Blue E

(c) Triphendioxazines

Triphendioxazine (V) was first discovered in 1879 by *G. Fischer* (J. pr. Chem., 1879, **19**, (2), 317), who prepared it by the oxidation of *o*-aminophenol with bleaching powder, but the structure was not reported until 1890 (*P. Seidel*, Ber., 1890, **23**, 182).

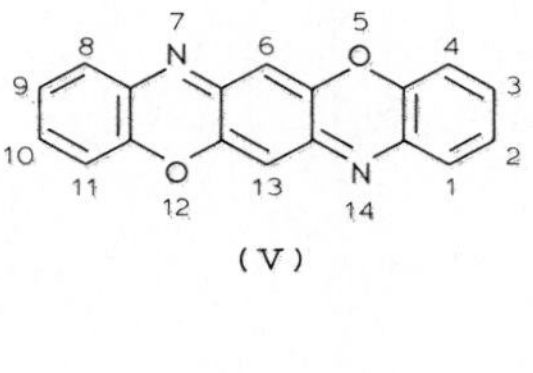

(V)

The synthesis of the commercially most important 6,13-dichlorotriphendioxazines (VII) was discovered in 1911 by the firm of *Meister, Lucius* and *Brüning* (G.P. 253091/1911). It involves the condensation of an arylamine with chloranil, chosen because of its ready availability, to give an intermediate (VI), often referred to as a "dianilide", which is converted into product (VII) in an oxidative cyclisation.

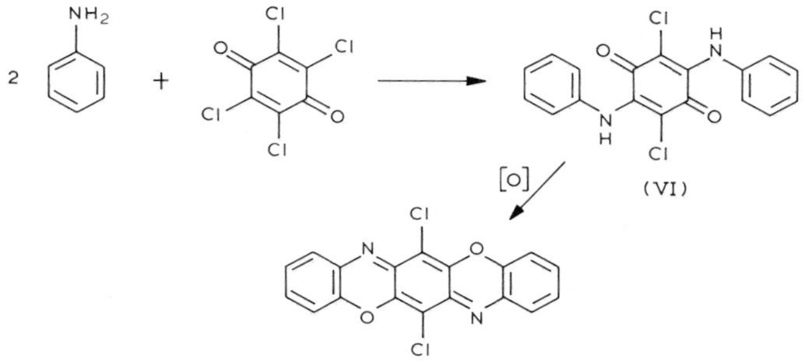

The initial condensation is conveniently carried out in alcoholic or aqueous alcoholic medium in the presence of a weak base such as magnesium oxide or sodium acetate. Oxidative cyclisation is effected by a variety of reagents, for example aluminium chloride or iron(III) chloride in a high boiling solvent such as nitrobenzene or by concentrated or fuming sulphuric acid (*Durand* and *Huguenin A.G.*, B.P. 197940/1924; 223481/1924; *I.G.*, B.P. 313094/1930). An alternative synthesis involving a non-oxidative cyclisation is condensation of an *o*-alkoxyarylamine with chloranil; the resultant 2,5-di-(*o*-alkoxyarylamino)-3,6-dichlorobenzoquinone (VIII) is cyclised by treatment with an acid chloride, *e.g.* benzoyl chloride or *p*-toluenesulphonyl chloride, in a high boiling solvent (*A. Brunner* and *K. Thiess*, U.S.P. 2,092,387/1937). This method is claimed to give better yields than the oxidative cyclisation.

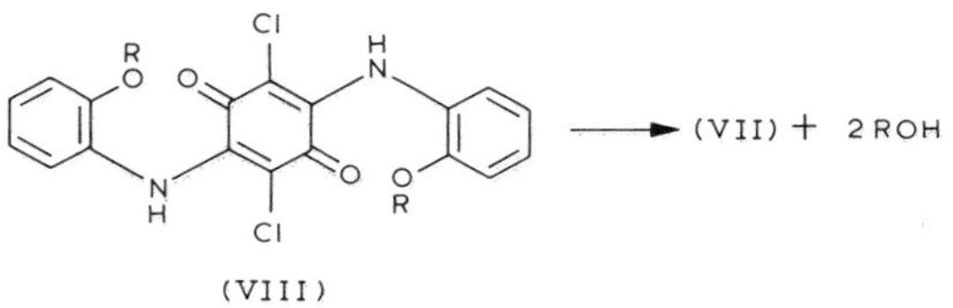

By modification of the chloranil, unsymmetrical triphendioxazines can be obtained. Treatment of chloranil with methanol and sodium acetate gives the 2-methoxy derivative which will react with arylamines under mild conditions to yield the corresponding 5-arylamino compound. This will then react at the 2-position, with a different arylamine under more vigorous conditions, and finally cyclise (*I.G.*, B.P. 416887/1934; H.E. Fierz-David, J. Brassell and F. Probst, Helv., 1939, **22**, 1348; G. Langbein, U.S.P. 2,229,099/1941; 2,267,741/1941; 2,288,194/1941; H. Greune and *Langbein*, U.S.P. 2,278,260/1942).

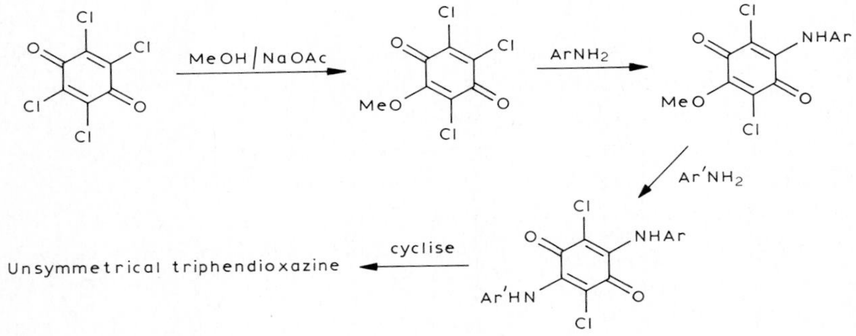

More recently mono-amino trichlorobenzoquinones have been used to prepare unsymmetrical products (*S.K. Jain, N.K. Goswami* and *R.R. Gupta*, Ann. Soc. Sci. Bruxelles Ser. 1, 1980, **94**, 63).

Triphendioxazines substituted at positions 6 and 13 by an acylamino group are prepared from 2,5-dihalogeno-3,6-bis(acylamino)-1,4-benzoquinones (*Geigy*, B.P. 1,039,926/1962; *Ciba*, B.P. 1,114,012/1966). Other less important methods of preparing triphendioxazines are given below.

(*i*) An *o*-aminophenol is oxidised to an aminophenoxazone which is condensed with a further molecule of *o*-aminophenol.

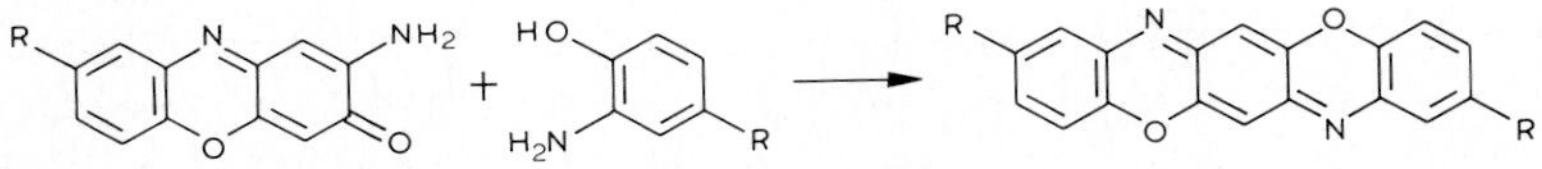

(R = H, Cl, CO_2H, etc.)

(*ii*) The condensation of 2-amino-4-nitrophenol with chloranil can be stopped at the trichloronitrophenoxazone stage; further reaction with *o*-aminophenol gives the nitrotriphendioxazine (*I.G.*, B.P. 411,132/1934).

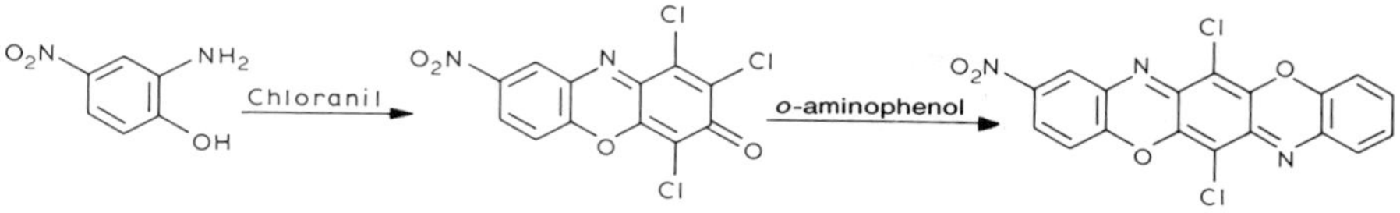

(*iii*) Reaction of 4,6-diaminoresorcinol sulphate with *o*-aminophenol affords triphendioxazine (*P. Seidel*, Ber., 1890, **23**, 182).

(*iv*) Condensation of 2,5-dihydroxybenzoquinones with *o*-aminophenols has been used to obtain a number of triphendioxazines (*F. Kehrmann*, Ber., 1893, **26**, 2375; 1896, **29**, 2076; Helv., 1926, **9** 866; 1928, **11** 1028).

(*v*) A special case is the synthesis of tetrabenzotriphendioxazine (*W. Burneleit*, U.S.P. 2,233,940/1941; G.P. 125,288/1942).

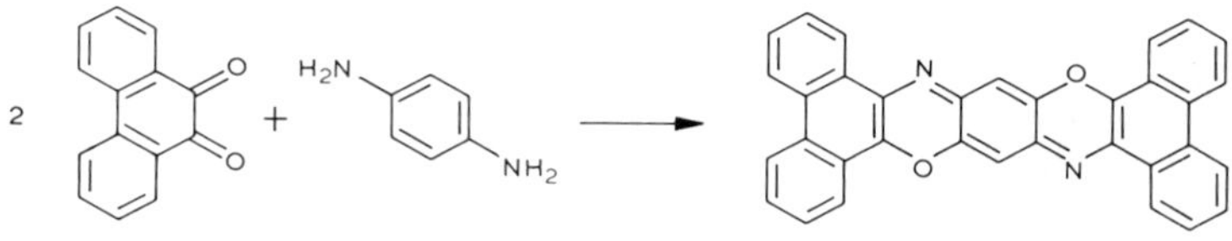

The triphendioxazines vary in colour from the red of the simple 6,13-dichloro derivatives through violet and blue to the greenish-blue of the more complex compounds, for example, Diamine Fast Blue F3GL (C.I. Direct Blue 109, 51310). They are well-defined crystalline compounds, the simple compounds having weakly basic properties. On reduction in organic solvents with, for example, phenylhydrazine, leuco compounds are obtained which readily reoxidise. Triphendioxazine itself can be nitrated (*Seidel, loc. cit.*). A peculiar reaction of 6,13-dichlorotriphendioxazine has been recorded: the product of its reaction with aniline is believed to be 6-chloro-3,10,13-triphenylaminotriphendioxazine (*Brassell* and *Probst, loc. cit.*).

Triphendioxazines are generally sparingly soluble in organic solvents and may therefore be used as pigments for paints, etc., examples being those from 3-amino-*N*-ethylcarbazole (B.I.O.S. 960, p. 75) and 4-aminodiphenyl (*R.W.G. Preston*, B.P. 646,099/1950). The sulphonated triphendioxazines are widely used as direct cotton dyestuffs and, as such, are noted for their brightness of shade and excellent fastness to light. These include Diamine Light Blue FFRL (C.I. Direct Blue 108, 51320) which is prepared from 3-amino-*N*-ethylcarbazole followed by sulphonation; Diamine Fast Blue F3GL (C.I. Direct Blue 109, 51310) from 1-aminopyrene followed by sulphonation and Diazol Light Pure Blue FF2JL (C.I. Direct Blue 106, 51300) the sulphonated derivative of the product from 4-aminodiphenylamine and chloranil.

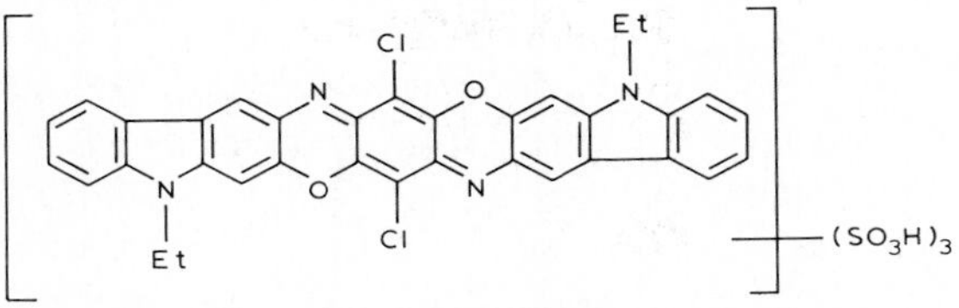

Diamine Light Blue FFRL

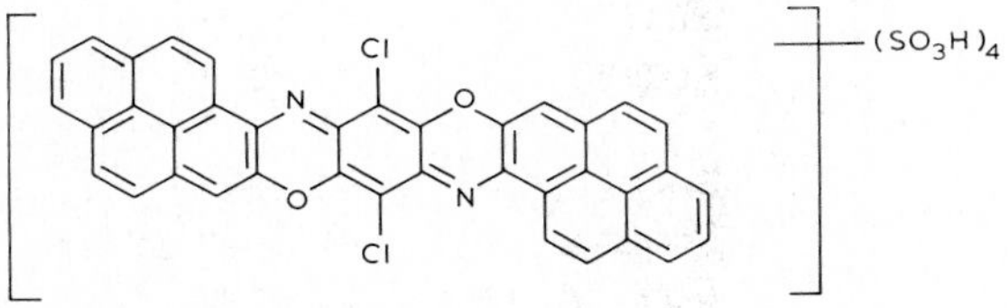

Diamine Fast Blue F3GL

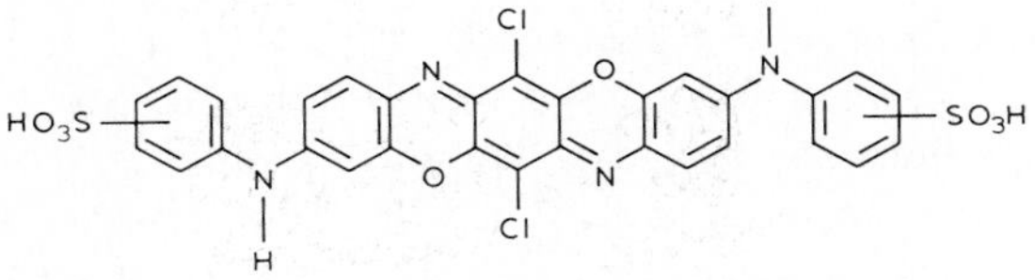

Diazol Light Pure Blue FF2JL

Commercial interest in triphendioxazine dyes and pigments has been maintained as shown by the publication of more than forty patents since

1970. Disclosures have been made concerning the application of triphendioxazines as acid dyes for nylon and wool (*B. Parton*, B.P. 1,353,604/1970), as direct dyes for cotton (*A.H.M. Renfrew*, E.P. 84,718/1982) and as cationic dyes for natural or synthetic fibres and for paper (*Ciba-Geigy*, E.P. 14,678/1979; E.P. 15,232/1979 and E.P. 15,233/1979; *B.A.S.F.*, E.P. 64,631/1981).

The bright blue fibre-reactive triphendioxazine dyestuffs for cotton, for example PROCION Blue H-EGN (C.I. Reactive Blue 198), have gained considerable commercial importance at the expense of the more traditional anthraquinonoid products. They are derived from 2,9-diamino-6,13-dichlorotriphendioxazine sulphonic acids in which the amino groups are substituted by a group bearing at least one fibre-reactive entity, for example a dichloro-*s*-triazinylamino group (*B. Parton*, B.P. 1,349,513/1972). For a comprehensive review of triphendioxazine textile dyestuffs see *Renfrew* (Rev. Prog. Color., 1985, **15**, 15).

4. Phenothiazine dyes*

Phenothiazine dyes are mostly salts of the oxidised form of 3,7-diaminophenothiazine with substituents on the amino groups and, although they have usually been represented as thionium salts, *i.e.* as *ortho*-quinonoid, they are undoubtedly best represented as *para*-quinonoid or as resonance hybrids as in (I) (p. 403, X = S).

The first phenothiazine to be made was 3,7-diaminophenothiazonium chloride, known as Lauth's Violet (C.I. Basic Dye, 52000) after its discoverer, who obtained it by oxidation of *p*-phenylenediamine with iron(III) chloride and sulphur (*Ch. Lauth*, Ber., 1876, **9**, 1035; Bull. Soc. chim. Fr., 1876, [ii], **26**, 422). Its constitution was determined by *A. Bernthsen* (Ann., 1885, **230**, 73., Ber., 1884, **17**, 611), who obtained it by nitration, reduction and subsequent oxidation of phenothiazine.

Methylene Blue (C.I. Basic Blue 9, 52015), 3,7-bis-(dimethylamino)phenothiazonium chloride is by far the most important phenothiazine dyestuff and is prepared by the oxidation of *N,N*-dimethyl-*p*-phenylenediamine with chromic acid in the presence of thiosulphate and dimethylaniline (*C. Roth*, B.P. 43/1886; *B.A.S.F.*, B.P. 10314/1888; *W.H. Claus* and *A. Rée*, B.P. 8221/1893; B.I.O.S. 1433, p. 11).

---

* K. Venkataraman, op. cit., p. 791.

In the manufacturing process dimethylaniline is nitrosated and the *p*-nitroso derivative is reduced with iron borings to *N,N*-dimethyl-*p*-phenylenediamine which is not isolated but treated, *in situ*, with sodium thiosulphate, sulphuric acid and sodium dichromate at 0° under carefully controlled conditions. Further dimethylaniline is then added together with more dichromate to form the green indaminethiosulphonic acid which is finally treated with copper(II) sulphate and a further quantity of dichromate at 60–70° to give Methylene Blue.

Leuco Methylene Blue

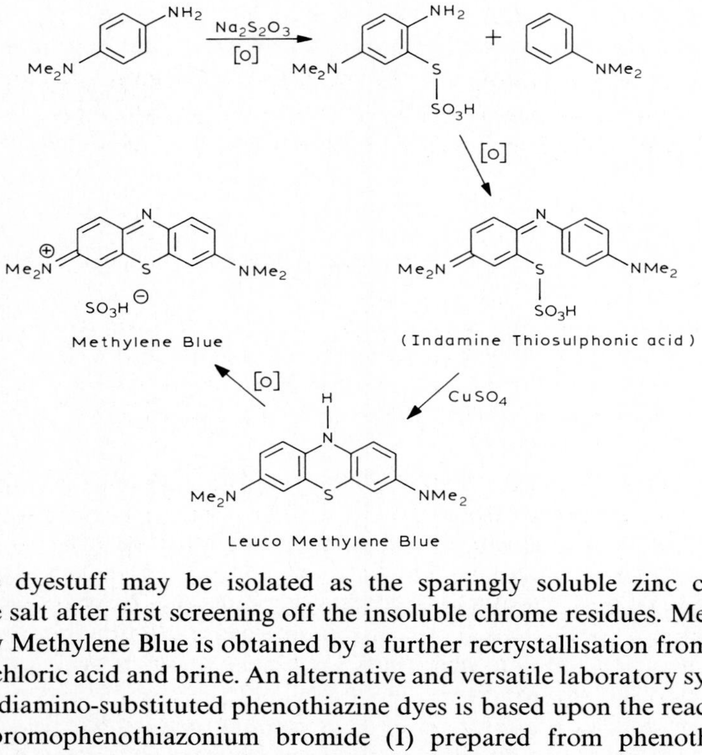

The dyestuff may be isolated as the sparingly soluble zinc chloride double salt after first screening off the insoluble chrome residues. Medicinal quality Methylene Blue is obtained by a further recrystallisation from dilute hydrochloric acid and brine. An alternative and versatile laboratory synthesis of 3,7-diamino-substituted phenothiazine dyes is based upon the reaction of 3,7-dibromophenothiazonium bromide (I) prepared from phenothiazine, with amines (*F. Kehrmann* and *L. Diserens*, Ber., 1915, **48**, 318; Ber. 1916, **49**, 53).

Phenothiazine (I)

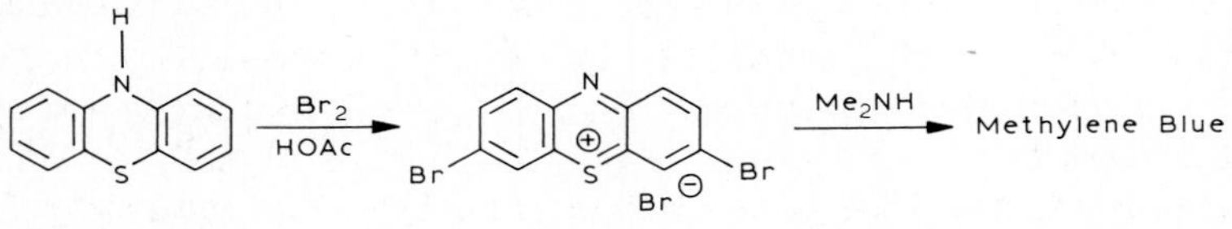

Other examples of phenothiazine dyestuffs which have been used are New Methylene Blue NSS (C.I. Basic Blue 24, 52030; *A. Weinberg*, B.P. 8407/1891), Toluidine Blue (C.I. Basic Blue 17, 52040; *Dandliker* and *Bernthsen*, B.P. 10314/1888) and Methylene Blue MT.

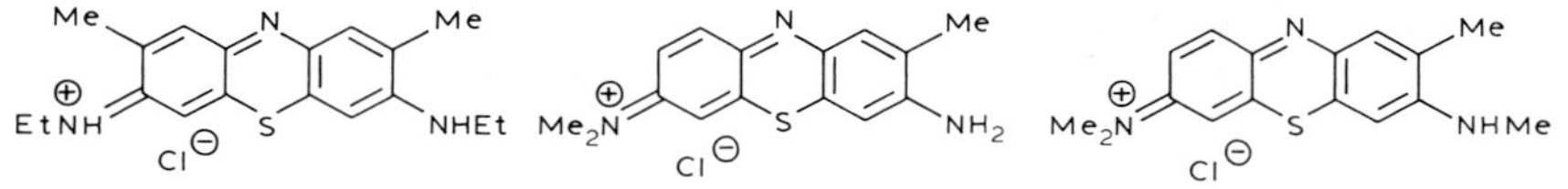

New Methylene Blue NXX   Toluidine Blue   Methylene Blue MT

Brilliant Alizarin Blue G (C.I. Mordant Dye, 52055) was made by condensation of 4-amino-N,N-dimethylaniline-3-thiosulphonic acid with 1,2-naphthaquinone-6-sulphonic acid (*M. Böniger*, B.P. 3886/1894):

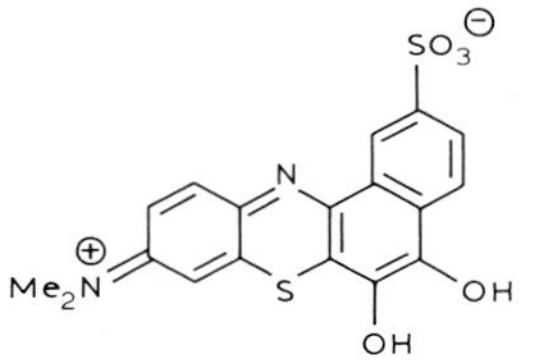

Brilliant Alizarin Blue G

The thiazine dyes of the Methylene Blue type form double salts with zinc chloride and are usually isolated as such because of the reduced solubility of these salts compared with the simple salts. The latter also form insoluble "lakes" with high molecular weight acids such as phosphotungstic and phosphomolybdic acids and as such are useful as pigments. Methylene Blue can be nitrated in the 4-position giving Methylene Green (C.I. Basic Green 5, 52020; *E. Ulrich*, B.P. 8992/1886, *R. Gnehm* and *E. Walder*, J. pr. Chem., 1907, [ii], **76**, 402).

Methylene Blue is used as a redox indicator in, for example, titanium-(III) chloride titrations (*E. Knecht*, J. Soc. Dy. Col., 1905, **21**, 9) or in place of starch in iodometric titrations; its insoluble perchlorate and dichromate can serve as the basis of a gravimetric determination (*F.A. Maurine* and *N. Deahl*, J. Amer. pharm. Assoc., 1943, **32**, 301). Its uses as a biological stain and in tests for tubercular infection in milk are well established and it is a weak antiseptic.

Leuco-Methylene Blue, a pale yellow water insoluble solid, m.p. 185°, may be obtained from the indaminethiosulphonic acid precursor by boiling an aqueous solution with dilute acid (*P. Landauer* and *H. Weil*, Ber., 1910, **43**, 198) or, more conveniently, by reduction of Methylene Blue in aqueous solution with sodium dithionite. It is readily re-oxidised to Methylene Blue on exposure to air and can be *N*-acylated. Thus *N*-benzoyl-leuco-Methylene Blue is obtained by benzoylation of leuco-Methylene Blue with benzoyl chloride in pyridine (*G. Cohn*, Ber., 1900, **33**, 1567), in aqueous alkaline medium (*National Cash Register Co.*, B.P. 725,275/1955), or in a variety of water-immiscible solvents (*S.R. Buk*, U.S.P. 2,909,520/1959).

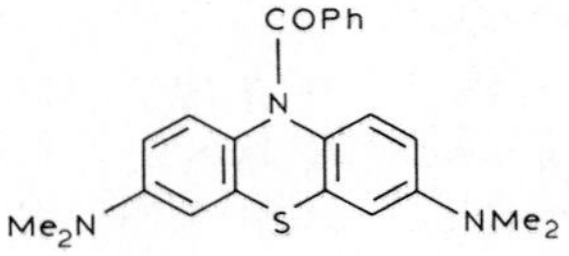

Benzoyl-leuco Methylene Blue

Benzoyl-leuco-Methylene Blue is used in carbonless, pressure-sensitive, copying papers. In these systems it is dissolved, along with other "colour formers", for example Crystal Violet Lactone, in a non-polar high boiling solvent and the solution is encapsulated in gelatin walled microcapsules. The suspension of capsules is applied to the underside of a sheet of paper. Application of pressure, by either writing or typing, to this paper causes the capsules to burst thereby transferring the colour former containing solution to a second, underlying, sheet of paper the upper surface of which is coated with an "acidic" material, generally an activated mineral such as silica or attapulgite clay. A Methylene Blue print is formed by hydrolysis of the *N*-benzoyl group on the active surface and subsequent aerial oxidation of the leuco-Methylene Blue so produced.

5. Sulphur dyes*

The sulphur dyes constitute a wide range of colouring matters cheaply produced by the introduction of sulphur into a variety of organic substances. Structural investigations are hindered by the presence of gross

---

* K. *Venkataraman*, op. cit., p. 1059 and Vol. VII, p. 1.

mixtures of non-crystalline products which are difficult to purify or degrade to identifiable compounds. They are, therefore, largely of unknown and variable composition.

The nature of the dyes produced by thionation is determined by: (a) the nature of the organic starting material, (b) the conditions of thionation, and (c) the method of work up and isolation employed. Since they do not fall into any well-defined chemical class they are, for convenience, classified in accordance with the chemical structure of the organic starting material.

The method and conditions of thionation form an integral part of a process resulting in a defined commercial product and are divided into four main types, *viz*.: (i) a sulphur bake, (ii) a sodium polysulphide bake, (iii) a sodium polysulphide melt in aqueous medium, and (iv) a sodium polysulphide melt in solvent medium.

Pressure autoclaves may be used for thionation temperatures above 100°. Other methods which are used, mainly in laboratory preparations, employ sulphur dichloride or chlorosulphonic acid as thionating agent. Following thionation the dyes are precipitated by aerial or chemical oxidation and/or acidification.

The sulphur colours are insoluble in water, acids and the common organic solvents unless sulphonic acid or basic groups are present. They are rather unstable to oxidation and, on storage, tend to deteriorate. In view of the lack of chemical stability almost any chemical treatment is liable to bring about some decomposition with consequent loss of sulphur, often as hydrogen sulphide; the dyes are thus affected by treatments involving hydrolysis, reduction or oxidation.

Sulphur dyes are still of very considerable commercial importance for the dyeing of cotton especially for heavy black shades where their cost advantage over other classes of dye is greatest. For application to the fibre they are first "vatted", that is reduced by and dissolved in an aqueous solution of sodium sulphide. The vat solution is then applied to the textile material and the colour is re-formed on the fibre by oxidation. A wide variety of shades are thus obtainable, particularly those of brown, blue and black; no true reds are produced although some dull red and red-brown hues are known. The shades obtained are generally considered to be dull in comparison with dyes of other chemical types but have generally excellent fastness to washing and to light.

The first sulphur dye, Cachou de Laval, was prepared by *E. Croissant* and *L.M.F. Bretanniére* (F.P. 98915/1873; B.P. 1489/1873) by baking sawdust with sodium polysulphide or sodium hydroxide and sulphur at temperatures above 200°. The green shade given on cotton is turned brown by treatment with dichromate. Vidal black (C.I. Sulphur Black 3, 53180, *R. Vidal* and *S.A. St. Denis*, B.P. 19880/1893) is, however, much more important and is prepared by heating either 4-aminophenol or *p*-phenylenediamine with sodium sulphide and sulphur at 180–210°. Vidal established the formation of diphenylamine intermediates in the sulphur melt and, to this day, diphenylamines are among the most important intermediates for the manufacture of sulphur dyes.

The first commercially successful product was Immedial Black (C.I. Sulphur Black 9, 53230) discovered by Kalischer (*L. Cassella and Co.*, B.P. 25234/1897) who prepared it by thionation of 2,4-dinitro-4'-hydroxydiphenylamine which was obtained by condensing 4-aminophenol with 2,4-dinitrochlorobenzene. Direct thionation of 2,4-dinitrophenol yields the important Sulphur Black T (C.I. Sulphur Black 1, 53185; *Vidal*, B.P. 16449/1896).

Another important class of intermediate for thionation are the indophenols. Hypochlorite oxidation of mixtures of either 4-aminophenol and diphenylamine or 4-aminodiphenylamine and phenol yields the indophenol (I), whilst (II) is prepared either from the reaction of 4-nitrosophenol with *N,N*-dimethylaniline or by the oxidation of a mixture of *N,N*-dimethyl-*p*-phenylenediamine and phenol.

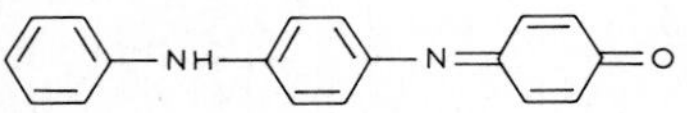

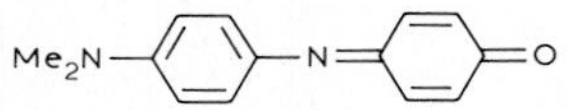

The insoluble blue indophenols are precipitated during the reaction. They are readily reduced to alkali-soluble leuco compounds which are then thionated by sodium polysulphide in boiling aqueous solution.

The sulphur dyes are mostly amorphous in appearance and examination of the dye from 4-hydroxydiphenylamine by X-ray diffraction supports this view (*W.N. Jones* and *E.E. Reid*, J. Amer. chem. Soc., 1932, **54**, 4393). Thiophorindigo CJ from the thionation of the indophenol (III) is reported to have been obtained in crystalline state from benzene (see *K. Venkataraman, op. cit.*, p. 1083):

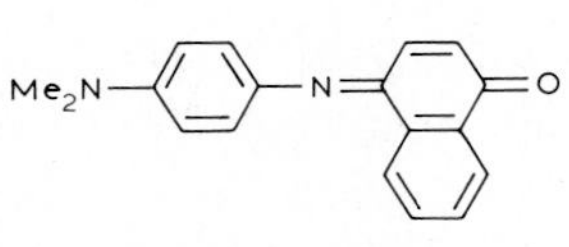

(a) Constitution of the sulphur dyes

Attempts to determine the constitution of the sulphur dyes have met with only limited success, mostly because they are invariably complex mixtures which are particularly difficult to resolve. However, the work of *R. Gnehm* and *E. Kaufler* (Ber., 1904, **37**, 2617, 3032), *R. Herz* (Ber., 1930, **63a**, 117), *Fierz-David et al.* (J. Soc. Dy. Col., 1935, **51**, 50) and *W. Zerweck, H. Ritter* and *M. Schubert* (Z. angew. Chem., 1948, **60a**, 141) has shown that the chromophores most commonly found in commercial products are ring structures of the thiazole (IV), thiazone (V), and thianthrene (VI) types.

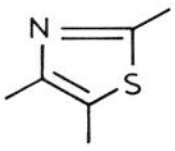

(IV) 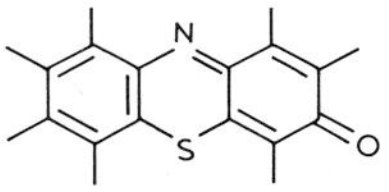   (V) 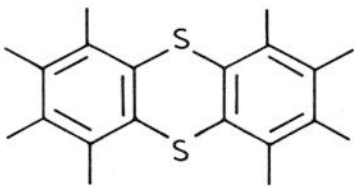   (VI)

A large proportion of the total sulphur is present either as polysulphide chains which may link aromatic nuclei or form part of a mercapto side chain or disulphoxide bridges. These side chains are essentially non-chromophoric but have an important role in the dyeing and fastness properties of the sulphur dye, for example, they may be reduced to water-soluble –SNa groups in the sodium sulphide vat and then form the final sulphides, *in situ*, upon oxidation.

Vidal formulated the production of his black dye in three stages, the molecule becoming progressively more complicated by the building up of thiazine rings from the *p*-phenylenediamine and 4-aminophenol employed. The first stage, involving diphenylamine formation, is followed by thionation to form (VII), (R = –OH or –NH$_2$) which results in a blue dye. After loss of ammonia the black dye is formed from the intermediate (VIII).

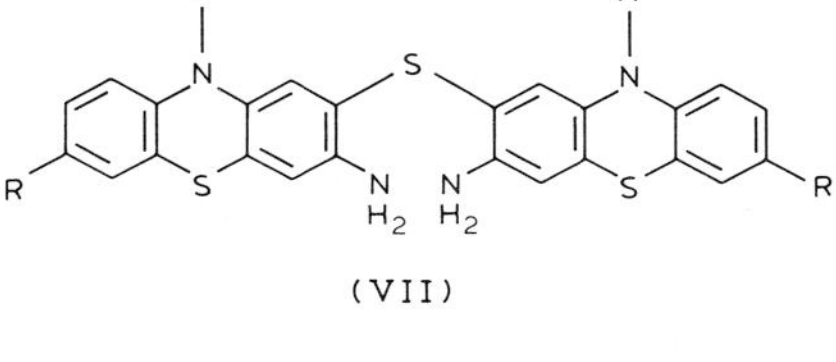

(VII)

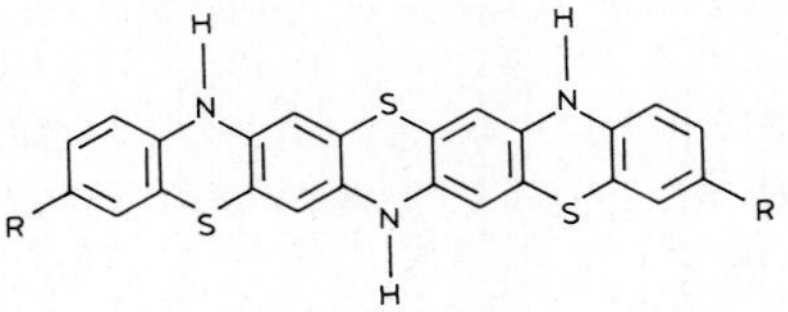

(VIII)

H.H. Hodgson (J. Soc. Dy. Col., 1924, **40**, 330; 1926, **42**, 76) examined the action of sulphur on aniline and showed the formation of disulphide links. Since these links would be expected to be relatively stable and the sulphur colours are, in fact, labile, *U. Perret* (Annali Chim. appl., 1926, **16**, 69) considers that polysulphide links must be present in the latter. In addition *T. Kubota* (J. chem. Soc. Japan, 1934, **55**, 569) in an examination of the reaction of 2,4,-diaminophenol with sodium polysulphide at 140° over varying periods of time found that the sulphur content of the dyestuff was reduced by prolonged treatment. He suggested the structure (IX) for the sulphur black intermediate: the polysulphide link S_x is said to be cleaved during vatting and not necessarily re-formed in the oxidation stage.

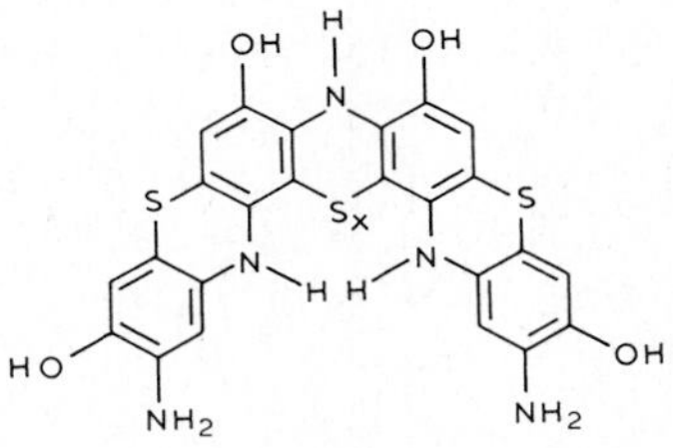

(IX)

W.N. Jones and *E.E. Reid* (J. Amer. chem. Soc., 1932, **54**, 4393) showed that extraneous sulphur, not contained in a ring system or as a thiophenol group, is removed by tin(II) chloride reduction and evolved as hydrogen sulphide. They argued for the presence of thiazone structures linked by polysulphide chains as in (X).

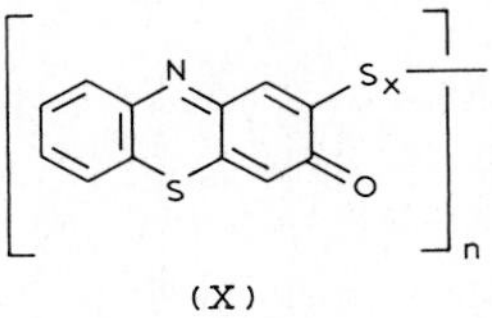

(X)

The presence of the thiazine ring in Immedial Pure Blue, which is usually prepared by reaction of 4-hydroxy-4'-dimethylaminodiphenylamine with sodium sulphide and sulphur at 115°, can be demonstrated by treating the dyestuff with potassium chromate and hydrogen bromide in a sealed tube. A tetrabromo derivative (XI) is formed which is identical with the product from the bromination of Methylene Violet (C.I. Basic Dye, 52041, *R. Gnehm* and *F. Kaufler*, Ber., 1904, **37**, 2617), which is itself prepared by the action of silver hydroxide on Methylene Blue.

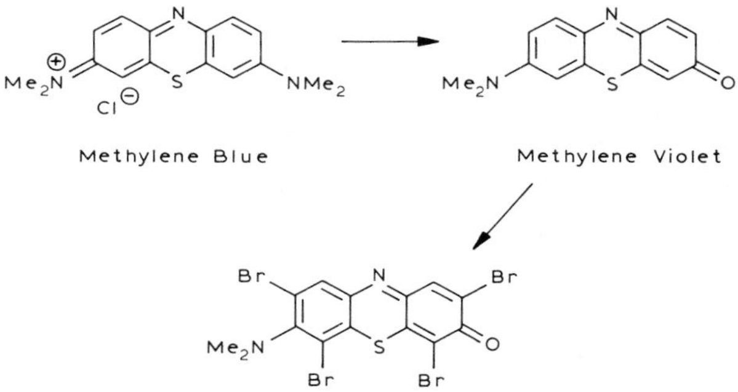

Methylene Blue → Methylene Violet

(XI)

H.E. Fierz-David et al. (J. Soc. Dy. Col., 1935, **51**, 50) have formulated Immedial Pure Blue as (XII), containing sulphoxide links.

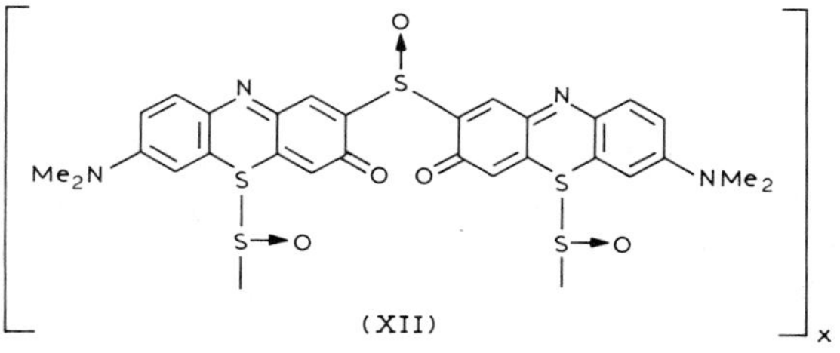

(XII)

The presence of the dimethylamino groups restricts the polynuclear structure of the molecule; when the intermediates used contain unsubstituted amino groups, thionation can lead to more condensed systems and black dyes.

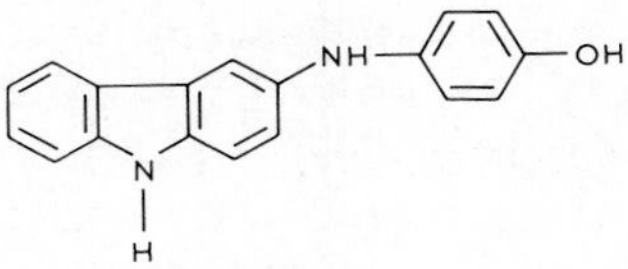

(XIII)

Another interesting example is that of Hydron Blue R (C.I. Vat Blue 43, 53630) which is prepared commercially by thionation of the hydroxyphenylamino carbazole (XIII). It may also be prepared from the trichloroindolophenothiazolone (XIV) in which chlorine atoms are successively replaced by sulphur atoms which then form links between two or more molecules. Substitution of all the halogens in this way gives a complex polymeric structure in which progressive oxidation of the sulphur to sulphoxide links may occur.

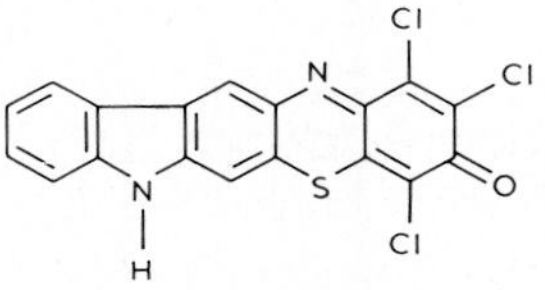

(XIV)

Introduction of blocking groups at any of the positions occupied by chlorine atoms in (XIV) substantially inhibits the satisfactory formation of a sulphur dye by thionation. The correctness of the proposed structure is demonstrated when Hydron Blue R is desulphurised with Raney nickel in morpholine whereupon the indolophenothiazone (XV) is isolated (*K.H. Shah, B.D. Tilak* and *K. Venkataraman*, J. Soc. Dy. Col., 1950, **66**, 333).

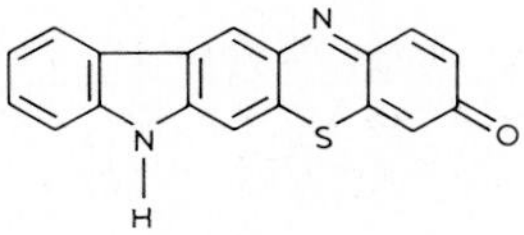

(XV)

Chapter 46

Quinazoline Alkaloids*

A. McKILLOP, M. SAINSBURY and B.P. SWANN

1. Simple quinazolines

A comprehensive review of the chemistry of the naturally occurring and biologically active quinazolines, with complete literature coverage through to 1965, has been published (see *W.L.F. Armarego*, "The Chemistry of Heterocyclic Compounds", ed. *A. Weissberger*; "Fused Pyrimidines", Part I, "Quinazolines", ed. *D.J. Brown*, Interscience Publishers, New York, 1967, pp. 490–518). More recent developments in this subject area are surveyed annually in the "Alkaloids", Specialist Periodical Reports, Royal Society of Chemistry and latterly in Natural Product Reports, Royal Society of Chemistry.

The very simple quinazolones, **glomerine**, $C_{10}H_{10}N_2O$, m.p. 200–210°, and **homoglomerine**, $C_{11}H_{12}N_2O$, m.p. 146–147°, have been isolated from the defensive secretions of the millipede *Glomaris marginata*, and identified, respectively, as the 2-methyl and the 2-ethyl derivatives of 1-methyl-4-quinazolone having structures (I) and (II) (*Y.C. Meinwald, J. Meinwald* and *T. Eisner*, Science, 1966, **154**, 390; Naturwiss., 1967, **54**, 1986; *K.H. Weiss* and *U. Maschwitz*, Z. Naturforsch., 1966, **21b**, 121).

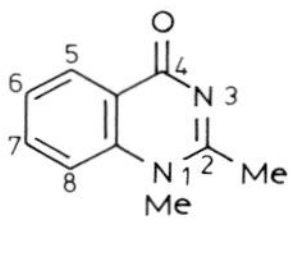

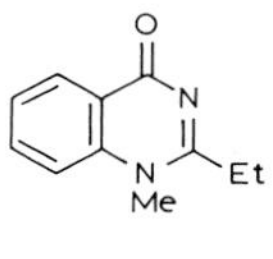

Several related alkaloids have been isolated from the Indian plant *Glycosmis arborea*. The most familiar of these is **arborine** (III), $C_{16}H_{14}N_2O$, m.p. 155–156°, *hydrochloride*, m.p. 215°; *picrate*, m.p. 172–173°. Alkaline

---

* This review covers the literature until end of December 1985.

hydrolysis of arborine gives *N*-methylanthranilic acid, phenylacetic acid and ammonia. Catalytic hydrogenation gives dihydroarborine, which yields phenylacetamide on hydrolysis. Consequently, structure (III) was proposed for arborine (*D. Chakravarti,* and *R.N. Chakravarti* and *S.C. Chakravarti,* J. chem. Soc., 1953, 3337; Experientia, 1953, **9**, 333). However, ozonolysis or oxidation of arborine with periodate gives benzaldehyde, which suggests that the compound can also exist in the tautomeric form (IV) (*A. Chatterjee*

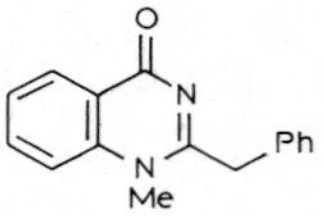

(III) 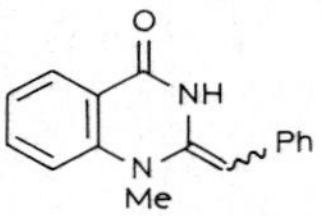   (IV)

and *S.G. Majumdar,* J. Amer. chem. Soc., 1953, **75**, 4365). A careful examination of spectral data showed such tautomerism to be possible, with form (III) being preferred in the solid state (*D.P. Chakravarti et al.,* Tetrahedron, 1961, **16**, 224).

The other minor alkaloids of *G. arborea* are **glycosmicine** (V), m.p. 270–271°, **glycorine** (VI); *hydrochloride,* m.p. 242° (dec.); *picrate,* m.p. 249° (dec.), **glycosminine** (VII), m.p. 249° (*S.C. Pakrashi et al., ibid.,* 1963, **19**, 1011), and **glycophymoline** (VIII), m.p. 165° (*M. Sarker* and *D.P. Chakraborty,* Phytochem., 1979, **18**, 694).

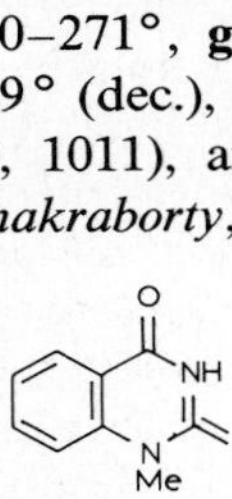

(V) 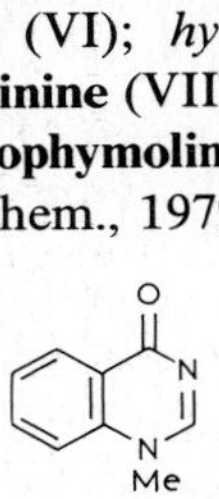   (VI) 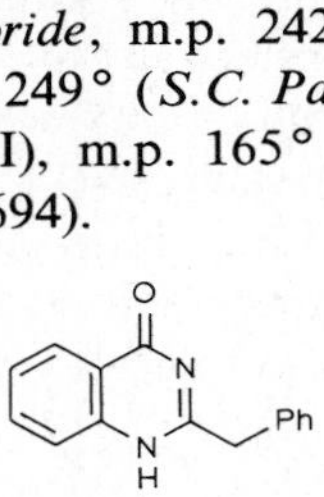   (VII) 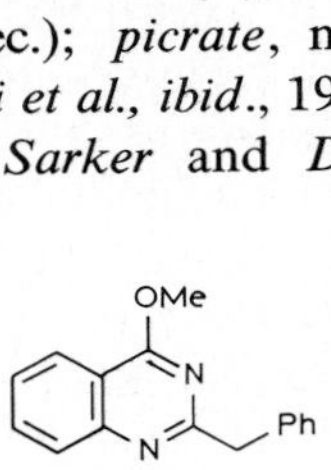   (VIII)

The structures of these compounds were deduced from their physical properties and confirmed by synthesis.

Sublimation of the hydrochlorides of glycorine and arborine above their m.ps. gives 4-quinazolone and glycosminine. Oxidation of arborine with chromic acid in acetic acid gives both glycosmicine and glycorine (*S.C. Pakrashi* and *J. Bhattacharyya, ibid.,* 1968, **24**, 1). The ultraviolet spectrum of each of the alkaloids exhibits characteristic maxima at λ 315 and 265 nm together with other bands which depend on the nature of the substituent at the nitrogen atoms (*Pakrashi, loc. cit.; Chakravarti et al., loc. cit.*).

The flower heads of *G. pentaphylla* (Retz.) D.C. [synonymous with *G. arborea* (Roxb.) D.C?] also yield glycosmicine and glycosminine, together with **glycophymine**, m.p. 254°, considered to be a tautomer of glycosminine

(*M. Sarkar* and *Chakraborty*, Phytochem., 1977, **16**, 2007). This suggestion is disputed by *Bhattacharyya* and *Pakrashi* (Heterocycles, 1979, **12**, 929) who have demonstrated that glycophymine and glycosminine are identical.

O-Methylation of glycosminine with dimethyl sulphate and sodium hydroxide affords glycophymoline (*Sarkar* and *Chakraborty, loc. cit.*). Interestingly, this plant also elaborates the amide (IX) and it has been shown that most of the alkaloids of this group can be synthesised from anilides of this type by heating them in xylene solution with ethyl carbamate and phosphorus pentoxide (*Bhattacharyya et al.*, Chem. Ind., 1978, 532).

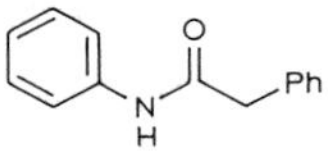

(IX)

E. Ziegler, W. Steiger and *Th. Kappe* (Monatsh. Chem., 1969, **100**, 949) showed that glomerine is produced when isatoic anhydride (X) is heated with thioacetamide. Similarly, if the *N*-methyl derivative of isatoic anhydride (XI) is heated with phenylthioacetamide arborine is obtained.

It is probable that imino ketones are intermediates in these reactions and some advantage is to be gained by substituting the isatoic anhydrides for sulphinamide anhydrides (*T. Kametani et al.*, J. chem. Soc. Perkin I, 1977, 2347; Heterocycles, 1977, **7**, 615; J. Amer. chem. Soc., 1977, **99**, 2306). Thus anthranilic acid when heated with thionyl chloride yields 3,2,1-benzoxathiazin-4(1*H*)-one *S*-oxide (XII) which reacts with the *O*-benzyllactimide (XIII, R^1 = CH(OBz)Me) to give 2-(1-benzyloxyethyl)quinazoline-4(3*H*)-one (XIV, R^2 = Bz). *O*-Debenzylation of this product gives (±)-**crysogine** (XIV, R^2 = H), m.p. 190–191°, the (−)-form of which has been isolated from the culture broth of *Penicillim chrysogenum* (*H. Hikino, M. Nabetani* and *T. Takemoto*, J. pharm. Soc., 1973, **93**, 619).

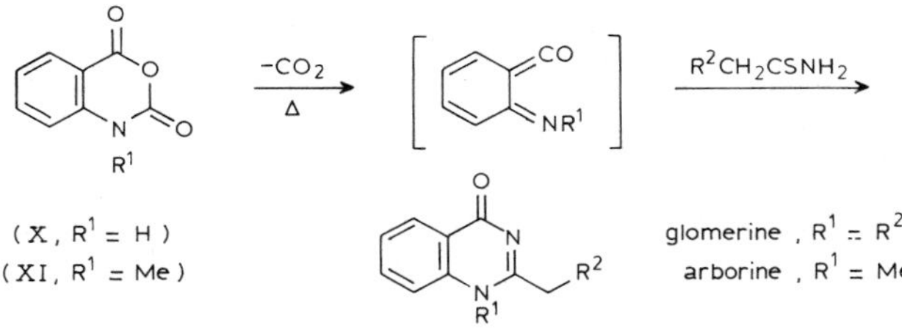

(X, R^1 = H)
(XI, R^1 = Me)

glomerine, R^1 = R^2 = H
arborine, R^1 = Me, R^2 = Ph

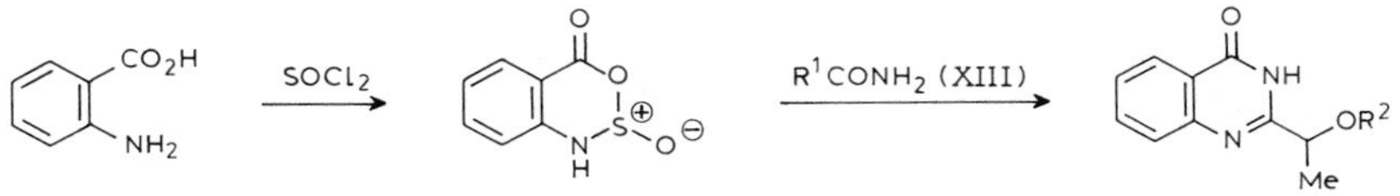

(XII)

(XIV)

Oxidation of the 2,3-dihydro-1-methyl-4(1*H*)-quinazolines (XV, R = H, Et or Bz) with mercury(II) ethylenediaminetetraacetate [Hg(II) EDTA] affords glycorine, homoglomerine and arborine, respectively. (*M. Möhrle* and *C.M. Seidel*, Arch. Pharm., 1976, **309**, 572).

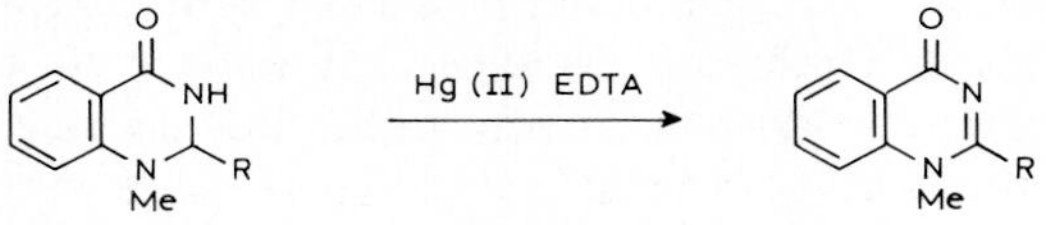

(XV)

Arborine and glycorine are present in *Glycosmis bilocularis* (*I.H. Bowen, K.P.W.C. Petera* and *J.R. Lewis*, Phytochem., 1978, **17**, 2125).

A series of compound structurally related to arborine have been prepared for pharmacological assessment, but none appear to have useful activity (*N.R. Naik, A.F. Amin* and *S.R. Patel*, J. Indian chem. Soc., 1979, **56**, 708).

2. Vasicine and related alkaloids

The Indian plant *Adhatoda vasica* Nees has long held an established position in folk medicine. The alkaloid **vasicine**, $C_{11}H_{12}N_2O$ was first isolated some 90 years ago and later work by *E. Späth* and collaborators (Ber., 1934, **67**, 45; 838) showed that a base, **peganine**, isolated from *Peganum harmala*, was identical with vasicine. The structure of the alkaloid was the subject of intensive investigations in the 1930s and these experiments have been fully summarised ("The Alkaloids", Vol. III, ed., *R.H.F. Manske*, 1953, Chapter 18). Vasicine was found to be an optically active tertiary base containing an alcoholic hydroxyl group. Oxidation with potassium permanganate in acidic solution gives 4-quinazolone (*T.P. Ghose*, J. Indian chem. Soc., 1927, **4**, 1; *Ghose et al.*, J. chem. Soc., 1932, 2740). Oxidation with alkaline permanganate gives an acid, $C_{10}H_8N_2O_3$, which yields anthranilic acid and glycine on alkaline hydrolysis, and on decarboxylation 3-methyl-4-quinazoline. Thus the acid is 4-quinazolone-3-acetic acid (I) (*Späth* and *E. Nikawitz*, Ber., 1934, **67**, 45) and the structure has been confirmed by the synthesis of the corresponding methyl ester (*Späth* and *F. Kuffner*, ibid., 1934, **67**, 1494).

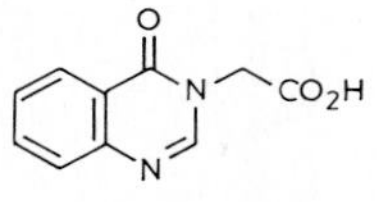

(I)

The reaction of the vasicine with phosphoryl chloride, followed by reduction with zinc and acid yields *deoxyvasicine*, $C_{11}H_{12}N_2$. This compound readily forms a benzylidene derivative, which led *Adams* to formulate deoxyvasicine as (II) since a similar reaction occurs with 2-alkylquinazolines (*W.E. Hanford* and *R. Adams*, J. Amer. chem. Soc., 1935, **57**, 921). This structure was soon confirmed by synthesis. 2-Nitrobenzyl chloride and methyl γ-aminobutyrate condense to form *N*-(2-nitrobenzyl)pyrrolidone which, on reduction and cyclisation, gives deoxyvasicine (*Späth et al.*, Ber., 1935, **68**, 497).

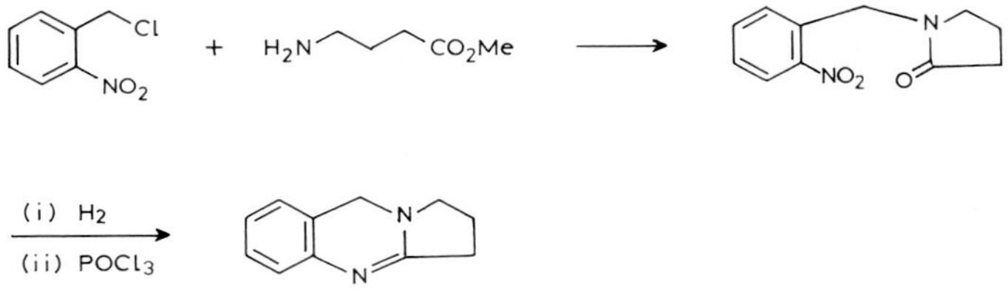

The position of the hydroxyl group in vasicine and the complete structure of the alkaloid follows from its synthesis (*idem, ibid.*, 1935, **68**, 699). Using a similar approach to that outlined above, 2-nitrobenzyl chloride was condensed with methyl γ-amino-α-hydroxybutyrate to give the corresponding pyrrolidone (III). Reduction of the nitro group of (III) with tin(II) chloride gave the amine which on distillation cyclised to (±)-vasicine. The base was resolved with tartaric acid.

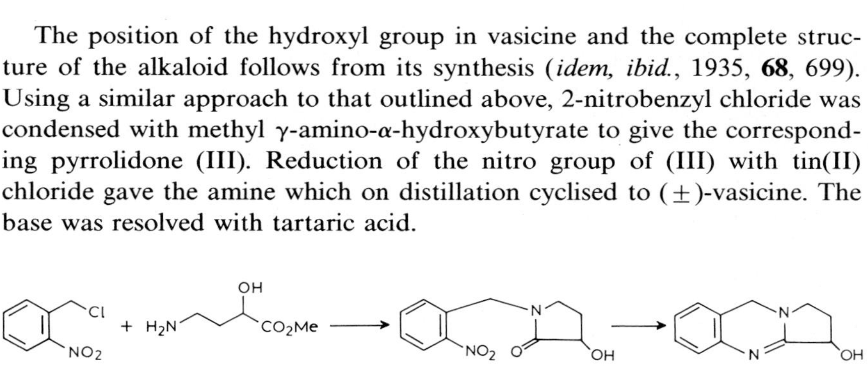

Other improved syntheses have been reported (*E. Späth* and *N. Platzer*, Ber., 1936, **69**, 255; *P.L. Southwick* and *J. Casanova Jr.*, J. Amer. chem. Soc., 1958, **80**, 1168). A synthesis starting from anthranilic acid (Scheme 1) yields vasicinone in 17% overall yield (*T. Onaka*, Tetrahedron Letters, 1971, 4387).

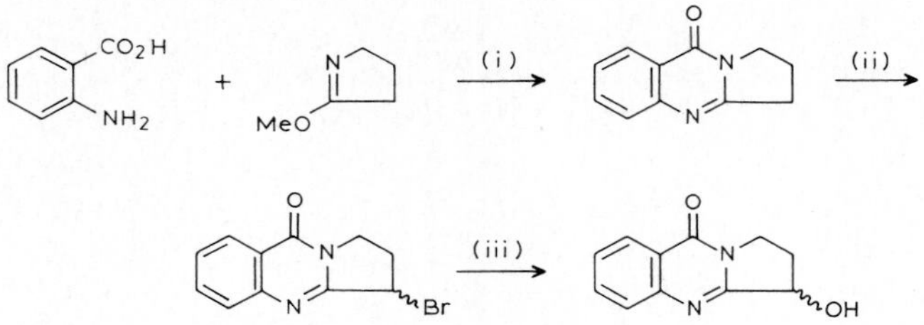

Reagents : (i) C$_6$H$_6$, reflux ; (ii) NBS/(PhCO)$_2$O$_2$/CCl$_4$; (iii) AcOH/AcONa/OH$^\ominus$

A modification of this route utilizes 3,2,1-benzoxathiazin-4(1*H*)-one *S*-oxide instead of anthranilic acid (*T. Kametani et al.*, J. Amer. chem. Soc., 1977, **99**, 2306). The benzoxathiazinone is probably a source of the iminoketene (IV), and this synthesis is then very similar to that used to prepare alkaloids of the simple quinazoline group (p. 439).

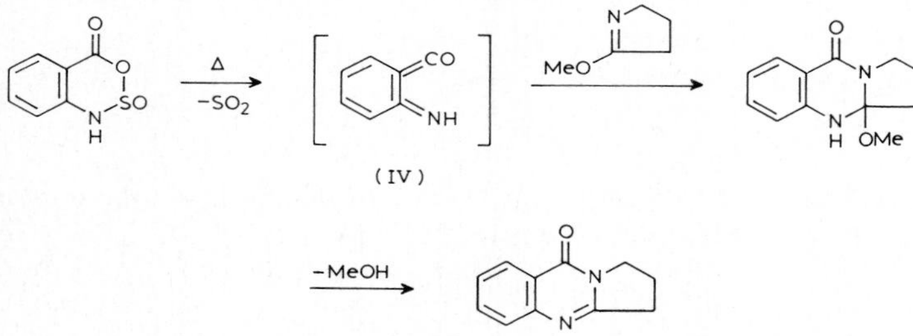

Condensation of anthranilic acid with 2-pyrrolidone, followed by reduction of the product with zinc and hydrochloric acid, gives desoxyvasicine (*Kh. Shakhidoyatov, A. Irishbaev* and *Ch.Sh. Kadyrov*, C.A., 1975, **82**, 86470).

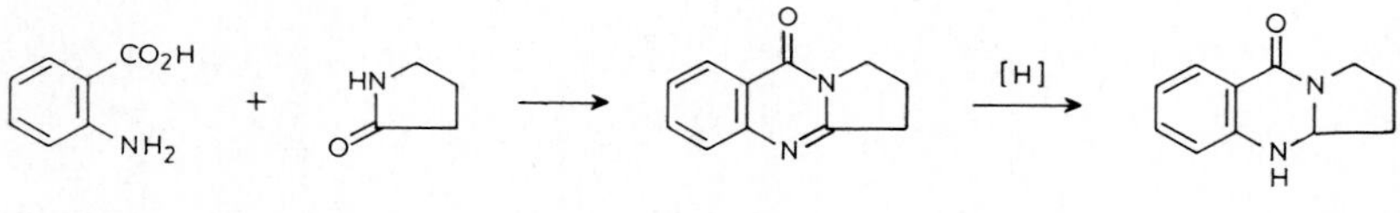

M. Möhrle and C.M. Seidel (Arch. Pharm., 1976, **309**, 503) have described another approach to the ring system. Thus the alkaloid vasicine may be prepared *via* the oxidative cyclisation of the amine (V) using mercury-(II) ethylenediaminetetraacetate [Hg(II) EDTA].

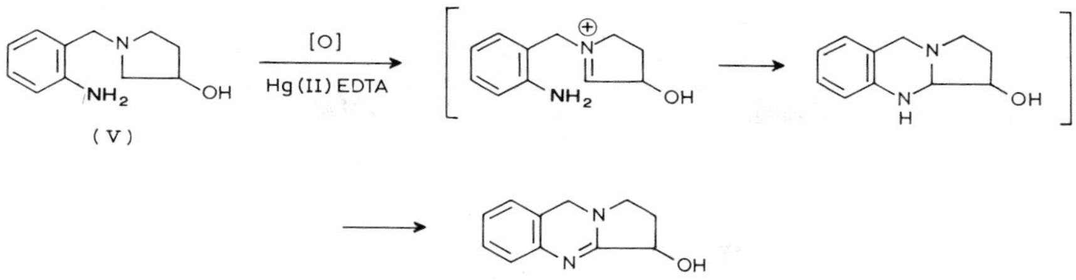

Some analogues of vascine have been made for biological testing (*R.L. Sharma et al.*, Indian J. Chem., 1979, **18B**, 449), but do not have advantageous pharmacological effects. Similarly some nitro derivatives have also been prepared (*M.P. Jain et al.*, Indian J. pharm. Soc., 1983, **45**, 178).

"Dimeric" deoxyvasicinone derivatives (VI) are obtained by the cyclocondensation of 6,6'-methylenedianthranilic acid with the appropriate lactams (VII) (*Shakhidoyatov* and *Kadyrov*, C.A., 1978, **88**, 630j), and thio-analogues of vascinone have been prepared by heating the alkaloid with phosphorus pentasulphide (*idem*, C.A., 1978, **88**, 105627e). The same authors have also condensed deoxyvasicinone with various aromatic aldehydes to give derivatives of the general structure (VIII) (C.A., 1978, **88**, 7166j; see also *M.N.A. Rao*, Indian J. pharm. Sci., 1982, **44**, 151).

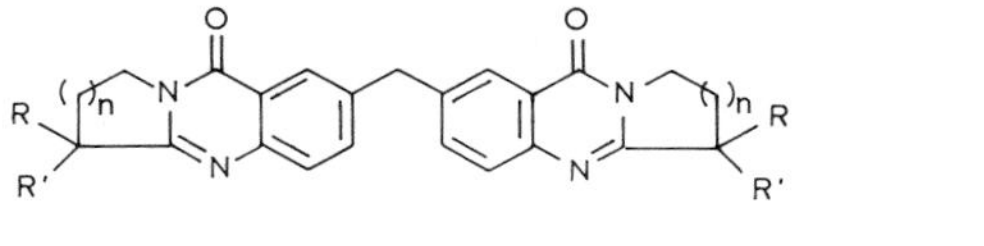

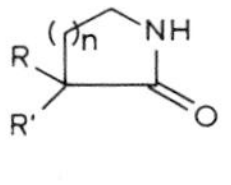

"Physiological" syntheses of deoxyvasicine (*C. Schöpf* and *F. Oechler*, Ann., 1936, **523**, 1) and vasicine (*N.J. Leonard* and *M.J. Martell Jr.*, Tetrahedron Letters, 1960, 2544) have also been described.

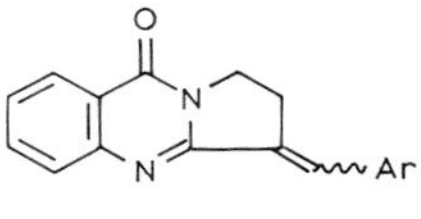

Both deoxyvasicine and vasicine readily undergo oxidation to the corresponding 4-quinazolones. Indeed, **vasicinone** has been claimed to be alkaloid in its own right (*D.R. Mehta, J.S. Naravone* and *R.M. Desai*, J. org.

Chem., 1963, **28**, 445). **Deoxyvasicinone** has also been isolated as a minor alkaloid from *Mackinlaya macrosciadea* (*N.K. Hart, S.R. Johns* and *J. Lamberton*, Aust. J. Chem., 1971, **24**, 223). The bronchial action of vasicine has been ascribed to its contamination by, or metabolism to, vasicinone (*G.W. Cambridge et al.*, Nature, 1962, **196**, 1217).

Vasicinone, vasicine and **vasicinol** (IX), m.p. 270–273° (dec.), $[\alpha]_D^{25}$ +2.5 (*c* 0.32, AcOH), have been isolated from *Sida cordifolia L.* (*S. Ghoshal et al.*, Phytochem., 1975, **14**, 830) and also from *S. acuta* Burm. and *S. rhombifolia L.* of Sri Lankan origin (*A.A.L. Gunnatilaka et al.*, Planta Med., 1980, **39**, 66). The same alkaloids are found in Indian plants of the same genus (*A. Prakash, R.K. Varma* and *S. Goshal, ibid.*, 1981, **43**, 384), and (±)-vasicine (peganine), vasicinone and deoxyvasicinone occur in *Peganum nigellastrum* (*D. Batsuren et al.*, C.A., 1978, **89** 126170). Five further alkaloids are claimed to occur in *Peganum harmala*. These are pegamine (X) and **peganidine** (XI) (*Kh.N. Khashimov et al.*, Khim. prir. Soedin., 1969, **5**, 599; 1970, **6**, 453) and **deoxypeganidine** (XII), **dipegine** (XIII) and **isopeganidine** (*B.Kh. Zharekeev, M.V. Telezhetskaya* and *S.Yu. Yunusov, ibid.*, 1973, **9**, 279; *Zharekeev, Khashimov* and *Telezhetskaya, ibid.*, 1974, **10**, 264). The structure of deoxypeganidine rests on spectroscopic evidence and the fact that upon oxidation it yields vasicinone. Dipegine is the first dimeric quinazoline alkaloid, and isopeganidine is described rather vaguely as "a racemic diastereoisomer of peganidine". *V.A. Snieckus* ("The Al-

TABLE 1

VASICINE ALKALOIDS

Alkaloid	$[\alpha]_D^0$	m.p. (°C)	Derivatives
Vasicine	−254 [a]	211–212	*Hydrochloride*, m.p. 205–207°
			Picrate, m.p. 190–191°
			Methiodide, m.p. 190°
Vasicinol	+2.5 [b]	270–273	
Vasicinone	−100 [a]	200–201	*Hydrochloride*, m.p. 232–234°
Vasicinolone	−	279	
Deoxyvasicine	−	96.5–97.5	*Picrate*, m.p. 205–206°
Hydroxyvasicine	−45.8 [b]	272–273	
Peganidine		189–190	*Semicarbazide*, m.p. 204–206°
Isopeganidine		169–170	*Picrate*, m.p. 177–178°
Deoxypeganidine		69–70	*Picrate*, m.p. 176°
Dipegine		221–223	
Nasicinolone		279	*Diacetyl, m.p.* 204°

(Solvents for optical rotation; a, chloroform; b, acetic acid.)

kaloids", Specialist Periodical Reports, The Chemical Society, Vol. 5, p. 108, 1975) has suggested, however, that both dipegine and isopeganidine are probably artefacts formed during the isolation procedure. There is reference in the Russian papers to yet another alkaloid **peganol**, which seemingly has not been fully characterized.

Vasicine, deoxyvasicinone are also present in the seeds of *P. harmala* (*A. Al-Shamma et al.*, J. nat. Prod., 1981, **44**, 745), and vasicinone occurs in *Nitraria sibirica* (*Z. Osmanov, A.A. Ibragimov* and *S. Yu Yunusov*, C.A., 1982, **97**, 20680c).

A further alkaloid isolated from *Adhatoda vasica* is **hydroxyvasicine** (XIV) (*E. Späth* and *F. Kesztler-Gandini*, Monatsh. Chem., 1960, **91**, 1150); its structure has been confirmed by synthesis (*F. Kuffner, G. Lennis* and *H. Bauer, ibid.*, 1960, **91**, 1152). The same alkaloid was also independently isolated by Indian workers (*A.K. Bhatnagar et al.*, Indian J. Chem., 1965, **3**, 524). **Nasicinolone** (XV) is also found in *A. vasica* and it is obtained synthetically by oxidation of vasicinol with hydrogen peroxide (*M.P. Jain* and *V.K. Sharma*, Planta Med., 1982, **46**, 250).

Physical data for some of these alkaloids are given in Table 1.

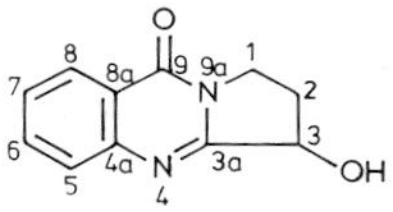

Vasicinone

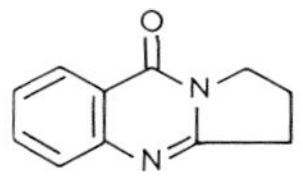

Deoxyvasicinone

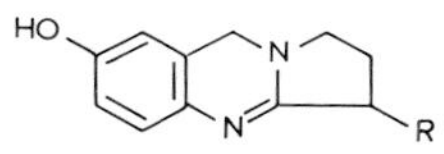

Vasicinol (IX, R = H)
Hydroxyvasicine (XIV, R = OH)

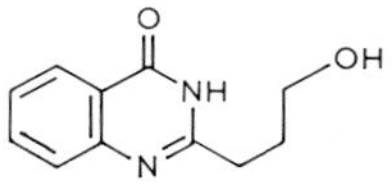

Pegamine (X)

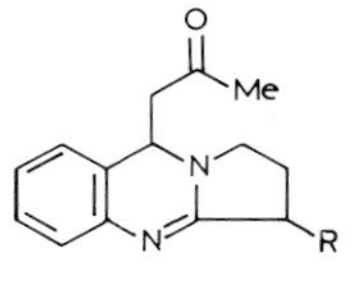

Peganidine (XI, R = OH)
Deoxypeganidine (XII, R = H)

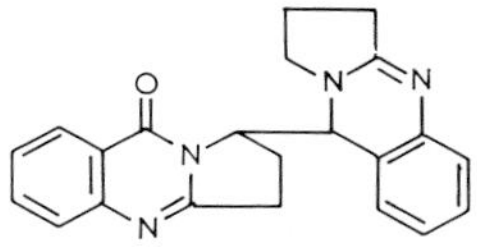

Dipegine (XIII)

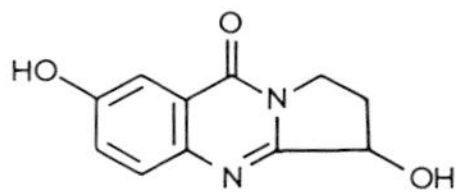

Nasicinolone (XV)

Vasicine is the subject of some interest because two distinct biosynthetic pathways to it have been established in higher plants. Both routes utilize anthranilic acid for one half of the alkaloid but differ as to which amino acid is used to construct the five-membered heterocycle. D.R. Liljergren (Phytochem., 1968, **7**, 1299; 1971, **10**, 2661), working with *Peganum harmala*, favours a sequence whereby ornithine is degraded to $^1\Delta$-pyrroline, which then combines with anthralinic acid:

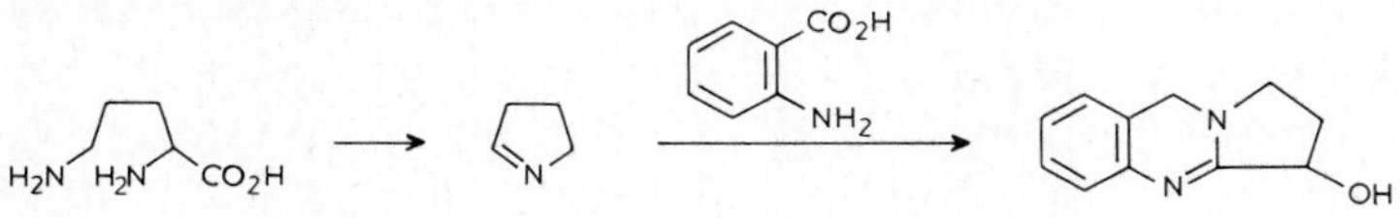

In *Adhatoda vasica*, however, aspartic acid is incorporated into vasicine, together with *N*-acetylanthranilic acid (*D. Gröger, Johne* and *K. Mothes*, Abhandl. Dent. Akad. Wiss. Berlin Klasse Chem. Geol. Biol., 1966, 581; Z. Pflanzenphysiol, 1969, **61**, 353).

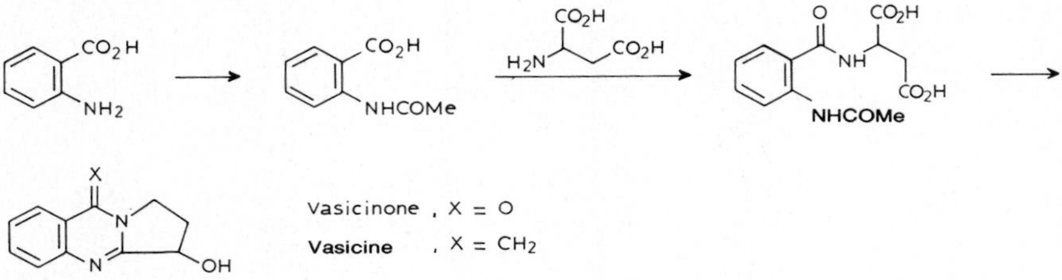

Vasicinone, X = O
Vasicine , X = CH_2

Interestingly a precursor-product relationship between vasicinone and vasicine could not be established (*Gröger* and *Mothes*, Arch. Pharm., 1960, **293**, 1049; *Gröger, Johne* and *Mothes*, Experientia, 1965, **21**, 13; *Johne, Gröger* and *G. Richter*, Arch. Pharm., 1968, **301**, 271).

The antibiotic trypthanthrin (XVI) is produced when the micro-organism *Candida lipolytica* is fed with *L*-tryptophan (*F. Schindler* and *H. Zahner*, Arch. Mikrobiol., 1971, **79**, 187). Its structure was deduced by *M. Brufani et al.* (Experienta, 1971, **27**, 1249) and the same alkaloid has been isolated from the higher plant *Strobilanthes cusia* O. Kuntze (*G. Honda* and *M. Tabata*, Planta Med., 1979, **36**, 85). This plant, cultivated primarily as a source of indigo dyestuff, has a long history in folk medicine as a cure for 'athlete's foot'.

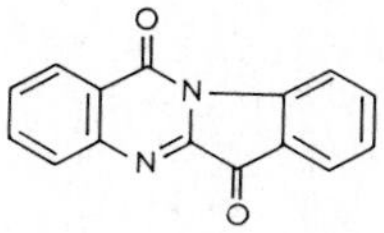

(XVI)

The seed husks of *Zanthoxylum arborescens* are a source of the quinazoline alkaloids (XVIII) and (XIX) which have m.ps. 100–102° and 133–134° respectively. The compound of structure (XVIII) is the more abundant and ^{13}C-n.m.r. spectroscopy showed it to contain a *N*-methyl rather than an *O*-methyl group, however, it was not known whether the phenylethyl substituent was attached to oxygen or to nitrogen. This argument was settled in favour of the latter formulation by a synthesis from *N*-methylisatoic anhydride and 2-phenylethylamine. In this the intermediate amide (XVII) was cyclised by treatment with methyl chloroformate (*D.L. Dreyer* and *R.C. Brenner*, Phytochem., 1980, **19**, 935).

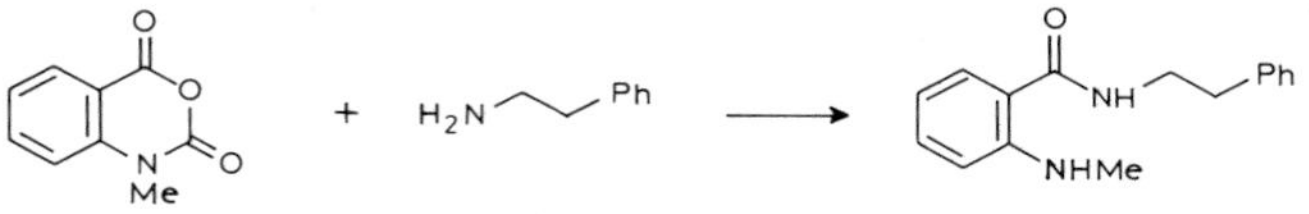

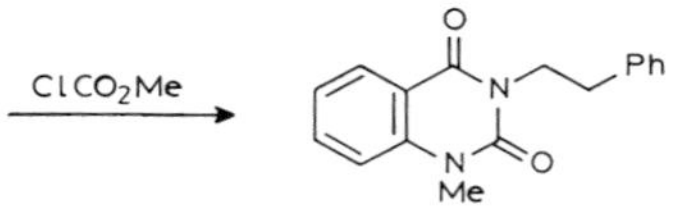

TABLE 2

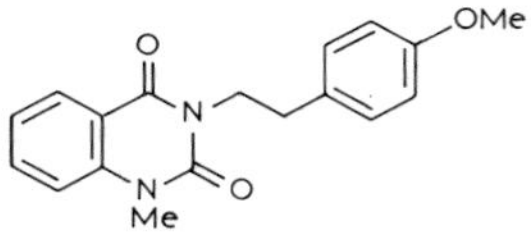

^{13}C-N.M.R. CHEMICAL SHIFTS DOWNFIELD FROM TMS FOR PEGANINE, VASICINONE AND DEOXYVASICINONE IN DEUTERIOCHLOROFORM SOLUTION

C Atom	Peganine	Vasicinone	Deoxyvasicinone
1	48.3	43.4	46.4
2	28.9	29.4	19.4
3	69.9	72.0	32.4
3a	163.9	160.3	159.4
4a	142.6	148.7	149.1
5	123.6	126.7	126.1
6	128.3	134.4	133.9
7	124.0	126.9	126.7
8	125.7	126.6	126.0
8a	118.9	120.4	120.4
9	47.1	160.6	160.6

TABLE 3

Name	Structure	R groups	m.p. (°C)	Ref.
Vasicoline	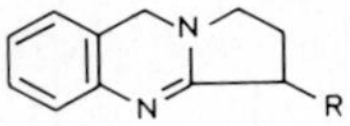	R 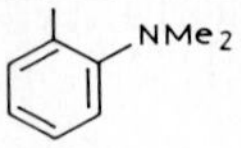 = 2-(NMe$_2$)-phenyl (o-tolyl with NMe$_2$)	135	1
Adhatodine		R 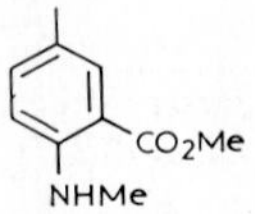 = 4-methyl-2-(NHMe)-phenyl with CO$_2$Me	183	1
Vasicolinone	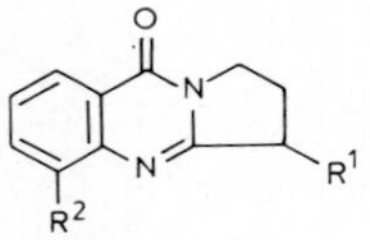	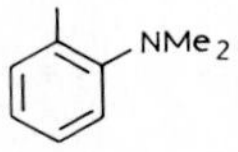 R^1 = 2-(NMe$_2$)-phenyl; R^2 = H	152	1
Deoxyaniflorine		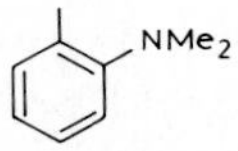 R^1 = 2-(NMe$_2$)-phenyl; R^2 = OMe	168–172	2
Anisotine		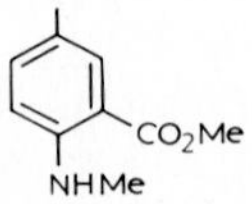 R^1 = 4-methyl-2-(NHMe)-phenyl with CO$_2$Me; R^2 = H	189–190	2
Anisessine		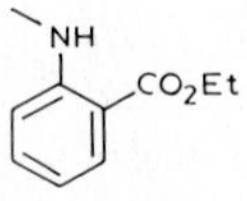 R^1 = 2-(NH-)-phenyl with CO$_2$Et; R^2 = H	170	2
Sessiflorine		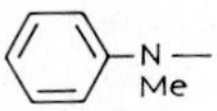 R^1 = phenyl-N(Me)–; R^2 = OMe	195–197	2,3
Aniflorine	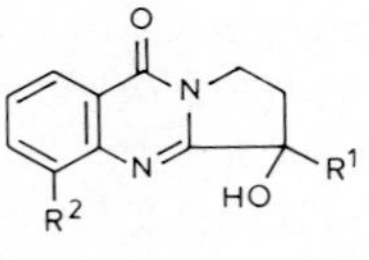	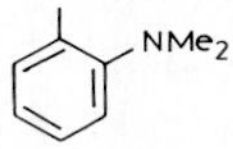 R^1 = 2-(NMe$_2$)-phenyl (with OH at ring carbon); R^2 = OMe	197	2

TABLE 3 (continued)

Unnamed (found in *A. vasica* only)	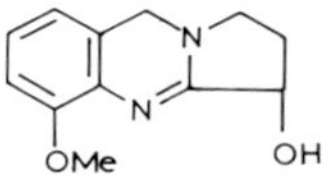 (structure: 8-methoxy-3-hydroxy-1,2,3,9-tetrahydropyrrolo[2,1-b]quinazoline)	224-225	4

1 S. *Johne*, D. *Gröger* and M. *Hesse*, Helv., 1971, **54**, 826.
2 R.R. *Arndt*, S.H. *Eggers* and A. *Jordaan*, Tetrahedron, 1967, **23**, 3521.
3 T. *Onaka* Tetrahedron Letters, 1971, 4387.
4 B.K. *Chowdhury* and P. *Bhattacharyya*, Phytochem., 1985, **24**, 3080.

Chemical shift data are of considerable value in the structural assignments of these alkaloids and S. *Johne et al.* (J. pr. Chem., 1977, **319**, 919) have analysed the ^{13}C-n.m.r. spectra of peganine, vasicinone, deoxyvasicinone and some of their derivatives. The chemical shift data for the alkaloids are summarised in Table 2.

A series of alkaloids based on the vasicine nucleus has been isolated from both *Adhatoda vasica* and *Anisotes sessiflorus*. Their structures, which were mainly deduced from ^{1}H-n.m.r. and mass spectrometry data, are summarised in Table 3.

3. Febrifugine and isofebrifugine

The powdered root of the plant *Dichroa febrifuga* has been used as a drug in China for a long time and has been found to possess antimalarial activity. Two alkaloids, **febrifugine** and **isofebrifugine** have been isolated from this source; both have the empirical formula $C_{16}H_{19}N_3O_3$. Difficulties in obtaining supplies of the raw material stimulated a search for alternative sources of these alkaloids, and the leaves of hydrangea were found to contain around 0.01% of febrifugine (*F. Ablondi et al.*, J. org. Chem., 1952, **17**, 14). The two bases are best separated by chromatography on alumina, followed by fractional crystallisation of their oxalates.

In the absence of spectroscopic techniques the structures of these alkaloids were elucidated by classical means but as the supply of the alkaloids was severely limited, many reactions were interpreted on the basis of comparison with model compounds (*B.R. Baker et al., ibid.*, 1952, **17**, 35;

52; 58). Febrifugine forms ketonic derivatives whereas isofebrifugine does not. Both bases absorb one equivalent of hydrogen to give different dihydro derivatives. Oxidation with permanganate gives 4-quinazolone-3-acetic acid (I) and 4-quinazolone (*T.Q. Chou et al.*, J. Amer. chem. Soc., 1948, **70**, 1765). Zinc dust distillation gives 3-acetonylquinazolone (II) (*J.H. Williams et al.*, J. org. Chem., 1952, **17**, 19). The presence of the quinazolone nucleus

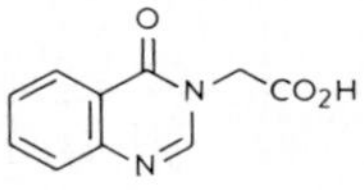

(I) 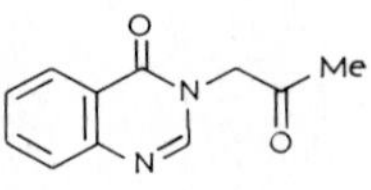   (II)

was confirmed both by u.v. studies and by the ready hydrolysis of the alkaloids to anthranilic acid. Febrifugine was also found to contain an alcoholic hydroxyl group and a secondary amino group. Formation of cyclic products on treatment with phenyl isothiocyanate suggested that this secondary amino function was β to the carbonyl group. The structure (III) for febrifugine was finally deduced by comparison of several synthetic compounds containing isomeric hydroxy-2-piperidyl or hydroxymethyl-2-pyrrolidyl groups attached to the side chain (*B.R. Baker et al., ibid.*, 1952, **17**, 68; 77; 97; 109; 116; 132) with the natural material. Synthetic ($\pm$)-3-(3-hydroxypiperidin-2-yl)-1-quinazol-4-on-3-yl-propan-2-one (III) had half the antimalarial activity of natural febrifugine. Several syntheses have subsequently been reported of optically active febrifugine (*idem, ibid.*, 1953, **18**, 178; *Baker* and *F.J. McEvoy, ibid.*, 1955, **20**, 136). The absolute configuration of natural febrifugine with respect to the piperidine ring was regarded as ($2'S$, $3'S$) and this stereochemistry was claimed to be necessary for antimalarial activity (*R.K. Hill* and *A.G. Edward*, Chem. Ind., 1962, 858).

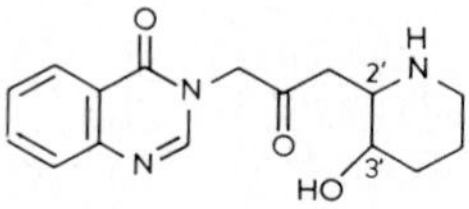

(III)

A reinvestigation of Baker's synthesis of febrifugine, however, showed that a key intermediate (IV) could readily isomerise to the *trans*-stereoisomer (V) at room temperature (*D.F. Barringer, Jr., et al.*, J. org. Chem., 1973, **38**, 1933). Further research into febrifugine itself revealed that a similar isomerisation could occur (*idem, ibid.*, 1973, **38**, 1937). Consequently, the stereochemistry at the 2'-position of febrifugine, and hence the absolute configuration based on this assignment, were both in error. Equilibration had occurred during the synthesis and although the final product was febrifugine, it did not relate stereochemically to the starting material. Thus, the correct absolute configuration of febrifugine is (2S, 3R) and its structure is as shown in formula (VI).

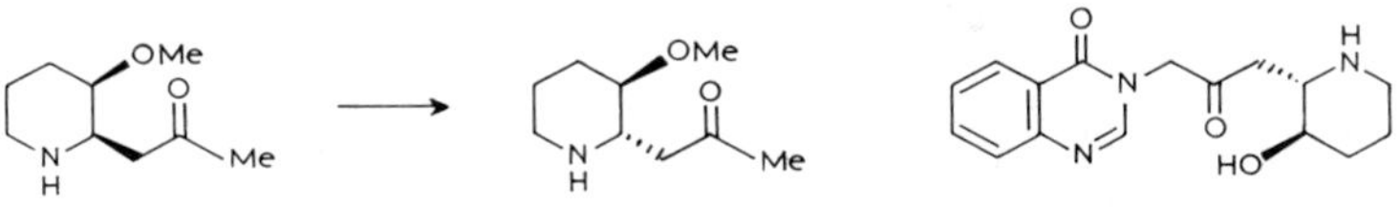

(IV)   →   (V)   (VI)

This stereochemistry correlates with the known absolute configuration of the hydroxyl group of δ-hydroxylysine and 5-hydroxypipecolic acid, both of which are plausible biosynthetic precursors (*Hill* and *Edwards, loc. cit.; B. Witkop*, Experientia, 1956, **12**, 372).

The structure of isofebrifugine was originally suggested to be a hemiacetal (*J.B. Koepfli, J.A. Brockman* and *J. Moffat*, J. Amer. chem. Soc., 1950, **72**, 3323), later confirmed by the absence of a carbonyl band in its i.r. spectrum (*Baker et al.*, J. org. Chem., 1953, **18**, 178). Thus isofebrifugine is one of the diastereoisomeric hemiacetals formed from febrifugine. Hydrogen bonding between the hydroxyl group and the carbonyl group of the quinazolone ring may account for the relative stability of isofebrifugine.

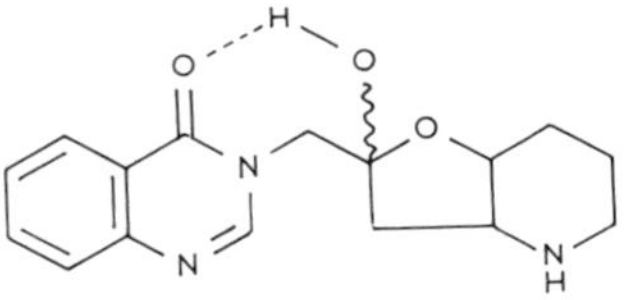

Isofebrifugine

Febrifugine has m.p. 139–140°, $[\alpha]_D$ +6° (chloroform); *hydrochloride* (B, HCl), m.p. 220°; *hydrochloride* (B, 2 HCl), m.p. 220–222° (dec.), *semicarbazone*, m.p. 187–188° (dec.). Isofebrifugine has m.p. 129–130°, $[\alpha]_D$ +131° (chloroform); *hydrochloride* (B, HCl) m.p. 210°; *oxalate*, m.p. 212–213° (dec.).

4. The indoloquinazoline alkaloids

These alkaloids form a small group related to both the β-carboline alkaloids and the more simple quinazoline alkaloids. They occur in the genera

Evodia, Hortia and *Euxylophora* (Rutaceae), and the dried fruits of *Evodia rutaecarpa* have been used in China as a drug for many years. From this source the two early members **evodiamine**, $C_{19}H_{17}N_3O$, m.p. 278°, $[\alpha]_D$ + 251° (acetone), and *rutaecarpine*, $C_{18}H_{13}N_3O$, m.p. 262°, $[\alpha]_D$ 0°, were isolated (*Y. Asahina* and *K. Kashiwaki*, J. pharm. Soc. Japan, 1915, **405**, 1293; *A.L. Chen* and *K.K. Chen*, J. Amer. pharm. Assoc., 1933, **22**, 716). Evodiamine is an optically active tertiary base which is decomposed by hot alkali to *N*-methylanthranilic acid and dihydronorharman. Hot aqueous acid hydrolysis gives the same acid and 3-β-aminoethylindole (*W.O. Kermack, W.H. Perkin* and *R. Robinson*, J. chem. Soc., 1921, **119**, 1602). This transformation proceeds *via* the intermediacy of "hydroxyevodiamine" (I) which itself has been reported to occur naturally (*A. Chatterjee, S. Bose,* and *A. Ghosh*, Tetrahedron, 1959, **7**, 257), although it may be an artefact (*M. Yamazaki* and *T. Kwana*, Yakagaku Zasshi, 1967, **87**, 608). Evodiamine is thus formulated as (II) and this structure has been confirmed by synthesis (*Y. Asahina* and *T. Ohta*, Ber., 1928, **61**, 319; 869).

(I)   (II)

The relationship between the two alkaloids was shown by pyrolysis of evodiamine hydrochloride, which resulted in decomposition to give methyl chloride and rutaecarpine (*idem*, J. pharm. Soc. Japan, 1926, **530**, 293). Alkaline hydrolysis of rutaecarpine yielded anthranilic acid and 3-β-aminoindole-2-carboxylic acid; hence rutaecarpine has structure (V). The alkaloid has been synthesised by several routes (*Asahina, Irie* and *Ohta, ibid.*, 1927, **543**, 51; *R.H.F. Manske* and *R. Robinson*, J. chem. Soc., 1927, 240; *Asahina, Manske* and *Robinson, ibid.*, 1708; *Asahina* and *Ohta*, J. chem. Soc. Japan, 1928, **48**, 313). A 'biological' synthesis was achieved in 70% yield by the reaction between an aqueous solution of 4,5-dihydro-3-carboline and 2-aminobenzaldehyde for 17 days at 25° and pH 6.9 (*C. Schöpf* and *H. Steur*, Ann., 1947, **558**, 124).

J. Bergman and *S. Bergman* (Heterocycles, 1981, **16**, 347; J. org. Chem., 1985, **50**, 1246) have synthesised rutaecarpine through the interaction of tryptamine and the 3,1-benzoxazinone (III), followed by cyclisation of the intermediate indole (IV) and elimination of trifluoromethane.

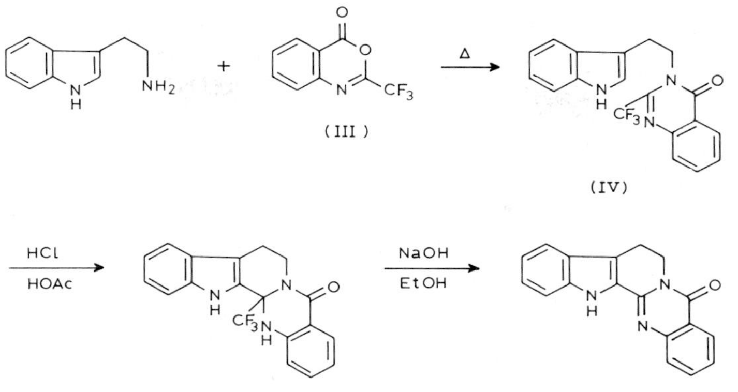

This is a modification of an earlier synthesis by *T. Kametani et al.* (J. Amer. chem. Soc., 1977, **99**, 2306) (see p. 439), but the overall yield is higher. For another synthesis of rutaecarpine, and also of vasicolinone (p. 448) see *C. Kaneko et al.* (Heterocycles, 1985, **23**, 1385).

Two more recently isolated members of this family, the alkaloids **hortiacine**, m.p. 215–252°, and **hortiamine**, m.p. 209°, contain a 10-methoxy substituent. The close similarity between hortiacine and rutaecarpine was soon established and it was shown that hortiacine is 10-methoxyruteacarpine (VI). The u.v. spectrum of hortiamine undergoes changes in protic solvents such as ethanol due to reversible hydration of the base to give adducts analogous to hydroxyevodiamine. Hydrogenation of hortiamine gives hortiacine and thus hortiamine is formulated as structure (VII) (*I.J. Pachter et al.*, J. Amer. chem. Soc., 1960, **82**, 5187; 1961, **83**, 635).

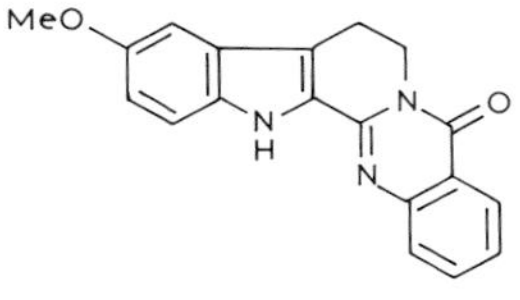

Another group of indoloquinazoline alkaloids has been isolated from the Brazilian tree *Euxylophora paraensis* Hub. (Rutaceae). They are all similar to the above alkaloids except that they contain an oxygenated quinazoline nucleus. These compounds are **euxylophorine A**, $C_{21}H_{19}N_3O_3$, m.p. 227–230° (VIII), **euxylophoricine A**, $C_{20}H_{17}N_3O_3$, m.p. 295–298° (IX),

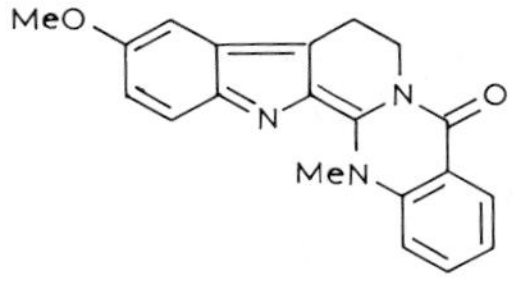

euxylophorine B, $C_{21}H_{17}N_3O_3$, m.p. 268–271° (dec.) (X), **euxylophoricine B**, $C_{20}H_{15}N_3O_3$, m.p. 310–312° (XI), **euxylophorine C**, $C_{20}H_{13}N_3O_3$, m.p. 207–209° (XII), **euxylophoricine C**, $C_{19}H_{11}N_3O_3$, m.p. 310–312° (XIII), **euxylophorine D**, $C_{22}H_{21}N_3O_4$, m.p. 256–260° C (XIV), **euxylophoricine D**, $C_{21}H_{17}N_3O_4$, m.p. 293–295° (XV), **euxylophoricine E**, $C_{22}H_{19}N_3O_4$, m.p. 290° (XVI), **euxylophoricine F**, $C_{19}H_{14}N_3O_3$, m.p. 226°, (XVII), **paraensine**, $C_{24}H_{21}N_3O_3$, m.p. 281–282° (XVIII) and 1-**hydroxyrutaecarpine**, m.p. 318–320° (XIX) (*L. Canonica et al.*, Tetrahedron Letters, 1968, 4865; *B. Danieli et al.*, Phytochem., 1972, **11**, 1833; Experientia, 1972, **28**, 249; Phytochem., 1973, **12**, 2523; 1974, **13**, 1603; 1976, **15**, 1095).

(VIII)

(IX)

(X)

(XI)

(XII)

(XIII)

(XIV)

(XV)

(XVI)

(XVII)

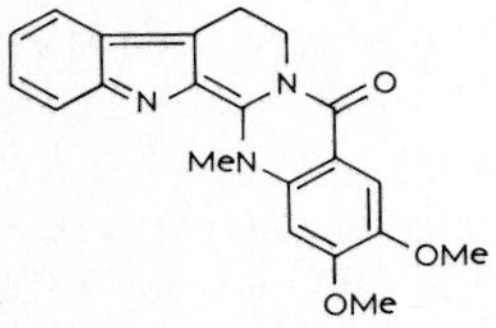

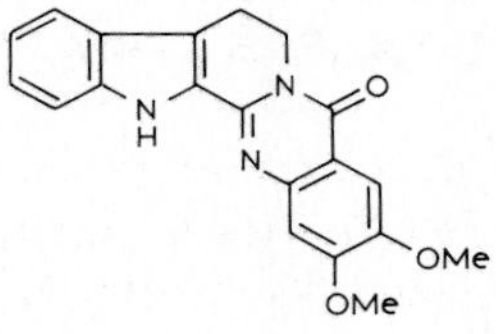

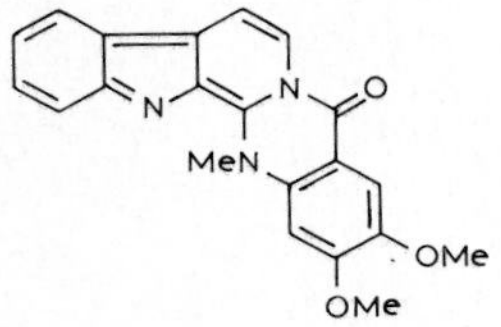

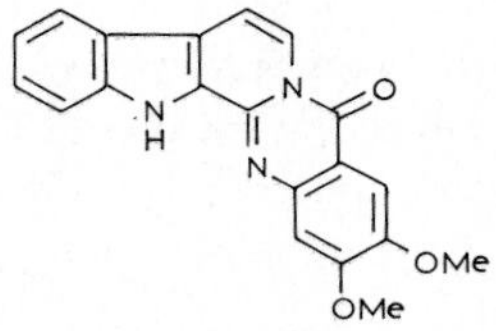

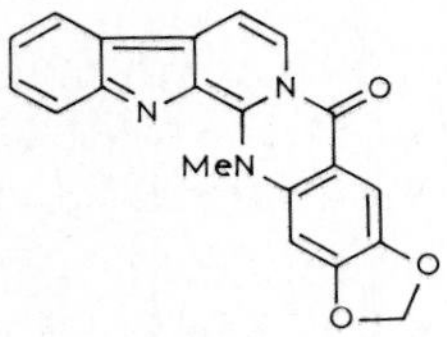

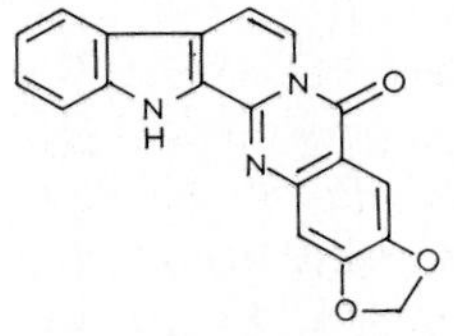

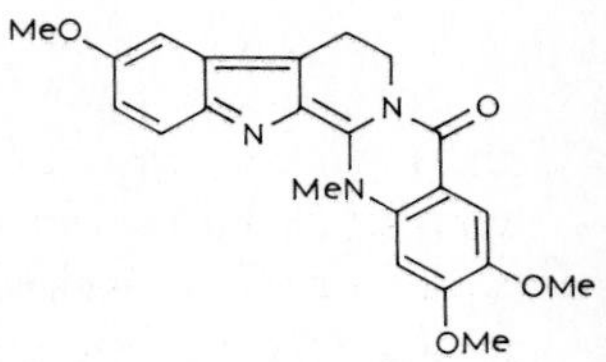

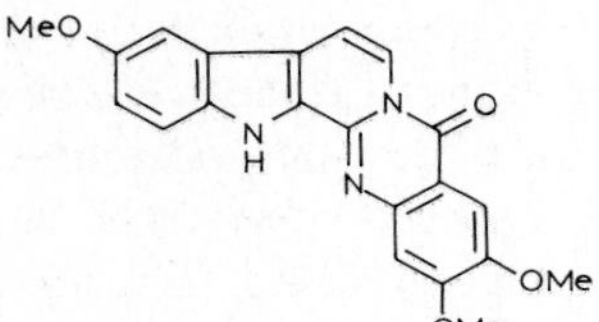

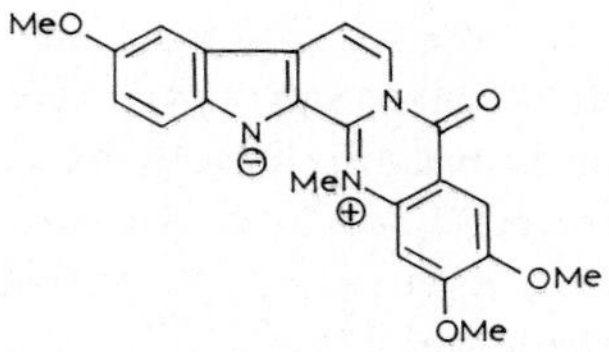

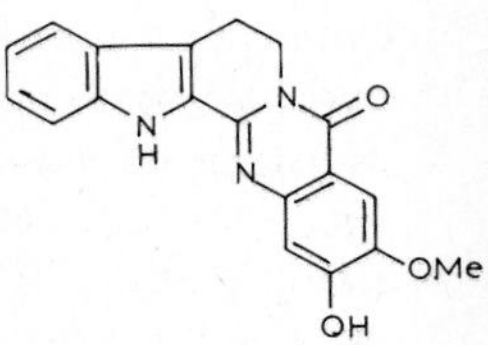

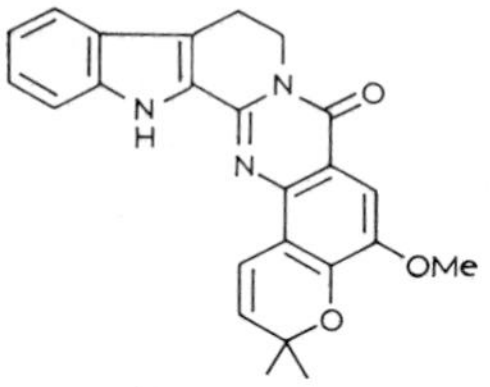

(XVIII) 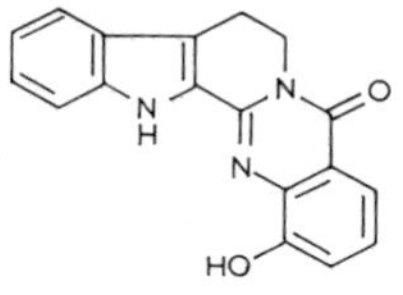   (XIX) 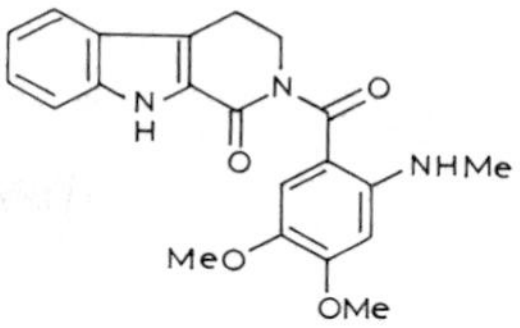   (XX)

Euxylophorine A (VIII) and its open-chain form (XX) exist in dynamic equilibrium in solution (*Danieli, G. Lesma* and *G. Palmisano*, Heterocycles, 1979, **12**, 353).

Evodia rutaecarpa is also a source of **dihydrorutaecarpine** (XXI, R = H) m.p. 214–216°, and its *N-formyl* derivative (XXI, R = CHO), m.p. 280–281° (*T. Kamikado* and *S. Murakoshi*, Agric. biol. Chem., 1978, **42**, 1515).

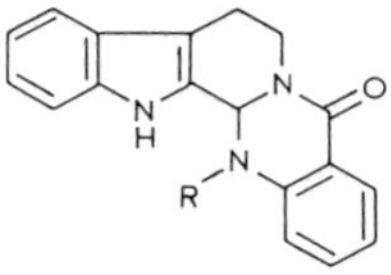

(XXI)

All these structures have been deduced by physical methods and from the results of hydrolysis experiments similar to those used to establish the identity of the earlier members of this series and a number of interconversions have been described, thus euxylophoricine B is formed by dehydrogenation of euxylophoricin A over selenium at 290°. Euxyloricine F may be degraded to 3-methoxyrutaecarpine by reaction with phenylisocyanate and triethylamine, followed by hydrogenation of the product over palladium on charcoal (*Danieli*, Phytochem., 1976, **15**, 1095). Euxylophorines A and B undergo similar pyrolytic rearrangements when heated at their melting points, thus euxylophorine B yields euxylophoricine B and its $N(13)$-methyl derivative (XXII) (*Danieli, Lesma* and *Palmisano*, Heterocycles, 1979, **12**, 1433). The methyl migration reactions appear to be intramolecular processes, although the mechanism is uncertain.

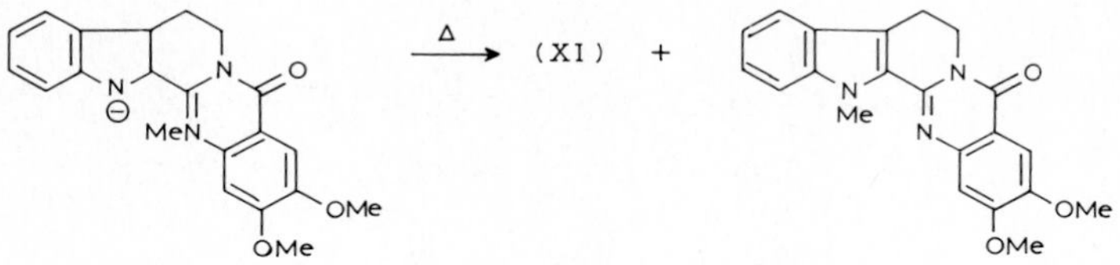

(XXII)

Although methyl iodide combines with rutaecarpine to give the $N(13)$-methyl derivative, other alkyl halides are much less reactive unless sodium carbonate is present in the reaction mixture (*Danieli* and *Palmisano*, Gazz., 1975, **105**, 45; J. heterocyclic Chem., 1977, **14**, 839).

Some of these alkaloids have been synthesised by condensation of tetrahydronorharmanone with the appropriate anthranilic acid, followed by cyclisation with phosphoryl chloride, and *Kametani* has extended his so-called "retro-mass" spectral synthesis (see p. 453) to the construction of rutaecarpine, evodiamine and the euxylophoricines A and C (J. Amer. chem. Soc., 1976. **98**, 618; 1977, **99**, 2306; J. chem. Soc. Perkin I, 1977, 2347; Heterocycles, 1977, **7**, 615; Chem. pharm. Bull., 1978, **26**, 1922).

Chapter 47

Compounds Containing a Six-Membered Ring with more than Two Hetero-Atoms

D.G. NEILSON and D. HUNTER

1. Triazines

The three possible classes of simple triazines are known. These are the 1,2,3-triazines (*v*-triazines). (I), the 1,2,4-triazines (*as*-triazines) (II) and the 1,3,5-triazines (*s*-triazines) (III).

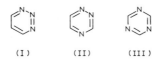

(a) *1,2,3-Triazines*

(i) Simple derivatives

Most work on simple 1,2,3-triazines has been undertaken within the last two decades (*R.J. Kobylecki* and *A. McKillop*, Adv. heterocyclic Chem., 1976, **19**, 215; *M.J. Hearn* and *F. Levy*, Org. Prep. Proced. Int., 1984, **16**, 201). By adaptation of a method first used to give fused systems, the parent compound (I), m.p. 69–71°, has been prepared *via* nickel peroxide oxidation of 1-aminopyrazole (*A. Ohsawa et al.*, Chem. Comm., 1981, 1174). The same workers (*ibid.*, 1980, 1182) had earlier prepared other members of this series, *e.g.* 4,5,6-*trimethyl*-1,2,3-*triazole*, m.p. 145° by oxidation of the correspondingly substituted 1-aminopyrazoles with lead tetraacetate—now the main reagent of choice (*H. Neunhoeffer et al.*, Ann., 1985, 1732). The first monocyclic compound to be isolated, 4,5,6-*triphenyl*-1,2,3-*triazine*, m.p. 276° (dec.), was however derived by thermolysis of 1,2,3-triphenyl-cyclopropenyl azide [(IV) → (V); R=Ph] and its structure proved by hydrolysis to 1,2,3-triphenylpropane-1,3-dione (*E.A. Chandross* and *G. Smolin-*

sky, Tetrahedron Letters, 1960, **13**, 19). This synthetic route, the most extensively used, has been adapted to give 4,5,6-tris(dialkylamino)-1,2,3-triazines (V, R=Me$_2$N, m.p. 93°) (*G. Seybold, U. Jersak* and *R. Gompper*, Angew. Chem. intern. Edn., 1973, **12**, 847) and 1,2,3-triazines with unsymmetrical substitution patterns, *e.g.* 4,6-*diphenyl*-5-(4-*tolyl*)-1,2,3-*triazine*, m.p. 222° (*H. Neunhoeffer, H.-D. Vötter* and *H. Ohl*, Ber., 1972, **105**, 3695), from precursors akin to (IV). In the case of compounds with different substituents, the bulkiest group is reported to occupy the 5-position (*R. Gompper* and *K. Schönafinger*, Ber., 1979, **112**, 1514).

Although 4,5,6-triphenyl-1,2,3-triazine yields 3,4,5-triphenylpyrazole on catalytic reduction (*E.A. Chandross* and *G. Smolinsky, loc. cit.*), 4,5,6-trimethyl-1,2,3-triazine with hydrogen over palladium gives its 2,5-*dihydro* derivative (VI), m.p. 54°, and, on oxidation with *m*-chloroperbenzoic acid, yields a mixture of the 1-*N-oxide*, m.p. 140°, and the 2-*N-oxide*, m.p. 112° (*H. Ohsawa et al.,* Chem. Comm., 1980, 1182). Lithium aluminium hydride also affords 2,5-dihydro reduction (*H. Neunhoeffer et al.*, Ann., 1985, 1732). 4,6-Dimethyl-1,2,3-triazine-2-oxide is reduced to the tetrahydro compound (VII) by sodium borohydride (*H. Arai et al.,* Heterocycles, 1982, 17) but is inert towards Grignard reagents (*A. Ohsawa, T. Kaihoh* and *H. Igeta*, Chem. Comm., 1985, 1370). The same workers report that Grignard reagents (*e.g.*, MeMgI) attack 1,2,3-triazine (I) at the 5-position yielding 5-methyl-2,5-dihydro-1,2,3-triazine along with crotonaldehyde formed by attack at the 4-position; 4,6-dimethyl-1,2,3-triazine undergoes attack at the 2-nitrogen along with reaction at the 5-position.

4,5,6-*Trichloro*-1,2,3-*triazine*, m.p. 111°, is attacked by nucleophiles preferentially at the 4- and then at the 6-position. 2-Methyl-1,2,3-triazinium iodides have been formed by the addition of methyl iodide to selected triazines (*R. Gompper* and *K. Schönafinger*, Ber., 1979, **112**, 1514; 1529; 1535).

Compound (V) (R=NMe$_2$) on pyrolysis affords 2,3,4-tris(dimethylamino)azacyclobutadiene (*G. Seybold, U. Jersak* and *R. Gompper*, Angew. Chem., Intern. Edn., 1973, **12**, 847), and 1,2,3-triazines acting as dienes in Diels-Alder reactions involving inverse electron demand yield substituted pyridines (*H. Neunhoeffer et al.,* Ann., 1985, 1732).

(ii) Fused 1,2,3-triazines

Although 4-(4-**methoxyphenyl**)-1,2,3-**benzotriazine**, m.p. 139–141°, formed by diazotisation of 2-amino-4′-methoxybenzophenone imine, was the first example of this class of compounds to be isolated, this preparative method has found little general application (*A.J. Nunn* and *K. Schofield*, J. chem. Soc., 1953, 716). **Benzo**-1,2,3-**triazine** (VIII, R=H), m.p. 119–120°, has been prepared by *C.W. Rees* and his co-workers (*B.M. Adger et al.*, J. chem. Soc. Perkin I, 1975, 31) by oxidation of 1-aminoindazole (IX, R=H) with lead tetraacetate in the presence of calcium oxide (2-aminoindazole reacts analogously). 4-*Phenyl*- (m.p. 156–157°), 4-*methoxy*- (m.p. 105–106°) and 4-*methyl*- (m.p. 120–121°) 1,2,3-*benzotriazine* have been prepared similarly from the respective precursors (IX, R=Ph, MeO, Me), the latter compound also arising by mercury(II) oxide oxidation of the azidohydrazone (X) or by lead tetraacetate oxidation of 2-aminoacetophenone hydrazone (XI) (*B.M. Adger et al., loc. cit.*).

(VIII)   (IX)   (X)   (XI)

1,2,3-Benzotriazine is susceptible to nucleophilic attack and is hydrolysed to 2-aminobenzaldehyde by dilute aqueous acid, but 4-substitution normally reduces reactivity.

1,2,3-**Benzotriazin**-4-**one** ("benzazimide"), m.p. 213° (dec.), and derivatives are well characterised. 1,2,3-Benzotriazin-4-one is prepared by diazotisation of either 2-aminobenzamide (*H. Finger*, J. pr. Chem., 1888, **37**, 431) or 2-aminobenzhydrazide with two molar equivalents of nitrous acid—one molar equivalent leading to the formation of 3-*amino*-3,4-*dihydro*-1,2,3-*benzotriazin*-4-*one*, m.p. 152–153° (dec.) (*G. Heller* and *A. Siller*, ibid., 1927, **116**, 9) which may also be prepared by the action of sulphamic acid on 1,2,3-benzotriazin-4-one (*J. Adamson et al.*, J. chem. Soc. C, 1971, 981). Benzotriazin-4-ones are soluble in aqueous base but are normally recovered on acidification. Hydrolysis in alkaline media or hot acid can give different products depending on the starting material but in many cases these are benzoic acid derivatives, *e.g.* anthranilic or salicyclic acids.

1,2,3-**Benzotriazine**-4-**thione** (XII), m.p. 187° (dec.), obtained from 2-aminothiobenzamide and nitrous acid (*A. Reissert* and *F. Grube*, Ber.,

1909, **42**, 3710) may be alkylated with alkyl halides in basic conditions to give 4-alkylthio derivatives, *e.g.* (XIII; R=Me, m.p. 100-101°; R=PhCH$_2$, m.p. 109-110°) (*B. Stanovnik* and *M. Tišler*, J. heterocyclic Chem., 1971, **8**, 785). Compounds (XII) and (XIII, R=Me) have acted as precursors for the formation of 4-*hydrazino*-1,2,3-*benzotriazine*, m.p. 188-189° (dec.) (*E.W. Parnell*, J. chem. Soc., 1961, 4930; *C. Grundman* and *H. Ulrich*, J. org. Chem., 1959, **24**, 272) which can be converted by nitrite in glacial acetic acid into 4-*azido*-1,2,3-*benzotriazine*, m.p. 118° (dec.) (*B. Stanovnik* and *M. Tišler, loc. cit.*).

(XII)   (XIII)   (XIV)   (XV)

4-**Amino**-1,2,3-**benzotriazine**, m.p. 266°, arises from the action of liquid ammonia in the presence of mercury(II) chloride on the thione (XII) and also from the action of nitrous acid on 2-aminobenzamidine dihydrochloride (*E.W. Parnell, loc. cit.*). Diazotisation of *N*-substituted 2-aminobenzamidines, *e.g. N*-2'-tolyl-2-aminobenzamidine, or cyclisation of 2-cyanophenyltriazenes (XIV) in boiling 70% ethanol leads to 4-imino-3,4-dihydro-1,2,3-benzotriazines (XV; *e.g.* R=*o*-tolyl, m.p. 100-101°) (*H.N.E. Stevens* and *M.F.G. Stevens*, J. chem. Soc. C, 1970, 765). The diazonium salt derived from 2-aminoacetophenone on treatment with an excess of methylamine and subsequently with neutral alumina yields 3-**methyl**-4-**methylene**-1,2,3-**benzotriazine**, m.p. 64-66°; the 3-phenyl compound prepared similarly is a red oil (*H. Fong* and *K. Vaughan*, Canad. J. Chem., 1975, **53**, 3714).

Whereas most 1,2,3-benzotriazines (VIII), when heated above 450°, break down into nitrogen, nitriles and benzynes, 4-aryl-1,2,3-benzotriazines yield 2-arylbenzazetes (as their dimers) on heating between 420 and 450° or on u.v. irradiation (*C.W. Rees, R.C. Storr* and *P.J. Whittle*, Chem. Comm., 1976, 411).

1*H*-Naphtho[1,8-de]triazine, (XVI, R=H) (decomposition point variously quoted between 230 and 260°) prepared by the diazotisation of 1,8-diaminonaphthalene (*H. Waldmann* and *S. Back*, Ann., 1940, **545**, 52) may be alkylated or arylated at the 1-position to give red compounds (XVI) and/or at the 2-position to give blue compounds (XVII) (*M.J. Perkins*, J. chem. Soc., 1964, 3005).

(XVI) (XVII)

(b) 1,2,4-Triazines

1,2,4-**Triazine** (II), m.p. 16–17°, b.p. 25–28°/0.5 mm, has been prepared directly from the interaction of glyoxal and formamidrazone (*H. Neunhoeffer* and *H. Henning*, Ber., 1968, **101**, 3952) and also by oxidation of 3-hydrazino-1,2,4-triazine with manganese dioxide (*W.W. Paudler* and *T.-K. Chen*, J. heterocyclic Chem., 1970, **7**, 767). *D.K. Krass* and *W.W. Paudler* (Synthesis, 1974, 351) have shown that better yields are obtained by decarboxylation of 3-carboethoxy-1,2,4-triazine, prepared from anhydrous glyoxal and ethyl oxalamidrazonate. Compound (XVIII) and its lower alkyl homologues are best stored below 0° in the dark. An X-ray structure determination of 5-(4-chlorophenyl)-1,2,4-triazine (*J.L. Atwood, D.K. Krass* and *W.W. Paudler*, J. heterocyclic Chem., 1974, **11**, 743) supported theoretical calculations (*M.J.S. Dewar* and *G.J. Gleicher*, J. chem. Phys., 1966, **44**, 759) allocating a higher contribution to the ground state of compound (XVIII) from the Kekule form (XVIIIa) than from structure (XVIIIb). For the π-deficient compound (XVIII), H–D exchange in basic or neutral media readily leads to the 3-deuterio derivative (*W.W. Paudler, J. Lee* and *T.-K. Chen*, Tetrahedron, 1973, **29**, 2495). 1,2,4-Triazines are relatively stable to acid but the lower aliphatic members undergo base hydrolysis. Alkyl and aryl monosubstituted derivatives are readily available from the condensation of the appropriate 1,2-dicarbonyl compound and an amidrazone. If a monosubstituted glyoxal is used, product (XIX) is the predominant isomer. 3-Trichloromethyl-5,6-diphenyl-1,2,4-triazine, prepared from the 3-methyl precursor by chlorination forms ylides (*S. Konno et al.*, Heterocycles, 1982, **19**, 1869).

3-*Methyl*-, 3-*phenyl*-, and 3,5,6-*triphenyl*-1,2,4-*triazines* have b.p. 64°/4 mm, m.p. 53° and m.p. 146°, respectively (*H. Neunhoeffer et al.*, Tetrahedron Letters, 1969, 3147). 5-*Methyl*- and 6-*methyl*-1,2,4-*triazines* boil at 90°/16 mm and 75°/15 mm respectively, and the 5-*phenyl*-, 6-*phenyl*-, and 5,6-*diphenyl* derivatives have m.p. 103°, 86°, and 117°, respectively (*H. Neunhoeffer* and *H. Henning, loc. cit.*).

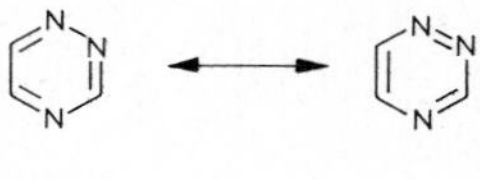

(XVIIIa)   (XVIIIb) 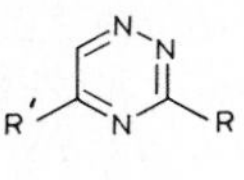   (XIX)

(i) Amino-1,2,4-triazines

Aminoguanidine condenses with dicarbonyl compounds, *e.g.* glyoxal and diacetyl yielding 3-**amino**-5,6-**dimethyl**- and 3-**amino**-1,2,4-**triazine**, m.p. 172° and 211–212° respectively (*J.G. Erickson*, J. Amer. chem. Soc., 1952, **74**, 4706). The 3-amino derivatives may also be prepared by nucleophilic displacement of halogen with ammonia or amines (*T. Sasaki et al.*, Tetrahedron, 1969, **25**, 1021; *A. Rykowski* and *H.C. van der Plas*, J. org. Chem., 1980, **45**, 881). 3-Amino-1,2,4-triazines can be acylated and react with nitrous acid to give the corresponding 1,2,4-triazin-3-ones (*T. Sasaki* and *K. Minamoto*, Chem. pharm. Bull. Tokyo, 1965, **13**, 1168). The 5-chloro atom of 3,5,6-trichloro-1,2,4-triazine can be replaced by reaction with ammonia or amines; subsequent hydrogenation of 5-amino-3,6-dichloro-1,2,4-triazine yields 5-**amino**-1,2,4-**triazine**, m.p. 232° (*A. Piskala, J. Gut* and *F. Sorm*, Chem. Ind., 1964, 1752). 6-Amino derivatives are not common; 6-*amino*-3,5-*diphenyl*-1,2,4-*triazine*, m.p. 243° (*H.G.O. Becker et al.*, J. pr. Chem., 1970, **312**, 669).

(ii) 1,2,4-Triazinecarboxylic acids

1,2,4-Triazine-3-carboxylates, *methyl ester*, m.p. 91–92° (*W.W. Paudler* and *R.E. Herbener*, J. heterocyclic Chem., 1967, **4**, 224); *ethyl ester*, m.p. 72–73° (*D.K. Krass* and *W.W. Paudler*, Synthesis, 1974, **6**, 351), are readily decarboxylated and act as electron-deficient dienes for Diels-Alder reactions.

Triethyl 1,2,4-*triazine*-3,5,6-*tricarboxylate*, b.p. 168–169°/1 mm, yields on saponification the tricarboxylic acid which is readily converted into the *anhydride* (XX), m.p. 175–183° (*R. Rätz* and *H.Schroeder*, J. org. Chem., 1958, **23**, 1931).

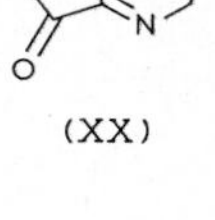

(XX)

(iii) Halogen-substituted 1,2,4-triazines

3-**Chloro**-5-**phenyl**-1,2,4-**triazine**, m.p. 121°, is obtained by the action of phosphorus oxychloride on 5-phenyl-1,2,4-triazin-3-one; its 3-*fluoro*-5-*phenyl* analogue has m.p. 93° (*A. Rykowski* and *H.C. van der Plas*, J. org. Chem., *loc. cit.*). 3,5,6-**Trichloro**-1,2,4-**triazine**, m.p. 56–58°, prepared by the action of phosphorus pentachloride on 6-bromo-1,2,4-triazine-3,5-dione in the presence of *N,N*-dimethylaniline (*B.A. Loving et al.*, J. heterocyclic Chem., 1971, **8**, 1095) is converted into the 3,5,6-trifluoro compound by vapour phase interaction with potassium fluoride (*M.G. Barlow et al.*, J. chem. Soc. Perkin I, 1980, 2254; 1982, 1245; 1251). The sequence of reaction of the chloro atoms is 5 > 6 > 3 for anionic nucleophiles and 5 > 3 ≈ 6 for neutral nucleophiles (*A Piskala* and *F. Sorm*, Coll. Czech. Chem. Comm., 1976, **41**, 465).

(iv) Hydroxy-1,2,4-triazines (triazinones)

1,2,4-Triazin-3-ones (3-hydroxy-1,2,4-triazines) arise from the condensation of a 1,2-dicarbonyl compound and a semicarbazide in the presence of base or acetic acid; both in solution and in the solid state the tautomer (XXI) predominates over (XXII, R=H). The use of appropriately substituted semicarbazides gives *N*-2- or *N*-4-substituted 1,2,4-triazin-3-ones.

1,2,4-*Triazin*-3-*one* as its potassium salt, m.p. 262–265° (*W.W. Paudler* and *J. Lee*, J. org. Chem., 1971, **36**, 3921); 5,6-*dimethyl*-1,2,4-*triazin*-3-*one*, m.p. 222–223° (*W. Seibert*, Ber., 1947, **80**, 494); 6-*phenyl*-1,2,4-*triazin*-3-*one*, m.p. 224–225° (*I. Lalezari et al.*, J. heterocyclic Chem., 1969, **6**, 403); 5-*phenyl*-1,2,4-*triazin*-3-*one*, m.p. 234° (*L. Wolff*, Ann., 1902, **325**, 129); 5,6-*diphenyl*-1,2,4-*triazin*-3-*one*, m.p. 225–226° (*H. Biltz* and *T. Arnd*, Ber., 1902, **35**, 344); 2,5,6-*trimethyl*-1,2,4-*triazin*-3-*one*, m.p. 86° (*W.W. Paudler* and *J. Lee, loc. cit.*). These compounds are, in general, stable to acids and bases and both *N*- and *O*-alkylation are feasible (*J. Daunis* and *C. Pigiere*, Bull. Soc. chim. Fr., 1973, 2818); 3-*methoxy*-1,2,4-*triazine*, (XXII, R=Me) m.p. 44–46° (*W.W. Paudler* and *T.-K. Chen, loc. cit.*).

(XXI)   (XXII)   (XXIII)

1,2,4-Triazin-5-ones (5-hydroxy-1,2,4-triazines) are synthesised from the reaction of amidrazones and α-ketocarboxylic acids. These compounds are base soluble and the main tautomeric form in solution is represented by formula (XXIII).

1,2,4-Triazin-5(2H)-one, its 2-*methyl* and 2,3,6-*trimethyl* derivatives have m.p. 196°, 125°, and 82°, respectively, the 4,6-*dimethyl*- and 3,4,6-*trimethyl*-1,2,4-*triazin*-5(4H)-*ones* melt at 100° and 139° respectively. (*J. Lee* and *W.W. Paudler*, J. heterocyclic Chem., 1972, **9**, 995). 5-*Methoxy*-3,6-*diphenyl*-1,2,4-*triazine* has m.p. 114° (*J. Daunis, R. Jacquier* and *G. Pigiere*, Tetrahedron, 1974, **30**, 3171).

5-**Methyl**- and 5-**phenyl**-4,5-**dihydro**-1,2,4-**triazin**-6-**ones**, m.p. 147–150° and 131° respectively, prepared by the interaction of an α-amino acid, dimethylformamide dimethylacetal and hydrazine, are readily oxidised (potassium permanganate in acetone/acetic acid) to 5-*methyl*-1,2,4-*triazin*-6-*one*, m.p. 110°, and its 5-*phenyl* analogue, m.p. 165° (*E.C. Taylor* and *J.E. Macor*, J. heterocyclic Chem., 1985, **22**, 409).

1,2,4-Triazine-3,5-diones arise from the condensation of α-ketocarboxylic acids and semicarbazide in basic conditions or by transformation of 3-thioxo-1,2,4-triazin-5-ones.

1,2,4-*Triazine*-3,5-*dione*, various quoted m.ps. 268–283° (*e.g. A. Novacek* and *J. Gut*, Coll. Czech. Chem. Comm., 1974, **37**, 3760); 6-*benzyl*- and 6-*phenyl*-1,2,4-*triazine*-3,5-*dione*, m.p. 208° and 262° respectively (*J. Bougault*, Compt. rend., 1914, **159**, 83); 2,4,6-*trimethyl*-1,2,4-*triazine*-3,5-*dione*, m.p. 101–103° (*R.H. Hall*, J. Amer. chem. Soc., 1958, **80**, 1145).

(v) 1,2,4-Triazine-N-oxides

3,5,6-Triphenyl-1,2,4-triazine on oxidation with peracetic acid yields both the 1-*oxide*, m.p. 207°, and the 2-*oxide*, m.p. 194° (*C.M. Atkinson et al.*, J. chem. Soc., 1964, 4209; *W.W. Paudler et al.*, J. org. Chem., 1977, **42**, 546; 1978, **43**, 2514). 1,2,4-Triazines unsubstituted in the 5-position tend to be oxidised to 1,2,4-triazin-5-ones rather than to N-oxides. 1,2,4-Triazine-4-oxides, *e.g.* 2,5,6-*trimethyl*-1,2,4-*triazine*-4-*oxide*, m.p. 93–94°, are prepared by the action of *ortho*-esters or imidates on α-oximinohydrazones (*V. Böhnisch, G. Burzer* and *H. Neunhoeffer*, Ann., 1977, 1713).

(vi) Triazinethiones

Phosphorus pentasulphide converts 1,2,4-triazin-3- or 5-ones and 3,5-diones into the related thiones or dithiones; 6-*amino*-3-*oxo*-1,2,4-*triazine*-5-*thione*, (XXIV) m.p. 242–244°, is isolated as an intermediate in the formation of 6-amino-1,2,4-triazine-3,5-dithione (*C.-C. Tzeng, N.C. Motola* and *R.P. Panzica*, J. org. Chem., 1983, **48**, 1271). Alkylation of thiones takes place almost exclusively at sulphur (*W.E. Taft* and *R.G. Shepherd*, J. med. Chem., 1967, **10**, 883); 5-*phenyl*-1,2,4-*triazine*-3-*thione*, m.p. 200° (*L. Wolff* and *H. Lindenhayn*, Ber., 1903, **36**, 4126); 3,6-*dimethyl*-1,2,4-*triazine*-5-*thione*, m.p. 228° (*J. Lee* and *W.W. Paudler*, J.

(XXIV)

heterocyclic Chem., 1972, **9**, 995); 6-*methyl*-3-*methylthio*-1,2,4-*triazine*, m.p. 43–44° (*J. Daunis* and *C. Pigiere*, Bull. Soc. chim. Fr., 1973, 2493).

(vii) Reduced 1,2,4-triazines

4,5-**Dihydro**-3-**methyl**-6-**phenyl**- and 4,5-**dihydro**-3,5,6-**trimethyl**-1,2,4-**triazine**, m.p. 134° and 115° respectively, are prepared by the action of hydrazine on the appropriate α-acylaminoketone and they can be oxidised to the parent compounds (*V. Sprio* and *P. Madonia*, Gazz. 1957, **87**, 992; *R. Metze*, Ber., 1958, **81**, 1863). 1,2,4-Triazin-3-ones are reduced by a wide range of reducing agents to 4,5-dihydro derivatives: 4,5-*dihydro*-1,2,4-*triazin*-3(2*H*)-*one*, m.p. 135–136° (*Y. Sanemitsu et al.*, J. heterocyclic Chem., 1981, **18**, 631; 1984, **21**, 639). Benzoin and semicarbazide condense to give 4,5-**dihydro**-5,6-**diphenyl**-1,2,4-**triazin**-3(2*H*)-**one**, m.p. 276° (*H. Biltz* and *C. Stellbaum*, Ann., 1905, **339**, 281). 1,4,5,6-Tetrahydro-1,2,4-triazines are synthesised by cyclisation of imidates or nitriles with 2-aminoethylhydrazines; 1,4,5,6-*tetrahydro*-3-(2-*phenylethyl*)-1,2,4-*triazine*, m.p. 114–115° (*D.L. Trepanier, K.L. Shriver* and *J.N. Eble*, J. med. Chem., 1969, **12**, 257). 4-**Amino**-2,6-**diphenyl**-**hydro**-1,2,4-**triazine**, m.p. 130°, formed by cyclisation of benzylidenephenacylhydrazine with formaldehyde, followed by removal of the 4-benzylidene group, reacts with nitrous acid to yield 2,6-*diphenyl*-2,3,4,5-*tetrahydro*-1,2,4-*triazine*, m.p. 160° (*M. Busch* and *K. Küspert*, J. pr. Chem., 1936, **144**, 273). Hexahydro-1,2-dibenzyl-1,2,4-triazine-3,6-dione, m.p. 113–116°, undergoes sequential hydrogenolysis to the 2-*benzyl* derivative, m.p. 198–200°, and then to *hexahydro*-1,2,4-*triazine*-3,6-*dione*, m.p. 222–226° (*T.J. Schwan*, J. heterocyclic Chem., 1983, **20**, 547).

(viii) 1,2,4-Benzotriazines (α-phenotriazines)

The title compounds may be prepared as follows.

(1) By cyclisation of formazans on treatment with sulphuric or acetic acid—when the formazan is not symmetrically substituted (XXV) one or other of the products (XXVI) or (XXVII) markedly predominates (*D. Jerchel* and *W. Woticky*, Ann., 1957, **605**, 191).

(2) By reduction of 1-acyl-2-(2-nitrophenyl)hydrazines (XXVIII) followed by oxidation (*e.g.* with potassium ferricyanide) of the dihydro intermediate.

(*3*) By cyclisation of 2-nitrophenylurea (XXIX; X=O) and related compounds (XXIX; X=S, NH) in alkali.

(*4*) By reduction of 1,2,4-benzotriazine-1-oxides (*G. Tennant*, J. chem. Soc. C, 1967, 2658).

1,2,4-Benzotriazines are in general stable, yellow/orange, crystalline substances whose reactions show obvious analogies with related simpler nitrogen heterocyclic systems. These 1,2,4-triazin-3-ones with phosphorus oxychloride yield 3-chloro derivatives which react readily with nucleophiles such as amines or hydrazines to form, *e.g.*, 3-amino-1,2,4-benzotriazines. 3-Amino-1,2,4-benzotriazines can be diazotised to yield 1,2,4-benzotriazin-3-ones.

1,2,4-**Benzotriazine** (α-phenotriazine), b.p. 235–240°/760 mm (*A. Bischler*, Ber., 1889, **22**, 2801), m.p. 74–75° (*E. Bamberger* and *E. Wheelright*, Ber., 1892, **25**, 3201), is obtained by method (1) utilising PhN : NC(COOEt) : NNHPh and by method (2) starting from compound (XXVIII; R=H). 3-*Methyl*-1,2,4-*benzotriazine*, m.p. 88–89° (*A. Bischler, loc. cit.*) or m.p. 97–98° (*B. Adger et al.,* Chem. Comm., 1971, 695); 3-*ethyl* analogue, m.p. 140–142° (*S. Kwee* and *H. Lund*, Acta Chem. Scand., 1969, **23**, 2711); 3-*benzyl*- (*G. Tennant, loc. cit.*) and 3-*phenyl*-1,2,4-*benzotriazines*, m.p. 87° and 126–127°, respectively (*R.A. Abramovitch* and *K. Schofield*, J. Chem. Soc., 1955, 2326). The 3-benzyl compound may be oxidised to 3-*benzoyl*-1,2,4-*benzotriazine*, m.p. 115°, and sodium dithionite reduction affords 1,2-dihydro-1,2,4-benzotriazines (*G. Tennant*, J. chem. Soc. C, 1967, 1279; 2658).

3-**Amino**-1,2,4-**benzotriazine**, m.p. 207°, and its 1-*oxide*, m.p. 269°, arise from 2-nitrophenylguanidine by methods (3) and (4) (*F. Arndt*, Ber., 1913, **46**, 3522). Oxidation (hydrogen peroxide in acetic acid) of 3-amino-1,2,4-benzotriazines at room temperature leads to the corresponding 2-oxides, *e.g.* 3-*amino*-1,2,4-*benzotriazine*-2-*oxide*, m.p. 187°, but prolonged oxidation at 50° gives rise to 1,4-dioxides (*J.C. Mason* and *G. Tennant*, J. chem. Soc. B, 1970, 911). Tin and hydrochloric acid converts 3-amino-1,2,4-benzotriazine-2-oxide into the corresponding 1,2-dihydro derivative (*F. Arndt* and *B. Rosenau*, Ber., 1917, **50**, 248). 3-*Hydrazino*-1,2,4-*benzotriazine* has m.p. 173–175° (*J. Jiu* and *G.P. Mueller*, J. org. Chem., 1959, **24**, 813).

1,2,4-**Benzotriazine**-3-**carboxylic acid** (subject to decomposition), its *ethyl ester*, m.p. 93°, and *hydrazide*, m.p. 207°, are prepared from *N*-(2-nitrophenyl)oxalamidrazonate (*R. Fusco* and *S. Rossi*, Gazz. 1956, **86**, 484). 1,2,4-Benzotriazine-3-acetic acids have been studied for their biological properties (*P.P. Mager et al.*, Pharmazie, 1981, **36**, 63).

3-**Chloro-** and 3-**bromo**-1,2,4-**benzotriazines**, m.p. 100–101° and 122°, respectively, are obtained via diazotisation of the 3-amino precursor in the presence of hydrogen chloride or potassium bromide (*F. Arndt* and *B. Rosenau, loc. cit.*). 3,7-**Dichloro**-1,2,4-**benzotriazine**, m.p. 140°, and its 1-*oxide*, m.p. 153–154°, are formed by reaction of the corresponding 1,2,4-benzotriazin-3-ones (below) with phosphorus oxychloride. Also 7-chloro-1,2,4-benzotriazin-3-one-1-oxide with phosphorus oxychloride/phosphorus pentachloride gives 3,7-dichloro-1,2,4-benzotriazine (*F.J. Wolf* and *K. Pfister*, U.S.P., 2,489,358; C.A., 1950, **44**, 3538b).

1,2,4-**Benzotriazin**-3-**one**, m.p. 209–210° (dec.), and its 1-*oxide*, m.p. 219° (dec.), are synthesised by method (3) using 2-nitrophenylurea or from their corresponding 3-amino derivatives with nitrous acid (*F. Arndt, loc. cit.*; *F. Arndt* and *B. Rosenau, loc. cit.*). 3-*Ethoxy*-1,2,4-*benzotriazine*, m.p. 74–76°, available from the action of sodium ethoxide on the 3-chloro compound, yields the 1-oxide on treatment with hydrogen peroxide/maleic anhydride (*T. Sasaki* and *M. Murata*, Ber., 1969, **102**, 3818). 1,2,4-Benzotriazin-3-one, on slow addition of diazomethane yields 3-*methoxy*-1,2,4-*benzotriazine*, m.p. 106°, whereas excess diazomethane results in the formation of 2-*methyl*-1,2,4-*benzotriazin*-3-*one*, m.p. 157–158° (dec.), and 4-*methyl*-1,2,4-*benzotriazin*-3-*one*, m.p. 202° (*L. Ergener*, C.A., 1950, **44**, 10718h). Alkylation (benzyl bromide) of the sodium salt of 1,2,4-benzotriazin-3-one-1-oxide yields 4-*benzyl*-4*H*-1,2,4-*benzotriazin*-3-*one*-1-*oxide*, m.p. 197° (*W.O. Foye, J.M. Kauffman* and *Y.H. Kim*, J. heterocyclic Chem., 1982, **19**, 497).

1,2,4-**Benzotriazine**-3-**thione**, m.p. 208–209°, is derived by reduction with zinc dust in alkali of its 1-*oxide*, m.p. 184° (methods 3 and 4 above) (*F. Arndt* and *B. Rosenau, loc. cit.*). Methylation (dimethyl sulphate) furnishes the corresponding 3-*methylthio* derivatives, m.p. 104° and 123°, respectively.

(ix) 1,2,4-Benzotriazines and related species with reduced ring systems

Dihydro-1,2,4-benzotriazines, often formulated as the 1,2-dihydro derivatives unless unambiguously having a 1,4-disubstitution pattern, are easily oxidised to the corresponding 1,2,4-benzotriazines. In addition to being intermediates in reaction (2) above they also arise by reduction of the related 1- or 2-oxide; 3-*benzyl*-1,2-*dihydro*-1,2,4-*benzotriazine*, m.p. 119° (*G. Tennant*, J. chem. Soc. C, 1967, 1279; 2658). Cycloalkane-1,2-diones condense with amidrazones to yield 1,2,4-triazines fused to cycloalkanes; *cyclohepta*-[e]- and *cyclooctano*-[e]-1,2,4-*triazines*, b.p. 64°/0.1 mm and b.p. 75°/0.1 mm respectively (*H. Neunhoeffer* and *K-H. Schnurrer*, in "Chemistry of 1,2,3-Triazines and 1,2,4-Triazines, Tetrazines and Pentazines", Wiley-Interscience, New York, 1978, p. 662) and 1,2,4-triazines are known fused to ring C of steroid systems (*S. Bergström* and *G.A.D. Haslewood*, J. chem. Soc., 1939, 540).

(x) Compounds containing 1,2,4-triazines fused to polycyclic and heterocyclic systems

1,2,4-Triazines fused to a very wide variety of polycyclic and heterocyclic systems have been reported, *e.g.* compounds (XXX–XXXII, XXXIV, XXXV).

(XXX)   (XXXI)

(XXXII)

Naphtho-[1,2-e]-1,2,4-triazine-2-carboxylic acid, m.p. 204–206°, *ethyl ester*, m.p. 158–160°, prepared from the precursor (XXXIII), decarboxylates to yield the parent compound (XXX), m.p. 140–141° (*R. Fusco* and *G. Bianchetti*, Gazz., 1957, **87**, 438).

(XXXIII)   (XXXIV)   (XXXV)

Naphtho-[1,2-e]-1,2,4-triazin-2-one, m.p. 272–273°, may be obtained by the action of nitrous acid on 2-*aminonaphtho*[1,2-e]-1,2,4-*triazine*, m.p. 200–201° (*R. Fusco* and *G. Bianchetti, ibid.*, 1957, **87**, 446), itself prepared by cyclisation of the aminoguanidyl derivative of 1-oximino-1,2-naphthoquinone (*F.L. Scott* and *F.J. Lalor*, Tetrahedron Letters, 1964, 641).
α-Naphthyldiphenylformazan in acid yields 3-**phenylnaphtho**-[1,2-e]-1,2,4-**triazine**, m.p. 145° (*F. Fichter* and *E. Schiess*, Ber., 1900, **33**, 747).

Naphtho-[2,1-e]-1,2,4-triazine (XXXI), m.p. 124–125°, arises from decarboxylation of its 3-*carboxylic acid*, m.p. 176–177°, *ethyl ester*, m.p. 150–151°, prepared from the analogue of species (XXXIII) (*R. Fusco* and *G. Bianchetti*, Gazz., 1957, **87**, 438).

9,10-Phenanthroquinone condenses with amidrazones to give 3-substituted phenanthro[9,10-e]-1,2,4-triazines (XXXII), *e.g.* aminoguanidine gives the 3-*amino* derivative, m.p. 262° (*J. Thiele* and *R. Bihan*, Ann., 1898, **302**, 299). Semicarbazide and thiosemicarbazide yield *phenanthro-*[9,10-e]-1,2,4-*triazin-3-one*, m.p. 285° (dec.) (*J. Schmidt et al.,* Ber., 1911, **44**, 276), and -3-*thione* (m.p. variable).

Isatin and aminoguanidine react to give a β-hydrazone which can be cyclised to 3-**amino**-1,2,4-**triazino**[5,6-b]**indole** (XXXIV), m.p. 350–354° (dec.) (*H. King* and *J. Wright*, J. chem. Soc., 1948, 2314). 5-Amino-6-cyano-3-ethoxy-1,2,4-triazine with guanidine yields 6,8-**diamino**-3-**ethoxypyrimido**-[4,5-e]-1,2,4-**triazine**, m.p. > 300°, which with ammonia gives the 3,6,8-triamino compound (*E.C. Taylor* and *S.F. Martin*, J. org. Chem., 1970, **35**, 3792). 5-Amino-4-hydrazinopyrimidines cyclise with *ortho*-esters or imidates to yield pyrimido-[5,4,-e]-1,2,4-triazines (XXXV) (*C. Temple, C.L. Kussner* and *J.A. Montgomery*, J. org. Chem., 1974, **39**, 2866).

(c) 1,3,5-Triazines

(i) Synthesis

1,3,5-Triazines (XXXVI) may be obtained as follows.

(*1*) By trimerisation of compounds of the type RCN where R may be H, halogen, alkyl, aryl, amino, perfluoroalkyl, *etc.*; mixtures of two nitriles have sometimes been employed.

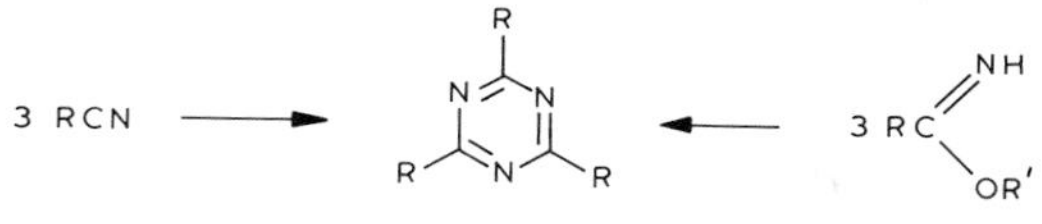

$$3\ RCN \longrightarrow \underset{(XXXVI)}{\text{triazine}} \longleftarrow 3\ RC\underset{OR'}{\overset{NH}{\diagup}}$$
(XXXVII)

High pressures and heat are often required, and acids, bases, and Lewis acids have all been used as catalysts for this reaction. When an alcohol is the solvent, imidate intermediates (XXXVII) have been proposed (*cf.* 2 below). Optimum conditions have been discussed (*D. Martin, M. Bauer* and *V.A. Pankratov*, Russ. chem. Rev., 1978, **47**, 975). The method is probably least effective for simple alkyl cyanides, and although with acids, aryl cyanates trimerise to 2,4,6-triaryloxy-1,3,5-triazines, alkyl cyanates first isomerise to isocyanates which in turn yield isocyanurates (*E. Grigat* and *R. Puetter*, Ber., 1964, **97**, 3012; *D. Martin* and *A. Weise*, Ber., 1967, **100**, 3747).

(2) By trimerisation of imidate bases (XXXVII) in the presence of acetic or trifluoroacetic acid (*F.C. Schaefer* and *G.A. Peters*, J. org. Chem., 1961, **26**, 2778; *G.A. Schvekhgeimer* and *A.P. Kryuchkova*, C.A., 1968, **68**, 78345q). The reaction may be modified to yield 2,4,6-unsymmetrically substituted 1,3,5-triazines by co-trimerisation of two imidates or by interaction of an amidine salt and a lower aliphatic imidate in which case the triazine (XXXVIII) has one substituent derived from the amidine and two from the imidate (*F.C. Schaefer*, J. org. Chem., 1962, **27**, 3362; 3608).

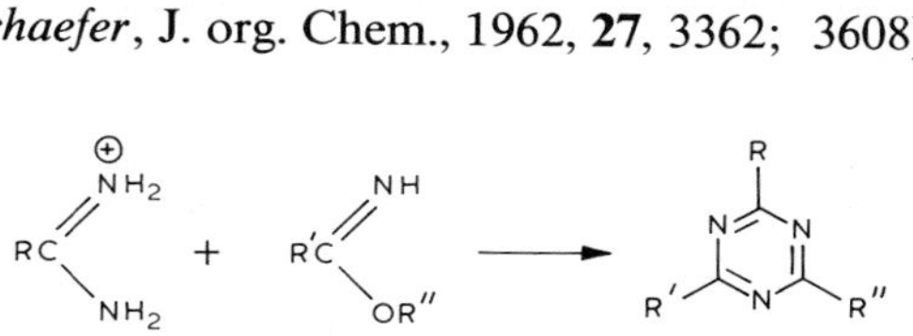

(XXXVIII)

Use of *N*-acylimidates permits the synthesis of 1,3,5-triazines with three different substituents but this reaction suffers from mixed products derived from transacylation reactions taking place (*H. Bader et al.*, J. org. Chem., 1965, **30**, 702; *P.-G. Baccar*, Compt. rend., 1967, **264C**, 352). Monosubstituted 1,3,5-triazines (XXXIX) result from the interaction of an imidate (or amidine) and 1,3,5-triazine (*F.C. Schaefer* and *G.A. Peters*, J. org. Chem., 1961, **26**, 2784; J. Amer. chem. Soc., 1959, **81**, 1470).

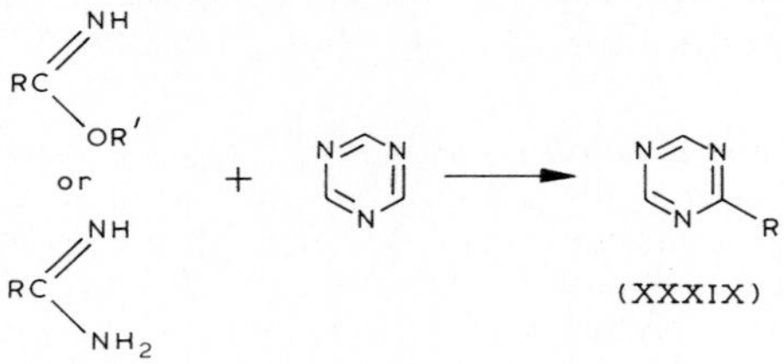

Imidates (and amidines) also react with 4,6-diaryl-1,2,3,5-oxathiadiazine dioxides to form 2,4,6-triaryl- or 2,4-diaryl-6-alkyl-1,3,5-triazines (*H. Weidinger* and *J. Kranz*, Ber., 1963, **96**, 2070).

(*3*) In addition to the two methods relating to amidines discussed in method (2), 1,3,5-triazines can be formed by trimerisation of formamidine or trichloroacetamidine — other amidines tending to give poor yields (*F.C. Schaefer et al.*, J. Amer. chem. Soc., 1959, **81**, 1466). Aryl amidines condense with acylating agents, *e.g.* phosgene (*T. Rappeport*, Ber., 1901, **34**, 1983) or ethyl formate (*H. Bredereck, F. Effenberger* and *A. Hoffmann*, Ber., 1963, **96**, 3265), in the latter case to give 2,4-disubstituted 1,3,5-triazines. Amidines along with cyanamides (*G.V. Boyd, P.F. Lindley* and *G.A. Nicolaou*, Chem. Comm., 1984, 1105) or *N*-cyanoamidines (*R.L.N. Harris*, Synthesis, 1980, 841; 1981, 907) condense with chloroiminium salts (XL) to yield 1,3,5-triazines of types (XLI) and (XLII) respectively.

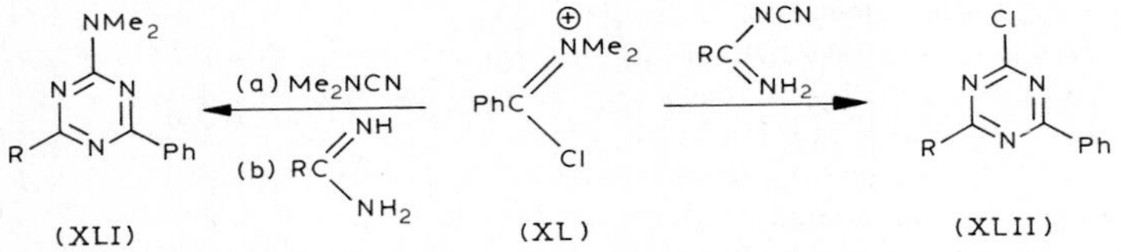

Amidines, guanidines, and isothioureas condense with *N,N*-dimethylformamide diethyl acetal (XLIII) to yield 1,3,5-triazines of type (XLIV) and (XLV) (*H. Bredereck et al.*, Angew. Chem., 1963, **75**, 825; 1964, **76**, 61; Ber., 1965, **98**, 3178).

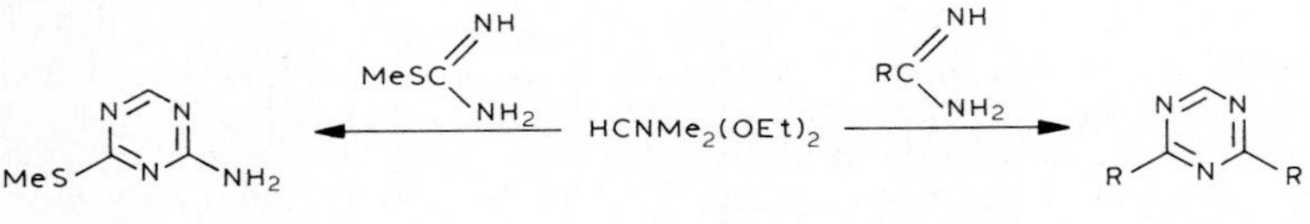

(*4*) By condensation of dicyanamide with, *e.g.*, nitriles, cyanamides, amines or urea to yield di- and tri-amino-substituted 1,3,5-triazines ("The Chemistry of Dicyanamide", American Cyanamid Co., New York, 1949).
(*5*) By the modification of cyanuric chloride (see below).

(ii) 1,3,5-Triazine and its alkyl- and aryl-substituted derivatives

1,3,5-**Triazine**, m.p. 86°, b.p. 114°/760 mm, was first prepared by *J.U. Nef* (Ann., 1895, **287**, 333) but its correct structure as a trimer of hydrogen cyanide was established much later by *C. Grundmann* and *A. Kreutzberger* (J. Amer. chem. Soc., 1954, **76**, 5646). 1,3,5-Triazine is volatile and sublimes; its Raman spectrum corresponds closely to that of benzene (*J. Goubeau et al.*, J. phys. Chem., 1954, **58**, 1078) and is in accord with its aromatic character. 1,3,5-Triazine is soluble in most organic solvents, is stable in anhydrous conditions but in aqueous acid forms ammonium formate or in aqueous alkali, formamidine (*L.E. Hinkel, E.E. Ayling* and *J.H. Beynon*, J. chem. Soc., 1935, 674; *J.U. Nef, loc. cit.*). Thus unlike many of its derivatives 1,3,5-triazine is highly reactive towards nucleophiles. 1,3,5-Triazine is formed in good yield by the action of hydrogen chloride on hydrogen cyanide *via* trimerisation of the initial adduct [HC≡NH]$^{\oplus}$Cl$^{\ominus}$ (*C. Grundmann* and *A. Kreutzberger, loc. cit.*; *A. Kreutzberger*, Fortschr. chem. Forsch., 1963, **4**, 273), or by base-catalysed condensation of formamidines (*F.C. Schaefer et al.*, J. Amer. chem. Soc., 1959, **81**, 1466) or benzyl formimidate (*F. Cramer, K. Pawelzik* and *J. Kupper*, Angew. Chem., 1956, **68**, 649). 1,3,5-Triazine heated in ethanol with ammonium chloride yields formamidine hydrochloride, and N,N'-disubstituted amidines are formed with primary amines (*C. Grundmann et al.*, J. org. Chem., 1956, **21**, 1037; J. Amer. chem. Soc., 1955, **77**, 6559). Diamines react similarly, ethylene diamine furnishing Δ^2-imidazoline and *o*-phenylenediamine, benzimidazole; substituted hydrazines yield formazans but can on occasions yield amidrazones (*C. Grundmann*, Angew. Chem. intern. Edn., 1963, **2**, 309). 1,3,5-Triazine reacts with imidates and amidines to yield mono-, di-, and tri-substituted derivatives, this being a useful route to these compounds (*F.C. Schaefer* and *G.A. Peters*, J. Amer. chem. Soc., 1959, **81**, 1470; J. org. Chem., 1961, **26**, 2784). Chlorination of 1,3,5-triazine at 140–200° yields 2,4-dichloro- and 2,4,6-trichloro-1,3,5-triazine (*C. Grundmann* and *A. Kreutzberger*, J. Amer. chem. Soc., 1955, **77**, 44).

2,4,6-**Trimethyl-** and 2,4,6-**triethyl**-1,3,5-**triazines**, m.p. 59–60° and 24–25° respectively, may be prepared by method 1 conducting the reaction under high pressure in a primary alcohol—pyrimidines can be formed as by-products (*T.L. Cairns et al.*, J. Amer. chem. Soc., 1952, **74**, 5633; *K. Yanagiya, M. Yasumoto* and *M. Kurabayashi*, Bull. chem. Soc. Japan, 1973, **46**, 2804). Halogen-substituted alkyl nitriles polymerise more readily; 2,4,6-*tris*(*trichloromethyl*)-1,3,5-*triazine*, m.p. 93° (*K. Wakabayashi, M. Tsunoda* and *Y. Suzuki*, Bull. chem. Soc. Japan, 1969, **42**, 2924); 2,4,6-*tris*(*perfluoroethyl*)-1,3,5-*triazine*, b.p. 121–122° (*W.L. Reilly* and *H.C. Brown*, J. org. Chem., 1957, **22**, 699).

2-**Methyl**-4,6-**diphenyl**-1,3,5-**triazine**, m.p. 110°, from benzamidine and acetic anhydride, gives on oxidation 4,6-*diphenyl*-1,3,5-*triazine*-2-*carboxylic acid*, m.p. 192° (dec.), which can be decarboxylated to 2,4-*diphenyl*-1,3,5-*triazine*, m.p. 75° (*F. Krafft* and *G. Koenig*, Ber., 1890, **23**, 2382).

2,4,6-**Triphenyl**-1,3,5-*triazine* (Cyaphenin), m.p. 232°, and 2,4,6-*tris*(*o*-, *m*-, and *p*-*tolyl*)-1,3,5-*triazines*, m.p. 110°, 152–153° and 278–279°, respectively, are readily available by method 1 (above), chlorosulphonic acid being a suitable catalyst (*A.H. Cook* and *D.G. Jones*, J. chem. Soc., 1941, 278). 2,4,6-Triaryl-substituted 1,3,5-triazines are stable to alkalis and acids and their benzene rings undergo electrophilic substitution reactions.

Mono- and di-substituted 1,3,5-triazines are available from amidines and ethyl formate or dimethylformamide diethyl acetal (method 3); 2,4-*dimethyl*-, 2,4-*diphenyl*- and 2,4-*dibenzyl*-1,3,5-*triazines* have m.p. 45–46°, 85–87° and 80–82°, respectively (*H. Bredereck et al.*, Angew. Chem. intern. Edn., 1963, **2**, 655).

(iii) Amino-substituted 1,3,5-triazines (melamines)

2,4,6-**Triamino**-1,3,5-**triazine** (melamine; XLVI, R=H) m.p. 354° (dec.), is obtained by polymerisation of dicyanamide in the presence of anhydrous ammonia in an autoclave (*E.C. Franklin*, J. Amer. chem. Soc., 1922, **44**, 486; *P.P. McClellan*, Ind. Eng. Chem., 1940, **32**, 1181), the process being of industrial importance for the preparation of melamine–formaldehyde thermosetting plastics. Trimerisation of cyanamide also gives the title compound but substituted cyanamides may either yield trisubstituted melamines (XLVI) or isomelamines (XLVII), *e.g.* phenylcyanamide at 20–80° cyclotrimerises to compound (XLVII, R=Ph) but at 200° yields species (XLVI, R=Ph).

(XLVI)   (XLVII)

The nature of the catalyst also influences the product (*V.V. Korshak et al.*, Izvest. Akad. Nauk S.S.S.R., Ser. Khim., 1973, 1408). 2-Amino-4,6-dihydroxy-1,3,5-triazine (ammelide) and 2,4-diamino-6-hydroxy-1,3,5-triazine (ammeline) obtainable along with melamine by heating dicyandiamide with aqueous ammonia or urea may also be synthesised by stepwise hydrolysis of melamine (*T.L. Davies et al.*, J. Amer. chem. Soc, 1921, **43**, 2230; 1922, **44**, 2595). Acylation of melamine is best carried out with acid anhydrides.

2,4,6-*Tris*(1-*aziridyl*)-1,3,5-*triazine*, m.p. 139° (dec.), is a cross-linking agent used in cancer chemotherapy (*V.P. Wystrach, D.W. Kaiser* and *F.C. Schaefer*, J. Amer. chem. Soc., 1955, **77**, 5915).

2,4-**Diamino**-1,3,5-**triazine** (formoguanamine), m.p. 329°, and 2,4-*diamino-6-phenyl*-1,3,5-*triazine*, m.p. 227–228°, are obtained by heating biguanide with ethyl formate and benzoyl chloride respectively (*K. Rackmann*, Ann., 1910, **376**, 163) and 2,4-*diamino-6-methyl*-1,3,5-*triazine* (acetoguanamine), m.p. 265°, from biguanide and acetic anhydride (*R. Andreasch*, Monatsh., 1927, **48**, 145).

(iv) Carbonyl derivatives and carboxylic acids

1,3,5-**Triazine**-2,4,6-**tricarboxylic acid**, m.p. 250° (dec.), can be obtained by alkaline hydrolysis of its *triethyl ester*, m.p. 169–170°, which is formed quantitatively by polymerising ethyl cyanoformate with hydrochloric acid (method 1). The *trimethyl ester* has m.p. 159–162°. The *acid chloride*, m.p. 54–56°, with benzene forms 2,4,6-**tribenzoyl**-1,3,5-**triazine**, m.p. 157–161°. Triazine aldehydes require special synthetic routes and are not available by many of the normal aldehyde syntheses (*C. Grundmann* and *E. Kober*, J. org. Chem., 1956, **21**, 1392; J. Amer. chem. Soc., 1958, **80**, 5547).

(v) Cyanuric acid (2,4,6-trihydroxy-1,3,5-triazine); trithiocyanuric acid

Cyanuric acid, first described by *C.W. Scheele* (1793), can be obtained by fusing urea and ammonium chloride at 250° for 15 minutes (*J.S. Mackay*, U.S.P. 2,527,316, 1950) or by hydrolysis of cyanuric chloride (see below). Polymerisation of cyanic acid vapour below 150° gives cyanuric acid as the main product (method 1). In the solid state cyanuric acid has been shown by Raman and infrared spectroscopy and X-ray analysis to exist in the keto form (XLVIII, R=H) rather than the hydroxy form (XLIX, R=H). In saturated aqueous solution at 20° there is present 5–6% of the enolic form which, however, predominates in highly alkaline solutions. Cyanuric acid crystallises as a dihydrate from water but may be obtained anhydrous. On heating, decomposition to cyanic acid occurs and mineral acid causes evolution of carbon dioxide. Treatment with phosphorus pentachloride at *ca*. 175° furnishes cyanuric chloride. Cyanuric acid forms two series of derivatives, the cyanurates in which alkyl or aryl groups are attached to oxygen (XLIX) and the isocyanurates with substituents on nitrogen (XLVIII). Both alkyl and aryl cyanurates may be prepared from cyanuric chloride (see below) but only the aryl derivatives (XLIX, R=aryl) can be obtained by the trimerisation of cyanates (method 1)—alkyl cyanates rearranging to isocyanates (*D. Martin, M. Bauer* and *V.A. Pankratov, loc. cit.*) which then give isocyanurates (*A.W. Hoffmann*, Ber., 1870, **3**, 269; 1885, **18**, 764). Alkyl cyanurates likewise isomerise to the corresponding isocyanurates (XLVIII). Trimethyl and *triphenyl cyanurates* have m.p. 135° and 225° respectively; *trimethyl* and *triphenyl isocyanurates*, m.p. 176° and 275° respectively.

(XLVIII)   (XLIX)

Trithiocyanuric acid (1,3,5-triazine-2,4,6-trithiol), dec. 360°, may be prepared by method 1 from thiocyanic acid. Alkyl thiocyanates trimerise similarly in the presence of acid catalysts (*D. Martin, M. Bauer* and *V.A. Pankratov, loc. cit.*). Triethyl trithiocyanurate, m.p. 27°, b.p. 150–152°/0.8 mm. **Triphenyl-** and **tris(*p*-tolyl)-trithiocyanurates**, m.p. 97° and 114° respectively, are prepared from cyanuric chloride and the appropriate sodium thiophenolate. 1,3,5-Trimethyl-1,3,5-*triazin-2-one*-4,6-*dithione*, m.p. 123° (*C.P. Joshua* and *S.K. Thomas*, Synthesis, 1982, 1070). 1-*Amino*-4,6-*diiminohexahydro*-1,3,5-*triazine*-2-*thione*, m.p. 170° (*C.P. Joshua* and *V.P. Rajan*, Austral. J. Chem., 1976, **29**, 1051).

(vi) Halogen-substituted 1,3,5-triazines—cyanuric halides

Whereas liquid cyanogen fluoride readily cyclotrimerises to 2,4,6-**triflu-**

oro-1,3,5-**triazine**, b.p. 74°/760 mm, and polymer at room temperature (*F. Fawcett* and *R.D. Lipscomb*, J. Amer. chem. Soc., 1964, **86**, 2576), trimerisation of cyanogen chloride and bromide requires the presence of catalysts, e.g. acids (method 1) (*D. Martin, M. Bauer* and *V.A. Pankratov, loc. cit.*). 2,4,6-*Trichloro*-1,3,5-*triazine*, cyanuric chloride, m.p. 145°, b.p. 198°/760 mm (*V.I. Mur*, Russ. chem. Rev., 1964, **33**, 92) also available from the action of chlorine on hydrogen cyanide in chloroform containing some alcohol (*O. Diels*, Ber., 1899, **32**, 691) undergoes successive nucleophilic displacement of the chlorine atoms by, e.g., amino, alkoxy or alkylthio groups. The reactivity of the remaining chlorine atoms diminishes as substitution proceeds and stepwise replacement with different groups is possible. Hydrolysis with 10% aqueous sodium hydroxide at 100° gives 2-chloro-4,6-dihydroxy-1,3,5-triazine but at 125° cyanuric acid is formed. Stepwise replacement of halogen occurs with aqueous alcohols in the presence of sodium bicarbonate and trialkyl cyanurates are formed from sodium alkoxides (*W.M. Pearlman et al.*, J. Amer. chem. Soc., 1949, **71**, 1128; 3248); partial or total replacement of halogen may also be made by reaction with sodium phenates (*F.C. Schaefer, J.T. Thurston* and *J.R. Dudley*, J. Amer. chem. Soc., 1951, **73**, 2990). Stepwise displacement with amines depends on the basicity of the amine and is a route to melamines, however weakly basic aromatic amines may give only disubstitution (*F. Naebe*, J. pr. Chem., 1916, **82**, 521). Complex synthetic dyestuffs capable of bonding to fabrics involve mono- and di-chloro-1,3,5-triazinyl residues (*J. Wegmann*, Tex. Rundsch., 1958, **13**, 323).

With cyanuric chloride, alkyl Grignard reagents yield 2-alkyl-4,6-dichloro-1,3,5-triazines and their aromatic counterparts, 2,4-diaryl-6-chloro-1,3,5-triazines (*R. Hirt, H. Nidecker* and *R. Berchtold*, Helv., 1950, **33**, 1365); Friedel-Crafts reactions are also possible (*H.E. Fierz-David* and *M. Matter*, J. Soc. Dy. Col., 1937, **53**, 424).

Cyanuric chloride is also a useful agent for converting alcohols (primary → tertiary) into alkyl chlorides, or iodides when sodium iodide is present (*S.R. Sandler*, J. org. Chem., 1970, **35**, 3967; Chem. Ind., 1971, 1416). Acid chlorides and acid fluorides may be prepared by treating a carboxylic acid with the appropriate cyanuric halide (*G.A. Olah, M. Nojima* and *I. Kerekes*, Synthesis, 1973, 487).

Cyanuric chloride also acts as a dehydrating agent, yielding nitriles from aldehyde oximes (*J.K. Chakrabarti* and *T.M. Hotten*, Chem. Comm., 1972, 1226), large ring lactones from ω-hydroxycarboxylic acids (*K. Venkataraman* and *D.R. Wagle*, Tetrahedron Letters, 1980, **21**, 1893) and isonitriles from *N*-monosubstituted formamides (*R. Wittmann*, Angew. Chem., 1961,

73, 219). β-Lactams may also be prepared by the action of cyanuric chloride on Schiff's bases and acetic acid derivatives (*M.S. Manhas, A.K. Bose* and *M.S. Khajavi*, Synthesis, 1981, 209).

Cyanuric bromide has m.p. 264–265°. 2-**Chloro**-4,6-**difluoro**- and 2,4-**dichloro**-6-**fluoro**-1,3,5-**triazine**, b.p. 113–114°/760 mm and 155°/760 mm respectively, are prepared by treating cyanuric chloride with SbF_3Cl or KSO_2F (*A.F. Maxwwell et al.*, J. Amer. chem. Soc., 1958, **80**, 548; *D.W. Grisely et al.*, J. org. Chem., 1958, **23**, 1802).

(vii) N-oxides and azides

1,3,5-Triazine-tri-*N*-oxide ("trifulmin") (explosive) and 2,4,6-*triphenyl*-1,3,5-*triazine-tri-N-oxide*, m.p. 125–130° (dec.) (*H. Wieland*, Ber., 1909, **42**, 803; 814). 1,3,5-Triazine azides, and in particular the 2,4,6-triazide, are very shock sensitive and have some use as explosives.

(viii) Compounds with a fully reduced 1,3,5-triazine ring

Hexamethylentetramine (hexamine) (L) discussed in Vol. I C, p. 41, is prepared by the action of formaldehyde on ammonia, monocyclic derivatives (LI) resulting when a primary amine replaces ammonia or when hexamine is heated with a primary amine (*L. Hortung*, J. pr. Chem., 1892, **46**, 1).

1,3,5-*Trimethylhexahydro*-1,3,5-*triazine*, b.p. 160–164°/760 mm, the *triethyl* and *triphenyl* analogues have b.p. 207–208°/760 mm and m.p. 141° respectively (*J. Graymore*, J. chem. Soc., 1924, **125**, 2283).

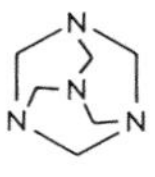

(L) 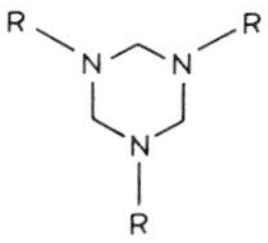   (LI) 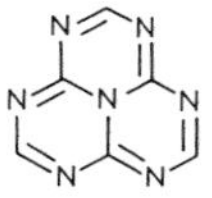   (LII)

(ix) Fused ring 1,3,5-triazines

1,3,5-Triazines fused to other heterocyclic systems are known. A series of fairly inert compounds of high heat stability and low solubility have been known since the work of *J. Liebig* (1834) but it fell to *L. Pauling* and *J.H. Sturdivant* (Proc. natl. Acad. Sci. USA, 1937, **23**, 615) to elucidate their structures based on compound (LII). The parent compound a 12π electron system (LII, cyamelurine) has been synthesised from methyl *N*-cyanoformimidate and 2,4-diamino-1,3,5-triazine. It is unstable to water (*R.S. Hosmane, M.A. Rossman* and *N.J. Leonard*, J. Amer. chem. Soc., 1982, **104**, 5497).

2. Thiadiazines

All six possible ring substitution patterns for this class of compounds are known. The scope is further widened by sulphur exhibiting variable valency.

(i) 1,2,3-Thiadiazines
First known only as their benzo derivatives, *e.g.* compound (I), m.p. 154° (dec.) (*E. Schrader*, J. pr. Chem., 1917, **96**, 180), monocyclic compounds, *e.g.* (II) m.p. 162°, have since been prepared by the action of hydrazines on 3-chloropropanesulphonyl chloride in basic media (*B. Helferich, R. Hoffmann* and *H. Mylenbusch*, J. pr. Chem., 1962, **19**, 56).

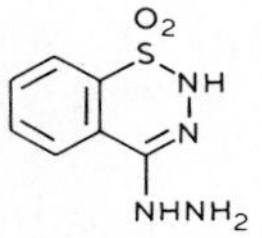

(ii) 1,2,4-Thiadiazines
1,2-Dibromoethane and thioureas under oxidative conditions ($KClO_3$) yield 1,2,4-thiadiazines; unsymmetrically substituted thioureas form single products (III) (*F. Kucera*, Monatsh., 1914, **35**, 137).

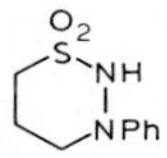

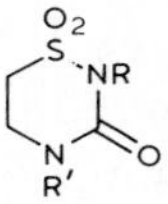

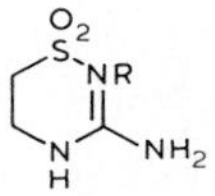

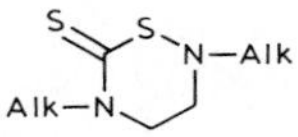

3-**Phenyl**-5,6-**dihydro**-2*H*-1,2,4-**thiadiazine**-1,1-**dioxide**, m.p. 214°, is derived from *N*-(2-chlorosulphonylethyl)benzimidoyl chloride and ammonia (*R. Winterbottom et al.*, J. Amer. chem. Soc., 1947, **69**, 1393) and amidines give related products, *e.g.* compound (IV), m.p. 180–184°, from guanidine with 2-chloroethanesulphonyl chloride (*A. Etienne, A. Le Berre* and *J.P. Giorgetti*, Bull. Soc. Chim. Fr., 1973, 985).

(iii) 1,2,5-Thiadiazines
The products of the interaction of carbon disulphide and N,N'-dialkyldiaminoethanes can be oxidised (I_2/KI in alkali) to tetrahydro-1,2,5-thiadiazine-6-thiones (V) (Alk = Et, m.p. 62°) (*R.A. Donia et al.*, J. org. Chem., 1949, **14**, 946).

(iv) 1,2,6-Thiadiazines
Sulphamides condense with 1,3-diketones to yield either 4*H*- or 2*H*-1,2,6-thiadiazines [(VI) and (VII) respectively] (*E.F. Deringer* and *J.E. Wilson*, J. org. Chem., 1952, **17**, 339; *J.B. Wright*, ibid., 1964, **29**, 1905).

3,5-Diamino-4H-1,2,6-thiadiazine has been prepared similarly from malononitrile (*H.A. Walter*, U.S.P., 2,454,262; C.A., 1949, **43**, 2648e). *G. Kresze* and *H. Grill* (Tetrahedron Letters, 1969, 4117) have prepared the S(IV) derivative (VIII), b.p. 28°/7 mm, from S(:NTs)$_2$ and propane-1,3-diamine, and *R.M. Acheson et al.* (Chem. Comm., 1983, 1002) have synthesised the 2,1,3-**benzothiadiazine** (IX), m.p. 189°, with its novel dithiosulphone group by heating 2-aminobenzamide and phosphorus pentasulphide in pyridine.

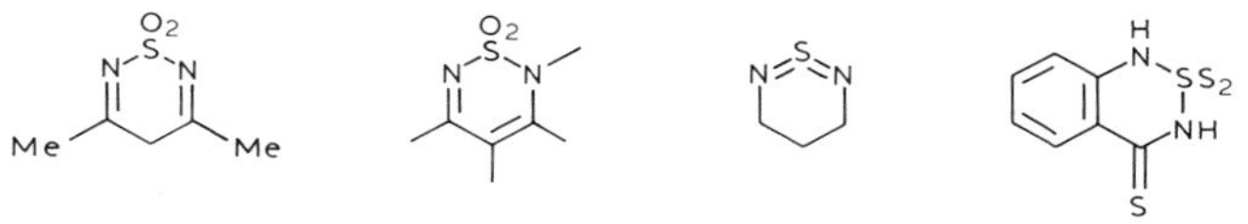

(VI)   (VII)   (VIII)   (IX)

(v) 1,3,4-Thiadiazines

Although α-haloketones condense with thiosemicarbazide or acid thiohydrazides to give 1,3,4-thiadiazines, confusion has arisen over aminothiazolines being ascribed thiadiazine structures. However, *H. Beyer* and co-workers (Ber., 1954, **87**, 1385; 1392; 1956, **89**, 1095) have shown that chloroacetone condenses with thiosemicarbazide in dilute hydrochloric acid to give compound (X). 2,5,6-*Triphenyl*-4H-1,3,4-*thiadiazine*, m.p. 192–193°, is the product of condensation in base of benzthiohydrazide and 2-chloro-1,2-diphenylethanal (*J. Sandstrom*, Arkiv. Kemi, 1960, **15**, 195). Treatment of acid N-2-hydroxyethylhydrazides with phosphorus pentasulphide gives compounds of type (XI) (*D.L. Trepanier et al.*, J. org. Chem., 1965, **30**, 2228) and the thiadiazolinone (XII) undergoes ring enlargement on attack by nucleophiles to give the title compounds, *e.g.* (XIII; Nu = NH$_2$) m.p. 77–78° (*H. Kristinsson, T. Winkler* and *M. Mollenkopf*, Helv., 1983, **66**, 2714).

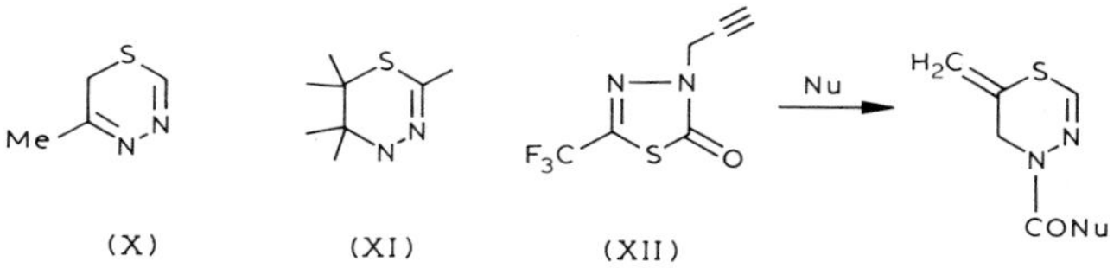

(X)   (XI)   (XII)   (XIII)

(vi) 1,3,5-Thiadiazines

Amines and formaldehyde unite with hydrogen sulphide at 0° to form tetrahydro-3,5-diaryl-1,3,5-thiadiazines (XIV) (*S.A. Hughes* and *E.B. McCall*, B.P., 943,273, 1963; C.A., 1964, **60**, 5528c) or with aryldithiocarbamic acids to give the related species (XV) (*G.M. Robinson* and *R. Robinson*, J. chem. Soc., 1923, **123**, 532). On treatment with BF_3/Et_2O and subsequently with alkali, a mixture of a nitrile, thioamide and an aldehyde leads to 4*H*-1,3,5-thiadiazines (XVI) (*C. Giordano, A. Belli* and *V. Bellotti*, Synthesis, 1975, 266) and 2*H*-derivatives (XVII) arise by the action of phosgene on $Me_2NCSNHC(:NH)SMe$ (*I. Iwataki*, Bull chem. Soc. Japan, 1972, **45**, 3218). *J. Goerdeler* and *W. Kunnes* (Ber., 1983, **116**, 2044) have condensed aldimines and ROCSNCS to obtain 2,3-dihydro-4*H*-derivatives (XVIII; R = Me, m.p. 155°; R = Ph, m.p. 153°).

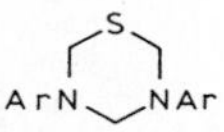

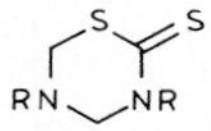

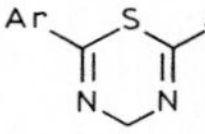

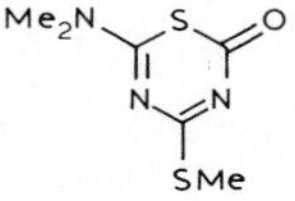

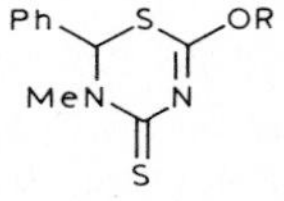

3. Oxadiazines

Examples of all six possible oxadiazine structures are known, although some are much more widely reported than others.

(i) 1,2,3-Oxadiazines

Early references to this system relate to benzo derivatives, 8-aminobenzo-4*H*-3,1,2-oxadiazin-4-one (I) being a reported product of the diazotisation of 2,3-diaminobenzoic acid (*P. Griess*, Ber., 1884, **17**, 603).

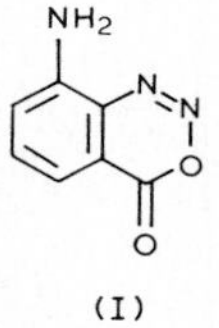

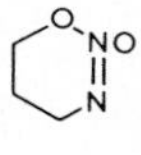

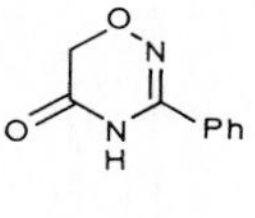

Recently, the 1,2,3-oxadiazine-2-oxide (II) has been formed by the action of base on the nitramide 4,3-$CH_3(O_2N)C_6H_3SO_2N(NO_2)(CH_2)_3Br$ at 10° (*O.A. Luk'yanov* and *T.V. Ternikova*, Izv. Akad. Nauk S.S.S.R. Ser. Khim., 1983, 667).

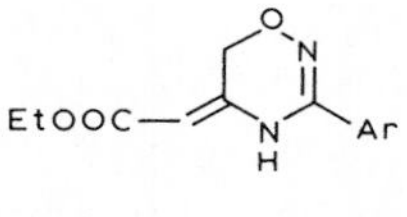

(ii) 1,2,4-Oxadiazines

Amidoximes condense at oxygen with α-halo-acids or -esters, e.g. benzamidoxime reacts with ethyl chloroacetate to give 5,6-**dihydro**-3-**phenyl**-4H-1,2,4-**oxadiazin**-5-**one** (III), m.p. 148° (*H. Koch*, Ber., 1889, **22**, 3161). γ-Haloacetoacetates react similarly to form the methylidene derivatives (IV) (*K. Tabei et al.*, Chem. pharm. Bull., 1982, **30**, 3987). 7-*Chloro*-3-*ethoxycarbonyl*-1,2,4-*benzoxadiazine*, m.p. 130–131°, arises from lead tetraacetate oxidation of amidoxime (V) (*C.W. Rees et al.*, Chem. Comm., 1975, 913; 914). 1,2,4-Benzoxadiazines undergo ring contraction to the corresponding benzoxazoles in refluxing chlorobenzene (*C.W. Rees et al., ibid.*, 962).

The **tetrahydro**-1,2,4-**oxadiazin**-3-**one** (VI; X = O), b.p. 100–105°/8 mm, prepared by the action of methylamine on the tosylate, $Me_2CHN(COOEt)OCH_2CH_2OTs$, is reduced by lithium aluminium hydride to the *tetrahydro-oxadiazine* (VI; X = H_2), b.p. 110°/760 mm (*F.G. Riddell et al.*, Heterocycles, 1978, **9**, 267; Tetrahedron, 1979, **35**, 1391).

(iii) 1,2,5-Oxadiazines

1,2-Diketone monoximes condense with aldehyde oximes to give 1,2,5-oxadiazines. 4-**Hydroxy**-3,4,6-**trimethyl**-1,2,5-**oxadiazine**, derived from diacetyl monoxime and acetaldehyde oxime, has m.p. 203° (dec.) (*O. Diels* and *R. van der Leeden*, Ber., 1905, **38**, 3357). The aminoketone oxime (VII) condenses with benzaldehyde to yield the 1,2,5-**oxadiazine** (VIII), m.p. 141°, whereas its geometric isomer gives an imidazoline-*N*-oxide (*H. Gnichtel* and *B. Möller*, Ber., 1981, **114**, 3170).

W. Kliegel and *G.-H. Franckenstein* (Ann., 1977, 956) have obtained perhydro-1,2,5-oxadiazines by ring enlargement of 1-alkyl-3-phenylimidazolidine-*N*-oxides.

(iv) 1,2,6-Oxadiazines

Diethyl acetonedicarboxylate on treatment with HNO_3/N_2O_3 and subsequently with sodium bisulphite yields the 1,2,6-oxadiazine (IX). Ethylation (Ag^+/EtI) gives in sequence the 2-ethyl and 4-ethoxy-2-ethyl derivative (*P. Henry* and *H. von Pechmann*, Ber., 1893, **26**, 997).

(v) 1,3,4-Oxadiazines

Hydrazines, and in particular acyl hydrazines, act as precursors of this system. *N*-Benzoyl-*N'*-phenylhydrazine reacts with chloroacetyl chloride to furnish the product (X), m.p. 96°,

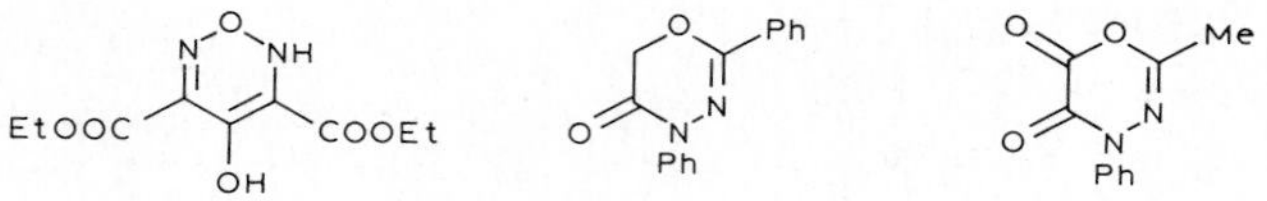

(IX)   (X)   (XI)

and the related 5,6-dione (XI), m.p. 141°, can be prepared via oxalyl chloride (*J. van Alphen*, Rec. Trav. chim., 1928, **47**, 673; 909). Cyclisation of PhCONHNHCH$_2$COPh in polyphosphoric/acetic acids at 140° yields 2,6-diphenyl-4*H*-1,3,4-oxadiazine (*K. Suhasini, T.V.P. Rao* and *V. Thirupathaiah*, Curr. Sci., 1983, **52**, 1133). 2-Alkyl-4,5-dimethyl-6-phenyltetrahydro-1,3,4-oxadiazines, prepared by the action of aldehydes on *N*-amino-l-ephedrine, are tautomeric with their open chain isomers γ-hydroxy-hydrazones PhCH(OH)CH(Me)N(Me)N = CHAlkyl (*L.C. Dorman*, J. org. Chem., 1967, **32**, 255; see also *A.A. Potekhin*, Zhur. org. Khim., 1971, **7**, 16). On the other hand, 2,3,4-*trimethyltetrahydro*-1,3,4-*oxadiazine*, b.p. 81°/15 mm, is stable (*I.J. Ferguson, A.R. Katritzky* and *D.M. Read*, J. chem. Soc. Perkin II, 1976, 1861). Azidodiketones condense with alkenes to yield 5,6-dihydro-4*H*-1,3,4-oxadiazines, *e.g.* compound (XII; R = Ph, m.p. 134°; or R = SEt, m.p. 114°) (*J. Firl* and *S. Sommer*, Tetrahedron Letters, 1971, 4193). Related work by *H.J. Hall* and *M. Wojciechowska* (J. org. Chem., 1978, **43**, 3348) utilises azidodicarboxylic esters.

(vi) 1,3,5-Oxadiazines

Isocyanates are common precursors of this system, methyl isocyanate trimerising in the presence of triethylphosphine to yield the tetrahydro-

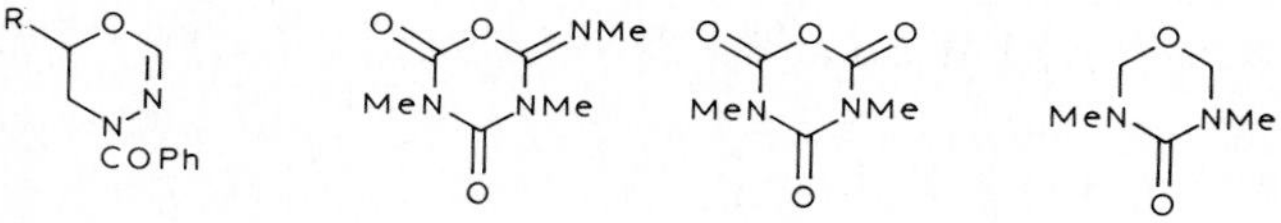

(XII)   (XIII)   (XIV)   (XV)

1,3,5-oxadiazine (XIII) (*C.K. Ingold* and *S.D. Weaver*, J. chem. Soc., 1925, **127**, 378). Alkyl isocyanates also condense with carbon dioxide in the presence of tributylphosphine to yield the related *trione* (XIV), m.p. 258°, but benzoyl isocyanate under similar conditions furnishes 2,6-diphenyl-1,3,5-oxadiazin-4-one (*A. Etienne et al.*, Bull. Soc. chim. Fr., 1972, 242, 251; 1974, 1497).

Acid condensation of urea or its derivatives with formaldehyde leads to tetrahydro-1,3,5-oxadiazin-4-ones, *e.g.* compound (XV), b.p. 75–76°/0.5 mm (*H. Kadowaki*, Bull. chem. Soc. Japan, 1936, **11**, 248) and the method has been adapted to accommodate higher aldehydes (*H. Petersen*, Angew. Chem., 1964, **76**, 909; Synthesis, 1973, 243). Compounds of type (XV) undergo hydrolysis and recyclisation in acid solutions yielding hexahydro-1,3,5-triazin-2-ones (*H. Petersen*, Synthesis, 1973, 243).

Hexafluoroacetone reacts with diethylcyanamide at 130° to yield 2,6-**bis(diethylamino)-4,4-bis(trifluoromethyl)-4H-1,3,5-oxadiazine**, m.p. 52° (*K. Burger* and *R. Simmerl*, Synthesis, 1983, 237), and related compounds, e.g. (XVI), have been prepared by [4 + 2] cycloaddition reactions of 4,4-bis(trifluoromethyl)-1-oxa-3-azabuta-1,3-dienes and α,β-unsaturated nitriles (*K. Burger, W. Schoentag* and *U. Wassmuth*, J. fluorine Chem., 1983, **22**, 99).

2,4,6-Triphenyl-1,3,5-oxadiazinium salts (diazapyrylium salts) (XVII) are synthesised from benzonitrile and benzoyl chloride in the presence of a Lewis acid (*R.R. Schmidt*, Ber., 1965, **98**, 334).

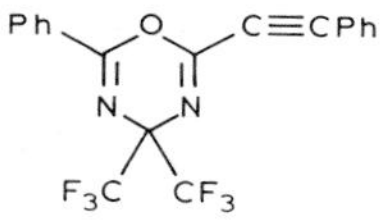

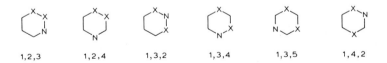

(XVI) 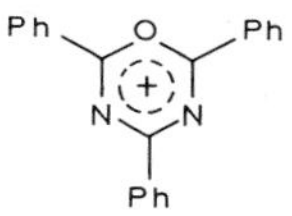   (XVII)

4. Dithiazines and dioxazines

The title compounds are represented by the following formulae, X = S and X = O respectively, but examples of certain structural types are as yet unknown.

1,2,3   1,2,4   1,3,2   1,3,4   1,3,5   1,4,2

(i) Dithiazines

Thiocarbamic esters and amides have been used to synthesise 1,2,4-dithiazines; for example sulphur dichloride condenses with $CH_2:CMeCH_2$-$NC_6H_{11}(cyclo)C(:S)OEt$ to yield compound (I) (*M. Muehlstaedt* and *R. Widera*, J. pr. Chem., 1978, **320**, 123) and compound (II), m.p. 195–199° (dec.), is formed by the action of acid at 0° on $PhNHCSNH(CH_2)_2S_2O_3Na$ (*J.B. Caldwell, B. Milligan* and *J.M. Swan*, J. chem. Soc., 1963, 2097).

Propane-1,3-disulphonyl chloride condenses with hydrazine salts in basic media to form **tetrahydro**-2-**amino**-1,3,2-**dithiazine**-1,1,3,3-**tetroxide** (III), m.p. 213° (*K.H. Linke, R. Bimczok* and *J. Lex*, Ber., 1975, **108**, 1087).

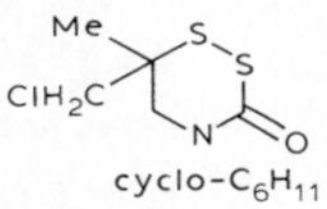

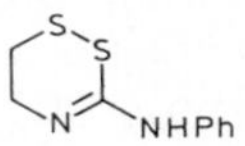

Thiobenzophenone undergoes photochemical cycloaddition reactions with imines to give reduced 1,3,5-dithiazines, *e.g.* *N*-methylbenzaldimine yields the dithiazine (IV) (*A. Ohno, N. Kito* and *T. Koizumi*, Tetrahedron Letters, 1971, 2421) and related syntheses have been carried out using thioketenes (*E. Schaumann, J. Ehlers* and *F.-F. Grabely*, Ber., 1980, **113**, 3010). Thioketones, thioamides and aldehydes (or ketones) condense together in boron trifluoride–etherate at room temperature to give 4*H*-1,3,5-dithiazines, the 2,2,6-*triphenyl* derivative having m.p. 130–132° (*C. Giordano* and *A. Belli*, Synthesis, 1977, 193, 476; 1978, 443).

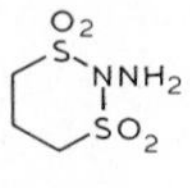

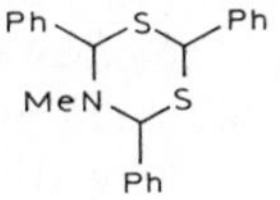

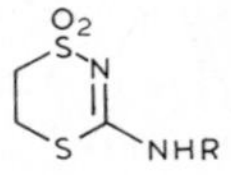

Vinyl sulphonamides react with alkyl and aryl isothiocyanates or with carbon disulphide in alkali to give 1,4,2-dithiazine-1,1-dioxides, *e.g.* compound (V; R = Me), m.p. 146–147°, *via* methyl isothiocyanate (*K. Nakahashi, S. Hirooka* and *K. Hasegawa*, Bull. chem. Soc. Japan, 1972, **45**, 3217; *K. Hasegawa, T. Sasaki* and *S. Hirooka, ibid.*, 1973, **46**, 696). The azide (VI) thermolyses to the isomeric 1,4,3-*dithiazines* (VII), m.p. 49–52°, and (VIII), m.p. 69–71°, which lose sulphur on further heating (*J. Nakayama et al.*, Tetrahedron Letters, 1983, **24**, 3729).

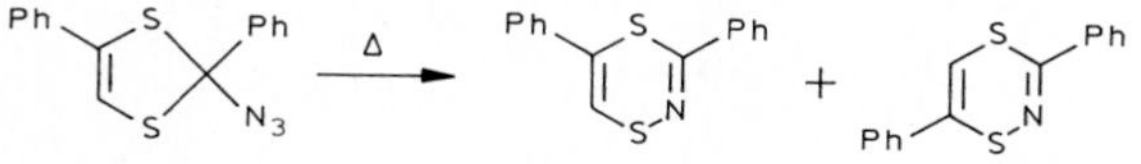

(ii) Dioxazines

A tentative claim to have prepared a tetrahydro-1,2,4-dioxazine which acts as a chemiluminescent peroxide has been made (*F. McCapra, Y.C. Chang* and *A. Burford*, Chem. Comm., 1976, 608). The 1,3,2-dioxazine (IX) has been synthesised by thermal cyclisation of MeOOCMe$_2$CN(OMe)O-(CH$_2$)$_3$OH in carbon tetrachloride (*V.G. Shtamburg et al.*, Iz. Akad. Nauk

S.S.S.R. Ser. Khim., 1980, 2669). The more accessible 1,4,2-system is derived in the main from hydroxamic acid derivatives, *e.g.* compound (X), m.p. 126–127°, from oxalyl chloride and *N*-benzoylhydroxylamine (*E.H. Burk* and *D.D. Carlos*, F.P. 1,543,701, 1968; C.A., 1970, **72**, 21506w) or the fully saturated analogue (XI) from benzilic acid, *N*-isopropylhydroxylamine and 1,1′-carbonyldiimidazole (*D. Geffken*, Arch. Pharm., 1980, **313**, 377). Formaldehyde condenses with amines or sulphonamides to give 1,3,5-dioxazines normally along with the related *s*-triazines and 1,3,5-oxadiazines. Compound (XII; $R = 4\text{-MeC}_6\text{H}_4\text{SO}_2$) has m.p. 139–140° and the analogue (XII; $R = \text{Me}_2\text{N(CH}_2)_3$) has b.p. 45°/0.01 mm.

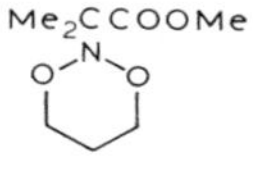

(IX) 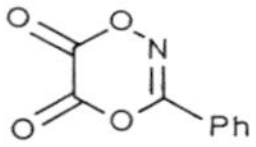 (X) 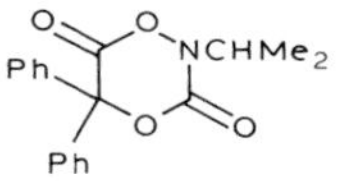 (XI) 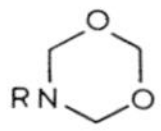 (XII)

5. Oxathiazines

Derivatives of the following ring systems have been noted (the representations are schematic and do not take account of double bonds). The 1,2,3- and 1,3,5-systems appear to have been the most extensively studied.

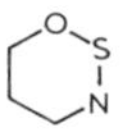

1,2,3 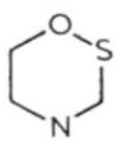   1,2,4 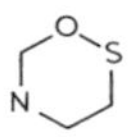   1,2,5 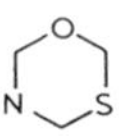   1,3,5 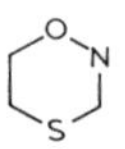   1,4,2 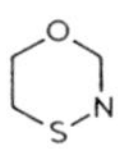   1,4,3

(i) 1,2,3-Oxathiazines

Chlorosulphonyl isocyanate reacts with alkynes (1:1 ratio) to yield 1,2,3-oxathiazine-2,2-dioxides, *e.g.* compound (I), m.p. 54–55°, which on treatment with lithium aluminium hydride gives its 3,4-dihydro derivative (II) (*E.J. Moriconi et al.*, Tetrahedron Letters, 1970, 27; J. org. Chem., 1972, **7**, 196). Compounds of type (I) hydrolyse in water or methanol but undergo nucleophilic displacement of chlorine with thiophenols (*J. Sander* and *K. Clauss*, Angew. Chem., 1980, **82**, 138).

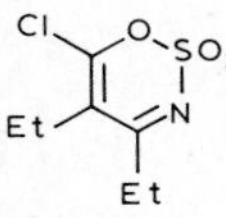

Related experiments but using excess halosulphonyl isocyanates on alkynes give rise to different products (III) (*e.g.* R = R' = Me, m.p. 108°) (*K. Clauss* and *H. Jensen*, Tetrahedron Letters, 1970, 119; *K.-D. Kampe*, ibid., 1970, 123). Such compounds (III) may also be derived by the action of chlorosulphonyl isocyanate in ether on ketones of the type $RCOCH_2R'$ (*J.K. Rasmussen* and *A. Hassner*, J. org. Chem., 1973, **38**, 2114; J. Amer. chem. Soc., 1975, **97**, 1451). Likewise, benzo-1,2,3-oxathiazine-2,2-dioxides may be prepared from 2-hydroxy-benzaldehydes or -acetophenones (*A. Kamal* and *P.B. Sattur*, Synthesis, 1981, 272).

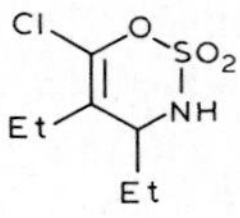

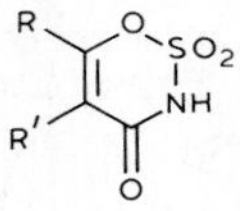

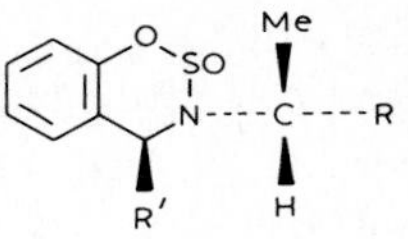

1,2,3-Oxathiazine-2-oxides (IV), useful as chiral auxiliaries, have been synthesised by the action of thionyl chloride on chiral *o*-hydroxybenzylamines. The oxathiazine (IV), after equilibration, is treated with phenylmagnesium bromide and then butyl lithium to give PhSOBu in high enantiomeric excess (*K. Hiroi, S. Sato* and *R. Kitayama*, Chem. Letters, 1980, 1595).

Further interest in these compounds has centred in their use as sweeteners.

(ii) 1,3,5-Oxathiazines

6*H*-1,3,5-Oxathiazin-6-ones arise *via* [4 + 2] cycloaddition reactions involving RCSNCO (R = EtO, Ph, *etc.*) and aldehydes or ketones, *e.g.* compound (V), m.p. 84°, from chloral (*A. Schulze* and *J. Goerdeler*, Tetrahedron Letters, 1974, 221; Ber., 1982, **115**, 3063; *K. Burger* and *H. Partscht*, Chem. Ztg., 1982, **106**, 303). Ketenes react similarly (*J. Goerdeler, R. Schimpf* and *M.-L. Tieot*, Ber., 1972, **105**, 3322).

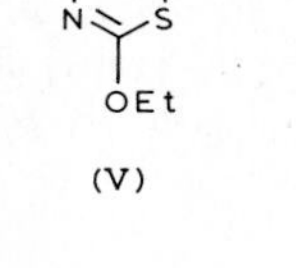

The title compounds may also be derived by the interaction of thioamides and aldehydes and ketones (VI; R = Me, m.p. 32°; R = Ph, m.p. 43°) (*C. Giordano et al.*, Synthesis, 1975, 789; 1979, 801; *K. Burger, R. Ottlinger* and *J. Albanbauer*, Ber., 1977, **110**, 2114). Some 1,3,5-oxathiazines on thermolysis form 1,3,5-thiadiazines, *e.g.* (VI) → (VII) (*K. Burger, loc. cit.*; *C. Giordano, A. Belli* and *L. Abis*, Tetrahedron Letters, 1979, 1537).

(iii) Other oxathiazine systems

Perhydro-1,4,2-oxathiazines (VIII) have been prepared by the condensation of ketones and the thiol $HS(CH_2)_2ONH_2$ (*L. Bauer, K.S. Suresh* and *B.K. Ghosh*, J. org. Chem., 1965, **30**, 949) and oximino chlorides condense with thiols to yield Δ-2,3-species (*S. Hoff* and *E. Zwanenburg*, Rec. Trav. chim., 1973, **92**, 929).

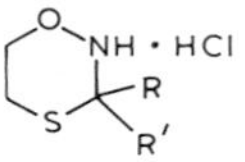

X-ray analysis confirms the structure of the 1,2,4-**oxathiazine** (IX), m.p. 173°, prepared by condensation of *N*-methylethanolamine and benzoyl isothiocyanate under oxidative conditions (*S. Solyom et al.*, Tetrahedron Letters, 1977, 4245)

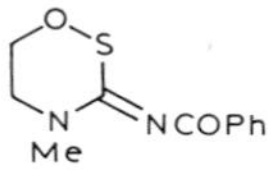

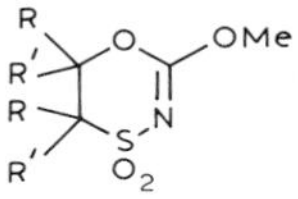

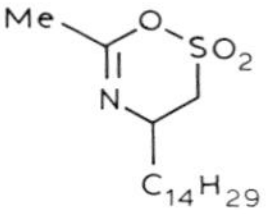

[4 + 2] Cycloreversion reactions of 1,4,3,5-oxathiadiazine-dioxides in the presence of alkenes give rise to 5,6-dihydro-1,4,3-oxathiazines of type (X) (*E.M. Burgess* and *W.M. Williams*, J. org. Chem., 1973, **38**, 1249) and 2,3-dihydro species have been derived from condensation in base of aldehyde imines and $PhCOCH_2SO_2Cl$ (*O. Tsuge* and *S. Iwanami*, Bull. chem. Soc. Japan, 1970, **43**, 3543).

Sulphur trioxide condenses with acetonitrile and hexadec-1-ene to give the 1,2,5-oxathiazine (XI) (*G.L. Broussalian*, C.A., 1966, **64**, 15902d) and also reacts with 4-methylphenyl isocyanate to yield 7-**methyl**-2,1,4-**benzoxathiazin**-3-**one**-1,1-**dioxide**, m.p. 183° (dec.) (*G. Balle, L. Born* and *D. Dieterich*, Angew. Chem., 1982, **84**, 872).

6. Trithianes, trioxanes, oxadithianes and dioxathianes

Six-membered ring compounds symmetrically substituted with either three oxygen or sulphur atoms are trimers of aldehydes (Vol. IC), thioaldehydes (Vol. IC) or thioketones (Vol. IC), respectively.

The naturally occurring 5-methylthio-1,2,3-trithiane (I) has been synthesised by selective methylation of propane-1,2,3-trithiol followed by treatment of the product with sulphur dichloride (*U. Anthoni et al.*, Tetrahedron, 1982, **38**, 2425) and the 1,2,4-trithiane (II) is the product of a Maillard reaction (*R.J.C. Kleipooc* and *A.C. Tas*, C.A., 1974, **81**, 63593p).

1,2,4-Trioxanes are the products of interaction (in air and in the absence of light) of arylamines and aldehydes, compound (III) having m.p. 80° (dec.) (*H. Yamamoto et al.*, J. chem. Soc. Perkin I, 1980, 2300).

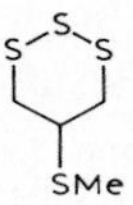

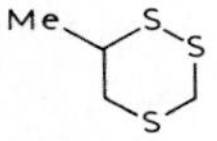

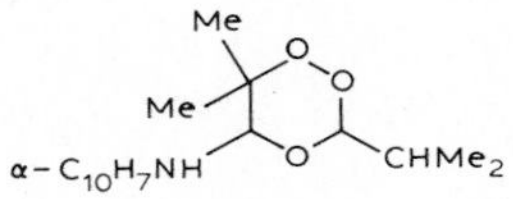

M.G. Gadzhieva and co-workers (Zhur. org. Khim., 1978, **14**, 2188) have prepared 1,3,5-oxadithianes (IV) from esters and sodium oxydimethylene dithiosulphate. Compound (IV) may be oxidised to its bis-dioxide. A 1,2,6-derivative (V) has also recently been reported (*F.E. Behr*, U.S.P. 4,329,478, 1982; C.A., 1982, **97**, 92993e).

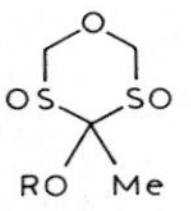

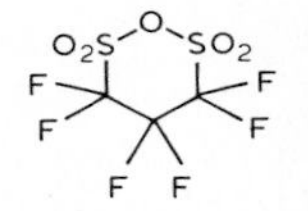

1,3,2-Dioxathiane-2-oxides (VI) arise from the action of thionyl chloride on 1,3-diols in base (*B.I. Mitsner et al.*, Zhur. org. Khim., 1973, **9**, 480; *L. Cazaux, G. Chassaing* and *P. Maroni*, Tetrahedron Letters, 1975, 2517) and the 1,3,4-derivative (VII), m.p. 178–179°, is one of the products of the reaction of sulphur dioxide on PhCOC(N$_2$)Ph (*J.B. Stothers, L.J. Danks* and *J.F. King*, Tetrahedron Letters, 1971, 2551).

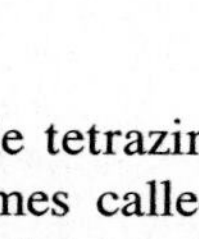

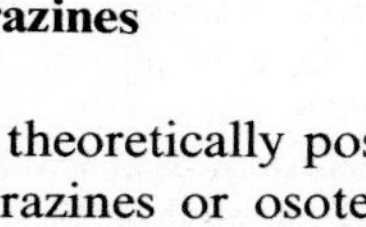

7. Tetrazines

Three classes of simple tetrazines are theoretically possible; (a) the 1,2,3,4-tetrazines (I) (sometimes called *v*-tetrazines or osotetrazines in the older literature), (b) the 1,2,3,5-tetrazines (II) (also known as *as*-tetrazines) and

(c) the 1,2,4,5-tetrazines (III) (commonly called symmetric or *s*-tetrazines). Of these three possible structural arrangements, derivatives of the 1,2,4,5-tetrazine system are by far the best known and indeed, 1,2,4,5-tetrazine is the only parent compound of the three to have been synthesised.

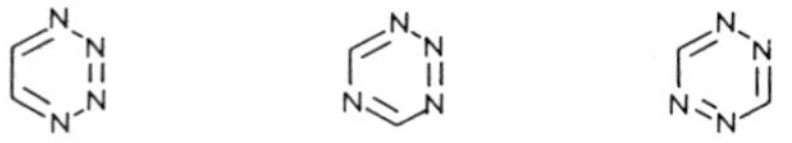

(a) 1,2,3,4-Tetrazines

All claims to have prepared derivatives of 1,2,3,4-tetrazines and which appear in the literature prior to the early 1970s are either clearly wrong or must be regarded with much scepticism as indeed must any later work not adequately backed by spectroscopic data and hence these early publications will not be referred to.

The first authenticated derivative of this series to be synthesised was 1,4-**dimethyl**-1,4,5,6-**tetrahydro**-1,2,3,4-**tetrazine** (IV; R = Me) b.p. 90°/10 mm, prepared in low yield by sodium hypochlorite oxidation of N,N'-diamino-N,N'-dimethylethylenediamine (*S.F. Nelson* and *R. Fibiger*, J. Amer. chem. Soc., 1972, **94**, 8497). Later a series of compounds based on structure (IV) was prepared in better yields (*D. Seebach et al.*, Angew. Chem. intern. Edn., 1973, **12**, 495) from lithium derivatives of alkyl nitrosamines. The intermediate 2-*oxide* (V; R = Me), m.p. 60°, for which an X-ray crystal structure was obtained, was reduced to the tetrazine (IV; R = Me) by lithium aluminium hydride (*D. Seebach et al.*, Helv., 1978, **61**, 1622).

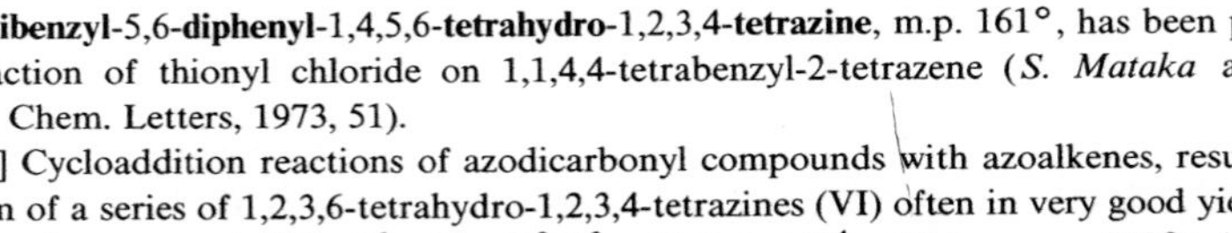

1,4-**Dibenzyl**-5,6-**diphenyl**-1,4,5,6-**tetrahydro**-1,2,3,4-**tetrazine**, m.p. 161°, has been prepared by the action of thionyl chloride on 1,1,4,4-tetrabenzyl-2-tetrazene (*S. Mataka* and *J.P. Anselme*, Chem. Letters, 1973, 51).

[4 + 2] Cycloaddition reactions of azodicarbonyl compounds with azoalkenes, result in the formation of a series of 1,2,3,6-tetrahydro-1,2,3,4-tetrazines (VI) often in very good yields. The structure of compound (VI; R^1 = Ph, R^2-R^3 = $(CH_2)_4$, R^4 = OMe; m.p. 132°) has been confirmed by an X-ray structure determination and by spectral evidence. Monocyclic derivatives of compound (VI) (*e.g.* $R^1 = R^2 = Me$, $R^3 = H$, $R^4 = Ph$; m.p. 134°) have been prepared (*S. Sommer* and *U. Schubert*, Angew. Chem. intern. Edn., 1979, **18**, 696).

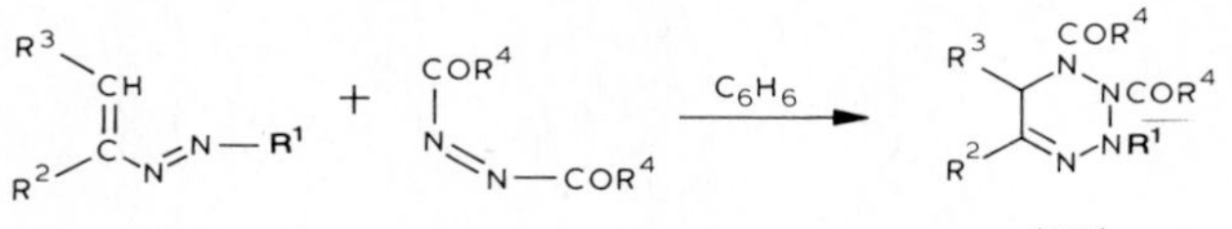

(VI)

1-**Methyl**-3*H*-**pyrazolo**[1,2-*a*]**benzotetrazin**-3-**one** (VII), a red solid, m.p. 188–189°, has been prepared and has been shown to decompose in hot ethanol (*A.M. Almerico* and *A.J. Boulton*, Chem. Comm., 1985, 204).

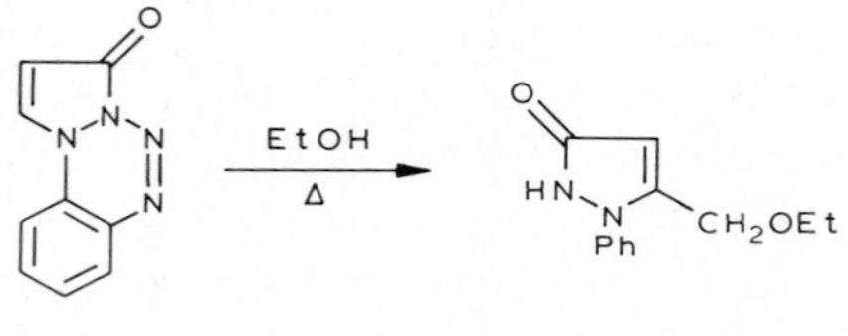

(VII)

(b) 1,2,3,5-Tetrazines

Derivatives of this ring structure are exceptionally rare and simple members are as yet unknown. Potassium salts of the tetrazines (VIII) and (IX) have been prepared by electrochemical oxidation of cyanamide using platinum anodes in aqueous potassium hydroxide. The potassium salts of the species (VIII) and (IX), on acidification, rearrange to *s*-triazine derivatives, *e.g.* compound (IX) produces melamine (*K. Kubo, T. Nonaka* and *K. Odo*, Bull chem. Soc. Japan, 1976, **49**, 1339).

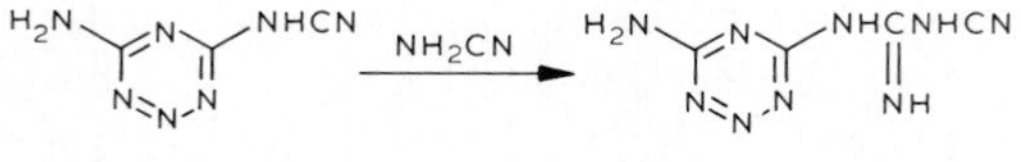

(VIII)   (IX)

A series of 1-aryl-4-dialkylamino-1,2,3,5-tetrazin-6-ones (XI) has been prepared from compounds of type (X) and their structure proved by X-ray analysis of compound (XI; Ar = Ph, R = Me). The tetrazinones (XI) all decompose at *ca.* 120° to give nitrogen, dimethylcyanamide and aryl isocyanates (*A.E. Baydar et al.*, J. chem. Soc. Perkin I, 1985, 415).

$$R_2N-C(NH_2)=N-N=NAr \quad (X) \longrightarrow R_2N-C(N=C=O)=N-N=NAr \longrightarrow R_2N-\text{(triazinone)} \quad (XI)$$

Conflicting theoretical evidence relating to the stability of 1,2,3,5-tetrazine (II) has been reviewed by *P.F. Wiley* ("Heterocyclic Compounds", Vol. 33, eds. *H. Neunhoeffer* and *P.F. Wiley*, Wiley, New York, 1978, p. 1296).

(c) 1,2,4,5-Tetrazines

(i) General preparations and properties

1,2,4,5-Tetrazines, normally orange-red to red-violet in colour, exhibit absorption maxima between 500 and 550 nm ($n \rightarrow \pi^*$ transitions) and between 250 and 300 nm ($\pi \rightarrow \pi^*$ transitions). The nitrogen atoms are non-basic and the ring system has only weakly developed aromatic properties resulting in many reactions leading to opening or transformation of the ring; hence 1,2,4,5-tetrazines with 3,6-diaryl or electron-withdrawing substituents can behave as dienes in Diels-Alder type reactions (*R.A. Carboni*, U.S.P., 2,817,662; C.A., 1958, **52**, 7360; *R.A. Carboni* and *R.V. Lindsey*, J. Amer. chem. Soc., 1959, **81**, 4342). With the exception of a number of the lower 3,6-dialkyl-1,2,4,5-tetrazines, which are somewhat unstable red oils, and the parent compound itself, 1,2,4,5-tetrazine (III), which although crystalline is unstable in air, 1,2,4,5-tetrazines are normally stable crystalline compounds.

1,2,4,5-Tetrazines are readily reduced to dihydro derivatives, sodium dithionite or zinc in acetic acid being commonly employed. The nature of these dihydro derivatives has been a source of much controversy with 1,2- and/or 1,4-dihydro structures (XII and XIII respectively) being assigned without much real evidence being put forward for either structure. The matter has also been further complicated by a rearrangement reaction of dihydro-1,2,4,5-tetrazines (XIII) under acid conditions which gives rise to the corresponding isomeric 4-amino-1,2,4-triazoles (XIV). However, n.m.r. studies suggest that the NH protons of dihydrotetrazines appear at lower field (δ 8.8–9.1) than the NH_2 protons of the 4-amino-1,2,4-triazoles (δ 3.4–6.2) (*R.A. Bowie et al.*, J. chem. Soc. Perkin I, 1972, 2395). X-ray evidence now clearly points to the 1,4-dihydrotetrazine structure (XIII) rather than the 1,2-dihydro form (XII) in the solid state (*D. Hunter, D.G. Neilson* and *T.J.R. Weakley*, J. chem. Soc. Perkin I, 1982, 1165; *A. Neugebauer et al.*, Ber., 1983, **116**, 2261). It is doubtful if any 1,2-dihydro structures exist in the solid state where for compound (XII) $R^1 = H$, although the derivatives $R^1 \neq H$ can be readily synthesised. The 1,4-dihydro-1,2,4,5-tetrazine ⇌ 1,2,4,5-tetrazine redox system is akin to the hydroquinone ⇌ quinone system and reduction potentials have been measured (*T. Troll*, Electrochim. Acta, 1982, **27**, 1311).

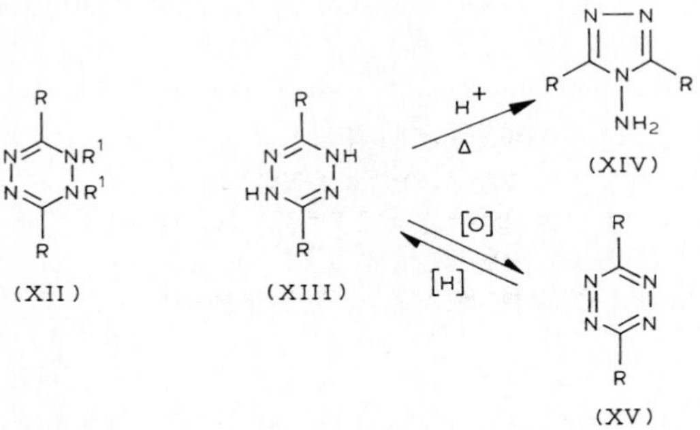

Hence the 1,4-dihydro derivatives (XIII) can be readily oxidised to the corresponding 1,2,4,5-tetrazines (XV), atmospheric oxygen, bromine or nitrous acid being commonly employed. Samples of 1,4-dihydro-1,2,4,5-tetrazines (normally off-white to straw in colour) appear relatively stable to air and light provided that they are pure and dry, otherwise oxidation is shown by the development of the red colouration of the parent tetrazine. Indeed, to obtain good samples of 1,4-dihydro-1,2,4,5-tetrazines (XIII) it is often advisable to prepare the tetrazine and reduce it back to its 1,4-dihydro derivative, *e.g.* with sodium dithionite.

The synthesis of 1,4-dihydro-1,2,4,5-tetrazines has been carried out by the action of hydrazine on nitriles (normally aryl) (*A. Charronat* and *P. Fabiani*, Compt. rend., 1955, **241**, 1783) or their derivatives, *e.g.* thioamides (*A. Junghahn*, Ber., 1898, **31**, 312), imidates (*A. Pinner*, Ber., 1893, **26**, 2126) or amidines (*J.L. Fahey et al.*, J. chem. Soc. C, 1970, 719). These preparations then yield the corresponding 1,2,4,5-tetrazines (XV) by mild oxidation of the dihydro intermediates (XIII).

$$RX + NH_2NH_2 \longrightarrow (XIII) \xrightarrow{[O]} (XV)$$

$$X = CN, CSNH_2, C(=NH)OR', C(=NH)NH_2$$

1,4-Dihydro-1,2,4,5-tetrazines, and subsequently the corresponding tetrazines by oxidation, may also be prepared by the action of hydrazine on the product of the reaction of diacyl hydrazines and phosphorus pentachloride (*R. Stollé* and *J. Laux*, Ber., 1911, **44**, 1127).

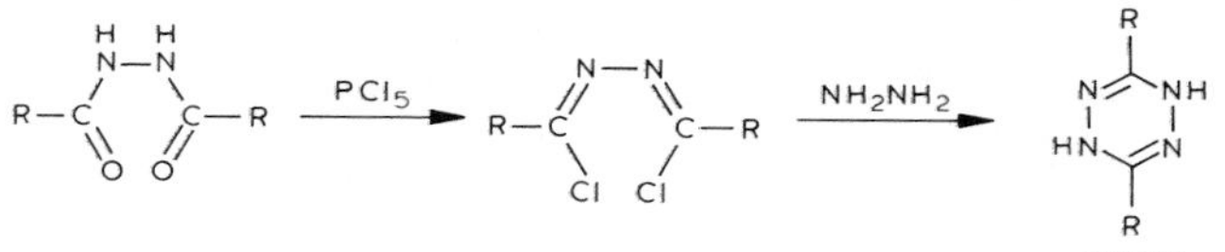

$$R-C(=O)-NH-NH-C(=O)-R \xrightarrow{PCl_5} R-C(Cl)=N-N=C(Cl)-R \xrightarrow{NH_2NH_2} \text{(XIII)}$$

(XIII)

Less commonly, 3,6-unsymmetrically-disubstituted-1,2,4,5-tetrazines have been synthesised by reacting a mixture of two different amidines (*R.A. Bowie et al.*, loc. cit.) or an amidine and an imidate (*S.A. Lang, B.D. Johnson* and *E. Cohen*, J. heterocyclic Chem., 1975, **12**, 1143), with hydrazine, and after oxidising, separating chromatographically the tetrazines formed.

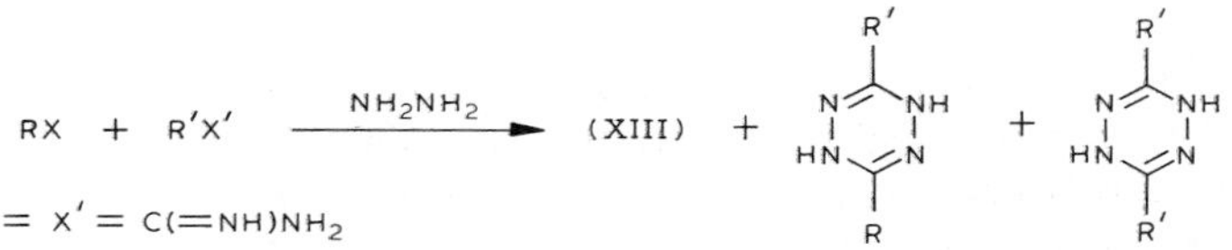

$$RX + R'X' \xrightarrow{NH_2NH_2} \text{(XIII)} + \text{[tetrazine with R' top, R bottom]} + \text{[tetrazine with R' top, R' bottom]}$$

X = X' = C(=NH)NH₂

X = C(=NH)NH₂ ;
X' = C(=NH)OR''

1,2,4,5-Tetrazines do not appear to yield *N*-oxides but instead peracids open the tetrazine ring and give 2,5-disubstituted-1,3,4-oxadiazoles, especially in alkaline media (*H.J. Haddadin, S.J. Firsan* and *B.S. Nader*, J. org. Chem., 1979, **44**, 629).

(ii) Reduced 1,2,4,5-tetrazines

Of the four possible isomeric dihydro-1,2,4,5-tetrazines having two hydrogens rather than substituents on nitrogen, the 1,2- and 3,6-dihydro series do not as yet appear to be known with certainty. 1,6-Dihydro derivatives, investigated as homoaromatic systems, have been reported to arise from the reduction of the parent tetrazine with sodium borohydride (*A. Counotte-Potman, H.C. van der Plas* and *B. von Veldhuizen*, J. org. Chem., 1981, **46**, 2138) or by oxidation (Pt/O₂) of hexahydro-1,2,4,5-tetrazines (*W. Skorianetz* and *E. sz. Kováts*, Helv., 1972, **55**, 1404). The 1,6-dihydro compounds also arise as intermediates in the formation of 6-alkylamino-3-aryl-1,2,4,5-tetrazines formed by attack of an alkylamine on a 3-aryl-1,2,4,5-tetrazine (*A. Counotte-Potman* and *H.C. van der Plas*, J. heterocyclic Chem., 1981, **18**, 123). 1,4-Dihydro-1,2,4,5-tetrazines are discussed above.

Few tetrahydro derivatives are known but *W. Skorianetz* and *E. sz. Kováts* (*loc. cit.*) investigated the formation of 1,2,3,6-tetrahydro-3,6-dimethyl-1,2,4,5-tetrazine by oxidation (HgO) of the corresponding hexahydro compound and they also isolated 1,2,3,4-tetrahydro-3,6-dimethyl-1,2,4,5-tetrazine by reduction (H_2S) of 1,4-dihydro-3,6-dimethyl-1,2,4,5-tetrazine.

Hexahydro-1,2,4,5-tetrazines, prepared from the interaction of hydrazines and aldehydes, are also known (*W. Skorianetz* and *E. sz. Kováts*, Helv., 1971, **54**, 1922) but more commonly, substituted hydrazines rather than hydrazine itself have been employed giving rise to more complex substitution patterns including fused systems (see later). Hexahydro-1,2,4,5-tetrazines are colourless compounds which undergo hydrolysis in both acidic and basic media.

(iii) Properties of 1,2,4,5-tetrazines

1,2,4,5-**Tetrazine**, purple rods, m.p. 99°, prepared by decarboxylation of its 3,6-dicarboxylic acid (below), was first characterised by *T. Curtius, A. Darapsky* and *E. Müller* (Ber., 1907, **40**, 84). 1,2,4,5-Tetrazine is volatile but unstable at ordinary temperatures and decomposes in air, hence the decarboxylation of the dicarboxylic acid is best carried out *in vacuo* (*C. Fridh et al.*, C.A., 1972, **77**, 113264). Metal salts (*e.g.* potassium) of somewhat uncertain composition have been reported (*D. Wood* and *F.W. Bergstrom*, J. Amer. chem. Soc., 1933, **55**, 3648). 1,2,4,5-Tetrazine can be reversibly reduced to 1,4-*dihydro*-1,2,4,5-*tetrazine*, yellow prisms, m.p. 125°, which may also be prepared by the action of hydrazine on ethyl formimidate, as above.

3,6-*Dimethyl*-1,2,4,5-*tetrazine* has m.p. 74° but by contrast most other members of the series substituted in the 3,6-positions with aliphatic groups of low molecular weight are somewhat unstable liquids for which boiling points are not quoted (*W. Skorianetz* and *E. sz. Kováts*, Tetrahedron Letters, 1966, 5067).

3,6-*Dibenzyl*-1,2,4,5-*tetrazine*, a stable solid, m.p. 74° (*A. Pinner*, Ber., 1897, **30**, 1889), decomposes in alkaline conditions giving, for example, on treatment with potassium hydroxide in methanol, 3-benzyl-7-methoxy-6-phenylimidazo[1,2-*b*]tetrazine among other products (*D.G. Neilson, K.M. Watson* and *T.J.R. Weakley*, J. chem. Soc. Perkin I., 1979, 333). 3,6-*Dibenzyl*-1,4-*dihydro*-1,2,4,5-*tetrazine*, m.p. 158°, isomerises readily, especially in acid conditions, to give 4-amino-3,5-dibenzyl-1,2,4-triazole (*R.A. Bowie et al., ibid.*, 1972, 2395).

Meso-, (±)- and optically active forms of 3,6-**bis**(1-**hydroxy**-1-**phenylethyl**)-1,2,4,5-**tetrazine** (m.p. 186–187°, 133–134° and 121°, respectively) and of its *p-methyl* derivative (m.p. 166–167°, 139–141° and 117–118°, respectively) have been prepared from the corresponding (±)- and (+)- or (−)-atrolact- and -*p*-methylatrolact-amidinium chlorides. C.D maxima were observed around 270–272 nm for the optically active compounds (*D.G. Neilson, S. Mahmood* and *K.M. Watson, ibid.*, 1973, 335).

3,6-**Diphenyl**-, 3,6-**bis**(*p*-**tolyl**)- and 3,6-**bis**(*m*-**tolyl**)-1,2,4,5-**tetrazines**, m.p. 198°, 232° and 150–152°, respectively are synthesised by oxidation of the corresponding 1,4-*dihydro* derivatives (m.p. 192°, 223° and 194°) (*A. Junghahn*, Ber., 1898, **31**, 312: *A. Pinner, ibid.*, 1893, **26**, 2126; 1894, **27**, 984, 3273; *R. Stollé*, J. pr. Chem., 1903, **68**, 464; 1906, **73**, 277). These fully aromatic compounds, among the most stable of the 1,2,4,5-tetrazines, are water insoluble and tend to have low solubility in many organic solvents. 3,6-Diaryl-1,2,4,5-tetrazines are hydrolysed to the corresponding hydrazones of the type ArCONHN = CHAr by treatment with boiling ethanolic solutions of potassium hydroxide (*A. Pinner*, Ber., 1894, **27**, 984). Despite its aromatic character, 3,6-diphenyl-1,2,4,5-tetrazine acts as a diene in Diels-Alder type reactions forming pyridazines and their 1,4-dihydro derivatives (*R.A. Carboni* and *R.V. Lindsey, loc. cit.*). Alkyl and aryl Grignard reagents react with 3,6-diphenyl-1,2,4,5-tetrazine attacking at a ring nitrogen to give rise to, *e.g.*, 1,4-*dihydro*-1-*methyl*-3,6-*diphenyl*-1,2,4,5-*tetrazine*, m.p. 159° (from MeMgI), and 1,4-*dihydro*-1,3,6-*triphenyl*-1,2,4,5-*tetrazine*, m.p. 124–125° (from PhMgBr) (*A. Neugebauer et al.*, Ber., 1983, **116**, 2261; *D. Hunter* and *D.G. Neilson*, J. chem. Soc. Perkin I, 1984, 2779).

1,4-*Dihydro*-3,6-*diphenyl*-1,2,4,5-*tetrazine*, m.p. 192°, is yellow in colour and non-basic. It can be hydrolysed in acid solution to hydrazine, the hydrazide (PhCONHNHCOPh) and benzoic acid. A solution of potassium hydroxide in methanol exposed to air hydrolyses the 1,4-dihydrotetrazine to 2,5-diphenyl-1,3,4-oxadiazole, the same reagent under nitrogen has no effect on the dihydrotetrazine (*D. Hunter* and *D.G. Neilson, ibid.*, 1985, 1081).

3-**Phenyl**-1,2,4,5-**tetrazine**, m.p. 125°, and 3-**methyl**-6-**phenyl**-1,2,4,5-**tetrazine**, m.p. 75–77°, are stable compounds derived by the action of hydrazine hydrate on benzamidine hydrochloride and formamidine acetate or acetamidine hydrochloride respectively (*R.A. Bowie et al., loc. cit.*).

3,6-*Dimethylthio*-1,2,4,5-*tetrazine*, m.p. 83°, (*A.W. Lutz et al.*, U.S.P. 3,155,488; C.A., 1965, **62**, 1676) undergoes successive replacement of its thiomethyl groups with amines, *e.g.* dimethylamine gives 3,6-**bis(dimethylamino)**-1,2,4,5-**tetrazine**, m.p. 121–123°.

3,6-**Diamino**-1,2,4,5-**tetrazine**, m.p. 300°, has been obtained from 1,2,4,5-tetrazine-3,6-dicarboxylic acid (see below) by way of the Curtius rearrangement of its diazide derivative and also from *S*-methyl isothiosemicarbazide (*C.-H. Lin, E. Lieber* and *J.P. Horwitz*, J. Amer. chem. Soc., 1954, **76**, 427).

1,2,4,5-**Tetrazine**-3,6-**dicarboxylic acid**, m.p. 148° (dec.), can be obtained by oxidation of its 1,4-dihydro derivative prepared by dimerisation of ethyl diazoacetate in base (*G.H. Spencer, P.C. Cross* and *K.B. Wiberg*, J. chem. Phys., 1961, **35**, 1939). It can readily be reduced back to its 1,4-dihydro form or decarboxylated to 1,2,4,5-tetrazine (see above). The dimethyl and diethyl esters are known (*T. Curtius* and *E. Rimele*, Ber., 1908, **41**, 3108) as is its 3,6-dicarboxamide (indefinite m.p.) which can be dehydrated to yield 3,6-dicyano-1,2,4,5-tetrazine (*E. Gryszkiewicz-Trochimowski* and *M. Bousquet*, Compt. rend., 1961, **253**, 2992).

3,6-**Dihydroxy**-1,2,4,5-**tetrazine** may be prepared by dehydrogenation of hexahydro-1,2,4,5-tetrazine-3,6-dione. The compound, m.p. 168° (dec.), exists as a tautomeric mixture (XVI ⇌ XVII) in solution but can be converted into 3,6-*dimethoxy*-1,2,4,5-*tetrazine*, m.p. 62–63°, by the action of diazomethane (*F.A. Neugebauer* and *H. Fischer*, Ann., 1982, 387).

(XVI) (XVII)

1,2,4,5-Tetraphenylhexahydro-1,2,4,5-tetrazine, m.p. 200°, results from the interaction of formaldehyde and hydrazobenzene; higher aldehydes give rise to related compounds with substituents in the 3,6-positions (*C.A. Bischoff*, Ber., 1898, **31**, 3250).

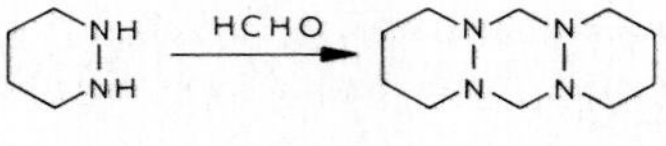

(XVIII)

Condensed systems are also known, *e.g.* formaldehyde condenses with cyclic hydrazines to give products of the type (XVIII), m.p. 170° (*R.A.Y. Jones et al.*, J. chem. Soc. Perkin II, 1974, 948), and bridged structures, *e.g.* compound (XIX), m.p. 144°, may also be derived *via* formaldehyde (*E. Schmitz* and *R. Ohme*, Ann., 1960, **635**, 82).

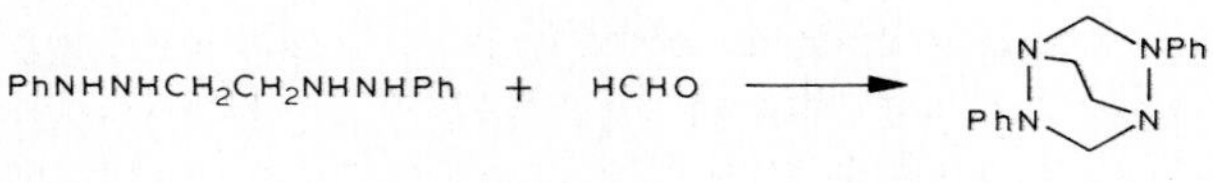

(XIX)

Early reports relating to the formation of hexahydro-1,2,4,5-tetrazine-3,6-dione (XX), also known are *p*-urazine, must be regarded with scepticism. However, *F.A. Neugebauer* and *H. Fischer* (*loc. cit.*) appear to have successfully synthesised the title compound by the reaction sequence below. Thermolysis over 200° converts the tetrazine (XX) into the triazolidine (XXI). However, the tetrazine (XX) may be dehydrogenated to give 3,6-dihydroxy-1,2,4,5-tetrazine.

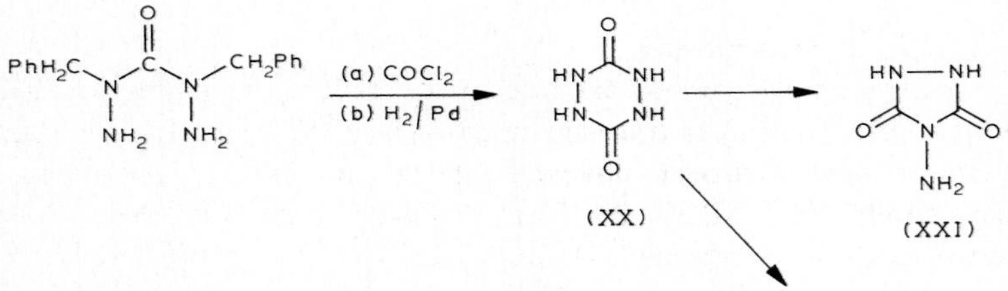

(XX) (XXI)

(XVI)

(iv) Verdazyls

Verdazyls (XXII) are intensely coloured (blue/green–black) stable free radicals, exhibiting e.s.r. spectra with nine lines and hyperfine splitting. Verdazyls, which are essentially derivatives of 2,3,4-trihydro-1,2,4,5-tetrazines have been much investigated by *Kuhn* and his co-workers (*R. Kuhn* and *H. Trischmann*, Angew. Chem., 1963, **75**, 294). Synthesis, normally starting from a 1,3,5-trisubstituted formazan, proceeds either by alkylation of the formazan in base followed by oxidation of the intermediate 1,2,3,4-

tetrahydro-1,2,4,5-tetrazine or alternatively by reduction of the cyclised product obtained from the condensation of an aldehyde and a 1,3,5-formazan.

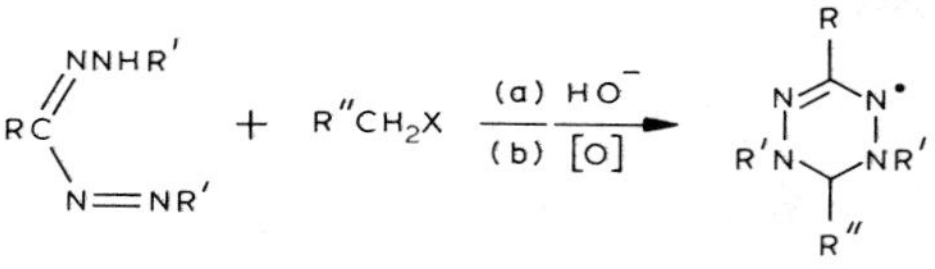

(XXII)

Reduction of a verdazyl, either catalytically (*F.A. Neugebauer* and *A. Mannschreck*, Tetrahedron, 1972, **28**, 2533) or chemically with sodium dithionite or zinc and acetic acid (*R. Kuhn* and *H. Trischmann*, Monatsch., 1964, **95**, 457) gives the corresponding 1,2,3,4-tetrahydro-1,2,4,5-tetrazine. Oxidation by oxygen over activated charcoal regenerates the formazan (*R. Kuhn, F.A. Neugebauer* and *H. Trischmann*, Monatsh, 1967, **98**, 726) and interaction with halogens gives verdazylium salts (*R. Kuhn* and *G. Fischer-Schwarz*, Monatsh., 1966, **97**, 517).

8. Thiatriazines and oxatriazines

Derivatives of 1,2,3,6-, 1,2,4,6- and 1,3,4,5-thiatriazine systems have been synthesised with the sulphur atom in various oxidation states.

(i) 2H-1,2,3,6-Thiatriazines

Compounds of type (I) arise from the action of hydrogen sulphide on $ArC(N_2Ph):C(N_2Ph)Ar$ but they rearrange to the related 1,2,3-triazoles on treatment with hydrogen chloride (*A. Spasov* and *B. Chemishev*, Dokl. Bulg. Akad. Nauk, 1970, **23**, 791; C.A., 1970, **73**, 120593z). Sulphur(IV) derivatives, *e.g.* compound (II) m.p. 114–116°, are obtained from [4 + 2] cycloaddition reactions of $ArSO_2NSO$ and N,N'-disubstituted diimides. Such compounds (II) are hydrolysed by aqueous acid and undergo ring contractions to Δ^3-1,2,3-thiadiazolines on treatment with trifluoroacetic acid (*S. Sommer* and *U. Schubert*, Ber., 1978, **111**, 1989).

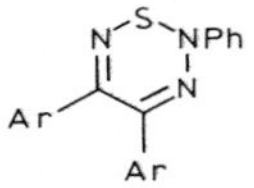

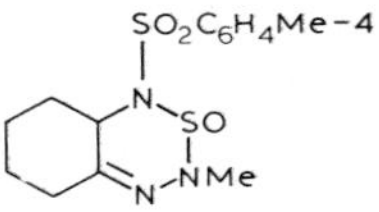

(ii) 1H-1,2,4,6-Thiatriazines

Amidines and their derivatives are useful starting materials for the title compounds; thus benzamidine reacts with $(CF_3)_2C:NSPh$ (*Y.G. Shermolovich*, Zhur. org. Khim., 1982, **18**, 2539), its *N*-bromo derivative with RSNa (*J. Goerdeler* and *D. Loeven*, Ber., 1954, **87**, 1079) and its *N*-cyano derivative with sulphur dichloride (*P.P. Kornuta, L.I. Derii* and *E.A. Romanenko*, C.A., 1978, **88**, 152576) to give compounds (III), (IV) and (V), respectively. Sulphur halogen atoms can be replaced by amines; likewise nucleophiles such as thiols and amines displace halogens on carbon (*W. Schram et al.*, Z. Chem., 1974, **14**, 471; 1975, **15**, 19, 57). 1-Chloro compounds yield free radicals (VI) on treatment with sodium or triphenylverdazyl (*L.N. Markovskii et al.*, Sulphur Letters, 1983, **1**, 143).

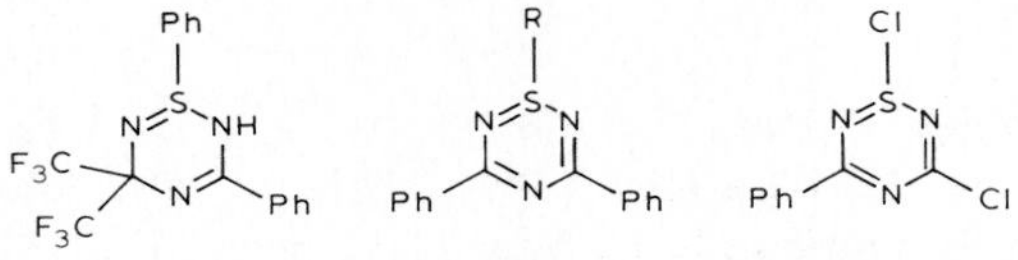

(iii) 2H-1,2,4,6-Thiatriazines

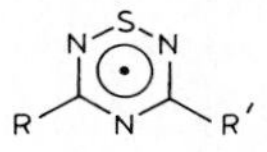

The N–S–N unit of this substitution pattern may be introduced by the action of formaldehyde and an amine on, *e.g.*, $(MeNH)_2SO_2$, compound (VII) (*B. Unterhalt, E. Seebach* and *D. Thamer*, Arch. Pharm., 1978, **311**, 47) or the N–C–N–C–N unit from biuret derivatives, *e.g.* compound (VIII) from bis-*O*-methylbiuret and $ArSO_2NSO$ (*E. Fischer, G. Rembarz* and *M. Teller*, J. pr. Chem., 1982, **324**, 920). Otherwise an N–C–N unit can be introduced by the condensation of amidines with $ClCON(Me)SO_2Cl$—2-aminopyridine leading to compound (IX) (*D. Bartholomew* and *I.T. Kay*, J. chem. Res.(S), 1977, 238)—or with chlorosulphonyl isocyanate [*e.g.* compound (X)] (*W. Friedrichsen, G. Moeckel* and *T. Debaerdemaeker*, Heterocycles, 1984, **22**, 63). The 3*H* of compound (XI) is reported to be replaced by chlorine on treatment with $PCl_5/POCl_3$—the halogen being susceptible to further nucleophilic displacement (*G. Hamprecht et al.*, Ger. Offen, 3,134,143 and 3,134,145; C.A., 1983, **98**, 198289r, 1983, **99**, 70774g).

(VII)

(VIII)

(IX)

(X)

(XI)

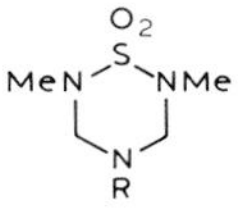

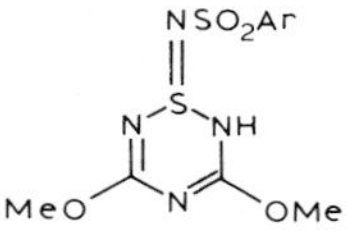

(iv) Other systems

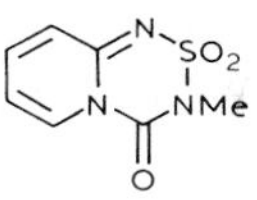

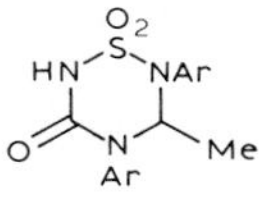

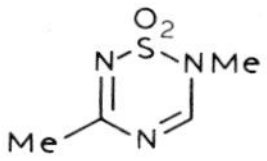

Bis-imines of the type RR′S(:NH)$_2$ react with *N*-chloromethylimidoyl halides to give compounds of type (XII) (*M. Haake, H. Fode* and *K. Ahrens*, Z. Naturforsch., 1973, **28B**, 539) and other S(VI) species, *e.g.* (XIII), have been derived by ring closure of the bis-acid fluoride FCONSF$_2$NCOF with amines (*I. Stahl, R. Mews* and *O. Glemser*, J. fluorine Chem., 1978, **11**, 455).

Plant protection products (XIV) have been prepared by the condensation of sulphamide and cyanates in base (*E. Grigat*, Ger. Offen., 2,026,625; C.A., 1972, **76**, 72564d).

(XII)

(XIII)

(XIV)

(XV)

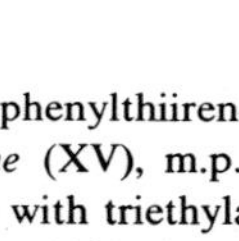

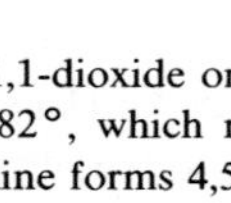

2,3-Diphenylthiirene-1,1-dioxide on treatment with lithium azide in acetonitrile yields the *thiatriazine* (XV), m.p. 182°, which may be ozonised to the 4-benzoyl derivative which on treatment with triethylamine forms 4,5-diphenyl-1,2,3-triazole (*B.B. Jarvis* and *G.P. Stahly*, J. org. Chem., 1980, **45**, 2604).

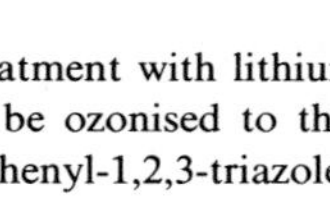

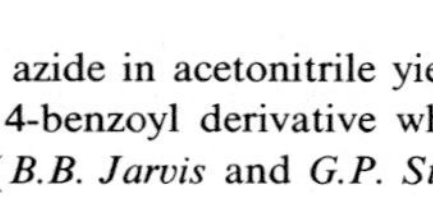

(v) *Oxatriazines*

2,4,6-**Triphenyl**-4*H*-1,3,4,5-**oxatriazine**, scarlet needles, m.p. 172°, and its 2,6-*diphenyl*-4-(*p-tolyl*) analogue, m.p. 140°, arise by photolysis of the corresponding triaryl-2*H*-1,2,3-triazole-*N*-oxides (*G.J. Gainsford* and *A.D. Woolhouse*, Austral. J. Chem., 1980, **33**, 2447).

9. Dithiadiazines, oxathiadiazines, dioxadiazines and dioxathiazines

(i) *Dithiadiazines*

Trichloromethyl sulphenyl chloride condenses with ammonia to give 3,6-**dichloro**-1,4,2,5-**dithiadiazine**, m.p. 201° (*A. Senning*, Angew. Chem., 1963, **75**, 450) and with organic bases, *e.g.* 2-aminopyridine, to furnish the reduced derivative (I), m.p. 130° (*J. Goerdeler* and *E.R. Erbach*, Ber., 1962, **95**, 1637).

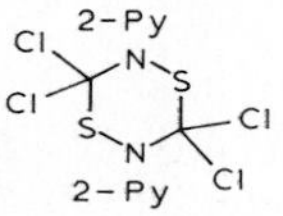

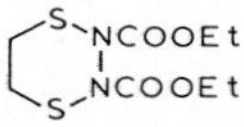

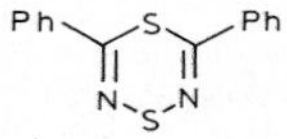

A related condensation of ethylene disulphenyl chloride with *N*,*N*'-dicarboethoxyhydrazine leads to a tetrahydro-1,4,2,3-dithiadiazine (II) (*K.H. Linke, R. Bimczok* and *H. Lingmann*, Angew. Chem. intern. Ed., 1971, **10**, 408).

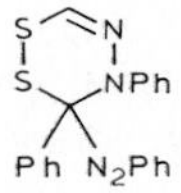

An X-ray structure analysis has shown the product of condensation of S_4N_4 and diphenylacetylene to be compound (III) (*S.T.A.K. Daley, C.W. Rees* and *D.J. Williams*, Chem. Comm., 1984, 57) and not 5,6-diphenyl-1,3,2,4-dithiadiazine as previously reported (*M. Tashiro* and *S. Mataka*, Heterocycles, 1976, **4**, 1243).

Thiohydrazides can be oxidised to give 1,2,4,5-dithiadiazines, compound (IV) prepared from PhCSNHNHPh having m.p. 120° (*D.H.R. Barton et al.*, J. chem. Soc. Perkin I, 1976, 38).

(ii) *Oxathiadiazines*

1,4,3,5-Oxathiadiazines (V) (R = R' = CCl_3, m.p. 146°; R = R' = CBr_3, m.p. 174°) are formed by the interaction of sulphur trioxide and haloalkyl nitriles (*A.A. Michurin, E.A. Sivenkov* and *E.N. Zil'berman*, C.A., 1970, **72**, 111430w). The reaction proceeds *via* a 1,3,2,4,5-dioxadithiazine intermediate which may be isolated and treated with a second nitrile (usually aryl or haloalkyl) to give unsymmetrically substituted compounds of type (V) along with the isomer (VI) (*I.V. Bodrikov, A.A. Michurin* and *V.L. Krasnov*, Zhur. org. Khim., 1975, **11**, 2217; 1977, **13**, 432; *E. Fischer, C. Mueller* and *G. Rembarz*, Z. Chem., 1977, **17**, 222). Compounds of type (VI; R = R') are also available from the action of sulphur trioxide on aromatic nitriles and such species (VI) have been found to be useful intermediates in the synthesis of heterocyclic compounds, reacting with amidines and related species to give 1,3,5-triazines, with hydrazines to give triazoles and with compounds with an active methylene group to yield pyrimidines (*H. Weidinger et al.*, Ber., 1963, **96**, 2070; Ann., 1968, **716**, 143).

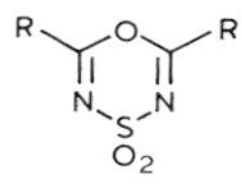

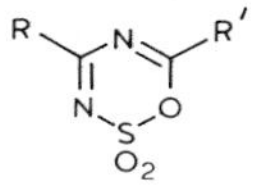

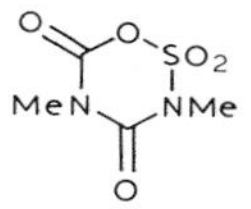

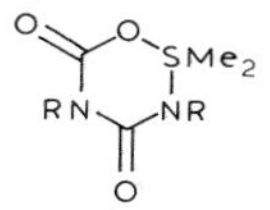

Entry to the 1,2,3,5-system is also possible by the action on isocyanates of either sulphur trioxide at 0° (*D. Arlt et al.*, Ger. Offen., 2,524,475, 1976); C.A., 1977, **86**, 106669r), compound (VII), or dimethyl sulphoxide at 80°, compound (VIII) (*Yu.I. Dergunov, N.N. Bochkareva* and *E.P. Trub*, Zhur. obshch. Khim., 1983, **53**, 2405).

(iii) Dioxadiazines

2-Azopyridine on heating in air with $PhCOC(N_2)Ph$ forms the 1,2,3,4-**dioxadiazine** (IX), m.p. 177°. Compound (IX) despite its peroxide link, appears to be stable but is hydrolysed by acid to azopyridine and benzilic acid (*M. Colonna* and *A. Risaliti*, Gazz., 1960, **90**, 1165). By contrast the 1,2,4,5-substitution pattern is less stable, compound (X) prepared by photo-oxidation of $Me_2C:NN:CMe_2$ having a half-life of around 20 minutes at 40° in CH_2Cl_2 (*P. Lechtken*, Z. Naturforsch., 1976, **31B**, 1436). Fully saturated 1,2,4,5-dioxadiazines have also been derived by the action of formaldehyde and hydrogen peroxide on N,N'-dialkylhydrazines (*E. Schmitz*, Ann., 1960, **635**, 73).

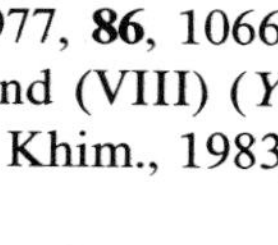

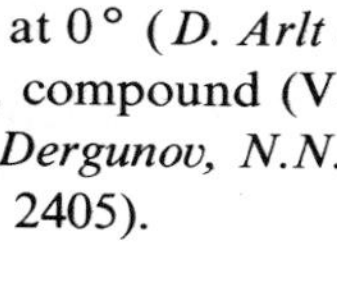

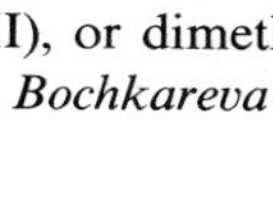

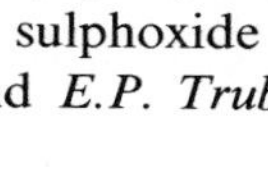

1,4,2,5-Dioxadiazines (XI; R = R') are nitrile oxide dimers, the dimerisation taking place with boron trifluoride (*S. Morrocchi et al.*, C.A., 1968, **69**, 77248u) or more usually in base (*W.J. Middleton*, J. org. Chem., 1984, **49**, 919; *F. De Sarlo*, J. chem. Soc. Perkin I, 1974, 1951). Related 3,6-unsymmetrically disubstituted compounds (XI) are derived from the interaction of amidoximes and oximino chlorides (*T. Sasaki et al.*, Bull. chem. Soc. Japan, 1970, **43**, 2991). The m.ps. of compounds (XI; R = R' = Ph and R = R' = 2,4,6-trimethylphenyl) are 102° and 172° respectively whereas (XI; R = R' = CF$_3$) has b.p. 62–63°.

On oxidation with nickel peroxide, the condensation product of methyl chlorformate and 1,2-bis(aminooxy)ethane forms the 1,4,2,3-**dioxadiazine-2,3-diester** (XII), m.p. 50° (dec.) (*D.K.W. Dixon, R.H. Weiss* and *W.M. Nelson*, Tetrahedron Letters, 1983, **24**, 4393).

(iv) Dioxathiazines

1,3,2,5-Dioxathiazines are reported to arise from the treatment of dimethyl sulphoxide with either acyl isocyanates in boron trifluoride–etherate (*B.A. Arbuzov, N.N. Zobova* and *O.V. Sofronova*, C.A., 1976, **84**, 31014v) or with hexafluoroacetone (*L.N. Kryukov et al.*, Zhur. org. Khim., 1980, **16**, 463).

10. Tetrathianes

Oxidation of *gem*-dithiols (*J. Jentzsch, J. Fabian* and *R. Meyer*, Ber., 1962, **95**, 1764) or of sodium *p*-diethylaminodithiobenzoate (*G. Cauquis* and *A. Deronzier*, Chem. Comm., 1978, 809) gives compounds of type (I) and (II) respectively.

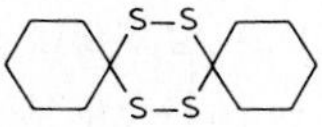

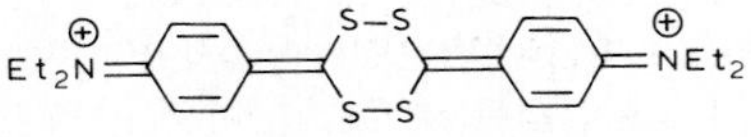

Sulphur is reported to add to tetrafluoroethene to give 5,5,6,6-**tetrafluoro**-1,2,3,4-**tetrathiane**, b.p. 59–61°/15 mm (*C.G. Krespan* and *W.R. Brasen*, J. org. Chem., 1962, **27**, 3995) and sulphur, sodium sulphide and formaldehyde in acidic conditions in dichloromethane yield 1,2,3,5-tetrathiane (*K. Morita, S. Kobayashi* and *H. Kinura*, Jap. P., 6,927,724, 1969; C.A., 1970, **72**, 21724r).

11. Pentazines

A claim by *F.D. Chattaway* and *G.D. Parkes* (J. chem. Soc., 1926, 113) to have prepared the **dihydro**-1,2,3,4,5-**pentazine** (I), m.p. 172° (dec.), by the action of nitrous acid on compound (II) was first challenged by *R. Stollé* (J. pr. Chem., 1926, **114**, 348) who suggested that the product was 1-(2,4-dibromoanilino)-5-phenyl-1,2,3,4-tetrazole (III), a fact later confirmed by *J.M. Burgess* and *M.S. Gibson* (Tetrahedron, 1962, **18**, 1001) using spectroscopic techniques.

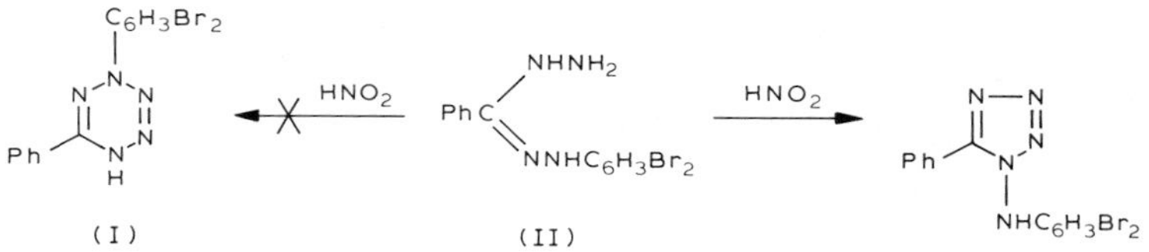

The parent compound has been the subject of much theoretical interest (*P.F. Wiley* in "Heterocyclic Compounds", Vol. 33, eds. *H. Neunhoeffer* and *P.F. Wiley*, Wiley, New York, 1978, p. 1298 and references therein) and it has been suggested on the basis of calculations that 1,2,3,4,5-pentazine itself cannot exist (*M.H. Palmer, A.J. Gaskell* and *R.H. Findlay*, Tetrahedron Letters, 1973, 4659; J. chem. Soc. Perkin II, 1974, 778).

12. Other ring systems with five hetero-atoms (N, O, S)

The only six-membered ring system with five identical hetero-atoms (N, O or S) known to exist is the pentathiane system, prepared from *gem*-dithiols and dichlorotrisulphane (ClSSSCl). H_2CS_5 is a relatively stable yellow compound, m.p. 95–96°, available from methane dithiol (*F. Fehér et al.*, Angew. Chem. intern. Ed., 1968, **7**, 301; Z. Naturforsch., 1979, **34B**, 1031).

1,3,2,4,6-Dithiatriazines arise from the interaction of ureas (*e.g.* compound (I), m.p. 157°, from N,N'-dimethylurea) and imido-bis(sulphonyl chlorides) (*M. Becke-Goehring* and *H.A. Slater*, Naturwiss., 1963, **50**, 353). Related syntheses utilise sulphur diimines; thus condensation of N,N'-bis(trimethylsilyl)sulphurdiimine in dichloromethane with chlorosulphonyl isocyanate gives the *dithiatriazine* (II), m.p. 87°, (*H.W. Roesky* and *B. Kuhtz*, Ber., 1974, **107**, 1; *R. Appel et al.*, Z. Naturforsch. 1974, **29B**, 799).

Alternatively, methyl *N*-chlorosulphonylcarbamate and dimethyl sulphone diimine react to yield the monobasic acid (III), m.p. 172°, (*M. Haake*, Angew. Chem. intern. Ed., 1971, **10**, 264).

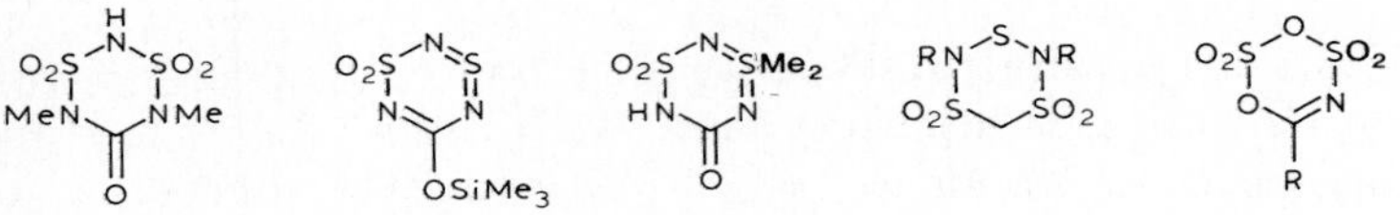

A 1,3,5,2,4-trithiadiazine (IV; $R = P_3N_3F_5$) has been prepared by the action of sulphur dichloride on $[P_3N_3F_5N(SnMe_3)SO_2]_2CH_2$ (*H.W. Roesky* and *M. Banek*, C.A., 1978, **89**, 43350u).

1,3,2,4,5-Dioxadithiazines (V) result from the addition of sulphur trioxide to haloalkyl or aryl nitriles (*N.P. Aktaev, V.A. Pashinin* and *G.A. Sokol'skii*, Zhur. org. Khim., 1974, **10**, 1428; *I.V. Bodrikov, A.A. Michurin* and *V.L. Krasnov*, ibid., 1975, **11**, 2217) (*cf.* 1,4,3,5-oxathiadiazines).

Guide to the Index

This index is constructed in a similar manner to the volume indexes of the first edition of the Chemistry of Carbon Compounds. However, to make the index easier to use, more descriptive entries have been made for the commonly occurring individual, and groups of chemicals.

The indexes cover primarily the chemical compounds mentioned in the text, and also include reactions and techniques, where named, and some sources of chemical compounds such as plant and animal species, oils, etc.

Chemical compounds have been indexed alphabetically under the names used by authors, editing being restricted to ensuring uniformity of entries under the same heading. In view of the alternative nomenclature that can often be used, a limited amount of cross-referencing has been done where it is considered to be helpful, but attention is particularly drawn to Convention 2 below.

For this and the succeeding volumes, the indexing conventions listed below have been adopted.

1. *Alphabetisation*

    (a) The following prefixes have not been counted for alphabetising:

n-	*o*-	*as*-	*meso*-	D-	*C*-
sec-	*m*-	*sym*-	*cis*-	DL-	*O*-
tert-	*p*-	*gem*-	*trans*-	L-	*N*-
	vic-		*endo*-		*S*-
		lin-	*exo*-		*Bz*-
					Py-

    Some prefixes and numbering have been omitted in the index, where they do not usefully contribute to the reference.

    (b) The following prefixes have been alphabetised:

Allo	Epi	Neo
Anti	Hetero	Nor
Cyclo	Homo	Pseudo
	Iso	

(c) A letter by letter alphabetical sequence is followed for entries, firstly for the main entry, followed by the descriptive entry. The only exception to this sequence is the placing of plural entries in front of the corresponding individual entries to prevent these being overlooked by a strict alphabetical sequence which could lead to a considerable separation of plural from individual entries. Thus "butanes" will come before n-butane, "butenes" before 1-butene, and 2-butene, etc.

2. *Cross references*

In view of the many alternative trivial and systematic names for chemical compounds, the indexes should be searched under any alternative names which may be indicated in the main body of the text. Only a limited amount of cross-referencing has been carried out, where it is considered that it would be helpful to the user.

3. *Esters*

In the case of lower alcohols esters are indexed only under the acid, e.g. propionic methyl ester, not methyl propionate. Ethyl is normally omitted *e.g.* acetic ester.

4. *Derivatives*

Simple derivatives are not normally indexed if they follow in the same short section of the text.

5. *Collective and plural entries*

In place of "– derivatives" or "– compounds" the plural entry has normally been used. Plural entries have occasionally been used where compounds of the same name but differing numbering appear in the same section of the text.

6. *Main entries*

The main entry of the more common individual compounds is indicated by heavy type. Where entries relate to sections of three pages or more, the page number is followed by "ff".

Index

Acetamidine, 168–170, 172, 173, 176, 179
Acetamidine acetate, reaction with hydrazine hydrate, 493
5-Acetamido-2-aminopyrimidine, 211
o-Acetamidobenzoic acid, 232
7-Acetamido-4-hydroxycinnoline, 62
5-Acetamidoluminol, 140
1-Acetamidophenazine, nitration, 379
β-Acetamidovinyl phenyl ketone, 183
Acetoguanamine (2,4-diamino-6-methyl-1,3,5-triazine), 472
Acetone 6-nitro-4-cinnolinyl hydrazone, 59, 69
Acetonylacetone, 2
3-Acetonylquinazolone, 450
4-Acetoxycinnoline, 82
4-Acetoxycinnoline-3-carboxylic acid, 84
5-Acetoxyiminopyrimidines, 209
4-Acetoxymethyl-2,6-dihydroxypyrimidine, 214
4-Acetoxymethyl-2-hydroxy-6-mercaptopyrimidine, 214
6-Acetoxymethyl-3-methoxypyridazine, 22
Acetylacetone, 155, 165, 170
2-(*N*-Acetylamino)-6-methylquinoxaline, 274
N-Acetylaminoquinoxaline, 274
N-Acetylanthranilic acid, 446
α-Acetyl-γ-butyrolactone, 173
3-Acetylcinnoline, 86, 89
2-Acetyl-1,2-dihydro-4-ethoxyphthalazine, 150
5-Acetyl-5,10-dihydrophenazine, 395
3-Acetyl-3,4-dihydrophthalazin-1(2*H*)-one, 149
5-Acetyl-2,4-dihydroxypyrimidine, 220
2-Acetyl-4-ethoxy-1,2-dihydrophthalazine, 150
5-Acetyl-4-ethyl-6-hydroxy-2-phenylpyrimidine, 187
5-Acetyloctahydrophenazine, 398
o-Acetylphenylurethane, 231
2-Acetylphthalazin-1(2*H*)-one, 152
1-Acetylpiperazine, 289
Acetylpyrazines, 255
2-Acetylpyrazine, 273
N-Acetyl-1,2,3,4-tetrahydroquinoxalines, 351, 353
2-Acetyl-3,5,6-trimethylpyrazine, 273
2-(*N*-Acylamino)quinoxalines, 261
N-Acylimidates, 469
1-Acyl-2-(2-nitrophenyl)hydrazines, 465
4-Acylpiperazine-1-carboxylates, 290
Acylpyrimidines, 190
1-Acyl-1,2,3,4-tetrahydroquinoxalines, 351, 353
Adhatoda vasica, 440, 445, 446
Adhatodine, 445
Aeruginsoin, 381
Alanine anhydride, 257
Aldoses, reactions with *o*-phenylenediamines, 342
3-Alkoxyacroleins, 167
2-Alkoxy-1-aminophenazines, 384
2-Alkoxy-6-aminopyrazines, synthesis from di(cyanomethyl)-*N*-nitrosoamines, 261
7-Alkoxy-1-aryl-4-(*p*-alkoxyphenyl)-1,2-dihydrocinnolines, synthesis from 1,1-bis(*p*-alkoxyphenyl)ethenes, 95
Alkoxycyanopyrazines, 274
4-Alkoxy-1,2-diphenyl-1,2,3,4-tetrahydrocinnolines, 95
α-Alkoxymethylene ketones, 169
1-Alkoxyphthalazines, melting points, 132
1-Alkoxyphthalazin-3-oxides, 124
2-Alkoxypyrazines, *N*-oxidation, 265
Alkoxypyrimidines, 192, 213, 217, 218
Alkoxyquinazolines, 237, 238
N-Alkylaminoacetonitriles, 250
6-Alkylamino-3-aryl-1,2,4,5-tetrazines, 492
Alkylaminopyrimidines, 209, 212
1-Alkyl-4-arylphthalazines, 117
Alkylcinnolines, synthesis and reactions, 47
–, u.v. spectra, 40, 42
3-Alkyl-4-cinnolinecarbonitriles, mass spectra, 90
N-Alkylcinnoline-3-carboxylic acids, esterification, 86
4-Alkylcinnoline-7-carboxylic acids, synthesis from 4-alkenyl-3-aminobenzoic acids, 91
2-Alkylcinnolinium-4-olates, photochemical rearrangement, 79
2-Alkyl-4,6-dichloro-1,3,5-triazines, 474

[505]

4-Alkyl-1,4-dihydro-6-methyl-1,4,2-oxathiazine-2,2-dioxides, 485
2-Alkyl-4,5-dimethyl-6-phenyltetrahydro-1,3,4-oxadiazines, 480
1-Alkyl-2,3-dimethylquinoxalinium salts, 316
N-Alkyl-N,N-diphenylacylamines, rearrangement, 278
2-Alkyl-1-hydroxy-1,2-dihydrophthalazine, 119
3-Alkyl-2-methylquinoxalines, 306
3-Alkyl-1-oxidophthalazinium betaines, synthesis, 133, 134
N-Alkylphenacylamines, rearrangement, 278
Alkylphenazines, synthesis, 366, 367
5-Alkylphenazinium salts, reduction, 393
1-Alkyl-3-phenylimidazolidin-N-oxides, ring expansion, 479
1-Alkyl-4-phenyl-1,2,3,4-tetrahydrocinnolines, reactions with aryldiazenium salts, 95
1-Alkylphthalazines, 117
2-Alkylphthalazinium hexacyanatoferrate iodides, 130
3-Alkyl-1(2H)-phthalazinium salts, 133
2-Alkylphthalazin-1(2H)-ones, 130
C-Alkylpiperidazines, 35
Alkylpyrazines, oxidation, 270
Alkylpyridazines, ^{1}H n.m.r. spectra, 10
Alkylpyrimidines, 190, 199, 200
N-Alkylpyrimidines, 217
1-Alkylpyrimidimium halides, 194
Alkylquinazolines, 233
N-Alkylquinazolines, 238
2-Alkylquinazolines, 441
4-Alkylquinazoline, 236
3-Alkylquinazol-4(3H)-ones, 79
N-Alkylquinoxalinones, 331, 332
5-Alkylrosindulines, 408
Alkylthiopyrimidines, 192
–, dealkylation, 224
–, oxidation, 224, 225
N-Alkylureas, 171
Alloxan, reactions with N,N-dimethyl-o-phenylenediamines, 303
–, – with o-phenylenediamines, 346
Alloxazines, 332
N-Allylanilines, dimerization, 291
8-Allyloxyquinazolines, 237
Amidines, 167, 173

–, reactions with N,N-dimethylformamide diethylacetal, 470
–, synthons for thiatriazines, 496
α-Amidino-α-aminoacetamides, 248
2-Amidinopyrazine, 273, 274
2-Amidoaminobenzo[f]cinnoline, 111
3-Amido-4-aminocinnoline, 87
4-Amidocinnolines, 89
4-Amidocinnoline, 85
3-Amido-4-hydrocinnoline, penicillin derivatives, 87
3-Amido-4-hydroxycinnolines, 87
3-Amido-4-hydroxycinnoline, 85
4-Amido-3-phenylcinnoline, 85
2-Amidrazonylpyrazine, 273, 274
α-Aminoacetaldehyde dimethylacetal, 267
Aminoacetamidine, 248
2-Aminoacetophenone hydrazone, 459
α-Aminoacids, 305
α-Aminoacid amides, 248
–, reactions, with α, β-dicarbonyl compounds, 262
Aminoacid nitriles, reactions with α,β-diketones, 263
2-Aminoanilinoacetonitrile, 325
2-Amino-3-arylquinoxalines, 326
p-Aminoazobenzene, 414
4-Aminoazobenzene hydrochloride, 407
N-Aminoazoliumphthalazines, 124
2-Aminobenzaldehyde, 232, 452, 459
2-Aminobenzamide, 459
–, reaction with phosphorus pentasulphide, 477
2-Aminobenzamidines, diazotisation, 460
2-Aminobenzamidine dihydrochloride, 460
4-Amino-6-benzamido-2-mercaptopyrimidine, 222
2-Aminobenzhydrazide, 459
1-Aminobenzo[c]cinnoline, 106
2-Aminobenzo[c]cinnoline, 106, 107
3-Aminobenzo[c]cinnoline, 106, 107
4-Aminobenzo[c]cinnoline, 106, 107
1-Aminobenzo[f]cinnoline-2-carbonitrile, 111
8-Aminobenzo-3,1,2-oxadiazin-4-one, 478
3-Amino-1,2,4-benzotriazines, 466
3-Amino-1,2,4-benzotriazine, 466
4-Amino-1,2,3-benzotriazine, 460
3-Amino-1,2,4-benzotriazin-1-oxide, 466
3-Amino-1,2,4-benzotriazin-2-oxide, 466
6-Amino-1-benzyl-4-thiouracil, 214

INDEX

6-Amino-1-benzyluracil, 214
4-Aminobiphenyl, 426
8-Amino-2,3-bis(phenylamino)-5-phenyl-phenazonium chloride, 414
3-Amino-6-bromopyridazine, 16
2-Amino-4-*tert*-butylanisole, oxidation, 367
2-Amino-5-carbamoylmethyl-4-hydroxy-pyrimidine, 222
2-Amino-3-carboxylic acids, decarboxylation 259
2-Amino-5-chlorocinnoline, 71
3-Amino-6-chlorocinnoline, 63
4-Amino-3-chlorocinnoline, 58, 63
4-Amino-6-chlorocinnoline, 63
4-Amino-7-chlorocinnoline, 63
3-Amino-6-chloro-4,5-dimethylpyridazine, 16
4-Amino-5-chloro-3-hydrazinopyridazine, 17
4-Amino-6-chloro-3-hydrazinopyridazine, 17
2-Amino-4-chloro-6-methyl-5-nitro-pyrimidine, 206
4-Amino-6-chloro-3-methylpyridazine, 16
6-Amino-3-chloro-4-methylpyridazine, 16
2-Amino-4-chloro-6-methylpyrimidine, 202
4-Amino-2-chloro-6-methylpyrimidine, 202
2-Amino-4-chloro-5-nitroacetophenone, 37
4-Amino-2-chloro-5-nitropyrimidine, 203
1-Amino-4-chlorophthalazine, 126, 127
3-Amino-6-chloropyridazine, 15, 16
4-Amino-5-chloropyridazine, 16
4-Amino-6-chloropyridazine-3(2H)-thione, 28
4-Amino-3-chloropyridazin-6(1H)-one, 23
4-Amino-6-chloropyridazin-3(2H)-one, 23
5-Amino-4-chloropyridazin-3(2H)-one, 23
5-Amino-6-chloropyridazin-3(2H)-one, 16
6-Amino-3-chloropyridazin-1-oxide. 18
4-Amino-6-chloropyrimidine, 191
2-Amino-3-chloroquinoxaline, 326
2-Amino-6-chloroquinoxaline, 324, 325
2-Amino-7-chloroquinoxaline, 325
Aminocinnolines, 62
–, diazotisation, 66
–, physical properties, 65
–, synthesis, 61
3-Aminocinnoline, 58, 61, 63, 65
4-Aminocinnoline, synthesis from phen-oxycinnolines, 82
4-Aminocinnoline, 57, 61, 63, 66, 68
5-Aminocinnolines, 60, 63, 64, 66
6-Aminocinnoline, 60, 63
7-Aminocinnoline, 61, 63
8-Aminocinnolines, 57, 60, 63, 66
5-Amino-6-cyano-3-ethoxy-1,2,4-triazine, 468
2-Aminocyclohexane-1,5-dione, self-dimerisation, 400
2-Amino-3,5-dibromopyrazine, 261
2-Amino-3,6-di(*sec*-butyl)pyrazine, 260
4-Amino-3,5-dichloropyridazine, 16
4(5)-Amino-3,6-dichloropyridazine, 15, 16
5-Amino-3,4-dichloropyridazine, 16
2-Amino-4,6-dichloropyrimidine, 203
4-Amino-2,6-dichloropyrimidine, 203
5-Amino-3,6-dichloro-1,2,4-triazine, 462
3-Amino-1,2-dihydro-1,2,4-benzotriazin-2-oxide, 466
3-Amino-3,4-dihydro-1,2,3-benzotriazin-4-one, 459
5-Amino-1,4-dihydrocinnoline, 61
4-Amino-1,2-dihydro-1-methyl-2-oxo-pyrimidine, 157
2-Amino-3,4-dihydroquinoxalines, 325
4-Amino-2,6-dihydroxy-5-(β-hydroxy-ethyl)pyrimidine, 179
2-Amino-4,6-dihydroxypyrimidine, 176
4-Amino-2,6-dihydroxypyrimidine, 179
6-Amino-2,3-dihydroxyquinoxaline, 333
2-Amino-4,6-dihydroxy-1,3,5-triazine (ammelide), 472
1-Amino-4,6-diiminohexahydro-1,3,5-triazine-2-thione, 473
5-Amino-3,6-dimethoxy-4-nitropyridazin-1-oxide, 19
4-Aminodimethylaniline, 418
4-Amino-*N,N*-dimethylaniline-3-thio-sulphonic acid, 429
2-Amino-4,6-dimethyl-5-nitropyrimidine, 206
5-Amino-6,7-dimethyl-3-phenylcinnoline, 63, 64
2-Amino-3,5-dimethylpyrazine, 260
2-Amino-5,6-dimethylpyrazine, 260
2-Amino-5,6-dimethylpyrazine-3-carbox-ylic acid, 274
4-Amino-3,6-dimethylpyridazine, 16
2-Amino-4,6-dimethylpyrimidine, 170

4-Amino-2,6-dimethylpyrimidine, 184, 210
2-Amino-6,7-dimethylquinoxaline, 325
5-Amino-2,3-dimethylquinoxaline, 325
6-Amino-2,3-dimethylquinoxaline, 325
3-Amino-5,6-dimethyl-1,2,4-triazine, 462
2-Aminodiphenylamine, 517
4-Aminodiphenylamine, 517
2-Amino-3,6-diphenylpyrazine, 260
2-Amino-5,6-diphenylpyrazine, 260
3-Amino-2,5-diphenylpyrazine, 260
2-Amino-5,6-diphenylpyrazine-3-carboxamide, 274
2-Amino-5,6-diphenylpyrazine-3-carboxylic acid, 274
6-Amino-2,3-diphenylquinoxaline, 325
4-Amino-2,6-diphenyl-2,3,4,5-tetrahydro-1,2,4-triazine, 465
6-Amino-3,5-diphenyl-1,2,4-triazine, 362
N-Amino-1-ephedrine, 480
4-Amino-5-(1′-ethoxyethylidene)amino-2-mercaptopyrimidine, 211
1-Amino-2-ethoxyphenazine, diazotisation, 384
3-Amino-N-ethylcarbazole, 426
3-β-Aminoethylindole, 452
3-Amino-6-ethylpyridazine, 16
2-Amino-3-formylpyrazine, 252
Aminoguanidine, reaction with isatin, 468
2-Aminohept-2-en-4-one, 171
4-Amino-5-hydrazinopyridazine, 17
α-Aminohydroxamic acids, reactions with α,β-dicarbonyl compounds, 265
6-Amino-4-hydroxy cinnoline, 63, 75
7-Amino-4-hydroxycinnoline, 63, 75
8-Amino-4-hydroxycinnoline, 63, 75
o-Amino-N-(β-hydroxyethyl) anilines, cyclodehydration, 35
2-Amino-4-hydroxy-5-(β-hydroxyethyl)pyrimidine, 210
4-Amino-6-hydroxy-2-mercaptopyrimidine, 179
2-Amino-4-hydroxy-6-methylpyrimidine, 173
4-Amino-6-hydroxy-2-methylthio-5-nitrosopyrimidine, 224
3-Amino-2-hydroxyphenazine, 382
5-Amino-4-hydroxyphthalazin-1(2H)-one (luminol), 139, 143
2-Amino-4-hydroxypyrimidine, 172
4-Amino-2-hydroxypyrimidine, 157, 177
6-Amino-6-hydroxypyrimidine, 179

2-Amino-2-hydroxyquinoxaline, diazotisation, 333
1-Aminoindazole, 459
2-Aminoindazole, 459
3-β-Aminoindole-3-carboxylic axid, 452
α-Aminoisobutyric acid, 297
N-Aminoisoindolinone, 146
α-Aminoketones, 249, 262
N-Aminomaleimides, 3
Aminomalonamidamidine, reaction with α,β-dicarbonyl compounds, 274
Aminomalonamides, reactions with α,β-dicarbonyl compounds, 262, 275
2-Amino-2-mercapto-6-methylpyrimidine, 179
4-Amino-2-mercaptopyrimidine, 222
2-Amino-4′-methoxybenzophenone imine, 459
8-Amino-6-methoxycinnoline, 38, 66
4-Amino-3-methoxy-6-methyl-5-nitropyridazine, 15
4-Amino-3-methoxy-6-methylpyridazine, 24
6-Amino-3-methoxy-4-methylpyridazine, 24
5-Amino-3-methoxy-4-nitropyridazin-1-oxide, 19
4-Amino-3-methoxypyridazine, 23
6-Amino-3-methoxypyridazine, 23
3-Amino-1-methylcinnoline, 61
3-Amino-4-methylcinnoline, 63
4-Amino-8-methylcinnoline, 63
8-Amino-4-methylcinnoline, 48
3-Amino-6-methyldithionracil, 214
Aminomethyleneacylurethane, 182
5-Amino-4-methyl-6-methylamino-2-methylthiopyrimidine, 188
6-Amino-3-methyl-5-methylaminouracil, 213
4-Amino-3-methyl-6-nitrocinnoline, 63
4-Amino-3-methyl-8-nitrocinnoline, 63
3-Amino-6-methyl-4-phenylpyridazine, 16
8-Amino-4-methylphthalazin-1(2H)-one, 131
2-Amino-2-methyl-1-propanol, cyclodehydration, 294
2-Amino-5-methylpyrazine, 260
2-Amino-6-methylpyrazine, 260
2-Amino-5-methylpyrazine-3-carboxamide, 274
2-Amino-6-methylpyrazine-3-carbox-

amide, 274
2-Amino-6-methylpyrazine-3-carboxylic acid, 274
3-Amino-5-methylpyridazine,, 15
3-Amino-6-methylpyridazine, 15, 17
4-Amino-3-methylpyridazine, 16
6-Amino-4-methylpyridazin-3(2H)-one, 23
2-Amino-3-methylquinoxaline, 325
2-Amino-5-methylquinoxaline, 325
2-Amino-6-methylquinoxaline, 325
2-Amino-7-methylquinoxaline, 325
2-Amino-8-methylquinoxaline, 325
5-Amino-7-methylquinoxaline, 325
6-Amino-2-methylquinoxaline, 325
6-Amino-7-methylquinoxaline, 325
6-Amino-8-methylquinoxaline, 325
1-Amino-3-methylquinoxalin-2-one, 332
2-Amino-α-(methylsulphinyl)- acetophenones, 37
2-Amino-3-methyluracil, 212
3-Amino-6-methyluracils, 214
6-Amino-1-methyluracil, 179
1-Amino-2-naphthol-6-sulphonic acid, 421
2-Aminonaphtho[1,2-e]-1,2,4-triazine, 468
α-Aminonitriles, reactions with α- oximinoketones, 259
5-Amino-4-nitroaminopyridazine, 17
2-Amino-2'-nitrobiphenyls, 100
4-Amino-3-nitrocinnoline, 63, 64, 67
4-Amino-6-nitrocinnoline, 63, 64
4-Amino-7-nitrocinnoline, 63
4-Amino-8-nitrocinnoline, 63
1-Amino-4-nitrophenazine, 378, 380
2-Amino-1-nitrophenazine, 380
6-Amino-1-nitrophenazine, 380
2-Amino-4-nitrophenol, 425
3-Amino-7-nitro-2-phenylquinoxalin-1- oxide, 337
5-Amino-4-nitropyridazin-1-oxide, 19
o-Aminonitrosobenzenes, reactions with phenylacetonitrile, 326
1-Aminooxindoles, 71
6-Amino-3-oxo-1,2,4-triazine-5-thione, 464
Aminophenanthraquinones, 406
3-Aminophenanthro[9,10-e]-1,2,4- triazine, 468
Aminophenazines, 381–384
–, synthesis from amino-1,2-benzo- quinones, 381

–, – from 6-amino-2-nitrodiphenylamines, 381
–, – from chlorophenazines, 383
–, – from 2,2'-diaminodiphenylamines, 382
–, – from 3,3-dimethylpentane-2,4-diones, 382
–, – from nitrophenazines, 384
–, – from o-phenylenediamines, 381, 382
–, – from polynitrodiphenylamines, 381
–, – from triaminobenzenes, 382
1-Aminophenazines, 361
1-Aminophenazine, 283, 378, 382, 384
2-Aminophenazine, 282–284
–, nitration, 379
2-Aminophenazinedioxides, synthesis from p-aminophenol, 371
Aminophenazine sulphonamides, 381
1-Aminophenazin-N-oxides, 373
2-Aminophenazin-N-oxides, 373
2-Aminophenol, 422
4-Aminophenol, 432
5-Amino-4-phenoxycinnoline, 83
7-Amino-4-phenoxycinnoline, 83
8-Amino-4-phenoxycinnoline, 83
5-Amino-3-phenylcinnoline, 63, 64
3-Amino-5-phenylphenazonium chloride (aposafranine), 407
2-Amino-4-phenylphthalazinium mesity- lenesulphonate, 124
3-Amino-6-phenylpyridazine, 16
4-Amino-5-phenylpyridazine, 16
2-Amino-4-phenylpyrimidine, 170
2-(p-Aminophenyl)quinoxaline, 318
2-Amino-4-phenyl-6-trifluoromethyl- pyrimidine, 170
Aminiphthalazines, synthesis, 126
1-Aminophthalazine, 126, 127
2-Aminophthalazinium mesitylene- sulphonate, 124
1-Aminophthalazinium salts, 127
4-Aminophthalazin-1(2H)-ones, 127
4-Aminophthalazin-1(2H)-one, 127
Aminopyrazines, 248, 259–261
–, ionisation constants, 261
–, reactions with nitrous acid, 263
–, synthesis from halopyrazines, 261
–, tautomerism, 261
–, u.v. spectra, 261
Aminopyrazine, 261
2-Aminopyrazine, 254, 258–260

510 INDEX

Aminopyrazinecarboxamides, 248
2-Aminopyrazine-3-carboxamides, 274
2-Aminopyrazine-3-carboxamide, 274
2-Aminopyrazine-3-carboxylic acids, decarboxylation, 260
2-Aminopyrazine-3-carboxylic acid, 274
2-Aminopyrazine-5-carboxylic acids, 261
Aminopyrazine esters, 248
Aminopyrazinium salts, physical data, 260
2-Aminopyrazinoic acids, reactions with nitrosylsulphuric acid, 274
2-Aminopyrazinoic acid, 272, 273
2-Amino-5-pyrazinoic acid, 274
2-Amino-6-pyrazinoic acid, 274
2-Aminopyrazin-1-oxides, 259
1-Aminopyrazoles, oxidation, 457
1-Aminopyrazole, 457
1-Aminopyrene, 426
Aminopyridazines, N-acylation, 16
–, N-sulphonylation, 16
3-Aminopyridazines, 19
3-Aminopyridazine, 15, 16
4-Aminopyridazine, 15
6-Aminopyridazine-3($2H$)-thione, 28
–, tautomerism, 28
4-Aminopyridazin-3($2H$)-one, 23
5-Aminopyridazin-3($2H$)-one, 23
6-Aminopyridazin-3($2H$)-one, 22
3-Aminopyridazin,-1-oxide, 19, 22
3-Aminopyridazin-2-oxide, 18
2-Aminopyridine, 496, 498
Aminopyrimidines, 209–211
–, acylation, 211, 212
–, tautomerism, 194
2-Aminopyrimidine, 167, 210, 213
4-Aminopyrimidine, 177
5-Aminopyrimidines, 191
6-Aminopyrimidine, 177
Aminopyrimidinones, 161
2-Aminopyrimidin-4(3H)-one, 161, 172
4-Aminopyrimidin-2(1H)-one, 157, 161
N-Aminopyrroles, 2
1-Aminopyrrolidines, ring expansion, 5
4-Aminoquinazoline, tautomerism, 230
2-Aminoquinazolin-4(3H)-one, methylation, 236, 237
5-Aminoquinazolinones, diazotisation, 233
Aminoquinoxalines, 324–327
2-Aminoquinoxalines, 319
–, synthesis, 309
2-Aminoquinoxaline, 325, 326

–, N-oxidation, 337
5-Aminoquinoxaline, 325
6-Aminoquinoxaline, 325
3-Aminoquinoxaline-2-carboxylic acid, 347
2-Aminoquinoxalin-1-oxide, 337
2-Aminoquinoxalin-N-oxides, 336
2-Aminotetrahydro-1,3,2-dithiazine-1,1,3,3-tetraoxide, 482
Aminothiazolines, 477
2-Aminothiobenzamide, 459, 460
Amino-1,2,4-triazines, 462
3-Amino-1,2,4-triazine, 462
5-Amino-1,2,4-triazine, 462
Amino-1,3,5-triazines (melamines), 472
6-Amino-1,2,4-triazine-3,5-dithione, 464
3-Amino-1,2,4-triazino[5,6-b]indole, 468
4-Amino-1,2,4-triazoles, 489
4-Amino-1,2,4-triazolidine-3,5-dione, 494
Amino trichlorobenzoquinones, 424
5-Amino-3,6,7-trimethylcinnoline, 63, 64
Ammelide (2-amino-4,6-dihydroxy-1,3,5-triazine), 472
Ammeline (2,4-diamino-6-hydroxy-1,3,5-triazine), 472
Ammonium anthranilate, 233
Anhydro-4-hydroxy-6-oxo-1,2,3-triphenylpyrimidine hydroxide, 176
Aniflorine, 448
Aniline, 160, 183
Aniline black (C.I. Pigment Black I, 50440), 516
2-Anilino-1-benzeneazonaphthalene, rearrangement, 368, 369
4-Anilinocinnoline, 63, 66
1-Anilino-4-phenylphthalazine, 124
4-Anilinovinylcinnoline, 48
Anisessine, 448
Anisotes sessiflora, 449
Anisotine, 448
Anthranilic acid, 439–442, 446, 450, 452, 456
Antrycide methosulphate, 68
Aposafranines, 405
Aposafranine (3-amino-5-phenylphenazonium chloride) 407
Aposafranone, 366, 407
3-(D-Arabotetrahydroxybutyl)cinnoline, 88
2-(D-Arabotetrahydroxybutyl)quinoxaline, 271, 342

INDEX

Arborine, 437, 440
4-Aroylpiperazine-1-carboxylates, 290
2-Arylbenzazetes, 460
4-Aryl-1,2,3-benzotriazines, 460
Arylcinnolines, 48
1-Aryl-4-dialkylamino-1,2,3,5-tetrazin-6-ones, 488
5-Aryl-2,4-diaminopyrimidines, 179
3-Aryl-1,2-dihydroquinoxalines, 349
1-Aryl-2,3-dimethylquinoxalinium salts, 316
3-Aryl-1-oxidophthalazinium betaines, 133
1-Aryloxyphthalazines, melting points, 132
N-Aryl-N-phenacylamines, dimerisation, 278
1-Aryl-4-phenyl-1,2,3,4-tetrahydrocinnolines, reactions with aryldiazenium salts, 95
1-Arylphthalazines, 118
3-Arylphthalazin-1(2H)-ium salts, 133
4-Arylphthalazin-1(2H)-ones, 130
Arylquinazolines, 233
Aspartic acid, 446
Aspergillic acid (6-sec-butyl-2-hydroxy-3-isobutylpyrazin-1-oxide), 231, 266
Atrolactamidinium chlorides, 492
Azanaphthalenes, 189
2-Azaprimaquine, 66
Azepino[1,2-a]quinoxalines, 314
α-Azido acids, reactions with o-phenylenediamine, 330
4-Azido-1,2,3-benzotriazine, 460
4-Azido cinnolin-N-oxides, 53
Azidodiketones, reactions with alkenes, 480
2-Azido-2,5-diphenyl-1,3-dithiole, 482
α-Azidoketones, 246, 25
2-Azidophenazine, 384
Azidopyrimidines, 204
6-Azidotetrazo[4,5-a]phthalazine, 128
4-Aziridinocinnoline, 63
4-Aziridino-6-nitrocinnoline, 63
Azo Carmine G, 407
Azo Carmine, synthesis, 407
Azodiphosphoric acid,
Azophenine, 414
2-Azopyridine, 499
2-Azopyrimidnine, 204

Balz-Schiemann reaction, 319
Bamberger-Ham reaction, 391

Barbiturates, 155
Barbituric acid, 160, 181, 196
–, tautomerism, 164
Beirut reaction, 308, 315, 337, 338, 359, 360, 370, 371, 397
Benzaldehyde, 438
Benzamide oxime, 173
Benzamidine, 168, 169, 179, 180
Benzamidine hydrochloride, reaction with hydrazine hydrate, 493
2-Benzamido-5-ethoxycarbonyl-4-methylpyrimidine, 168
Benzamidoxime, reaction with ethyl chloroacetate, 479
Benzazimide, (1,2,3-benzotriazin-4-one), 459
6-Benzenazo-4-aminocinnoline, 67
1-Benzenazo-2-hydroxynaphthalene-8-carboxylic acid, 113
1-Benzenazo-4-hydroxynaphthalene-8-carboxylic acid, 113
1-Benzenesulphonamidobenzo[c]-cinnoline, reduction and ring cleavage, 107
2-Benzenesulphonamidobiphenyl, 107
5-Benzenesulphonyloctahydrophenazine, 398
Benzil, 286, 302, 318
Benzimidazole, 471
Benzo[c]cinnolines, nomenclature, 99
–, reduction, 101
–, synthesis, 100-103
Benzo[c]cinnoline, 27, 100–103
–, chemical reactions, 103
–, halogenation, 105,106
–, nitration, 107
–, photolysis, 103
–, physical properties, 102
1H-Benzo[d,e]cinnoline, 110
–, synthesis from 8-formylnaphthalene, 113
Benzo[f]cinnolines, synthesis by photochemical methods, 111
Benzo[f]cinnoline, 110
Benzo[g]cinnoline, 110
Benzo[h]cinnoline, 110
Benzo[c]cinnoline-2-carboxylic acid, 108
Benzo[c]cinnoline-3-carboxylic acid, 108
Benzo[c]cinnoline-2,9-dicarboxylic acid, 108
Benzo[c]cinnoline-3,8-dicarboxylic acid,

108, 109
Benzo[c]cinnoline-4,7-dicarboxylic acid, 108
Benzo[c]cinnoline-1,10-dimethanol monoxide, 109
Benzo[c]cinnoline-5,6-dioxide, 100, 104
Benzo[c]cinnoline-5,6-malonide, 109
Benzo[c]cinnolin-5-oxide, 100, 104
Benzo-1,3-diazines (quinazolines), 227–234
Benzofuranones, reactions with enolate anions, 339, 340
Benzofuroxans, reactions with enolate anions, 337–339
–, – with phenolate anions, 370, 371
–, – with o-quinones, 371
Benzofuroxan, reactions with anils, 360
–, – with pyrrolidino enamines, 397
Benzoin, reaction with o-phenylenediamine, 348
–, – with N-phenylethylenediamine, 283
Benzophenazines, 367–369
Benzo[a]phenazine (1,2-benzophenazine), 367
Benzo[b]phenazine, 369
Benzo[a]phenazinedioxide, 360
Benzophenosafranines, 405
Benzopyrazines (quinoxalines), 301–349
Benzo[c]pyridazo[1,2,a]cinnolinium salts, 103
1,4-Benzoquinone dichloroimines, 406
1,2-Benzoquinone dioxime, reactions with dicarbonyl compounds, 340
2,1,3-Benzothiadrazine, 477
1,2,3-Benzotriazine, 459
1,2,4-Benzotriazines (α-phenotriazines), 465
–, reduced forms, 467
1,2,4-Benzotriazine, 466
1,2,4-Benzotriazine-3-acetic acids, 466
1,2,4-Benzotriazine-3-carboxylic acid, 466
1,2,3-Benzotriazine-4-thione, 459
1,2,4-Benzotriazine-3-thione, 467
1,2,3-Benzotriazin-4-one(benzazimide), 459
1,2,4-Benzotriazin-3-ones, 466
1,2,4-Benzotriazin-3-one, 467
1,2,4-Benzotriazin-3-on-1-oxide, 467
1,2,4-Benzotriazin-1-oxides, 466
Benzoxadiazepines, 342
1,2,4-Benzoxadiazines, ring contraction, 479
1,2,3-Benzoxathiazine-2,2-dioxides, 484
3,2,1-Benzoxathiazin-4(1H)-on-S-oxide, 439, 442
Benzoylacetone, 171, 185
3-Benzoyl-1,2,4-benzotriazine, 466
1-Benzoyl-1,2-bisdimethylaminomethylene, 169
2-Benzoyl-2-cyano-1,2-dihydrophthalazine, 149
5-Benzoyl-5,10-dihydrophenazine, 395
Benzoylguanidine, 168
N-Benzoylhydroxylamine, reaction with oxalyl chloride, 483
Benzoyl-leuco-Methylene Blue, 430
2-Benzoyl-3-methylquinoxaline, reaction with acetophenone, 361, 367
N-Benzoyl-N'-phenylhydrazine, reaction with chloroacetyl chloride, 479
3-Benzoyl-2-phenylquinoxaline, 338
1-Benzoylphthalazine, 122
1-Benzoylpiperazine, 289
1-Benzoyl-3,3,3-trifluroacetone, 170
Benzthiohydrazide, reaction with 2-chloro-1,2-diphenylethanal, 477
3-Benzylamino-6-methylpyridazine, 17
3-Benzylaminopyridazine, 17
3-Benzyl-1,2,4-benzotriazine, 466
4-Benzyl-1,2,4-benzotriazin-3(4H)-on-1-oxide, 467
Benzyl 4-carbamyl-1-piperazinecarboxylates, 291
3-Benzyl-1,2,4-benzotriazine, 467
3-Benzyl-2,5-dihydroxy-6-methylpyrazine, 269
1-Benzyl-3,5-dimethylpiperazinone, 295
N-Benzyl-N,N-diphenylacylamine, reaction with benzylamine, 276, 278
N-Benzylethylenediamine, 286
2-Benzylhexahydro-1,2,4-triazine-3,6-dione, 465
3-Benzyl-2-hydroxyquinoxaline, 329
Benzylic acid, 499
α-Benzylideneacetylacetone, 185
N-Benzylideneaminoisatin, 38
N-Benzylideneamino-N-phenyloxamoyl chloride, 38
Benzylideneoctahydrophenazines, epoxidation, 400
O-Benzyllactimides, 439
3-Benzyl-7-methoxy-6-phenylimidazo[1,2-

b]tetrazine, 492
N-Benzyl-1-methylethylenediamine, reaction with α-bromopropionic acid, 295
1-Benzyl-4-methylpiperazine-2,6-dione, 300
1-Benzyl-3-methylquinoxalin-2-one, 332
1-Benzyloxycarbonylpiperazine, 291
2-(1-Benzyloxyethyl)quinazolin-4(3*H*)-one, 439
6-Benzyloxypyrimidine, 217
6-Benzyl-1,2,2,5,5-pentamethyltetrahydropyrazine, 286
1-Benzylphenazine, 368
2-Benzylphenazine, 368
2-Benzyl-3-phenylquinoxaline, 312
1-Benzylphthalazine, 122
1-Benzylphthalazin-2-oxide, 123
1-Benzylpiperazine, 290
Benzylpyrazine, 257
N-Benzylpyrazinium salts, reduction, 288
2-Benzylquinoxaline, 310, 312
Benzyl sodiocyanoacetate, 238
Benzylthiopyrimidines, 192
2-Benzyl-3-(*p*-tolyl)quinoxaline, 312
6-Benzyl-1,2,4-triazine-3,4-dione, 464
Benzynes, 460
4,4′-Bicinnolinyl, 91, 94
Bicyclomycin, 296
Biginelli reaction, 185
Biguanide, reaction with acetic anhydride, 472
Bindschedlers Green 404
Bioluminescence, 241
Bipyrimidines, 207
5,5′-Bipyrimidine, 208
2,2′-Biquinoxalinyl, 311
3,6-Bis(acylamino)-2,5-dihalogeno-1,4-benzoquinones, 424
1,2-Bis(aminooxy)ethane, reaction with methyl chloroformate, 500
2,3-Bis(bromomethyl)quinoxaline, 317
2,3-Bis(bromomethyl)quinoxaline, 345
Bischler synthesis, of quinazolines, 231
2,3-Bis(dibromomethyl)quinoxaline, 317
2,6-Bis(diethylamino)-4,4-bis(trifluoromethyl)-4*H*-1,3,5-oxadiazine, 481
Bis(diethylamino)ethene, 291
3,7-Bis-(dimethylamino)phenothiazonium chloride (methylene blue), 427
2,4-Bis(dimethylamino)pyrimidine, 216
3,6-Bis(dimethylamino)-1,2,4,5-tetrazine, 493
Bis(dimethylamino)trimethinium salts, 170
2,3-Bis(ethoxycarbonyl)quinoxalin-1-oxide, 340
1,4-Bis(fluoroalkyl)hexahydropthalazines, 153
3,6-Bis(1-hydroxy-1-phenylethyl)-1,2,4,5-tetrazine, 492
4,4′-Bis(methylamino)-6,6′-azocinnoline, 67
1,4-Bis(methylthio)phthalazine, 142
2,5-Bis(phenylamino)-1,4-bis(phenylimino)benzene, 414
3,6-Bis(3-tolyl)-1,4-dihydro-1,2,4,5-tetrazine, oxidation, 492, 493
3,6-Bis(4-tolyl)-1,4-dihydro-1,2,4,5-tetrazine, oxidation, 492, 493
3,6-Bis(3-tolyl)-1,2,4,5-tetrazine, 492
1,4-Bis(trifluoromethyl)-4a,5,6,7,8,8a-hexahydrophthalazine, 153
4,4-Bis(trifluoromethyl)-1-oxa-3-azabuta-1,3-dienes, reactions with α,β-unsaturated nitriles, 481
2,4-Bis(trimethylsilyloxy)pyrimidine, 216
N,*N*′-Bis(trimethylsilyl)sulphurdiimine, 501
Bis(triphenylphosphine)palladium dichloride, 206
Blasticidin S, 209
Blue I, 407, 411
Blue II, 411
Borsche synthesis, cinnolines, 37
Bredereck synthesis, 185
Brilliant Alizarin Bleu G. (C.I. Mordant Dye 52055), 429
ω-Bromoacetophenone, aminolysis, 246
1-Bromobenzo[*c*]cinnoline, 106
2-Bromobenzo[*c*]cinnoline, 106
4-Bromobenzo[*c*]cinnoline, 106
1-Bromobenzo[*c*]cinnolin-*N*-oxides, 106
2-Bromobenzo[*c*]cinnolin-6-oxide, 106
4-Bromobenzo[*c*[cinnolin-6-oxide, 106
3-Bromo-1,2,4-benzotriazine, 467
6-Bromo-3-carboxy-4-hydroxy-2-methylcinnolinium hydroxide, 78
5-Bromo-4-carboxy-2-methylpyrimidine, 169
3-Bromo-4-chlorocinnoline, 55
5-Bromo-4-chloropyridazin-3(2*H*)-one, 23
5-Bromo-2-chloropyrimidine, 202

3-Bromocinnolines, displacement reaction with copper(I) cyanide, 87
3-Bromocinnoline, 54–56,61
4-Bromocinnoline, 54
6-Bromocinnoline, 55
5-Bromo-2-diethoxycarbonylmethylpyrimidine, 206
6-Bromo-1,4-dihydro-4-oxocinnolinecarboxylic acid, alkylation, 78
5-Bromo-5,6-dihydro-2,4,6-trihydroxypyrimidine, 196
6-Bromo-2,3-dihydroxy-1-methylquinoxaline, 333
5-Bromo-2,4-dimethoxypyrimidine, 207
6-Bromo-2,3-dimethylquinoxaline, 320
2-Bromo-3,6-diphenylpyrazine, 258
4-Bromo-3-ethoxycarbonylcinnoline, 98
2-Bromo-3-ethylquinooxaline, 320
5-Bromo-2-fluoropyrimidine, 206
3-Bromo-4-hydrazinocinnoline, 69
6-Bromo-4-hydrazinocinnoline, 69
3-Bromo-4-hydroxycinnoline, 56
6-Bromo-4-hydroxycinnoline-3-carboxylic acid, 85
5-Bromo-2-hydroxy-3-phenylpyrazine, 263
5-Bromo-4-hydroxy-6-phenyl pyrimidine, 206
3-Bromo-2-hydroxypyrazines, 263
2-Bromo-5-hydroxypyrazine, 297
α-Bromoketones, 249
β-Bromolaevulic acid, reaction with ammonia, 253
4-Bromo-1-methoxyphenazine, 377
3-Bromo-4-methylcinnoline, 55
1-Bromomethylphenazine, 367
2-Bromomethylphenazine, 367
3-Bromo-6-methylpyridazine, 12
2-(Bromomethyl)quinoxaline, 313, 344
6-(Bromomethyl)quinoxaline, 318
Bromomucic acid, 169
1-Bromophenazine, 376
2-Bromophenazine, 376
1-Bromophenazin-N-oxides, 373
2-Bromophenazin-N-oxides, 373
5-Bromo-2-phenylpyrimidine, 207
1-Bromophthalazine, 125, 126
Bromopyrazines, 257, 258
–, cyanation, 270, 271, 273
Bromopyrazine, 257
2-Bromopyrazine, 258
3-Bromopyridazine, 12

6-Bromopyridazine-3($2H$)thione, 28
5-Bromopyrimidin-2-amine, 161
2-Bromopyrimidine, 207
5-Bromopyrimidines, 196
5-Bromopyrimidine, 195, 205, 208
2-Bromoquinoxaline, 319, 320
6-Bromoquinoxaline, 320
6-Bromoquinoxaline-2,3-dicarboxylic acid, 248
Bromouracil, 196
5-Bromouracil, 160
E-5-(2-Bromovinyl)-2′-deoxyuridine, 226
Busch reaction, 207
6-Butyl-5,6-dihydro-3,6-dimethylpyridazine, 32
5-Butyl-2-dimethylamino-4-hydroxy-6-methylpyrimidine, 215
3-Butyl-3,6-dimethyl-2,3,4,5-tetrahydropyridazine, 34
6-sec-Butyl-2-hydroxy-3-isobutylpyrazine (deoxyaspergillic acid), 266
6-sec-Butyl-2-hydroxy-3-isobutylpyrazin-1-oxide(aspergillic acid), 266
4-tert-Butyl-3-methoxy-6-phenyl-4,5-dihydropyridazine, 32
4-tert-Butyl-8-nitrocinnoline, 60
3-Butyloxy-3-isopropyl-2-phenylacrylonitrile, 179
2-tert-Butylphenazine, 368
3-Butylpyridazine, 10, 27
5-Butyluracil, 164

Cachou de Laval, 431
Candida lipolytica, 446
Capri Blue GON(C.I. Basic Dye 51015, 418
2-Carbamoyl-4-hydroxy-6-methylpyrimidine, 219
3-Carbamoyl-2-hydroxypyrazines, 262
1-Carboxamido-5,10-dihydrophenazine, 395
1-Carboxy-5,10-dihydrophenazine, 395
2-Carboxy-5,10-dihydrophenazine, 395
4-Carboxy-2,6-dihydroxypyrimidine, 220
5-Carboxy-2,4-diphenylpyrimidine, 219
1-Carboxy-5-methylphenazinium salts, 390
4-Carboxy-2-methylphthazin-($2H$)-thione, 145
2-Carboxy-5-methylpyrazine, decarboxylation, 253
3-Carboxymethyluracils, 189

INDEX 515

1-Carboxyphenazin-N-oxides, 373
4-Carboxyphthalazin-1($2H$)-one, 145
Carboxypyrimidines, 200, 218–220
–, esterification, 219
Carboxyquinazolines, 236
2-Carboxyquinazolin-N-oxides, 336
Carboxyureides, hydrolysis, 329
Catechols, reactions with o-phenylenediamines, 366
Catechol, reaction with o-phenylenediamine, 309
4-Chloroacetylcinnoline (4-cinnolinyl chloromethyl ketone), 90
5-Chloroarylquinazolines, 234
1-Chlorobenzo[c]cinnoline, 106
2-Chlorobenzo[c]cinnoline, 106
3-Chlorobenzo[c]cinnoline, 106
4-Chlorobenzo[c]cinnoline, 106
3-Chloro-1,2,4-benzotriazine, 467
7-Chloro-1,2,4-benzotriazin-3-on-1-oxide, 467
1-Chloro-2-(β-chloroethylamino)-cyclohexane, cyclisation, 353
3-Chlorocinnoline, 55, 56
–, reduction, 94
4-Chloro cinnoline, 39, 57, 61, 62, 66, 68, 125
–, reduction, 94
6-Chlorocinnoline, 55
7-Chlorocinnoline, 55
8-Chlorocinnoline, 55
6-Chlorocinnolin-4($1H$)-one, reaction with methyltosylate, 79
3-Chlorocinnolin-1-oxide, 50
3-Chloro-4-cyano-6-methylpyridazine, 28
4-Chloro-2-cyano-6-methylpyrimidine, 219
3-Chloro-4-cyanopyridazine, 28
3-Chloro-6-cyanopyridazine, 28
Chlorocyclizine, 289
6-Chloro-3,4-diaminocinnoline, 67
4-Chloro-3,5-diaminopyridazine, 16
5-Chloro-3,4-diaminopyridazine, 16
6-Chloro-3,4-diaminopyridazine, 16
6-Chloro-2,4-diaminopyrimidine, 203
2-Chloro-3,6-di(sec-butyl)pyrazine, 257, 258
2-Chloro-4,6-difluoro-1,3,5-triazine, 475
3-Chloro-5,6-dihydrobenzo[h]cinnoline, 111
3-Chloro-8,9-dihydro-6-methoxy-$1H$-benzo[d,e]cinnoline, 113

6-Chloro-2,3-dihydroxyquinoxaline, 333
2-Chloro-4,6-dihydroxy-1,3,5-triazine, 474
4-Chloro-3,6-dimethoxypyridazine, 22
6-Chloro-3,4-dimethoxypyridazine, 24
7-Chloro-4-(4′-dimethylamino-1′-methylbutylamino)cinnoline, 62
2-Chloro-3,6-dimethylpyrazine, 257, 258
3-Chloro-4,6-dimethylpyridazine, 12
4-Chloro-3,6-dimethylpyridazine, 12
6-Chloro-3,4-dimethylpyridazine, 10, 12
4-Chloro-3,6-dimethylpyridazin-1-oxide, 18
6-Chloro-3,4-dimethylpyridazin-1-oxide, 18
2-Chloro-4,6-dimethylpyrimidine, 202
5-Chloro-2,3-dimethylquinoxaline, 320
6-Chloro-2,3-dimethylquinoxaline, 320
6-Chloro-2,3-dimethyl-1,4,9,10-tetrazaphenanthrene, 67
4-Chloro-2,2′-dinitrodiphenylamine, 359, 375
3-Chloro-4,6-diphenylpyridazine, 12
6-Chloro-3,4-dipheylpyridazine, 12
7-Chloro-3-ethoxycarbonyl-1,2,4-benzoxadiazine, 479
4-Chloro-6-ethoxy-5-nitropyrimidine, 204
α-(β-Chloroethylamino)-α-phenylacetophenone, 285
3-Chloro-4-ethyl-6-methylpyridazine, 12
N-(β-Chloroethyl)-o-nitroanilines, reductive cyclisation, 352
3-3-Chloro-4-ethyl-6-phenylpyridazine, 12
3-Chloro-6-ethylpyridazine, 12
2-Chloro-6-ethylquinoxaline, 320
4-Chloro-6-fluorocinnoline, 55
2-(4-Chloro-2-fluorophenyl)-5,6,7,8-tetrahydrocinnolin-3-one, herbicidal activity, 98
3-Chloro-4-hydrazinocinnoline, 69
6-Chloro-4-hydrazinocinnoline, 69
3-Chloro-6-hydrazino-4-methylpyridazine, 17
6-Chloro-3-hydrazino-4-methylpyridazine, 17
6-Chloro-3-hydrazinopyridazine, 17
3-Chloro-4-hydroxycinnoline, 75
5-Chloro-4-hydroxycinnoline, 75
6-Chloro-3-hydroxycinnoline, 71
6-Chloro-4-hydroxycinnoline, 75
7-Chloro-4-hydroxycinnoline, 75
8-Chloro-4-hydroxycinnoline, 75, 77

6-Chloro-4-hydroxycinnoline-3-carboxylic acid, 85
–, esterification, 85
4-Chloro-6-hydroxy-2-methylpyrimidine, 205
6-Chloro-4-hydroxy-2-methylpyrimidine, 205
6-Chloro-2-hydroxyphenazine, 386
5-Chloro-4-hydroxy-3-phenylcinnoline, synthesis from 6-chloro-2-hydrazinobenzoic acid, 74
5-Chloro-2-hydroxypyrazine, 257
4-Chloro-6-hydroxypyrimidine, 205
6-Chloro-2-hydroxyquinoxaline, 329
7-Chloro-2-hydroxyquinoxaline, 329
Chloroiminium salts, reactions with amidines, 470
2-Chloro-8-iodophenezine, 376
4-Chloro-5-methoxycinnoline, 80
4-Chloro-6-methoxycinnoline, 80
4-Chloro-7-methoxycinnoline, 80
4-Chloro-8-methoxycinnoline, 80
6-Chloro-4-methoxycinnoline, 77
3-Chloro-6-methoxy-4-methylpyridazine, 23
6-Chloro-3-methoxy-4-methylpyridazine, 23
4-Chloro-1-methoxyphenazine, 386
1-Chloro-4-methoxyphthalazine, 126
3-Chloro-6-methoxypyridazine, 22
6-Chloro-3-methoxypyridazine, 23
6-Chloro-3-methoxypyridazin-1-oxide, 18
2-Chloro-4-methoxypyrimidine, 204
3-Chloro-4-methylcinnoline, 55
4-Chloro-3-methylcinnoline, 47
6-Chloro-4-methylcinnoline, 48, 55
7-Chloro-4-methylcinnoline, 48
8-Chloro-4-methylcinnoline, 55
6-Chloro-1-methylcinnolin-4-one, 66, 84
6-Chloro-2-methylcinnolin-4($1H$)one betaine, adduct formation with dimethyl acetylenedicarboxylate, 79
N-Chloromethylimidoyl halides reactions with bis-imines, 497
3-Chloro-6-methyl-4-nitropyridazin-1-oxide, 19
6-Chloro-1-methyl-4-oxocinnoline-3-carboxylic acid, decarboxylation, 84
3-Chloro-6-methyl-4-phenylpyridazine, 12
1-Chloro-4-methylphthalazine, 125, 120
4-Chloro-1-methylphthalazine, 122

4-Chloro-2-methylphthalazin-1($2H$)-one, 126
3-Chloro-4-methylpyridazine, 12
–, ^{1}H n.m.r. spectrum, 13
3-Chloro-5-methylpyridazine, 12
–, ^{1}H n.m.r. spectrum, 13
3-Chloro-6-methylpyridazine, 12, 16, 17
6-Chloro-3-methylpyridazine, 27
3-Chloro-4-methylpyridazin-6($1H$)-one, 23
6-Chloro-4-methylpyridazin-3($2H$)-one, 23
3-Chloro-4-methylpyridazin-1-oxide, 18
3-Chloro-4-methylpyridazin-2-oxide, 18
3-Chloro-6-methylpyridazin-1-oxide, 18
4-Chloro-3-methylpyridazin-1-oxide, 18
4-Chloro-5-methylpyridazin-1-oxide, 18
5-Chloro-3-methylpyridazin-1-oxide, 18
5-Chloro-3-methylpyridazin-2-oxide, 18
2-Chloro-3-methylquinoxaline, 320
2-Chloro-5-methylquinoxaline, 320
2-Chloro-6-methylquinoxaline, 320
2-Chloro-7-methylquinoxaline, 320
6-Chloro-2-methyl-7-sulphonamidophthalazin-1-one, reduction 146
4-Chloro-3-nitrocinnoline, 56
4-Chloro-6-nitrocinnoline, 61
Chloronitropyrimidines, 204
2-Chloro-5-nitroyrimidine, 202
Chlorophenazines, synthesis from phenazine oxides, 372
1-Chlorophenazine, 364, 376, 377
–, reaction with sodium azide, 383
2-Chlorophenazine, 259, 372, 375, 376
–, ammonolysis, 383
–, reaction with sodium azide, 383
2-Chlorophenazine-5,10-dioxide, 386
1-Chlorophenazin-N-oxides, 373
2-Chlorophenazin-N-oxides, 373
3-Chloro-4-phenoxycinnoline, 83
5-Chloro-4-phenoxycinnoline, 83
6-Chloro-4-phenoxycinnoline, 83
7-Chloro-4-phenoxycinnoline, 83
8-Chloro-4-phenoxycinnoline, 83
4-Chloro-3-(phenylethynyl)cinnoline, 59
1-Chloro-4-phenylphthalazine, 123, 125, 126
3-Chloro-6-phenylpyridazine, 12
3-Chloro-5-phenyl-1,2,4-triazine, 463
5-(4-Chlorophenyl)-1,2,4-triazine, 461
1-Chlorophthalazine, 117, 124–126, 142

–, reductive dechlorination, 116
4-Chlorophthalazin-1(2H)-one, 126, 128
Chloropyrazines, 258, 259
Chloropyrazine, 257, 260
3-Chloropyridazine, 12
–, ^{1}H n.m.r. spectium, 13
6-Chloropyridazine-3-carboxylic acid, 27
6-Chloropyridazine-3(2H)thione, 28
6-Chloropyridazin-3(2H)-one, 23
6-Chloropyridzin-3(2H)one-4-carboxylic acid, 27
3-Chloropyridazin-1-oxide, 18
3-Chloropyridazin-2-oxide, 18
Chloropyrimidines, 211
2-Chloropyrimidine, 202, 204
5-Chloropyrimidine, 186
2-Chloroquinazoline, ammonolysis, 238
3-Chloroquinazoline, 125
6-Chloroquinazoline, 234
–, ammonolysis, 238
Chloroquinoxalines, 341
2-Chloroquinoxalines, 319
–, reactions with ammonia, 324
2-Chloroquinoxaline, 319–321
5-Chloroquinoxaline, 320
6-Chloroquinoxaline, 320
3-Chloroquinoxaline-2-carboxylic acid, 347
6-Chloroquinoxaline-2-carboxylic acid 1,4-dioxide, 334
6-Chloroquinoxaline-2,3-dicarboxylic acid, 348
6-Chloroquinoxaline-2,3(1H,4H)-dione, 333
2-Chloroquinoxalin-N-oxides, 336
6-Chloroquinoxalin-1-oxide, 341
Chlororaphine, 390
N-(2-Chlorosulphonylethyl)benzimidoyl chloride, reaction with ammonia, 476
Chlorosulphonyl isocyanate 496, 501
3-Chloro-5,6,7,8-tetrahydrocinnoline, 97
3-Chloro-5,6,7,8-tetrahydrocinnolin-1-oxide, 51
6-Chloro-1,2,3,4-tetrahydroquinoxaline, 306, 351
5-Chloro-3,10,13-triphenylaminotriphendioxazine, 425
Chromazurin Blue E (C.I. Mordant Blue 59, 51125), 422
Chromobacerium iodinium 355
Chrysoidine, 406

Cinnolines, 35
–, bioactivity, 46
–, ^{13}C n.m.r. spectra data, 44
–, chromatography, 45
–, electrochemical reduction, 94
–, electrophilic substitution, 52
–, electrosynthesis, 50
–, ^{1}H n.m.r. spectral data, 43
–, i.r. spectra, 42
–, reactions, 45
–, reduction, 92
–, synthesis, 36
–, synthesis from monosaccharides, 88
–, u.v. spectra, 42
Cinnoline, mass spectrum, 121
–, physical properties, 39
–, reduction to 1-aminoindoles, 92
–, synthesis, 39
–, u.v. spectrum, 41
Cinnoline-3-carbonitrile, 90
Cinnoline-4-carbonitriles, reaction with Grignard reagents, 91
Cinnoline-4-carbonitrile, 85, 90
Cinnoline-4-carbonylchloride, 85, 90
Cinnoline-4-carboxaldehyde oxime, 48
Cinnoline-3-carboxylates, 85, 86
Cinnolinecarboxylic acids, preparation, 83–87
Cinnoline-3-carboxylic acids, from 4-chlorocinnolinecarboxylates, 84
Cinnoline-3-carboxylic acid, 85
Cinnoline-4-carboxylic acids, decarboxylation, 88
Cinnoline-4-carboxylic acid, 39
–, esterification, 89
Cinnoline-4-carboxylic acid N-oxide, 89
Cinnoline-4,6-dicarbonitrile, 85
Cinnoline-1,2-dicarboxylate, synthesis from styrene and diethyl azodicarboxylate, 91
Cinnoline-1,2-dioxide, 49
4-Cinnolinethiol, 57
Cinnolin-3(2H)-one silver salt, glucosidation, 73
Cinnolin-N-oxides, 49, 51
–, biological activity, 52
–, nitration, 59
–, reduction, 53
–, spectroscopy, 52
Cinnolin-1-oxide, 49
Cinnolin-2-oxide, 49, 54

4-Cinnolinyl methyl ketones, 91
Cinoxacin, biological action, 87
Claisen condenstion, 201
Copper tetrapyrazinotetraazaporphorin tetrahydrate, 274
Crysogine, 439
Crystal Violet lactone, 430
Curtius rearrangement, 493
Cyamelurine, 475
Cyanoacetaldehyde diethylacetal, 177
2-Cyano-γ-butyrolactone, 179
3-Cyanocinnolines, 86, 87
5-Cyanocinnoline, 66
5-Cyano-1-cyclohexyluracil, 219
5-Cyanocytosine, 177
5-Cyano-4,6-diamino-2-phenylpyrimidine, 180
3-Cyano-5,10-dihydro-5-methylphenazine, 365
4-Cyano-5,6-dimethylpyridazin-3(2H)-one, 10, 28
4-Cyano-5-5,6-diphenylpyridazin-3(2H)-one, 28
1-Cyanoethylcinnolin-4-one, 78
Cyanogen, reaction with o-phenylenediamine, 309
N-Cyanomethyl-o-phenylenediamine, reaction with hydroxylamine, 306
4-Cyano-5-methylpyridazin-3(2H)-one, 25
4-Cyano-6-methylpyridazin-3(2H)-one, 28
1-Cyano-2-nitrophenazine, 380
1-Cyanophenazine, 391
2-Cyanophenazine, 381, 391
1-Cyano-4-phenylphthalazine, 124
2-Cyanophenyltriazenes, cyclisation, 460
4-Cyanophthalazin-1(2H)-one, 145
Cyanopyrazinamides, 271
Cyanopyrazines, 258, 273, 274
Cyanopyrazine, 273
Cyanopyridazines, synthesis, 28
3-Cyanopyridazine, 28
4-Cyanopyridazine, 28
4-Cyanopyridazin-3(2H)-one, 28
5-Cyanopyrimidines, 206
2-Cyanoquinoxalin-N-oxides, 336
4-Cyano-5,6,7,8-tetrahydrocinnolin-3(2H)-ones, synthesis, 99
Cyanotetrahydroquinoxalines, 305
Cyanurates, 473
Cyanuric acid (2,4,6-trihydroxy-1,3,5-triazine), 473

Cyanuric bromide, 475
Cyanuric chloride, 473–475
Cyaphenin (2,4,6-triphenyl-1,3,5-triazine), 471
Cyclic hydrazines, reactions with formaldehyde, 494
Cyclizine, 289
Cycloheptano[e]-1,2,4-triazine, 467
3-Cyclohexyl-4-hydroxycinnolines, methoxy derivatives, 74
1-Cyclohexyluracil, 219
Cyclooctano[e]-1,2,4-triazine, 467
Cytosine, 157, 161, 177, 197, 198, 209
–, mass spectrum, 164
–, tautomerism, 164

Dakin-West reaction, 246
Decahydrocinnoline, 95
Decahydrophenazine, 400, 401
Decahydroquinoxalines, 353, 354
Decahydroquinoxaline, 311
cis-Decahydroquinoxaline, 354
trans-Decahydroquinoxaline, 353
Decahydroquinoxalinones, reduction, 353, 354
Delphine Blue (C.I. Morant Blue 56, 51120), 422
Deoxyaniflorine, 448
Deoxyaspergillic acid, (6-sec-butyl-2-hydroxy-3-isobutylpyrazine), 266
Deoxypeganidine, 444
Deoxyvasicine, 441
–, oxidation, 443
Deoxyvasicinone, 444, 447, 449
–, reactions with aldehydes, 442
Deoxyvasicinone "dimers", 443
Desoxyvasicine, 442
Diacetyl, 247, 315
3,3'-Diacetyl-4,4'-diaminoazobenzene, 67
1,4-Diacetyl-3,6-dibenzylpiperazine-2,5-dione, 299
5,10-Diacetyl-5,10-dihydrophenazine, 379, 394, 395
1,4-Diacetyl-1,4-dihydropyrazine, 277
1,4-Diacetyl-2,3-diphenylpyrazine, 278
1,4-Diacetyl-2,3-diphenyl-1,4,5,6-tetrahydropyrazine, 278
N,N'-Diacetylluminol, 140
N,O-Diacetylluminol, 140
Diacetylmonoxime, reduction, 253
Diacetylvasicolinone, 444

5,10-Diacetyloctahydrophenazine, 398
2,3-Diacetyl-1,2,3,4-tetrahydrophthalazine, 150
1,4-Diacyl-1,2,3,4-tetrahydroquinoxalines, 351, 353
2,5-Di-(2-alkoxyarylamino)-3,6-dichlorobenzoquinones, 423
1,4-Dialkoxyphthalazines, 124
Dialkoxypyrazines, 266, 267
2,5-Dialkoxypyrazines, photo-oxygenation, 269
2-(β-Dialkylaminoethyl)quinoxalines, 312
Dialkyl azodicarboxylates, adducts with dienes, 4
5,10-Dialkyl-5,10-dihydrophenazines, 393
1,2-Dialkyl-2,5-diphenyl-1,2-dihydropyrazines, 276
N,N'-Dialkylhydrazines, 499
N,N-Dialkyl-p-phenylenediamine dyestuff (F.I.A.T. 1313, II, 383), 411
3,3-Dialkyltetrahydrogquinoxalines, 305
3,6-Dialkyl-1,2,4,5-tetrazines, 489
2,3-Dialkynylquinoxalines, 323
Dialuric acid, 160
α,β-Diamines, condenstion with α,β-diketones, 247
Diamine Fast Blue F3 GL (C.I. Direct Blue 109, 51310), 425,426
Diamine Light Blue FFRL (C.I. Direct Blue 108, 51320), 426
2,5-Di(ω-aminoalkyl)pyrazines, 251
4,4'-Diamino-6,6'-azocinnoline, 67
3,8-Diaminobenzo[c]cinnoline, 107
2,3-Diaminobenzoic acid, diazotisation, 478
4,6-Diamino-1-benzoyl-1,2-dihydro-2-thiopyrimidine, 222
4,6-Diamino-2-benzoylthiopyrimidine, 222
α,β-Diaminocarboxylic acids, 248
2,4-Diamino-5(4-chlorophenyl)-6-ethylpyrimidine, 155
2,3-Diamino-6-chloroxquinoxaline, 325
Diaminocinnolines, 64
3,4-Diaminocinnoline, 63, 67
4,6-Diaminocinnoline, 63, 64
4,7-Diaminocinnoline, 63
4,8-Diaminocinnoline, 63
4,6-Diaminocinnoline hydrochloride, 68
N^1,N^3-Di(4-amino-6-cinnolyl)guanidine, 68
1,2-Diaminocyclohexane, 354

–, reductive deamination, 402
2,9-Diamino-6,13-dichlorotriphendioxazine, 427
2,4-Diamino-5,6-dihydroxypyrimidine, 198
4,7-Diamino-1,10-dimethylbenzo[c]-cinnoline, 105, 107
N,N'-Diamino-N,N'-dimethylenediamine, oxidation, 487
3,6-Diamino-3,6-dimethyloctane, 34
2,2'-Diaminodiphenylamine, oxidative cyclisation, 363
4,4'-Diaminodiphenylamine, 14
6,8-Diamino-3-ethoxypyrimidino[4,5-e]-1,2,4-triazine, 468
2,4-Diamino-6-hydroxy-1,3,5-triazine (ammeline), 472
2,4-Diamino-6-isopropyl-5-phenylpyrimidine, 179
Diaminomalonitrile, 247–250, 268, 270, 273
4,6-Diamino-2-mercaptopyrimidine, 180, 227
4,5-Diamino-3-methoxy-6-methylpyridazine, 24
2,3-Diamino-6-methylquinoxaline, 325
2,4-Diamino-6-methyl-1,3,5-triazine (acetoguanamine), 472
3,5-Diamino-4-nitropyridazine, 15
1,2-Diaminophenazine, 383, 384
1,3-Diaminophenazine, 383
1,4-Diaminophenazine, 383
2,3-Diaminophenazine, 382, 383, 412
2,7-Diaminophenazine, 838
2,8-Diaminophenazine, 282, 283
2,4-Diaminophenol, 434
3,7-Diaminophenothiazines, oxidised forms, 427
3,7-Diaminophenothiazonium chloride (Lauths' violet), 427
2,8-Diamino-3-phenylamino-5-phenylphenazonium chloride, 414
2,4-Diamino-6-phenyl-1,3,5-triazine, 472
1,4-Diaminophthalazine, 126, 127
1,4-Diaminophthalazine, 121
1,2-Diaminopropane, 247
α,β-Diaminopropionic acid, reaction with glyoxal, 270
2,6-Diaminopyrazine, 261
3,6-Diaminopyridazine, 15
2,5-Diaminopyrimidine, 211

4,5-Diaminopyrimidine, 211
4,6-Diaminopyrimidine, 211
5,6-Diaminopyrimidines, 188
4,6-Diaminopyrimidine-2(1H)-thione, 180
2,3-Diaminoquinoxalines, hydrolysis, 309
2,3-Diaminoquinoxaline, 323, 325–327
2,7-Diaminoquinoxaline, 309
3,6-Diaminoquinoxaline, 325
4,6-Diaminoresorcinol sulphate, 425
3,6-Diamino-1,2,4,5-tetrazine, 493
3,5-Diamino-4H-1,2,6-thiadiazine, 477
2,4-Diamino-1,3,5-triazine (formoguanamine), 472, 475
"Dianilides", 423
Dianils, cycloadditions with dimethylketene, 284
2,4-Diaryl-6-chloro-1,3,5-triazines, 474
1,4-Diaryl-2,5-dimethylpiperazines, 291
4,6-Diaryl-1,2,3,5-oxathiadrazinedroxides, reactions with imidates, 470
3,5-Diaryltetrahydro-1,3,5-thiadrazines, 478
3,6-Diaryl-1,2,4,5-tetrazines, 489, 492, 493
Diaveridine, 209
Diazanaphthalenes, protonation, 40
4,9-Diazapyrene, 109
Diazapyrylium salts (2,4,6-triphenyl-1,3,5-oxadiazinium salts), 481
1,4-Diazobicyclo[2.2.2]octane (DABCO), 292
Diazobicyclo[3.2.1]octanes, 269
Diazo compounds, adducts with cyclopropenes, 5
α-Diazoketones, 246
Diazol Light Pure Blue FF2JL (D.I. Direct Blue 106,51300), 426
Diazomethyl ketones, reduction, 257
5-Diazo-3-methyluracil, 212
1-Diazophenazin-2(1H)-one, 384
3-Diazophenyl-3,4-dihydro-3,4-diphenyl-1,2,4,5-dithiadiazine, 498
5,10-Dibenzenesulphonyloctahydrophenazine, 398
2,2'-Dibenzimidazoyl, 334
Dibenzophenazines, 367–369
Dibenzo[a,h]phenazine, 363, 369
Dibenzo[a,i]phenazine, 369
Dibenzo[a,j]phenazine, 369
Dibenzo phenosafranines, 405
Dibenzopyrazines, 354–402

3,4-Dibenzoylcinnoline, 85, 90
1,2-Dibenzoylcinnoline-4-carbonitrile, 91
N,N'-Dibenzoylethylenediamine, 279
3,4-Dibenzylcinnoline, oxidation, 90
1,4-Dibenzyl-1,4-dihydro-2,6-diphenylpyrazine, 278
5,10-Dibenzyl-5,10-dihydrophenazine, 395
3,6-Dibenzyl-1,4-dihydro-1,2,4,5-tetrazine, 492
1,4-Dibenzyl-5,6-diphenyl-1,4,5,6-tetrahydro-1,2,3,4-tetrazine, 487
1,2-Dibenzylhexahydro-1,2,4-triazine-3,6-dione, 465
2,5-Dibenzyloxy-2,5-dihydropyrazine, 282
2,4-Dibenzyloxy-6-fluoro-5-methylpyrimidine, 204
1,4-Dibenzylphenazine, 368
1,4-Dibenzylpiperazine-2,5-dione, 297
2,5-Dibenzylpyrazine, 257
3,6-Dibenzyl-1,2,4,5-tetrazine, 492
2,4-Dibenzyl-1,3,5-triazine, 472
1-(2,4-Dibromoanilino)-5-phenyl-1,2,3,4-tetrazole, 501
1,4-Dibromobenzo[c]cinnoline, 106
1,4-Dibromobenzo[c]cinnolin-6-oxide, 106
3,4-Dibromocinnoline, 58
2,3-Dibromo-5,6-dimethylpyrazine, 258
2-(Dibromomethyl)-4,6-dimethylpyrimidine, 189
1-(Dibromomethyl)phenazine, 367
3,6-Dibromo-4-methylpyridazine, 12
2-(Dibromomethyl)quinoxaline, 313
1,4-Dibromophenazine, 376
2,7-Dibromophenazines, synthesis from benzenesulphenanilides, 375
2,7-Dibromophenazine, 376
2,8-Dibromophenazine, 376
3,7-Dibromophenothiazonium bromide, 428
1,4-Dibromophthalazine, 125, 126
2,3-Dibromopyrazine, 258
2,5-Dibromopyrazine, 258
2,6-Dibromopyrazine, 255, 261, 273
3,6-Dibromopyridazine, 12
5,5'-Dibromopyrimidines, 196
2,3-Dibromoquinoxalines, 334
2,3-Dibromoquinoxaline, 320, 322
5,8-Dibromo-5,6,7,8-tetrahydrocinnolin-3(2H)-one, 98
3,6-Di(sec-butyl)-2,5-dichloropyrazine,

257
3,6-Di-(sec-butyl)-2-hydroxypyrazine, 262
2,7-Di-*tert*-butylphenazine, 362, 367, 368
2,3-Dicarboxy-5,6-dimethylpiperazine, decarboxylation, 253
4,5-Dicarboxy-2,6-diphenylpyrimidine, 218, 219
N-Dichloroacetylphenylalanine, 299
3,7-Dichloro-1,2,4-benzotriazine, 467
3,7-Dichloro-1,2,4-benzotriazin-1-oxide, 467
2,4-Dichloro-5-(α-chlorovinyl)-pyrimidine, 220
2,4-Dichloro[c]cinnoline, 106
3,4-Dichlorocinnoline, 55, 56, 58
3,7-Dichlorocinnoline, 56
4,6-Dichlorocinnoline, 55
4,7-Dichlorocinnoline, 39, 55
4,8-Dichlorocinnoline, 55
5,6-Dichloro-5,6-dicyanopiperazine-2,3-dione, 296
1,1-Dichlorodifluoroethene, 225
6,7-Dichloro-2,3-dihydroxyquinoxaline, 333
2,3-Dichloro-5,6-dimethylpyrazine, 258
2,5-Dichloro-3,6-dimethylpyrazine, 257
3,6-Dichloro-4,5-dimethylpyridazine, 10
4,6-Dichloro-2,5-dimethylpyrimidine, 200
2,5-Dichloro-3,6-diphenylpyrazine, 258
3,6-Dichloro-4,5-diphenylpyridazine, 12
3,6-Dichloro-1,4,2,5-dithiadiazine, 498
3,6-Dichloro-5-ethyl-4-methylpyridazine, 12
2,3-Dichloroethylpropenoates, 173
2,4-Dichloro-6-fluoro-1,3,5-triazine, 475
3,6-Dichloro-4-hydrazinopyridazine, 17
3,6-Dichloro-4-hydroxycinnoline, 58
5,8-Dichloro-4-hydroxycinnoline, 75
6,7-Dichloro-4-hydroxycinnoline, 37
6,8-Dichloro-4-hydroxycinnoline, 75
7,8-Dichloro-4-hydroxycinnoline, 55, 75
6,7-Dichloro-2-hydroxyquinoxaline, 329
2,4-Dichloro-1-methoxyphenazine, 386
3,6-Dichloro-4-methoxypyridazin-1-oxide, 18
3,6-Dichloro-4-methoxypyridazin-2-oxide, 19
3,6-Dichloro-4-methyl-5-phenylpyridazine, 12
2,3-Di(chloromethyl)pyrazine, 254
3,6-Dichloro-4-methylpyridazine, 10, 12,

27
2,4-Dichloro-6-methylpyrimidine, 202, 205
4,6-Dichloro-2-methylpyrimidine, 205
2,4-Dichloro-5-nitropyrimidine, 203
4,6-Dichloro-5-nitropyrimidine, 202, 204
1,9-Dichloro-1,2,3,4,6,7,8,9-octahydrophenazin-5-oxide, 399, 400
1,2-Dichloraphenazine, 376
1,3-Dichlorophenazine, 376
1,4-Dichlorophenazine, 359, 364, 376, 377
1,6-Dichlorophenazine, 376
1,9-Dichlorophenazine, 376
2,3-Dichlorophenazine, 376
2,6-Dichlorophenazine, 372
2,7-Dichlorophenazines, synthesis from benzenesulphenanilides, 375
2,7-Dichlorophenazine, 376
2,8-Dichlorophenazine, 376
2,7-Dichlorophenazin-N-oxides, 373
2,7-Dichlorophenazin-5-oxide, 372
5,6-Dichloro-4-phenoxycinnoline, 83
3,6-Dichloro-4-phenylpyridazine, 12
1,4-Dichlorophthalazine, 124–126, 134
–, reductive dechlorination, 116
2,3-Dichloropyrazine, 257
3,5-Dichloropyrazine, 257
2,6-Dichloropyrazine, 257
2,6-Dichloropyrazin-3-ones, 250
3,6-Dichloropyridazine, 12
–, ^{1}H n.m.r. spectrum, 13
3,6-Dichloropyridazine-4-carboxylic acid, 27
4,5-Dichloropyridazinones, 4
3,4-Dichloropyridazin-6(1H)-one, 23
4,5-Dichloropyridazin-3(2H)-one, 14, 23
4,6-Dichloropyridazin-3(2H)-one, 14, 23
5,6-Dichloropyridazin-3(2H)-one, 14
3,6-Dichloropyridazin-1-oxide, 18
2,4-Dichloropyrimidine, 166, 202, 204
4,6-Dichloropyrimidine, 191, 204, 211
2,4-Dichloroquinazoline, 234, 238
6,8-Dichloroquinazoline, 234
2,3-Dichloroquinoxalines, 334
–, synthesis, 322
2,3-Dichloroquinoxaline, 268, 315, 320, 322, 323, 326
2,6-Dichloroquinoxaline, 320
2,7-Dichloroquinoxaline, 320
5,8-Dichloroquinoxaline, 320
6,7-Dichloroquinoxaline, 320
1,4-Dichloro-5,6,7,8-tetrahydro-

phthalazine, 152
6,13-Dichlorotriphendioxanes, 423
6,13-Dichlorotriphendioxazine, 425
Dichlorotrisulphane, 501
Dichroa febrifua, 499
4,4'-Dicinnolinyl, 39
Dicinnolinylguanidines, 68
Dicinnolinylthioureas, 68
Dicinnolinylureas, 68
Dicyanamide, 471
–, polymerisation, 472
2,9-Dicyanobenzo[c]cinnoline, 108
3,4-Dicyanocinnoline, 58
5,6-Dicyano-2,3-dihydroxypyrazine, 268
2,3-Dicyano-5,6-dimethylpyrazine, 273
2,3-Dicyano-2,3-diphenylpiperazine, 294
2,3-Dicyano-2,3-diphenylpyrazine, 279
2,3-Dicyano-5,6-diphenylpyrazine, 279
2,3-Dicyano-5-methylpyrazine, 250
1,4-Dicyanophthalazine, 145
5,6-Dicyanopiperazine-2,3-dione, 296
2,3-Dicyanopyrazines, 247, 270, 274
2,3-Dicyanopyrazine, 273
2,5-Dicyanopyrazine, 247, 273
2,6-Dicyanopyrazine, 273
3,6-Dicyano-2,2,5,5-tetramethylpiperazine, 294
3,6-Dicyano-1,2,4,5-tetrazine, 493
Dicyclohexano[a,d]-1,2,4,5-hexahydrotetrazine, 494
1,3-Dicyclohexylbarbituric acid, 175
N,N'-Dicyclohexylcarbodiimide, 175
3,8-Di(dimethylamino)benzo[c]-cinnoline, 107
Diels Alder reaction, 256
1,2-Diethoxycarbonyl-1,2-dihydro-3,6-diphenylpyridazine, 31
1,4-Diethoxycarbonyl-5,10-dihydrophenazine, 395
1,4-Diethoxycarbonyl-1,4-dihydropyrazine, 288
1,2-Diethoxycarbonyl-1,2-dihydropyridazine, 31
1,2-Diethoxycarbonylhexahydropyridazine, 31
N,N'-Diethoxycarbonylhydrazine, reaction with ethene-1,2-disulphenyl chloride, 498
4,6-Diethoxycarbonyl-5-hydroxy-1,2,6-oxadiazine, 479
2,3-Diethoxycarbonyltetrahydro-1,4,2,3-dithiadiazine, 498
1,4-Diethoxycarbonyl-1,2,3,4-tetrahydrophenazine, 397
1,2-Diethoxycarbonyl-1,2,3,6-tetrahydropyridazine, 33, 35
1,4-Diethoxycarbonyl-1,2,3,4-tetrahydroquinoxaline, 353
2,5-Diethoxy-3,6-dimethylpyrazine, photooxygenation, 269
2-Diethoxymethyl-2-formyl-3-(3',4',5'-trimethoxyphenyl)proprionitrile, 118
2,6-Diethoxy-1-methylpyrimidinium iodide, 218
2,5-Diethoxypyrazines, 251, 268, 269
2,6-Diethoxypyrazine, 267
2,4-Diethoxypyrimidine, 218
4,6-Diethoxypyrimidine, 204
Diethylacetylene dicarboxylate, reaction with o-phenylenediamine, 328
3,8-Diethylaminobenzo[c]cinnoline, 107
β-(Diethylamino)ethylamine, 322
3-(N,N-Diethylamino)-6-nitrosophenol, 418
1-Diethylaminopropyne, reactions with pyrazines, 256
3,8-Diethylbenzo[c]cinnoline, 107
1-Diethylcarbamyl-4-methylpiperazine, 291
1-Diethylcarbamylpiperazine, 291
Diethylcyanamide, reaction with hexafluoroacetone, 481
Diethyl 1,4-dihydrobenzo[c]cinnoline-1,2-dicarboxylate, by electrochemical carboxylation, 109
1,4-Diethyl-2,3-dihydro-5,6-diphenylpyrazine, 286
5,10-Diethyl-5,10-dihydrophenazine, 395
Diethyl 3,6-dimethylpyrazine-2,5-dicarboxylate, 270
3,6-Diethyl-3,6-dimethyl-3,4,5,6-tetrahydropyridazine, 34
Diethyl 3,6-diphenyl-1,2,3,6-tetrahydropyridazine-1,2-dicarboxylate, 27
Diethyl ethoxymethylenemalonate, 172
Diethyl fumarate, reactions with dihydropyrazines, 279
3,6-Diethyl-2-hydroxy-5-pyrazinoic acid, 275
Diethyl 3-methyl-6-phenyl-1,2,3,4-tetrahydropyridazine-1,2-dicarboxylate, 27
Diethyl 1,2,3,5,6,7,8,8a-octahydrocinno-

line-1,2-dicarboxylate, 95
Diethyl phthalazine-1,4-dicarboxylate, 144
2,5-Diethylpyrazine, 253
1,4-Diethylquinoxaline, reduction, 350
2,3-Diethylquinoxaline, 312
1,4-Diethyl-1,2,3,4-tetrahydroquinoxaline, 311
2,3-Difluoroquinoxalines, 333
1,4-Diformyl-1,2,3,4-tetrahydroquinoxaline, 311
1,4-Dihydrazinophthalazines, synthesis, 128
1,4-Dihydrazinophthalazine, 128
3,6-Dihydrazinopyridazine, 17
1,2-Dihydro-1,2-(alkylmalonyl)-4-phenylcinnolines, 92
Dihydroarborine, 438
1,2-Dihydro-1,2-(arylmalonyl)-4-phenylcinnolines, 92
5,6-Dihydrobenzo[c]cinnolines, radical cations, 109
5,6-Dihydrobenzo[c]cinnoline, 100, 109
5,6-Dihydrobenzo[h]cinnoline, 112
5,10-Dihydrobenzo[b]phenazine, 369
1,2-Dihydro-1,2,4-benzotriazines, 466
4,5-Dihydro-3-carboline, 452
1,2-Dihydrocinnoline, 92
1,4-Dihydrocinnolines, synthesis from ring enlargement of 1-aminoindoles, 96
1,4-Dihydrocinnoline, 39, 45, 92, 93
–, – synthesis from 2-(o-hydroxyaminophenyl)ethylamine, 95
1a,7b-Dihydro-1H-cyclopropa[c]cinnolines, thermal rearrangement, 86
5,10-Dihydro-1,6-dihydroxy-3,8-dimethoxyphenazine, 389
1,4-Dihydro-5,10-dihydroxy-4,4-dimethylbenzo[g]cinnolin-3(2H)-one, 112
1,2-Dihydro-3,6-dihydroxy-1,2-diphenyl-4-ethylpyridazine, 31
1,2-Dihydro-3,6-dihydroxy-1,2-diphenyl-4-methylpyridazine, 31
1,2-Dihydro-3,6-dihydroxy-1,2-diphenyl-4-propylpyridazine, 31
2,5-Dihydro-3,6-di(2-hydroxyphenyl)pyrazine, 282
5,10-Dihydro-5,10-di(methoxymethyl)phenazine, 364, 393–395
5,6-Dihydro-1,10-dimethylbenzo[c]cinnoline, 105, 109
2,5-Dihydro-2,5-dimethyl-3,6-diphenylpyrazine, 281
2,5-Dihydro-3,6-dimethyl-2,5-diphenylpyrazine, 281
2,6-Dihydro-3,5-dimethyl-1,3,5-oxadiazin-4-one, 480
1,2-Dihydro-1,4-dimethyl-2-oxo-6-propylpyrimidine, 171
1,2-Dihydro-1,6-dimethyl-2-oxo-4-propylpyrimidine, 171
5,10-Dihydro-5,10-dimethylphenazine, 394, 395
2,5-Dihydro-3,6-dimethylpyrazine, 281
1,2-Dihydro-1,3-dimethylquinoxalinone, 314
5,6-Dihydro-1,4-dimethyl-1,2,3,4-tetrazine, 487
1,4-Dihydro-3,6-dimethyl-1,2,4,5-tetrazine, 492
5,6-Dihydro-1,4-dimethyl-1,2,3,4-tetrazin-2-oxide, 487
2,3-Dihydro-3,9-dioxo-2-phenylbenzo[d,e]cinnoline, 113
1,2-Dihydro-2,3-diphenylcinnoline, 54
1,4-Dihydro-1,4-diphenylcinnoline, reaction with trans-stilbene, 94
1,2-Dihydro-1,4-diphenyl-6-methyl-2-oxopyrimidine, 171
1,2-Dihydro-1,6-diphenyl-4-methyl-2-thiopyrimidine, 171
5,10-Dihydro-5,10-diphenylphenazine, 394, 395
1,2-Dihydro-2,4-diphenylphthalazine, 147
3,4-Dihydro-4,4-diphenylphthalazin-1(2H)one, synthesis from Crystal Violet lactone, 149
2,3-Dihydro-2,3-diphenylpiperazine, 294
2,3-Dihydro-2,3-diphenylpyrazine, 280
–, isomers, 280
2,3-Dihydro-5,6-diphenylpyrazine, 279, 286
2,5-Dihydro-3,6-diphenylpyrazine, 281
5,6-Dihydro-2,3-diphenylpyrazine, 278, 279
2,3-Dihydro-5,6-diphenylpyrazine-2-carboxamide, 279
1,2-Dihydro-3,6-diphenylpyridazine, 31
1,2-Dihydro-2,3-diphenylquinoxaline, 349
1,4-Dihydro-3,6-diphenyl-1,2,4,5-tetrazine, oxidation, 492, 493
4,5-Dihydro-5,6-diphenyl-1,2,4-triazin-3(2H)-one, 465

5,10-Dihydro-5,10-di(trimethylsilyl)-
  phenazine, 363
1,4-Dihydro-1,4-di(trimethylsilyl)-
  pyrazine, 277
1,4-Dihydro-3-ethoxycarbonylcinnoline,
  98
5,10-Dihydro-5-ethoxycarbonylphenazine,
  395
2,3-Dihydro-3-ethoxymethylene-5-methyl-
  2-oxofuran, 174
5,10-Dihydro-5-ethylphenazine, 366, 395
1,2-Dihydro-4-ethylphthalazine, 150
3,4-Dihydro-3-ethylphthalazin-1(2H)-one,
  150
2,5-Dihydro-2,2,3,5,5,6-hexamethyl-
  pyrazine, 281
5,6-Dihydro-3-hydrazinobenzo[h]-
  cinnolinium chloride, 111
9,10-Dihydro-3-hydrazino-6-methoxy-1H-
  benzo[d,e]cinnoline, 113
8,9-Dihydro-3-hydroxy-6-methoxy-1H-
  benzo[d,e]cinnoline, 113
5,10-Dihydro-1-hydroxyphenazine, 395
5,10-Dihydro-2-hydroxyphenazine, 395
2,3-Dihydro-1-hydroxy-3-phenyl-
  phthalazinyl-4-acetic acid, 135
1,2-Dihydro-1-hydroxy-1,2,4-triphenyl-
  phthalazine, 147
3,4-Dihydro-4-imino-1,2,3-benzotriazines,
  460
1,3-Dihydroisoindoles, synthesis from reduction of 1-chloro-4-alkylphthalazines,
  126
5,10-Dihydro-1-methoxy carbonyl-
  phenazine, 395
5,10-Dihydro-5-methoxycarbonyl-
  phenazine, 364, 394, 395
5,10-Dihydro-2-methoxyphenazine, 395
2,3-Dihydro-5-methoxyquinoxaline, 333
1,4-Dihydro-4-methylcinnoline, 46, 92
1,2-Dihydro-1-methyl-2-oxopyrazine, 264,
  277
5,10-Dihydro-1-methylphenazine, 390, 395
5,10-Dihydro-2-methylphenazine, 395
5,10-Dihydro-5-methylphenazine, 365, 395
1,2-Dihydro-1-methylphthalazine, 117
1,2-Dihydro-2-methylphthalazine, 146, 149
1,2-Dihydro-4-methylphthalazine, 146
5,10-Dihydro-5-methyl-10-phenylpyrazine,
  394, 395
4,5-Dihydro-3-methyl-6-phenyl-1,2,4-
  triazine, 465
2,3-Dihydro-1-methyl-4(1H)-quinazo-
  lines, 440
1,2-Dihydro-3-methylquinoxaline, 349
1,4-Dihydro-2-methylquinoxaline, 349
4,5-Dihydro-5-methyl-1,2,4-triazin-6-one,
  464
1,4-Dihydro-5-nitro cinnoline, 61
5,10-Dihydro-1-nitrophenazine, 395
5,10-Dihydro-2-nitrophenazine, 395
Dihydronorharman, 452
5,6-Dihydro-4H-1,3,4-oxadiazines, 480
5,6-Dihydro-4H-1,2,3-oxadiazin-2-oxide,
  478
5,10-Dihydro-5-oxo-3-phenylbenzo[g]-
  cinnoline, 112
5,6-Dihydro-1,4,3-oxathiazines, 485
5,10-Dihydrophenazines, 358, 392–397
–, synthesis, 392–394
5,10-Dihydrophenazine, 363, 395
–, chemical reactions, 394
5,10-Dihydrophenazine-phenazine molecular complex, 396
5,6-Dihydro-5-phenylbenzo[c]cinnoline,
  103
10a,10b-Dihydro-5-phenylbenzo[c]-
  cinnolinium cation, 109
5,10-Dihydro-3-phenylbenzo[g]cinnolin-
  5-one, 112
1,2-Dihydro-4-phenylcinnoline, 92, 93
1,4-Dihydro-3-phenylcinnoline, 89
1,4-Dihydro-4-phenylcinnoline, 92, 93
1,2-Dihydro-4-phenylcinnoline-1,2-
  phthalide, 92
5,6-Dihydro-3-phenyl-4H-1,2,4-oxadiazin-
  5-one, 479
5,10-Dihydro-5-phenylphenazine, 395
1,2-Dihydro-2-phenylphthalazine, 149
Dihydro-2-phenylquinoxalines, 318
1,2-Dihydro-3-phenylquinoxaline, 249
5,6-Dihydro-3-phenyl-2H-1,2,4-thia-
  diazine-1,1-dioxide, 476
4,5-Dihydro-5-phenyl-1,2,4-triazin-6-one,
  464
1,2-Dihydrophthalazines, oxidation, 148
–, synthesis, 145
1,2-Dihydrophthalazine, 132, 149
–, synthesis, 146
1,4-Dihydrophthalazine, 150
3,4-Dihydrophthalazin-1(2H)-ones, 148
–, synthesis from phthalazinones, 145

3,4-Dihydrophthazin-1(2H)-one, 131, 149
Dihydropyocyanine, 386
Dihydropyrazines, dehydrogenation, 247, 251
–, synthesis, 247–251
1,2-Dihydropyrazines, 275–277
1,4-Dihydropyrazines, 277–279
2,3-Dihydropyrazines, 279–280
2,5-Dihydropyrazines, 280–282
–, photo-oxygenation, 269
2,6-Dihydropyrazines, cycloaddition reactions, 269
Dihydropyridazines, 2
1,2-Dihydropyridazines, 30
1,4-Dihydropyridazines, 4
–, synthesis, 31
1,6-Dihydropyridazines, synthesis, 31, 32
3,6-Dihydropyridazines, 33
4,5-Dihydropyridazines, synthesis, 32
5,6-Dihydropyridazines, 32
4,5-Dihydropyridazin-3(2H)-ones, 1
Dihydropyrimidines, 197, 210
Dihydroquinoxalines, 349, 350
1,4-Dihydroquinoxaline, 311, 349
5,12-Dihydroquinoxalo[2,3-b]phenazines (fluorindines), 412, 413
Dihydrorutaecarpine, 455
5,6-Dihydro-1,5,6,10-tetramethylbenzo-[c]cinnoline, 105
2,3-Dihydro-2,2,5,5-tetramethylpiperazine, 294
2,3-Dihydro-2,2,5,5-tetramethylpyrazine, 281
2,5-Dihydro-2,2,5,5-tetramethylpyrazine, 281
1,4-Dihydro-1,2,4,5-tetrazines, 489–491
1,4-Dihydro-1,2,4,5-tetrazine, 492
1,4-Dihydro-1,2,4,5-tetrazine-3,6-dicarboxylic acid, 493
2,5-Dihydro-4,5,6-trimethyl-1,2,3-triazine, 458
4,5-Dihydro-3,5,6-trimethyl-1,2,4-triazine, 465
4,5-Dihydro-1,2,4-triazin-3(2H)-one, 465
1,4-Dihydro-1,3,4-triphenylcinnoline, 94
3,8-Dihydroxybenzo[c]cinnoline, 108
3,4-Dihydroxycinnolines, 71
3,6-Dihydroxycinnoline, 71, 73
1,6-Dihydroxy-3,8-dimethyoxyphenazine, 387, 389
2,3-Dihydroxy-1,4-dimethyl-6-methoxyquinoxaline, 333
2,5-Dihydroxy-3,6-dimethylpyrazine, 267
2,4-Dihydroxy-5,6-dimethylpyrimidine, 200
2,3-Dihydroxy-6,7-dinitroquinoxaline, 333
2,3-Dihydroxydioxane, 303, 310
2,5-Dihydroxy-3,6-diphenylpyrazine, 267
2,6-Dihydroxy-3,5-diphenylpyrazine, 267, 268
2,4-Dihydroxy-6-formyl-5-methylpyrimidine, 200
4,6-Dihydroxy-2-mercaptopyrimidine, 176
2,3-Dihydroxy-5-methyl-6-phenylpyrazine, 267
2,3-Dihydroxy-5-methylpyrazine, 267
2,4-Dihydroxy-5-methylpyrimidine, 172
2,4-Dihydroxy-6-methylpyrimidine, 173, 174, 205
2,3-Dihydroxy-N-methylquinoxaline, 334
2,3-Dihydroxy-6-methylquinoxaline, 333
2,6-Dihydroxymorpholine, reaction with hydroxylamine, 252
2,3-Dihydroxynaphthalene, reaction with o-phenylenediamine, 369
2′,5′-Dihydroxy-2-nitrodiphenylamine reductive cyclisation, 386
2,3-Dihydroxy-5-nitroquinoxaline, 333
2,3-Dihydroxy-6-nitroquinoxaline, 333
1,2-Dihydroxyphenazines, 381, 388
1,3-Dihydroxyphenazine, 388
1,4-Dihydroxyphenazine, 388
1,6-Dihydroxyphenazine, 388
1,7-Dihydroxyphenazine, 388
1,8-Dihydroxyphenazine, 388
1,9-Dihydroxyphenazine, 388
2,3-Dihydroxyphenazine, 384, 388
2,7-Dihydroxyphenazine, 388
2,8-Dihydroxyphenazine, 387, 388
2,3-Dihydroxy-5-phenylpyrazine, 267
Dihydroxypyrazines, 266–269
2,3-Dihydroxypyrazines, 263
–, alkylation, 268
2,3-Dihydroxypyrazine, 257, 267
2,5-Dihydroxypyrazines, cycloaddition reactions, 269
2,3-Dihydroxypyrazine-5,6-dicarboxylic acid, 268
4,6-Dihydroxypyrimidine, 175, 181
2,3-Dihydroxyquinoxalines (quinoxaline-2,3(1H,4H)diones), 309, 322–334
2,3-Dihydroxyquinoxaline, 311, 334

Dihydroxytartaric acids, reactions with *o*-phenylenediamines, 348
3,6-Dihydroxy-1,2,4,5-tetrazine, 493, 494
Diiminosuccinonitrile, 284, 296, 327
2,7-Diiodophenazine, 376
1,4-Diiodophthalazine, 126
2,6-Diiodopyridazine, 12
3,6-Diisobutyl-2-hydroxypyrazine (flavacol), 266
N,N-Diisobutylpiperidazine, 34
3,6-Diisobutylpyrazine, 266
2,5-Diisopropylpyrazine, 253
Diketene, 174
α,β-Diketones, condensation with α,β-diamines, 247
2,5-Diketopiperazine, alkylation, 282
2,4-Dimercaptopyrimidine, 221
5,5-Dimethoxybarbituric acid, 175
3,8-Dimethoxybenzo[*c*]cinnoline, 108
4,4-Dimethoxybutan-2-one, 168
1,2-Dimethoxycarbonyl-3,6-dimethyl-1,2,3,6-tetrahydropyridazine, 33
1,2-Dimethoxycarbonyl-3,6-diphenyl-1,2,3,6-tetrahydropyridazine, 33
1,4-Dimethoxycarbonylpiperazine-2,3-dione, 296
1,4-Dimethoxycarbonylpiperazinone, 296
2,6-Dimethoxycarbonylpyrazine, 276
2,3-Dimethoxycarbonyltetrahydro-1,4,2,3-dioxadiazine, 500
2,5-Dimethoxy-2,5-dihydrofurans, 5
1,1-Dimethoxy-4-(*N,N*-dimethylamino)-but-3-en-2-one, 169
2,3-Dimethoxy-5,6-dimethylpyrazine, 267
2,5-Dimethoxy-3,6-dimethylpyrazine, 267
3,6'-Dimethoxy-2,2'-dinitrodiphenylamine, cyclisation, 387
2,3-Dimethoxy-5,6-diphenylpyrazine, 267
2,5-Dimethoxy-3,6-diphenylpyrazine, 267
2,6-Dimethoxy-4-hydroxypyrimidine, 170
4,6-Dimethoxy-3-methylpyridazin-1-oxide, 19
4-Dimethoxymethylpyridmidine, 169
1,2-Dimethoxyphenazine, 387
1,3-Dimethoxyphenazine, 388
1,4-Dimethoxyphenazine, 388
1,6-Dimethoxyphenazine, 387, 388
1,8-Dimethoxyphenazine, 388
1,9-Dimethoxyphenazine, 388
2,3-Dimethoxyphenazine, 388
2,7-Dimethoxyphenazine, 388
2,8-Dimethoxyphenazine, 388
3,4-Dimethoxy-*o*-phenylenediamine, reaction with 1,2-cyclohexanadione, 387
2,3-Dimethoxypyrazine, 266, 267
2,5-Dimethoxypyrazine, 266, 267
3,4-Dimethoxypyridazine, 24
3,6-Dimethoxypyridazine, 24
–, rearrangement, 26
3,6-Dimethoxypyridazin-1-oxide, 19
2,4-Dimethoxypyrimidine, 204
2,3-Dimethoxyquinoxaline, 334
6,7-Dimethoxy-1,2,3,4-tetrahydrophenazine, 397
6,8-Dimethoxy-1,2,3,4-tetrahydrophenazine, 397
6,9-Dimethoxy-1,2,3,4-tetrahydrophenazine,
7,8-Dimethoxy-1,2,3,4-tetrahydrophenazine, 397
3,6-Dimethoxy-1,2,4,5-tetrazine, 493
Dimethyl acetylenedicarboxylate, 269
–, adduct with 2-methylquinoxaline, 313, 314
–, cycloaddition with 2,3-dimethylquinoxaline, 315
–, Diels-Alder reactions with pyrazines, 256
p-(Dimethylamino)anils, reactions with *o*-phenylenediamine, 326
3,8-Dimethylaminobenzo[*c*]cinnoline, 107
3-Dimethylaminocinnoline, 61, 63
4-Dimethylaminocinnoline, 63
5-Dimethylamino-2,4-diphenylpyrimidine, 169
2-*N,N*-Dimethylaminoethylpiperazine-1-carboxylic acid, 290
4'-Dimethylamino-4-hydroxydiphenylamine, 435
4-Dimethylamino-2-methylcinnolinium halides, 66
5-[3-(Dimethylamino)-propyl]-1,3,5-oxadiazine, 483
2,4-Dimethylaniline, oxidation, 363
1,3-Dimethylbarbituric acid, 175
1,8-Dimethylbenzo[*c*]cinnoline, 105
1,10-Dimethylbenzo[*c*]cinnoline, 105
–, nitration, 104
2,4-Dimethylbenzo[*c*]cinnoline, 105
2,9-Dimethylbenzo[*c*]cinnoline, 105
3,8-Dimethylbenzo[*c*]cinnoline, 105

INDEX

4,7-Dimethylbenzo[c]cinnoline, 105
2,9-Dimethylbenzo[c]cinnoline-dioxide, 105
3,8-Dimethylbenzo[c]cinnolin-1-oxide, 105
3,5-Dimethyl-2-cyanopyrazine, 273
5,6-Dimethyl-2-cyanopyrazine, 273
4a,9a-Dimethyldecahydrophenazine, 401
3,6-Dimethyl-4,5-dihydropyridazine, 34
5,7-Dimethyl-3,4-dihydroxycinnoline, 71
1,4-Dimethyl-2,3-dihydroxyquinoxaline, 334
Dimethyl α,α-dimethoxymalonate, 175
1,3-Dimethyl-2,4-dioxo-6-methylamino-5-nitroso-1,2,3,4-tetrahydro pyrimidine, 209
3,6-Dimethyl-2,5-diphenylpyrazine, 257
3,3-Dimethyl-1,3,2,4,6-dithiatriazin-5(6H)-on-1,1-dioxide, 502
4,6-Dimethyl-1,3,2,4,6-dithiatriazin-5(2H)-on-1,1,3,3-tetroxide, 501, 502
1,6-Dimethyl-5-ethoxycarbonyl-2-hydroxy-4-methylpyrimidine, 223
1,4-Dimethyl-2-ethoxycarbonyl-3-phenyl-1,4,5,6-tetrahydropyrazine, 283
Dimethyl ethoxymethylenemalonate, 173
3,6-Dimethyl-2-ethoxypyrazine, 265
4,6-Dimethyl-2-ethylaminopyrimidine, 170
N,N-Dimethylethylenediamine, 291
3,3-Dimethyl-4-ethylpiperazin-2-one, 296
3,6-Dimethyl-4-ethylpyridazine, 10
N,N-Dimethylformamide diethylacetal, 370
N,N-Dimethylformamide dimethylacetal, 211
Dimethylglyoxime, hydrogenation, 253
1,2-Dimethylhexahydropyridazine, 35
5,6-Dimethyl-2-hydroxy-3-phenylpyrazine, 262
5,6-Dimethyl-2-hydroxy-3-phenylpyrazin-1-oxide, 265
3,5-Dimethyl-2-hydroxypyrazine, 262
3,6-Dimethyl-2-hydroxypyrazine, 262
5,6-Dimethyl-2-hydroxypyrazine, 262
5,6-Dimethyl-2-hydroxypyrazine-3-carboxamide, 275
3,6-Dimethyl-2-hydroxy-5-pyrazinoic acid, 275
3,6-Dimethyl-2-hydroxypyrazin-1-oxide, 266
4,6-Dimethyl-2-hydroxypyrimidine, 170, 198, 200
3,5-Dimethyl-2-hydroxyquinoxaline, 329
3,6-Dimethyl-2-hydroxyquinoxaline, 329
3,7-Dimethyl-2-hydroxyquinoxaline, 329
2,6-Dimethyl-4-iodopyrimidine, 206
Dimethylketene, 284
4,6-Dimethyl-2-mercaptopyrimidine, 170
3,5-Dimethyl-2-methylimino-1,3,5-oxadiazine-4,6-dione, 480
2,5-Dimethyl-5-methylpyrimidine, 168
6,9-Dimethyl-2-methylthiopurine, 188
1,10-Dimethyl-4-nitrobenzo[c]cinnoline, 104
1,10-Dimethyl-7-nitrobenzo[c]cinnoline, 104
3,4-Dimethyl-8-nitrocinnoline, 60
1,2-Dimethyl-5-nitrophthalazinium iodide, 122
1,3-Dimethyl-5-nitrophthalazinium iodide, 122
3,6-Dimethyl-4-nitropyridazin-1-oxide, 19
2,3-Dimethyl-5-nitroquinoxaline, 324
2,3-Dimethyl-6-nitroquinoxaline, 324
3,5-Dimethyl-1,3,5-oxadiazine-2,4,6-trione, 480
7,7-Dimethyl-5-oxo-3-phenyl-5,6,7,8-tetrahydrocinnoline, 64, 97
Dimethylphenazhydrin, 396
1,3-Dimethylphenazine, 368
1,6-Dimethylphenazine, 368
1,7-Dimethylphenazine, 367, 368
1,8-Dimethylphenazine, 367, 368
1,9-Dimethylphenazine, 368
2,3-Dimethylphenazine, 368
2,7-Dimethylphenazine, 367, 368
2,3-Dimethylphenazin-N-oxides, 373
2,2-Dimethyl-3-phenyl-2H-azirine, 286
N,N-Dimethyl-p-phenylenediamine, 419, 427, 428
N,N'-Dimethyl-o-phenylenediamine, reaction with cyclohexane-1,2-dione, 393
N,N-Dimethyl-p-phenylenediamine-phenol, oxidation, 432
4,6-Dimethyl-2-phenylpyrimidine, 165
3,6-Dimethyl-3-phenyl-2,3,4,5-tetrahydropyridazine, 34
1,4-Dimethylphthalazine, 122
2,3-Dimethylphthalazine-1,4-dione, reduction, 150
2,5-Dimethylpiperazines, 293

1,2-Dimethylpiperazine, 35
1,4-Dimethylpiperazine, reaction with ethylenechlorohydrin, 292, 293
2,3-Dimethylpiperazine, dehydrogenation, 253
3,6-Dimethylpiperazine-2,5-dione, 257
3,3-Dimethylpiperazin-2-one, 294
–, reductive N-alkylation, 296
1,4-Dimethylpiperazin-N-oxides, 292
2,3-Dimethylpyrazine, 253, 254, 352
2,5-Dimethylpyrazine, 254, 260, 263
2,6-Dimethylpyrazine, 250, 253
3,6-Dimethylpyrazine-2,5-dicarboxylic acid, 272
5,6-Dimethylpyrazine-2,3-dicarboxylic acid, 272
3,4-Dimethylpyridazine, 10
3,5-Dimethylpyridazine, 10
3,6-Dimethylpyridazine, 2, 10
4,5-Dimethylpyridazine, 10
3,6-Dimethylpyrazine-2-carboxylic acid, 272
3,6-Dimethylpyridazine-4,5-dicarboxylic acid, 27
1,2-Dimethylpyridazine-3,6-dione, 25
3,6-Dimethylpyridazine-1,2-dioxide, 19
2,6-Dimethylpyridazin-3($2H$)-one, 25
3,4-Dimethylpyridazin-6($1H$)-one, 23
4,6-Dimethylpyridazin-3($2H$)-one, 23
5,6-Dimethylpyridazin-3($2H$)-one-4-carboxylic acid, 27
3,4-Dimethylpyridazin-1-oxide, 18
3,6-Dimethylpyridazin-1-oxide, 18, 22
4,6-Dimethylpyrimidin-2-amine, 161
4,5-Dimethylpyrimidine, 200
4,6-Dimethylpyrimidine, 159
4,6-Dimethylpyrimidin-1-oxide, 227
2,4-Dimethylquinazoline, 236
2,3-Dimethylquinoxaline, 312, 315
–, addition reactions, 315, 316
–, alkylation, 316
–, photo-adduct with tetrahydrofuran, 317
2,6-Dimethylquinoxaline, 317
2,7-Dimethylquinoxaline, 317
6,7-Dimethylquinoxaline, 317
6,7-Dimethylquinoxaline-2,3-dicarboxylic acid, 348
2,3-Dimethylquinoxaline-1,4-dioxide, 308, 338, 361
–, deoxygenation, 315
–, reaction with benzil, 367

–, reactions with α-diketones, 372
1,3-Dimethylquinoxalin-2-one, 332
2,3-Dimethylquinoxalin-N-oxides, 336
2,3-Dimethyl-5,6,7,8-tetrabromoquinoxaline, 317
3,5-Dimethyl-5,6,7,8-tetrahydrocinnoline, 97
Dimethyl 4,5,6,7-tetrahydroisoindole-1,3-decarboxylate, 152
Dimethyl 5,6,7,8-tetrahydrophthalazine-1,4-dicarboxylate, 152
1,4-Dimethyltetrahydropyrazine, 291
2,3-Dimethyl-1,2,3,4-tetrahydroquinoxalines, 351, 353
2,3-Dimethyl-1,2,3,4-tetrahydroquinoxaline, 317
1,4-Dimethyl-1,4,5,6-tetrahydro-1,2,3,4-tetrazine, 487
3,6-Dimethyl-1,2,3,6-tetrahydro-1,2,4,5-tetrazine, 492
3,6-Dimethyl-1,2,4,5-tetrazine, 492
3,6-Dimethylthio-1,2,4,5-tetrazine, 493
2,4-Dimethyl-1,3,5-triazine, 472
4,6-Dimethyl-1,2,3-triazine, 458
3,6-Dimethyl-1,2,4-triazine-5-thione, 464
4,6-Dimethyl-1,2,4-triazin-5($4H$)-one, 464
5,6-Dimethyl-1,2,4-triazin-3-one, 463
4,5-Dimethyl-1,2,3-triazin-2-oxide, reduction, 458
5,7-Dimethyl-o-triazolo[1,5-a]pyrimidine, 189
3,6-Dimethyluracil, 183
N,N'-Dimethylurea, 168, 175, 501
1,4-Dimorpholinophthalazine, 138
Dimroth rearrangement, 177, 194, 212, 213, 236, 237
2,3-Di-(β-naphthyl)quinoxaline, 312
3,4-Dinitroaminopyridazine, 17
1,3-Dinitrobenzene, 193
1,4-Dinitrobenzo[c]cinnoline, 106
1,10-Dinitrobenzo[c]cinnoline, 104, 107
2,4-Dinitrochlorobenzene, 432
2,2'-Dinitrodiphenylamines, 359
–, intramolecular cyclisation, 375
2,4-Dinitrofluorobenzene, reactions with α-phenylacetamidines, 337
2,3-Dinitro-4'-hydroxydiphenylamine, 432
2,6-Dinitro-4'-methoxydiphenylamine, 379
6,8-Dinitro-5-methoxyquinoxaline, 324
1,6-Dinitrophenazine, 378, 380

1,9-Dinitrophenazine, 378, 380
2,4-Dinitrophenol, dithionation, 432
5,6-Dinitroquinoxaline, 310
5,6-Dinitroquinoxaline, 323
6,7-Dinitroquinoxaline, 324
6,7-Dinitroquinoxaline-2,3-dione, 317
Dioxadiazines, 498, 499
1,2,3,4-Dioxadiazines, 499
1,2,4,5-Dioxadiazines, 499
1,4,2,5-Dioxadiazines, 500
1,3,2,4,5-Dioxadithiazines, 498, 499
1,3,2,6,5-Dioxadithiazines, 502
Dioxathianes, 485, 486
1,3,2-Dioxathian-2-oxides, 486
Dioxathiazines, 498
1,3,2,5-Dioxathiazines, 500
Dioxazines, 481–483
1,6-Dioxo-1,2,3,4,6,7,8,9-octahydro-phenazine, 400
Dioxosuccinates, reactions with o-phenylenediamines, 348
Dipegine, 444
Diphenylacetylene, 498
N,N'-Diphenylbenzamidine, 176
Diphenylbipyrimidine, 207
1,4-Diphenyl-1,3-butadiene, 27, 33
1,3-Diphenylcinnolin-4($1H$)-one, 77
Diphenylcyclopropenone, 173
–, cycloaddition with 2-methylquinoxaline, 314
3,6-Diphenyl-4,5-dihydropyridazine, 32
3,6-Diphenyl-1,4,2-5-dioxadiazine, 500
5,6-Diphenyl-1,3,2,4-dithiadrazine, incorrect structure allocation, 498
2,5-Diphenyl-1,4,3-dithiazine, 482
2,6-Diphenyl-1,4,3-dithiazine, 482
1,6-Diphenyl-3-ethoxycarbonyl-4-hydroxy-pyridazine, 30
3,4-Diphenyl-6-hydrazinopyridazine, 17
4,6-Diphenyl-3-hydrazinopyridazine, 17
5,6-Diphenyl-2-hydroxy-3-methylpyrazine, 262
3,5-Diphenyl-2-hydroxypyrazine, 262
3,6-Diphenyl-2-hydroxypyrazine, 264
5,6-Diphenyl-2-hydroxypyrazine, 262, 264
5,6-Diphenyl-2-hydroxypyrazine-3-carboxamide, 275
3,5-Diphenyl-2-hydroxypyrazin-1-oxide, 266
Diphenylketene, 284
–, reaction with dihydropyrazines, 279

3,6-Diphenyl-5-methoxy-1,2,4-triazine, 464
1,4-Diphenyl-2-methylpiperazine, 293
4,6-Diphenyl-2-methyl-1,3,5-triazine, 471
2,3-Diphenyl-6-nitroquinoxaline, 324
2,6-Diphenyl-4,H-1,3,4-oxadiazine, 480
1,4-Diphenyl-1,3,4-oxadiazin-5($6H$)-one, 480
2,6-Diphenyl-1,3,5-oxadiazin-4-one, 480
2,5-Diphenyl-1,3,4-oxadiazole, 493
3,4-Diphenyl-1-oxidophthalazinium betaine, 149
1,4-Diphenyloxyphthalazine, 127
1,3-Diphenylphenazine, 361, 367, 368
2,3-Diphenylphenazine, 367, 368
2,7-Diphenylphenazine, 368
2,3-Diphenylphenazin-N-oxides, 369
1,4-Diphenylphthalazine, 122
–, synthesis from 1,4-diphenyl-sym-tetrazine, 118
4,4-Diphenylphthalazin-1-one, 149
1,4-Diphenylphthalazin-2-oxide, 124
1,4-Diphenylpiperazine, 292
2,3-Diphenylpiperazine, 294
1,4-Diphenylpiperazine-2,3-dione, 296
1,4-Diphenylpiperazine-2,6-dione, 300
1,4-Diphenylpiperazine-1,4-dioxides, 292
2,3-Diphenylpyrazine, 257, 278
2,5-Diphenylpyrazine, 246, 257
2,6-Diphenylpyrazine, 246, 257
5,6-Diphenylpyrazine-2-carboxamide, 279
3,4-Diphenylpyridazine, 10
3,5-Diphenylpyridazine, 10
3,6-Diphenylpyridazine, 10-31
4,5-Diphenylpyridazine, 10
3,4-Diphenylpyridazin-6($1H$)-one, 23
4,6-Diphenylpyridazin-3($2H$)-one, 23
5,6-Diphenylpyridazin-3($2H$)-one-4-carboxylic acid, 27
2,4-Diphenylpyrimidine, 219
1,3-Diphenyl-2H-pyrrolo[3,4-c]cinnoline, oxidation, 90
2,3-Diphenylquinazolin-4($3H$)-one, 233
2,3-Diphenylquinoxaline, 303, 312, 318
–, reduction, 349
2,3-Diphenylquinoxalin-N-oxides, 336
2,3-Diphenylquinoxalin-1-oxide, photorearrangement, 342
2,6-Diphenyl-4-stilbenyl-1,3,4,5-thiatriazine-1,1-dioxide, 497
1,4-Diphenyl-1,2,3,4-tetrahydrocinnoline,

94
2,3-Diphenyl-1,2,3,4-tetrahydrophthalazine, 151
3,6-Diphenyl-3,4,5,6-tetrahydropyridazines, 34
1,3-Diphenyl-1,4,5,6-tetrahydropyridazine, 34
3,6-Diphenyl-1,4,5,6-tetrahydropyridazine, 34
1,4-Diphenyl-1,2,3,4-tetrahydropyrazine-2,3-dione, 296
2,3-Diphenyl-1,2,3,4-tetrahydroquinoxalines, 351, 353
2,6-Diphenyl-2,3,4,5-tetrahydro-1,2,4-triazine, 465
3,6-Diphenyl-1,2,4,5-tetrazine, 492
–, behaviour as diene, 493
2,3-Diphenylthiirene-1,1-dioxide, 497
2,4-Diphenyl-1,3,5-triazine, 472
5,6-Diphenyl-1,2,4-triazine, 461
2,4-Diphenyl-1,3,5-triazine-2-carboxylic acid, 471
4,6-Diphenyl-1,3,5-triazine-2-carboxylic acid, 471
5,6-Diphenyl-1,2,4-triazin-3-one, 463
4,5-Diphenyl-1,2,3-triazole, 497
5,6-Diphenyl-3-trichloromethyl-1,2,4-triazine, 461
2,6-Diphenyl-4-(4-tolyl)-4H-1,3,4,5-oxatriazine, 498
4,6-Diphenyl-5(4-tolyl)-1,2,3-triazine, 458
3-(4'-Diphenylyl)-2-phenylquinoxaline, 312
2-(4'(4'-Diphenylyl)quinoxaline, 312
2,3-Dipropylquinoxaline, 316
1,4-Dipyrazin-4'-on-1'-ylphthalazines, 128
1,2-Di(pyrazinyl)ethane, 255
2,5-Distyrylpyrazine, 255
3,6-Distyrylpyridazin-1-oxide, 22
N,N'-Disubstituted piperazines, stereochemistry, 292
Dithiadrazines, 498
1,3,2,4,6-Dithiatriazines, 501
Dithiazines, 481, 482
1,2,4-Dithiazines, 481, 482
1,4,2-Dithiazine-1,1-dioxides, 482
N,N'-Ditosyl-o-phenylenediamines, 352
3,6-Di-(trifluoromethyl)-1,4,2,5-dioxadiazine, 500
3,6-Di-(2,4,6-trimethylphenyl)-1,4,2,5-dioxadiazine, 500

1,2,3,4,5,6,7,8,9,10,11,12-Dodecahydrobenzo[c]cinnoline, 110

Ethano-bridged tetrahydroquinoxalines, 353
Ethanolamine, 155
3-Ethoxy-1,2,4-benzotriazine, 467
3-Ethoxy-1,2,4-benzotriazin-1-oxide, 467
3-Ethoxycarbonyl-1,2,4-benzotriazine, 466
3-Ethoxycarbonylcinnolin-4(1H)-thione, 98
1-(2'-Ethoxycarbonylhydrazino)-phthalazine, 129
5-Ethoxycarbonyl-4-hydroxy-2-methylpyrimidine, 173
5-Ethoxycarbonylmethyl-4-hydroxy-2-mercaptopyrimidine, 222
2-Ethoxycarbonylnaphtho[1,2-e]-1,2,4-triazine, 468
1-Ethoxycarbonylpiperazine, 289
2-Ethoxycarbonylpiperazines, 291
6-Ethoxycarbonylpyridazin-3(2H)-one, 27
1-Ethoxycarbonyltetrahydropyridazine, 31
1-Ethoxycarbonyl-1,2,3,4-tetrahydroquinoxalines, 353
3-Ethoxycarbonyl-1,2,4-triazine, 461, 462
4-Ethoxycinnoline, 77
3-Ethoxy-2-methylacrolein, 168
α-(Ethoxymethylene)acetonitrile, 178
α-(Ethoxymethylene)acetophenone, 169
4-Ethoxy-6-methylpyrimidine-2-carbonitrile, 227
4-Ethoxy-6-methylpyrimidin-N-oxide, 227
Ethoxymethylquinoxaline, 271
4-Ethoxyoxalylmethylpyrimidine, 201
2-Ethoxyphenazine, dialkylation, 286
2-Ethoxypyrazine, 265
4-Ethoxypyridazine, 23
4-Ethoxyquinazoline, 239
4-Ethoxyquinazolin-1-oxide, 239
Ethyl 3-aminocrotonate, 183
Ethyl 3-amino-2-cyanoacrylate, 184
Ethyl 4-amino-2-methylpyrimidine-5-carboxylate, 184
2-Ethylbenzonitrile, 64
3-Ethyl-1,2,4-benzotriazine, 466
Ethyl α-benzoyl-α-chloroacetate, reaction with N,N-dimethylethylenediamine, 283
Ethyl cinnoline-3-carboxylate, 85
Ethyl cinnoline-4-carboxylate, 85

Ethyl cyanoacetate, 179
3-Ethyl-2-cyanopyrazine, 273
Ethylenediamine, 247
Ethyl ethoxymethylenecyanoacetate, 177
Ethyl 6-ethyl-4-hydroxy-3-cinnolyl-
  carboxylate, radiolabelled, 83
Ethyl formimidate, reaction with
  hydrazine, 492
Ethyl formylacetate, 172
Ethyl α-formylacetoacetate, 168
Ethyl guanidine, 170
1-Ethyl-4-hydroxy-3-methanesulphonyl-
  cinnoline, reaction with potassium cyanide, 84
Ethyl 4-hydroxy-6-phenylcinnoline-3-
  carboxylate, 84
3-Ethyl-2-hydroxyquinoxaline, 329
Ethyl 2-mercaptoethylacetoacetate, 174
Ethyl 2-methoxy(thiocarbonyl)acetate, 176
4-Ethyl-3-methylpyridazine, 10
5-Ethyl-3-methylpyridazine, 10
5-Ethyl-4-methylpyridazine, 10
6-Ethyl-3-methylpyridazine, 10
4-Ethyl-6-methylpyridazin-3($2H$)-one, 23
Ethyl 1-methylthiophthalazine-4-carboxylate, 145
Ethyl oxalamidrazinate, 461
Ethyl α-oximinoacetoacetate, reaction with
  o-phenylenediamine, 328
2-Ethylphenazine, 368
10-Ethylphenazin-1($10H$)-one, 366
10-Ethylphenazin-2($10H$)-one, 365
Ethylphenazinium ethylsulphate, 365
Ethylphenazinium iodide, 364
5-Ethylphenazyl radical, 366
Ethyl 3-phenylcinnoline-4-carboxylate, 85
1-Ethylphthalazine, 122
1-Ethylphthalazinium salts, 122
Ethyl phthalazin-1($2H$)-thione-4-carboxylate, 145
Ethyl piperazinoate, 291
Ethyl propynoate, 173
Ethylpyrazines, oxidation, 255
3-Ethylpyridazine, 10
6-Ethylpyridazin-3($2H$)-one, 23
Ethyl 3-pyridazinylpyruvate, 11
Ethyl 4-pyridazinylpyruvate, 11
2-Ethylquinoxaline, 312
Ethyl sodioformylacetate, 172
6-Ethyl-3-sulphanilamidopyridazine, 30

3-Ethylthio-6-methylthiopyridazine, 29
5-Ethyl-3,4,6-trimethylpyridazine, 10
Ethyl β-ureidocrotonic acid, 173
Eurhodines, 381, 405–409
Eurhodols, 405–407
Euxylophora paraensis, 453
Euxylophoricine A, 453, 455, 456
Euxylophoricine B, 455
Euxylophoricine C, 456
Euxylophoricine D, 454
Euxylophoricine E, 454
Euxylophoricine F, 454
Euxylophorine A, 453, 455
Euxylophorine B, 453
Euxylophorine D, 454
Evodiamine, 452
Evodia rutecarpa, 452, 455

Febrifugine, 228, 449–451
–, salts, 451
Flavacol (3,6-diisobutyl-2-hydroxy-
  pyrazine), 266
Flavazoles, 344
Flaviinduline O, 404, 406
Fluoflavin, 323
Fluorindines (5,12-Dihydroquinoxalo[2,3-
  b]phenazines), 412, 413, 415
Fluorindine, 413
8-Fluoro-4-hydroxycinnoline, 56
1-Fluorophenazine, 376
2-Fluorophenazine, 376
3-Fluoro-5-phenyl-1,2,4-triazine, 463
Fluoropyrazines, 258, 259
2-Fluoropyrazine, 258
–, reaction with ammonia, 260
2-Fluoropyrazin-1-oxide, 265
2-Fluroquinoxaline, 319, 320
5-Fluoro-2,4,6-triaminopyrimidine, 212
5-Fluoro-2,4,6-trisheptafluoroisopropyl-
  pyrimidine, 207
5-(Fluorovinyl)pyrimidines, 225
Fluorubin, 311, 323
Formamide, 181
Formamidine, 175, 471
–, trimerisation, 470
Formamidine acetate, reaction with
  hydrazine hydrate, 493
Formamidrazone, 461
1,3,5-Formazans, 494, 495
Formoguanamine (2,4-diamino-1,3,5-
  triazine), 472

532 INDEX

Formylacetic acid, 172
Formylacrylic acids, 4
Formylacrylic esters, 4
Formylbromopyruvic acid, 169
3-Formylcinnoline, 85, 88
4-Formylcinnoline, 85, 89
N-Formyldihydrorutaecarpine, 455
1-Formylphenazine, 367, 391
2-Formylphenazine, 391
2-Formylphenazin-N-oxides, 373
Formylpyrimidines, 200, 220
2-Formylquinoxalin-N-oxides, 336
5-Formyluracil, 190
Frankland-Kolbé synthesis, 184
Furan-2-carboxamidine, 167
Furanoquinoxalines, 321
2-Furylglyoxalic acid, 343
2-Fur-2′-yl-5-nitropyrimidine, 167
2-(2-Furyl)quinoxaline, 343
Fusel oils, 241

Gallamide, 420
Gallamine Blue (C.I. Mordant Dye 51060), 420
Gallanil Violet (C.I. Mordant Blue 45, 51065), 420, 421
Gallanilide, 420
Gallo Blue E (C.I. Mordant Violet 54, 51040), 420
Gallocyanines, 420–422
Gallocyanine, 420
Glomaris marginata, 437
Glomerine, 437, 439
D-Glucose, reaction with ammonia, 253
Glyceraldehyde, 250
Glycine, 440
Glycine amide, reaction with cyclohexane-1,2-dione, 354
Glycomis arborea, 437, 438
Glycomis bilocularis, 440
Glycomis pentaphylla, 438
Glycophymine, 438, 439
Glycophymoline, 438, 439
Glycorine, 438, 440
Glycosmicine, 438
Glycosminine, 438
Glyoxal, 2, 247, 302, 303, 340, 461
–, reactions with o-phenylenediamines, 334
Gougerotin, 209
Grimmel synthesis, of quinazolines, 232

Guanidines, 173, 180
–, reactions with N,N-dimethylformamide diethylacetal, 470
Guanidine, 167, 170, 172, 175, 176, 178, 179
–, reaction with 2-chloroethanesulphonyl-chloride, 476
Guanidine carbonate, 170

α-Halogenoacid halides, 249
Halogenobenzo[c]cinnolines, 107
Halogenobenzo[c]cinnolin-N-oxides, deoxygenation, 107
α-Halogenocarbonyl compounds, reactions with o-phenylenediamines, 304
Halogenocinnolines, 54–56
Halogenonitro-4-aziridinocinnolines, 62
Halogeno-2-nirodiphenylamines, cyclisation, 375
p-Halonitrosobenzenes, cyclodimerisation, 375
Halogenophenazines, 375–378
–, reactions with nucleophiles, 378
Halogenophthalazines, synthesis, 124, 125
Halogenopyrazines, 254, 257–259
–, hydrolysis, 263
Halogenopyridazin-N-oxides, 14
Halogenopyrimidines, 201–208
5-Halogenopyrimidines, 193, 195, 196
Halogenoquinoxalines, 319–323
Halogeno-1,2,4-triazines, 463
Halogeno-1,3,5-triazines (cyanuric halides), 473, 474
Heptan-2,4-dione, 171
Hetrazin, 291
3,4,5,6,7,8-Hexachlorocinnoline, 55
Hexachlorophthalazine, 126
Hexachloroquinoxaline, 320
Hexafluoroacetone, 305
3,4,5,6,7,8-Hexafluorocinnoline, 55
1,4,5,6,7,8-Hexafluorophthalazine, 125, 126
Hexafluoroquinoxaline, 320
1,2,3,4,5,6-Hexahydrobenzo[f]cinnoline, 110
2,3,4,4a,5,10-Hexahydrobenzo[g]cinnolinones catalytic hydrogenation, 112
1,2,3,5,6,10b-Hexahydrobenzo[f]cinnolin-2-ones, 110
Hexahydrocinnolines, 97
4,4a,5,6,7,8-Hexahydrocinnolines, oxida-

INDEX 533

tion, 96
4,4a,5,6,7,8-Hexahydrocinnolin-3(3H)-ones, conversion into azamorphinans, 97
–, reduction to decahydrocinnolines, 97
1,4,5,6,7,8-Hexahydrocinnolin-5-ones, synthesis, 99
2,3,4,4a,5,6-Hexahydro-3-oxo-5-phenylbenzo[h]cinnoline, 111
2,3,4,4a,5,6-Hexahydro-5-phenylbenzo[h]cinnolin-3-one, 111
4a,5,6,7,8,8a-Hexahydrophthalazines, 153
Hexahydropyrazines (piperazines), 287–294
Hexahydropyridazines, (piperidazines), 34
Hexahydro-1,2,4,5 tetrazines, 491
Hexahydro-1,2,4,5-tetrazine-3,6-dione, 493, 494
Hexahydro-1,3,5-triazin-2-ones, 481
1,4,5,6,7,8-Hexamethoxyphthalazine, 126
2,3,4,7,8,9-Hexamethylbenzo[c]cinnoline, 105
Hexamethylenetetramine (hexamine), 475
1,2,2,5,5,6-Hexamethyltetrahydropyrazine, 286
Hexamine (hexamethylenetetramine), 475
Hilbert-Johnson reaction, 217, 218, 238
Hinsberg reaction, 302
Hofmann reaction, 219, 231, 259, 272, 324
Homoglomerine, 437, 44
Hortiacine, 453
Hortiamine, 453
Hydantoins, 186, 187
3-Hydrazino-1,2,4-benzotriazine, 466
4-Hydrazino-1,2,3-benzotriazine, 460
Hydrazinocinnolines, synthesis, 68
4-Hydrazinocinnoline, 68, 69
1-Hydrazino-3-hydroxyphthalazine, 130
4-Hydrazino-6-methoxycinnoline, 69
4-Hydrazino-3-methylcinnoline, 69
3-Hydrazino-6-methylpyridazine, 17
4-Hydrazino-6-nitrocinnoline, 69
4-Hydrazino-7-nitrocinnoline, 69
4-Hydrazino-3-phenylcinnoline, 69
3-Hydrazino-6-phenylpyridazine, 17
Hydrazinophthalazines, 127
1-Hydrazinophthalazines, acylation, 129
–, nitrosation, 129
1-Hydrazinophthalazine, 128
–, reaction with ethyl chloroformate, 130
4-Hydrazinophthalazin-1(2H)-one, 128
3-Hydrazinopyridazine, 17
4-Hydrazinopyridazine, 17
Hydrazinopyrimidines, 203
4-Hydrazino-3,6,8-triaminopyrimidino-[4,5-e]-1,2,4-triazine, 468
3-Hydrazino-1,2,4-triazine, 461
Hydrazobenzene, reaction with formaldehyde, 494
Hydron Blue R (C.I. Vat Blue 43, 53630), 436
3-Hydroxy-1-(2-aminobenzyl)pyrrolidine, 442
1-Hydroxyaminophthalazines, synthesis from alkoxyphthalazines, 127
3-(o-Hydroxyaryl) quinoxalin-1-oxides, 339
2-Hydroxybenzo[c]cinnoline, 106
3-Hydroxybenzo[c]cinnoline, 106, 108
1-Hydroxybenzo[c]cinnoline-10-carboxylic acid lactone, 108
3-Hydroxybenzo[c]cinnolin-6-oxide, 108
Hydroxycinnolines, 62
–, electrophilic substitution, 77, 78
–, phenolic properties, 81, 82
–, physical properties, 70
–, spectra, 82
3-Hydroxycinnoline, 56, 65, 70–72
–, reaction with phosphorus halides, 74
–, reduction, 73, 93
4-Hydroxycinnolines, N- vs. O-acetylation, 79
–, acetyl derivatives, 75
–, bromination, 56
–, reactions, 77
–, synthesis, 56, 74, 76
–, tautomerism, 36
4-Hydroxycinnoline, 70, 77
–, alkylation, 78
–, mass spectrum, 79
–, nitration, 59
–, reduction, 77, 93
5-Hydroxycinnolines, synthesis and physical properties, 80
6-Hydroxycinnolines, synthesis and physical properties, 80
7-Hydroxycinnolines, synthesis and physical properties, 80
8-Hydroxycinnolines, synthesis and physical properties, 80
8-Hydroxycinnoline, 81

4-Hydroxycinnoline-6-carbonitrile, synthesis from ethyl β-(2-amino-4-ethoxycarbonylbenzoyl) propionate, 91
4-Hydroxycinnoline-3-carboxylic acids, 38, 83
–, decarboxylation, 74–76
4-Hydroxycinnoline-3-carboxylic acid, 36, 85
–, attempted esterification, 85, 86
8-Hydroxycinnoline-4-carboxylic acid, 81
4-Hydroxycinnolinium cation, 78
6-Hydroxycytosine, 179
2-Hydroxydihydropyrimidines, 185
Hydroxydihydroquinoxalines, synthesis, 309
4-Hydroxy-3,6-dimethyl-5-nitropyridazin-1-oxide, 19
4-Hydroxy-2,6-dimethylpyrimidine, 173
4-Hydroxydiphenylamine, 432
2-Hydroxy-3,6-diphenylpyrazine, 264
2-Hydroxyethylamine, 288
N-(2-Hydroxyethyl)ethylenediamine, 288
1-(β-Hydroxyethyl)piperazine, 290, 292
2-(β-Hydroxyethyl)quinoxaline, 311
Hydroxyevodiamine, 453
α-Hydroxyiminoketones, 247
4-Hydroxy-3-iodocinnoline, 56
4-Hydroxy-8-iodocinnoline, 56
2-Hydroxy-3-isobutylpyrazine, 265
δ-Hydroxylysine, 451
4-Hydroxy-2-mercapto-6-methylpyrimidine, 173
4-Hydroxy-6-mercapto-2-methylpyrimidine, 176, 205
4-Hydroxy-5-methoxycinnoline, 75, 80
4-Hydroxy-6-methoxycinnoline, 75, 80
4-Hydroxy-7-methoxycinnoline, 75, 80
4-Hydroxy-8-methoxycinnoline, 75, 80
4-Hydroxy-6-methoxycinnoline-3-carboxylic acid, 81, 85
1-Hydroxy-6-methoxyphenazine, 384
2-Hydroxy-6-methoxyphenazine, 386
3-Hydroxy-4-methylcinnoline, 72, 73
4-Hydroxy-3-methylcinnoline, 75
4-Hydroxy-4-methylcinnoline, 81
4-Hydroxy-7-methylcinnoline, 75
4-Hydroxy-8-methylcinnoline, 75
8-Hydroxy-4-methylcinnoline, 80
4-Hydroxy-2-methylcinnolinium hydroxide, 78
2-(Hydroxymethyl)-5-methylpyrazine, 271

4-Hydroxy-5-methyl-2-methylthiopyrimidine, ammonolysis, 223, 224
2-Hydroxy-6-methylphenazine, 386
1-Hydroxy-5-methylphenazinium methylsulphate, 386
2-Hydroxy-3-methyl-5-phenylpyrazine, 262
6-Hydroxy-4-methyl-5-phenylpyridazin-3(2H)-one, 24
2-Hydroxy-3-methyl-5-phenylpyrazin-1-oxide, 266
2-Hydroxy-3-methylpyrazine, 262
2-Hydroxy-5-methyl-3-pyrazinoic acid, 275
3-Hydroxymethylpyridazine, 27
6-Hydroxy-2-methylpyridazin-3-one, 25
2-Hydroxy-5-methylpyrimidine, 168
4-Hydroxy-2-methylpyrimidine, 172
2-Hydroxymethylquinoxaline, 312
2-Hydroxy-3-methylquinoxaline, 329
2-Hydroxy-5-methylquinoxaline, 329
2-Hydroxy-7-methylquinoxaline, 329
5-Hydroxymethyluracils, 214
5-Hydroxymethyluracil, 190
5-Hydroxy-6-methyluracil, 215
4-Hydroxy-3-nitrocinnoline, 56, 67, 75
4-Hydroxy-5-nitrocinnoline, 75
4-Hydroxy-6-nitrocinnoline, 75, 77
–, chlorination, 61
4-Hydroxy-7-nitrocinnoline, 75
4-Hydroxy-8-nitrocinnoline, 50, 75, 77
2-Hydroxy-1-nitrophenazine, 380
2-Hydroxy-5-nitro-3-phenylpyrazine, 264
2-Hydroxy-5-nitropyrimidine, 195
2-Hydroxy-6-nitroquinoxaline, 329
Hydroxyphenanthraquinones, 406
Hydroxyphenazines (phenazinols), 384–390
–, methylation, 385, 386
–, synthesis from 2-aminoindophenols, 387
–, – from 1-aminophenazines, 386
–, – from o-benzoquinones, 385
–, – from 6-methoxy-2-nitrodiphenylamines, 386
–, – from o-phenylenediamines, 385, 386
–, – from pyrogallol, 386
–, tautomerism, 385
1-Hydroxyphenazine, 384–388
2-Hydroxyphenazine, 370, 384, 386–388
2-Hydroxyphenazine-5,10-dioxide, 370, 386
1-Hydroxyphenazin-N-oxides, 360, 373

2-Hydroxyphenazin-N-oxides, 373
Hydroxyphenaylaminocarbazoles, 436
3-Hydroxy-4-phenylcinnoline, 73
4-Hydroxy-3-phenylcinnoline, 74, 77, 88
–, synthesis from 3-phenylcinnoline, 77
4-Hydroxy-4-phenylcinnoline, 81
1-Hydroxy-3-phenyl-3,4-dihydro-
    phthalazinyl-4-acetic acid, 134
4-Hydroxy-2-phenylphthalazin-1($2H$)-
    one, 115, 134, 147
2-Hydroxy-2-phenylpyrazine, bromination,
    263
2-Hydroxy-3-phenylpyrazine, 262, 264
2-Hydroxy-3-phenylquinoxaline, 329
2-Hydroxy-4-phenylpyrimidine, 169
Hydroxyphthalazine, synthesis, 130
1-Hydroxyphthalazin-1($2H$)ones, acetyl
    and methyl derivatives, 139
–, alkylation, 138
–, oxidation and reduction, 138
–, physical properties, 137, 138
–, synthesis from N-aminophthalimide,
    137
4-Hydroxyphthalazin-1($2H$)-ones (1,4-
    dihydroxyphthalazines or 1,4-
    phthalazindiones), synthesis, 136, 137
4-Hydroxyphthazin-1($2H$)-one, 115, 124,
    126, 137, 142, 150
5-Hydroxypipecolic acid, 451
3-(3-Hydroxypiperidin-2-yl)-1-quinazol-4-
    on-3-yl-propan-2-one, 450
3-Hydroxy-2-polyhydroxyalkylquinoxalines,
    343
Hydroxypyrazines, 248, 261–265
–, alkylation, 264
–, electrophilic subotitation, 263
–, i.r. spectra, 261
–, phenylation, 264
–, tautomerism, 261
2-Hydroxypyrazines, 258
–, N-alkylation, 277
–, N-oxidation, 265
–, synthesis, 249
2-Hydroxypyrazine, 257, 262, 264, 265
2-Hydroxypyrazine,-3-carboxamide, 275
Hydroxypyrazine carboxylic acids,
    decarboxylation, 263
2-Hydroxy-3-pyrazinoic acid, 274
2-Hydropyrazin-N-oxides, 265, 266
Hydroxypyrazolo[b]pyrazines, 274
Hydroxypyridazines, tautomerism, 24

6-Hydroxypyridazine-3($2H$)-thione, 28
–, tautomerism, 28
6-Hydroxypyridazin-3($2H$)-one, 24
Hydroxypyrimidines, 192, 213–218
–, amination, 216
–, i.r. spectra, 164
–, tautomerism, 194
2-Hydroxypyrimidine, 195, 210
Hydroxyquinoxalines, (quinoxalin-2($1H$)-
    ones), 303, 327–334
–, hydroxylated derivatives, 335
2-Hydroxyquinoxalines, chemical reac-
    tions, 331
–, synthesis, 328
2-Hydroxyquinoxaline, 329, 341
–, hydroxylation, 333
–, O-silylation, 331
–, tautomerism, 327, 328
2-Hydroxyquinoxaline-3-carboxylic acid,
    333
3-Hydroxyquinoxaline-2-carboxylic acids,
    345, 347
2-Hydroxyquinoxalin-4-oxides, synthesis
    from o-nitroamiles, 330
2-Hydroxyuiqnoxalin-4-oxide, 341, 342.
1-Hydroxtyrutaecarpine, 454.
8-Hydroxy-4-styrylcinnoline, 81
4-Hydroxy-5,6,7,8-tetrahydrophthalazin-
    1($2H$)-one, 152
Hydroxy-1,2,4-triazines, (triazinones),
    463
4-Hydroxy-3,4,6-trimethyl-1,2,5-oxa-
    diazine, 479
2-Hydroxy-3,5,6-trimethylpyrazine, 262
2-Hydroxy-3,5,6-trimethylpyrazin-1-oxide,
    265
4-Hydroxy-2,5,6-triphenylpyrimidine, 173
Hydroxyvasicine, 444, 445

Imidazoles, 186
Δ^2-Imidazoline, 471
Imidazopyrimidines, 189
Imido-bis(sulphonyl chlorides), 501
Imidoyl chlorides, reactions with o-
    phenylenediamines, 305, 330
β-Iminobutyronitrile, 179
Immedial Black (C.I. Sulphur Black 9,
    53230), 432
Immedial Pure Blue, 435
Indamine dyes, 404
Indaminethiosulphonic acids, 430

Indazole, 49
11-Indeno[1,2-c]cinnoline-11-one, 90
Indigo dyestuff, 446
Indolophenothiazines, 436
Indoloquinazoline alkaloids, 451–456
Indulines, 414-416
Induline 3 B, 414, 415
Induline 6 B, 414, 415
Induline Scarlet (C.I. Basic Dye 50080), 408
Induline Spirit Soluble (C.I. Solvent Blue 7, 50400), 415
Iodinin, 355, 369, 384, 387
3-Iodo-4-chlorocinnoline, 55
3-Iodocinnoline, 55, 56
6-Iodo-2,3-dimethylquinoxaline, 320
1-Iodophenazine, 376
2-Iodophenazine, 376
3-Iodo-6-phenylpyridazine, 12
1-Iodophthalazine, 126
Iodopyrazines, 258, 259
2-Iodopyrazine, 258
2-Iodoquinoxaline, 319, 320
6-Iodoquinoxline, 320
Isatin, reaction with aminoguanidine, 468
Isatoic anhydride, 439
Isocyanates, 189
Isocyanic acid, 269
Isocytosine, 161, 172
Isofebrifugine, 449, 451
Isomelamines, 472
Isopeganidine, 444
2-Isopropyl-4-methyl tetrahydro-1,2,4-oxadiazine, 479
1-Isopropylphenazine, 368
2-Isopropylphenazine, 368
2-Isopropylquinoxaline, 312
Isothiocyanates, 189
Isothioureas, reactions with N,N-dimethylformamide dicthylacetal, 470

α-Ketogluconic acids reaction with o-phenylenediamine, 343
Ketopiperazines (piperazinones), 294
β-Ketosulphones, 248
β-Ketosulphoxides, reations with o-phenylenediamines, 305

β-Lactams, 284
Lauth's Violet (C.I. Basic Dye 52000), 427
Leuco-Methylene Blue, 429

5-Lithiopyrimidines, 225
Lumazines, hydrolysis, 260, 274
Luminol (5-amino-4-hyroxyphthalazin-1(2H)-one), 139
–, chemiluminescence, 139, 140, 141

Mackinlaya macroscidea, 444
Maillard reaction, 486
Malediamide, 181
Maleic anhydride, 3
–, adduct with 2,3-dimethylquinoxaline, 315
Malic acid, 172
Malondialdehyde tetraethyl acetal, 166
Malondiamide, 181
Malondiamidines, 181, 182
Malonic acid, 2
Malonnitriles, 180
Malononitrile, 188
–, reaction with dihydropyrazines, 279, 280
Malonyl dichloride, 175
Mannich reaction, 201, 236
Mauveine (C.I. 50245), 409
Melamines (amino-1,3,5-triazines), 472, 488
Meldola's Blue (C.I. Basic blue 6, 51175), 417, 419
Meldola's synthesis, 420
6-Mercapto-4-methyl-3(2H)-pyridazinethione, 28
2-Mercapto-4-methylpyrimidine, 169
2-Mercapto-5-methylpyrimidine, 168
4-Mercapto-6-methylpyrimidine, 223
Mercaptophthalazines, synthesis, 141
4-Mercaptophthalazin-1(2H)-thione, 139, 142
6-Mercaptopyridazine-3(2H)-thione, 28
–, tautomerism, 28
Mercaptopyrimidines, 192, 221
–, acylation, 222
–, dealkylation, 224
–, desulphurisation, 223
–, oxidation, 223
–, tautomerism, 194
2-Mercaptopyrimidine, 166, 222
Mercuric ethylenediaminetetraacetate, 440, 442
Mesoxalates, reactions with o-phenylenediamine, 328
Mesoxalic acids, reactions with o-phenyl-

enediamines, 345
Metallopyrimidines, 225, 226
3-Methanesulphonyl-4-hydroxycinnolines, 38
Methaqualone, 228
2-Methoxybenzo[c]cinnoline, 101, 106, 108
3-Methoxybenzo[c]cinnoline, 108
4-Methoxybenzo[c]cinnoline, 106
6-Methoxy-1H-benzo[d,e]cinnoline, synthesis from 1-formyl-8-hydroxy-4-methoxynaphthalene, 113
3-Methoxy-1,2,4-benzotriazine, 467
4-Methoxy-1,2,3-benzotriazine, 459
N-Methoxycarbonyl-N-methylguanidine, 168
1-Methoxycarbonylpiperazine, 290
3-Methoxycarbonyl-1,2,4-triazine, 462
3-Methoxycinnoline, 58, 72
4-Methoxycinnoline, 72, 77
–, reduction, 93, 94
5-Methoxycinnoline, 80
6-Methoxycinnoline, 80
7-Methoxycinnoline, 80
8-Methoxycinnoline, 80
4-Methoxycinnoline-1-oxide, 50
8-Methoxy-1,2,3,4,5,6-hexahydrobenzo-[f]cinnoline, 110
8-Methoxy-4-methylcinnoline, 80
3-Methoxy-6-methyl-4-nitropyridazin-1-oxide, 19
3-Methoxy-4-methyl-6-phenylpyridazine, 23
3-Methoxy-4-methylpyridazine, 23
3-Methoxy-6-methylpyridazine, 23, 25
3-Methoxy-6-methylpyridazin-1-oxide, 19, 22
4-Methoxy-3-methylpyridazin-1-oxide, 19
6-Methoxy-2-methylpyridazin-3-one, 25
4-Methoxy-6-methylpyrimidine, 159
4-Methoxy-6-nitrocinnoline, 77
2-Methoxy-1-nitrophenazine, 380
7-Methoxy-1-nitrophenazine, 379
1-Methoxy-5-nitrophthalazine, 133
6-Methoxy-5-nitroquinoxaline, 324
5-Methoxy-2-nitrosophenol, deoxygenation, 389
5-Methoxy-1-oxotetralin-8-carboxylic acid, 113
1-Methoxyphenazine, 385, 386, 388
–, bromination, 377

–, nitration, 379
2-Methoxyphenazine, 388
–, bromination, 377
–, dealkylation, 386
2-Methoxyphenazine-1-carboxylic acid, 390
6-Methoxyphenazine-1-carboxylic acid, 390
1-Methoxyphenazin-N-oxides, 373
2-Methoxyphenazin-N-oxides, 373
4-(4-Methoxyphenyl)-1,2,3-benzotriazine, 459
3-Methoxy-6-phenylpyridazine, 23, 32
1-Methoxyphthalazine, mass spectrum, 121
1-Methoxyphthalazin-1(2H)-one, 137
3-Methoxypyridazine, 23
4-Methoxypyridazine, 23
6-Methoxypyridazine-3(2H)-thione, 28
6-Methoxypyridazin-3(2H)-one, 24, 25
3-Methoxypyridazin-1-oxide, 19, 22
4-Methoxypyridazin-1-oxide, 19
4-Methoxypyridazin-2-oxide, 19
4-Methoxypyridmidine, 159, 217
Methoxyquinoxalines, nitration, 324
5-Methoxyquinoxaline, 324
6-Methoxyquinoxaline, 324
3-Methoxyrutaecarpine, 455
10-Methoxyrutaecarpine, 453
p-Methoxystyrene, 284
3-Methoxy-5,6,7,8-tetrahydrocinnoline, 97
6-Methoxy-1,2,3,4-tetrahydrophenazine, 397
Methoxy-4-toluenesulphonylhydrazino-cinnoline hydrochlorides, 80
3-Methoxy-1,2,4-triazine, 463
Methyl γ-aminobutyrate, 441
4-Methylaminocinnoline, 63, 64
N-Methyl-4-aminocinnolinium iodides, 67
Methyl γ-amino-α-hydroxybutyrate, 441
2-Methylamino-5-nitroaniline, reaction with butyl glyoxylate, 303
2-(N-Methylamino) pyrimidine, 161
2-Methylaminoquinazolin-4(3H)-one, 23
N-Methylanthranilic acid, 438, 452
1-Methylbenzo[c]cinnoline, 105
2-Methylbenzo[c]cinnoline, 105
3-Methylbenzo[c]cinnoline, 105
3-Methyl-1H-benzo[d,e]cinnoline, synthesis from 8-acetylnaphthalene, 113
4-Methylbenzo]c]cinnoline, 105

2-Methylbenzo]c]cinnoline-9-carboxylic acid, 108
3-Methyl-1,2,4-benzotriazine, 466
4-Methyl-1,2,3-benzotriazine, 459
2-Methyl-1,2,4-benzotriazin-3-one, 467
4-Methyl-1,2,4-benzotriazin-3-one, 467
7-Methyl-2,1,4-benzoxathiazin-3-one-1,1-dioxide, 485
3-Methylcinnoline, 47, 88
4-Methylcinnoline, 45, 46, 48, 54
–, oxidation to 4-formylcinnoline, 89
–, reduction, 93, 94
–, X-ray structure, 40
2-Methylcinnolinium perchlorate, 59
1-Methylcinnolin-4-one, 78
2-Methylcinnolin-3($2H$)-one, 72
4-Methylcinnolin-1-oxide, 50
–, photolysis, 54
4-Methylcinnolin-2-oxide, photolysis, 54
Methyl N-cyanoforminidate, 475
3-Methyl-2-cyanopyrazine, 273
2-Methylcyclohexanone, oxidative dimerisation, 401
1-Methylcytosine, 157
3-Methylcytosine, 157
3-Methylcytosine-5-carboxylic acid, 177
4-Methyl-1,4-dihydrocinnoline, 93
1-Methyl-3,4-diphenylcinnoline, 93
2-Methyl-5,6-diphenylpyrazine, 257
1-Methyldithiouracil, 215
Methylene Blue (C.I. Basic Blue 9, 52015), 427–430, 435
6,6′-Methylenedianthranilic acid, cyclocondensations with lactams, 443
Methylene Green (C.I. Basic Green 5, 52020), 429
4-Methylene-3-(4-nitrophenyl)-1-oxidophthalazinium betaine, 135
Methylene Violet (C.I. Basic Dye 52041), 435
Methyl 2-ethoxycarbonyldithioacetate, 176
N-Methyleuxylophoricine B, 455
Methyl gallate, 420
3-Methylindazole, 54
3-Methylindole, 54
N-Methylisatoic anhydride, 447
Methylisocyanate, 183, 232
–, trimerisation, 480
1-Methylisoindole, 125
β-Methylmalic acid, 172
Methyl 3-methoxy-2-methoxymethylene-propionate, 173
4-Methyl-3-methoxy-6-methylthio-pyridazine, 29
4-Methyl-6-methoxy-3-methylthio-pyridazine, 29
3-Methyl-4-methylene-1,2,3-benzo-triazines, 460
3-Methyl-4-methylene-1,2,3-benzotriazine, 64
6-Methyl-3-methylthiopyridazine, 29
6-Methyl-3-methylthio-1,2,4-triazine, 465
3-N-Methylnitroaminopyridazine, 17
4-Methyl-8-nitrocinnoline, 48, 60
7-Methyl-8-nitrocinnoline, 60
4-Methyl-3-nitrocinnolin-1-oxide, 50
3-Methyl-5-nitrocytosine, 177
3-Methyl-6-nitro-4-phenoxycinnoline, 83
3-Methyl-8-nitro-4-phenoxycinnoline, 83
4-Methyl-3-(p-nitrophenyl)-1-oxido-phthalazinium betaine, 135
2-Methyl-5-nitrophthalazinium iodide 133
3-Methyl-5-nitrophthalazinium iodide, 133
2-Methyl-5-nitrophthalazin-1($2H$)-one, 133
2-Methyl-8-nitrophthalazin-1($2H$)-one, 133
4-Methyl-5-nitrophthalazin-1($2H$)-one, 133
4-Methyl-8-nitrophthalazin-1($2H$)-one, 122, 133
–, reduction, 131
3-Methyl-5-nitropyridazin-2-oxide, 19
4-Methyl-1-oxido-3-phenylphthalazinium betaine, 135
3-Methyl-1-oxidophthalazinium betaine, adducts with alkynes, 133, 134
Methylphenazhydrin, 396
1-Methylphenazine, 367, 368, 396
2-Methylphenazine, 367, 368
Methylphenazinium methosulphate, 364, 387
–, reaction with nucleophiles, 365
Methylphenazinium salts reactions with carbanions, 366
5-Methylphenazin-1($5H$)-one (pyocanine), 365
10-Methylphenazin-2($10H$)-one, 365, 387
1-Methylphenazin-N-oxides, 373
2-Methylphenazin-N-oxides, 373
6-Methyl-3-phenoxy-4-phenylpyridazine, 23

1-Methyl-4-phenyl-1,3-butadiene, 27
3-Methyl-4-phenylcinnoline, 48, 49
6-Methyl-3-phenylcinnoline, 49
4-Methylphenylisocyanate, reaction with sulphur trioxide, 485
4-Methyl-3-phenyl-1-oxidophthalazinium betaine, 135
1-Methyl-4-phenylphthalazine, 123
–, Synthesis from bis(diethoxycarbonyl)methyl-4-phenylphthalazine, 123
3-Methyl-6-phenylpyridazine, 10, 27
4-Methyl-3-phenylpyridazine, 10
4-Methyl-5-phenylpyridazine, 10
5-Methyl-3-phenylpyridazine, 10
4-Methyl-5-phenylpyridazine-3-carboxylic acid, 27
6-Methyl-5-phenylpyridazine-3,4-dicarboxylic acid, 27
4-Methyl-6-phenylpyridazin-3(2H)-one, 23
6-Methyl-4-phenylpyridazin-3(2H)-one, 23
2-Methyl-4-phenylpyrimidine, 183
4-Methyl-6-phenylpyrimidine, 185
2-Methyl-3-phenylquinoxaline, 312
1-Methyl-3-phenylquinoxalin-2-one, 332
3-Methyl-1-phenylquinoxalin-2-one, 332
5-Methyl-3-phenyl-5,6,7,8-tetrahydrocinnoline, 97
3-Methyl-6-phenyl-1,2,4,5-tetrazine, 493
1-Methylphthalazine, 117, 118, 122, 149
–, mass spectrum, 121
2-Methylphthalazinium iodide, 119
–, reduction, 146
1-Methylphthalazinium salts, 122
4-Methylphthalazin-1(2H)-one, 115, 146
2-Methylphthalazin-1-one-4-carboxylic acid, 145
1-Methylphthalazin-2-oxide, 123
1-Methylphthalazin-3-oxide, 123
2-Methylphthalazin-1-thione-4-carboxylic acid, 145
1-Methylpiperazine, 290
2-Methylpiperazine, 293
–, dehydrogenation, 253
4-Methylpiperazine-2,6-dione, 300
Methylpyrazines, 255
Methylpyrazine, 253
–, oxidation, 271
–, phenylation, 257

–, reaction with 1-diethylaminopropyne, 256
3-Methylpyrazine-2-carboxamide, 273
5-Methylpyrazine-2,3-dicarboxylic acid, 271, 272, 313
3-Methyl-2-pyrazinoic acid, 272
5-Methyl-2-pyrazinoic acid, 271, 272
6-Methyl-2-pyrazinoic acid, 271, 272
1-Methylpyrazolo[1,2,-a]benzotetrazin-3-one, 488
Methylpyridazines, mass spectra, 11
3-Methylpyridazine, 9, 10
4-Methylpyridazine, 10, 28
3-Methylpyridazine-1,2-dioxide, 19
6-Methylpyridazine-3(2H)-thione, 28
4-Methylpyridazin-3(2H)-one, 23
5-Methylpyridazin-3(2H)-one, 23
6-Methylpyridazin-3(2H)-one, 23, 25
5-Methylpyridazin-3(2H)-one-4-carboxylic acid, 27
6-Methylpyridazin-3(2H)-one-4-carboxylic acid, 27
6-Methylpyridazin-3(2H)-on-1-oxide, 19
3-Methylpyridazin-1-oxide, 18
3-Methylpyridazin-2-oxide, 18
4-Methylpyridazin-1-oxide, 18
4-Methylpyridazin-2-oxide, 18
N-Methylpyridone, 143
4-Methylpyrimidin-2-amine, 161
Methylpyrimidines, 201
4-Methylpyrimidine, 159, 199–201
5-Methylpyrimidines, 199
5-Methylpyrimidine-2,4(1H,3H)-dione, 172
1-Methylpyrimidin-2(1H)-one, 167
4-Methylpyrimidin-2(1H)-one, 159
6-Methylpyrimidin-4-ylmethylacetate, 227
N-Methylpyrrolid-2-one, 258
2-Methylquinazolines, condensation with benzaldehyde, 235
2-Methylquinazoline, 230
–, reaction with chloral, 235
3-Methyl-4-quinazoline, 440
4-Methylquinazolines, condensation with benzaldehyde, 235
4-Methylquinazoline, 230
3-Methylquinazolin-2(3H)-one, 232
2-Methylquinoxaline, 311, 312, 349
–, cycloaddition reactions 313, 314
–, N-methylation, 314
–, reactions with araldehydes, 345

4-Methylquinoxaline, 321
5-Methylquinoxaline, 317
–, oxidation, 271
6-Methylquinoxaline, 317, 318
3-Methylquinoxaline-2-carboxaldehyde, 344
3-Methylquinoxaline-2-carboxylic acid, 347
6-Methylquinoxaline-2,3-dicarboxylic acid, 348
1-Methylquinoxalin-2-one, 332
2-Methylquinoxalin-N-oxides, 336
5-Methyl-2-styrylpyrazine, 255
6-Methyl-3-sulphanilamidopyridazine, 30
3-Methylsulphinyl-4($1H$)-cinnoline diacetate, 79
3-Methylsulphinylcinnolin-4($1H$)-ones, 38
1-Methylsulphinylphthalazine, 142
2-Methylsulphinylpyrimidine, 224
3-Methylsulphonylcinnolines, 86
1-Methylsulphonylphthalazine, 117, 142
2-Methylsulphonylpyrimidine, 224
2-Methyl-1,2,3,4-tetrahydrocinnoline, 93
3-Methyl-5,6,7,8-tetrahydrocinnoline, 96, 97
4-Methyl-1,2,3,4-tetrahydrocinnoline, 93
2-Methyl-1,2,3,4-tetrahydrocinnolin-3-one, 72
1-Methyl-1,2,3,4-tetrahydrophenazine, 397
2-Methyl-1,2,3,4-tetrahydrophenazine, 397
7-Methyl-1,2,3,4-tetrahydrophenazine, 397
2-Methyl-1,2,3,4-tetrahydroquinoxalines, 353
2-Methyl-1,2,3,4-tetrahydroquinoxaline, 351
6-Methyl-1,2,3,4-tetrahydroquinoxaline, 351
3-Methylthio-1,2,4-benzotriazine, 467
1-Methylthiophthalazine, 142
1-Methylthiophthalazine-4-carboxylic acid, 145
3-Methylthiopyridazine, 29
4-Methylthiopyridazine, 29
2-Methylthiopyrimidine, 224
5-Methylthiopyrimidine, 205
5-Methylthio-1,2,3-trithiane, 486
1-Methyl-4-thiouracil, 215
N-Methylthiourea, 178
3-Methyl-1,2,4-triazine, 461
5-Methyl-1,2,4-triazine, 461
6-Methyl-1,2,4-triazine, 461

2-Methyl-1,2,3-triazinium iodides, 458
2-Methyl-1,2,4-triazin-5($2H$)-one, 464
5-Methyl-1,2,4-triazin-6-one, 464
3-Methyl-1,2,4-triazolo[4,5-a]-phthalazine, 129
5-Methyl-4-trichloromethylpyridine, 200
5-Methyl-2,4,6-trifluoropyrimidine, 204
N-Methyluracil, 195
5-Methyluracil, 172
Methylureas, 171
N-Methylurea, 167, 179
O-Methylurea, 176
N-Methylurethane, 182
Michler's hydrol, 419
1-Morpholino-6-oximinocyclohexene, reaction with hydrogen chloride, 399
Mucochloric acid, 4
Myxin, 355, 369, 284, 387

1,2-Naphthaquinone-6-sulphonic acid, 429
Naphtho[1,2-c]cinnoline, 105
Naphtho[1,2-c]-7,8-diazaanthracen-2,9,14-trione, 112
2-Naphthol, 360, 417, 418
1,2-Naphthoquinone, reaction with o-phenylenediamine, 357, 367
1,2-Naphthoquinone-4-sulphonic acid, 421
$1H$-Naphtho[1,8-d,e]triazine, 460
Naphtho[2,1-e]-1,2,4-triazine, 468
Naphtho[2,1-e]-1,2,4-triazine-2-carboxylic acid, 468
Naphtho[2,1-e]-1,2,4-triazin-2-one, 468
1-Naphthylamine, 369, 418
2-Naphthylamine, oxidation, 363
–, oxidative dimerisation, 369
α-Naphthyldiphenylformazan, 468
3-(α-Naphthyl)-2-phenylquinoxaline, 312
2-(α-Naphthyl)quinoxaline, 312
2-(β-Naphthyl)quinoxaline, 312
Nasicinolone, 444, 445
New Blue B (C.I. Basic Blue 10, 51190), 419
New Blue F (C.I. Basic Dye 51200), 419
New Methylene Blue GG (C.I. Basic Dye 51195), 419
New Methylene Blue NSS (C.I. Basic Blue 24, 52030), 429
Niementowski synthesis, of quinazolines, 232
Nigrosines, 414–416
Nigrosine Spirit Soluble (C.I. Solvent

Black 5, 50415), 415
Nile Blue A (C.I. Basic Blue 12, 51180), 418, 420
Nitramide, 478
Nitraria Sibirica, 445
ω-Nitroacetophenones, reactions with *o*-phenylenediamines, 304, 305
3 Nitroaminopyridazine, 16
4-Nitrominopyridazine, 17
α-(2-Nitroanilino)carboxylic acids, reductive cyclisation, 306, 330
4-Nitroanthranilic acid
6-Nitro-4-aziridinocinnoline, anti-tumour activity, 62
4-Nitrobenzaldehyde, 196, 198
Nitrobenzo[*c*]cinnolines, reduction, 107
1-Nitrobenzo[*c*]cinnoline, 106, 107
2-Nitrobenzo[*c*]cinnoline, 106
3-Nitrobenzo[*c*]cinnoline, 106
4-Nitrobenzo[*c*]cinnoline, 107
1-Nitrobenzo[*c*]cinnolin-6-oxide, 104
2-Nitrobenzo[*c*]cinnolin-6-oxide, 104
4-Nitrobenzotriazole, 324
2-Nitrobenzylchloride, reaction with methyl γ-aminobutyrate, 441
–, reaction with methyl γ-amino-α-hydroxybutyrate, 441
N-(2-Nitrobenzyl)pyrrolidone, 441
Nitrocinnolines, reduction, 60
–, synthesis, 59
3-Nitrocinnoline, 60
5-Nitrocinnoline, 45, 60, 61
6-Nitrocinnoline, 60
7-Nitrocinnoline, 60
8-Nitrocinnoline, 45, 60, 61
4-Nitrocinnolin-1-oxide, 61
8-Nitrocinnolin-2-oxide, 50
2-Nitrodiphenylamines, intramolecular cyclisation, 370
2-Nitrodiphenylamine, cyclisation, 363
2-(4-Nitrofuryl)quinoxaline, 343
7-Nitroindazole, 50
4-(Nitromethyl)cinnoline, 48
3-Nitro-4-methylcinnoline, 60
Nitrophenazines, 378–381
–, photochemical reactions with amines, 380
–, reactions with nucleophiles, 379, 380
–, reduction, 379
1-Nitrophenazine, 378, 380
2-Nitrophenazine, 379, 380

4-Nitrophenazin-1(5*H*)-one, 378
1-Nitrophenazin-5-oxide, 372
2-Nitrophenazin-10-oxide deoxygenation, 379
–, reactions with carbanions, 372
3-Nitrophenazin-5-oxide, 372
3-Nitro-4-phenoxycinnoline, 83
5-Nitro-4-phenoxycinnoline, 83
6-Nitro-4-phenoxycinnoline, 64, 83
7-Nitro-4-phenoxycinnoline, 83
8-Nitro-4-phenoxycinnoline, 83
2-(*o*-Nitrophenyl)glyoxaline, 318
5-(4-Nitrophenylhydroxymethyl)uracil, 196
N-(2-Nitrophenyl)oxalamidrazonate, 466
2-(*p*-Nitrophenyl)quinoxaline, 318
2-Nitrophenylurea, 467
5-Nitrophthalazines, synthesis by photocyclisation of acylhydrazines, 119
5-Nitrophthalazine, 121, 133
Nitrophthalazinium salts, nitration, 133
5-Nitrophthalazin-1(2*H*)-one, 133
8-Nitrophthalazin-1(2*H*)-one, 133
5-Nitrophthalazin-1(2*H*)-one-4-carboxylic acid, 145
7-Nitrophthalazin-1(2*H*)-one-4-carboxylic acid, 145
Nitropyridazines, 14
3-Nitropyridazin-1-oxide, 19
4-Nitropyridazin-1-oxide, 19
5-Nitropyridazin-1-oxide, 19
3-Nitropyridine, 159
Nitropyrimidines, 208, 209
2-Nitropyrimidine, 194
5-Nitropyrimidines, 191, 192
5-Nitropyrimidine, 202
5-Nitropyrimidin-2-amine, 161
7-Nitroquinazolin-2,4(1*H*,3*H*)-dione, 232
6-Nitroquinazoline, 234
Nitroquinoxalines, 323, 324
5-Nitroquinoxaline, 310, 323, 324
6-Nitroquinoxaline, 324
6-Nitroquinoxaline-2,3-dicarboxylic acid, 348
6-Nitroquinoxaline-2,3-dione, 317
Nitrosobenzenes, dimerisation, 370
p-Nitrosodiethylaniline, 417, 418
1-Nitroso-2-naphthylamine, 369
–, reaction-4-nitrosophenol, with *N,N*-dimethylaniline, 432
Nitrosopyrimidines, 208, 209

5-Nitrosopyrimidines, 191, 192
5-Nitro-1,2,3,4-tetrahydroquinoxaline, 351
6-Nitro-1,2,3,4-tetrahydroquinoxaline, 351
Nitrotriphendioxazines, 425
5-Nitrouracil, 160

Octafluorophenazine, 376
1,2,3,5,6,7,8,8a-Octahydrocinnoline,
1,2,3,4,4a,5,10,10a-Octahydro-10,10-dimethyl-8-methoxybenzo[g]cinnolin-2-ones, 113
5,5a,6,7,8,9,9a,10-Octahydro-4,9a-dimethyl-6-methylene-2-phenylbenzo[g]-3($2H$)-one, 112
Octahydrophenazines, 398–400
–, dehydrogenation, 363
–, reduction, 401
–, – synthesis from 2-chlorocyclohexanones, 399
–, – from 1-chlorocyclohexene oxides, 399
–, – from hydroxycyclohexanones, 399
cis-1,2,3,4,4a,5,10,10a-Octahydrophenazine, 397, 398
$trans$-1,2,3,4,4a,5,10,10a-Octahydrophenezine, 398
1,2,3,4,6,7,8,9-Octahydrophenazine, 399
3,4,4a,5,6,7,8,8a-Octahydrophthalazin-1($2H$)-one, 152
Ornithine, 446
Orotic acid, 186, 220
Oxadiazines, 188, 478–481
1,2,3-Oxadiazines, 478
1,2,4-Oxadiazines, 479
1,2,5-Oxadiazines, 479
1,2,5-Oxadiazine, 479
1,2,6-Oxadiazines, 479
1,3,4-Oxadiazines, 479, 480
1,3,5-Oxadiazines, 480, 481
1,3,4-Oxadiazoles, 491
Oxadithianes, 485, 486
1,3,5-Oxadithianes, 486
Oxalates, reactions with o-phenylenediamines, 333
Oxalic acid derivatives, 303
Oxathiadiazines, 498, 499
1,4,3,5-Oxathiadiazines, 502
Oxathiadiazine, dioxides, 485, 499
Oxathiazines, 483–485
1,2,3-Oxathiazines, 483, 484
1,2,5-Oxathiazines, 485
1,3,5-Oxathiazines, 484, 485

1,2,3-Oxathiazine-2,2-dioxides, 483
1,3,5-Oxathiazin-6($2H$)-ones, 484
1,2,3-Oxathiazin-2-oxides, 484
Oxatriazines, 495, 498
Oxazines, 187
Oxazine dyes, 403
Oxazolinones, 330
1-Oxido-3-phenylphthalazinium betaine, 135, 136
Oxidiphthalazinium betaines, 130
1-Oxidophthalazinium betaines, spectral data, 134, 135
Oximes, rearrangement, 246
Oximinoacetone, 253
α-Oximinohydrazines, 464
α-Oximinoketones, reactions with α-aminonitriles, 259
1-Oximino-1,2-naphthoquinone, aminoguanidinyl derivative, 468
5-Oxo-5,6,7,8-tetrahydrocinnolines, 64
3-Oxo-1,2,3,4-tetrahydrocinnoline, 93
4-Oxo-1,2,3,4-tetrahydrocinnoline, 94
4-Oxo-1,2,3,4-tetrahydrocinnolinium chloride, 93
5-Oxo-5,6,7,8-tetrahydrocinnolinium bromide, 97

Paraensine, 454
Pegamine, 445
Peganidine, 444, 445
Peganine, 440, 447, 449
Peganol, 445
$Peganum\ harmala$, 440, 445, 446
$Peganum\ nigellastrum$, 444
$Penicillin\ chrysogenum$, 439
1,5,6,7,8-Pentafluoro-4-methoxyphthalazine, 126
4,5,6,7,8-Pentafluorophthalazin-1($2H$)-one, 126
1,2,2,5,5-Pentamethyl-6-phenyltetrahydropyrazine, 286
Pentazines, theoretical considerations, 501
Pentoses, reactions with o-phenylenediamines, 342, 342
Perfluoro-1,2,6-oxadithiane-2,2,6,6-tetroxide, 486
Perfluoropropene, 225
Perhydro-1,2,5-oxadiazines, 479
Perhydro-1,4,2-oxathiazines, 485
N-Phenacylbenzylamine hydrobromide, self condensation, 276

2-Phenacylcyclohexane-1,3-diones, 64
Phenanthrene, 54
9,10-Phenanthroquinone, 468
Phenanthro[9,10,e]-1,2,4-triazine-3-thione, 468
Phenanthro[9,10e]-1,2,4-triazin-3-one, 468
Phenazhydrin, 396
Phenazines (dibenzopyrazines), 354–367
–, nitration, 378
–, physical and spectroscopic properties, 356, 357
–, reduced forms, 392–402
–, reduction, 393, 394
–, synthesis, 357–363
–, – from o-aminodiphenylamine, 262
–, – from 2-amino-4-butylanisole, 362
–, – from anilines, 359
–, – from arylamines, 363
–, – from benzenesulphenanilides, 362
–, – from benzofuroxans, 359, 360
–, – from cyclohexane-1,2-diones, 361
–, – from 2,2'-diaminodiphenylamines, 361
–, – from enamines, 360
–, – from 2-nitrodiphenylamines, 358, 359
–, – from phenolate anions, 360
–, – from 1,2-quinones, 357, 358
–, – from quinoxalines, 361
Phenazines-dihydrophenazines molecular adducts, 396
Phenazine, 271, 363–365, 396
–, dyestuffs, 355, 356, 381
–, herbicides, 355
–, nomenclature, 354
–, oxidation, 348
Phenazinecarboxylic acids, 390, 391
–, synthesis from o-aminodiphenylamines, 390
–, – anthranilic acids, 391
–, – from cyanophenazines, 391
–, – from methylphenazines, 391
–, – from o-phenylenediamines, 39
–, – from o-quinones, 390
Phenazine-1-carboxylic acid, 390, 392
Phenazine-2-carboxylic acid, 392
Phenazine-1,4-dicarboxylic acid, 392
Phenazine-1,6-dicarboxylic acid, 391, 392
Phenazine-1,8-dicarboxylic acid, 392
Phenazine-1,9-dicarboxylic acid, 392
Phenazine-2,7-dicarboxylic acid, 391, 392

Phenazine-5,10-dioxide, 364
–, acetylation, 374
Phenazine methosulphate, 364
Phenazinesulphonic acids, 378–381
Phenazine-2-sulphonic acid, 380
Phenazinium salts, 364–366
1-Phenazinol dioxides, 371
Phenazin-1(5H)-one, 385
Phenazin-2-(10H)-one, 385
Phenazin-N-oxides, 369–375
–, deoxygenation, 374, 375
–, nitration, 372
–, reduction, 374, 375
Phenazin-5-oxide, 364, 372, 373
–, photochemical irradiation, 374
Phenazinyl radicals, 380
Phenocyanine VS (C.I. Mordant Dye 51140), 421
Phenosafranines, 405
Phenosafranine, 355
–, deamination, 366
Phenothiazine, 428
Phenothiazine dyes, 427–430
α-Phenotriazines (1,2,4-benzotriazines), 465
Phenoxazine dyes, 403, 404, 417–420
4-Phenoxycinnolines, synthesis, 82
4-Phenoxycinnoline, 57, 62, 82, 83
Phenylacetamide, 438
Phenylacetic acid, 438
Phenylaposafrannines, 405
3-Phenylazirine, dimerisation, 281
p-Phenylazobenzene rearrangement, 367
4-Phenylazo-1-naphthylamine, reaction with o-phenylenediamine, 409
2-Phenylbenzimidazole, 318
N-Phenylbenzimidoyl chloride, 233
1-Phenylbenzo[c]cinnoline, 105
2-Phenylbenzo[c]cinnoline, 105
3-Phenylbenzo[c]cinnoline, 105
3-Phenyl-1H-benzo[d,e]cinnoline, synthesis from 8-benzoylnaphthalene, 113
5-Phenylbenzo[c]cinnolin-1-one, 103, 104
N-Phenyl-1,4-benzoquinodiimine, 516
3-Phenyl-1,2,4-benzotriazine, 466
4-Phenyl-1,2,3-benzotriazine, 459
3-Phenylcinnoline, 49, 50, 64, 88
–, reduction, 46
4-Phenylcinnoline, 45, 48, 49
–, oxidation, 46
–, reduction, 93

3-Phenyl-4-cinnolinecarbonylchloride, 90
3-Phenylcinnoline-4-carboxylic acid, 85
–, decarboxylation, 88
–, electrochemical reduction, 89
1-Phenylcinnolin-4(1H)-ones, general synthesis, 76
α-Phenyl-4-cinnolinylacetonitrile, 91
Phenylcyanamide, cyclotrimerisation, 472
3-Phenyl-2-cyanopyrazine, 273
3-Phenyl-1,4,2-dioxazin-5,6-dione, 483
Phenylenediamines, 160
o-Phenylenediamine, reaction with 1,4-benzoquinone, 413
–, – with α,β-dicarbonylcompounds, 302–304
p-Phenylenediamine, 414, 432
–, oxidation of aniline mixture, 409
Phenylethylamine, 447
3-(2-Phenylethyl)-1,4,5,6-tetrahydro-1,2,4-triazine, 465
3-(Phenylethynyl)cinnoline, 58
Phenylglycine anhydride, 298
Phenylgloxal, 318
3-Phenyl-4-hydroxycinnolin-1-oxide, 50
Phenylisothiocyanate
Phenylketene, 284
N-Phenylmaleimide, adduct with 2,3-dimethylquinoxaline, 315
3-Phenylnaphtho[1,2-e]-1,2,4-triazine, 468
2-Phenyl-3-(β-naphthyl)quinoxaline, 312
3-Phenyl-1-oxidophthalazinium betaine, 135
1-Phenylphenazine, 367, 368
2-Phenylphenazine, 367, 368
5-Phenylphenazinium chloride, 366
5-Phenylphenazinium salts, 372
1-Phenylphenazin-N-oxides, 373
2-Phenylphenazin-N-oxides, 373
5-Phenylphenazinium chloride, 407
1-Phenylphthalazines, 122
1-Phenylphthalazine, 118, 122
–, Mass spectrum, 121
2-Phenylphthalazin-1-one, 147
3-Phenylphthalazin-1(2H)-one, 146
2-Phenylphthalazin-1(2H)-one-4-carboxylic acid, 145
1-Phenylphthalazin-3-oxides, 123
2-Phenylpiperazine, 44, 293
3-Phenylpiperazine-2,6-dione, 300
3-Phenylpiperazinone, 294

Phenylpyrazine, 257
3-Phenylpyrazine-2-carboxamide, 273
5-Phenylpyrazinoic acid, 270
6-Phenylpyrazinoic acid, 270
3-Phenylpyridazine, 10
4-Phenylpyridazine, 10
6-Phenylpyridazine-3-carboxylic acid, 27
5-Phenylpyridazine-4-carboxylic acid, 27
5-Phenylyridazine-3,4-dicarboxylic acid, 27, 46
6-Phenylpyridazin-3(2H)-one, 23
2-Phenylpyrimidine, 168
4-Phenylpyrimidine, 201
2-Phenylquinoxaline, 49, 312, 318, 349
3-Phenylquinoxaline-2-carboxylic acid, 347
2-Phenylquinoxaline-1,4-dioxide, 338
3-Phenylquinoxalin-2(1H)-one, 304
6-Phenyl-3-sulphanilamidopyridazine, 30
3-Phenyl-5,6,7,8-tetrahydrocinnoline, 96, 87
2-Phenyl-1,2,3,4-tetrahydroquinoxaline, 351
3-Phenyl-1,2,3,4-tetrahydroquinoxalin-2-one, 304
3-Phenyl-1,2,4,5-tetrazine, 493
Phenylthioacetamide, 439
3-Phenyl-1,2,4-triazine, 461
5-Phenyl-1,2.4-triazine, 461
6-Phenyl-1,2,4-triazine, 461
6-Phenyl-1,2,4-triazine-3,5-dione, 464
5-Phenyl-1,2,4-triazine-3-thione, 464
5-Phenyl-1,2,4-triazin-3-one, 463
5-Phenyl-1,2,4-triazin-6-one, 464
6-Phenyl-1,2,4-triazin-3-one, 463
N-Phenyltriflamide, 249
N-Phenylurea, 171
Phosphomolybdic acid, insoluble "lakes", 429
Phosphotungstic acid insoluble "lakes", 429
Phthalazines, 35
–, nucleophilic displacement reactions, 117
–, quaternisation with alkylhalides, 119
–, reactions with Grignard reagents, 117
–, Reissert compounds, 118, 121
–, synthesis, 114, 115
Phthalazine, 116, 119
–, adduct with maleic anhydride, 122
–, chemical reactions, 121

–, photoreduction, 146
–, physical properties, 119, 120
–, reduction, 122
–, Reissert compounds, 146
–, spectral data, 120, 121
–, ylides, 122
Phthalazine, carboxylic acids, 144, 145
Phthalazine-1-carboxylic acid, 144
Phthalazine-1,4-dione, [4+2] cycloaddition reaction, 143, 144
Phthalazinethiones, synthesis, 141
Phthalazine-1(2H)-thione, 142
–, physical data, 142
Phthalazinium salts, physical properties, 119
Phthalazinones, reduction, 131
–, spectral properties, 131
–, synthesis, 130
Phthalazin-1(2H)-ones, physical data, 132, 133
–, quaternization, 133
–, synthesis, 114, 115, 130, 130
Phthalazin-1(2H)-one, 124, 125, 133
–, reductive ring-contraction to phthalimidine, 132
–, synthesis from phthalaldehydic acid, 114
–, tautomerism, 131
Phthalazin-1(2H)-one-4-carboxylic acid, 145
Phthalazin-N-oxides, physical properties, 123
–, reduction, 118
–, synthesis, 123
Phthalazin-2-oxide, 123
2-(1-Phthalazinyl)phthalazin-1(2H)-one, 125
Pinner synthesis, 170
Piperazines (hexahydro pyrazines), 254, 287–294
–, conformation, 287
–, pharmacology, 287
–, synthesis, 288
Piperazine, 288
–, adducts with alkenes, 289
Piperazine-1-acetic acid, 290
Piperazine-2,5-diones, alkylation, 298
–, condensation with araldehydes, 298, 299
–, ethylation, 269
–, natural occurrence, 296

–, reactions with triethyloxonium tetrafluoroborate, 251
Piperazine-2,5-dione, 257
Piperazine-2,6-dione, 300
Piperazine-2,5-dithione, 299
Piperazinetetraone, 300
Piperazinetrione, 300
Piperazinones (ketopiperazines), 294
Piperazinone, 294
β-1-Piperazinopropionic acid, 290
Piperidazine (hexahydropyridazine), 34
Plasmodium gallimaceum, 62
Polyhydroxyalkylquinoxalines, 342, 344
Primaquine, 66
PROCION Blue H-EGN (C.I. Reactive Blue 198), 427
PROCION Blue MX-7RX (C.I. Reactive Blue 161 I.C.I.), 412
Propane-1,2,3-trithiol, methylation, 486
Propargyl alcohols reactions with *o*-phenylenediamines, 306
3-Propylpyridazine, 10
Pseudomaonas aeruginosa, 355
Pseudomonas chlororaphis, 390
Pteridines, 189
–, oxidative cleavage, 252
Puffer fish, 228
Pulcherrimin, 241
Pyocanine (5-methylphenazin-1(5H)-one), 355, 365
Pyocyanine, 384, 386
–, biosynthesis, 390
Pyrazinamide, 242, 259, 271, 272
Pyrazine (1,4-diazines), 241–300
–, alkylation, 254
–, amination, 260
–, basicity, 242
–, ^{13}C n.m.r. spectra, 245
–, electrophilic substitution, 253
–, homolytic substitution, 254
–, mass spectra, 245
–, natural occurrence, 241
–, nucleophilic substitution, 254
–, oxidation, 243, 254
–, physical and spectroscopic properties, 243–245
–, quaternisation, 242, 243
–, reduction, 254, 288
–, synthesis, 245–252
–, – from α-aminocarbonyl compounds, 245, 246

Pyrazine, 261
–, electrochemical reduction, 278
–, electronic configuration, 243, 244
–, halogenation, 254
–, ^{1}H n.m.r. spectrum, 244, 245
–, infrared spectrum, 244
–, ionisation constants, 244
–, Raman spetrum, 244
–, reaction with 1-diethylaminopropyne, 256
–, synthesis, 252, 253
–, ultraviolet spectrum, 244
Pyrazine-2-carboxamides, 259
Pyrazinecarboxylic acids, 248, 259, 270–272
–, decarboxylation, 271
Pyrazinecarboxylic acid, decarboxylation, 252, 253
Pyrazine(p-diazine), 252
Pyrazine-2,3-dicarboxamides, 259
Pyrazine-2,3-dicarboxyamide, 272
Pyrazine-2,6-dicarboxamide, 272, 273
Pyrazine-2,5-dicarboxyhydrazide, 272
Pyrazine-2,3-dicarboxylic acid, 271, 272, 311
Pyrazine-2,5-dicarboxylic acid, 271, 272
Pyrazine-2,6-dicarboxylic acid, 271, 272
Pyrazine-2,5-diisocyanate, 272
Pyrazine-1,4-dioxide, 257
Pyrazine-2,5-diurethane, 272
Pyrazinesulphonates, 259
Pyrazinetetracarboxylic acid, 171, 272
Pyrazinetricarboxylic acid, 171, 272, 348
Pyrazinium, bromide, 257
Pyrazinoic acid, 271, 272
Pyrazinones, 261
Pyrazin-N-oxides, 257
–, synthesis, 254
Pyrazin-1-oxide, 257
Pyrazinylcarbinols, 259
Pyrazolocinnoline, 59
Pyrazylmethylsodium, 255
Pyridazines, 189
–, synthesis from cyanoacetic acid, 3
–, – from 1,2-diazepines, 5
–, – from 1,2-diketones, 2
–, – from 1,4-diketones, 2
–, – from isoxazoles, 5
–, – from pyrroles, 5
–, – from monosaccharides, 6
–, – from pyrans, 5

–, – from pyridines, 5
Pyridazine, 6, 159
–, chemical reactions, 9
–, ^{1}H n.m.r. data, 7
–, mass spectrum, 121
–, physical properties, 6
–, reduction, 9
–, spectra, 8
–, sulphur compounds, 28
Pyridazinecarboxylic acids, synthesis, 26
Pyridazine-3-carboxylic acid, 27
Pyridazine-4-carboxylic acid, 27
Pyridazine-4,5-dicarboxylic acid, 27, 121
Pyridazine-1,2-dioxides, ^{1}H n.m.r. spectral data, 21
Pyridazine-1,2-dioxide, 19
Pyridazine-3,4,5,6-tetracarboxylic acid, 27
Pyridazinethiones, basity, 28
Pyridazine-3($2H$)-thione, 28
–, X-ray structure, 28
Pyridazine-4($1H$)-thione, 28
Pyridazinones, acylation, 26
–, alkylation, 24
–, ^{1}H n.mr. spectral data, 25
–, mass spectra, 24
–, synthesis, 22
Pyridazin-3($2H$)-ones, 1,3,23
Pyridazin-4($1H$)-one, 23
Pyridazin-N-oxides, 11, 14, 18
–, ^{1}H n.m.r. spectral data, 20
Pyridazin-1-oxide, 18
β-4-Pyridazinylacaylic acid, 28
N-Pyridazinylpyrazoles, 17
Pyridoazepines, 256
3-Pyridylcinnolines, 49
Pyrimethamine, 155
Pyrimidinamines, 160, 191, 209
Pyrimidin-2-amine, 161
Pyrimidin-5-amine, 160
Pyrimidines, 155–227
–, by products from 1,3,5-triazine, syntheses, 471
–, general properties, 193, 194
–, ionisation contants, 159
–, oxidation, 198
–, photochemical reactions, 197
–, physical data, 199
–, reactions with electrophiles, 195–197
–, – with nucleophiles, 197
–, reduction, 198
–, spectroscopy, 162–164

–, sulphur containing derivatives, 221
–, synthesis, 165, 193
–, – from β-aldehydoesters, 172
–, – from β-aldehydoketones, 168
–, – β-aldehydronitriles, 177
–, – from azoles, 189
–, – from β-dialdehydes, 166
–, – from β-diesters, 175
–, – from β-diketones, 170
–, – from β-dinitriles, 180
–, – from β-ester nitriles, 179
–, – from imidazoles, 186
–, – from β-ketoesters, 173
–, – from β-ketonitriles, 179
–, – from purines, 188
–, – from pyrroles, 186
–, – from thiazines, 188
Pyrimidine, 199
–, structure, 158, 159
–, tautomerism, 164
Pyrimidine-2-carbaldehydes, 190
Pyrimidine-4-carbaldehydes, 190
Pyrimidine-2-carboxylate, 190
Pyrimidine carboxylic acids, 159, 160, 191
Pyrimidine-2,4-diamine, 161
Pyrimidine-4,5-diamine, 161
Pyrimidine-4,5-dicarboxylic acid, 235
Pyrimidine, nucleosides, 218, 225
Pyrimidine-2-sulphonic acids, 160
Pyrimidinethiones, 159, 160
–, tautomerism, 164
Pyrimidine-2(1H)-thione, 166
Pyrimidin-5-ols, 159, 192
Pyrimidinones, 159, 160
–, alkylation, 215
–, tautomerism, 164
Pyrimidino[5,4-e]-1,2,4-triazines, 468
Pyrimidin-N-oxides, 161, 193, 196, 227
Pyrimidin-5-ylcopper compounds, 226
Pyrimidin-5-ylpalladium 226
2-Pyrimidin-2′-ylpyrimidine, 207
Pyrimid-4-yl acetate, 227
Pyrimidylaminocinnolines, quaternary salts, 68
Pyrroles, 186
Pyrrolidines, 188
2-Pyrrolidone, 442
1△-Pyrroline, 446
Pyrrolopyrazines, 256
Pyruvaldehyde 340,
Pyruvohydroxamic acid, reaction with aminoacetone, 266

Quinazolinamines, 236
Quinazolines, 227–234, 437
–, general properties, 229, 230
–, reactions, 234
–, synthesis, 231–234
Quinazoline, hydrogenation, 235
Quinazoline alkaloids, 228, 437–456
Quinazoline carbonitriles, 236
Quinazoline-4-carbonitrile, 233
Quinazoline-2,4(1H,3H)-dione, silylation, 237
Quinazoline sulphonamides, 236
Quinazoline-2-thiones, 233
Quinazoline-2(1H)-thione, 234
Quinazoline-4-thiones, 233, 238
Quinazolinols, alkylation, 237
Quinazolinones reaction with phosphorus pentasulphide, 237
Quinazolin-2-one, tautomerism, 230
Quinazolin-4-ones, 232
–, tautomerism, 230
Quinazolin-4(3H)-one, 232, 235, 438, 440, 450
Quinazol-4-one-3-acetic acid, 440, 450
Quinazolin-N-oxides, 234, 238, 239
Quinol, 315
3-Quinolinylcinnolines, 49
o-Quinone diimides cycloaddition reactions, 310
Quinoxalines (benzopyrazines), 301–349
–, oxidative cleavage, 252
–, physical properties, 301, 302
–, spectroscopy, 301, 302
–, synthesis, 302, 310
Quinoxaline, 301–303, 310–312, 318
–, addition reactions, 311
–, oxidation, 270
–, substitution reactions, 310
Quinoxaline carboxaldehydes, 344, 345
Quinoxaline-2-carboxaldehydes, 343
Quinoxaline-2-carboxaldehyde, 344
Quinoxaline-2-carboxaldehyde 1,4-dioxide, 344
Quinoxaline-2-carboxamides, 324
Quinoxaline-2-carboxylates, 347
Quinoxaline-2-carboxylic acids, 343–347
Quinoxaline-2-carboxylic acid, 347
Quinoxaline-2-carboxylic acid 1,4-di-N-oxide, 301

Quinoxaline-2,3-dicarboxaldehyde, 345
Quinoxaline-2,3-dicarboxylates, 348
Quinoxaline-2,3-dicarboxylic acids, 348
Quinoxaline-2,3-dicarboxylic acid, 271, 303, 348, 364
Quinoxaline-2,3-dicarboxylic anhydride, 349
Quinoxaline-2,3(1H,4H)-diones (2,3-dihydroquinoxalines), 333, 334
Quinoxaline-2,3-(1H,4H)-dione, 333
Quinoxaline-1,4-dioxide, 311, 322
Quinoxaline, ureides, 346
Quinoxalino[2,3-e]cinnoline, 58
Quinoxalinones, reduction, 332
Quinoxalin-2(1H)-ones (hydroxyquinoxalines), 327–334
Quinoxalino[2,3-a]phenazine, 398
Quinoxalin-N-oxides, 334–342
–, chemical reactions, 341, 342
–, deoxygenation, 307
–, synthesis from benzofuroxan, 308
–, – from dioximes, 308
–, – from enaminones, 307
–, – from o-nitroacetanilides, 307
Quinoxalin-1-oxide, 311
β-Quinoxalinylpropionamide, 313
β-Quinoxalinylpropionic acid, 313

Rhoduline Heliotrope 3b (C.E. Basic Violet 6, 50055), 410
Rinkes synthesis, 181, 231
Ris reaction, 358, 366
Rosindones, 405
Rosindulines, 405
Rutaecarpine, 452, 456

Safranines, 355, 409
Safranine (phenosafranine or S.I. Basic Dye 50200), 409
Safranine MN (C.I. Basic Violet 8, 50210), 410
Safranine-T.(C.I. Basic Red 2, 50240; F.I.A.T. 1313, II, 367) 410
Sessiflorine, 448
Sida acuta, 444
Sida cordifolia, 444
Sida rhombifolia, 444
Sodionitromalondialdehyde, 167
Sodium 1-benzeneazo-1,2-dihydro-2-oxonaphthalene-1-sulphonate, 134
Sodium oxydimethylenedithiosulphate, 486
Sphoerides phyrens, 228
Sphoerides rubripes, 228
Stilbene, 54
Streptomyces ambrofaciens, 301, 334
Strobilanthes cusia, 446
Styrene oxides, reactions with o-phenylenediamines, 305
2-Styrylazobenzene, 54
4-Styrylcinnolines, 53
4-Styrylcinnoline, 48
Styrylpyrazines, 255
Styrylpyrimidine, 199
Styrylquinoxalines, oxidation, 270
2-Styrylquinoxalines, 345
2-Styrylquinoxaline, 311
1,2-Succinylpiperidazine, 35
Sulfalene, 242
Sulphadiazine, 155
Sulphamides, reactions with 1,3-diketones, 476
Sulphamide, reactions with cyanates, 497
3-Sulphanilamidopyridazine, 30
Sulphinamide anhydrides, 439
N-Sulphinylaniline, 292
Sulphonylated safranines, 411
1-Sulphonyl-1,2,3,4-tetrahydroquinoxalines, 351
Sulphur Black T (C.I. Sulphur Black 1, 53185; Vidal B.P. 16449/1896), 432
Sulphur diimines, 501
Sulphur dyes, 430–436

Tannic acid, 420
Tetra-O-acetyl-α-D-glucopyranosyl bromide, 22
2-(Tetra-O-acetyl-α-D-glucopyranosyl)-cinnolin-3(2H)-one, 73
2,4,5,6-Tetraaminopyrimidine, 212
Tetrabenzotriphendioxazine, 425
1,1,4,4-Tetrabenzyl-2-tetrazene, reaction with thionyl chloride, 487
1,4,6,9-Tetrabromophenazine, 376
Tetrachloro-o-benzoquinone, cycloaddition with 2-methylquinoxaline
3,4,7,8-Tetrachlorocinnoline, 55, 56
1,2,3,4-Tetrachlorophenazine, 376
1,4,6,8-Tetrachlorophenazine, 376
1,4,6,9-Tetrachlorophenazine, 376, 377
2,3,7,8-Tetrachlorophenazine, 376
Tetrachloropyrazine, 257, 258

INDEX

3,4,5,6-Tetrachloropyridazine, 12
Tetracyanoquinodimethane, adduct with 5,10-dihydro-5,10-dimethylphenazine, 394
Tetracyano-*p*-quinodimethane charge transfer complexes with dihydrophenazines, 396
Tetradecahydrophenazines, 401, 402
1,1,3,3-Tetraethoxypropane, 166
Tetraethyl 4-(1,2-dicarboxyhydrazino)-1,2,3,4-tetrahydrocinnoline-1,2-dicarboxylate, 95
5,6,7,8-Tetrafluorocinnoline, 55
1,2,3,4-Tetrafluorophenazine, 376
Tetrafluoropyrazine, 258
1,4,5,6-Tetrafluoropyridazine, 12
1,4,5,6-Terafluoropyrimidine, 207
5,5,6,6-Tetrafluoro-1,2,3,4-tetrathiane, 500
1,2,3,4-Tetrahydrobenzo[*c*]cinnoline, 109, 110
2,3,5,6-Tetrahydrobenzo[*h*]cinnolin-3-one, 111
Tetrahydrocinnolines, mass spectra, 96
1,2,3,4-Tetrahydrocinnolines, electrochemical oxidation, 95
1,2,3,4-Tetrahydrocinnoline, 39, 77, 92, 93
5,6,7,8-Tetrahydrocinnoline, 96
1,2,3,4-Tetrahydrocinnoline-1,2-phthalide, 92
5,6,7,8-Tetrahydrocinnolin-3(2*H*)-ones, reactions, 98
–, reduction, 98
–, reductive rearrangement to 1-aminooxindole, 73
–, synthesis, 98
1,2,3,4-Tetrahydro-2,4-dimethylbenzo[*c*]cinnoline, 109
1,2,3,4-Tetrahydro-4,6-dimethylbenzo[*c*]cinnoline, 109
1,2,3,4-Tetrahydro-2,3-dimethylphthalazine, 150
1,2,3,4-Tetrahydro-2,3-diphenylphthalazine, 151
5,6,7,8-Tetrahydro-1-iodophthalazine, 151
1,2,3,4-Tetrahydro-8-methoxybenzo[*c*]-cinnoline, 109
1,2,3,4-Tetrahydro-1-methylphthalazine, 146
Tetrahydronorharmanone, 456
Tetrahydro-1,2,4-oxadiazin-3-one, 479
1,2,3,4-Tetrahydro-1-oxo-3-phenyl-naphthalene-2-acetic acid, 111
1,2,3,4-Tetrahydrophenazines, 397, 398
1,2,3,4-Tetrahydrophenazine, 360, 397
–, dehydrogenation, 363
–, reduction, 401
1,2,3,4-Tetrahydrophenazine-9,10-dioxide, 397
1,2,3,4-Tetrahydro-3-phenylcinnoline-1,2-phthalide, 94
1,2,3,4-Tetrahydrophthalazines, synthesis, 150
5,6,7,8-Tetrahydrophthalazines, synthesis, 151
1,2,3,4-Tetrahydrophthalazine, 150, 151
–, chemical reactions, 151
–, physical properties, 151
5,6,7,8-Tetrahydrophthalazine, 151, 152
4a,5,8,8a-Tetrahydrophthalazin-1(2*H*)-one, 153
–, synthesis, 153
5,6,7,8-Tetrahydrophthalazin-1(2*H*)-one, 152
Tetrahydropyrazines, 283, 287
1,2,3,6-Tetrahydropyrazin-2-ones, 286, 287
1,2,34,6-Tetrahydropyridazines, 4, 33
1,4,5,6-Tetrahydropyridazines, 5, 33, 34
2,3,4,5-Tetrahydropyridazines, 34
Tetrahydroquinazolines, 235
Tetrahydroquinoxalines, 332, 350
1,2,3,4-Tetrahydroquinoxalines, 311
–, conformation, 352, 353
1,2,3,4-Tetrahydroquinoxaline, 309, 350, 351
5,6,7,8-Tetrahydroquinoxalines, 352
2,3,4,5-Tetrahydro-2,2,5,5-tetramethylpyrazine, 285
1,2,3,6-Tetrahydro-1,2,3,4-tetrazines, 487
1,2,3,4-Tetrahydro-1,2,4,5-tetrazines, 494, 495
Tetrahydro-1,2,5-thiadiazine-6-thiones, 476
1,4,5,6-Tetrahydro-1,2,4-triphenylpyrazine, 283
Tetrahydroxybutylquinoxalines, 345
2-Tetrahydroxybutylquinoxalines, synthesis from osone hydrazones, 343
1,2,3,4-Tetrahydroxyphenazine, 387
Tetrakis(triphenylphosphine)palladium, 258, 259
2,3,8,9-Tetramethylbenzo[*c*]cinnoline, 105

1,3,6,8-Tetramethylphenazine, 363, 367, 368
2,2,5,5-Tetramethylpiperazines, 294
2,3,5,6-Tetramethylpiperazines, 293
2,2,5,5-Tetramethylpiperazine, 294
2,3,5,6-Tetramethylpiperazine, 293
3,3,6,6-Tetramethylpiperazine-2,5-dione, 297
Tetramethylpyrazine, 253
–, oxidation, 271
3,4,5,6-Tetramethylpyridazine, 10
2,3,6,7-Tetramethylquinoxaline, 317
Tetraphenyl azodiphosphate, 4
1,2,4,5-Tetraphenylhexahydro-1,2,4,5-tetrazine, 494
2,3,5,6-Tetraphenylpiperazines, 294
Tetraphenylpyrazine, 257
Tetrathianes, 500
1,2,3,4-Tetrathiane, 500
Tetra(trifluoromethyl)pyrazine, 271
Tetrazines, 486–494
1,2,3,4-Tetrazines (v-tetrazines or osotetrazines), 486–488
1,2,3,5-Tetrazines (as-tetrazines), 486, 488, 489
1,2,4,5-Tetrazines (s-tetrazines), 31, 486, 489–49
–, reduced forms, 491
1,2,4,5-Tetrazine, 492
Tetrazine-1,4-dicarboxylate, cycloaddition with benzyne, 144
1,2,4,5-Tetrazine-3,6-dicarboxylic acid, 492, 493
1,2,3,5-Tetrazin-6-ones, decomposition to dimethylcyanamide, 488, 489
Tetrazolo[4,5-a]phthalazine, 129
Tetrazolo[1,5-b]pyridazines, 17
Tetrazolopyrimidines, 204
Tetrodotoxin, 228
Thiadiazines, 476–478
1,2,3-Thiadiazines, 476
1,2,4-Thiadazines, 476
1,2,5-Thiadiazines, 476
1,2,6-Thiadiazines, 476
1,3,4-Thiadiazines, 477
1,3,5-Thiadiazines, 477, 478
Δ^3-1,2,3-Thiadiazolines, 495
Thiadiazolinones, ring expansion, 477
Thiatriazines, 495–497
$1H$-1,2,4,6-Thiatriazines, 496
$2H$-1,2,3,6-Thiatriazines, 495

$2H$-1,2,4,6-Thiatriazines, 496
Thiazine dyes, 403
Thiobenzophenone, photochemical additions to imines, 482
Thiocarbamates, 481
Thiohydrazides, synthons for dithiadiazines, 498
Thiophorindigo C.T., 432
Thiopropionamide, 187
2-Thiopyrimidines, 215, 221
Thiouracil, 172, 182, 215
Thioureas, 173, 176
Thiourea, 168, 169, 172, 175, 179, 180, 205, 221
Thymine, 172, 197, 198
–, mass spectrum, 163
–, tautomerism, 164
5-(4-Toluenesulphonyl)-1,3,5-oxadiazine, 483
Toluidine Blue (C.I. Basic Blue 17, 52040) 429
4-p-Tolylcinnoline, 48
p-Tolyl-D-isoglucosamine, reaction with hydrazine, 343
1-Tosyloxyphenazine, 372
2,4,5-Triaminopyrimidine, 211
2,4,6-Triaminopyrimidine, 180
2,4,6-Triamino-1,3,5-triazine (melamine), 472
2,4,6-Triaryloxy-1,3,5-triazines, 469
Triaryl-$2H$-1,2,3-triazol-N-oxides, photolysis, 498
Triazines, 457–475, 483
1,2,3-Triazines, 457
–, fused ring systems, 459
1,2,4-Triazines, 461–468
–, fused ring systems, 467
–, reduced forms, 465
1,3,5-Triazines, 468–475
–, fused ring systems, 475
1,2,4-Triazine, 461
1,3,5-Triazine, 471
1,2,4-Triazinecarboxylic acids, 462
1,3,5-Triazinecarboxylic acids, and related compounds, 472, 473
1,2,4-Triazine-5,6-dicarboxylic acid anhydride, 462
1,2,4-Triazine-3,5-dione, 464
Triazinethiones, 464, 465
1,3,5-Triazine-2,4,6-triazide, 475
1,2,4-Triazine-3,5,6-tricarboxylic acid, 462

1,3,5-Triazine-2,4,6-tricarboxylic acid, 472
1,3,5-Triazine-tri-*N*-oxide (trifumin), 475
1,3,5-Triazine-2,4,6-trithiol (trithiocyanuric acid), 473
Triazinones (hydroxy-1,2,4-triazines), 463
1,2,4-Triazin-5-ones, 463, 464
1,2,4-Triazin-3-one, 463
1,2,4-Triazin-5(2*H*)-one, 464
2*H-asym*-Triazino[3,4-*a*]phthazin-3,4-dione, 130
1,2,4-Triazin-*N*-oxides, 464
1,2,3-Triazoles, 495
sym-Triazolo[3,4-*a*]phthalazines, 129
1,2,4-Triazolo[4,5-*a*]phthalazine, 129
Triazolopyrimidines, 189
Tri-*O*-benzoylribosylamine, 183
2,4,6-Tribenzoyl-1,3,5-triazine, 473
2-(Tribromomethyl)quinoxaline, 313
Tricarbonyl compounds, reactions with *o*-phenylenediamines, 346
Trichloroacetaldehyde, 225
Trichloroacetamidine, trimerisation, 470
3,4,6-Trichlorocinnoline, 55, 58
3,4,7-Trichlorocinnoline, 56
4,6,7-Trichlorocinnoline, 55
4,7,8-Trichlorocinnoline, 55
3,7,8-Trichloro-4-hydroxycinnoline, 55
Trichloroindolophenothiazolones, 436
3-Trichloromethyl-1,2,4-triazolo[4,5-*a*]phthalazine, 129
Trichloronitrophenoxazones, 425
1,2,3-Trichlorophenazine, 316
1,2,9-Trichlorophenazine, 316
1,4,6-Trichlorophenazine, 376
1,4,7-Trichlorophenazine, 376
Trichloropyridazines, 14
3,4,5-Trichloropyridazine, 12, 16
4,4,6-Trichloropyridazine, 12
2,4,6-Trichloropyrimidine, 203
2,4,6-Trichloro-1,3,5-triazine, 474
3,5,6-Trichloro-1,2,4-triazine, 462
4,5,6-Trichloro-1,2,3-triazine, 458
Tricyanomethane, 180
2,3,5-Tri(dimethylamino)azacyclobutadiene, 458
3,5,6-Triethoxycarbonyl-1,2,4-triazine, 462
1,3,3-Triethoxypropene, 167
1,3,5-Triethylhexahydro-1,3,5-triazine, 475
2,4,6-Triethyl-1,3,5-triazine, 471
2-Trifluoromethyl-3,1-benzoxazin-4-one, 452, 453

3-Trifluoromethyl-1,2,4-triazolo[4,5-*a*]-phthalazine, 129
2,4,6-Trifluoro-1,3,5-triazine, 473
3,5,6-Trifluoro-1,2,4-triazine, 463
Trifulmin (1,3,5-triazine-tri-*N*-oxide), 475
2,3,4-Trihydro-1,2,4,5-tetrazines, 494
2,4,6-Trihydro-1,3,5,2,6-trithiadiazine-3,3,5,5-tetroxides, 502
2,3,7-Trihydroxyphenazine-1,6-dicarboxylic acid, 390
2,4,6-Trihydroxy-1,3,5-triazine (cyanuric acid), 473
Trimethoprim, 178, 209
2,3,7-Trimethoxyphenazine, 384
1,2,4-Trimethylbenzo[*c*]cinnoline, 105
1,3,5-Trimethylhexahydro-1,3,5-triazine, 475
Trimethyloxonium borofluoride, 9
3,3,4-Trimethylpiperazin-2-one, 296
Trimethylpyrazines, 253
2,3,5-Trimethylpyrazine, 247, 253
3,5,6-Trimethylpyrazine-2-carboxamide, 273
3,4,6-Trimethylpyridazine, 10
2,4,6-Trimethylpyridimine, 170
2,3,6-Trimethylquinoxaline, 317
5-Trimethylsilyl-1,3,2,4,6-dithiatriazine-1,1-dioxide, 502
2,3,4-Trimethyltetrahydro-1,3,4-oxadiazine, 480
2,4,6-Trimethyl-1,3,5-triazine, 471
4,5,6-Trimethyl-1,2,3-triazine, 458
2,4,6-Trimethyl-1,2,4-triazine-3,5-dione, 464
2,3,6-Trimethyl-1,2,4-triazin-5(2*H*)-one, 464
3,4,6-Trimethyl-1,2,4-triazin-5(4*H*)-one, 464
1,3,5-Trimethyl-1,3,5-triazin-2-one-4,6-dithione, 473
2,5,6-Trimethyl-1,2,4-triazin-4-oxide, 464
4,5,6-Trimethyl-1,2,3-triazin-1-oxide, 458
4,5,6-Trimethyl-1,2,3-triazin-2-oxide, 458
4,5,6-Trimethyl-1,2,3-triazole, 457
Trioxanes, 485, 486
1,2,4-Trioxanes, 486
2-Trioxanylquinoxaline, 344
Triphenazineoxazines, 415
Triphendioxazines, 412, 422–427
Triphendioxazine, 422
–, nitration, 425

1,3,4-Triphenylcinnoline, 93
1,4,4-Triphenylcinnoline, 93
3,5,6-Triphenyl-2-cyanopyrazine, 273
1,2,3-Triphenylcyclopropenyl azide, thermolysis, 457
1,3,5-Triphenylhexahydro-1,3,5-triazine, 475
2,4,6-Triphenyl-1,3,5-oxadiazinium salts (diazapyrylium salts), 481
2,4,6-Triphenyl-4H-1,3,4,5-oxatriazine, 498
1,2,3-Triphenylpiperazine, 283
1,2,3-Triphenylpropane-1,3-dione, 457
3,4,5-Triphenylpyrazole, 458
3,4,6-Triphenylpyridazine, 10
2,3,6-Triphenylquinoxaline, 312
2,5,6-Triphenyl-4H-1,3,4-thiadrazine, 477
2,4,6-Triphenyl-1,3,5-triazine (cyaphenin), 471
3,5,6-Triphenyl-1,2,4-triazine, 461
–, oxidation, 464
2,4,6-Triphenyl-1,3,5-triazinetri-N-oxide, 475
3,5,6-Triphenyl-1,2,4-triazin-1-oxide, 464
4,5,6-Triphenyl-1,2,3-triazole, 457, 458
Triphenylverdazyl, 496
2,4,6-Tris(aziridyl)-1,3,5-triazine, 472
4,5,6-Tris(dialkylamino)-1,2,3-triazines, 458
2,4,6-Tris(perfluoroethyl)-1,3,5-triazine, 471
2,4,6-Tris(2-tolyl)-1,3,5-triazine, 471
3,4,6-Tris(3-tolyl)-1,3,5-triazine, 471
2,4,6-Tris(4-tolyl)-1,3,5-triazine, 471
2,4,6-Tris(trichloromethyl)-1,3,5-triazine, 471
Trithianes, 485, 486
1,2,4-Trithiane, 486
Trithiocyanurates, 473
Trithiocyanuric acid (1,3,5-triazine-2,4,6-trithiol), 473
Trypanosome congolense, 67, 68
Tryptamine, 452
L-Tryptophan, 446

Tryphthanthrin, 446

Ullmann reaction, 207
Uracils, 159
Uracil, 159, 181, 182, 196–198, 216
–, lithio derivatives, 225
–, mass spectrum, 163
–, tautomersim, 164
Ureas, 173
Urea, 174, 179, 185
Urea-quinoxaline, molecular complex, 310
Uric acid, 155
Uridines, 159

Vascinone thioanalogues, 443
Vasicine (Peganie), 440-442, 444-446
–, biosynthesis 445
–, oxidation, 443
Vasicinol, 444, 445
Vasicinone, 441, 443–447, 449
Vasicoline, 448
Vasicolinone, 448, 453
Verdazyls, 494, 495
Verdazylium salts, 495
Vidal black (X.I. Sulphur Black 3, 53180), 432
Vilsmeier reaction, 201
5-Vinyl-2'-deoxyuridine, 226
2-Vinylfuran, 284
Vitamin B1, 155
Vivian reaction, 363, 375, 381

Whitehead synthesis, 177
Widman-Stoermer synthesis, cinnolines, 36
Wohl-Aue reaction, 359, 367, 370, 375, 386, 391
Wool Fast Blue BL (C.I. Acid Blue 59; 50315; F.I.A.T. 1313, II, 380), 411
2-(m-4-Xylyl)quinoxaline, 312

Zanthoxylum arborescens, 447